Edited by
Claudio Tomasi, Sandro Fuzzi, and
Alexander Kokhanovsky

Atmospheric Aerosols

Wiley Series in Atmospheric Physics and Remote Sensing

Series Editor: Alexander Kokhanovsky

Wendisch, M. / Brenguier, J.-L. (eds.)

Airborne Measurements for Environmental Research

Methods and Instruments

2013

Coakley Jr., J. A. / Yang, P

Atmospheric Radiation

A Primer with Illustrative Solutions

2014

Stamnes, K. / Stamnes, J. J.

Radiative Transfer in Coupled Environmental Systems

An Introduction to Forward and Inverse Modeling

2015

Tomasi, C. / Fuzzi, S. / Kokhanovsky, A. (eds.)

Atmospheric Aerosols

Life Cycles and Effects on Air Quality and Climate

2016

Forthcoming:

Kokhanovsky, A. / Natraj, V.

Analytical Methods in Atmospheric Radiative Transfer

Huang, X. / Yang, P.

Radiative Transfer Processes in Weather and Climate Models

North, G. R. / Kim, K.-Y.

Energy Balance Climate Models

Davis, A. B. / Marshak, A.

Multi-dimensional Radiative Transfer

Theory, Observation, and Computation

Minnis, P. *et al.*

Satellite Remote Sensing of Clouds

Zhang, Z. *et al.*

Polarimetric Remote Sensing

Aerosols and Clouds

Weng, F.

Satellite Microwave Remote Sensing

Fundamentals and Applications

Edited by Claudio Tomasi, Sandro Fuzzi, and Alexander Kokhanovsky

Atmospheric Aerosols

Life Cycles and Effects on Air Quality and Climate

WILEY-VCH
Verlag GmbH & Co. KGaA

The Editors

Prof. Claudio Tomasi
Institute of Atmospheric Sciences & Climate ISAC
Area della Ricerca del CNR
Via Gobetti 101
40129 Bologna
Italy

Prof. Sandro Fuzzi
Institute of Atmospheric Sciences & Climate ISAC
Area della Ricerca del CNR
Via Gobetti 101
40129 Bologna
Italy

Dr. Alexander Kokhanovsky
EUMETSAT
Remote sensing Products
EUMETSAT-Allee 1
64295 Darmstadt
Germany

and

Moscow Engineering Physics Institute (MEPhI)
National Research Nuclear University
Kashirskoe Str. 31
115 409 Moscow
Russia

A book of the Wiley Series in Atmospheric Physics and Remote Sensing

The Series Editor

Dr. Alexander Kokhanovsky
EUMETSAT
Remote sensing Products
EUMETSAT-Allee 1
64295 Darmstadt
Germany

and

Moscow Engineering Physics Institute (MEPhI)
National Research Nuclear University
Kashirskoe Str. 31
115 409 Moscow
Russia

Cover
NASA

Library of Congress Card No.: applied for

British Library Cataloguing-in-Publication Data
A catalogue record for this book is available from the British Library.

Bibliographic information published by the Deutsche Nationalbibliothek
The Deutsche Nationalbibliothek lists this publication in the Deutsche Nationalbibliografie; detailed bibliographic data are available on the Internet at <http://dnb.d-nb.de>.

Print ISBN: 978-3-527-33645-6
ePDF ISBN: 978-3-527-33643-2
ePub ISBN: 978-3-527-33641-8
Mobi ISBN: 978-3-527-33642-5
oBook ISBN: 978-3-527-33644-9

Cover Design Grafik-Design Schulz
Typesetting SPi Global, Chennai, India
Printing and Binding Markono Print Media Pte Ltd, Singapore

Printed on acid-free paper

Contents

List of Contributors

Alessandra Bonazza
Institute of Atmospheric Sciences and Climate (ISAC)
National Research Council of Italy (CNR)
Via Gobetti 101
40129 Bologna
Italy

Julien Boulon
Université Paul Sabatier
Institut de Recherche en Astrophysique et Planetologie (IRAP)
CNRS-UMR5277
14 avenue E. Belin
31400 Toulouse
France

Maurizio Busetto
Institute of Atmospheric Sciences and Climate (ISAC)
National Research Council of Italy (CNR)
Via Gobetti 101
40129 Bologna
Italy

Marina Camatini
University of Milano Bicocca
POLARIS Research Centre
Department of Environmental Science
Piazza della Scienza 1
20126 Milano
Italy

Paola De Nuntiis
Institute of Atmospheric Sciences and Climate (ISAC)
National Research Council of Italy (CNR)
Via Gobetti 101
40129 Bologna
Italy

Sandro Fuzzi
Institute of Atmospheric Sciences and Climate (ISAC)
National Research Council of Italy (CNR)
Via Gobetti 101
40129 Bologna
Italy

Stefania Gilardoni
Institute of Atmospheric
Sciences and Climate (ISAC)
National Research Council of
Italy (CNR)
Via Gobetti 101
40129 Bologna
Italy

Maurizio Gualtieri
Italian National Agency for New
Technologies
Energy and Sustainable
Economic Development
(ENEA)-SSPT-MET-INAT
POLARIS Research Centre
40129 Bologna
Italy

Alexander A. Kokhanovsky
EUMETSAT
Remote Sensing Products
EUMETSAT Allee 1
D64295 Darmstadt
Germany

and

National Research Nuclear
University
Moscow Engineering Physics
Institute (MEPhI)
Kashirskoe Str. 31
115 409 Moscow
Russia

Christian Lanconelli
Institute of Atmospheric
Sciences and Climate (ISAC)
National Research Council of
Italy (CNR)
Via Gobetti 101
40129 Bologna
Italy

Kwon H. Lee
Gangneung-Wonju National
University (GWNU)
Department of Atmospheric and
Environmental Sciences
7 Jukheon-gil
Gangneung, Gwangwondo 25457
Republic of Korea

Angelo Lupi
Institute of Atmospheric
Sciences and Climate (ISAC)
National Research Council of
Italy (CNR)
Via Gobetti 101
40129 Bologna
Italy

Paolo Mandrioli
Institute of Atmospheric
Sciences and Climate (ISAC)
National Research Council of
Italy (CNR)
Via Gobetti 101
40129 Bologna
Italy

Mauro Mazzola
Institute of Atmospheric
Sciences and Climate (ISAC)
National Research Council of
Italy (CNR)
Via Gobetti 101
40129 Bologna
Italy

Sonoyo Mukai
The Kyoto College of Graduate
Studies for Informatics
7 Tanaka Monzencho
Sakyo
Kyoto 606-8225
Japan

Makiko Nakata
Kindai University
3-4-1 Kowakae
Higashi-Osaka
Osaka 577-8502
Japan

Walter Di Nicolantonio
CGS S.p.A – Compagnia
Generale per lo Spazio
OHB SE Company
Department of Satellites and
Missions
via Gallarate 150
20151 Milan
Italy

Boyan H. Petkov
Institute of Atmospheric
Sciences and Climate (ISAC)
National Research Council of
Italy (CNR)
Via Gobetti 101
40129 Bologna
Italy

Cristina Sabbioni
Institute of Atmospheric
Sciences and Climate (ISAC)
National Research Council of
Italy (CNR)
Via Gobetti 101
40129 Bologna
Italy

Giulio Sancini
University of Milano Bicocca
POLARIS Research Centre
School of Medicine and Surgery
20900 Monza
Italy

Itaru Sano
Kindai University
3-4-1 Kowakae
Higashi-Osaka
Osaka 577 – 8502
Japan

Karine Sellegri
Université Blaise Pascal
Laboratoire de Météorologie
Physique (LaMP)
CNRS-UMR 6016
4 avenue Blaise Pascal
63170 Aubière
France

Claudio Tomasi
Institute of Atmospheric
Sciences and Climate (ISAC)
National Research Council of
Italy (CNR)
Via Gobetti 101
40129 Bologna
Italy

Preface

Atmospheric aerosol, also commonly called airborne particulate matter (PM), is a subject of extensive research that, since the beginning of the 1980s, has received increased attention from the atmospheric science community. This is in part due to the enormous advances in measurement technologies from that period, which have allowed for an increasingly accurate understanding of the chemical composition and of the physical properties of atmospheric aerosol and its life cycle in the atmosphere.

The growing scientific interest in atmospheric aerosol is also due to its high importance for environmental policy. In fact, PM constitutes one of the most challenging problems both for air quality and for climate change policies. Atmospheric aerosol affects air quality and, in turn, human and ecosystem well-being and also have an important role in the Earth's climate system. Understanding of aerosol nucleation, emission, deposition, transport, and life cycle is probably the most pressing issue in air quality regulation worldwide, and at the same time it represents one of the biggest sources of uncertainty in current climate simulations.

This book, which collects contributions from an international team of scientists with different backgrounds, is aimed at providing an overall interdisciplinary picture of the aerosol lifecycle, from nucleation and emission to atmospheric processing. Also, the measurement techniques and the aerosol environmental effects are discussed.

The first chapter of the book provides an overview of the main sources of atmospheric aerosol, both primary and secondary and of natural and anthropogenic origin.

Chapters 2 and 3 describe the atmospheric processing and removal of aerosol: new particle formation, coagulation and condensation processes, wet and dry deposition, and aerosol–cloud interaction.

Chapter 4 provides an overview of the chemical climatology of atmospheric aerosol and of some emerging issues in aerosol science such as the nature and processes of organic aerosol and the chemical properties of aerosol affecting their cloud condensation nuclei (CCN) and ice nuclei (IN) ability.

Chapter 5 introduces the basic elements of aerosol optics which are needed for the understanding of aerosol radiation transfer modelling (Chapter 6) and the remote sensing of atmospheric aerosol both from ground-based measurements and from airborne and space platforms (Chapter 7). This latter chapter also provides useful examples of international aerosol remote sensing experiments carried out in different regions of the world.

Chapter 8 addresses the role of aerosol on climate, both in terms of direct (i.e., radiation scattering and absorption properties of aerosol particles) and indirect (i.e., the effect of aerosol on cloud structure and radiative properties acting as CCN and IN) effects.

Chapter 9 examines the effects of aerosol on air quality and the interactions between air quality and climate, the retrieval of aerosol data from satellite measurements, and a characterization of mass concentration and optical properties of desert dust in different areas of the Earth.

Chapter 10 then describes the effects of PM on human health, both in terms of epidemiology and toxicology, the quantification of human exposure to particulate pollutants, and the mechanisms and effects of the interaction between PM and the human body.

The last chapter addresses an aspect rarely discussed in books of aerosol science, that is, the effects of aerosols that cause damage to heritage materials, both indoor and outdoor: monumental complexes, archaeological sites, and heritage objects. The chapter describes the monitoring of the damages; the physical, chemical, and biological mechanisms of interaction between atmospheric aerosols and cultural heritage material; and the guidelines for the conservation of cultural heritage.

We wish to thank all colleagues who have contributed to the book and have provided such a wide and interdisciplinary overview on atmospheric aerosol: their origin, processing in the atmosphere, measurement techniques, and environmental effects. We also wish to thank Prof. Teruyuki Nakajima for having kindly agreed to write the foreword to this book.

Bologna
August 1, 2016

Claudio Tomasi
Sandro Fuzzi
Alexander Kokhanovsky

Foreword

Atmospheric aerosol composed of small liquid and solid particles suspended in air is an important atmospheric constituent that has always attracted scientific research. It impacts various phenomena on earth, including atmospheric chemistry, cloud and rain, climate, health, and biological activity among others. In the 1950s–1960s, urban air pollution, causing phenomenon such as the Los Angeles City smog, posed a serious social problem. Hence, understanding the atmospheric chemical reactions and aerosol properties was an urgent requirement for the research community to focus on. During this period, basic atmospheric chemistry was conceptualized. In the 1980s–1990s, aerosols received growing attention from the climate researchers owing to global warming. In this period, climate and aerosol researchers collaborated to develop climate models that take into account aerosols and other chemical species. They found that aerosols affect the earth's climate directly through their radiative effect and indirectly through their interactions with the cloud system. The net radiative forcing of aerosols was estimated to be negative, that means cooling of the earth system, but the estimated uncertainty was significantly large. Satellite remote sensing also flourished during this period, and a global perspective of the aerosol phenomena was conceived, such as continental-scale transportation of dust and other particles and change in the cloud droplet size on interacting with aerosols.

The current era is known as the era of human beings or Anthropocene, as coined by the Nobel laureate Paul Crutzen. This era is marked by human activities changing the earth on a scale observable from the space. The United Nation's Sustainable Development Goals (SDGs) have outlined the prioritization scheme for the maximization of the benefits of symbiotic coexistence with nature. In this era, aerosol science can provide a solution for various aerosol-related problems by linking various processes of different scales, such as microphysical, chemical, mesoscale, cloud, and global processes. High-resolution modelling of the weather, cloud, and rain considers aerosols as the cloud condensation nuclei (CCN) and radiative forcing agent to improve the weather forecasting for disaster prevention. A data assimilation system has been developed to utilize the data from ground-based networks and satellites along with model simulations for providing useful information on the air quality to regulate and evaluate its public health impact.

As described above, aerosol-related sciences have become an important research field. In this regard, I am pleased to see that this book is being published. It includes the widespread effects of aerosols, which is otherwise difficult to fully understand without a well-organized textbook. This book provides a comprehensive knowledge of aerosol science. It covers almost everything from the basic concepts to various applications and from theoretical bases to observations with well-organized equations, figures, and tables to provide the readers with a quantitative approach to the subject.

I believe this book will provide the readers a great opportunity to explore the wonders of aerosol science.

Teruyuki Nakajima
Earth Observation Research Center
JAXA, Tsukuba, Japan

Acknowledgments

Claudio Tomasi and Angelo Lupi gratefully acknowledge the colleagues Cristina Sabbioni, Paolo Mandrioli, and Alessandra Bonazza (Institute of Atmospheric Sciences and Climate, ISAC – C.N.R., Bologna, Italy); Emanuela Molinaroli (Department of Environmental Sciences, University of Venice, Italy); Angelo Ibba (Department of Earth Sciences, University of Cagliari, Italy); Annie Gaudichet (LISA, Paris University, France); and Hélène Cachier (LSCE/IPSL, Gif-sur-Yvette, France) for providing the scanning electron microscopy (SEM) and transmission electron microscopy (TEM) images of aerosol particles shown in Figures 1.4, 1.7, 1.8, and 1.11 of Chapter 1.

Sandro Fuzzi and Stefania Gilardoni acknowledge the support of the European Projects ACCENT (Atmospheric Composition Change: the European Network) and ACCENT-Plus (Atmospheric Composition Change: the European Network-Policy Support and Science) for the preparation of Chapter 4 and part of Chapter 9.

Alexander Kokhanovsky acknowledges support of the excellence center for applied mathematics and theoretical physics within MEPhI Academic Excellence Project (contract No. 02.a03.21.0005, 27.08.2013). He is grateful to his former colleagues at the Institute of Physics in Minsk (Belarus) and also at the Institute of Environmental Physics (Bremen, Germany) for discussions on various aspects of radiative transfer, satellite atmospheric optics, and atmospheric aerosol. A lot of insights in aerosol optics, image transfer, and radiative transfer have been gained via joint work with Eleonora Zege, Vladimir Rozanov, Wolfgang von Hoyningen-Huene, Gerrit de Leeuw, Teruyuki Nakajima, Reiner Weichert, Alan Jones, and John Burrows.

Claudio Tomasi, Boyan H. Petkov, Christian Lanconelli, Maurizio Busetto, Mauro Mazzola, Angelo Lupi, Alexander Kokhanovsky, and Kwon H. Lee acknowledge the colleagues G. P. Gobbi and F. Angelini (Institute of Atmospheric Sciences and Climate, ISAC – C.N.R., Rome Tor Vergata, Italy), for providing the automated Vaisala LD-40 ceilometer data set collected at the Torre Sarca (University of Milano-Bicocca) station in the center of Milan (Italy) during the QUITSAT field campaigns of summer 2007 and winter 2008, and the colleagues Ezio Bolzacchini and Luca Ferrero (Dept. of Earth and Environmental Sciences, University of Milano-Bicocca, Milan, Italy) for providing the OPC (1.108 "Dustcheck" GRIMM model) data set collected at the Torre Sarca (Milano-Bicocca University) station in the center of Milan (Italy) during the three years from 2005 to 2008, which have been examined in Chapter 7. K.H. Lee's work was funded by the Korean Meteorological Administration Research and Development Program under Grant KMIPA2015-2012.

Marina Camatini, Maurizio Gualtieri, and Giulio Sancini of the POLARIS Research Centre (University of Milano-Bicocca, Milan, Italy) acknowledge Dr. Laura Capasso (POLARIS Research Centre), who has realized some figures shown in Chapter 10.

1
Primary and Secondary Sources of Atmospheric Aerosol

Claudio Tomasi and Angelo Lupi

1.1
Introduction

Atmospheric aerosols are suspensions of any substance existing in the solid and/or liquid phase in the atmosphere (except pure water) under normal conditions and having a minimum stability in air assuring an atmospheric lifetime of at least 1 h. Generated by natural sources (i.e., wind-borne dust, sea spray, volcanic debris, biogenic aerosol) and/or anthropogenic activities (i.e., sulfates and nitrates from industrial emissions, wind-forced mineral dust mobilized in areas exploited for agricultural activities, fossil fuel combustion, and waste and biomass burning), aerosol particles range in size from a few nanometers to several tens of microns. As a result of internal cohesive forces and their negligible terminal fall speeds, aerosol particles can first assume sizes appreciably larger than the most common air molecules and subsequently increase to reach sizes ranging most frequently from less than 10^{-3} to no more than 100 μm (Heintzenberg, 1994). Particles with sizes smaller than 20–30 Å (1 Å $= 10^{-10}$ m) are usually classified as clusters or small ions, while mineral and tropospheric volcanic dust particles with sizes greater than a few hundred microns are not considered to belong to the coarse aerosol class, since they have very short lifetimes. Aerosol particles grown by condensation to become cloud droplets are not classified as aerosols, although a cloud droplet needs a relatively small aerosol particle acting as a condensation nucleus for its formation under normal atmospheric conditions. Similarly, precipitation elements such as rain droplets, snowflakes, and ice crystals are not classified as aerosols (Heintzenberg, 1994). Although present in considerably lower concentrations than those of the main air molecules, aerosol particles play a very important role in numerous meteorological, physical, and chemical processes occurring in the atmosphere, such as the electrical conductivity of air, condensation of water vapor on small nuclei and subsequent formation of fog and cloud droplets, acid rains, scattering, and absorption of both incoming solar (shortwave) radiation and thermal terrestrial (longwave) radiation. The interaction processes between atmospheric aerosols and the downwelling and upwelling radiation fluxes of solar

Atmospheric Aerosols: Life Cycles and Effects on Air Quality and Climate, First Edition.
Edited by Claudio Tomasi, Sandro Fuzzi, and Alexander Kokhanovsky.

and terrestrial radiation at the surface play a major role in defining the radiation budget of our planet and, hence, the Earth's climate (Chylek and Coakley, 1974).

To give an idea of the shape of an aerosol particle suspended in dry air, a schematic representation of a particle originating from the aggregation of various kinds of particulate matter fragments is shown in Figure 1.1. It consists of several small unit structures of different chemical composition and origin (soluble acid substances, sodium chloride crystals of marine origin, ammonium sulfates, insoluble carbonaceous matter, insoluble mineral dust, and insoluble organic substances), held together by interparticle adhesive forces in such a way that an aerosol particle behaves as a single unit in suspension. Thus, the same particle often contains distinct homogeneous entities, which are internally mixed to form aggregates of different components.

The insoluble carbonaceous and organic substances often consist of gas-borne particulate matter pieces from incomplete combustion, which predominantly contain carbon and other combustion-produced materials. When the surrounding air relative humidity (RH) increases to reach values higher than 65–70%, the same particle (containing soluble substances) grows gradually by condensation of water vapor to become a water droplet in which pieces of insoluble matter are suspended, as can be seen in the (b) of Figure 1.1 (see also Hänel, 1976), while the various soluble materials reach different solution states as a result of their appreciably differing deliquescence properties. In this way, an internally mixed particle evolves assuming the characteristics of an aggregate consisting of different particulate phases. Figure 1.1 also shows that dry aerosol particles can often exhibit irregular shapes, which can considerably differ from the spherical

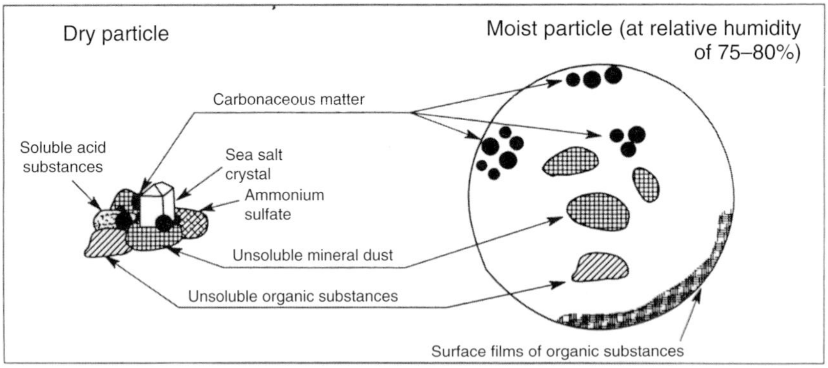

Figure 1.1 Schematic representation of an aerosol particle for dry air conditions (left) and humid air (for relative humidity (RH) = 75–80%) conditions (right), consisting of particulate matter pieces of soluble (i.e., soluble acid substances, sea-salt crystal, ammonium sulfates) and insoluble substances (carbonaceous matter, mineral dust, organic substances), which remain suspended inside the moist particle gradually growing by condensation until becoming a water droplet with soluble salts, acids, and organic compounds. (Adapted from a draft presented by Gottfried Hänel in a seminar given in 1985 at the FISBAT-CNR Institute, Bologna, Italy.)

one. Thus, the size of each real aerosol particle is generally evaluated in terms of an "equivalent" diameter a, for which the volume of such an ideal spherical particle is equal to that of the real particle.

Aerosol particles cover a size range of more than five orders of magnitude, with "equivalent" sizes ranging from 5×10^{-3} to 2.5 μm for fine particles and greater than 2.5 μm for coarse particles (Hinds, 1999). The fine particles include both (i) the so-called Aitken nuclei, having sizes mainly ranging from 5×10^{-3} to 5×10^{-2} μm, and (ii) the so-called "accumulation" particles having sizes ranging from 5×10^{-2} to about 2 μm. In this classification, it is worth mentioning that (i) the nuclei constitute the most important part of the so-called ultrafine particles (which have sizes $<10^{-1}$ μm) and mainly form through condensation of hot vapors during combustion processes and/or nucleation of atmospheric gaseous species to form fresh particles and (ii) the accumulation particles are mainly generated through coagulation of small particles belonging to the nuclei class and condensation of vapors onto existing particles, inducing them to grow appreciably. Consequently, the particle concentration within this size subrange increases, and the accumulation mode becomes gradually more evident, so named because the particle removal mechanisms are poorly efficient in limiting the concentration of such an intermediate-size class of particles. Therefore, such particles have longer residence times than the nuclei, and their number concentration tends to increase through "accumulation" of these particles within such a size class. Among the coarse particles, those having sizes ranging from 10 μm to the previously established upper limit of 100 μm are often called "giant" particles. They mainly contain man-made, sea-salt, and natural dust aerosols, being subject to sufficiently high sedimentation velocities and, hence, very efficiently removed in rather short times.

As shown in Figure 1.2, aerosols with diameters ranging from 10^{-3} to 2×10^{-1} μm can play an important role in cloud and precipitation physics, because water and ice aerosols form cloud droplets and ice crystals with diameters varying mainly from about 2×10^{-2} to more than 10^3 μm. These growth processes lead to the incorporation of particulate matter into cloud droplets during the formation of precipitation and hence contribute to removing aerosols from the atmosphere through the so-called wet deposition processes.

Aerosols also play a fundamental role in enhancing the electricity characteristics of the atmosphere, mainly due to molecular aggregates carrying an electric charge. These particles are called ions and are divided into (i) small ions, with sizes varying from 3×10^{-4} to no more than 10^{-3} μm, and (ii) large ions, with sizes varying from 10^{-3} to about 5×10^{-1} μm. The presence of these ions determine the electrical conductivity of air. Therefore, their increase in concentration can change the magnitude of the fair weather atmospheric electric field. In the lower atmosphere, ions are mainly produced by cosmic rays and, to a lesser extent, by ionization due to crustal radioactive materials within the surface layer of the atmosphere. Ions are removed from the atmosphere through the combination of ions of opposite sign. Small ions are not much larger than molecules and have electrical mobility (defined as their velocity in an

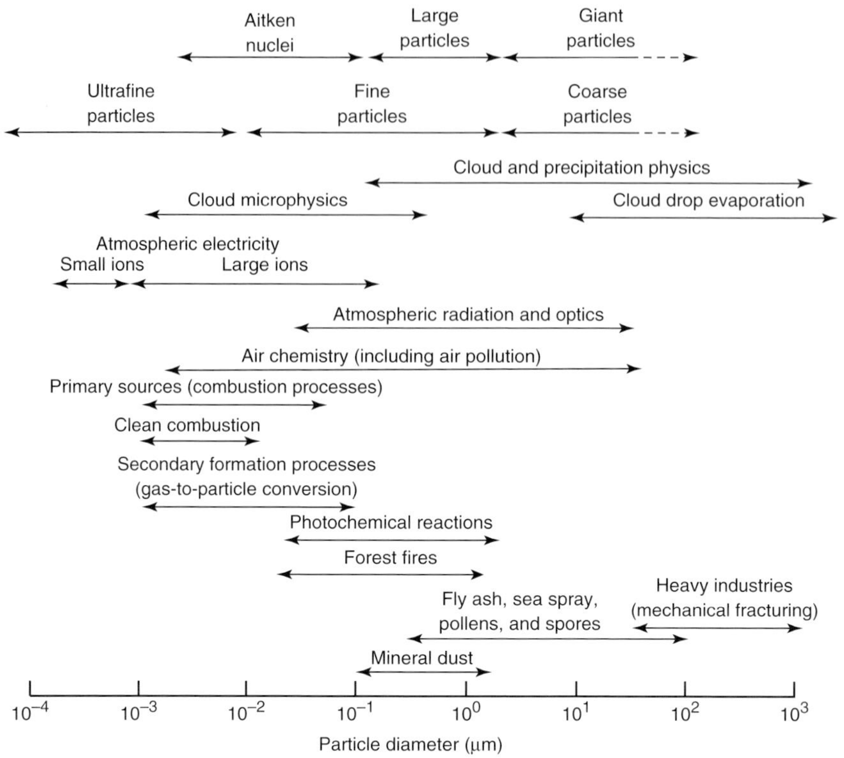

Figure 1.2 Size range of aerosol particles in the atmosphere and their role in atmospheric physics and chemistry.

electric field equal to $1\,V\,m^{-1}$) ranging from about 1 to $2 \times 10^4\,m\,s^{-1}$ at normal temperature and pressure (NTP) conditions. Conversely, the electrical mobility of large ions is very low, generally varying from 3×10^{-8} to $8 \times 10^{-7}\,m\,s^{-1}$. Thus, the concentration of small ions usually varies from about 40 to $1500\,cm^{-3}$ at sea level, and that of large ions from about $200\,cm^{-3}$ in maritime air to more than $8 \times 10^5\,cm^{-3}$ in the most polluted urban areas. Electrical conductivity of the air is proportional to the product of ion mobility by ion concentration, so it is generally produced by small ions in unpolluted areas. Conversely, the concentration of small ions in polluted urban areas tends to decrease as a result of their capture by both large ions and uncharged aerosols, which all exhibit very high concentrations in highly polluted areas. Consequently, the electrical conductivity of air associated with fair weather atmospheric conditions assumes the lowest values for the highest concentrations of large ions. In this view, the decrease of at least 20% in the electrical conductivity of the air, as observed over the Northern Atlantic Ocean during the twentieth century, is currently attributed to a doubling in the concentration of particles with sizes ranging from 0.02 to 0.2 μm, resulting from the increase in background pollution conditions measured in North America and Europe (Wallace and Hobbs, 2006).

Combustion aerosols produced by forest fires have sizes ranging for the major part from 10^{-3} to 10^{-1} μm, while mineral dust particles generated by soil erosion and wind-forced mobilization present sizes mainly varying from 10^{-1} to no more than 5 μm. Figure 1.2 shows that fly ash, sea spray, pollens, and spores cover all together the size range from 5×10^{-1} to 10^2 μm, while industrial man-made aerosols fall within the range from 5×10^1 to more than 10^2 μm. Chemical processes involve particles mainly generated by air pollution processes, with sizes in general varying from 10^{-3} to 10^1 μm. More precisely, the aerosol polydispersions of different origins usually cover the following size intervals:

- From less than 10^{-3} to 5×10^{-2} μm, for aerosols generated by primary combustion processes
- From 10^{-3} to 10^{-2} μm, for aerosols produced by clean combustion
- From 10^{-3} to 10^{-1} μm, for secondary aerosols formed through gas-to-particle (g-to-p) conversion processes
- From 5×10^{-2} to about 2 μm, for aerosols originated by photochemical reactions

Aerosol radiative effects on solar and terrestrial radiation are produced more efficiently by particles with sizes ranging from 5×10^{-2} to 5×10^1 μm, which are able to cause marked scattering and absorption of incoming shortwave radiation at wavelengths varying from 0.4 to 2.2 μm (Charlson *et al.*, 1991). As a result of these interactions, nonabsorbing aerosol layers generally produce significant cooling effects, especially when poorly absorbing particles are suspended above low-reflectance surfaces, such as those of the oceanic regions (Bush and Valero, 2002). By contrast, appreciable warming effects can be induced near the surface by strongly absorbing particle layers suspended above bright surfaces, such as those covered by glaciers and snow fields in Greenland and Antarctica (Chylek and Coakley, 1974). The most intense radiative effects are mainly induced directly through scattering and absorption of incoming solar radiation, but appreciable exchanges of infrared radiation between the surface and the atmosphere can occur in the presence of dense aerosol layers near the surface, usually causing rather marked cooling effects within the ground layer. Besides these direct effects induced by aerosol particles on the shortwave and longwave radiation budget of the surface–atmosphere system (Charlson *et al.*, 1992; Penner, Dickinson, and O'Neill, 1992), aerosols exert an important influence on climate inducing various indirect effects, which can appreciably modify the size-distribution curves of cloud droplets and ice crystals, enhance the liquid water content (LWC) of clouds, favor longer cloud life, and strongly influence the heterogeneous chemistry of the atmosphere (Schwartz *et al.*, 1995; Jensen and Toon, 1997; Lohmann and Lesins, 2002).

The preceding remarks clearly indicate that aerosol is unique in its complexity among the atmospheric constituents and strongly influences the Earth's climate system. Airborne particulate matter is not only generated by particle direct emission mechanisms but can also form from emissions of certain gases that either condense as particles directly or undergo chemical transformations to gaseous species, which subsequently become particles by condensation. The variety of the morphological, optical, and chemical composition properties of airborne aerosols closely depends on the formation processes of particulate matter and their subsequent aging processes occurring in the atmosphere. Considering only the formation processes of primary and secondary aerosols, the present study describes the various physicochemical processes acting as sources of marine, wind-borne (dust), volcanic, biological, combustion and anthropogenic and/or industrial aerosols, and the chemical reactions leading to the formation of secondary aerosols of both natural and anthropogenic origin. This chapter is divided into the following five sections:

Section 1.1, presenting the primary sources of natural aerosols (mineral dust, sea salt, tropospheric volcanic dust, biogenic aerosols, and forest fire and biomass burning smokes generated by natural processes).

Section 1.2, describing the formation of secondary aerosols of natural origin, like sulfate particles in the troposphere from natural SO_2 and sulfur compounds, natural nitrates from tropospheric nitrogen oxides, organic aerosols from biogenic volatile organic compounds (VOCs), and stratospheric sulfates formed from SO_2 of volcanic or marine origin.

Section 1.3, dealing with the primary sources of anthropogenic aerosols (industrial dust, fossil fuel combustion particles, including carbonaceous (soot) substances, and waste and biomass burning particulate matter).

Section 1.4, describing the main chemical processes forming secondary anthropogenic aerosols (mainly sulfates from SO_2, nitrates from NO_x, and organic aerosols).

Section 1.5, providing the most reliable estimates of the global annual emission fluxes of particulate matter associated with the various primary and secondary formation processes. The estimates were in part taken from the literature of the past 20 years and in part calculated by assuming that they agree with the most realistic evaluations of the global atmospheric mass burdens of the various types of natural and anthropogenic particles.

1.2
A General Classification of Aerosol Sources

Airborne aerosol particles are directly generated by surface sources or through a combination of physical and chemical and sometimes biological processes occurring in the atmosphere and in the adjacent reservoirs. Among these processes, three general types of sources are commonly distinguished:

1. "Bulk-to-particle (b-to-p) conversion," leading to the production of (i) mineral dust particles when the Earth's crust provides the solid base material; (ii) maritime (sea-salt) particles when the liquid base material is constituted by the natural marine water reservoirs; and (iii) biological aerosols when

the particulate solid material is furnished by plants (mainly plant debris and pollens) and animals. It is evident that a variety of physical, chemical, and biological precursors are necessary in all these b-to-p conversion processes for the division of the bulk material into particles before its emission into the atmosphere.

2. "G-to-p conversion" in which condensable vapors lead to either the nucleation of new particles or the condensational growth of existing particles. In these cases, both physical and chemical processes are necessary for the accretion of precursors (most frequently molecules), which are by themselves too small to be initially counted as particles.
3. "Combustion" processes which are assumed to constitute a third typological class of particle sources, even though they are, strictly speaking, a combination of the first two types of formation processes. The main difference between the combustion processes and the b-to-p and g-to-p conversions lies in the high temperatures at which the combustion processes take place, which facilitate the formation of such particles presenting shapes and composition features that cannot be achieved solely through the first two source processes mentioned earlier.

Emitted directly as particles (primary aerosol) through b-to-p conversion processes or originating in the atmosphere through g-to-p conversion processes (secondary aerosol), atmospheric aerosols of both natural and anthropogenic origin present composition characteristics closely dependent on their formation processes, with number concentration generally decreasing rapidly as their sizes gradually increase (Junge, 1963).

1.3 Primary Aerosols of Natural Origin

Significant natural surface sources of primary aerosol particles include the emission of sea spray, release of soil and rock debris (mineral dust) and biogenic aerosols, emission of biomass burning smoke, and injection of volcanic debris at tropospheric altitudes by violent eruptions. A negligible contribution to the overall atmospheric aerosol loading is also given by space, in the form of cosmic aerosols, but these fine particles are deemed to exert only a very weak influence on the aerosol characteristics of the high-altitude atmospheric regions, where particle concentration is always very low. Thus, cosmic rays do not substantially alter the air properties of the low stratosphere and the human environment conditions observed in the troposphere. The aforementioned primary mechanisms that generate the different types of particles formed at the terrestrial surface are each characterized by well-diversified morphological features, chemical composition, optical properties, and deposition patterns. They are described in detail in the following subsections.

1.3.1
Sea-Salt Particles

The oceans constitute the main source of sea-salt aerosols. They are estimated to originate a very large amount of particulate matter per year, including the coarse particles (having sizes $a \geq 2.5\,\mu m$), which are generally be transported over short distances because of their rapid removal due to gravitational settling. Abundance of sea-salt particles is second only to mineral dust in contributing to the overall global particulate mass content of the troposphere (Andreae and Rosenfeld, 2008), in which hygroscopic salts, such as NaCl, KCl, $CaSO_4$, and $(NH_4)_2SO_4$ provide about 3.5% of the overall tropospheric water mass. Relatively high percentage mass concentrations of various salt ions are estimated to furnish relative mass percentages equal to 55.0% Cl^-, 30.6% Na^+, 7.7% $SO_4{}^{2-}$, 3.7% Mg^{2+}, 1.2% Ca^{2+}, 1.1% K^+, and 0.7% due to other ions. Pósfai *et al.* (1995) collected a large variety of sea-salt aerosols during the Atlantic Stratocumulus Transition Experiment/Marine Aerosol and Gas Exchange (ASTEX/MAGE) field campaign undertaken in June 1992 over North Atlantic and studied their morphological characteristics using transmission electron microscopy (TEM) techniques. They found that oceanic aerosols may have different composition features in clean, intermediate, and dirty samples. The major species present in clean samples included NaCl molecules with mixed cations (Na^+, Mg^{2+}, K^+, and Ca^{2+}), sulfate ions, and to a lesser extent $NaNO_3$, presenting uniform composition features of sea-salt mode particles. The excess in sulfate and nitrate concentrations is reasonably due to the oxidation of SO_2 in the sea-salt aerosol water and the reactions of NO_x with NaCl. The same compounds were also found to be present in intermediate samples in which compositional groups characterized by low and high losses of Cl^- ions from sea salt were distinguished, with most Cl^- losses compensated by $NaNO_3$ formation. Several compositional groups were found in the dirty samples, including Na_2SO_4 (with minor contents of Mg, K, and Ca), $(NH_4)_2SO_4$, and silicates, in addition to the particle types present in clean and intermediate samples. The distinct compositional groups monitored in the dirty samples revealed that long-range transport of continental air masses has favored the mixing of aerosols, while ozone oxidation and cloud processing could have contributed to the formation of excess sulfate in such samples.

During the Aerosol Characterization Experiment 2 (ACE-2) conducted in summer 1997 over the North Atlantic, Li, Anderson, and Buseck (2003a) found that the major maritime aerosol types include fresh and partly or completely reacted sea salt consisting of NaCl, mixed cations (Na, Mg, K, and Ca), sulfate (Na_2SO_4), and nitrate ($NaNO_3$), confirming the Pósfai *et al.* (1995) evaluations. In addition to the aforementioned marine components, particles of industrial origin, including $(NH_4)_2SO_4$, soot, fly ash, silica, Fe oxide, and $CaSO_4$, were found in the samples, together with minor mineral dust contents. Li, Anderson, and Buseck (2003a) also pointed out that (i) only a sea-salt mass fraction of 0–30% remains unreacted along the Atlantic Ocean coasts of southern Portugal with the anthropogenic aerosol transported from Europe – while the rest was partly reacted or converted to sulfates and nitrates – and (ii) the sea-salt mass fraction sampled at Punta del Hidalgo (Canary Islands) was much less affected by industrial pollution, with only 5% of the particles that were completely reacted, demonstrating that the dilution of pollution varies considerably as a function of the distance of samplings from sources.

More generally, in the most remote areas of our planet, just above the ocean surface, sea salts are generally found to dominate the mass of both submicrometer and supermicrometer particles. Sea-salt aerosols are generated by various physical processes, especially the rising of entrained air bubbles to the sea surface and the subsequent bursting of such bubbles during whitecap formation, through effectiveness features that strongly depend on wind speed (Blanchard and Woodcock, 1957). These aerosol particles are often the dominant cause of solar light scattering and the main contributor of cloud nuclei in the atmosphere above the most remote oceanic regions, provided that wind speed is high enough and the other aerosol sources are weak (O'Dowd *et al.*, 1997). In fact, sea-salt particles can grow considerably as a function of RH due to their usually high hygroscopic properties (Pósfai *et al.*, 1998) and often act as very efficient cloud condensation nuclei (CCNs), creating major cloud nucleating effects (O'Dowd *et al.*, 1997). Therefore, the characterization of the maritime aerosol production processes occurring at the oceanic surface is of great importance to achieve correct evaluations of their indirect chemical effects in the marine atmosphere, especially those induced by particles with diameters $a < 200$ nm (Leck and Bigg, 2005).

The maritime particles are ejected into the air through the bubble bursting mechanism occurring at the ocean surface during whitecap formation (Monahan, Spiel, and Davidson, 1986), as can be seen looking at the schematic sequence presented in Figure 1.3. It shows that bubbles with $a \geq 2$ mm first reach the ocean surface (in parts (1)–(3)), each of them ejecting 100–200 film droplets into the air when the upper portion of the air bubble film bursts (see part (4)). These small "film droplets" subsequently evaporate, leaving behind sea-salt particles with $a \leq 0.3$ μm (as can be seen in part (5)). One to five larger drops break away from each jet that forms when a bubble bursts (as shown in part (6)), and these jet drops are thrown about 15 cm up into the air. Some of these drops subsequently evaporate and leave behind sea-salt particles with $a > 2$ μm, containing not only sea salts but also organic compounds and bacteria that are already present in the surface layer of the ocean. This is due to the fact that the surface microlayer of the ocean is enriched in microorganisms, viruses, and extracellular biogenic material, which can enter the atmosphere through such a bubble bursting mechanism. Consequently, sea-salt particles usually contain about 10% organic matter (OM) (Middlebrook, Murphy, and Thomson, 1998), but currently it is not well known whether these biogenic constituents are internally mixed with sea salt or whether they also form agglomerate pools of externally mixed organic particles (Bigg and Leck, 2008).

As a result of the mechanisms described in Figure 1.3, sea-salt particles cover a wide size range from about 0.05 to 10 μm, presenting in general bimodal size-distribution curves with a first mode centered at $a_c \approx 0.1$ μm and consisting of

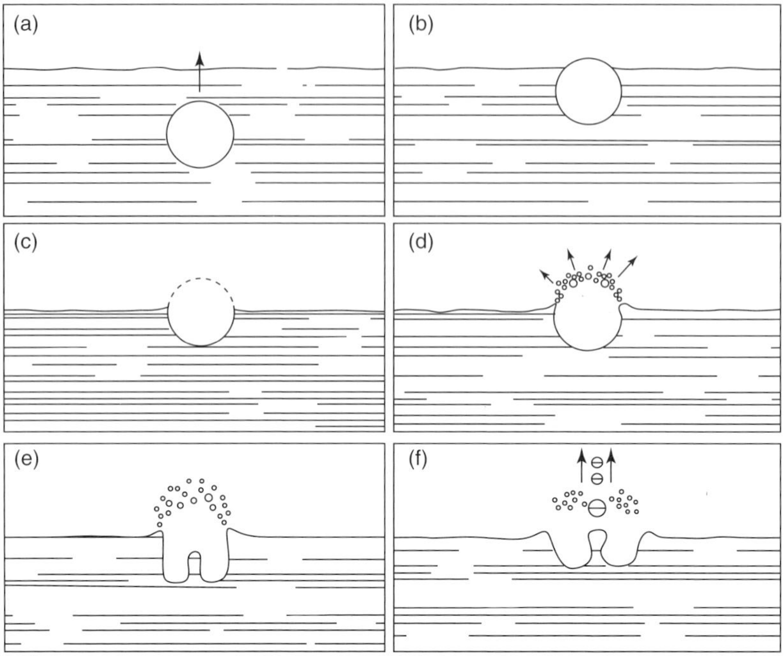

Figure 1.3 Schematic sequence of the various phases through which the film droplets and jet drops are produced when an air bubble bursts at the sea surface: (a) the air bubble is coming to sea surface; (b) the air bubble reaches the surface; (c) the sea water film starts to break; (d) droplets of ~5–30 μm diameter form when the upper portion of the bubble film bursts; (e) film droplets start to evaporate leaving sea-salt particles and other materials in the air; and (f) when the bubble bursts, 1–5 large drops (of sizes equal to about 15% the air bubble diameter) break away from the jet formed during the bubble burst. The time between phases (c) and (f) is ~2 ms.

particles originated from film drops and the second mode centered at $a_c \approx 2.5\,\mu m$ and containing particles forming from the jet drops (Mårtensson *et al.*, 2003). These particles exhibit a wide range of lifetime Δt_L, depending on the large variety of sea-salt particle sizes. In fact, the largest droplets fall rapidly to the ground within their area of origin, while only the smallest aerosol particles formed at the ocean surface play a major role in determining maritime aerosol properties on a large scale. The particles are originated by bubble bursting and have sizes ranging approximately from 0.1 to 1 μm, therefore having residence times in the atmosphere long enough to allow sampling in high concentrations even at continental sites (Sinha *et al.*, 2008).

The average global value of sea-salt particle production in the oceanic regions is estimated to be close to $100\,cm^{-2}\,s^{-1}$, including the large drops formed from windblown spray and foam. As mentioned earlier, the coarse sea-salt particles have rather short lifetimes Δt_L in the air, due to their large sizes. Some scanning electron microscopy (SEM) images of maritime aerosol particles consisting of pure sea-salt (halite) crystals or containing NaCl and other sea-salt cubic crystals alone

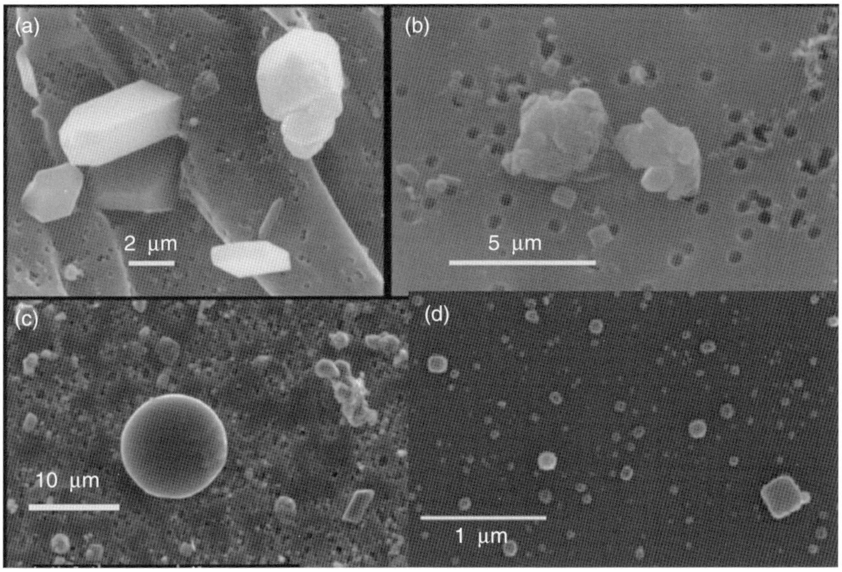

Figure 1.4 SEM images of sea-salt and other maritime particles: (a) particles sampled off the shore of Sardinia (Italy); (b) aggregates of sea-salt (halite) particles and numerous small sea-salt crystals of about 1 μm sizes, with a large-size (∼4 μm) particle on the left, containing sea-salt crystals and mineral dust, and a large-size (∼4 μm) particle on the right, consisting of various sea-salt crystals; (c) sea-salt and anthropogenic particles sampled off the shore of Malta (Mediterranean Sea), including some sea-salt crystals of cubic shape and a larger irregular-shaped particle (on the right) formed by aggregation of marine particles, together with a large-size spherical particle (having a diameter of ∼14 μm) in the middle, presumably due to coal combustion (from the intense ship traffic in the Sicily Channel, near Malta); and (d) several submicron sea-salt cubic crystals sampled near the island of Malta. (Reproduced with permission of Alessandra Bonazza, ISAC-CNR Institute, Bologna, Italy.)

or aggregated with sulfate and nitrate particles are shown in Figure 1.4, as obtained by examining some particulate samples collected at various sites in Sardinia (Italy) and in the Sicily Channel, near the island of Malta. Interesting SEM images of sea-salt cubic crystals of various sizes and aged sea-salt particles have been shown by Sinha *et al.* (2008), obtained for samples collected in the surroundings of Mainz (Germany), that is, in an area very far from the Atlantic Ocean. SEM images of sea-salt particles have also been provided by Li, Anderson, and Buseck (2003a), sampled at Sagres (southern Portugal) and Punta del Hidalgo (Canary Islands, Spain) during the ACE-2, which show evidence of the morphological characteristics of (i) a sea-salt particle consisting of euhedral NaCl with tabular Na_2SO_4, (ii) a particle of euhedral NaCl with mixed-cation sulfate rims, and (iii) a completely converted sea-salt crystal consisting of Na_2SO_4 and $NaNO_3$.

The dry sea-salt particles transported away by winds can very easily form solution droplets in all cases where RH exceeds 65–70%. Ambient gases (e.g., SO_2 and CO_2) are also taken up by these droplets, changing their ionic composition.

For example, the reaction of OH (gaseous phase) with sea-salt particles generates OH^- ions in the liquid droplets and can lead to an increase in the production of SO_4^{2-} ions (in liquid phase) by aqueous-phase reactions and a concomitant reduction in the concentration of Cl^- ions (in liquid phase). Consequently, the ratio of Cl^- to Na^+ ions in sea-salt particles collected in the atmosphere is generally much lower than in seawater. The excess SO_4^{2-} ions (in liquid phase) over those of bulk seawater is referred to as "nonsea salt (nss) sulfate."

The oxidation of Br^- and Cl^- ions in liquid phase, occurring in solutions of sea-salt particles for both these ions, can produce BrO_x and ClO_x species. But catalytic reactions involving BrO_x and ClO_x, similar to those observed also in the stratosphere, turn out to destroy O_3 through a mechanism postulated to explain the depletion of O_3 from about 40 to less than 0.5 ppbv, as occasionally observed over periods of hours to days in the Arctic boundary layer, usually starting at polar sunrise and continuing during the early spring period.

In order to analyze the emissions and atmospheric distribution of the maritime aerosol components and evaluate the mechanisms forming sea-salt aerosols and the atmospheric content of such particles, it is necessary to use size-resolved models. A semiempirical formulation was proposed by Gong *et al.* (1998) to establish a relationship between the size-segregated surface emission rate of sea-salt aerosols and the wind field intensity over the particle diameter range from 0.06 to 16 μm. On the average, the complex mechanism described in Figure 1.3 leads to a number concentration of sea-salt particles with diameters $a > 0.2$ μm varying from 0.1 to 25 cm^{-3} near the surface. As can be seen, the sea-salt particle formation process generates a broad range of particle sizes. Consequently, the particle production rate is strongly size dependent, as demonstrated by Monahan, Spiel, and Davidson (1986), who proposed the following parameterization for a size-dependent production rate $J_{s\text{-}s}$ (expressed in $m^{-2}\ \mu m^{-1}\ s^{-1}$) of sea-salt particles at the surface level:

$$J_{s-s} = 1.373\ \text{WS}^{3.41}(a/2)^{-3}[(1 + 0.057(a/2)^{1.05}]10^{1.19\exp(B)}, \tag{1.1}$$

where a is measured in micrometer at RH = 80%, WS is the wind speed measured in meter per second at the altitude $z = 10$ m above sea level (a.m.s.l.), and $B = \{[0.38 - \ln(d/2)]/0.65\}^2$.

The dependence features of total mass concentration $M_{s\text{-}s}$ of sea-salt particles produced by winds on the surface-level wind speed was parameterized by Jaenicke (1988) in terms of the following two empirical formulas defined over different ranges of height z and wind speed WS:

$$M_{s-s} = 4.26\exp(0.160\text{WS}) \tag{1.2}$$

over the $5 \leq z \leq 15$ m and $1 \leq \text{WS} \leq 21\ m\ s^{-1}$ ranges and

$$M_{s-s} = 2.82\exp(0.152\ \ \text{WS}) \tag{1.3}$$

over the $10 \leq z \leq 600$ m and $5 \leq \text{WS} \leq 35\ m\ s^{-1}$ ranges.

For the present-day climate, estimates of the overall global annual emission flux Φ_e of sea-salt particles from ocean to atmosphere have been made over the past three decades. Flux Φ_e was evaluated to be of 300 Tg per year by SMIC (1971) over the range $a < 20\,\mu m$ and to (i) range from 5×10^2 to 2×10^3 Tg per year by Jaenicke (1988) over the whole size range, (ii) vary from 10^3 to 3×10^3 Tg per year by Erickson and Duce (1988), (iii) be equal to 5.9×10^3 Tg per year (Tegen *et al.*, 1997), and (iv) vary from 10^3 to 10^4 Tg per year by Seinfeld and Pandis (1998, 2006), assuming an average global value of 1.3×10^3 Tg per year for sea-salt coarse particles only. Partial flux values were proposed by the IPCC (2001) equal to 54 Tg per year for fine particles with $a < 1\,\mu m$ and 3.29×10^3 Tg per year for coarse particles over the $1 \le a \le 16\,\mu m$ range, leading to a value of $\Phi_e = 3344$ Tg per year, which was very close to the value of 3.3×10^3 Tg per year assumed by Jaenicke (2005), while higher evaluations were proposed by Gong, Barrie, and Lazare (2002) ($\Phi_e = 1.01 \times 10^4$ Tg per year) and Tsigaridis *et al.* (2006) ($\Phi_e = 7.804 \times 10^3$ Tg per year). Taking these estimates into account, Andreae and Rosenfeld (2008) provided a reasonable range of Φ_e for sea-salt particles from 3×10^3 to 2×10^4 Tg per year, as reported in Table 1.1.

1.3.2
Mineral Dust

Mineral dust originates mainly from desert and semiarid land surfaces as a result of the wind forces that mobilize the soil particles. The main dust mobilization regions include the Sahara desert and other desert regions that constitute the dust belt, a chain of arid regions extending not only over North Africa but also over South Africa, and the Middle East Asia and China (Gobi Desert), besides some wide high-altitude desert regions in South America. In addition, dry lakes and lakebeds and other once-wet areas act also as particularly efficient sources of atmospheric dust (Prospero, 1999). Consequently, dust particles constitute the major component of the atmospheric aerosol content in the subtropical regions of the planet. This is due to the fact that all the aforementioned arid and semiarid regions of the Earth acting as dust sources occupy about one-third of the global land area, desert dust being the dominant particle type even in air masses thousands of kilometers away from the source. Estimates of the annual emission flux Φ_e of mineral dust made over the global scale vary from 10^3 to 5×10^3 Tg per year (Duce, 1995), presenting marked spatial and temporal variations from one region of the Earth to another, because such dust source regions also include semiarid desert fringes and dry land areas, where the vegetation cover was seriously disturbed by human activities.

Particle reactions and internal mixing during transport of mineral dust can substantially change the composition of the original aerosol. For instance, mineral particles become internally mixed with sea-salt components, perhaps through cloud processing (Trochkine *et al.*, 2003). In other cases, Saharan minerals were associated with sulfur and OM transported from both urban and agricultural pollution areas (Falkovich *et al.*, 2001) and with organics contained in combustion smokes and anthropogenic particulate matter (Gao, Anderson, and Hua, 2007),

Table 1.1 Estimates of the annual emission fluxes (measured in teragram per year (being $1\ \mathrm{Tg\,yr^{-1}} = 10^6\ \mathrm{ton\,yr^{-1}}$)) of natural aerosols on a global scale from various sources, as found in the literature over the past 15 years.

Natural particles	Annual global emission flux ($\mathrm{Tg\,yr^{-1}}$)
Sea salt (total, sizes <16 μm)	3344 (IPCC, 2001)
Sea salt (sizes <1 μm)	54 (IPCC, 2001)
Sea salt (1–16 μm size range)	3290 (IPCC, 2001)
Sea salt (overall)	10 100 (Gong, Barrie, and Lazare, 2002), 3300 (Jaenicke, 2005), 7804 (Tsigaridis *et al.*, 2006), ranging from 3000 to 20 000 $\mathrm{Tg\,y^{-1}}$ (Andreae and Rosenfeld, 2008)
Mineral (soil) dust (total, sizes <20 μm)	2150 (IPCC, 2001)
Mineral (soil) dust (sizes <1 μm)	110 (IPCC, 2001)
Mineral (soil) dust total (1–2 μm size range)	290 (IPCC, 2001)
Mineral (soil) dust total (2–20 μm size range)	1750 (IPCC, 2001)
Mineral dust (0.1–10 μm size)	1000–2150 (average = 1490) (Zender, Brian, and Newman, 2003)
Mineral dust (overall)	2000 (Jaenicke, 2005), 1704 (Tsigaridis *et al.*, 2006), ranging from 1000 to 2150 $\mathrm{Tg\,y^{-1}}$ (Andreae and Rosenfeld, 2008)
Volcanic dust (coarse particles only)	30 (Seinfeld and Pandis, 1998)
Sulfates from volcanic SO_2	10 (Hobbs, 2000)
Volcanic sulfates (as NH_4HSO_4)	21 (IPCC, 2001)
Volcanic SO_2	9.2 (Tsigaridis *et al.*, 2006)
Cosmic dust in the upper mesosphere	3×10^{-2} to 1.1×10^{-1} (Plane, 2012)
Cosmic dust in the middle atmosphere	2×10^{-3} to 2×10^{-2} (Plane, 2012), 1.5×10^{-4} to 4×10^{-2} (Gardner *et al.*, 2014)
Biogenic aerosol	1000 (Jaenicke, 2005)
Biogenic sulfate (as NH_4HSO_4)	57 (IPCC, 2001)
Biogenic carbonaceous aerosol (sizes > 1 μm)	56 (IPCC, 2001)
Biogenic primary organic aerosol	15–70 (Andreae and Rosenfeld, 2008)
Biogenic VOC compounds	16 (IPCC, 2001)
Secondary organic aerosol from biogenic VOC	11.2 (Chung and Seinfeld, 2002)
Secondary organic aerosol	2.5–83 (Andreae and Rosenfeld, 2008)
Sulfates (from all the natural primary and secondary sources)	107–374 (Andreae and Rosenfeld, 2008)
Nitrates (overall, from natural primary and secondary sources)	12–27 (Andreae and Rosenfeld, 2008)
Secondary sulfates from DMS	12.4 (Liao *et al.*, 2003), 18.5 (Tsigaridis *et al.*, 2006)
Carbonaceous aerosols from biomass burning (sizes < 2 μm)	54 (IPCC, 2001)
Primary organic aerosol	44.4 (Tsigaridis *et al.*, 2006)
Biomass burning organic	26–70 (Andreae and Rosenfeld, 2008)
Total natural particles over the whole size range	5875 (IPCC, 2001) 4200–22 800 (Andreae and Rosenfeld, 2008), including the United Nations (1979) estimate of volcanic debris

suggesting that mineral dust is a potential cleansing agent for organic pollutants (Pósfai and Buseck, 2010).

Dust deflation occurs in a source region when the frictional wind speed at the surface exceeds a threshold value, which is a function of surface roughness elements, grain size, soil moisture, and surface geological characteristics. The mobilized desert dust can be then transported by winds over long distances, even thousands of kilometers from their source areas. Crusting of soil surfaces and limitation of particle availability can contribute efficiently to reduce the dust release from a source region. In addition, the disturbance of such surfaces by human activities can strongly enhance the dust mobilization potentiality of these land areas. Up to 50% of the current atmospheric dust load has been estimated to originate from disturbed soil surfaces and should therefore be considered anthropogenic in origin (Tegen and Fung, 1995). Dust deflation can change in response to currently occurring natural climate events: for instance, Saharan dust transport to Barbados was estimated to increase during the El Niño years (Prospero and Nees, 1986), and dust export to the Mediterranean and North Atlantic was found to be closely correlated with the North Atlantic Oscillation (Moulin *et al.*, 1997). Further information on dust mobilization, transport, and interannual variability are available in the literature, as given by various models over regional global scales (Marticorena *et al.*, 1997; Tegen and Miller, 1998).

As shown in Table 1.1, the estimate of the average range of annual global emission flux Φ_e made by Zender, Brian, and Newman (2003, 2004) for mineral dust over the whole size range from 0.1 to 10 μm is from 1000 to 2150 Tg per year, giving an average value of 1490 Tg per year. This estimate substantially agrees with the maximum estimate of 1800 Tg per year made by Jaenicke (1988) and the soil dust range of 1000–5000 Tg per year given by Duce (1995) and that of 1000–3000 Tg per year proposed by Seinfeld and Pandis (1998, 2006), with an average value of 1500 Tg per year defined for the coarse particles of soil dust. Values of annual global emission flux Φ_e of mineral dust were evaluated by the IPCC (2001) to be equal to 2150 Tg per year over the whole range of particle diameter range $a < 20\,\mu m$, of which 110 Tg per year was attributed to the range $a < 1\,\mu m$, 290 Tg per year to the range $1 \leq a \leq 2\,\mu m$, and 1750 Tg per year to the range $2 \leq a \leq 20\,\mu m$. More recent evaluations of Φ_e for mineral dust were provided over the last decade, yielding values of 2000 Tg per year (Jaenicke, 2005) and 1704 Tg per year (Tsigaridis *et al.*, 2006), which are very close to the upper limit of 2150 Tg per year determined by Zender, Miller, and Tegen (2004) and subsequently confirmed by Andreae and Rosenfeld (2008).

Large uncertainties exist in explaining the mineral dust emission processes, which arise not only from the complexity of the processes raising dust into the atmosphere under wind forcing but also on the nature of the arid and semiarid surfaces and the atmospheric turbulence fields capable of involving the mobilized dust and transporting it over long distances. Soil particles are mobilized by wind forcing. The threshold value of the frictional wind speed at the ground is estimated to be equal to $\sim 0.2\,m\,s^{-1}$ for particles with equivalent diameter a ranging from 50 to 200 μm and for soils containing 50% clay or tilled soils, because smaller particles adhere better to the surface and cannot be mobilized (Mullins *et al.*, 1992). To reach a frictional speed of 0.2 m s^{-1}, the wind speed WS must be higher than several meters per second at the height of a few meters above the ground. A major source of relatively smaller particles (with a ranging from ~10 to 100 μm) is saltation, in which the larger grains become airborne, fly a few meters, and then land on the ground, creating a burst of smaller dust particles, as shown in the schematic representation of Figure 1.5.

The formation of crustal particles can be considered an important b-to-p conversion process that generates sand particles from crustal material through a two-stage sequence, the first consisting of the physicochemical erosion processes dividing the bulk material into small grains and the second causing the wind-forced ejection of such particles into the atmosphere:

1. Two physical models represent the first stage. The first assumes slow processes, which divide each grain randomly into two different parts, leading to a size distribution that tends to assume a lognormal analytical form after many repetitions. The second model assumes faster processes, which divide each grain randomly into a certain number of parts, giving an overall exponential mass size-distribution curve, resulting from the envelope of different unimodal size distributions, as shown in the example of Figure 1.6 suggested by Junge (1979). It is worth noting that the first model is a special case of the second one and both these modeled processes are expected to act simultaneously on the natural sands.
2. The second stage of the desert dust formation process consists of particle ejection. It is much harder to model this physical mechanism. For grains

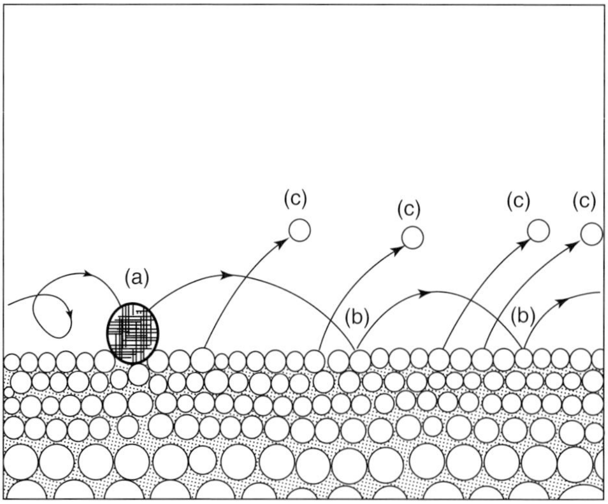

Figure 1.5 Schematic representation of the saltation mechanism through which sand particles are mobilized by wind: a large particle (a) is lifted by wind and then lands on the ground (b), creating a burst of smaller dust particles (c).

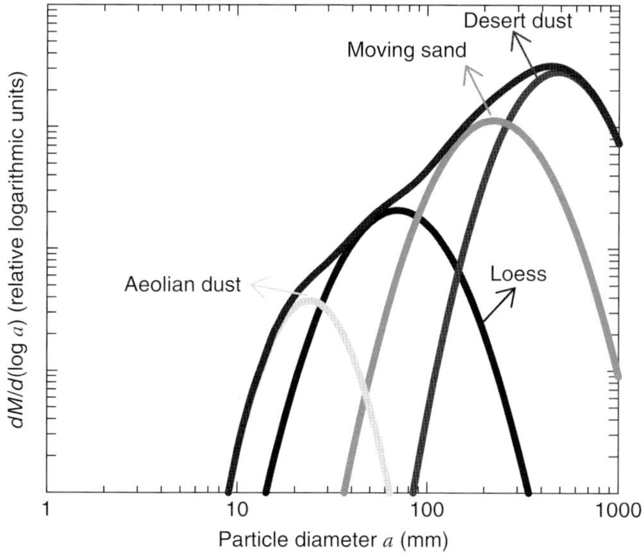

Figure 1.6 Schematic picture of the mineral dust particle mass size distributions generated in different ground sands, characterized by multimodal features associated with the various mobilization processes. (Adapted from a graph of Junge (1979).)

with $a > 100\,\mu m$, critical flow velocities for the ejection have been proposed in the literature, establishing that all grains become airborne for wind speed $WS > 1\,m\,s^{-1}$ at the ground. Consequently, airborne particle size distributions are independent of grain sizes, and, hence, all grain sizes are involved in such mobilization processes. On this matter, Jaenicke (1988) evaluated that the airborne mass concentration M_c of crustal material near the surface varies as a function of wind speed WS according to the following mean relationship:

$$M_c = 52.77 \exp(0.30\ \ WS), \tag{1.4}$$

assumed to be valid over the $0.5\,m\,s^{-1} < WS < 18\,m\,s^{-1}$ range.

Closely depending on the geological characteristics of the mobilization areas, the morphological features of mineral dust particles can vary widely, presenting features that change appreciably with the chemical composition and mineralogical characteristics of the bulk material forming such crustal aerosols. The main species found in soil dust are quartz, clays, calcite, gypsum, and iron oxides, with optical properties depending on the relative abundance of the various minerals. Some SEM images of wind-forced dust particles and soil particles consisting of smectite, illite, and gypsum and sampled at different localities in Italy are shown in Figure 1.7. More details on the composition of such soil dust samples collected in the Mediterranean area are presented by Molinaroli *et al.* (1999). Numerous SEM and TEM images of mineral dust particles are available in the literature, such as (i) those sampled by Sinha *et al.* (2008) in central Europe (at

Figure 1.7 Scanning electron microscopy (SEM) images of mineral (aeolian) dust particles consisting of (a) quartz, (b) dolomite, (c) kaolinite, (d) palygorskite of Saharan origin, (e) smectite, (f) illite, and (g), and (h) gypsum (calcium sulfate dihydrate) in both cases. (Reproduced with the permission from Emanuela Molinaroli, Ca' Foscari University, Dept. of Environmental Sciences, Venice, Italy.)

Mainz, Germany), consisting of silicates, primary and secondary gypsum, calcium carbonate, and iron oxide; (ii) those sampled by Li, Anderson, and Buseck (2003a) at Sagres (southern Portugal), during the June–July 1997 ACE-2 campaign conducted over the North Atlantic Ocean, regarding an aluminosilicate sphere mixed with Na_2SO_4 crystals, an aggregate of euhedral magnetite crystals, a quartz crystal, a smectite particle, and a kaolinite particle coated with ammonium sulfate; (iii) some silicate mixed particles with small NaCl and Na_2SO_4 aggregated crystals, sampled by Li, Anderson, and Buseck (2003a) at Punta del Hidalgo (Canary Islands, Spain) during the ACE-2; and (iv) and those sampled by Li *et al.* (2003b) in mineral dust collected for hazy conditions in South Africa, showing muscovite particles with aggregated small tar balls and a quartz particle aggregated with smectite and organic particles.

Most dust particles are not soluble in water. Therefore, major changes in the shape and morphological characteristics of such aerosols due to high RH conditions of air are rarely observed. However, the mineral core of such airborne particles can be covered by a water or ice shell when transported at high altitudes within highly moist air masses. These changes in the structural features of dust particles can lead to marked modifications in the radiative properties of dust particles, with lower values of the real part of the refractive index than those usually found for dust particles suspended in dry air.

Significant variations in the optical properties of mineral dust can also be caused by changes in the chemical composition of these particles often occurring when mineral dust is mixed with aerosols having different composition features during their transport over long distances. Examples of these variations have been, for instance, documented in cases where (i) calcite and halite particles relatively close to the source partially react to sulfate during dust storms (Okada and Kai, 2004); (ii) mineral dust types containing a significant fraction of carbonate minerals in certain desert areas act as effective sinks for nitric acid to form nitrates (Krueger *et al.*, 2004); (iii) carbonate minerals convert to sulfate particles or mixed (sulfate and nitrate) particles (Sobanska *et al.*, 2003) or nitrates only (Matsuki *et al.*, 2005); (iv) dust particles mixed internally with sea-salt aerosols, when transported within clouds over oceans (Trochkine *et al.*, 2003); (v) desert dust is associated with sulfur compounds and OM from surrounding urban and agricultural polluted areas (Falkovich *et al.*, 2001); (vi) desert dust particles mixed with organics from combustion products over oceans, acting as cleansing agent for organic pollutants (Gao, Anderson, and Hua, 2007); and (vii) mineral dust contributing to stimulate phytoplankton growth (Cwiertny *et al.*, 2008), playing an important role in the global biogeochemical cycle of iron, which is a significant limiting nutrient in many oceanic ecosystems (Jickells *et al.*, 2005).

The mineral dust particles mobilized in desert regions can be transported over long distances, as frequently observed over the Mediterranean by means of satellite-based observations made by TOMS, SEVIRI, MODIS, MISR, and other sensors, which gather information on such transport episodes from North Africa toward Europe (Karam *et al.*, 2010). More intense desert dust transport episodes, during which dust particles can be transported thousands of kilometers away from their sources, are frequently observed in various areas of the planet: (i) from Sahara-Sahel toward Atlantic Ocean and the Caribbean region (Prospero, 1999), (ii) from mid-eastern Asia toward the Persian Gulf and Indian Ocean, and (iii) from the Gobi Desert and Central Asia toward the Northern Pacific Ocean, as shown by the POLDER/ADEOS-1 observations performed by Deuzé *et al.* (1999). It is also important to mention that plumes of Asian dust are often detected over Alaska by means of satellite observations and in the Arctic region by means of ground-based sun-photometer measurements performed mainly in spring (Stone *et al.*, 2007). Therefore, mineral dust is expected to cause important direct and indirect effects on the terrestrial climate system.

The transport of ineral dust particles is in general accompanied by sharp and marked changes in the multimodal size-distribution curves of columnar aerosol polydispersions, especially within the size range typical of coarse particles, associated with significant changes in the optical properties of these aerosols. Gravitational settling and wet deposition act as the major removal processes of mineral dust from the atmosphere, limiting the atmospheric lifetimes Δt_L of such particles to no more than 2 weeks on average. It is obvious that the lifetime Δt_L of mineral dust particles depends closely on particle size, since large particles are quickly removed from the atmosphere by gravitational settling, while submicron particles can remain suspended in air over periods of several weeks.

1.3.3 Biogenic Aerosols

Primary biogenic aerosols are solid and liquid particles released into the atmosphere from plants and animals. They consist of plant debris (cuticular waxes, leaf fragments), microbial particles (living and dead viruses, bacterial cells, fungi, spores, pollens, algae, seeds, etc.), insects, humic matter, and other biogenic debris, such as marine colloids and pieces of animal skins. Due to their highly different origins, biological aerosol particles exhibit a large variety of shapes and cover a wide size range from less than 0.1 μm to at least 250 μm. For instance, fungal spores have sizes mainly ranging from 1 to 30 μm, while (i) most pollens present sizes varying usually from 20 to 60 μm, and (ii) bacteria, algae, protozoa, fungi, and viruses have sizes generally smaller than a few micrometers, those of viruses varying in general from 0.1 to 0.3 μm and those of bacteria from 0.2 μm to no more than 8.0 μm.

Figure 1.8 shows some SEM images of aerodiffuse biological particles sampled at rural sites in the Po Valley (Italy) by the ISAC-CNR. The images show some pollen samples, with nearly spherical shapes and sizes varying mainly from 15 to 60 μm, and a spherical-shaped fungus spore, with a diameter of ∼15 μm. TEM images of biogenic particles sampled in Southern Africa in the Madikwe Game Reserve were shown by Pósfai *et al.* (2003a), presenting a group of brochosomes (C-, O-, and Si-bearing particles secreted by leaf-hopping insects) showing spherical shapes with a equal to ∼250 nm. Other interesting TEM images of biogenic particles have been presented by Pósfai and Buseck (2010), including an atypical

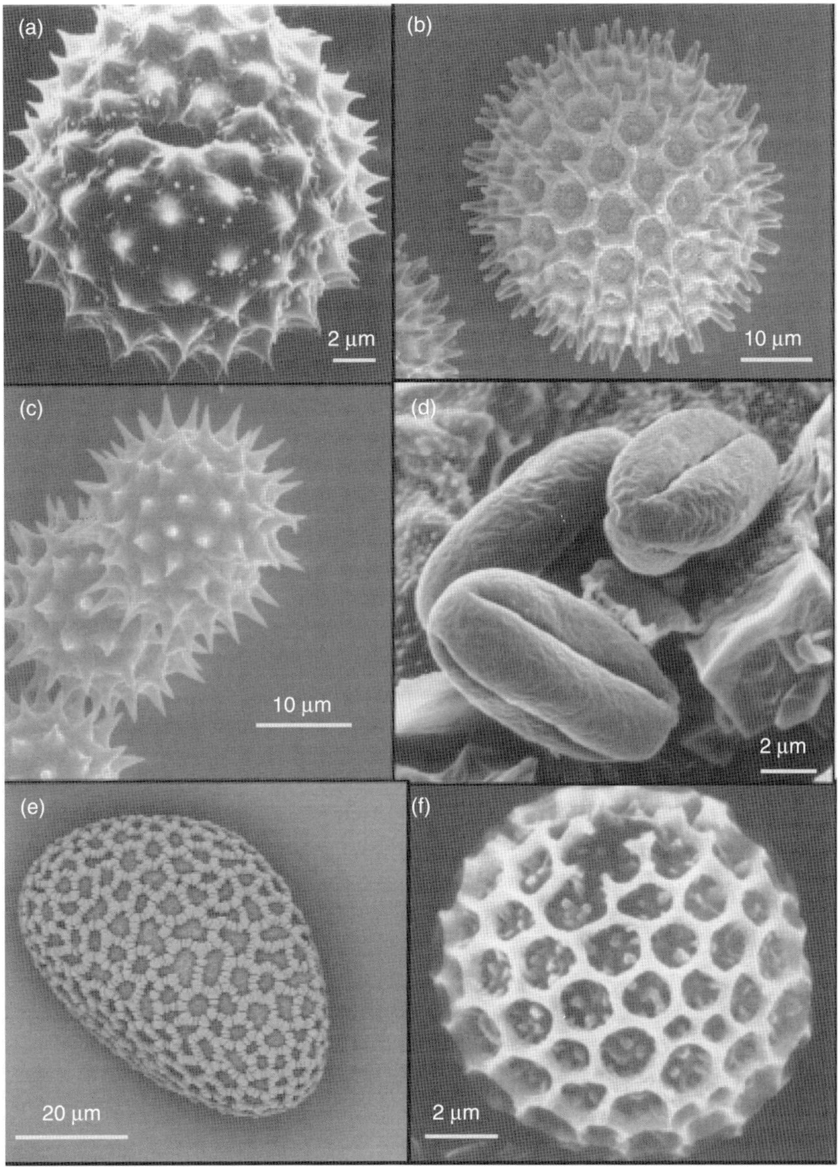

Figure 1.8 SEM images of aerodiffuse biological particles sampled at rural sites of the Po Valley (northern Italy): (a) pollen of *Ambrosia* sp. (Po Valley), (b) pollen of Convolvulaceae sp. (Po Valley), (c) pollen of Asteraceae (Helian.) (Po Valley), (d) pollen of *Castanea sativa* sp. (Po Valley), (e) pollen of Liliaceae sp. (Po Valley), and (f) fungus spore of *Ustilago* sp. (Po Valley). (Reproduced with the permission from Paolo Mandrioli, ISAC-CNR Institute, Bologna, Italy.).

oceanic microorganism sampled at Cape Grim with sizes of ~1 μm and biogenic debris consisting of several plant fragments collected at a rural site in Hungary, with sizes varying from 2 to 10 μm.

As a result of the large variability of sources and typologies, the number density concentrations of biogenic aerosol particles are close to 1% of the total aerosol concentration in remote oceanic regions and equal to around 2–3% in continental environments (Pósfai *et al.*, 2003b; Winiwarter *et al.*, 2009), but can reach percentage values of ~25% in continental aerosol (Matthias-Maser and Jaenicke, 1994) and 35% in Amazonia (Elbert *et al.*, 2007), presenting wide variations related to the source and particle type. These remarks show that the relative contribution of biological aerosols can be very high in densely vegetated regions, particularly in the moist tropical regions (Simoneit, Cardoso, and Robinson, 1990). Conversely, biological aerosols suspended in an urban midlatitude temperate environment can contribute to the total aerosol mass concentration by percentages of 10–30% over the submicron and supermicron size ranges (Matthias-Maser and Jaenicke, 1995). The number density concentration of biological aerosols can vary appreciably with season, location, and altitude. Low number concentrations are usually measured at midlatitudes in mountain locations during winter, and very low concentrations are recorded usually at polar latitudes in all seasons. Typical concentrations of biological particles in the midlatitude regions were found to be (i) higher than 200 m^{-3} for grassy pollens, (ii) ranging from about 100 to 400 m^{-3} for fungal spores (in water), (iii) equal to ~0.5 m^{-3} for bacteria over remote oceans (Pósfai *et al.*, 2003b), (iv) varying from 80 to 800 m^{-3} for bacteria in the urban New York City area, and (v) close to about 10^4 m^{-3} for bacteria over sewage treatment plants. Jaenicke (2005) estimated that the average annual global emission flux Φ_e from biospheric sources is approximately equal to 1000 Tg per year for atmospheric primary particles, compared to the estimates of 2000 Tg per year for mineral dust and 3300 Tg per year for sea-salt particles (see Table 1.1).

Biogenic aerosols can sometimes occupy up to 30% of the total atmospheric aerosol volume at a given location, especially in the remote tropical continental areas covered by pluvial forests, and exhibit at least three times lower concentrations in the remote oceanic regions (Elbert *et al.*, 2007). Nevertheless, inland biogenic aerosols can travel very long distances owing to their low density, often being collected thousand kilometers from their origin area. The occurrence of such long-range transport episodes is commonly used for the identification of the origin of a certain air mass at a given location (Mandrioli *et al.*, 1980). The biogenic aerosols are in general widespread and occupy the whole troposphere, presenting highly variable number concentrations with altitude. For instance, spores of a number of molds were identified at 11 km altitude in the atmosphere (Cadle, 1966), suggesting that the primary biological particles can efficiently participate in the cloud and ice nucleation processes. Many problems in atmospheric science are related to the primary biological particles, including the spread of pathogens and the roles of microorganisms in cloud and ice nucleation (Morris *et al.*, 2011). Biogenic aerosols can easily penetrate the respiratory systems of animals and humans, while pollens can cause serious allergologic pathologies affecting various biological functions of humans. Viruses and bacteria can be attached to other particles (e.g., dust grains, pollens, and spores) and transported over large distances by them and by the cloud particles that were grown on biogenic nuclei. Some bacteria are also suspected to be active in cloud droplets, governing the biological activities occurring inside these droplets (Sattler, Puxbaum, and Psenner, 2001).

In evaluating the optical effects induced by biological aerosols, it is important to take into account that most of these particles are not spherical (Pósfai *et al.*, 2003a; Niemi *et al.*, 2006; Pósfai and Buseck, 2010). In particular, many bacteria are rod shaped and are not characterized by just one size. Bacteria also have an internal structure and cannot be considered homogeneous objects in light scattering studies. More generally, bioaerosols often have quite complex shapes and variable internal structures. Therefore, it is very difficult to simulate their optical characteristics, even using advanced computers. In addition, the refractive index of these particles needed for the theoretical modeling of their optical effects is poorly known. Due to the fact that only left-handed amino acids and right-handed sugars exist in nature, it is clear that the biological aerosols are mostly chiral. This implies that the refractive index depends on the sense of rotation of incident electromagnetic circularly polarized waves (Kokhanovsky, 2003). The presence of humic-like substances makes these aerosol particles light absorbing, especially in the ultraviolet (UV)-B wavelength range (Havers *et al.*, 1998). It is currently known that many biological particles not only absorb and scatter light but can also fluoresce when zapped with a beam of UV light. Due to their relatively low background number concentrations, however, biological particles are deemed to produce in general rather low direct radiative effects on the Earth's climatic system. As mentioned earlier, there is evidence that primary biogenic particles can efficiently act as cloud droplets and ice nuclei (Schnell and Vali, 1976). Thus, biogenic particles are expected to play a not negligible role in inducing indirect climatic effects since they favor the formation of cloud particles. Quantitative estimates of these effects currently deserve closer investigation.

1.3.4 Forest Fire Smoke

Biomass burning smoke is emitted into the atmosphere by intense fires of forests, savannah grass, and other types of vegetation, occurring most frequently in the tropical regions during drought periods. Emissions from burning vegetation include elemental carbon (EC) and organic carbon (OC) as well as other particulate substances, together with gases such as CO_2, CO, NO_x, CH_4, and nonmethane hydrocarbons (NMHCs). The term EC refers to the most refractory part of the carbonaceous aerosol particles, which oxidize above a certain threshold in combustion experiments. However, it is often called free carbon" or "graphitic" and most frequently "black carbon" (BC), being a combustion product capable of strongly absorbing the incoming solar radiation over the whole visible wavelength range. The acronyms BC and EC are inappropriate and it is better to use the term "soot" to indicate the primary particles produced by combustion processes (Andreae and Gelencsér, 2006; Bond and Bergstrom, 2006). These particles have a distinctive structure consisting of graphene-like layers that are wrapped into (i) spherules having diameters equal to tens of nanometer and resembling nano-onions and (ii) aggregates of spherules into branching or compact clusters. Numerous examples of such complex layered structures and aggregates of soot spherules are available in the literature illustrated by means of both SEM and TEM images made for particle samples collected during different fire episodes by (a) Li *et al.* (2003b) during the Madikwe Game Reserve fire (South Africa) on August 20, 2000, showing SEM images of a tar ball (arrowed) in young smoke and soot aggregates in young smoke; (b) Li *et al.* (2003b) in the Timbavati (South Africa) region during the fire of September 7, 2000, showing TEM images of (i) tar balls from aged smoke, collected downwind from the Timbavati fire, together with potassium-salt particles having organic coatings and most of the KCl crystals in the young smoke that were transformed into potassium sulfate and nitrate with the aging of the smoke, and (ii) potassium salts that have formed inclusions within organic particles; and (c) Pósfai *et al.* (2003a) in the Kruger National Park (South Africa) during the fire of August 17, 2000, observed in the presence of both smoke and haze particles, showing TEM images of (i) a typical portion of a sample of young smoke from a smoldering and flaming fire, in which most particles are carbonaceous (organic) with inorganic K-sulfate inclusions, (ii) individual spherules consisting of graphitic layers, and (iii) a few nanometer-sized graphite nucleus within a soot globule.

Three distinct types of carbonaceous particles were mostly found by Pósfai *et al.* (2003a) in the smoke samples collected at the Kruger National Park in August 2000: (i) particles with inorganic (K-salt) inclusions; (ii) the so-called tar balls, which constitute a new and distinctive type of particles produced by biomass burning emissions (Pósfai *et al.*, 2003b, 2004); and (iii) soot particles. It is worth mentioning that the large variety of soot particles, carbonaceous particulate aggregates, graphitic nanoparticles, and the so-called tar balls are all generated by forest and savannah fires, which can be clearly observed using high-resolution transmission electron microscopy (HRTEM) techniques (Pósfai *et al.*, 1999), since this instrument can provide useful diagnoses for soot identification in complex internally mixed particles. Pósfai *et al.* (2003a) showed that the relative number density concentrations of organic particles assumed the highest values in young smoke, whereas tar balls were dominant in slight smoke from smoldering fire, with $\Delta t_L = {\sim}1$ h. Flaming fires were evaluated to emit relatively more soot particles than smoldering fires, but soot was a minor constituent of all such studied plumes. Further aging effects were observed to cause the accumulation of sulfate on organic and soot particles, as indicated by the large number of internally mixed organic/sulfate particles recorded in the haze cases. Externally mixed ammonium sulfate particles were found to dominate in the boundary layer hazes, whereas organic/sulfate particles were the most abundant type in the upper hazes. The measurements also showed that the elevated haze layers were affected more strongly by biomass smoke and those in the boundary layer less apparently. Particles of all the aforementioned kinds are usually found in high concentrations for stable meteorological conditions persisting in the atmospheric ground layer over the regions involved by forest and savannah fires.

The analysis of smoke particles sampled by Li *et al.* (2003b) during the SAFARI 2000 experiment conducted in South Africa provided evidence that potassium salts and organic particles were the predominant species, whereas more K_2SO_4 and KNO_3 particles were found in aged smoke. This difference indicates that with the aging of the smoke, KCl particles from the fires were converted to K_2SO_4 and KNO_3 through reactions with sulfur- and nitrogen-bearing species from biomass burning as well as other sources. More soot was found in smoke from flaming grass fires than bush and wood fires, probably due to the prevailing flaming combustion in grass fires. The high abundance of organic particles and soluble salts was deemed to affect the hygroscopic properties of biomass burning aerosol, thereby also influencing the potentiality of such particles to act as CCNs.

The effect of gaseous emissions from biomass burning on atmospheric chemistry and climate is critically important. Biomass burning is one of the major sources of atmospheric aerosols, mainly consisting of small smoke particles containing primarily organic compounds and EC, mainly having submicron sizes. In fact, combustion processes generate in general (i) a very high number concentration of small particles, with $a < 0.2\,\mu m$; (ii) a high number of accumulation mode particles with sizes ranging from 0.2 to 1 μm; (iii) a relatively low number of coarse particles, with sizes varying from 1 to 2 μm; and (iv) only a few coarse-mode particles, with sizes greater than 2 μm. This implies that a large fraction of smoke particles can easily penetrate the respiratory system of humans causing various health problems. Most of these submicron aerosols are produced by large-scale savannah burning and forest fires occurring during the dry season in Africa, the Amazon Basin, and southwestern Asia, which can also induce sharp increases in the surface-level ozone concentration on a regional scale, to such an extent as to approach the O_3 concentration level measured in urban industrialized regions and observed to vary often from 60 to 100 ppbv. Such forest fire particles are rapidly uplifted to high altitudes and transported over long distances, with major effects on air quality, biogeochemical distribution of nutrients, and climate. The climatic effects are mainly associated with an appreciable decrease in surface reflectance and a decrease in the single scattering albedo of columnar aerosols, which both appreciably contribute to cause local warming effects in the atmosphere.

Vegetation is predominantly composed of cellulose, hemicellulose, and lignin and constitutes the major fuel consumed in natural biomass burning. Therefore, the combustion of biomass organic components involves a complex sequence of physical transformations and chemical reactions, including pyrolysis, depolymerization, water elimination, fragmentation, oxidation, char formation, and volatilization. During the earlier stages of combustion, known as flaming combustion, hydrocarbons volatilized from the decomposing biomass are rapidly oxidized in a flame upon mixing with air. When the flux of these combustible volatile compounds falls below a critical level, flaming expires and smoldering or glowing combustion starts to occur. Taking place at such lower temperatures, these combustion processes involve a gradual gas–solid phase reaction between oxygen and the remaining reactive char, thus emitting large amounts of incompletely oxidized pyrolysis products. Many of these compounds have sufficiently low vapor pressures to be found in the particulate phase when released into the ambient atmosphere. Therefore, OM often constitutes the major fraction of smoke aerosols, sometimes accounting for more than 90% of the total aerosol mass (Graham *et al.*, 2002), while the fraction of water-soluble organic compounds consists of a complete mixture of oxygenated substances derived primarily from biomass burning.

Smoke aerosols may cause a number of spectacular optical atmospheric effects, such as the blue moon and Sun. As mentioned earlier, smoke aerosols contain a large fraction of soot substances. These particles consist in part of aggregates and in part of small aerosols, of which (i) the aggregates generally have sizes greater than 1 μm, and are formed by coalescence of ultimate or primary small particles having sizes mainly varying from 50 to 100 nm, and (ii) small aerosols exhibit sizes that are mainly smaller than 0.2 μm. As mentioned previously, soot particles are often assembled in chain-like structures. This makes it impossible to use spherical particle models to estimate their radiative properties, because the complex refractive index of soot varies as a function of its structure and production chain, presenting relatively high values of its real part at visible and near-infrared wavelengths varying mainly from 1.74 to 1.83 and values of its imaginary part usually close to 0.4–0.5, therefore considerably higher than those typical of maritime, mineral dust, and secondary water-soluble aerosols.

Several million grams of particles can be released by the burning of one tropical forest hectare (10^4 m^2). On a global scale, SMIC (1971) estimated that forest fires and slash-burning debris contributed to give a global annual average production flux of aerosol particle mass varying from 3 to 150 Tg per year, with the fraction of forest fire smoke particles with $a < 20$ μm that provides a value of Φ_e not exceeding 50 Tg per year (United Nations, 1979). Accordingly, flux Φ_e including the whole smoke atmospheric load from forest fires was estimated to range from 5 to 150 Tg per year by Jaenicke (1988). On the basis of most recent evaluations, it was estimated by the IPCC (2001) that about 54 Tg per year of carbonaceous aerosols with $a < 2$ μm is released into the atmosphere each year by natural biomass burning, of which 28 Tg per year in the northern hemisphere and 26 Tg per year in the southern hemisphere. This estimate of emission flux Φ_e is relatively small in comparison to that of overall aerosols formed from all the natural and anthropogenic sources. A fraction of Φ_e equal to about 10% and, hence, to ~6 Tg per year, is attributed to carbonaceous particles, consisting mainly of EC, which does not usually volatilize below the temperature of 550 °C. As reported in Table 1.1, Tsigaridis *et al.* (2006) calculated that overall flux Φ_e of primary organic aerosol from forest fires is equal to 44.4 Tg per year, while the flux Φ_e for biomass burning organic particles was estimated by Andreae and Rosenfeld (2008) to range from 26 to 70 Tg per year. This evaluation agrees very well with that of OM different from EC, equal to about 48 Tg per year on a global scale.

Aerosols emitted from forest fires consist mainly of submicron particles giving form to an accumulation particle mode, usually peaked at $a \approx 0.1$ µm, and containing relatively high concentrations of EC that may have major light absorption effects. Smoke has important local effects on the environment: (i) provoking serious diseases in humans, animals, and plants; (ii) affecting the Earth albedo through deposition of soot particles at the surface, at the same time causing a marked lowering of the overall single scattering albedo of airborne aerosols, which may consequently induce pronounced warming effects in the atmosphere; (iii) causing the decrease of local and regional visibility because of their light scattering and absorption properties; (iv) acting as efficient CCNs, which can considerably modify the cloud droplet size distribution and the cloud reflectance characteristics; and (v) efficiently changing the cloud microphysical processes, leading to major variations in precipitation patterns. On a global scale, the combined direct and indirect radiative effects induced by these aerosols are estimated to be responsible for a net radiative forcing comparable to that induced by sulfate aerosols. In particular, boreal forest fires are estimated to contribute as much as 20–50% to the observed BC concentrations in the Arctic during summer. When deposited at the ground in these polar regions covered by sea ice, snow fields, and glaciers, smoke particles originated from summer forest fires in North America and Siberia are often transported toward Northern Canada, Greenland, the Svalbard islands, and other Arctic land regions, causing considerable changes in the surface albedo (Stohl *et al.*, 2006). For instance, over 5 million hectares of boreal forests burned in Alaska and Canada during summer 2004, and an enormous transport of smoke particles toward the Arctic followed. During this episode, values of aerosol optical thickness higher than 0.2–0.3 at the visible wavelengths were measured at Summit (Greenland) and Ny-Ålesund (Svalbard) in June and July 2004 (Tomasi *et al.*, 2007), showing some very high peaks exceeding the unit value at Barrow (Alaska) in June 2004 (Stone *et al.*, 2008). The combined mechanism of transport of the forest fire aerosols toward higher latitudes (Paris *et al.*, 2009) and the subsequent deposition on high-albedo surfaces is estimated to cause a reduction of the snow-covered areas in the Arctic and a not negligible decrease in the surface albedo of such polar regions, leading to significant variations in the radiation budget of the surface–atmosphere system (Flanner *et al.*, 2011).

1.3.5
Volcanic Dust in the Troposphere

Volcanoes emit both particles and gases into the troposphere during their moderate eruptions and inject considerable amounts of gases into the lower stratosphere during the most violent episodes, at altitudes most frequently between 15 and 25 km. In these often dramatic circumstances, volcanic eruptions can play an important role in stratospheric chemistry, causing chemical processes that are described in detail in Section 1.4.4 dealing with the formation of natural marine secondary and volcanic sulfate aerosols in the stratosphere. Two components of volcanic emissions contribute mainly to the formation of volcanic aerosol particles at tropospheric altitudes, primarily consisting of dust and sulfates formed from gaseous sulfur. Estimates of the global annual flux Φ_e of volcanic dust injected into the troposphere were made in the 1990s by Jones, Charlson, and Rodhe (1994), who proposed values ranging from 4 to 10^4 Tg per year (evaluated for extremely explosive eruption episodes). An average estimate of the annual volcanic aerosol emission flux Φ_e equal to ~33 Tg per year was proposed by Andreae (1995), which agrees closely with other evaluations of Φ_e for volcanic debris, giving a range from 25 to 150 Tg per year (SMIC, 1971; United Nations, 1979) and a range from 4 to 90 Tg per year (Jaenicke, 1988), for which an average value of 30 Tg per year was determined by Kiehl and Rodhe (1995). On the basis of these evaluations, all reported in Table 1.1, we assumed that the tropospheric volcanic dust emission rate Φ_e ranged from 4 to 150 Tg per year over the past 50 years, with an average value of 30 Tg per year.

Volcanic sources contribute significantly to the sulfate aerosol burden in the upper troposphere, where they might take part to the formation of ice particles, therefore representing an important potential for inducing strong indirect radiative effects (Graf, Feichter, and Langmann, 1997). Support for this contention lies in the evidence of cirrus cloud formation from volcanic aerosols (Sassen, 1992) and arises from the results achieved by Song *et al.* (1996), indicating that the interannual variability of high-altitude clouds is currently associated with explosive eruption episodes. Graf, Feichter, and Langmann (1997) estimated that the radiative effects of volcanic sulfates are only slightly weaker than those induced by anthropogenic sulfates, even though the anthropogenic SO_2 source intensity is about five times greater. This surprising impact capability on climate arises from the higher efficiency of volcanic sulfur in producing sulfate aerosols, evaluated to be 4.5 times higher than that of anthropogenic sulfur. This is because SO_2 molecules released from volcanoes at high tropospheric altitudes have a longer residence time, mainly due to their appreciably lower dry deposition rate with respect to that typical of anthropogenic SO_2 emitted at the surface level (Benkovitz *et al.*, 1994).

The coarse solid matter particles (like powder-size matter, volcanic ash, or other igneous rock materials) blown into the air by an erupting volcano have generally short residence times in the troposphere, not exceeding a few weeks, as observed during the Eyjafjallajökull eruption in Iceland during April 2010 and the subsequent spread over Europe and North Atlantic in the second half of April and early May (Campanelli *et al.*, 2012). In this view, the term "volcanic dust" is here more appropriately referred to powder-size material than to fly ash particles most commonly having very large sizes, although both powder-size and fly ash particles usually cover the size range from 0.3 to more than 30 μm, having mass density close to 2.5×10^6 g m^{-3}. Some TEM images of volcanic ash particles of various sizes were shown by Schumann *et al.* (2011), as sampled inside the plume emitted by the Eyjafjallajökull volcano in April 2010 and penetrating the North Atlantic on May 2, 2010. They exhibit irregular shapes with jagged edges and sizes ranging from 0.7 to more than 20 μm. Silicates and mixed particles with $a < 500$ nm were found to constitute the most abundant groups. Such sampling measurements also provided evidence of sulfur-containing particles inside the Eyjafjallajökull plume, clearly indicating that the aerosol content in the size range $a < 500$ nm was dominated by secondary particles mainly consisting of ammonium sulfates and nitrates. Numerous submicron particles of ammonium sulfate were also sampled in very high concentration, together with supermicron particles having irregularly prismatic shape (with sizes of ~2 μm), and small droplets of sulfuric acid. In addition, numerous agglomerates were observed, consisting of (i) rather large silicate particles having sizes of ~10 μm and containing predominantly mixtures of various minerals, feldspars, amphiboles/pyroxenes, and pure SiO_2 minerals and (ii) smaller ammonium sulfate particles having sizes $a < 1$ μm, often crystallized out as droplets.

The residence times of the fly ash particles (mainly with $a < 1$ μm) forming drifting volcanic clouds can range from days to weeks, while the ash particles with sizes mainly ranging from 1 to 20 μm can remain suspended at atmospheric altitudes only for short times, varying from a few hours to 2 days. Due to these relatively long residence times in the atmosphere, both fine and coarse ash particles can produce locally relevant optical effects, which can be realistically evaluated using the Mie (1908) theory applied to scattering by spherical particles. Being insoluble in water, the larger volcanic ash particles exhibit in general sizes ranging from 20 μm to about 1 mm, presenting most frequently irregular shapes with sharp and jagged edges, as shown by Muñoz *et al.* (2004) for particles sampled during the volcanic eruptions of Mt St Helens on May 18, 1980, Mt Redoubt in 1989 and 1990, and Mt Spurr in August and September 1992.

The ash particles having sizes equal to several tens of micrometer fall not far from the volcano, having residence times of no more than 1–2 h. Therefore, they are estimated to produce only negligible radiative effects on climate. Volcanic ash particles of giant sizes look like grained fragments with sizes of 1–2 mm and consist mostly of broken glass shards and variable amounts of broken crystal and lithic fragments. However, having lifetimes of no more than a few hours and sizes greater than 1 mm, they cannot be classified as aerosols, being largely beyond the upper size limit commonly defined for aerosols (Rose and Durant, 2009). Most of the particles ejected from volcanoes in the form of dust and ash consist of water-insoluble substances (minerals, silicates, and metallic oxides such as SiO_2, Al_2O_3, and Fe_2O_3) and remain for the most part confined to the troposphere.

Finally, it is important to emphasize that volcanic eruptions can play an important role in modifying the sulfate aerosol load of the upper troposphere, where they act as condensation nuclei for ice particles. Therefore, they can significantly favor cirrus cloud formation, thus contributing to produce relevant indirect radiative forcing effects. Satellite-based observations of volcanic aerosol and clouds confirm that an interannual variability link exists between explosive volcanic eruptions and high-level cloud formation and cloudiness index (Graf, Feichter, and Langmann, 1997).

1.3.6 Cosmic Dust

Cosmic dust in the solar system is mainly constituted by small particles having sizes ranging from a few molecules to 0.1 μm. In the solar system, cosmic dust is formed from comet dust, asteroidal dust, Kuiper belt dust, and interstellar dust, the last component causing the zodiacal light. Examining spacecraft measurements of the cosmic dust flux, it was estimated by Zook (2001) that as many as 4×10^4 t of cosmic dust reach the Earth's surface every year, with nearly all of the meteoritic mass residing in grains with masses ranging from 10^{-19} to 10^{-7} g. Less than 25% of this mass is believed to derive from asteroids. Data from the spacecrafts of Pioneer 8 and 9 missions gave evidence for a flux of submicron dust grains leaving the solar system under radiation pressure at the rate of about 10 tons s^{-1}, while a mass flux of interstellar grains equal on average to 3×10^{-19} g s^{-1} has been estimated to pass through the solar system on the basis of the measurements taken on board the spacecrafts of Galileo and Ulysses missions. Kwok and Zhang (2011) reported that cosmic dust contains complex OM (consisting of amorphous organic solids with a mixed aromatic–aliphatic structure) that could be created naturally, and rapidly, by stars. More recently, examining a set of zodiacal cloud observations and measurements made with a spaceborne dust detector, Plane (2012) found that the daily mass input of interplanetary dust particles entering the Earth's atmosphere ranges from 100 to 300 t, in agreement with the accumulation rates of cosmic-enriched elements (Ir, Pt, Os, and superparamagnetic Fe) evaluated in polar ice cores and deep-sea sediments. In contrast, it was estimated by Plane (2012) from measurements in the middle atmosphere conducted with radar, LIDAR, high-flying aircraft, and satellite remote sensing techniques that the annual input flux Φ_e varies between 5 and 50 tons per day. The previous evaluations lead to obtain values of cosmic dust Φ_e ranging from 3×10^{-2} to 1.1×10^{-1} Tg per year at around 100 km altitude and from 2×10^{-3} to 2×10^{-2} Tg per year at the 50 km altitude. These estimates are two to three orders of magnitude smaller than those given by the weaker surface-level sources, as can be seen in Table 1.1. It is important to take into account that cosmic dust particles enter the atmosphere at high speeds and undergo significant ablation passing through the atmosphere. The resulting metals injected into the atmosphere are involved in a diverse range of phenomena, including (i) the formation of layers of metal atoms and ions; (ii) the nucleation of noctilucent clouds, which are a sensitive marker of climate change; (iii) impacts on stratospheric aerosols and ozone chemistry, which need to be considered against the background of a cooling stratosphere and geoengineering plans to increase sulfate aerosol; and (iv) fertilization of the ocean with bioavailable Fe, which has potential climate feedbacks. Gardner *et al.* (2014) estimated that the global influx of cosmic dust ranges from 0.4 to 110 t per day and therefore from

1.5×10^4 to 4×10^{-2} Tg per year, for which the average global flux of cosmic dust entering the Earth's atmosphere can be evaluated as equal to 2×10^{-2} Tg per year.

1.4
Secondary Aerosols of Natural Origin

Natural secondary aerosols originate in the atmosphere as a result of *in situ* g-to-p conversion processes. They are composed mostly of sulfates and nitrates formed through condensation of sulfur- and nitrogen-containing gases and may (i) condense onto existing particles, thereby increasing the mass (but not the number) of particles, such a process being favored when the surface area of existing particles is high and the supersaturation of the gases is low, or (ii) form new particles, with sizes smaller than 0.01 μm in general. The amount of aerosols produced by g-to-p conversions is comparable to that generated by direct emissions of natural aerosols and exceeds that due to direct emissions of anthropogenic aerosols. Three major families of chemical species can be identified in such g-to-p conversion processes, involving natural gaseous compounds containing sulfur, nitrogen, and organic and carbonaceous molecules. In fact, various organic substances (originating, e.g., from gases emitted by plants) can additionally furnish a large contribution to the total aerosol mass (Seinfeld and Pandis, 1998). In particular, SO_2 is oxidized to H_2SO_4 and the rate of conversion is influenced by the presence of heavy metal ions (e.g., Fe, Mn, V). Among the proposed reactions, the following six are commonly estimated to be the most important:

$$SO_3 + H_2O \rightarrow H_2SO_4 \tag{1.5}$$

$$NO + O_2 + H \rightarrow HNO_3 \tag{1.6}$$

$$SO_2 + C_2H_2 + \text{allene} \rightarrow C_3H_4S_2O_3 \tag{1.7}$$

$$NO_2 + \text{hydrocarbons} + \text{photochemistry} \rightarrow \text{organic nitrates} \tag{1.8}$$

$$SO_2 + \text{alkanes} \rightarrow \text{sulfinilic acid} \tag{1.9}$$

$$O_3 + \text{olefines} \rightarrow \text{organic particles} \tag{1.10}$$

The particles forming through these reactions generally exhibit spherical shapes, with number density concentrations usually ranging from 3×10^3 to 7×10^3 cm^{-3}, but sometimes reaching values exceeding 1.5×10^4 cm^{-3} for the heaviest pollution conditions occurring in urban areas. The size distributions of these aerosols are in general realistically represented by lognormal curves with geometrical standard deviations assuming values close to 0.70. The mass concentration of these particles can vary considerably as a function of RH conditions of the surrounding air. Due to their relatively high number concentration at locations all over the world, secondary particles play an important role in the global aerosol budget, also due to the remarkable contribution from anthropogenic gaseous emissions.

1.4.1
Natural Sulfate Particles from Tropospheric SO_2 and Sulfur Compounds

Sulfate aerosols are produced by chemical reactions in the atmosphere involving gaseous precursors, with the exception of sea-salt sulfate and gypsum dust particles. Together with those formed from anthropogenic sources through secondary processes (which are described in Section 1.6.1), sulfate particles have compositions ranging from sulfuric acid droplets to ammonium sulfates and constitute a major aerosol type in the troposphere. The key controlling variables for the production of sulfate aerosols from its precursors are (i) the source strength of the precursor substance, (ii) the fraction of precursors removed before conversion to sulfate, and (iii) the chemical transformation rates along with the gas-phase and aqueous chemical pathways for sulfate formation from SO_2. The atmospheric burden of sulfate aerosol is then regulated by the interplay of production, transport, and wet and dry deposition processes. Sulfate aerosols nucleate homogeneously or form on existing particles from gaseous precursors. Natural aerosols are formed from SO_2 emitted by volcanoes into the atmosphere and from dimethyl sulfide (i.e., CH_3SCH_3, or referred to as DMS) emitted by biogenic sources, especially marine plankton.

Natural sulfur is mainly assimilated by living organisms and is then released into the atmosphere as one of the final metabolism products. The most important atmospheric sulfur gases are sulfur dioxide (SO_2), hydrogen sulfide (H_2S), DMS, carbonyl sulfide (COS), and carbon disulfide (CS_2). Among them, the three following reduced sulfur gases of biogenic origin are the most important in the formation of sulfate aerosols:

1. H_2S, which is produced by sulfate-reducing bacteria and is mainly emitted from swamps and sediments. Here, microbial degradation of dead OM, occurring through biogenic reactions in soils, marshland, and plants, contributes together with halocarbons to release H_2S in the air, which is then rapidly converted to SO_2.
2. DMS, which is emitted from marine phytoplankton together with dimethyl disulfide and is then oxidized to SO_2 and sulfate aerosols.
3. COS, which is released in much smaller amounts from the biosphere but remains in the atmosphere during much longer lifetimes of 44 years on average.

All three gaseous species serve as precursor gases for sulfate aerosol production. The regions of the ocean with high organic content and biological productivity (i.e., upwelling regions, coastal waters, and salt marshes) constitute the major sources of CS_2 and COS, where various biogenic reactions in the oceans act as sources of DMS, COS, and CS_2, primarily due to phytoplankton. Released into the atmosphere, DMS and COS are first oxidized to SO_2. In particular, DMS is oxidized by free radicals that contain oxygen (such as NO_3^- and OH^-). Since OH^- forms during the day via photochemical processes, the reaction with OH^- during sunlight hours is the primary process leading to DMS oxidation. Intermediate oxidation products are methanesulfonic acid (CH_3SO_3H), dimethyl sulfoxide, and dimethyl sulfone. DMS is not subject to dry deposition and can therefore be converted to SO_2 far enough from the ground to avoid large deposition losses. For this reason, DMS is the main precursor of SO_2 in remote oceanic air. In polluted atmospheres, NO_3 can build up at night and lead to the DMS + NO_3 reaction, from which SO_2 is formed near the ground. The reactivity of COS in the troposphere is low, allowing COS to enter the stratosphere by diffusion. At these low stratospheric levels, oxidation of COS dominates the production of stratospheric sulfate aerosol during periods of weak volcanic activity, when direct injection of SO_2 into the stratosphere is negligible.

Sulfur compounds exist in both reduced and oxidized states, with oxidation numbers from −2 to +6. When these reduced sulfur gases are released into the oxygen-rich atmosphere, they are mainly oxidized by more than 65% to the +6 oxidation state of H_2SO_4 (i.e., oxidized to SO_4^- with S(VI)), while the remainder of the SO_2 is oxidized to the +4 oxidation state of SO_2 (i.e., S(IV)), and a small percentage is removed by dry deposition. In fact, the +6 oxidation state is the stable form of sulfur in the presence of oxygen. Oxidation of sulfur compounds illustrates an effect that often applies to other compounds, for which the more oxidized species generally have a high affinity for water (e.g., sulfuric acid). Consequently, such oxidized species are removed very rapidly and efficiently from the atmosphere by wet processes.

As mentioned earlier, the principal natural sources of SO_2 are the oxidation processes of DMS and H_2S. For example, H_2S reacts with OH to form H_2O and HS and then forms SO_2 through the following sequence of reactions:

$$HS + O_3 \rightarrow HSO + O_2 \tag{1.11}$$

$$HS + NO_2 \rightarrow HSO + NO \tag{1.12}$$

$$HSO + O_3 \rightarrow HSO_2 + O_2 \tag{1.13}$$

$$HSO_2 + O_2 \rightarrow HO_2 + SO_2 \tag{1.14}$$

Volcanoes, biomass burning, and forest fires are also natural sources of tropospheric SO_2, although the strongest source of SO_2 all over the world is fossil fuel combustion, estimated to yield an annual average global flux $\Phi_e = 75$ Tg(S) per year of sulfur, against $\Phi_e = 10$ Tg(S) per year of volcanic sulfur, and $\Phi_e = 3$ Tg(S) per year of sulfur from biomass burning, as can be seen in Table 1.2, where the estimates of atmospheric content of sulfur compounds (H_2S, SO_4^{2-}, SO_2, CS_2, DMS, and COS) are given together with those of the emission and removal annual global fluxes from the principal sources and sinks of sulfur-containing

Table 1.2 Principal sources and sinks of sulfur-containing species on a global scale, defining the sulfur reservoirs of the surface–troposphere system where continents contain 2×10^{10} Tg(S) and oceans 1.6×10^{9} Tg(S) (Hobbs, 2000) (being 1 Tg = 10^{6} t).

Reservoir	Atmospheric mass content (Tg(S))	Entering annual global flux (Tg(S) yr^{-1})	Outgoing annual global flux (Tg(S) yr^{-1})
H_2S	0.03	7 (from soils and marshlands)	7 (toward the SO_2 reservoir)
SO_4^{2-}	1.20[a]	3 (volcanoes)	10 (dry deposition)
		80 (SO_2 reservoir)	73 (wet deposition)
SO_2	0.30	7 (H_2S reservoir)	80 (toward the SO_4^- reservoir)
		3 (biomass burning)	40 (dry deposition)
		10 (volcanoes)	—
		75 (anthropogenic activities and fossil fuel combustion)	—
		25 (DMS reservoir)	—
		0.5 (CS_2 reservoir)	—
CS_2	0.10	0.7 (soils, plants and industry)	0.5 (toward the SO_2 reservoir)
		0.3 (biogenic processes in oceans)	0.5 (toward the COS reservoir)
DMS	0.05	25 (biogenic processes in oceans)	25 (toward the SO_2 reservoir)
COS	2.50	0.3 (biogenic processes in oceans)	0.9 (dry deposition)
		0.3 (soils)	0.2 (toward the stratosphere)
		0.5 (CS_2 reservoir)	—

a) 40–320 Tg(S) of SO_4^{2-} are involved in the sea-salt cycle (emission and removal in atmosphere–ocean exchanges).

species, respectively (Hobbs, 2000). Estimates of the entering global flux Φ_e of SO_2 defined by the IPCC (2001) are presented in Table 1.4 for comparison. The value of $\Phi_e = 75$ Tg(S) per year of SO_2 proposed by Hobbs (2000) as due to fossil fuel combustion is in close accordance with the value of $\Phi_e = 76$ Tg(S) per year provided by Benkovitz *et al.* (1996), while the estimate of Φ_e made by Hobbs (2000) for Φ_e for the sulfur from biomass burning was appreciably higher than that estimated by Spiro, Jacob, and Logan (1992) around 2.2 Tg(S) per year. The entering global flux of particulate matter sulfur from volcanoes was estimated by Andres and Kasgnoc (1998) to be 9.3 Tg(S) per year (including H_2S) and, hence, slightly lower than the Hobbs (2000) value. A total entering annual global flux of SO_2 equal to 88 Tg(S) per year was therefore assumed by the IPCC (2001), including a small contribution of 0.06 Tg(S) per year from aircraft traffic (Penner *et al.*, 1998). The entering annual global flux of DMS from biogenic processes in the oceans was evaluated by Hobbs (2000) to be 25 Tg(S) per year, in accordance with the overall estimate of IPCC (2001) including also H_2S, consisting of 24 Tg(S) per year from oceans (Kettle and Andreae, 2000) and 1.0 Tg(S) year from land biota and soils (Bates *et al.*, 1992; Andreae and Jaeschke, 1992), as reported in Table 1.4.

In the gas phase, SO_2 is oxidized through the following sequence of three reactions:

$$SO_2 + OH + M \rightarrow HOSO_2 + M, \tag{1.15}$$

where M is a third molecule to which the exothermal energy is transferred;

$$HOSO_2 + O_2 \rightarrow HO_2 + SO_3, \tag{1.16}$$

yielding SO_3, which then reacts with water in the final reaction; and

$$SO_3 + H_2O \rightarrow H_2SO_4, \tag{1.17}$$

which may also occur in cloud water to form a sulfuric acid molecule.

The main sources of H_2S are emissions from soils, marshlands, oceans, and volcanoes. The only significant sink for H_2S is oxidation to SO_2, through the four reactions in Eqs 1.11–1.14. Biological reactions in the oceans involving phytoplankton emit several sulfur gases, of which DMS exhibits the largest emission rate, being then removed from the atmosphere primarily by its reaction with OH to produce SO_2. The sulfur gas with the highest concentration in the unpolluted atmosphere is COS (~0.5 ppbv), due to biogenic sources and oxidation of CS_2 by OH. Molecules of COS are very stable in the troposphere and can be transported into the stratosphere, where they constitute the dominant source of sulfate particles during the quiescent periods between two consecutive violent and strong volcanic eruptions, which are capable of injecting sulfur gases into the stratosphere through the mechanisms described in Section 1.4.4.

The preceding remarks state that various sulfur gases (e.g., H_2S, CS_2, COS, DMS) can be oxidized to SO_2, which is then oxidized to sulfate ($SO_4{}^{2-}$), the dominant gas-phase routes being the reactions defined in Eqs. 1.15–1.17. It is estimated that H_2S oxidizes to SO_2 within about 2 days on average at tropospheric levels, during the sulfate particle formation processes.

An important role is played by sulfuric acid in the formation of sulfates by ammonia (NH_3), which is the principal basic gas in the atmosphere originating from soils, animal waste, fertilizers, and industrial emissions, having residence time in the lower troposphere of ~10 days. Ammonia neutralizes acid species through reactions such as

$$2NH_3 + H_2SO_4 \rightarrow (NH_4)_2SO_4. \tag{1.18}$$

This implies that the primary removal mechanisms for NH_3 involve its conversion to ammonium-containing aerosols, which are then removed from the atmosphere and transported to the ground by wet and dry deposition. However, on a global scale, heterogeneous reactions of SO_2 in cloud water dominate the conversion of SO_2 to ammonium-containing droplets. Over the oceans, sulfates derived from DMS contribute to the growth of existing particles. Sulfates are also produced in and around clouds, the subsequent evaporation of cloud water releasing these sulfate particles into the air, as reported in Table 1.2.

The chemical pathway of precursors converted to sulfate is important because it changes the radiative effects. Most SO_2 is converted to sulfate either in the gas phase or in cloud droplets that later evaporate. Model calculations suggest that aqueous-phase oxidation is dominant over the global scale. Both processes produce sulfate mostly in submicron aerosols, which act as efficient light scatterers, but the size distribution of sulfate in aerosols is different for gas phase and aqueous production. The size distribution of sulfate particles formed in the gas-phase process also depends on the interplay of nucleation, condensation, and coagulation processes. Two types of chemical interaction have been recognized to reduce the radiative impact of sulfate aerosols by causing some to condense onto larger particles with lower scattering efficiencies and shorter atmospheric lifetimes: (i) heterogeneous reactions of SO_2 on mineral aerosols (Andreae and Crutzen, 1997) and (ii) oxidation of SO_2 to sulfate in sea-salt-containing cloud droplets and deliquesced sea-salt aerosols. The second process forms a substantial fraction of nss sulfate particles that numerically predominate over that of coarse sea-salt particles, especially under conditions where the rate of photochemical H_2SO_4 production is low and the amount of sea-salt aerosol surface available is high (Sievering *et al.*, 1992). Sulfate in aerosol particles is present as sulfuric acid, ammonium sulfate, and intermediate compounds, depending on the availability of gaseous ammonia to neutralize the sulfuric acid formed from SO_2. In a modeling study, Adams, Seinfeld, and Koch (1999) estimated that the global mean NH_4^+/SO_4^{2-} mole ratio is close to unit, in good agreement with experimental data. This not only increases the mass of sulfate aerosol by some 17% but also changes the hydration behavior and refractive index of the aerosol. The overall effects are of the order of 10%, which is relatively low if compared with the uncertainties discussed earlier.

On this matter, an interesting hypothesis on the existence of a DMS global climate feedback formulated by Charlson *et al.* (1987), called the CLAW hypothesis from the initials of the four coauthors' surnames, is worth mentioning. This theory states that biogenic DMS production by marine phytoplankton influences global climate through a feedback mechanism involving sulfate aerosols and clouds through the following 10 steps:

1. DMS diffuses from the sea surface to the atmosphere.
2. DMS is oxidized to a sulfate aerosol.
3. Sulfate aerosols act as the main source of CCNs over the oceans.
4. Cloud albedo increases.
5. Global temperature is lowered.
6. Phytoplankton production decreases.
7. DMS emissions decrease.
8. Atmospheric load of sulfate aerosols decreases.

9. Cloud albedo decreases.
10. Global temperature rises, increasing both phytoplankton production and DMS emissions.

1.4.2 Natural Nitrate Particles from Tropospheric Nitrogen Oxides

Nitrogen molecules (N_2) constitute more than 99.99% of the nitrogen present in the atmosphere and have an overall atmospheric content of 2×10^9 Tg(N). Nitrous oxide (N_2O) is one of the most important greenhouse gases and contributes to provide most of the remaining nitrogen, giving a considerably lower atmospheric content of 1.3×10^3 Tg(N). The other nitrogen species in the atmosphere are therefore present in very low trace concentrations, yielding an overall content of 0.75 Tg(N). Nonetheless, they are of crucial importance in atmospheric chemistry, since they participate in important aerosol formation processes. Ammonium (NH_3), for example, is the only basic gas in the atmosphere responsible for neutralizing acids produced by the oxidation of SO_2 and NO_2. The ammonium salts of sulfuric and nitric acid formed through these processes become atmospheric aerosols, mainly through the reactions involving nitrogen oxides NO and NO_2 in both tropospheric and stratospheric chemistry processes. All of the nitrogen-containing gases in the air are involved in biological nitrogen fixation and denitrification. "Fixation" refers to the reduction and incorporation of nitrogen from the atmosphere into living biomass. This transformation is accomplished by bacteria that have special enzymes for this task, usually producing NH_3. The term "fixed nitrogen" refers to nitrogen contained in chemical compounds that can be used by plants and microorganisms. Under aerobic (i.e., oxygen-rich) conditions, other specialized bacteria can oxidize ammonia to nitrite (NO_2^-) ions and then to nitrate (NO_3^-) ions through a process referred to as "nitrification." Most plants use nitrate taken through their roots to satisfy their nitrogen needs. Some of the nitrate undergoes bacterial reduction to N_2 and N_2O (termed "denitrification"), which returns fixed nitrogen from the biosphere to the atmosphere. In this case, nitrate acts as the oxidizing agent. Therefore, denitrification generally occurs under anaerobic conditions, where oxygen is not available. Fixed nitrogen can also be returned from plants to the atmosphere in the form of N_2O. Biomass burning returns fixed nitrogen to the atmosphere such as N_2, NH_3, N_2O, and NO_x.

The oxides of nitrogen, that is, nitric oxide (NO) and nitrogen dioxide (NO_2), which together are referred to as NO_x, play a very important role in atmospheric chemistry, contributing to determine the entering and outgoing global annual fluxes of nitrogen estimated in Table 1.3, because they take part in all the formation and removal processes of nitrogen-containing compounds occurring in the atmosphere. The nitrogen oxides are produced by fossil fuel combustion and biomass burning and originated from soils, lightning, NH_3 oxidation, aircraft emissions, and transport from the stratosphere. Nitrogen oxides are emitted into the troposphere primarily as NO, but during the day NO rapidly establishes an

Table 1.3 Principal sources and sinks of nitrogen-containing species on a global scale, defining the nitrogen reservoirs of the surface–troposphere system where continents contain 9×10^9 Tg(N) and oceans 1.1×10^7 Tg(N) (Hobbs, 2000) (being 1 Tg = 10^6 t).

Reservoir	Atmospheric mass content (Tg(N))	Entering annual global flux (Tg(N) yr^{-1})	Outgoing annual global flux (Tg(N) yr^{-1})
N_2	2×10^9	100 (denitrification at the ocean surface)	100 (fixation at the ocean surface)
		140 (denitrification by bacteria in plants, soils, and water)	140 (fixation by bacteria in plants, soils, and water)
N_2O	1.3×10^3	2 (from oceans)	9 (toward the stratosphere)
		0.5 (biomass burning)	—
		4 (fossil fuel consumption and industrial activities)	—
		6 (soil emissions due to nitrification, denitrification, and fertilizers)	—
NO_x	0.15	6 (soil emissions due to nitrification, denitrification, and fertilizers)	20 (dry removal of NO_x)
		12 (biomass burning)	25 (wet removal of NO_x)
		20 (fossil fuel combustion)	—
		3 (reservoir of NH_3 and NH_4^+)	—
		5 (lightning in clouds)	—
NH_3 and NH_4^+	0.60	7 (from oceans)	15 (dry removal of NO_3 and NH_4^+)
		2 (biomass burning)	32 (wet removal of NO_3 and NH_4^+)
		12 (soil emissions including fertilizer loss)	3 (toward the NO_x reservoir)
		26 (decomposition of proteins), of which	—
		23 in urea from domestic animals and	—
		3 in urea from wild animals	—

equilibrium with NO_2 through the null cycle consisting of the following two reactions:

$$NO + O_3 \rightarrow NO_2 + O_2 \tag{1.19}$$

$$NO_2 + O_2 + M + h\nu \rightarrow NO + O_3 + M \tag{1.20}$$

where M represents an inert molecule, capable of absorbing the excess molecular energies. Once NO is converted to NO_2, a number of reaction paths are available. During the night, NO_x is present only as NO_2 due to the reaction represented in Eq. (1.19), while in the daytime the principal sink for NO_x is given by the following reaction:

$$NO_2 + OH + M \rightarrow HNO_3 + M. \tag{1.21}$$

Nitric acid (HNO_3) is removed in about 1 week by dry and wet deposition. During the night hours, NO_2 is oxidized by O_3 to NO_3, and the NO_3 then reacts with NO_2 to produce N_2O_5, which reacts subsequently with water on particles and cloud droplets to form HNO_3. The evaporation of cloud water releases these nitrate particles into the air, so that the resulting residence time of NO_2 is of about 1 day.

Examining the data given in Table 1.3, Hobbs (2000) estimated that the overall entering annual global flux of NO_x was equal to 46 Tg(N) per year, while the IPCC (2001) estimated a flux of ~41 Tg(N) per year, of which:

1. 5.5 Tg(N) per year is due to soil emissions associated with nitrification and denitrification as well as to the effects arising from the use of fertilizers (Yienger and Levy, 1995), such an overall flux being subdivided into ~2.2 Tg(N) per year coming from agricultural soils and ~3.2 Tg(N) per year from natural soils, while Hobbs (2000) evaluated this overall flux as equal to 6 Tg(N) per year.
2. 6.4 Tg(N) per year is given by biomass burning (Liousse *et al.*, 1996) against the estimate of 12 Tg(N) per year made by Hobbs (2000).
3. 21 Tg(N) per year is due to fossil fuel combustion (Benkovitz *et al.*, 1996) in place of the Hobbs (2000) estimate equal to 20 Tg(N) per year.
4. 7.0 Tg(N) per year (Price, Penner, and Prather, 1997) is due to lightning against the estimate of 5.0 Tg(N) per year made by Hobbs (2000).

All these estimates are given in Table 1.4, while those of Hobbs (2000) are presented in Table 1.3. The main discrepancies resulting from such a comparison are due to the fact that the IPCC (2001) neglected the contribution made by the reservoir of NH_3 and NH_4^+, while Hobbs (2000) assumed a value of 3 Tg(N) per year for evaluating such a contribution and neglected the aircraft traffic contribution, which was evaluated by Penner *et al.* (1999) to be equal to 58 Tg(N) per year. The overall entering annual global flux of nitrogen from NH_3 was estimated by the IPCC (2001) to be 54 Tg(N) per year (Bouwman *et al.*, 1997), while Hobbs (2000) assumed a total value of 47 Tg(N) per year. The partial contributions were attributed by Bouwman *et al.* (1997) to (i) domestic animals (21.6 Tg(N) per year),

Table 1.4 Estimates of the annual emission fluxes of natural and anthropogenic aerosol precursors (measured in Tg(S), Tg(N), and Tg(C) per year) on a global scale, as estimated by the IPCC (2001) and determined by various authors over the past two decades.

Emissions from natural precursors	Estimates and references
DMS and H_2S from oceans	24 Tg(S) yr^{-1} (Kettle and Andreae, 2000)
DMS and H_2S from land biota and soils	1.0 Tg(S) yr^{-1} (Andreae and Jaeschke, 1992)
NH_3 (natural overall)	10.7 Tg(N) yr^{-1} (IPCC, 2001)
NO_x from natural soils and lightning	10.2 Tg(N) yr^{-1} (IPCC, 2001)
VOCs from isoprene	467.5 Tg(C) yr^{-1} (Tsigaridis *et al.*, 2006)
VOCs from monoterpenes	120.8 Tg(C) yr^{-1} (Tsigaridis *et al.*, 2006)
Aromatic VOCs	14.3 Tg(C) yr^{-1} (Tsigaridis *et al.*, 2006)
Overall VOCs	250.9 Tg(C) yr^{-1} (Tsigaridis *et al.*, 2006)
Emissions from anthropogenic precursors	
SO_2 (anthropogenic overall)	78.3 Tg(S) yr^{-1} (IPCC, 2001)
NO_x (anthropogenic overall)	30.2 Tg(N) yr^{-1} (IPCC, 2001)
NH_3 (anthropogenic overall)	42.8 Tg(N) yr^{-1} (IPCC, 2001)

(ii) agriculture (12.6 Tg(N) per year), (iii) humans (2.6 Tg(N) per year), (iv) biomass burning (5.7 Tg(N) per year), (v) fossil fuel combustion and industry (0.3 Tg(N) per year), (vi) natural soils (2.4 Tg(N) per year), (vii) wild animals (0.1 Tg(N) per year), and (viii) oceans (8.2 Tg(N) per year).

Aerosol nitrate is closely tied to the relative abundances of ammonium and sulfate. If ammonia is available in excess of the amount required to neutralize sulfuric acid, nitrate can form small-size radiatively efficient aerosol particles. In the presence of an accumulation mode of aerosols containing sulfuric acid, nitric acid is deposited on alkaline mineral or sea-salt particles having larger sizes. Because coarse-mode particles are less efficient in causing light scattering per unit mass, this process reduces the radiative impact of nitrate particles in practice. The global burden of ammonium nitrate aerosols was estimated by Andreae (1995) to be 0.24 Tg from natural sources and 0.40 Tg (as NH_4NO_3) from anthropogenic sources, finding that anthropogenic nitrates cause only 2% of the total direct aerosol-induced forcing on a global scale. Similar estimates of global burden were derived by Jacobson (2001), who estimated that the global mean direct radiative forcing induced by anthropogenic nitrates is equal to $-2.4 \times 10^{\times 2}$ W m^{-2} on average. Adams, Seinfeld, and Koch (1999) obtained an even lower value of 0.17 Tg (as NO_3^- ions) for the global nitrate burden. Part of the difference with respect to the Andreae's (1995) evaluation was plausibly attributed to the fact that the Adams, Seinfeld, and Koch (1999) model did not include nitrate deposition on sea-salt aerosol. Estimates obtained from other studies indicate that direct radiative forcing due to ammonium nitrate is about 1/10 of the sulfate forcing. However, the importance of aerosol nitrate will probably increase appreciably over the twenty-first century, because it was forecast that NO_x emissions will more than triple over such a period, while the emissions of SO_2 will decline slightly. Thus, assuming an increase in agricultural emissions of ammonia, it is conceivable that direct forcing by ammonium nitrate could become comparable in magnitude to that induced by sulfates within the next 50 years (Adams *et al.*, 2001).

1.4.3
Organic Aerosols from Biogenic Volatile Organic Compounds

Organic and carbonaceous aerosols are formed by g-to-p conversion involving gases released from the biosphere and VOCs, such as crude oil leaking to the Earth's surface. As pointed out in Section 1.3.5, natural carbonaceous particles are emitted directly into the atmosphere, mainly during the biomass combustion processes occurring in forest and savannah areas. Therefore, they contain both EC and OC. In addition to biomass burning products, the main natural sources of carbonaceous aerosols are atmospheric oxidation of biogenic VOCs, which mainly consist of isoprene and terpene emitted by terrestrial plants (Claeys *et al.*, 2004), as reported in Table 1.4. Photooxidation of VOCs reduces their vapor pressure, while organic gases condense onto existing aerosol particles. Polymerization continues in the condensed phase and results in semi- or nonvolatile organic species (Gelencsér, 2004). Several thousands of VOCs are emitted not only by plants but also by numerous other biogenic sources into the atmosphere, providing a contribution which is globally predominant over that provided by anthropogenic sources (Tsigaridis and Kanakidou, 2007). In fact, it was estimated in the IPCC (2001) report that the overall volatile organic emissions are equal to 236 Tg(C) per year, of which 127 Tg(C) per year is from terpenes (i.e., ~54%) (Guenther *et al.*, 1995) and 10^9 Tg(C) per year from anthropogenic sources (equivalent to ~46%) (Piccot, Watson, and Jones, 1992), the second sources being mainly constituted by (i) motor vehicles that are in practice the primary source of VOCs in the most polluted areas of our planet, because of the incomplete combustion and vaporization of fuel producing hydrocarbons, and (ii) evaporation of solvents, evaluated to be the second most important anthropogenic source of VOCs in the world. Some of the more important VOCs are isoprene (C_5H_8), ethane (C_2H_4), and monoterpenes, yielding the global annual emission fluxes reported in Table 1.4, according to Tsigaridis *et al.* (2006). Isoprene accounts for about 50% of the NMHC. The photooxidation of isoprene can produce compounds having vapor pressures so low as to condense onto preexisting particles. This process could account for about 5–20% of the annual secondary organic aerosol (SOA) production from biogenic sources. Terpenes constitute a class of hydrocarbons that evaporate from leaves. About 80% of these emissions oxidize to organic aerosols within about 1 h. Emissions from vegetation constitute a significant source of hydrocarbons, which can react photochemically with NO and NO_2 to produce O_3, thereby playing a central role in atmospheric chemistry. Estimates of the annual emission fluxes of these aerosol precursors on a global scale are given in Table 1.4, as determined for both natural and anthropogenic sources.

1.4.4
Sulfate Particles from Marine and Volcanic SO_2 Formed in the Stratosphere

Stratospheric aerosols are composed of an aqueous solution of 60–80% sulfuric acid for temperatures varying from −80 to −45 °C, respectively (Shen, Wooldrige, and Molina, 1995). The source of the globally distributed background stratospheric aerosol, formed in the absence of perturbations caused by violent volcanic eruptions, is given by the emission of COS at the ocean surface, as a result of metabolism in marine areas, producing a high content of COS molecules near the surface, which are subsequently transported across the tropopause and then oxidized at stratospheric altitudes. Besides this sulfur compound, the transport from the troposphere to the stratosphere also involves other chemical species, such as (i) chlorine (Cl), whose major natural source is methyl chloride (CH_3Cl), which derives in part from biological activities in seawater, wood molds, and biomass burning, and (ii) halogen compounds (e.g., chlorine and bromine species) produced by biological activities in the oceans. COS is chemically inert and water insoluble and therefore has a long tropospheric residence time. For this reason, it diffuses into the stratosphere, where it is photolytically dissociated by solar UV radiation to form sulfuric acid through the following reaction:

$$\mathrm{COS} + h\nu \rightarrow \mathrm{CO} + \mathrm{S}, \qquad (1.22)$$

where the incoming photons $h\nu$ belong to the UV radiation spectral range from 200 to 260 nm (Lin, Sim, and Ono, 2011), followed by these reactions:

$$\mathrm{S} + \mathrm{O_2} \rightarrow \mathrm{SO} + \mathrm{O} \qquad (1.23)$$

and

$$\mathrm{O} + \mathrm{COS} \rightarrow \mathrm{SO} + \mathrm{CO} \qquad (1.24)$$

$$\mathrm{SO} + \mathrm{O_2} \rightarrow \mathrm{SO_2} + \mathrm{O} \qquad (1.25)$$

or

$$\mathrm{SO} + \mathrm{O_3} \rightarrow \mathrm{SO_2} + \mathrm{O_2}, \qquad (1.26)$$

both leading to oxidation of SO into SO_2. The sequence of reactions (1.22–1.26) leads to the net result:

$$2\mathrm{COS} + \mathrm{O_2} + \mathrm{O_3} + (\mathrm{UV})\ h\nu \rightarrow 2\mathrm{CO} + 2\mathrm{SO_2} + \mathrm{O}, \qquad (1.27)$$

which contributes efficiently to form stratospheric SO_2, from which natural aerosols are formed at these altitudes. Alternatively, but with less efficiency, COS can react with a hydroxyl OH to yield SO_2.

As a result of the overall conversion defined in Eq. (1.27), the mixing ratio of COS decreases with height, passing from about 0.4 ppbv at the tropopause level to 0.02 ppbv at around 30 km altitude. Consequently, the SO_2 concentration remains roughly constant with height, assuming rather stable values of ~0.05 ppbv. The production of stratospheric sulfate aerosols at these altitudes primarily occurs through oxidation of SO_2 by OH, through the reactions described in Eqs. (1.15) and (1.16) to form HO_2 and SO_3. In the stratosphere, an alternative path for the formation of SO_3 is also given by the reaction

$$SO_2 + O + M \rightarrow SO_3 + M, \quad (1.28)$$

in which the exothermal energy is transferred to the third molecule M. Subsequently, SO_3 is combined with H_2O in the reaction of Eq. (1.17) leading to the formation of sulfuric acid molecules. The subsequent conversion of the H_2SO_4 vapor to liquid H_2SO_4 can occur because of the following two main mechanisms: (i) the combination of molecules of H_2SO_4 and H_2O (i.e., *homogeneous bimolecular nucleation*) and/or the combination of H_2SO_4, H_2O, and HNO_3 to form new (primarily sulfuric acid) droplets (i.e., *homogeneous heteromolecular nucleation*) and (ii) vapor condensation of H_2SO_4, H_2O, and HNO_3 onto the surfaces of preexisting particles with sizes >0.30 μm (i.e., *heterogeneous heteromolecular nucleation*).

Model calculations indicate that the second mechanism is the more likely route in the stratosphere. The tropical stratosphere is probably the major region where the nucleation processes occur, and the aerosols are then transported to higher latitudes by large-scale atmospheric motions. As a result of the aforementioned chemical processes, the concentration of H_2SO_4 liquid droplets assumes a peak at around 20 km. The previous explanation based on the conversion of COS in SO_2 and the subsequent formation of H_2SO_4 is supported by the fact that other sulfur-containing species, such as SO_2, DMS, and CS_2, do not persist long enough in the troposphere to be transported to the stratosphere. Therefore, their role in the formation of stratospheric sulfuric acid droplets is wholly negligible. The background amount of stratospheric sulfate is mainly due to UV photolysis of organic COS forming SO_2, although the direct contribution of tropospheric SO_2 injected in the stratosphere across the tropical tropopause region plays a significant role in forming sulfates at the lower stratospheric levels and accounts for about one-third of total stratospheric sulfate mass content. This quantity is currently estimated to be equal to ∼0.15 Tg(S) during volcanically quiet periods, therefore accounting for no more than 15% of the overall sulfate content in the troposphere and stratosphere.

Sulfur direct emissions in the stratosphere occur mainly in the form of SO_2, even though other sulfur species may be present in the volcanic plume, predominantly SO_4^{2-} ions and H_2S, which were evaluated to assume percentage concentrations of around 1% of the total amount and only in some cases to exceed percentages of around 10–20%. In these less frequent cases, H_2S oxidizes to SO_2 in about 2 days at the tropospheric levels and within 10 days at the lower stratospheric levels. The state of unperturbed background stratospheric aerosol was relatively rare during the past decades because several very violent volcanic eruptions took place during the past 50 years, in which significant amounts of SO_2 were directly injected into the lower and middle stratosphere, as occurred during the series of Gunung Agung eruptions in 1963–1964 in Bali (Indonesia), the El Chichón volcano eruption of April 1982 in Mexico, and Mt Pinatubo eruption in the

Philippines (June 1991). The observations of these volcanic eruption events indicated that the associated transient climatic effects may be very important in such circumstances, showing that the trends in the frequency of volcanic eruptions can constitute an efficient climatic factor capable of strongly influencing the average surface temperature, as was understood by analyzing the well-documented data set recorded during the evolution of the Pinatubo plume, from which valuable information was achieved by the scientific community on the climate effects that can be caused by a violent volcanic eruption.

In all these episodes, a few months after the direct injection of large amounts of SO_2 and lower concentrations of minor gases into the stratosphere, small particles were formed at relatively low stratospheric altitudes consisting of sulfuric acid (~75%) and liquid water (~25%). These droplets were transported globally during the subsequent months, producing rather high values of stratospheric aerosol optical depth with respect to the globally average background conditions of this optical parameter, causing pronounced cooling effects on the tropospheric and stratospheric temperatures (McCormick, Thomason, and Trepte, 1995). To give a measure of the strong changes in the aerosol optical thickness caused by the formation of Mt Pinatubo aerosol layers in the stratosphere, Figure 1.9 shows the time patterns of the daily mean values of aerosol optical thickness $\tau_a(\lambda)$ measured at various visible and near-infrared wavelengths and the corresponding atmospheric turbidity parameters α and β (Ångström, 1964) obtained from the sun-photometer measurements performed at the CNR Pyramid Laboratory (5050 m a.m.s.l.) in the Himalayan region (Nepal) on several clear-sky days during summer 1991 and summer 1992 (Tomasi, Vitale, and Tarozzi, 1997). These spectral dependence patterns were found to be very similar to those measured by Russell *et al.* (1993) at the Mauna Loa Observatory during the same months. The results recorded at the CNR Pyramid Laboratory provided evidence of the sharp increase in $\tau_a(\lambda)$ at all wavelengths λ and the corresponding decrease in α and the simultaneous increase in β due to the arrival of the stratospheric volcanic aerosol layers on July 27 above this high-altitude observation site. Evaluations of the vertical mass loading M_S of volcanic aerosols suspended in the low stratosphere, made using the average multimodal size distributions defined by Pueschel *et al.* (1993) from measurements performed at the Mauna Loa Observatory (Hawaii), indicated that M_S increased from less than 0.04 to more than 0.07 g m^{-2} within a few days in late July 1991, due to the arrival of the Mt Pinatubo stratospheric particles.

Figure 1.9 shows that the decay of aerosol extinction features recorded 1 year later caused a decrease of τ_a(0.50 μm) from more than 0.20 in summer 1991 to around 0.17 in summer 1992, presenting decay features similar to those observed by Pueschel *et al.* (1993) at the Mauna Loa Observatory. These variations in the Himalayan region occurred with (i) the gradual decrease of α from about 0.8 to 0.5, caused by the growth of the coarse particles; (ii) the decrease of β from more than 0.12 to less than 0.10, due to the decreasing trend of $\tau_a(\lambda)$; and (iii) the appreciable decrease of M_S from 0.07 g m^{-2} in summer 1991 to less than 0.05 g m^{-2} in summer 1992. These findings clearly showed that the columnar particulate mass content M_s decreased from more than 0.03 g m^{-2} in summer 1991 to about 0.02 g m^{-2} in

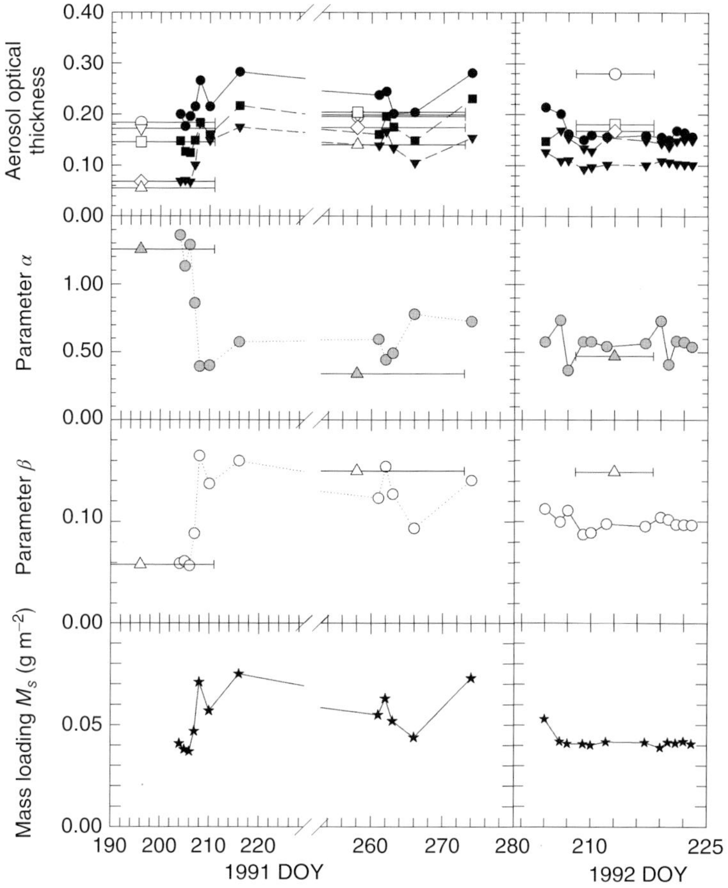

Figure 1.9 Upper part: time patterns of the daily mean values of aerosol optical thickness $\tau_a(\lambda)$ measured by us at the 380 nm (solid circles), 500 nm (solid squares), and 875 nm (solid triangles) wavelengths (Tomasi, Vitale, and Tarozzi, 1997) at the Pyramid Laboratory (5050 m a.m.s.l.) in Nepal, during summer 1991 (left-hand side) and summer 1992 (right-hand side), and of the monthly mean values of $\tau_a(\lambda)$ measured by Pueschel *et al.* (1993) at the Mauna Loa Observatory (Hawaii) in July and September 1991 (left-hand side) and in July and August 1992 (right-hand side) at the 382 nm (open circles), 451 nm (open down triangles), 528 nm (open squares), 865 nm (open up triangles), and 1060 nm (open diamonds) wavelengths. Middle part: time patterns of the daily mean values of Ångström's (1964) exponent α (solid circles) and atmospheric turbidity parameter β (open circles), determined at the Pyramid Laboratory during summer 1991 (left-hand side) and summer 1992 (right-hand side) and compared with the monthly mean values of α and β (open triangles) determined by Pueschel *et al.* (1993) in July and September 1991 at the Mauna Loa Observatory. Lower part: time patterns of the daily mean values of the Pinatubo aerosol mass loading M_s, as determined by us from the spectral values of $\tau_a(\lambda)$ measured during summer 1991 and summer 1992, in terms of variable linear combinations of background aerosol model and bimodal extinction model defined by Pueschel *et al.* (1993) for volcanic particles a few months old (summer 1991) and the trimodal model of Pueschel *et al.* (1993) for an aged population of Pinatubo particles (summer 1992).

summer 1992, indicating that the mean half-time was considerably longer than one year. The time patterns of $\tau_a(\lambda)$ shown in Figure 1.9 give a measure of the sharp increase in aerosol extinction within the stratospheric region below the 25 km altitude a few months after the eruption. Such an increase in $\tau_a(\lambda)$ arose from the formation of stratospheric particles mainly due to the injection of sulfur-containing volcanic gases and the subsequent reactions defined in Eqs. (1.15–1.17). The following increase in M_s occurred as a result of the gradual aging of the Mt Pinatubo sulfate aerosols at altitudes ranging from 8 to 26 km.

These pronounced evolutionary patterns in the size-distribution parameters of volcanic particles were observed at all latitudes of the two hemispheres over a period of 2–3 years after the eruption, as shown by the latitudinal and temporal patterns of the stratospheric aerosol depth $\tau_a(1.02\ \mu m)$ derived by Thomason *et al.* (2008) from the SAM II, SAGE, SAGE II, and SAGE III satellite-borne measurements performed over the period from January 1985 to the end of 2005 and completed with supplementary LIDAR data. This study provided evidence of (i) the gradual decrease in the atmospheric extinction observed in 1985 and early 1986 after the El Chichón eruption of 1982 and (ii) the formation of the Mt Pinatubo volcanic particle cloud in 1991 a few months after the eruption, followed by a slow decrease from 1991 to 1996. They confirm that the total removal period of the Pinatubo aerosol loading in the stratosphere lasted about 4 years, leading to an average estimate of the residence time of these aerosol particles in the low stratosphere equal to around 2.5 ± 1.2 years. It was evaluated that the Mt Pinatubo eruption of June 15, 1991, ejected very large amounts of ash and gases into the atmosphere so strongly that the volcano's plume penetrated into the stratosphere.

The aerosol mass loading ejected by this violent volcanic eruption was estimated to be about 1.5×10^7 t of SO_2 into the stratosphere, where it reacted with hydroxyl and water to form sulfuric acid droplets, yielding an overall stratospheric aerosol mass content of about 30 Tg on a global scale. The Mt Pinatubo eruption caused an abrupt increase in stratospheric aerosol optical depth by a factor of 10–100 times that measured before the eruption. Three months after the eruption, it was found that the stratospheric region at midlatitudes near the Philippines had warmed suddenly by 2.5–3.0 °C due to the increased aerosol concentration. In the following few months, a huge stratospheric load of sulfuric acid droplets was observed to accumulate, leading to the formation of a dense layer of volcanic particles around the planet. This abrupt increase was gauged to cause a decrease of about 5 °C in the temperature of the midlatitude tropospheric region above the Philippines and an overall average decrease in the global surface temperature of about −0.5 °C over the subsequent 2 years. A value of 2 years was therefore assumed to be a reliable estimate of the stratospheric relaxation time to background conditions (McCormick, Thomason, and Trepte, 1995). Volcanic stratospheric particles can persist for up to a few years since they are not efficiently removed by wet deposition, having very low fall speeds at those altitudes. Bearing in mind that only one strong volcanic eruption occurs every 10 years on average, it can be stated that stratospheric aerosol is only rarely in a totally unperturbed state by volcanic emissions.

Volcanic eruptions can have a strong impact on the stratospheric load of aerosol particles. The lifetime of coarse particles (dust and ash) directly injected into the stratosphere is relatively short (not longer than 1–2 months) due to the efficient removal of settling coarse particles. About 3 months of posteruptive aging are needed for chemical and microphysical processes to form the stratospheric layer of sulfate particles, causing the sharp increase in volcanic aerosol mass and aerosol optical thickness shown in Figure 1.9 at the end of July 1991 (209th DOY). The analysis of satellite observations made by Thomason *et al.* (2008) estimated that (i) the average stratospheric $\tau_a(1.02\,\mu m)$ was equal to 0.1 over the entire global scale (for latitudes from 80°N to 80°S) during the first 6 months of 1992 and (ii) a total time of ~4 years was needed to return to an average background value of $\tau_a(1.02\,\mu m) = 3 \times 10^{-3}$. An analysis of these observations indicated that the Pinatubo stratospheric particles produced a mean radiative cooling forcing of about $-4\,W\,m^{-2}$ over a global scale at the beginning of 1992, that is, about 6–8 months after the eruption. This average radiative forcing effect decayed exponentially over the subsequent years until reaching a value of around $-0.1\,W\,m^{-2}$ at the end of 1996.

The trend observed in the frequency of volcanic eruptions during the past 50 years has been estimated to have a major impact on the temperature conditions of the low stratosphere, because the sulfur emissions from volcanoes can produce a strong increase in the stratospheric load of aerosol particles. Considering that the volcanic aerosols efficiently reflect the shortwave solar radiation and absorb the longwave terrestrial radiation, the formation and growth of volcanic particles cause an increase in the planetary albedo, giving rise to important climatic effects: satellite measurements revealed a 1.4% increase in solar radiation reflected from the atmosphere toward space over a period of several months after the Mt Pinatubo eruption.

Such emissions occur mainly in the form of SO_2, even though other sulfur species may be present in the volcanic plume, predominantly sulfate ions and H_2S, the latter being the second molecular species that is oxidized to SO_2 in about 10 days at stratospheric altitudes. Estimates of the emission of sulfur-containing species from quiescent degassing and eruptions vary on average from 7 to 14 Tg(S) per year. The historical record of SO_2 emissions by erupting volcanoes shows that over 100 Tg of SO_2 can be emitted in a single event, such as the Mt Tambora (Sumbawa, Indonesia) eruption in 1815, which presumably led to very strong transient cooling effects evaluated to range from −0.1 to at least −0.3 °C. Estimates of the global annual flux Φ_e of sulfate volcanic aerosols formed in the stratosphere were recently made by Toohey *et al.* (2011), who determined a value of 17 Tg per year for the Mt Pinatubo eruption in June 1991. Therefore, considering also the years with only weak eruptions, a reasonable range from 2 to 17 Tg per year was assumed for the events occurring during the past 20 years, with an average value equal to about half the Mt Pinatubo value.

Within the stratospheric layers of sulfuric acid droplets, the concentration of Aitken nuclei is higher at the low stratospheric altitudes and generally decreases slowly with height. In contrast, particles with sizes ranging from about 0.2 to 4 μm reach a maximum concentration of ~0.1 cm^{-3} at altitudes of around 17–20 km at midlatitudes. This part of the low stratosphere, where background layers of sulfate particles can persist over long periods, is called the "stratospheric sulfate layer" or also the "Junge layer," to honor the memory of Christian Junge who discovered these thin aerosol layers in the late 1950s. As mentioned earlier, the most significant source of SO_2 in the stratosphere is given by the major volcanic eruptions, which inject SO_2 into the stratosphere. This molecule is then converted to H_2SO_4, mainly through the reactions defined in Eqs. (1.15–1.17), in which SO_2 is oxidized to SO_3 and SO_3 is then combined with water to form sulfuric acid. As mentioned previously, the oxidation of SO_2 to form SO_3 may also occur through the reaction of Eq. (1.28) in the presence of an external molecule. Finally, H_2SO_4 vapor is then converted to liquid H_2SO_4 through the homogeneous heteromolecular nucleation and heterogeneous heteromolecular nucleation processes, among which the heterogeneous processes are evaluated to be more efficient at stratospheric levels.

The enhancement of the stratospheric sulfate layer originated from volcanic eruptions can cause depletions in stratospheric O_3, due to the fact that H_2SO_4 droplets act to modify the distribution of catalytically active free radicals. For example, the eruption of Mount Pinatubo caused dramatic depletions in stratospheric ozone, which caused an indirect radiative forcing estimated to be equal to about −0.1 W m^{-2} and hence considerably smaller than the direct aerosol-induced radiative forcing. The observed stratospheric load of sulfate particles is estimated to be ~0.14 Tg(S) during volcanically quiet periods. Calculations with global climate models suggest that the radiative forcing of volcanic sulfate particles is only slightly smaller than that induced by anthropogenic sulfates suspended in the troposphere, even though the anthropogenic SO_2 source intensity is about five times greater. The main reason is that SO_2 is released from volcanoes at higher altitudes and therefore has a longer residence time than anthropogenic sulfates. Relatively dense stratospheric layers of sulfate aerosols were estimated to cause general cooling effects on the climate system. Theoretical studies have been developed on the possibility of artificial production of sulfate aerosols in the stratosphere with the aim of increasing the planetary albedo and hence slowing down the current global warming trends (Crutzen, 2006). The benefits and disadvantages of such efforts need to be carefully discussed before entering the field of artificial modifications of climate, but the idea first proposed by Budyko (1977) is itself very stimulating.

1.5 Primary Anthropogenic Aerosols

A significant fraction of tropospheric aerosol particles is anthropogenic and consists of both primary particles (mainly due to diesel exhaust and vehicular traffic dust) and secondary particles formed from the gases emitted by urban and industrial activities. Anthropogenic aerosol emissions arise primarily from the following four source categories: (i) fuel combustion, (ii) industrial processes, (iii) nonindustrial fugitive sources (roadway dust from paved and unpaved roads, wind erosion of cropland, construction, etc.), and (iv) transportation activities with cars, ships, airplanes, and other vehicles. Therefore, anthropogenic aerosols contain sulfate, ammonium, nitrate, trace metals, carbonaceous material, and water. The carbonaceous fraction of this particulate matter consists of both EC and OC, the first emitted directly into the atmosphere from combustion processes and the second emitted directly or formed through atmospheric condensation of low-volatility organic gases.

Anthropogenic aerosols certainly contribute to the total tropospheric aerosol mass loading by more than 10%. The influence of this relatively small but increasing contribution on the climate system is not exactly known and needs to be assessed in future model studies and experiments. In their coupled aerosol and gas-phase chemistry transport model, Tsigaridis *et al.* (2006) used emission values relative to the year 1990 equal to (i) 73 Tg per year for anthropogenic SO_2, (ii) 45 Tg per year for NO_x, (iii) 44.1 Tg per year for NH_3, (iv) 1051.8 Tg per year for carbon monoxide, (v) 7.5 Tg per year for BC, (vi) 250.9 Tg per year for VOCs, (vii) 44.4 Tg per year for primary organic aerosols, (viii) 19.2 Tg per year for CH_2O, and (ix) 14.3 Tg per year for aromatic VOCs. These values are low in comparison with the annual emission flux values assumed by Tsigaridis *et al.* (2006) for natural dust (1704 Tg per year) and sea salt (7804 Tg per year) in their coupled model (see Table 1.1) but indicate that a considerable influence is exerted by humankind emissions on the current atmospheric composition. Use of biological materials by humans results in the emission of many chemical species into the atmosphere, such as (i) CO_2, CO, NO_x, N_2O, NH_3, SO_2, and hydrogen chloride (HCl) arising from the combustion of oil, gas, coal, and wood; (ii) hydrocarbons from vehicles, refineries, paints, and solvents; (iii) H_2S and DMS from paper mills and oil refineries; (iv) COS from natural gas; and (v) chloroform $CHCl_3$ from combustion of petroleum, bleaching of woods, and solvents. The change in trace gases can have negative effects on the stability of the climate system, favoring global warming, and nonreversible processes in the Earth–atmosphere system. Therefore, it is of basic importance to monitor the columnar contents of trace gases and the aerosol concentration along the vertical atmospheric path on a global scale using satellite measurements. Although the increased gaseous concentrations lead to stronger atmospheric absorption effects and, therefore, to a more pronounced warming of the atmosphere, it is worth considering that airborne aerosol load could increase or decrease the overall albedo of the Earth–atmosphere system, to an extent closely depending on the ground albedo. For the low albedo typical of ocean and dark vegetation in the visible, additional anthropogenic aerosols induce an increase in the planetary albedo and therefore favor a cooling of the atmosphere. Aerosols with a high soot content appear dark and hence reduce planetary albedo, especially in the polar regions presenting bright snow surfaces. Consequently, additional warming is expected to occur due to current efforts to clean polluted areas of the planet, because this could reduce the pollution load but increase the risks related to heat waves (Meehl and Tebaldi, 2004).

The global input of anthropogenic particles into the atmosphere is ~20% (by mass) of that from natural sources. For particles with diameters $a > 5\,\mu m$, direct emissions from anthropogenic sources dominate over secondary aerosols, which mainly originate in the atmosphere through g-to-p conversion of anthropogenic gases. The opposite case applies to most small particles, for which g-to-p conversion is the largely prevailing source favoring the increase in the concentration of anthropogenic aerosols. About 35% of the number concentration of atmospheric aerosols is estimated to be given by sulfates formed through oxidation of SO_2. The worldwide direct emission into the atmosphere of primary anthropogenic particles with sizes smaller than 10 μm was estimated to be ~350 Tg per year (Jaenicke, 1988), excluding the secondary particles formed through g-to-p conversion. On the basis of these remarks, the primary anthropogenic aerosols can be subdivided into the three following classes: (i) industrial dust, (ii) particles from fossil fuel combustion that include an important fraction of carbonaceous (soot) aerosols, and (iii) particles from waste and biomass burning.

1.5.1 Industrial Dust

The main anthropogenic sources of industrial dust are given by (i) transportation vehicular traffic and dust from roads, (ii) coal and fuel combustion in industrial processes, (iii) cement manufacturing, (iv) metallurgy, and (v) waste incineration. Wind erosion of tilled land can contribute to yield supplementary amounts of dust. In fact, three broad categories account in general for nearly all of the potential fugitive emissions including mineral dust and food agricultural products. Primary metals are also included in the first category and have a considerable impact considerably on environmental quality. Industrial dust has been widely monitored and regulated in developed countries, verifying that such an impact of fugitive dust emissions is relatively limited because they consist mostly of giant particles, which settle a short distance from the source. Fugitive dust sources are most efficient in rural areas. Relatively low nonindustrial fugitive dust emissions are caused by traffic entrainment of dust from paved and unpaved roads, agricultural activities, building construction, and fires.

During the twentieth century, the emission of particles from anthropogenic sources into the atmosphere was a small fraction of the overall particulate mass emitted from natural sources. Currently, the aforementioned aerosol sources are responsible for the most conspicuous impact of anthropogenic aerosols on environmental quality and were monitored and regulated over the past 20 years in North America and Europe. The result was that the emission of industrial dust aerosols has decreased significantly in these areas during the past two decades, obtaining a general improvement of air quality. On the other hand, the anthropogenic emission of aerosols has considerably increased over the past 50 years in world areas where industrialization has grown without stringent emission controls, especially in Asia. It is estimated that such a growth in industrialization could lead to a considerable increase in the industrial dust emission flux until reaching values of particulate mass global emission flux higher than 300 Tg per year by 2040 (Wolf and Hidy, 1997) with respect to evaluations of industrial dust emission estimated in the 1990s to vary from about 100 Tg per year excluding carbonaceous matter (Kiehl and Rodhe, 1995) to 200 Tg per year (Wolf and Hidy, 1997), as also shown in Table 1.5. The previous limitations are probably not of crucial importance for the current climatic conditions, considering the source strength of these emissions and the fact that many industrial dust particles have sizes <1 μm and hence can cause only relatively weak extinction effects on the incoming solar radiation.

1.5.2
Anthropogenic Aerosols from Fossil Fuel Combustion and Carbonaceous (Soot) Particles

Particulate matter emissions worldwide have been dominated by fossil fuel combustion (primarily coal) and biomass burning. These emissions are projected to double by the year 2040, largely due to anticipated increases in fossil fuel combustion, which are predicted to occur mainly in China and India. The combustion of all carbon-containing particles leads to the formation of EC particles, partially combusted fuel components, and products of the high-temperature reactions in the flame. These primary EC particles are very small (mostly with sizes smaller than 2 μm), but they rapidly aggregate through coagulation, starting already in the flame. In the case of solid fuels, there are several pathways leading to a variety of particle sizes, shapes, and composition. To give an idea of the variety of physicochemical processes taking place during the combustion of pulverized coal, a schematic representation of the pathways of particle formation during the combustion of a pulverized coal particle of 70 μm size is shown in Figure 1.10 (Okazaki, 1993).

It can be seen that the particle may be subject to two initial pathways of swelling or shrinking: (a) in the first case, the particle breaks into various fragments after swelling, forming a large number of particles with sizes ranging mainly from 0.5 to 30 μm; and (b) in the second case, the shrinking of the particle leads to (i) vaporization and condensation, until surface enrichment of very small particles is obtained; (ii) vaporization followed by nucleation and coalescence (and also coagulation, after Heintzenberg (1994)), to form a large number of particles with sizes ranging from 0.02 to 0.2 μm; (iii) direct agglomeration and coagulation, producing particles that may grow up to 30 μm sizes; (iv) expansion and quenching, yielding particles of 10–90 μm sizes; and (v) expansion and disintegration, generating particles with sizes $a < 30$ μm. The high-temperature fragmentation of burning coal and other solid fuels is a special case of the b-to-p conversion process. Supersaturated vapors are formed in all combustion processes described earlier. Therefore, the g-to-p conversion processes are also expected to occur during combustion.

Some SEM images of industrial aerosol particles formed through fossil fuel combustion (oil, coal, and distilled oil) are shown in Figure 1.11, presenting nearly spherical shapes. These images present a characterization of combustion particles

Table 1.5 Estimates of the annual emission fluxes of anthropogenic aerosols on a global scale from various sources and their precursors, as found in the literature of the past two decades.

Anthropogenic direct emission fluxes	Estimates (in teragram per year, being $1\ Tg\ yr^{-1} = 10^6\ ton\ yr^{-1}$) and references
Sulfates from industrial activities and fossil fuel combustion	75 (Hobbs, 2000)
Anthropogenic sulfates (as NH_4HSO_4)	122 (IPCC, 2001)
Secondary sulfates from SO_2	48.6 (in $Tg(S)\ yr^{-1}$) (Liao *et al.*, 2003)
Anthropogenic SO_2	73.0 (Tsigaridis *et al.*, 2006)
Nitrates from N_2O (fossil fuel combustion and industrial activities)	10 (Hobbs, 2000)
Nitrates from NO_x (fertilizers and fossil fuel combustion)	26 (Hobbs, 2000), 45 (Tsigaridis *et al.*, 2006)
Nitrates from NH_3 (fertilizers and agricultural activities)	12 (Hobbs, 2000)
Nitrates from NH_3 (breeding activities)	23 (Hobbs, 2000)
Anthropogenic nitrates (as NO_3^- ions)	14.2 (IPCC, 2001)
Secondary nitrates from NO_x (NO_3^- ions only)	21.3 (Liao *et al.*, 2004)
Secondary nitrates from NH_3	44.1 (Tsigaridis *et al.*, 2006)
Carbonaceous aerosols from biomass burning (sizes < 2 μm)	54 (IPCC, 2001)
Carbonaceous aerosols from fossil fuel (sizes < 2 μm)	28.4 (IPCC, 2001)
Fossil fuel organic matter (sizes < 2 μm)	6.60 (IPCC, 2001)
Biomass burning black carbon (sizes < 2 μm)	5.7 (IPCC, 2001)
Anthropogenic organic compounds	0.6 (IPCC, 2001)
Carbonaceous aerosols from aircraft	0.006 (IPCC, 2001)
Industrial dust, etc. (sizes > 1 μm)	100 (IPCC, 2001)
Industrial dust	40–130 (Andreae and Rosenfeld, 2008)
Biomass burning black carbon (sizes < 2 μm)	5.6 (IPCC, 2001)
Black carbon	7.5 (Tsigaridis *et al.*, 2006)
Black carbon (soot)	Ranging from 8 to 14 $Tg\ yr^{-1}$ (Andreae and Rosenfeld, 2008)
Subtotal (anthropogenic)	440 (IPCC, 2001)
Total (natural + anthropogenic)	6315 (IPCC, 2001)

sampled in the Po Valley (Italy) area, as emitted by an oil fuel-fired power plant (Bacci *et al.*, 1983), a coal-fueled power plant (Del Monte and Sabbioni, 1984), and a domestic heating unit fueled by distilled oil (Sabbioni and Zappia, 1992). An analysis of these samples indicated that (a) oil-fired aerosol particles have spherical shapes presenting a porous surface, internal cavities, and a spongy structure,

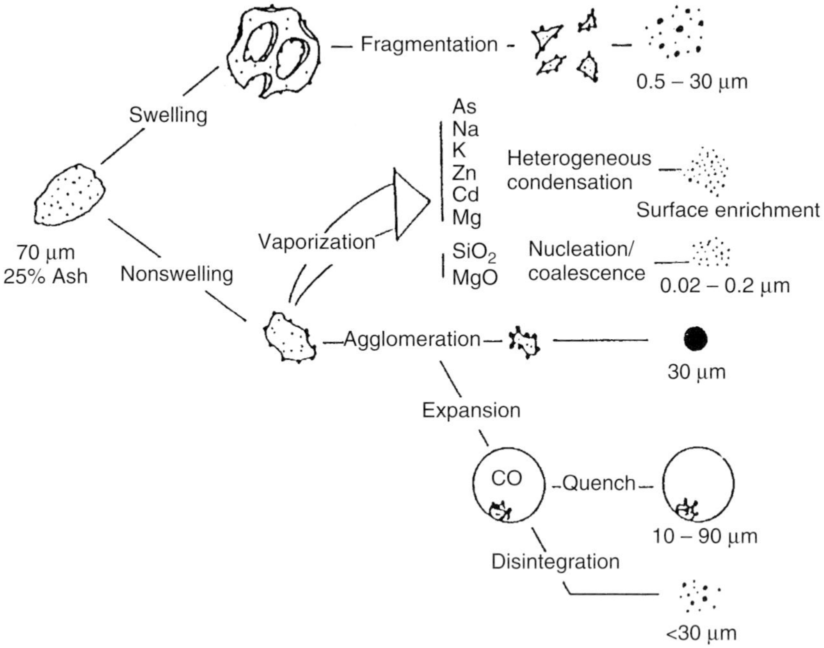

Figure 1.10 Pathways of particle formation during the combustion of pulverized coal through different combinations of swelling, shrinking, fragmentation, vaporization, condensation, nucleation, coagulation, expansion, quenching, and disintegration processes. (Adapted from a graph of Okazaki (1993).)

as can be seen in Figure 1.11a and b, which often contain sodium and vanadium oxide hydrate crystals having prismatic shapes and presenting relatively high elemental concentrations for all size classes over the 0.01 – 50 μm range (mainly with traces of Na, Al, Mg, S, K, V, Fe, Ni, Cu, Zn, and Pb elements); (b) fly ash particles emitted by a coal-fired power plant, such as that shown in Figure 1.11c, which may include different components of particulate matter, such as (i) glassy aluminosilicates of variable composition, (ii) spongy carbonaceous matter, (iii) spherical metallic particles containing different iron oxide phases (magnetite, hematite, maghemite), (iv) spherical rutile particles, (v) spherical lime particles, and (vi) mineral formless particles, containing mainly quartz and mullite; and (c) particles generated by domestic heating units fueled by distilled oil exhibit different morphological typologies (irregular, rounded to spherical, spherical with circular pores, smooth spherical particles, and spherical agglomerated particles, as presented in the example of Figure 1.11d); they often host crystals with different structural shapes, such as "rosettes" composed of twinned crystals with hexagonal habit or elongated crystals, in which the elemental composition is mainly given by S, Fe, Pb, Ti, V, Mn, Cu, Zn, Cd, Ni, Co, Ca, Si, and Cr.

Figure 1.11 clearly indicates that these soot particles change appreciably in shape passing from fractal-like fresh diesel soot particles to more compact aggregates of spherical shape soot particles. Schnaiter *et al.* (2005) pointed out that

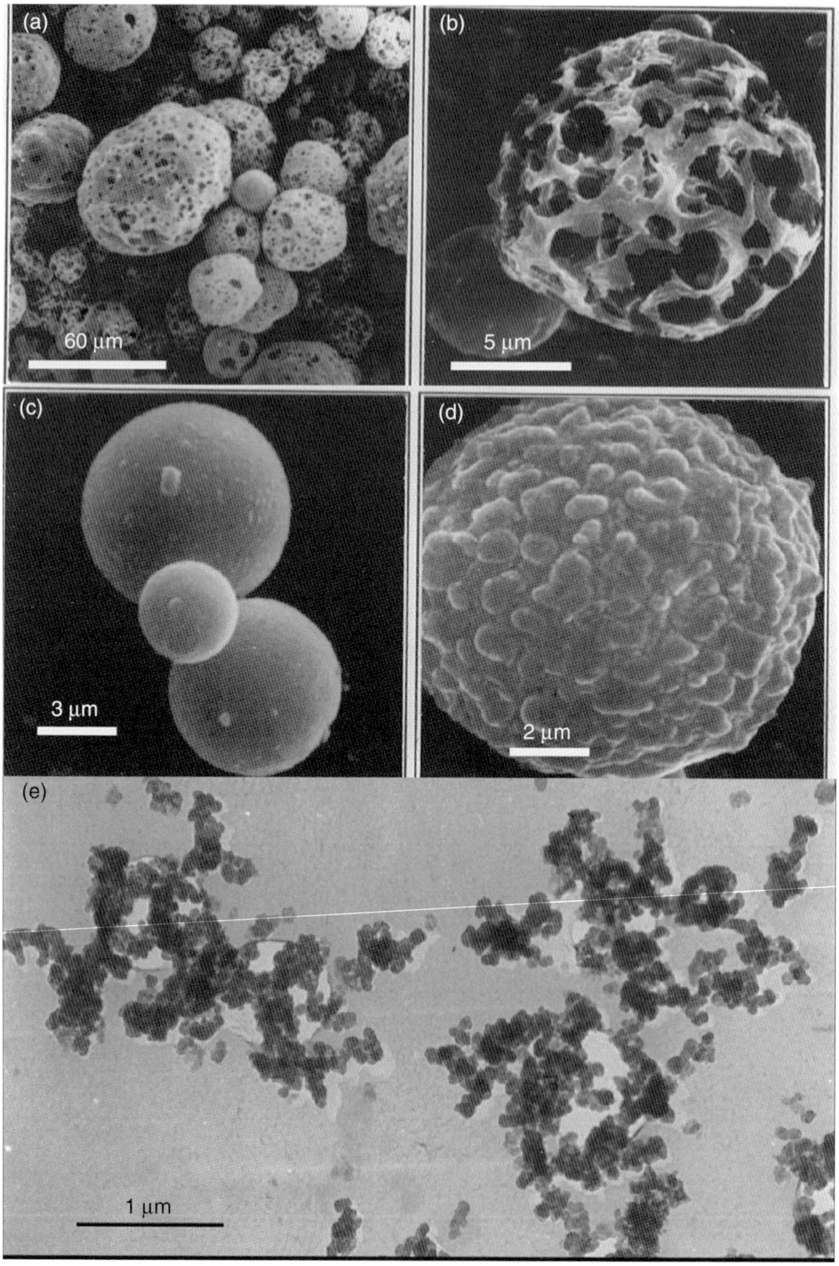

Figure 1.11 SEM images of industrial aerosol particles: (a, b) particles formed from fuel oil combustion, (c) soot particles from coal combustion, (d) particle from distilled oil fuel combustion, and (e) atmospheric diesel soot particles. ((a)–(d) Reproduced with permission and courtesy of Cristina Sabbioni, ISAC-CNR Institute, Bologna, Italy. (e) Reproduced with permission and courtesy of Annie Gaudichet (LISA, Paris, France) and Hélène Cachier.)

particle growth after the initial aggregate restructuring was reflected by a clear decrease in Ångström's (1964) exponent and in the hemispheric backscattering ratio measured in the laboratory, in accordance with the Mie (1908) theory. Considering that these combustion particles exhibit nearly spherical shapes in most cases, it seems not difficult to calculate the radiative effects induced by these particles applying the Mie (1908) theory directly to spherical particle polydispersions, provided that realistic values of the real and imaginary parts of complex refractive index are known. On this matter, the most reliable estimates available in the literature indicate that the real part of the refractive index for soot particles is equal to 1.74 at 0.30 μm wavelength, 1.75 at wavelengths increasing from 0.40 to 1.06 μm and slowly increasing at the longer wavelengths, assuming values equal to 1.76 at 1.30 μm and 1.83 at 2.5 μm. Correspondingly, the imaginary part was estimated to assume values decreasing from 0.47 at 0.30 μm to 0.43 at 0.694 and 0.860 μm wavelengths and then increasing from 0.44 at 1.06 μm to 0.51 at 2.50 μm.

The problem of BC influence on the planetary radiative budget (in terms of solar radiation absorption by aerosols and subsequent direct aerosol-induced radiative forcing effects) is a hot topic in modern research. In particular, aerosols transported from highly polluted areas in Europe toward the Arctic can lead to a decrease in the planetary albedo over areas where these soot aerosols are deposited, rendering the snow and ice surfaces dirtier and more polluted. These changes in the surface albedo lead to variations in the local radiation budget that can be associated with a more intense absorption of solar radiation due to such absorbing aerosol particles (Hansen and Nazarenko, 2004). Among the atmospheric aerosols of natural and anthropogenic origin, those emitted by combustion sources such as industrial plants, heating units, and vehicular traffic can cause major damage to buildings, bridges, monuments, and other human constructions through deposition on surfaces and the subsequent formation of substances, most commonly gypsum (calcium sulfate dihydrate), calcite (calcium carbonate), and noncarbonated carbon (Sabbioni, 1992).

As mentioned before, the carbonaceous fraction of ambient particulate matter consists of EC and various organic compounds (mainly OC): (i) EC is also often called BC and has a chemical structure similar to impure graphite, being emitted directly into the atmosphere during combustion processes, and (ii) OC can be emitted directly by primary sources or formed *in situ* by condensation of low-volatility products formed through photooxidation of hydrocarbons (secondary OC). It refers to only a part of the organic material mass, the rest being given by hydrogen, oxygen, nitrogen, and so on. Additional small quantities of particulate carbon may exist either as carbonates (e.g., $CaCO_3$) or CO_2 adsorbed onto particulate matter such as soot. Therefore, atmospheric carbonaceous particles consist of two major components, that is, BC (which is the most abundant light-absorbing aerosol species in the atmosphere and sometimes referred to as EC or free carbon) and OM. EC can be produced only in combustion processes and is therefore solely primary. Organic particulate matter is a complex mixture of many classes of compounds either directly emitted from sources or produced through atmospheric reactions involving gaseous organic precursors. A major reason for the study of OM is the possibility that such compounds pose a health hazard. Specifically, certain fractions of particulate OM, especially those containing polycyclic aromatic hydrocarbons (PAHs), were found to be carcinogenic in animals and mutagenic in *in vitro* bioassays.

Carbonaceous compounds make up a large but highly variable fraction of atmospheric aerosols. Organics are the largest single component of biomass burning aerosols, which can contribute to light scattering at least as much as the sulfate component. They are also important constituents of aerosols suspended in the upper troposphere. The presence of polar functional groups, particularly carboxylic and dicarboxylic acids, makes many of the organic compounds in aerosols water soluble and allows them to play a role in cloud droplet nucleation, acting as efficient cloud nuclei and hence having a major indirect effect on climate. Carbonaceous particles are produced by the combustion of liquid or gaseous fuels and consist of both EC and OC, presenting a wide variety of shapes assumed by soot, fly ash, and OC particles of anthropogenic (industrial and urban) origins. Various SEM and TEM images of such complex particle shapes are available in the literature. Pósfai *et al.* (1999) showed some TEM images of particles collected in the boundary layer of the North Atlantic Ocean (Azores islands), in which ammonium sulfate particles include small soot particles and fly ash spheres, and some TEM images of a typical chain-like soot aggregate, in which a carbon film connects individual small spherules within the aggregate and fly ash spheres consisting of amorphous silica, having sizes of 1 − 2 μm, clearly indicating that soot particles are mainly agglomerates of small nearly spherical carbonaceous particles forming typical chains. Other TEM and SEM images were shown by Li, Anderson, and Buseck (2003a) during the ACE-2 conducted at Sagres in southern Portugal, in which soot particles aggregated to form particles of different composition, such as silica or sea salt: (i) a TEM image showing, for instance, a typical chain-like soot aggregate of soot spheres having sizes of more than 10 μm, (ii) a SEM image showing a soot aggregate mixed with sea-salt crystals, (iii) a high-resolution TEM image showing an example of typical discontinuous onion-like structure of graphitic particle layers, and (iv) a TEM image illustrating an aggregate of rounded amorphous carbonaceous particles, sampled at Sagres, providing evidence under high magnification that these carbonaceous particles sometimes do not show the graphitic layers characterizing soot particles. Some SEM images of OC and fly ash particles sampled in the urban area of Mainz (Germany) have been also shown by Sinha *et al.* (2008), such organic aerosols being mainly formed during *in cloud* processes.

On the basis of the previous analyses, it can be stated that the industrial and urban aerosols are in general agglomerates of small roughly spherical elementary carbonaceous particles, in which the size and morphology of the clusters may vary widely. The small spherical elementary particles are remarkably consistent from one to another and have sizes ranging from 20 to 30 nm. They can cluster with each other, forming straight or branched chains, which can then agglomerate to form soot particles having sizes up to a few micrometers. Soot formed in combustion processes is not a unique substance, since it consists mainly of carbon atoms but also contains up to 10% moles of hydrogen as well as traces of other elements. The composition of a soot particle aged at the high temperatures of a flame is typically C_8H (Flagan and Seinfeld, 1988). Soot particles usually contain more hydrogen earlier in the flame. Subsequently, they absorb organic vapors when the combustion products cool down and frequently accumulate significant mass fractions of organic compounds. Thus, soot substances consist in general of a mixture of EC, OC, and other elements such as oxygen, nitrogen, and hydrogen incorporated in its graphitic structure (Chang, Brodzinsky, and Gundel, 1982). The EC found in atmospheric particles is not usually present in the large-size particles of highly structured pure graphite but tends to form three-dimensional arrays of carbon with small fractions of other elements, containing a high number of crystallites having sizes of 2–3 nm, all consisting of several carbon layers presenting the hexagonal structure of graphite. Therefore, the density of these soot particles is around $2\ \mathrm{g\,cm^{-3}}$.

Soot forms in a flame as the result of a sequence of events starting with the oxidation and/or pyrolysis of the fuel into small molecules, such as those of acetylene, C_2H_2, and PAHs, which are considered the main molecular intermediates for soot formation and growth (McKinnon and Howard, 1990). These particles grow through the initial formation of soot nuclei and the subsequent rapid effects of surface reactions (Harris and Weiner, 1983a,b). However, the soot nuclei represent only a small fraction of the overall soot mass produced. The formation of soot particles depends strongly on the fuel composition, because the rank ordering of the sooting tendency of fuel components is naphthalenes → benzenes → aliphatics, and the order of sooting tendencies of the aliphatics (alkanes, alkenes, alkynes) varies dramatically depending on the flame type (Flagan and Seinfeld, 1988), and soot formation depending critically on the carbon–oxygen ratio in the hydrocarbon–air mixture (Wagner, 1981).

The typical layered structure of soot spherules can be successfully investigated using the HRTEM techniques for identifying the soot components in complex internally mixed particles, as evidenced by Pósfai *et al.* (1999), since the nanostructural features of primary soot spherules reflect their distinct sources (from oil boilers, jet aircraft, and diesel engines) (Hays and Vander Wal, 2007). In addition, the fractal properties of soot aggregates are influenced by fuel type and combustion conditions, this allowing a source apportionment in some cases from the structure variations and fractal parameters of atmospheric soot (Wentzel *et al.*, 2003; van Poppel *et al.*, 2005).

Together with wood burning fires, diesel vehicles are effective sources of EC, estimated to emit particulate matter consisting of more than 90% submicron particles, generally distributed within a mode centered at diameter $a_c \approx 0.1\ \mu m$. EC mass concentration usually varies from 0.2 to $2.0\ \mu g\,m^{-3}$ in rural and remote areas and from 1.5 to $20\ \mu g\,m^{-3}$ in urban areas, where values often higher than $10\ \mu g\,m^{-3}$ can be found in the PM_{10} samples. Conversely, EC concentration is very low over the remote oceans, where values ranging mainly from 5 to $20\ ng\,m^{-3}$ are usually measured, and assumes values not exceeding a few nanograms per cubic meter in the coastal areas of Antarctica (Wolff and Cachier, 1998). The organic component of ambient particles in both polluted and remote areas is a complex mixture of hundreds of organic compounds (including *n*-alkanes, *n*-alkanoic acids, *n*-alkanals, aliphatic dicarboxylic acids, diterpenoid acids and retene, aromatic polycarboxylic acids, PAHs, polycyclic aromatic ketones and quinones, steroids, N-containing compounds, regular steranes, pentacyclic triterpanes, and iso- and anteiso-alkanes), as pointed out by Seinfeld and Pandis (1998, 2006). Most studies have assumed in the past decades that OC concentration was simply given by the concentration of carbon measured in microgram(C) per cubic meter, therefore without including the contribution to the aerosol mass of the other elements, namely, oxygen, hydrogen, and nitrogen, present in the organic aerosol compounds. It was suggested in the early 1990s that measured OC values should therefore be multiplied by a factor of 1.5 to calculate the total organic mass associated with OC, while alternative values of such a conversion factor varying between 1.2 and 2.0 were proposed in the literature. The mass size distributions of OC particles sampled in urban polluted areas usually present bimodal features, with the first mode centered at an aerodynamic diameter of around 0.1 μm and the second mode centered over the 0.5 – 1 μm aerodynamic diameter range. In remote marine areas, the ambient aerosol OC concentration was found to vary from 0.1 $\mu g\,m^{-3}$ (in the Samoa archipelago and New Zealand) to 0.8 $\mu g\,m^{-3}$ (Enewetak Atoll in the Pacific Ocean). Primary carbonaceous particles are produced by combustion (pyrogenic), chemical (commercial products), geologic (fossil fuel), and natural (biogenic) sources. Annual global primary OC emissions were estimated to vary from 15 to 80 Tg per year, with open burning contributions around 75% of the total, while fossil fuel combustion was estimated to provide a partial contribution of about 7%. Several hundred organic compounds have been identified in primary organic aerosol emissions, due to a variety of sources: the major constituents of the resolved urban aerosol mass are normal alkanoic acids (with annual average concentrations of 0.25 – 0.30 $\mu g\,m^{-3}$), aliphatic dicarboxylic acids (0.20 – 0.30 $\mu g\,m^{-3}$), and aromatic polycarboxylic acids (~0.10 $\mu g\,m^{-3}$).

1.5.3
Anthropogenic Aerosols from Waste and Biomass Burning

Although the controlled technological burning of fossil fuels emits almost pure soot, a much greater variety of carbonaceous particles results from biofuel, agricultural, and uncontrolled biomass fires (Reid *et al.*, 2005). Biomass burning produces atmospheric particles in amounts and concentrations that can affect both the regional and global climate (Forster *et al.*, 2007). In the past 10 years, individual particle TEM studies have greatly improved our knowledge of particle formation from biomass fires and revealed that agricultural fires severely affect atmospheric composition and air quality, even in developed regions such as Europe (Niemi *et al.*, 2006) and North America (Hudson *et al.*, 2004). In addition to soot particles, the major types of particulate matter generated by waste and biomass burning include organic particles, many of which also contain inorganic salt inclusions (Martins *et al.*, 1998; Li *et al.*, 2003b) and the so-called tar balls (Pósfai *et al.*, 2004), which are spherical and have sizes greater than individual soot spherules. They not only consist mainly of carbon and oxygen but sometimes also contain nitrogen and traces of sulfur, silicate, and other elements. The distribution of carbon is homogeneous, and oxygen is enriched in the ~30 nm outer layer of the spherules (Hand *et al.*, 2005) where the organic compositions resemble those of atmospheric humic-like substances (Tivanski *et al.*, 2007).

The intentional burning of biomass is the major source of combustion products suspended in the atmosphere. Most of these burning episodes take place in the tropics where emissions from burning vegetation arise from any uncontrolled combustion process and include both EC and OM, besides CO_2, CO, NO_x, CH_4, and NMHCs. The quantity and type of emissions from an intentional biomass fire depend not only on the type of vegetation but also on its moisture content, ambient temperature, RH conditions, and local wind speed. Other sources of biomass burning aerosols are the human-induced burning, such as wood or agricultural wastes, which together with intentional forest fires are estimated to produce about 40% of carbonaceous particulate matter emissions over a global scale. The overall annual emissions from fossil fuel and biomass burning are gauged to be 10–30 Tg per year and 45–80 Tg per year, respectively (IPCC, 2001). Combustion processes constitute the dominant source of BC, which contribute to emit the prevailing particulate mass fraction in the submicron size fraction and furnish an overall annual emission of 12–17 Tg per year (with an average value of 14 Tg per year on the global scale), of which 6–9 Tg per year is mainly from intentional biomass burning and 6–8 Tg per year from fossil fuel combustion (IPCC, 2001).

1.6 Secondary Anthropogenic Aerosols

Secondary liquid particles are usually quite small and exhibit shapes that can be correctly approximated by spheres in radiative transfer calculations. However, many anthropogenic aerosols show irregular shapes, since they originate and grow through two basic processes which both regulate the mass transfer from the gas phase to the particulate phase (g-to-p conversion) where (i) preexisting particles may grow through material condensation from the gas phase and (ii) new particles may form through homogeneous nucleation. In these g-to-p conversion processes, three major families of chemical species are involved, containing sulfur or nitrogen or organic and carbonaceous substances. The most important role in these processes is played by the following two reactions:

$$SO_3 + H_2O \rightarrow H_2SO_4, \quad (1.29)$$

$$NO + O_2 + H \rightarrow HNO_3, \quad (1.30)$$

forming molecules of sulfuric acid and nitric acid, respectively. The subsequent conversion of the H_2SO_4 vapor to liquid H_2SO_4 occurs through the following mechanisms, already considered in Section 1.4.4 to describe the formation of volcanic sulfuric acid droplets in the stratosphere: (i) the combination of H_2SO_4 and H_2O molecules in *homogeneous bimolecular nucleation* processes and/or the combination of H_2SO_4, H_2O, and HNO_3 in *homogeneous heteromolecular nucleation* processes to form new primarily sulfuric acid droplets and (ii) the vapor condensation of H_2SO_4, H_2O, and HNO_3 onto the surfaces of preexisting particles with sizes no greater than 0.3 μm in *heterogeneous heteromolecular nucleation* processes. These *in situ* chemical reactions are classified as *homogeneous* or *heterogeneous* according to the following concepts: (i) a homogeneous reaction involves reactants that are all in the same phase, and (ii) a heterogeneous reaction involves reactants that are in two or more phases, such as that occurring in a mixing of an inorganic aerosol (like sulfuric acid or nitric acid) with organic compounds (like aldehydes).

The formation of SOA particles is in part due to the VOCs emitted from anthropogenic sources and in part to sulfur- and nitrogen-containing gases. The nucleation of nanometer-sized sulfate particles has been observed in rural, urban, and coastal environments (Kulmala *et al.*, 2004), probably enhanced by the presence of aromatic acids (Zhang *et al.*, 2004). This suggests that organics are already present in the nucleation of small sulfate particles. In particular, SOA/sulfate mixed particles were formed through a further condensation of organics onto freshly nucleated particles (Adachi and Buseck, 2008; Smith *et al.*, 2008). In general, SOA particles are amorphous and composed of light elements, so that it is very difficult to study their structural characteristics using the TEM and SEM techniques and adopting the related methodologies (Pósfai and Buseck, 2010).

Secondary anthropogenic aerosols are transported by the airflows they encounter during their life in the atmosphere, until covering the global scale. In fact, aerosols formed through g-to-p conversion processes can be transported over long distances, since the time required for g-to-p conversion and the relatively small sizes of the particles formed during such a process lead to long atmospheric residence times for such aerosols. This is, for instance, the case of acidic aerosols such as sulfates and nitrates, which contribute to the so-called acid rains: it was found in the 1980s that SO_2 emitted from power plants in the United Kingdom were most intensely deposited as sulfate particles far inland in continental Europe.

1.6.1
Secondary Particles from SO_2

About 35% of the number density concentration of airborne aerosols consists of sulfates forming through oxidation of anthropogenic SO_2 emissions, in large part originated from fossil fuel combustion (primarily coal and oil). Therefore, these emissions can lead to dramatic pollution effects in urban areas where the consumption of coal and oil is very strong, causing high local concentrations of aerosols that are commonly termed "smog." The term *smog* derives from *smoke* and *fog* and was originally coined to refer to heavily polluted air that can form in cities (generally in winter under calm, stable, and moist conditions of air) due to emissions of SO_2 and aerosols from fossil fuel burning. The term is now applied to all forms of severe air pollution, particularly to those occurring in urban areas and causing a strong reduction of visibility.

Prior to the introduction of air pollution abatement laws in the latter part of the twentieth century, many of the most populated cities in Europe and North America regularly suffered from severe and frequent smog episodes. The London event was notorious and known as *London smog*: in the presence of smog, where particles swelled in size under high air RH conditions and acted in part as condensation nuclei on which fog droplets formed. In these cases, sulfur dioxide gas dissolves in the fog droplets, in which it is oxidized to form sulfuric acid. After the London smog episode, laws were passed in the United Kingdom and elsewhere banning the use of coal in open fires for domestic heating and the emission of black smoke and requiring industries to switch to cleaner burning fuels. Nevertheless, pollution is still a serious problem in some cities in Europe and in the United States as well as in many large cities of developing countries (China, India) due to the burning of coal and wood and the lack of strict air pollution controls.

Sulfuric acid is important in aerosol formation for humid air conditions, as emphasized in a number of studies (Kiang *et al.*, 1973; Mirabel and Katz, 1974; Kulmala and Laaksonen, 1990). The efficiency of this process can be first studied considering the simplified binary *sulfuric acid–water system* before discussing more complicated atmospheric aerosol systems. Sulfuric acid is very hygroscopic and absorbs significant amounts of water even at extremely low values of air RH. If RH exceeds 50%, the H_2SO_4 concentration in solution will be lower than 40% by mass, and the H_2SO_4 mole fraction smaller than 0.1. For a temperature of 20 °C, this corresponds to equilibrium vapor pressures lower than 10–12 mm Hg. Under all conditions, the concentration of H_2SO_4 in the gas phase is much less than the aerosol sulfate concentration. For particles having sizes <0.1 μm, the H_2SO_4 mole fraction in the droplet is highly dependent on particle size. Thus, for each fixed small droplet size, the water concentration increases as RH increases.

The reaction

$$H_2SO_4\ (g) \rightleftharpoons H^+ + HSO_4 \tag{1.31}$$

can often be neglected in atmospheric aerosol calculations in all cases where the vapor pressure of $H_2SO_4(g)$ is practically null over atmospheric particles. The whole system is therefore described by the bisulfate dissociation reaction

$$HSO_4^- \rightleftharpoons H^+ + SO_4^{2-} \tag{1.32}$$

with an equilibrium constant at 298 K. Since the molar ratio of HSO_4^- to SO_4^{2-} is proportional to the hydrogen ion concentration, it is expected that HSO_4^- ions will be present in acidic particles.

In the *sulfuric acid–ammonia–water system*, aerosol composition depends mainly on air temperature, air RH, and concentrations of ammonia and sulfuric acid. However, a series of compounds may exist in the aerosol phase, including both solids (like letovicite ($(NH_4)_3H(SO_4)_2$), $(NH_4)_2SO_4$, and NH_4HSO_4) and aqueous solutions of NH_4^+, SO_4^{2-}, HSO_4^-, and NH_3 (aq.). For environments in which NH_3 is poorly available, sulfuric acid exists in the aerosol phase in the form of H_2SO_4. As the NH_3 availability increases, H_2SO_4 is converted to HSO_4^- and its salts and finally, if there is an abundance of NH_3, to SO_4^{2-} and its salts. For total (gas and aerosol) ammonia concentrations lower than 0.8 μg m^{-3}, the aerosol phase consists mainly of H_2SO_4 (aq.), while some molecules of NH_4HSO_4 (s) are also present. A significant amount of water accompanies the H_2SO_4 (aq.) even for low RH conditions. This regime is characterized by an ammonia/sulfuric acid molar ratio lower than ~0.5. For higher values of this ratio, the following aerosol composition features are found: (i) when the ammonia/sulfuric acid molar ratio varies from 0.5 to 1.25, NH_4HSO_4 (s) is the dominant aerosol component for this system; (ii) when the ratio reaches unity (for ammonia concentration equal to 1.8 μg m^{-3}), the salt $(NH_4)_3H(SO_4)_2$ (s) (letovicite) is formed in the aerosol phase, gradually replacing NH_4HSO_4 (s); (iii) for a molar ratio of 1.25, $(NH_4)_3H(SO_4)_2$ (s) is the dominant aerosol component, and for a value equal to 1.5 the aerosol phase consists almost exclusively of letovicite; (iv) for even higher ammonia concentrations, $(NH_4)_2SO_4$ starts to form; (v) for an ammonia concentration equal to 3.6 μg m^{-3} (molar ratio 2), the aerosol consists of only $(NH_4)_2SO_4$ in the solid state; (vi) when the total ammonia concentration is lower than 3.6 μg m^{-3}, aerosol almost completely exists in the form of sulfate salts; and (vii) further increases in the available ammonia do not change the aerosol composition, but the excess ammonia remains in the gas phase as NH_3 (g).

It is expected that an increased availability of NH_3 would result in a monotonic increase in the total aerosol mass. This is not the case observed for ammonia-poor environmental conditions, that is, for ammonia/sulfuric acid molar ratio lower than unit. An increase in NH_3 concentration results in a reduction of the H_2SO_4 (aq.) and water. The overall aerosol mass decreases mainly because of the water loss, reaching a minimum for an ammonia concentration of 1.8 μg m^{-3}. A further increase in the ammonia concentration leads to a corresponding increase in the overall aerosol mass.

At the 298 K temperature, the deliquescence RH values for the NH_4HSO_4 and $(NH_4)_3H(SO_4)_2$ salts are 40% and 69%, respectively. A system characterized by higher RH (equal to 75%, for instance) is that the aerosol consists of a liquid solution of H_2SO_4, HSO_4^-, SO_4^{2-}, and NH_4^+ over the usual ammonia concentration range in which the formation of these salts is favored. It is important to bear in mind that (i) H_2SO_4 dominates for ammonia/sulfuric acid molar ratios lower than 0.5, (ii) HSO_4^- is the main aerosol component for values of the above ratio ranging from 0.5 to 1.5, and (iii) sulfate is formed for higher ammonia concentrations. When there is enough ammonia to completely neutralize the available sulfate forming $(NH_4)_2SO_4$, the liquid aerosol loses its water content and becomes solid. The deliquescence RH for $(NH_4)_2SO_4$ is equal to 80% at this temperature. Therefore, assuming it is on the deliquescence branch of the hysteresis curve, the aerosol will exist as solid matter only. These changes in the particulate chemical composition and the accompanying phase transformation significantly modify the hygroscopic characteristics of aerosol particles, resulting in a monotonic decrease of their water mass. The net result is a significant reduction of the overall aerosol mass with increasing ammonia for this system and an abrupt change accompanying the phase transition from liquid to solid. Summarizing, in very acidic atmospheres (i.e., for NH_3/H_2SO_4 molar ratios lower than 0.5), aerosol particles exist primarily as H_2SO_4 solutions, while in acidic atmospheres (in which the NH_3/H_2SO_4 molar ratio ranges from 0.5 to 1.5), the particles consist mainly of bisulfate. If there is a sufficient ammonia concentration to neutralize the available sulfuric acid, $(NH_4)_2SO_4$ or NH_4^+ and SO_4^{2-} solution constitutes the prevailing composition of the aerosol phase. For acidic atmospheres, all the available ammonia is taken up by the aerosol phase, and only for NH_3/H_2SO_4 molar ratios higher than 2 can ammonia also exist in the gas phase.

Numerous SEM and TEM images of ammonium sulfate aerosols with different sizes can be found in the literature, mixed with smaller soot particles or nearly spherical silica particles and with soot inclusions or surrounded by numerous tiny satellite droplets. For instance, Li, Anderson, and Buseck (2003a) showed images of ammonium sulfate particles mixed with small soot particles and silica spheres sampled at Sagres (Portugal) during the ACE-2 campaign and ammonium sulfate particles with soot inclusions as well as a particle consisting of NaCl with a mixed-cation sulfate rim, both sampled at Punta del Hidalgo (Canary Islands). The morphologic characteristics of some ammonium sulfate particles surrounded by numerous tiny satellite droplets were described by Kojima, Buseck, and Reeves (2005). Ammonium sulfate particles collected in the free troposphere near Pietersburg (South Africa) were shown by Li *et al.* (2003b), exhibiting both round and elongated shapes, in some cases aggregated with soot when larger in size. Various ammonium sulfate particles sampled in the urban environment of Mainz (Germany) were shown by Sinha *et al.* (2008), providing evidence of their mixed composition in such a polluted ambient.

The nearly spherical shape characteristics of sulfate ammonium particles facilitate the calculations based on the Mie (1908) theory of the radiative effects produced by these aerosol particles that scatter and absorb solar radiation. In fact, ammonium sulfate particles are highly hygroscopic and therefore gradually tend to assume more spherical shapes as the RH of the surrounding air increases to exceed 70–75% (Hänel, 1976). In order to better study the radiative effects of sulfate particles, the evaluations of the sulfate particulate matter refractive index proposed by Toon, Pollack, and Khare (1976) and Shettle and Fenn (1979) can be conveniently used. On the basis of these data, the real part of refractive index was estimated to have a stable value of 1.53 over the 0.30–0.70 μm wavelength range and to decrease with wavelength until reaching a value of 1.42 within the 2.0–2.5 μm spectral interval, while the imaginary part was in general very low at all the visible and near-infrared wavelengths, assuming values varying from 5×10^{-3} to 8×10^{-3} over the 0.3–0.7 μm wavelength range and slightly higher values in the near-infrared 0.86–2.50 μm range, varying from 8×10^{-3} to 2×10^{-2}, with a maximum at 1.30 μm and a pronounced minimum at around 2.00 μm.

Anthropogenic emissions constitute the main source of SO_2 over continents and in polluted marine air masses, where most sulfur compounds are lost by deposition before oxidation. Therefore, anthropogenic aerosols are mainly formed from SO_2 emitted by industrial activities, domestic heating, and vehicular traffic. As shown in Table 1.2, they predominate in concentration over those formed in the troposphere from natural SO_2. Being mostly generated by fossil fuel burning, the source distribution and magnitude of anthropogenic SO_2 are fairly well known, and recent estimates differ one from the other by no more than 30% (Lelieveld *et al.*, 1997). Anthropogenic SO_2 emissions were estimated to have decreased by ~25% from 1980 to 2000 when the major emitting regions shifted from Europe and North America to Southeast Asia (Liu, Penner, and Herzog, 2005).

1.6.2
Secondary Particles from NO_x

Nitrate particles generally exhibit larger sizes than sulfate particles suspended in marine air. Because seawater contains negligible nitrate concentrations, the nitrate in these particles must derive from the condensation of gaseous HNO_3, possibly through g-to-p conversion in the liquid phase. Nitrates are also common in continental aerosols, mainly formed from agricultural activities and deriving in part from the condensation of HNO_3 onto larger and more alkaline mineral particles, with sizes predominantly ranging from about 0.2 to 20 μm. The *in situ* chemical reactions involving and generating many important atmospheric trace constituents lead to the formation of nitrate aerosols. Most of such gaseous reactions are initiated by photolysis involving radicals and occur with one to three different molecules. These reactions can be classified as *homogeneous* or *heterogeneous*, according to the aforementioned definitions.

One of the most important homogeneous reactions is the following:

$$NO_2\ (g) + O_3\ (g) \rightarrow NO_3\ (g) + O_2\ (g), \quad (1.33)$$

which constitutes in practice the major source of the nitrate radical (NO_3) in the atmosphere. Mixing an inorganic aerosol, like nitric acid (HNO_3), with organic compounds (aldehydes) which can appreciably increase the rate of aerosol growth is an example of a heterogeneous reaction.

Odd nitrogen (NO_y), or total reactive nitrogen as it is sometimes called, refers to the sum of NO_x plus all the compounds produced by the atmospheric oxidation of NO_x, including HNO_3, NO_3, dinitrogen pentoxide (N_2O_5), and peroxyacetyl nitrate (PAN for short). Nitric acid can form from N_2O_5 in cloud water, and subsequently the evaporation of cloud water releases nitrate particles into the air. The main sources of NO_y are anthropogenic emissions, but away from pollution sources, soils and lightning can also play dominant roles. Interactions between NO_y and NMHC can lead to photochemical smogs in cities. On regional and global scales, interactions of NO_y with odd hydrogen have a strong influence on

OH concentrations. Even with rigorous pollution controls, increasing numbers of cars in urban centers can produce high concentrations of NO, which can then be converted to NO_2 through the following reaction:

$$2NO + O_2 \rightarrow 2NO_2. \tag{1.34}$$

The NO_2 can then be involved in the production of O_3. Because the rate coefficient of the reaction represented in Eq. (1.34) increases as temperature decreases, the production of NO_2 can become significant in urban areas during cold winters.

In the *ammonia–nitric acid–water system*, NH_4NO_3 is found in the aqueous state for air RH higher than the deliquescence point. In fact, ammonia and nitric acid react in the atmosphere to form ammonium nitrate (NH_4NO_3):

$$NH_3\ (g) + HNO_3\ (g) \rightarrow NH_4NO_3\ (s). \tag{1.35}$$

Ammonium nitrate is formed in areas characterized by high ammonia and nitric acid concentrations in the presence of low sulfate concentrations. Depending on the ambient RH, ammonium nitrate may exist as a solid or aqueous solution of NH_4^+ and NO_3^-. Equilibrium concentrations of gaseous NH_3 and HNO_3 and the resulting concentration of solid or aqueous NH_4NO_3 can be calculated on the basis of fundamental thermodynamic principles using the method of Stelson and Seinfeld (1982a,b). For low air RH conditions, at which NH_4NO_3 is in the solid state, the dissociation constant is equal to the product of the partial pressures of NH_3 and HNO_3. This parameter increases almost linearly as a function of temperature and therefore is quite sensitive to temperature changes, varying by more than two orders of magnitude for typical ambient conditions. Lower temperatures correspond to lower values of the dissociation constant and, hence, to lower equilibrium values of the NH_3 and HNO_3 gas-phase concentrations. Thus, lower temperatures shift the equilibrium of the system toward the aerosol phase, inducing an increase in the aerosol mass concentration of NH_4NO_3.

In ammonium nitrate solutions for RH higher than that of deliquescence, NH_4NO_3 is found in the aqueous state. The corresponding dissociation reaction is

$$NH_3\ (g) + HNO_3\ (g) \rightleftharpoons NH_4^+ + NO_3^-. \tag{1.36}$$

Solution concentrations with 8–26 molalities can be expected in wetted atmospheric aerosol. These concentrations depend not only on the aerosol nitrate and ammonium but also on the amount of water in the aerosol phase. Therefore, calculation of the aerosol solution composition requires estimation of the aerosol water content.

In the *ammonia–nitric acid–sulfuric acid–water* system, gaseous components (NH_3, HNO_3, H_2SO_4, H_2O) are present together with aqueous-phase components (NH_4^+, H^+, HSO_4^-, SO_4^{2-}, NO_3^-, and H_2O) and solid components (NH_4HSO_4, $(NH_4)_2SO_4$, NH_4NO_3, $(NH_4)_2SO_4 \cdot 2NH_4NO_3$, $(NH_4)_2SO_4 \cdot 3NH_4NO_3$, and $(NH_4)_3H(SO_4)_2$). Two remarks are useful to determine a priori the composition of the aerosol existing in such a system: (i) sulfuric acid possesses an extremely low vapor pressure, and (ii) $(NH_4)_2SO_4$ in the solid or aqueous state is the preferred form of sulfate. The first note implies that the amount of sulfuric acid in the gas phase will be negligible, and the second means that, if possible, each mole of sulfate will remove two moles of ammonia from the gas phase. Based on these observations, two regimes of interest can be considered that are ammonia poor and ammonia rich, respectively:

1. In the ammonia-poor case, there is insufficient NH_3 to neutralize the available sulfate, and therefore the aerosol phase will be acidic. The vapor pressure of NH_3 will be low, and the sulfate will tend to drive the nitrate to the gas phase. Since the NH_3 partial pressure will be low, the NH_3-HNO_3 partial pressure product will also be low, so that ammonium nitrate levels tend to be low or null, and sulfate may exist as bisulfate.
2. In the ammonia-rich case, there is an excess of ammonia, so that the aerosol phase will be largely neutralized. The ammonia that does not react with sulfate is thus available for reacting with nitrate until producing NH_4NO_3.

Therefore, at very low ammonia concentrations, sulfuric acid and bisulfate constitute the aerosol composition. As ammonia increases, ammonium nitrate becomes a significant aerosol constituent. The aerosol LWC varies nonlinearly as a function of these variations, reaching a minimum close to the transition between the two regimes. The formation of ammonium nitrate is often limited by the availability of one of the reactants. Rural areas with significant ammonia emissions are often in this regime, where NH_4NO_3 formation is nitric acid limited. The boundary between the nitric acid-limited and ammonia-limited regimes depends on temperature, air RH, sulfate concentrations, and concentrations of the other major inorganic aerosol components.

Reductions in the sulfate concentration affect the inorganic particle concentration in two different ways: (i) part of the sulfate may be replaced by nitric acid, leading to an increase in the ammonium nitrate content of the aerosol, while sulfate decrease frees up ammonia to react with nitric acid and transfers it to the aerosol phase; (ii) if the particles are aqueous, there is also a second effect, for which the reduction of the sulfate ions in solution increases the equilibrium vapor pressure product of the ammonia and nitric acid in the particles, shifting the ammonium nitrate partitioning toward the gas phase. In this second scenario for aqueous particles, the substitution of sulfate by nitrate is in general relatively small: sulfate reductions are accompanied by an increase in the aerosol nitrate and a reduction of the aerosol ammonium, water, and total mass. However, such a reduction of the mass is in general nonlinear. For example, a reduction of total sulfate from 30 to $10\,\mu g\,m^{-3}$ results in a net reduction of the dry aerosol mass (excluding water) of only $12.9\,\mu g\,m^{-3}$, because of the increase in the aerosol nitrate concentration by $10\,\mu g\,m^{-3}$, while ammonium decreases by only $2.9\,\mu g\,m^{-3}$.

These g-to-p conversion processes caused dramatic episodes of urban pollution during the second half of the twentieth century, when emissions of vehicle exhausts constituted an important source of pollutants in many urban areas. For sunlight and stagnant meteorological conditions, the combination of chemical species in the strongly polluted urban air can lead to *photochemical smog* cases, which are also called *Los Angeles-type smogs*. They are characterized by high concentrations of a wide variety of pollutants, such as nitrogen oxides, O_3, CO, hydrocarbons, aldehydes (and other eye-irritant materials), and sometimes sulfuric acid. The chemical reactions leading to photochemical smog are very complex, due to the interactions of a large number of organic pollutants (e.g., hydrocarbons such as ethylene and butane) with nitrogen oxides. The reactions start with the photolysis process

$$NO_2 + h\nu \rightarrow NO + O, \tag{1.37}$$

induced with rate coefficient j by the incoming solar radiation with wavelength $\lambda < 430$ nm, which is followed by the reaction

$$O + O_2 + M \rightarrow O_3 + M \tag{1.38}$$

forming O_3 molecules with rate coefficient k_1. However, O_3 is then depleted by the rapid reaction

$$O_3 + NO \rightarrow NO_2 + O_2, \tag{1.39}$$

characterized by a high rate coefficient k_2. If there are no other reactions, the reactions in Eqs (1.37 – 1.39) are expected to lead to a steady-state concentration of O_3 given by

$$[O_3] = j[NO_2]/k_2[NO]. \tag{1.40}$$

Equation (1.40) predicts O_3 concentrations in urban-polluted air of only ~0.03 ppmv, whereas typical values are well above this concentration and can exceed 0.5 ppmv. Therefore, other chemical reactions need to be involved to explain the photochemical smog occurrence. Most effective are reactions that oxidize NO to NO_2 without consuming O_3, since this would allow a net O_3 production, favoring a buildup of ozone concentrations during the day. The reactions leading to photochemical smog and due to organic compounds are examined in the next section dedicated to the SOAs.

In marine air, the main contributors to the mass of inorganic aerosols are the ions Na^+, Cl^-, Mg^{2+}, SO_4^{2-}, K^+, and Ca^{2+}. Apart from SO_4^{2-}, these compounds are contained primarily in particles with sizes of no more than a few micrometers, because they originate from sea-salt particles derived from bubble bursting. Sulfate mass concentrations exhibit the highest values over the 0.1 – 1 μm size range. As mentioned before, nitrates constitute the particles with sizes larger than those of sulfates in marine air. Therefore, seawater contains negligible concentrations of nitrate-containing particles, which are formed by condensation of gaseous HNO_3, mainly through g-to-p conversion in the liquid phase. For this reason, nitrate particles are common in continental aerosols, formed through condensation of HNO_3 onto larger and more alkaline mineral particles, assuming sizes varying mainly from 0.2 to 20 μm.

Considering other *inorganic aerosol species*, aerosol particles exhibit considerably high concentrations in regions close to seawater, consisting mainly of Na^+ and

Cl^- ions. Sodium and chloride interact with several aerosol components, generating a variety of particles formed during these reactions, including ammonium chloride, sodium nitrate, sodium sulfate, and sodium bisulfate, while HCl(g) may be released to the gas phase. The addition of NaCl to an urban aerosol can have a series of interesting effects, including the following reaction of NaCl with HNO_3:

$$\mathrm{NaCl\ (s) + HNO_3\ (g) \rightleftharpoons NaNO_3\ (s) + HCl\ (g).} \tag{1.41}$$

As a result of this reaction (see, for instance, De Bock, Van Maldered, and Van Grieken (1994)), more nitrate is transferred to the aerosol phase and is associated with the coarse sea-salt particles. At the same time, hydrochloric acid is liberated and the aerosol particles appear to be chloride deficient. As found by Pakkanen (1996) examining particulate samples collected in the urban area of Helsinki (Finland), high atmospheric concentrations of coarse nitrate particles can be formed in the reaction of nitric acid with sea- and soil-derived coarse particles. In particular, the reaction with sea-salt particles resulted in the evaporation of chlorine ions as HCl. The presence of nss sulfate ions indicates that also H_2SO_4 and/or SO_2 can react with sea salt and contribute to the loss of chlorine ions. In fact, such a chloride deficiency substantially results from the following pair of reactions:

$$\mathrm{2NaCl\ (s) + H_2SO_4\ (g) \rightleftharpoons Na_2SO_4\ (s) + 2HCl\ (g)} \tag{1.42}$$

and

$$\mathrm{NaCl\ (s) + H_2SO_4\ (g) \rightleftharpoons NaHSO_4\ (s) + HCl\ (g).} \tag{1.43}$$

The overall percentage of evaporated chlorine ions was found to decrease as the particle size increases, so that Cl^- evaporation was found to be almost complete over the $1-2\,\mu m$ size range, indicating that all major sea-derived chlorides (NaCl, $MgCl_2$, $CaCl_2$, and KCl) may react with acidic species in the atmosphere (Pakkanen, 1996).

1.6.3
Secondary Organic Aerosols

SOA particles originate in the atmosphere through the mass transfer of low-pressure products given by the oxidation of organic gases to the aerosol phase. Carbonaceous particles emitted directly into the atmosphere derive mainly from biomass fires. Organic and carbonaceous aerosols are produced by g-to-p conversion of gases released from the biosphere and from volatile compounds, such as crude oil leaking to the Earth's surface. The organic gases are oxidized in the gas phase by species such as the hydroxyl radical (OH), ozone (O_3), and the nitrate radical (NO_3), so that such oxidation products gradually accumulate. Some of these products have low volatilities and condense on the available particles in an effort to establish equilibrium between the gas and aerosol phases. The mass transfer flux of these products to the aerosol phase is proportional to the difference between their gas-phase concentration and their concentration in the gas phase at the particle surface. Thus, there are two separate steps involved in the production of SOAs: (i) the organic aerosol compound is produced in the gas phase during the reaction of parent organic gases, with rates closely depending on the gas-phase chemistry of the organic aerosol precursors, and (ii) the organic compound partitions between the gas and particulate phases form SOA, occurring through variable interactions among the various compounds present in both phases.

In the formation of photochemical smog, the OH radical can initiate a chain reaction, which attacks the hydrocarbon pollutants in urban air, through one of the following three reactions:

$$OH + CH_4 \rightarrow H_2O + CH_3 \tag{1.44}$$

$$OH + CO \rightarrow H + CO_2 \tag{1.45}$$

or

$$OH + CH_3CHO \rightarrow H_2O + CH_3CO. \tag{1.46}$$

The resulting radicals – CH_3 in Eq. (1.44), H in Eq. (1.45), and CH_3CO in Eq. (1.46) – are then involved in reactions that oxidize NO to NO_2 and regenerate OH. For example, CH_3 formed through the reaction given in Eq. (1.44) can initiate the following sequence of reactions:

$$CH_3 + O_2 \rightarrow CH_3O_2 \tag{1.47}$$

$$CH_3O_2 + NO \rightarrow CH_3O + NO_2 \tag{1.48}$$

$$CH_3O + O_2 \rightarrow HCHO + HO_2 \tag{1.49}$$

$$HO_2 + NO \rightarrow NO_2 + OH \tag{1.50}$$

to form OH and oxidize NO to NO_2.

The net result of these four reactions together with the reaction described in Eq. (1.44) is given by

$$CH_4 + 2O_2 + 2NO \rightarrow H_2O + 2NO_2 + HCHO, \tag{1.51}$$

where the oxidation of CH_4 results in the oxidation of NO to NO_2 without consuming O_3.

The reaction defined in Eq. (1.51) produces formaldehyde (HCHO), which is an eye irritant and constitutes an efficient source of HO_x, as can be seen in the sequence of the two following reactions producing HO_2:

$$HCHO + h\nu \rightarrow H + HCO \tag{1.52}$$

and

$$HCO + O_2 \rightarrow HO_2 + CO. \tag{1.53}$$

Similarly, the acetyl radical (CH_3CO) formed in the reaction described in Eq. (1.46) is involved in a series of reactions leading to the formation of the methyl radical CH_3 and the peroxyacetyl radical (CH_3COO_2). The methyl radical oxidizes NO through the reactions defined in Eqs. (1.47–1.50), and the peroxyacetyl radical reacts with nitrogen dioxide, as follows:

$$CH_3COO_2 + NO_2 \rightarrow CH_3COO_2NO_2. \tag{1.54}$$

The chemical species on the right-hand side of the reaction represented in Eq. (1.54) is the vapor of a colorless and dangerously explosive liquid called PAN, an important component of photochemical smog and another major eye irritant. Other alkenes oxidize NO to NO_2 without consuming O_3 and regenerate OH, doing this much faster than the previously reported reactions. The concentrations of some of the major components of photochemical smog in Los Angeles vary typically throughout the day. For these high-pollution conditions, ozone precursors, such as NO_x and hydrocarbons, build up during the morning rush hour, and aldehydes, O_3, and PAN peak in the early afternoon.

The role of PAHs in air pollution and public health was first studied in the early 1940s following the discovery that organic extracts of particles from polluted air (e.g., benzo[*a*]pyrene ($C_{20}H_{12}$), a five-ring PAH) produced cancer in laboratory experiments on animals, its metabolites being mutagenic and highly carcinogenic. The PAHs are emitted by diesel and gasoline engines, coal-fired electric power-generating plants, biomass burning, and cigarettes and are present in air as volatile particulates and gases. Reactions initiated by OH in the day and NO_3 during the night convert gaseous PAHs to nitro-PAH derivatives, which are responsible for about 50% of the mutagenic activity of respirable airborne particles monitored in southern California.

1.7
Concluding Remarks on the Global Annual Emission Fluxes of Natural and Anthropogenic Aerosol Mass

The previous sections showed that significant contributions are made by primary and secondary sources of natural and anthropogenic particulate matter to the global airborne aerosol loading. Table 1.1 presents the estimates of the average annual emissions of the various aerosol types on a global scale, due to the primary and secondary natural sources, chosen among those proposed in the literature over the past decades. Similarly, Table 1.5 provides the estimates of the average global annual emissions from the primary and secondary sources of anthropogenic aerosols evaluated since 2000. Examining the data reported in Table 1.1, it can be stated that the global annual emission flux Φ_e of natural primary particulate matter into the atmosphere varies:

1. over the $3 \times 10^3 - 2.0 \times 10^4$ Tg per year range (Andreae and Rosenfeld, 2008) for the sea-salt particles formed by wind forcing on the oceanic surfaces, with

average annual values of 1.01×10^4 Tg per year (Gong, Barrie, and Lazare, 2002), 3.3×10^3 Tg per year (Jaenicke, 2005), and 7.8×10^3 Tg per year (Tsigaridis *et al.*, 2006), for which an arithmetic mean value of about 7×10^3 Tg per year was reported in Table 1.6;

2. over the $1.0 \times 10^3 - 2.15 \times 10^3$ Tg per year range (Andreae and Rosenfeld, 2008) for the mineral dust mobilized by winds in desert regions and more generally on arid soil and rock debris, with average annual values of 1.49×10^3 Tg per year (Zender, Brian, and Newman, 2003), 2×10^3 Tg per year (Jaenicke, 2005), and 1.7×10^3 Tg per year (Tsigaridis *et al.*, 2006), for which an arithmetic mean value of 1.73×10^3 Tg per year was reported in Table 1.6;
3. from 15 to 70 Tg per year (Andreae and Rosenfeld, 2008) for the primary organic aerosol emitted from the various biological sources, with an average value of 50 Tg per year estimated by Kiehl and Rodhe (1995) for biological debris;
4. from 26 to 70 Tg per year (Andreae and Rosenfeld, 2008) for the biomass burning organic aerosol, with an average annual value of 44.4 Tg per year proposed by Tsigaridis *et al.* (2006) in good agreement with the IPCC (2001) estimate equal to 56 Tg per year;
5. from 4 to 90 Tg per year (Jaenicke, 1988) for primary and secondary aerosol fluxes injected into the troposphere by volcanic eruptions, with an average annual value of 30 Tg per year (Kiehl and Rodhe, 1995) (see Table 1.1), this quantity varying largely from one year to another, since it depends closely on the number frequency and intensity of volcanic eruptions;
6. from 4 to 10^4 Tg per year (Seinfeld and Pandis, 1998) for volcanic aerosol formed in the stratosphere from volcanic SO_2, with an average annual estimate of 20 Tg per year (Kiehl and Rodhe, 1995) comparable to those of (i) 21 Tg per year (IPCC, 2001), (ii) 30 Tg per year (Seinfeld and Pandis, 1998) – which are considerably higher than the value of 9.2 Tg per year proposed by Tsigaridis *et al.* (2006) – and (iii) the average value of 10 Tg per year derived by Toohey *et al.* (2011) over the last 20 years (see Table 1.1); and
7. from 1.5×10^{-4} to 4×10^{-2} Tg per year (Gardner *et al.*, 2014) for the global flux of cosmic dust entering the Earth's atmosphere, which are several orders of magnitude smaller than those attributed to sea salt and mineral dust sources, these evaluations implying an estimate of the average global flux of cosmic dust equal to 2×10^{-2} Tg per year, as reported in Table 1.6.

In addition, significant emission fluxes of secondary aerosol particles of natural origins need to be taken into account:

1. The overall sulfate aerosol flux Φ_e was estimated to vary from 107 to 374 Tg per year by Andreae and Rosenfeld (2008), including the flux of sulfate aerosols formed from natural DMS, which was estimated to be equal to 12.4 Tg per year by Liao *et al.* (2003) and 18.5 Tg per year by Tsigaridis *et al.* (2006). In agreement with the evaluations of Andreae and Rosenfeld

Table 1.6 Ranges and average values of global emission fluxes (measured in teragram per year, being 1 Tg yr^{-1} = 10^6 ton yr^{-1}) of the most important aerosol types due to natural, anthropogenic, and overall mixed (natural + anthropogenic) sources, as derived from the data given in Tables 1.1 and 1.5 for natural and anthropogenic particles and derived from the evaluations available in the literature of the past 20 years.

Aerosol particle type	Natural sources		Anthropogenic sources		Overall mixed sources	
	Range	Average	Range	Average	Range	Average
Sea salt	3000–20 000	7068	—		3000–20 000	7068
Mineral dust	1000–2150	1731	50–250	150	1050–2400	1881
Biogenic primary organic	15–70	50	—	—	15–70	50
Biomass burning organic	26–70	44	12–270	90	38–340	134
Tropospheric volcanic	4–90	30	—	—	4–90	30
Stratospheric volcanic	2–17	10	—	—	2–17	10
Cosmic dust in the middle atmosphere	2×10^{-4} to 4×10^{-2}	2×10^{-2}	—	—	2×10^{-4} to 4×10^{-2}	2×10^{-2}
Sulfates	107–374	165	50–122	73	157–496	238
Nitrates	12–27	25	90–118	95	102–145	115
Overall biogenic from isoprene, monoterpenes, and VOCs	835–1000	900	—	—	835–1000	900
Biogenic secondary organic	3–83	11	—	—	3–83	11
Industrial dust	—	—	40–130	100	40–130	100
Carbonaceous from hydrocarbons and VOCs	—	—	15–90	29	15–90	29
Black carbon (soot)	—	—	8–14	12	8–14	12

(2008), an average annual value of 165 Tg per year was assumed in Table 1.6, according to the SMIC (1971) evaluations.

2. The overall nitrate aerosol flux was estimated by Andreae and Rosenfeld (2008) to range from 12 to 27 Tg per year, with an average annual value of 25 Tg per year determined by the IPCC (2001) and reported in Table 1.6.
3. From about 3 to 83 Tg per year (Andreae and Rosenfeld, 2008) for the SOA, mainly given by the SOA formed from biogenic VOCs, estimated to provide an average global annual flux of about 11 Tg per year (Chung and Seinfeld, 2002). In reality, the overall emission of biogenic aerosol was estimated to be equal to 8.35×10^2 Tg per year by Tsigaridis *et al.* (2006), due to isoprenes, monoterpenes, and nonaromatic and aromatic VOCs, thus indicating that the overall annual flux of biological aerosol should be with good approximation equal to 10^3 Tg per year, as estimated by Jaenicke (2005) (see also Table 1.6).

For all the above average annual global estimates of natural primary and secondary particles, the overall annual emission of natural aerosol on the global scale was evaluated to be 5875 Tg per year on average (IPCC, 2001) but to vary from one year to another between 4185 and more than 2.2×10^4 Tg per year if the evaluations of Andreae and Rosenfeld (2008) are taken into account and the volcanic debris estimates proposed by the United Nations (1979) are included. More than 80% of the overall global annual emission flux of particulate mass turns out to be given by sea-salt particles.

The global annual emissions of particulate matter due to man-made activities arise primarily from four source categories that are (i) fuel combustion, (ii) industrial processes, (iii) nonindustrial fugitive sources (such as roadway dust from paved and unpaved roads, wind erosion of cropland, construction, etc.), and (iv) transportation sources (cars, ships, airplanes, etc.). Fugitive particles are those not emitted from a definable point such as a stack. Industrial fugitive dust emissions result from wind erosion of storage piles and unpaved plant roads and from vehicular traffic over plant roads. Fugitive process emissions are related to industrial activities, such as materials handling, loading, and transfer operations. Transportation source emissions occur mainly as vehicle exhaust and vehicle-related particles from tire, clutch, and brake wear. Engine-related particulate emissions are composed primarily of lead halides, sulfates, and carbonaceous matter and have sizes mostly smaller than 1 μm. About 40% of particles from tire wear are less than 10 μm and about 20% less than 1 μm, being primarily composed of carbon. Particles from brake linings are less than 1 μm and are composed mainly of carbon and asbestos.

The estimate of emissions of a certain aerosol species from a source is based on a technique using "emission factors" based on source-specific emission measurements evaluated as a function of activity level (e.g., amount of annual production at an industrial facility) with regard to each source. For example, to sample a power plant's emissions of SO_2 or NO_x at the stack, knowing the plant's boiler design and its consumption rate, the sulfur and nitrogen content of fuel burned can be used to calculate an emission factor of kilograms of SO_2 or NO_x emitted per metric ton (Mg) of fuel consumed. Emission factors currently in use are developed from only a limited sampling of the emissions source population for any given category, and the values reported are an average of those limited samples and might not be statistically representative of populations. The formulation of emission factors for mobile sources, the major sources of VOCs and NO_x, is based on rather complex emission estimation models used in conjunction with data from laboratory testing of representative groups of motor vehicles.

Anthropogenic activities emit rather high number concentrations of particles into the atmosphere, both directly and through g-to-p conversion. For particles with $a \geq 2.5\,\mu m$ (coarse particles), human activities worldwide are estimated to produce ~15% of natural emissions, with industrial processes, fuel combustion, and g-to-p conversion accounting for ~80% of the anthropogenic emissions. However, anthropogenic sources in urban areas are much more important. For particles with $a < 2\,\mu m$ (fine particles), human activities produce ~20% of the overall natural emissions, with g-to-p conversion accounting for ~90% of the anthropogenic emissions. Table 1.5 summarizes the estimates of the annual global emissions of particulate matter due to anthropogenic activities, as estimated through various studies reported in the literature. From these data, it can be stated that the annual global emissions of particulate mass due to anthropogenic activities vary:

1. from 50 to 250 Tg per year (SMIC, 1971) for the dust from soil erosion and wind-forced mobilization associated with agricultural activities, with an average annual value of 150 Tg per year;
2. from 12 to 270 Tg per year (Liousse *et al.*, 1996; IPCC, 2001) of carbonaceous aerosols from deforestation activities and biomass burning, with an average annual value of 90 Tg per year (Seinfeld and Pandis, 1998);
3. from about 50 to 122 Tg per year (IPCC, 2001; Liao *et al.*, 2003) for anthropogenic sulfates, with an average annual value of 73 Tg per year (Tsigaridis *et al.*, 2006);
4. from 40 to 118 Tg per year (IPCC, 2001) for anthropogenic nitrates, with an average annual value of 44 Tg per year (Tsigaridis *et al.*, 2006);
5. from 40 to 130 Tg per year (Seinfeld and Pandis, 1998) for industrial dust, with an average annual value of 100 Tg per year (Kiehl and Rodhe, 1995);
6. from 15 to 90 Tg per year (SMIC, 1971; Liousse *et al.*, 1996) of carbonaceous aerosols from hydrocarbons, anthropogenic VOCs, and industrial activities, with an average annual value of 29 Tg per year (IPCC, 2001); and
7. from 8 to 14 Tg per year (Andreae and Rosenfeld, 2008) for BC (soot) aerosols, with an average annual value of 12 Tg per year (Liousse *et al.*, 1996).

For these average estimates of the annual global emissions of various anthropogenic aerosol types, an overall emission of ~580 Tg per year can be assumed on average for the anthropogenic particulate matter. Thus, for the overall contribution of natural aerosols determined previously, the total (natural + anthropogenic) annual emissions over a global scale provide approximatively an overall average annual flux of about 6.5×10^3 Tg per year, which does not differ appreciably from the estimate of 6.3×10^3 Tg per year established by the IPCC (2001).

Abbreviations

BC	black carbon
CCN	cloud condensation nuclei
EC	elemental carbon
HRTEM	high-resolution transmission electron microscope
LWC	liquid water content
NTP	normal temperature and pressure
OC	organic carbon
PAHs	polycyclic aromatic hydrocarbons
SEM	scanning electron microscope
TEM	transmission electron microscope
VOCs	volatile organic compounds

List of Symbols

Latin

Δt_L	atmospheric lifetime of aerosol particles
a	"equivalent" diameter of an aerosol particle assumed to have a spherical shape
a_c	mode diameter of an aerosol size-distribution curve
B	shape parameter used by Monahan *et al.* (1986) in their size-dependent production rate formula
h	Planck's constant, defining the quantum of action, that is, the proportionality constant between the energy of a photon and its frequency ν: It is estimated to be equal to 6.626×10^{-34} J s
J_{s-s}	production rate of sea salt particles at the surface level (measured in $m^{-2}\ \mu m^{-1}\ s^{-1}$)
M_c	airborne mass concentration of crustal material near the surface
M_S	vertical mass loading M_S of volcanic aerosols in the stratosphere
$M_{s\text{-}s}$	total mass concentration of sea salt particles, measured in $\mu g\ m^{-3}$ and produced by wind forcing on the sea surface
RH	relative humidity of air (measured in %)
WS	wind speed (measured in $m\ s^{-1}$)
z	altitude above mean sea level (measured in m a.m.s.l.)

Greek

ν	frequency of the incoming solar irradiance
$\tau_a(\lambda)$	aerosol optical thickness at wavelength λ

Φ_e annual global emission flux of aerosol particulate mass (measured in Tg per year)

α Ångström exponent (i.e., first atmospheric turbidity parameter given by the Ångström's formula)

β second atmospheric turbidity parameter (giving the best-fit value of $\tau_a(\lambda)$ at the wavelength $\lambda = 1$ μm)

References

Adachi, K. and Buseck, P.R. (2008) Internally mixed soot, sulfates, and organic matter in aerosol particles from Mexico City. *Atmos. Chem. Phys.*, **8** (21), 6469–6481. doi: 10.5194/acp-8-6469-2008

Adams, P.J., Seinfeld, J.H., and Koch, D.M. (1999) Global concentrations of tropospheric sulfate, nitrate and ammonium aerosol simulated in a general circulation model. *J. Geophys. Res.*, **104** (D11), 13791–13823. doi: 10.1029/1999JD900083

Adams, P.J., Seinfeld, J.H., Koch, D.M., Mickley, L., and Jacob, D. (2001) General circulation model assessment of direct radiative forcing by the sulfate-nitrate-ammonium-water inorganic aerosol system. *J. Geophys. Res.*, **106** (D1), 1097–1111. doi: 10.1029/2000JD900512

Andreae, M.O. (1995) Climatic effects of changing atmospheric aerosol levels, in *World Survey of Climatology: Future Climates of the World*, vol. **16** (ed. A. Henderson-Sellers), Elsevier, Amsterdam, pp. 341–392.

Andreae, M.O. and Crutzen, P.J. (1997) Atmospheric aerosols: biogeochemical sources and role in atmospheric chemistry. *Science*, **276** (5315), 1052–1056. doi: 10.1126/science.276.5315.1052

Andreae, M.O. and Gelencsér, A. (2006) Black carbon or brown carbon? The nature of light-absorbing carbonaceous aerosols. *Atmos. Chem. Phys.*, **6** (10), 3131–3148. doi: 10.5194/acp-6-3131-2006

Andreae, M.O. and Jaeschke, W.A. (1992) Exchange of sulphur between biosphere and atmosphere over temperate and tropical regions, in *Sulphur Cycling on the Continents: Wetlands, Terrestrial Ecosystems and Associated Water Bodies, SCOPE 48* (eds R.W. Howarth, J.W.B. Stewart, and M.V. Ivanov), John Wiley & Sons, Ltd, Chichester, pp. 27–61.

Andreae, M.O. and Rosenfeld, D. (2008) Aerosol-cloud-precipitation interactions. Part 1. The nature and sources of cloud-active aerosols. *Earth Sci. Rev.*, **89** (1-2), 13–41. doi: 10.1016/j.earscirev.2008.03.001

Andres, R.J. and Kasgnoc, A.D. (1998) A time-averaged inventory of subaerial volcanic sulfur emissions. *J. Geophys. Res.*, **103** (D19), 25251–25261. doi: 10.1029/98JD02091

Ångström, A. (1964) The parameters of atmospheric turbidity. *Tellus*, **16** (1), 64–75. doi: 10.1111/j.2153-3490.1964.tb00144.x

Bacci, P., Del Monte, M., Longhetto, A., Piano, A., Prodi, F., Redaelli, P., Sabbioni, C., and Ventura, A. (1983) Characterization of the particulate emission by a large oil fuel fired power plant. *J. Aerosol Sci.*, **14** (4), 557–572. doi: 10.1016/0021-8502(83)90011-3

Bates, T.S., Lamb, B.K., Guenther, A., Dignon, J., and Stoiber, R.E. (1992) Sulphur emissions to the atmosphere from natural sources. *J. Atmos. Chem.*, **14** (1-4), 315–337. doi: 10.1007/BF00115242

Benkovitz, C.M., Berkowitz, C.M., Easter, R.C., Nemesure, S., Wagner, R., and Schwartz, S.E. (1994) Sulfate over the North Atlantic and adjacent continental regions: evaluation of October and November 1986 using a three-dimensional model driven by observation-derived meteorology. *J. Geophys. Res.*, **99** (D10), 20725–20756. doi: 10.1029/94JD01634

Benkovitz, C.M., Scholtz, M.T., Pacyna, J., Tarrasón, L., Dignon, J., Voldner, E.C., Spiro, P.A., Logan, J.A., and Graedel, T.E. (1996) Global gridded inventories of anthropogenic emissions of sulfur and nitrogen. *J. Geophys. Res.*, **101** (D22), 29239–29253. doi: 10.1029/96JD00126

Bigg, E.K. and Leck, C. (2008) The composition of fragments of bubbles bursting at the ocean surface. *J. Geophys. Res.*, **113**, D11209. doi: 10.1029/2007JD009078

Blanchard, D.C. and Woodcock, A.H. (1957) Bubble formation and modification in the sea and its meteorological significance. *Tellus*, **9** (2), 145–158. doi: 10.1111/j.2153-3490.1957.tb01867.x

Bond, T.C. and Bergstrom, R.W. (2006) Light absorption by carbonaceous particles: an investigative review. *Aerosol Sci. Technol.*, **40** (1), 27–67. doi: 10.1080/02786820500421521

Bouwman, A.F., Lee, D.S., Asman, W.A.H., Dentener, F.J., Van Der Hoek, K.W., and Olivier, J.G.J. (1997) A global high-resolution emission inventory for ammonia. *Global Biogeochem. Cycles*, **11** (4), 561–588. doi: 10.1029/97GB02266

Budyko, M.I. (1977) *Climatic Changes*, American Geophysical Union, Waverly Press, Inc., Washington, DC (USA), p. 261.

Bush, B.C. and Valero, F.P.J. (2002) Spectral aerosol radiative forcing at the surface during the Indian Ocean Experiment (INDOEX). *J. Geophys. Res.*, **107** (D19), 8003. doi: 10.1029/2000JD000020

Cadle, R.D. (1966) *Particles in the Atmosphere and Space*, Reinhold, New York.

Campanelli, M., Estelles, V., Smyth, T., Tomasi, C., Martìnez-Lozano, M.P., Claxton, B., Muller, P., Pappalardo, G., Pietruczuk, A., Shanklin, J., Colwell, S., Wrench, C., Lupi, A., Mazzola, M., Lanconelli, C., Vitale, V., Congeduti, F., Dionisi, D., Cardillo, F., Cacciani, M., Casasanta, G., and Nakajima, T. (2012) Monitoring of Eyjafjallajökull volcanic aerosol by the new European Skynet Radiometers (ESR) network. *Atmos. Environ.*, **48**, 33–45. doi: 10.1016/j.atmosenv.2011.09.070

Chang, S.G., Brodzinsky, R., Gundel, L.A., and Novakov, T. (1982) Chemical and catalytic properties of elemental carbon, in *Particulate Carbon: Atmospheric Life Cycle* (eds G.T. Wolff and R.L. Klimsch), Plenum Press, New York, pp. 158–181.

Charlson, R.J., Langner, J., Rodhe, H., Leovy, C.B., and Warren, S.G. (1991) Perturbation of the northern hemisphere radiative balance by backscattering from anthropogenic sulfate aerosols. *Tellus B*, **43** (4), 152–163. doi: 10.1034/j.1600-0870.1991.00013.x

Charlson, R.J., Lovelock, J.E., Andreae, M.O., and Warren, S.G. (1987) Oceanic phytoplankton, atmospheric sulphur, cloud albedo and climate. *Nature*, **326** (6114), 655–661. doi: 10.1038/326655a0

Charlson, R.J., Schwartz, S.E., Hales, J.M., Cess, R.D., Coakley, J.A. Jr.,, Hansen, J.E., and Hofmann, D.J. (1992) Climate forcing by anthropogenic aerosols. *Science*, **255** (5043), 423–430. doi: 10.1126/science.255.5043.423

Chung, S.H. and Seinfeld, J.H. (2002) Global distribution and climate forcing of carbonaceous aerosols. *J. Geophys. Res.*, **107**, D19, AAC 14-1–AAC 14-33. doi: 10.1029/2001JD001397

Chylek, P. and Coakley, J.A. Jr., (1974) Aerosols and climate. *Science*, **183** (4120), 75–77. doi: 10.1126/science.183.4120.75

Claeys, M., Graham, B., Vas, G., Wang, W., Vermeylen, R., Pashynska, V., Cafmeyer, J., Guyon, P., Andreae, M.O., Artaxo, P., and Maenhaut, W. (2004) Formation of secondary organic aerosols through photooxidation of isoprene. *Science*, **303** (5661), 1173–1176. doi: 10.1126/science.1092805

Crutzen, P. (2006) Albedo enhancement by stratospheric sulfur injections: a contribution to resolve a policy dilemma? *Clim. Change*, **77** (3-4), 211–220. doi: 10.1007/s10584-006-9101-y

Cwiertny, D.M., Baltrusaitis, J., Hunter, G.J., Laskin, A., Scherer, M.M., and Grassian, V.H. (2008) Characterization and acid-mobilization study of iron-containing mineral dust source materials. *J. Geophys. Res.*, **113**, D05202. doi: 10.1029/2007JD009332

De Bock, L.A., Van Maldered, H., and Van Grieken, R.E. (1994) Individual aerosol particle composition variations ion air masses crossing the North Sea. *Environ. Sci. Technol.*, **28** (8), 1513–1520. doi: 10.1021/es00057a021

Del Monte, M. and Sabbioni, C. (1984) Morphology and mineralogy of fly ash from a coal-fueled power plant. *Arch. Meteorol. Geophys. Bioclimatol., Ser. B*, **35** (1-2), 93–104. doi: 10.1007/BF02269412

Deuzé, J.L., Herman, M., Goloub, P., Tanré, D., and Marchand, A. (1999) Characterization of aerosols over ocean from POLDER/ADEOS-1. *Geophys. Res. Lett.*, **26** (10), 1421–1424. doi: 10.1029/1999GL900168

Duce, R. (1995) Distributions and fluxes of mineral aerosol, in *Aerosol Forcing on Climate* (eds R.J. Charlson and J. Heintzenberg), John Wiley & Sons, Ltd, Chichester, pp. 43–72.

Elbert, W., Taylor, P.E., Andreae, M.O., and Pöschl, U. (2007) Contribution of fungi to primary biogenic aerosols in the atmosphere: wet and dry discharged spores, carbohydrates, and inorganic ions. *Atmos. Chem. Phys.*, **7** (17), 4569–4588. doi: 10.5194/acp-7-4569-2007

Erickson, D.J. III, and Duce, R.A. (1988) On the global flux of atmospheric sea salt. *J. Geophys. Res.*, **93** (c011), 14079–14088. doi: 10.1029/JC093ic011p14079

Falkovich, A.H., Ganor, E., Levin, Z., Formenti, P., and Rudich, Y. (2001) Chemical and mineralogical analysis of individual mineral dust particles. *J. Geophys. Res.*, **106** (D16), 18029–18036. doi: 10.1029/2000JD900430

Flagan, R.C. and Seinfeld, J.H. (1988) *Fundamentals of Air Pollution Engineering*, Prentice-Hall, Englewood Cliffs, NJ.

Flanner, M.G., Shell, K.M., Barlage, M., Perovich, D.K., and Tschudi, M.A. (2011) Radiative forcing and albedo feedback from the Northern Hemisphere cryosphere between 1979 and 2008. *Nat. Geosci.*, **4** (3), 151–155. doi: 10.1038/ngeo1062

Forster, P., Ramaswamy, V., Artaxo, P., Berntsen, T., Betts, R., Fahey, D.W., Haywood, J., Lean, J., Lowe, D.C., Myhre, G., Nganga, J., Prinn, R., Raga, G., Schulz, M., and Van Dorland, R. (2007) Changes in atmospheric constituents and radiative forcing, in *Climate Change 2007: The Physical Science Basis. Contribution of Working Group I to the Fourth Assessment Report of the Intergovernmental Panel on Climate Change* (eds S. Solomon, D. Qin, M. Manning, Z. Chen, M. Marquis, K.B. Averyt, M. Tignor, and H.L. Miller), Cambridge University Press, Cambridge, pp. 129–234.

Gao, Y., Anderson, J.R., and Hua, X. (2007) Dust characteristics over the North Pacific observed through shipboard measurements during the ACE-Asia experiment. *Atmos. Environ.*, **41** (36), 7907–7922. doi: 10.1016/j.atmosenv.2007.06.060

Gardner, C.S., Liu, A.Z., Marsh, D.R., Feng, W., and Plane, J.M.C. (2014) Inferring the global cosmic dust influx to the Earth's atmosphere from lidar observations of the vertical flux of mesospheric Na. *J. Geophys. Res.*, **119** (9), 7870–7879. doi: 10.1002/2014JA020383

Gelencsér, A. (2004) *Carbonaceous Aerosol*, Springer-Verlag, New York and Berlin, p. 350.

Gong, S.L., Barrie, L.A., Blanchet, J.-P., and Spacek, L. (1998) Modeling size-distributed sea salt aerosols in the atmosphere: an application using Canadian climate models, in *Air Pollution Modeling and Its Applications XII*, NATO, Challenges of Modern Society, vol. **22** (eds S.-E. Gryning and N. Chaumerliac), Plenum Press, New York, pp. 337–345. doi: 10.1007/978-1-4757-9128-0_35

Gong, S.L., Barrie, L.A., and Lazare, M. (2002) Canadian Aerosol Module (CAM): a size-segregated simulation of atmospheric aerosol processes for climate and air quality models. 2. Global sea-salt aerosol and its budgets. *J. Geophys. Res.*, **107** (D24), 4779. doi: 10.1029/2001JD002004

Graf, H.-F., Feichter, J., and Langmann, B. (1997) Volcanic sulfur emissions: estimates of source strength and its contribution to the global sulfate distribution. *J. Geophys. Res.*, **102** (D9), 10727–10738. doi: 10.1029/96JD03265

Graham, B., Mayol-Bracero, O.L., Guyon, P., Roberts, G.C., Decesari, S., Facchini, M.C., Artaxo, P., Maenhaut, W., Köll, P., and Andreae, M.O. (2002) Water-soluble organic compounds in biomass burning aerosols over Amazonia. *J. Geophys. Res.*, **107** (D20), 8047. doi: 10.1029/2001JD000336

Guenther, A., Hewitt, C.N., Erickson, D., Fall, R., Geron, C., Graedel, T., Harley, P., Klinger, L., Lerdau, M., McKay, W.A., Pierce, T., Scholes, B., Steinbrecher, R., Tallamraju, R., Taylor, J., and Zimmerman, P. (1995) A global model of natural volatile organic compound emissions. *J. Geophys. Res.*, **100** (D5), 8873–8892. doi: 10.1029/94JD02950

Hand, J.L., Malm, W.C., Laskin, A., Day, D., Lee, T., Wang, C., Carrico, C., Carrillo, J., Cowin, J.P., Collett, J. Jr.,, and Iedema, M.J. (2005) Optical, physical, and chemical properties of tar balls observed during the Yosemite Aerosol Characterization Study. *J. Geophys. Res.*, **110**, D21210. doi: 10.1029/2004JD005728

Hänel, G. (1976) The properties of atmospheric aerosol particles as functions of the relative humidity at thermodynamic equilibrium with the surrounding moist air. *Adv. Geophys.*, **19**, 73–188. doi: 10.1016/S0065-2687(08)60142-9

Hansen, J.E. and Nazarenko, L. (2004) Soot climate forcing via snow and ice albedos. *Proc. Natl. Acad. Sci. U.S.A.*, **101** (2), 423–428. doi: 10.1073/pnas.2237157100

Harris, S.J. and Weiner, A.M. (1983a) Surface growth of soot particles in premixed ethylene/air flames. *Combust. Sci. Technol.*, **31** (3-4), 155–167. doi: 10.1080/00102208308923637

Harris, S.J. and Weiner, A.M. (1983b) Determination of the rate constant for soot surface growth. *Combust. Sci. Technol.*, **32** (5-6), 267–275. doi: 10.1080/00102208308923661

Havers, N., Burba, P., Lambert, J., and Klochow, D. (1998) Spectroscopic characterization of humic-like substances in airborne particulate matter. *J. Atmos. Chem.*, **29** (1), 45–54. doi: 10.1023/A:1005875225800

Hays, M.D. and Vander Wal, R.L. (2007) Heterogeneous soot nanostructure in atmospheric and combustion source aerosols. *Energy Fuels*, **21** (2), 801–811. doi: 10.1021/ef060442h

Heintzenberg, J. (1994) The life cycle of the atmospheric aerosol, in *Topics in Atmospheric and Interstellar Physics and Chemistry*, Vol. **1**, Chapter XII (ed. F. Boutron), Les Editions de Physique, Sciences, Les Ulis, France ERCA, pp. 251–270.

Hinds, W.C. (1999) *Aerosol Technology: Properties, Behavior, and Measurement of Airborne Particles*, 2nd edn, John Wiley & Sons, Inc., New York, pp. 504.

Hobbs, P.V. (2000) *Introduction to Atmospheric Chemistry*, Cambridge University Press, New York, p. 182.

Hudson, P.K., Murphy, D.M., Cziczo, D.J., Thomson, D.S., de Gouw, J.A., Warneke, C., Holloway, J., Jost, H.-J., and Hübler, G. (2004) Biomass-burning particle measurements: characteristics composition and chemical processing. *J. Geophys. Res.*, **109**, D23S27. doi: 10.1029/2003JD004398

IPCC (Intergovernmental Panel on Climate Change) (2001) *Climate Change 2001. The Scientific Basis* (eds J.T. Houghton, Y. Ding, D.J. Griggs, M. Noguer, P.J. van der Linden, X. Dai, K. Maskell, and C.A. Johnson), Cambridge University Press, Cambridge and New York, p. 881.

Jacobson, M.Z. (2001) Global direct radiative forcing due to multicomponent anthropogenic and natural aerosols. *J. Geophys. Res.*, **106** (D2), 1551–1568. doi: 10.1029/2000JD900514

Jaenicke, R. (1988) Aerosol physics and chemistry, in *Landolt-Börnstein Numerical Data and Functional Relationship in Science and Technology*, New Series Group V, Geophysics and Space Research, Meteorology, vol. **4** (ed. G. Fischer), Springer-Verlag, Heidelberg, pp. 391–457.

Jaenicke, R. (2005) Abundance of cellular material and proteins in the atmosphere. *Science*, **308**, 73. doi: 10.1126/science.1106335

Jensen, E.J. and Toon, O.B. (1997) The potential impact of soot particles from aircraft exhaust on cirrus clouds. *Geophys. Res. Lett.*, **24** (3), 249–252. doi: 10.1029/96GL03235

Jickells, T.D., An, Z.S., Andersen, K.K., Baker, A.R., Bergametti, C., Brooks, N., Cao, J.J., Boyd, P.W., Duce, R.A., Hunter, K.A., Kawahata, H., Kubilay, N., la Roche, J., Liss, P.S., Mahowald, N., Prospero, J.M., Ridgwell, A.J., Tegen, I., and Torres, R. (2005) Global iron connections between desert dust, ocean biogeochemistry, and climate. *Science*, **308** (5718), 67–71. doi: 10.1126/science.1105959

Jones, P.R., Charlson, R.J., and Rodhe, H. (1994) Aerosols, in *Climate Change 1994: Radiative Forcing of Climate Change and An Evaluation of the IPCC 1992 Emission Scenarios* (eds J.T. Houghton,

L.G. Meira Filho, J.P. Bruce, H. Lee, B.A. Callander, and E.F. Haites), Cambridge University Press, Cambridge, pp. 131–162.

Junge, C.E. (1963) *Air Chemistry and Radioactivity*, Academic Press, New York, p. 382.

Junge, C. (1979) The importance of mineral dust as an atmospheric constituent, in *Saharan Dust: Mobilization, Transport, Deposition*, SCOPE Report No. 14 (ed. C. Morales), John Wiley & Sons, Ltd, Chichester, pp. 49–60.

Karam, D.B., Flamant, C., Cuesta, J., Pelon, J., and Williams, E. (2010) Dust emission and transport associated with a Saharan depression: February 2007 case. *J. Geophys. Res.*, **115** (D4), D00H27. doi: 10.1029/2009JD012390

Kettle, A.J. and Andreae, M.O. (2000) Flux of dimethylsulfide from the oceans: a comparison of updated data sets and flux models. *J. Geophys. Res.*, **105** (D22), 26793–26808. doi: 10.1029/2000JD900252

Kiang, C.S., Stauffer, D., Mohnen, V.A., Bricard, J., and Vigla, D. (1973) Heteromolecular nucleation theory applied to gas-to-particle conversion. *Atmos. Environ.*, **7** (12), 1279–1283. doi: 10.1016/0004-6981(73)90137-6

Kiehl, J.T. and Rodhe, H. (1995) Modeling geographical and seasonal forcing due to aerosols, in *Aerosol Forcing of Climate* (eds R.J. Charlson and J. Heintzenberg), John Wiley & Sons, Inc., New York, pp. 281–296.

Kojima, T., Buseck, P.R., and Reeves, J.M. (2005) Aerosol particles from tropical convective systems: 2. Cloud bases. *J. Geophys. Res.*, **110**, D09203. doi: 10.1029/2004JD005173

Kokhanovsky, A.A. (2003) *Polarization Optics of Random Media*, Springer-Praxis, Chichester, p. 224.

Krueger, B.J., Grassian, V.H., Cowin, J.P., and Laskin, A. (2004) Heterogeneous chemistry of individual mineral dust particles from different dust source regions: the importance of particle mineralogy. *Atmos. Environ.*, **38** (36), 6253–6261. doi: 10.1016/j.atmosenv.2004.07.010

Kulmala, M. and Laaksonen, A. (1990) Binary nucleation of water sulfuric acid system. Comparison of classical theories with different H_2SO_4 saturation vapor pressures. *J. Chem. Phys.*, **93** (1), 696–701.

Kulmala, M., Vehkamäki, H., Petäjä, T., Dal Maso, M., Lauri, A., Kerminen, V.-M., Birmili, W., and McMurry, P.H. (2004) Formation and growth rates of ultrafine atmospheric particles: a review of observations. *J. Aerosol Sci.*, **35** (2), 143–176. doi: 10.1016/j.jaerosci.2003.10.003

Kwok, S. and Zhang, Y. (2011) Mixed aromatic–aliphatic organic nanoparticles as carriers of unidentified infrared emission features. *Nature*, **479** (7371), 80–83. doi: 10.1038/nature10542

Leck, C. and Bigg, E.K. (2005) Source and evolution of the marine aerosol - A new perspective. *Geophys. Res. Lett.*, **32** (L19803), 1–4. doi: 10.1029/2005GL023651

Lelieveld, J., Roelofs, G.J., Ganzeveld, L., Feichter, J., and Rodhe, H. (1997) Terrestrial sources and distribution of atmospheric sulphur. *Philos. Trans. R. Soc. London, Ser. B*, **352**, 149–158. doi: 10.1098/rstb.1997.0010

Li, J., Anderson, J.R., and Buseck, P.R. (2003a) TEM study of aerosol particles from clean and polluted marine boundary layers over the North Atlantic. *J. Geophys. Res.*, **108** (D6), 4189. doi: 10.1029/2002JD002106

Li, J., Pósfai, M., Hobbs, P.V., and Buseck, P.R. (2003b) Individual aerosol particles from biomass burning in southern Africa: 2. Compositions and aging of inorganic particles. *J. Geophys. Res.*, **108** (D13), 8484. doi: 10.1029/2002JD002310

Liao, H., Adams, P.J., Seinfeld, J.H., Mickley, L.J., and Jacob, D.J. (2003) Interactions between tropospheric chemistry and aerosols in a unified GCM simulation. *J. Geophys. Res.*, **108** (D1), 4001. doi: 10.1029/2001JD001260

Liao, H., Seinfeld, J.H., Adams, P.J., and Mickley, L.J. (2004) Global radiative forcing of coupled tropospheric ozone and aerosols in a unified general circulation model. *J. Geophys. Res.*, **109**, D16207. doi: 10.1029/2001JD002004

Lin, Y., Sim, M.S., and Ono, S. (2011) Multiple-sulfur isotope effects during photolysis of carbonyl sulfide. *Atmos. Chem. Phys.*, **11** (19), 10283–10292. doi: 10.5194/acp-11-10283-2011

Liousse, C., Penner, J.E., Chuang, C., Walton, J.J., Eddleman, H., and Cachier, H. (1996) A global three-dimensional model study of carbonaceous aerosols. *J. Geophys. Res.*, **101** (D14), 19411–19432. doi: 10.1029/95JD03426

Liu, X., Penner, J.E., and Herzog, M. (2005) Global modeling of aerosol dynamics: model description, evaluation, and interactions between sulfate and nonsulfate aerosols. *J. Geophys. Res.*, **110**, D18206. doi: 10.1029/2004JD005674

Lohmann, U. and Lesins, G. (2002) Stronger constrains on the anthropogenic indirect aerosol effect. *Science*, **298** (5595), 1012–1015. doi: 10.1126/science.1075405

McCormick, M.P., Thomason, L.W., and Trepte, C.R. (1995) Atmospheric effects of the Mt. Pinatubo eruption. *Nature*, **373** (6513), 399–404. doi: 10.1038/373399a0

McKinnon, J.T. and Howard, J.B. (1990) Application of soot formation model: effects of chlorine. *Combust. Sci. Technol.*, **74** (1-6), 175–197. doi: 10.1080/00102209008951687

Mandrioli, P., Negrini, M.G., Scarani, C., Tampieri, F., and Trombetti, F. (1980) Mesoscale transport of Corylus pollen grains in winter atmosphere. *Grana*, **19** (3), 227–233. doi: 10.1080/00173138009425007

Mårtensson, E.M., Nilsson, E.D., de Leeuw, G., Cohen, L.H., and Hansson, H.-C. (2003) Laboratory simulations and parameterization of the primary marine aerosol production. *J. Geophys. Res.*, **108** (D9), 4297. doi: 10.1029/2002JD002263

Marticorena, B., Bergametti, G., Aumont, B., Callot, Y., N'Doumé, C., and Legrand, M. (1997) Modeling the atmospheric dust cycle. 2. Simulation of Saharan dust sources. *J. Geophys. Res.*, **102** (D4), 4387–4404. doi: 10.1029/96JD02964

Martins, J.V., Hobbs, P.V., Weiss, R.E., and Artaxo, P. (1998) Sphericity and morphology of smoke particles from biomass burning in Brazil. *J. Geophys. Res.*, **103** (D24), 32051–32057. doi: 10.1029/98JD01153

Matsuki, A., Iwasaka, Y., Shi, G., Zhang, D., Trochkine, D., Yamada, M., Kim, Y.-S., Chen, B., Nagatani, T., Miyazawa, T., Nagatani, M., and Nakata, H. (2005) Morphological and chemical modification of mineral dust: observational insight into the heterogeneous uptake of acidic gases. *Geophys. Res. Lett.*, **32**, L22806. doi: 10.1029/2005GL024176

Matthias-Maser, S. and Jaenicke, R. (1994) Examination of atmospheric bioaerosol particles with radii > 0.2 μm. *J. Aerosol Sci.*, **25** (8), 1605–1613. doi: 10.1016/0021-8502(94)90228-3

Matthias-Maser, S. and Jaenicke, R. (1995) The size distribution of primary biological aerosol particles with radii > 0.2 μm in an urban-rural influenced region. *Atmos. Res.*, **39** (4), 279–286. doi: 10.1016/0169-8095(95)00017-8

Meehl, G.A. and Tebaldi, C. (2004) More intense, more frequent, and longer lasting heat waves in the 21st century. *Science*, **305** (5686), 994–997. doi: 10.1126/science.1098704

Middlebrook, A.M., Murphy, D.M., and Thomson, D.S. (1998) Observations of organic material in individual marine particles at Cape Grim during the First Aerosol Characterization Experiment (ACE 1). *J. Geophys. Res.*, **103** (D13), 16475–16483. doi: 10.1029/97JD03719

Mie, G. (1908) Beiträge zur Optik trüber Medien, speziell kolloidaler Metallösungen. *Ann. Phys., Vierte Folge*, **25** (3), 377–445.

Mirabel, P. and Katz, J.L. (1974) Binary homogeneous nucleation as a mechanism for the formation of aerosols. *J. Chem. Phys.*, **60** (3), 1138–1144. doi: 10.1063/1.1681124

Molinaroli, E., Pistolato, M., Rampazzo, G., and Guerzoni, S. (1999) Geochemistry of natural and anthropogenic fall-out (aerosol and precipitation) collected from the NW Mediterranean: two different multivariate statistical approaches. *Appl. Geochem.*, **14** (4), 27–36. doi: 10.1016/S0883-2927(98)00062-6

Monahan, E.C., Spiel, D.E., and Davidson, K.L. (1986) Model of marine aerosol generation via whitecaps and wave disruption, in *Oceanic Whitecaps and their Role in Air-Sea Exchange Processes* (eds E.C. Monahan and G. McNiocaill), D. Reidel Publishing, Dordrecht, pp. 167–174.

Morris, C.E., Sands, D.C., Bardin, M., Jaenicke, R., Vogel, B., Leyronas, C., Ariya, P.A., and Psenner, R. (2011) Microbiology and atmospheric processes: research challenges concerning the impact of airborne micro-organisms on the atmosphere and climate. *Biogeosciences*, **8** (17-25), 91–212. doi: 10.5194/bg-8-17-2011

Moulin, C., Lambert, C.E., Dulac, F., and Dayan, U. (1997) Control of atmospheric export of dust from North Africa by the North Atlantic Oscillation. *Nature*, **387** (6634), 691–694. doi: 10.1038/42679

Mullins, M.E., Michaels, L.P., Menon, V., Locke, B., and Ranade, M.B. (1992) Effect of geometry on particle adhesion. *Aerosol Sci. Technol.*, **17** (2), 105–118. doi: 10.1080/02786829208959564

Muñoz, O., Volten, H., Hovenier, J.W., Veihelmann, B., van der Zande, W.J., Waters, L.B.F.M., and Rose, W.I. (2004) Scattering matrices of volcanic ash particles of Mount St. Helens, Redoubt, and Mount Spurr Volcanoes. *J. Geophys. Res.*, **109**, D16201. doi: 10.1029/2004JD004684

Niemi, J.V., Saarikoski, S., Tervahattu, H., Mäkelä, T., Hillamo, R., Vehkamäki, H., Sogacheva, L., and Kulmala, M. (2006) Changes in background aerosol composition in Finland during polluted and clean periods studied by TEM/EDX individual particle analysis. *Atmos. Chem. Phys.*, **6** (12), 5049–5066. doi: 10.5194/acp-6-5049-2006

O'Dowd, C.D., Smith, M.H., Consterdine, I.E., and Lowe, J.A. (1997) Marine aerosol, sea-salt, and the marine sulphur cycle: a short review. *Atmos. Environ.*, **31** (1), 73–80. doi: 10.1016/S1352-2310(96)00106-9

Okada, K. and Kai, K. (2004) Atmospheric mineral particles collected at Qira in the Taklamakan Desert, China. *Atmos. Environ.*, **38** (40), 6927–6935. doi: 10.1016/j.atmosenv.2004.03.078

Okazaki, K. (1993) Submicron particle formation in pulverized coal combustion. *J. Aerosol Res. Jpn.*, **7**, 289–291.

Pakkanen, T.A. (1996) Study of formation of coarse particle nitrate aerosol. *Atmos. Environ.*, **30** (14), 2475–2482. doi: 10.1016/1352-2310(95)00492-0

Paris, J.-D., Stohl, A., Nédélec, P., Arshinov, M.Y., Panchenko, M.V., Shmargunov, V.P., Law, K.S., Belan, B.D., and Ciais, P. (2009) Wildfire smoke in the Siberian Arctic in summer: source characterization and plume evolution from airborne measurements. *Atmos. Chem. Phys.*, **9** (23), 9315–9327. doi: 10.5194/acp-9-9315-2009

Penner, J.E., Bergmann, D.J., Walton, J.J., Kinnison, D., Prather, M.J., Rotman, D., Price, C., Pickering, K.E., and Baughcum, S.L. (1998) An evaluation of upper tropospheric NO_x with two models. *J. Geophys. Res.*, **103** (D17), 22097–22113. doi: 10.1029/98JD01565

Penner, J., Dickinson, R., and O'Neill, C. (1992) Effects of aerosols from biomass burning on the global radiation budget. *Science*, **256** (5062), 1432–1434. doi: 10.1126/science.256.5062.1432

Penner, J.E., Lister, D., Griggs, D., Docken, D., and MacFarland, M.M. (eds) (1999) *Aviation and the Global Atmosphere*, Intergovernmental Panel on Climate Change, Special Report, , Cambridge University Press, Cambridge, p. 373.

Piccot, S.D., Watson, J.J., and Jones, J.W. (1992) A global inventory of volatile organic compound emissions from anthropogenic sources. *J. Geophys. Res.*, **97** (D9), 9897–9912. doi: 10.1029/92JD00682

Plane, J.M.C. (2012) Cosmic dust in the earth's atmosphere. *Chem. Soc. Rev.*, **41** (19), 6507–6518. doi: 10.1039/C2CS35132C

Pósfai, M. and Buseck, P.R. (2010) Nature and climate effects of individual tropospheric aerosol particles. *Annu. Rev. Earth Planet. Sci.*, **38**, 17–43. doi: 10.1146/annurev.earth.031208.100032

Pósfai, M., Anderson, J.R., Buseck, P.R., and Sievering, H. (1995) Compositional variations of sea-salt-mode aerosol particles from the North Atlantic. *J. Geophys. Res.*, **100** (D11), 23063–23074. doi: 10.1029/95JD01636

Pósfai, M., Anderson, J.R., Buseck, P.R., and Sievering, H. (1999) Soot and sulfate aerosol particles in the remote marine troposphere. *J. Geophys. Res.*, **104** (D17), 21685–21693. doi: 10.1029/1999JD900208

Pósfai, M., Gelencsér, A., Simonics, R., Arató, K., Li, J., Hobbs, P.V., and Buseck, P.R. (2004) Atmospheric tar balls: particles from biomass and biofuel burning. *J. Geophys. Res.*, **109**, D06213. doi: 10.1029/2003JD004169

Pósfai, M., Simonics, R., Li, J., Hobbs, P.V., and Buseck, P.R. (2003a) Individual aerosol particles from biomass burning in southern Africa: 1. Compositions and size distributions of carbonaceous particles. *J. Geophys. Res.*, **108** (D13), 8483. doi: 10.1029/2002JD002291

Pósfai, M., Li, J., Anderson, J.R., and Buseck, P.R. (2003b) Aerosol bacteria over the Southern Ocean during ACE-1. *Atmos. Res.*, **66** (4), 231–240. doi: 10.1016/S0169-8095(03)00039-5

Pósfai, M., Xu, H., Anderson, J.R., and Buseck, P.R. (1998) Wet and dry sizes of atmospheric aerosol particles: an AFM-TEM study. *Geophys. Res. Lett.*, **25** (11), 1907–1910. doi: 10.1029/98GL01416

Price, C., Penner, J., and Prather, M. (1997) NO_x from lightning. 1. Global distribution based on lightning physics. *J. Geophys. Res.*, **102** (D5), 5929–5941. doi: 10.1029/96JD03504

Prospero, J.M. (1999) Long-range transport of mineral dust in the global atmosphere: impact of African dust on the environment of the southeastern United States. *Proc. Natl. Acad. Sci. U.S.A.*, **96** (7), 3398–3403. doi: 10.1073/pnas.96.7.3396

Prospero, J.M. and Nees, R.T. (1986) Impact of the North African drought and El Niño on mineral dust in the Barbados trade winds. *Nature*, **320** (6064), 735–738. doi: 10.1038/320735a0

Pueschel, R.F., Kinne, S.A., Russell, P.B., Snetsinger, K.G., and Livingston, J.M. (1993) Effects of the 1991 Pinatubo volcanic eruption on the physical and radiative properties of stratospheric aerosols, in IRS '92: Current Problems in Atmospheric Radiation, Proceedings of the International Radiation Symposium, Tallinn (Estonia), 2-8 August 1992 (eds S. Keevallik and O. Kärner), A. Deepak Publishing, Hampton, VA, pp. 183–186.

Reid, J.S., Koppmann, R., Eck, T.F., and Eleuterio, D.P. (2005) A review of biomass burning emissions. Part II: Intensive physical properties of biomass burning particles. *Atmos. Chem. Phys.*, **5** (3), 799–825. doi: 10.5194/acp-5-799-2005

Rose, W.I. and Durant, A.J. (2009) Fine ash content of explosive eruptions. *J. Volcanol. Geotherm. Res.*, **186** (1-2), 32–39. doi: 10.1016/j.jvolgeores.2009.01.010

Russell, P.B., Livingston, J.M., Dutton, E.G., Pueschel, R.F., Reagan, J.A., DeFoor, T.E., Box, M.A., Allen, D., Pilewskie, P., Herman, B.M., Kinne, S.A., and Hofmann, D.J. (1993) Pinatubo and pre-Pinatubo optical-depth spectra: mauna Loa measurements, comparisons, inferred particle size distributions, radiative effects, and relationship to Lidar data. *J. Geophys. Res.*, **98** (D12), 22969–22985. doi: 10.1029/93JD02308

Sabbioni, C. (1992) Characterization of atmospheric particles on monuments by scanning electron microscopy/energy dispersive X-ray analyses. Electron Microscopy 92. Material Sciences, Proceedings of 10th European Congress on Electron Microscopy (EUREM 92), Vol. 2, September 1992, Granada, Spain.

Sabbioni, C. and Zappia, G. (1992) Characterization of particles emitted by domestic heating units fueled by distilled oil. *Atmos. Environ.*, **26** (18), 3297–3304. doi: 10.1016/0960-1686(92)90346-M

Sassen, K. (1992) Evidence for liquid-phase cirrus cloud formation from volcanic aerosols: climatic implications. *Science*, **257** (5069), 516–519. doi: 10.1126/science.257.5069.516

Sattler, B., Puxbaum, H., and Psenner, R. (2001) Bacterial growth in supercooled cloud droplets. *Geophys. Res. Lett.*, **28** (2), 239–242. doi: 10.1029/2000GL011684

Schnaiter, M., Linke, C., Möhler, O., Naumann, K.-H., Saathoff, H., Wagner, R., Schurath, U., and Wehner, B. (2005)

Absorption amplification of black carbon internally mixed with secondary organic aerosol. *J. Geophys. Res.*, **110**, D19. doi: 10.1029/2005JD006046

Schnell, R.C. and Vali, G. (1976) Biogenic ice nuclei: Part I. Terrestrial and marine sources. *J. Atmos. Sci.*, **33** (8), 1554–1564. doi: 10.1175/1520-0469(1976)033<1554:BINPIT>2.0.CO;2

Schumann, U., Weinzierl, B., Reitebuch, O., Schlager, H., Minikin, A., Forster, C., Baumann, R., Sailer, T., Graf, K., Mannstein, H., Voigt, C., Rahm, S., Simmet, R., Scheibe, M., Lichtenstern, M., Stock, P., Rüba, H., Schäuble, D., Tafferner, A., Rautenhaus, M., Gerz, T., Ziereis, H., Krautstrunk, M., Mallaun, C., Gayet, J.-F., Lieke, K., Kandler, K., Ebert, M., Weinbruch, S., Stohl, A., Gasteiger, J., Groß, S., Freudenthaler, V., Wiegner, M., Ansmann, A., Tesche, M., Olafsson, H., and Sturm, K. (2011) Airborne observations of the Eyjafjalla volcano ash cloud over Europe during air space closure in April and May 2010. *Atmos. Chem. Phys.*, **11** (5), 2245–2279. doi: 10.5194/acp-11-2245-2011

Schwartz, S.E., Arnold, F., Blanchet, J.-P., Durkee, P.A., Hofmann, D.J., Hoppel, W.A., King, M.D., Lacis, A.A., Nakajima, T., Ogren, J.A., Toon, O.B., and Wendisch, M. (1995) Group report: connections between aerosol properties and forcing of climate, in *Aerosol Forcing of Climate* (eds R.J. Charlson and J. Heintzenberg), John Wiley & Sons, Inc., New York, pp. 251–280.

Seinfeld, J.H. and Pandis, S.N. (1998) *Atmospheric Chemistry and Physics, from Air Pollution to Climate Change*, 1st edn, John Wiley & Sons, Inc., New York, p. 1326.

Seinfeld, J.H. and Pandis, S.N. (2006) *Atmospheric Chemistry and Physics, from Air Pollution to Climate Change*, 2nd edn, John Wiley & Sons, Inc., Hoboken, NJ, p. 1225.

Shen, T.L., Wooldrige, P.J., and Molina, M.J. (1995) Stratospheric pollution and ozone depletion, in *Composition, Chemistry, and Climate of the Atmosphere* (ed. H.B. Singh), Van Nostrand Reinhold, New York, pp. 394–442.

Shettle, E.P. and Fenn, R.W. (1979) Models for the Aerosols of the Lower Atmosphere and the Effects of Humidity Variations on their Optical Properties. Environmental Research Papers, No. 676, Air Force Geophys. Lab., AFGL-Techn. Rep. 79-0214, Hanscom AFB, MA, 94 pp.

Sievering, H., Boatman, J., Gorman, E., Kim, Y., Anderson, L., Ennis, G., Luria, M., and Pandis, S. (1992) Removal of sulphur from the marine boundary layer by ozone oxidation in sea-salt aerosols. *Nature*, **360** (6404), 571–573. doi: 10.1038/360571a0

Simoneit, B.R.T., Cardoso, J.N., and Robinson, N. (1990) An assessment of the origin and composition of higher molecular weight organic matter in aerosols over Amazonia. *Chemosphere*, **21** (10-11), 1285–1301. doi: 10.1016/0045-6535(90)90145-J

Sinha, B.W., Hoppe, P., Huth, J., Foley, S., and Andreae, M.O. (2008) Sulfur isotope analyses of individual aerosol particles in the urban aerosol at a central European site (Mainz, Germany). *Atmos. Chem. Phys.*, **8** (23), 7217–7238. doi: 10.5194/acp-8-7217-2008

SMIC (1971) Inadvertent Climate Modification, Report of the Study of Man's Impact on Climate (SMIC), MIT Press, Cambridge, MA, pp. 185–232.

Smith, J.N., Dunn, M.J., VanReken, T.M., Iida, K., Stolzenburg, M.R., McMurry, P.H., and Huey, L.G. (2008) Chemical composition of atmospheric nanoparticles formed from nucleation in Tecamac, Mexico: evidence for an important role for organic species in nanoparticle growth. *Geophys. Res. Lett.*, **35**, L04808. doi: 10.1029/2007GL032523

Sobanska, S., Coeur, C., Maenhaut, W., and Adams, F. (2003) SEM-EDX characterisation of tropospheric aerosols in the Negev desert (Israel). *J. Atmos. Chem.*, **44** (3), 299–322. doi: 10.1023/A:1022969302107

Song, N., Starr, D.O.'C., Wuebbles, D.J., Williams, A., and Larson, S.M. (1996) Volcanic aerosols and interannual variation of high clouds. *Geophys. Res. Lett.*, **23** (19), 2657–2660. doi: 10.1029/96GL02372

Spiro, P.A., Jacob, D.J., and Logan, J.A. (1992) Global inventory of sulfur emissions with 1° x 1° resolution. *J. Geophys. Res.*, **97** (D5), 6023–6036. doi: 10.1029/91JD03139

Stelson, A.W. and Seinfeld, J.H. (1982a) Relative humidity and temperature dependence of the ammonium nitrate dissociation constant. *Atmos. Environ.*, **16** (5), 983–992. doi: 10.1016/0004-6981(82)90184-6

Stelson, A.W. and Seinfeld, J.H. (1982b) Thermodynamic prediction of the water activity, NH_4NO_3 dissociation constant, density and refractive index for the system NH_4NO_3-$(NH_4)_2SO_4$-H_2O at 25 °C. *Atmos. Environ.*, **16** (10), 2507–2514. doi: 10.1016/0004-6981(82)90142-1

Stohl, A., Andrews, E., Burkhart, J.F., Forster, C., Herber, A., Hoch, S.W., Kowal, D., Lunder, C., Mefford, T., Ogren, J.A., Sharma, S., Spichtinger, N., Stebel, K., Stone, R., Ström, J., Tørseth, K., Wehrli, C., and Yttri, K.E. (2006) Pan-Arctic enhancements of light absorbing aerosol concentrations due to North American boreal forest fires during summer 2004. *J. Geophys. Res.*, **111**, D22214. doi: 10.1029/2006JD007216

Stone, R.S., Anderson, G.P., Andrews, E., Dutton, E.G., Shettle, E.P., and Berk, A. (2007) Incursions and radiative impact of Asian dust in northern Alaska. *Geophys. Res. Lett.*, **34**, L14815. doi: 10.1029/2007GL029878

Stone, R.S., Anderson, G.P., Shettle, E.P., Andrews, E., Loukachine, K., Dutton, E.G., Schaaf, C., and Roman, M.O. III, (2008) Radiative impact of boreal smoke in the Arctic: observed and modeled. *J. Geophys. Res.*, **113**, D14S16. doi: 10.1029/2007JD009657

Tegen, I. and Fung, I. (1995) Contribution to the atmospheric mineral aerosol load from land surface modification. *J. Geophys. Res.*, **100** (D9), 18707–18726. doi: 10.1029/95JD02051

Tegen, I., Hollrig, P., Chin, M., Fung, I., Jacob, D., and Penner, J. (1997) Contribution of different aerosol species to the global aerosol extinction optical thickness: estimates from model results. *J. Geophys. Res.*, **102** (D20), 23895–23915. doi: 10.1029/97JD01864

Tegen, I. and Miller, R. (1998) A general circulation model study of the interannual variability of soil dust aerosol. *J. Geophys. Res.*, **103** (D20), 25975–25995. doi: 10.1029/98JD02345

Thomason, L.W., Burton, S.P., Luo, B.-P., and Peter, T. (2008) SAGE II measurements of stratospheric aerosol properties at non-volcanic levels. *Atmos. Chem. Phys.*, **8** (4), 983–995. doi: 10.5194/acp-8-983-2008

Tivanski, A.V., Hopkins, R.J., Tyliszczak, T., and Gilles, M.K. (2007) Oxygenated interface on biomass burn tar balls determined by single particle scanning transmission X-ray microscopy. *J. Phys. Chem. A*, **111** (25), 5448–5458. doi: 10.1021/jp070155u

Tomasi, C., Vitale, V., and Tarozzi, L. (1997) Sun-photometric measurements of atmospheric turbidity variations caused by the Pinatubo aerosol cloud in the Himalayan region during the summer periods of 1991 and 1992. *Il Nuovo Cimento C*, **20** (1), 61–88.

Tomasi, C., Vitale, V., Lupi, A., Di Carmine, C., Campanelli, M., Herber, A., Treffeisen, R., Stone, R.S., Andrews, E., Sharma, S., Radionov, V., von Hoyningen-Huene, W., Stebel, K., Hansen, G.H., Myhre, C.L., Wehrli, C., Aaltonen, V., Lihavainen, H., Virkkula, A., Hillamo, R., Ström, J., Toledano, C., Cachorro, V., Ortiz, P., de Frutos, A., Blindheim, S., Frioud, M., Gausa, M., Zielinski, T., Petelski, T., and Yamanouchi, T. (2007) Aerosols in polar regions: a historical overview based on optical depth and in situ observations. *J. Geophys. Res.*, **112**, D16205. doi: 10.1029/2007JD008432

Toohey, M., Krüger, K., Niemeier, U., and Timmreck, C. (2011) The influence of eruption season on the global aerosol evolution and radiative impact of tropical volcanic eruptions. *Atmos. Chem. Phys.*, **11** (23), 12351–12367. doi: 10.5194/acp-11-12351-2011

Toon, O.B., Pollack, J.B., and Khare, B.N. (1976) The optical constants of several atmospheric aerosol species: ammonium sulfate, aluminum oxide, and sodium chloride. *J. Geophys. Res.*, **81** (33), 5733–5748. doi: 10.1029/JC081i033p05733

Trochkine, D., Iwasaka, Y., Matsuki, A., Yamada, M., Kim, Y.-S., Nagatani, T., Zhang, D., Shi, G.-Y., and Shen, Z. (2003) Mineral aerosol particles collected in Dunhuang, China, and their comparison with chemically modified particles collected over Japan. *J. Geophys. Res.*, **108** (D23), 8642. doi: 10.1029/2002JD003268

Tsigaridis, K. and Kanakidou, M. (2007) Secondary organic aerosol importance in the future atmosphere. *Atmos. Environ.*, **41** (22), 4682–4692. doi: 10.1016/j.atmosenv.2007.03.045

Tsigaridis, K., Krol, M., Dentener, F.J., Balkanski, Y., Lathière, J., Metzger, S., Hauglustaine, D.A., and Kanakidou, M. (2006) Change in global aerosol composition since preindustrial times. *Atmos. Chem. Phys.*, **6** (12), 5143–5162. doi: 10.5194/acp-6-5143-2006

United Nations (1979) *Fine Particulate Pollution*, Pergamon Press, New York.

van Poppel, L.H., Friedrich, H., Spinsby, J., Chung, S.H., Seinfeld, J.H., and Buseck, P.R. (2005) Electron tomography of nanoparticle clusters: implications for atmospheric lifetimes and radiative forcing of soot. *Geophys. Res. Lett.*, **32**, L24811. doi: 10.1029/2005GL024461

Wagner, H.G. (1981) Soot formation - an overview, in *Particulate Carbon Formation During Combustion* (eds D.C. Siegla and G.W. Smith), Plenum Press, New York, pp. 1–29.

Wallace, J.M. and Hobbs, P.V. (2006) *Atmospheric Science: An Introductory Survey*, 2nd edn, Elsevier Academic Press, Burlington, MA, pp. 153–208.

Wentzel, M., Gorzawski, H., Naumann, K.H., Saathoff, H., and Weinbruch, S. (2003) Transmission electron microscopical and aerosol dynamical characterization of soot aerosols. *J. Aerosol Sci.*, **34** (10), 1347–1370. doi: 10.1016/S0021-8502(03)00360-4

Winiwarter, W., Bauer, H., Caseiro, A., and Puxbaum, H. (2009) Quantifying emissions of primary biological aerosol particle mass in Europe. *Atmos. Environ.*, **43** (7), 1403–1407. doi: 10.1016/j.atmosenv.2008.01.037

Wolf, M.E. and Hidy, G.M. (1997) Aerosols and climate: anthropogenic emissions and trends for 50 years. *J. Geophys. Res.*, **102** (D10), 11113–11121. doi: 10.1029/97JD00199

Wolff, E.W. and Cachier, H. (1998) Concentrations and seasonal cycle of black carbon in aerosol at a coastal Antarctic station. *J. Geophys. Res.*, **103** (D9), 11033–11041. doi: 10.1029/97JD01363

Yienger, J.J. and Levy, H. II, (1995) Empirical model of global soil-biogenic NO_x emissions. *J. Geophys. Res.*, **100** (D6), 11447–11464. doi: 10.1029/95JD00370

Zender, C.S., Brian, H., and Newman, D. (2003) Mineral Dust Entrainment And Deposition (DEAD) model: description and 1990s dust climatology. *J. Geophys. Res.*, **107** (D24), 4416. doi: 10.1029/2002JD002775

Zender, C.S., Miller, R.L., and Tegen, I. (2004) Quantifying mineral dust mass budgets: terminology, constraints, and current estimates. *Eos Trans. Am. Geophys. Union*, **85** (48), 509–512. doi: 10.1029/2004EO480002

Zhang, R., Suh, I., Zhao, J., Zhang, D., Fortner, E.C., Tie, X., Molina, L.T., and Molina, M.J. (2004) Atmospheric new particle formation enhanced by organic acids. *Science*, **304** (5676), 1487–1490. doi: 10.1126/science.1095139

Zook, H.A. (2001) Spacecraft measurements of the cosmic dust flux, in *Accretion of Extraterrestrial Matter Throughout Earth's History* (eds B. Peucker-Ehrenbrink and B. Schmitz), Springer US, Kluwer Academic/Plenum Publishers, New York, pp. 75–92. doi: 10.1007/978-1-4419-8694-8_5

2
Aerosol Nucleation in the Terrestrial Atmosphere

Karine Sellegri and Julien Boulon

2.1
Introduction

The quantification of aerosol particle emission sources is essential for the prediction of anthropogenic impacts on the atmospheric composition. Aerosols can be directly emitted from primary sources or be formed in the atmosphere from nucleation of gaseous precursors, that is, by formation of a new particulate phase from the clustering of gaseous molecules. A large fraction, yet not properly estimated, of the atmospheric particle number concentration is formed by gas-to-particle conversion through nucleation. New nanoparticles formed by nucleation will grow to climate relevant sizes by condensation and coagulation and then contribute to the total cloud condensation nuclei (CCN) number burden of the atmosphere, hereby influencing climate (Figure 2.1).

Observations of the formation of new particles by nucleation in the atmosphere have multiplied in the last 20 years thanks to the evolution of the instrumentation toward aerosol detection limits reaching the below 10 nm size range. The difficulty in estimating the number concentration of particles formed by nucleation lies in the fact that the nucleation process is a complex issue occurring at scales that are intermediate between the gas phase and the condensed phase. Limitations exist both in the experimental techniques allowing to investigate nanometric sizes and in the theoretical approaches that cannot be directly validated. In this chapter, we will provide an overview of the theoretical approaches used to describe the nucleation process and subsequent growth of clusters. We will then present the instrumental techniques used to investigate the physical and chemical properties of clusters and newly formed nanoparticles and the most common metrics used to characterize the process from observations. The main findings resulting from observations reported from different environments and from chamber experiments will be summarized, and at last, the impact of nucleation in the atmosphere will be evaluated both from the experimental and modeling points of view.

Atmospheric Aerosols: Life Cycles and Effects on Air Quality and Climate, First Edition.
Edited by Claudio Tomasi, Sandro Fuzzi, and Alexander Kokhanovsky.

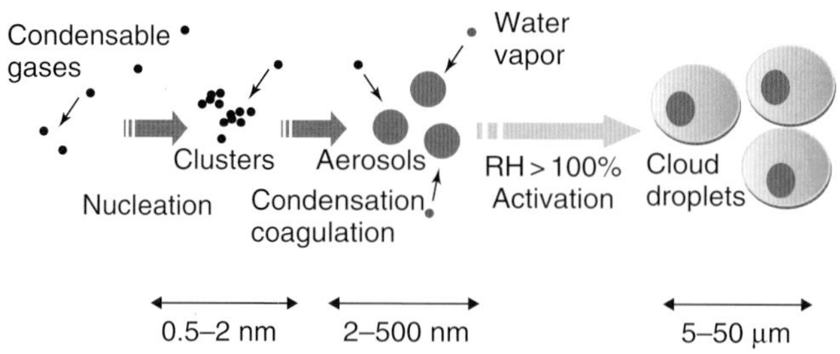

Figure 2.1 Schematic representation of nucleation, growth, and activation of aerosol particles.

2.2 Theoretical Basis of Nucleation and Growth of New Particles in the Atmosphere

Atmospheric new particle formation (NPF) consists in a complicated set of processes that includes the production of nanometer-size clusters from gaseous vapors and the growth of these clusters to detectable sizes. In order to introduce the complexity of the nucleation, the most relevant models that describe atmospheric nucleation will be presented in this section. Nucleation is defined as the formation of new particles by attachment of monomers until reaching a critical size at which the particle is thermodynamically stable. During this process, the surface formation is the main energy barrier. Once this barrier overcame, the process of nucleation itself has occurred, and the particle will spontaneously grow through condensation (sticking collisions between monomers and the particle) and coagulation (sticking collision between two particles). In atmospheric conditions, coagulation is negligible compared to condensation in terms of particle size evolution and will not be detailed in this section.

2.2.1 Introduction to Nucleation Theories Useful in Atmospheric Sciences

Most of theoretical works that have been conducted to date are derived from the classical nucleation theory (CNT) established by Becker and Döring (1935). In atmospheric sciences, the nucleation and growth of nanoparticle is considered as an isothermal process, meaning that the temperature of the cluster and the vapor is assumed to be the same. This assumption implies that between two cluster–monomer collisions, the cluster is thermalized by a buffer species. A deep analysis of nonisothermal nucleation can be found in the theoretical work of Feder *et al.* (1966). This analysis was completed later to account for phase transitions due to internal energy variation during the growth of particles in cold vapors by L'Hermite (2009). This aspect is still not accounted for in descriptions of atmospheric NPF while its importance has been underlined by many authors (Virtanen *et al.*, 2010; Koop *et al.*, 2011; Perraud *et al.*, 2012) since the aerosol state phase plays a major role in aerosol chemistry. In the following, isothermal classical nucleation models commonly used in the atmospheric science community will be presented. These models are based on the capillarity approximation, a common framework of mesoscopic physics which mainly relies on surface-to-volume ratio increasing as the droplet size decreases (see Oxtoby, 1992; Ford, 1997 for more details).

2.2.1.1 The Unary System Model

The unary system model (USM) is the "simplest" nucleation model as it describes the nucleation of one component in a medium composed by its own gas phase. Despite its nonrelevance for atmospheric aerosol purposes, it is a fundamental element to build up more atmospheric representative models. As a consequence of size decreasing, more molecules of a particle are localized on the surface than within the volume. These surface molecules have a lower binding energy compared to inner molecules which have more surroundings. At a given temperature, surface molecules tend to evaporate more than the inner ones. As the size decrease, the surface-to-volume ratio increases, meaning that the equilibrium vapor pressure increases as well. The vapor saturation ratio S is defined as

$$S = \frac{P_X(T)}{P_{\mathrm{sat},X}(T)}, \tag{2.1}$$

where $P_X(T)$ is partial pressure of the species X and $P_{\mathrm{sat},X}(T)$ the saturation vapor pressure of X in equilibrium with its liquid phase over a flat surface at temperature T. $S < 1$ for unsaturated vapor, $S = 1$ for saturated vapor, and $S > 1$ for supersaturated vapor.

In supersaturated conditions, there is a size for which the evaporation rate of the clusters is equal to the sticking collision rate. Below this critical size, the cluster evaporates, while above this critical size the cluster spontaneously grows since the sticking collision rate is greater than the evaporation rate. This later case is the so-called NPF regime. From the thermodynamic point of view, the change in the Gibbs free energy budget ΔG must be negative to allow the particle (surface) formation, meaning that the energy is minimized when forming a particle of a radius r. The variation of the Gibbs free energy associated with the particle formation of a given radius r is defined by Eq. (2.2):

$$\Delta G = \frac{4\pi r^3}{3v_{\mathrm{part}}} \times (g_{\mathrm{part}} - g_{\mathrm{vap}}) + 4\pi\sigma r^2, \tag{2.2}$$

where $4\pi r^3/3v_{\mathrm{part}}$ is the number of molecule X in the particle (r is the particle radius and v_{part} is the volume of a molecule X in the particle) and g_{vap} and g_{part} are, respectively, the Gibbs free energy for a molecule X in the gas phase and in the particle. The term $4\pi\sigma r^2$ is the free energy associated with an interface with radius of curvature r and surface tension σ.

The term $(g_{\mathrm{part}} - g_{\mathrm{vap}})$ is linked to the surface formation and can be derived from the Kelvin equation, assuming an ideal gas and equilibrium as follows:

$$(g_{\mathrm{part}} - g_{\mathrm{vap}}) = -k_{\mathrm{B}}T\ln(S), \tag{2.3}$$

where k_B is the Boltzmann constant, T the absolute temperature, and S the saturation ratio (Eq. (2.1)). By combining Eqs (2.2) and (2.3), the Gibbs free energy change during nucleation can be expressed as follows:

$$\Delta G = 4\pi\sigma r^2 - \frac{4\pi}{3} \times \frac{k_B T \ln(S)}{v_{drop}} r^3. \tag{2.4}$$

The free energy needed to form a new cluster (Eq. (2.4)) is composed of a surface term (proportional to r^2) and a volume term (proportional to r^3) of opposite signs as long as $S > 1$. At a given temperature, for high enough S, the volume term will eventually overcome the surface term, leading to a negative free energy of formation. As an example, the variation of ΔG as a function of the radius of the particle for pure water and for different values of the saturation ratio S is reported in Figure 2.2.

The critical radius r^* is the radius for which the condensation of any additional molecule will lead to a larger diameter which has a lower δG which in turn will lead to the condensation of more molecules and hence a spontaneous growth. Radius r^* is obtained when ΔG reaches a maximum and hence is defined by Eq. (2.5):

$$r^* \text{ for } \frac{\delta(\Delta G)}{\delta r} = 0 \Leftrightarrow r^* = \frac{2\sigma v_{part}}{k_B T \ln(S)}. \tag{2.5}$$

The nucleation rate J can be derived not only from the thermodynamic point of view developed earlier but also from a kinetic approach. In the kinetic approach the cluster population balance is controlled by the sticking collision rate and by the evaporation rate assuming a steady state. The two approaches provide different but equivalent expressions of the nucleation rate J. Contrary to the thermodynamic approach, the kinetic approach provides an easy-to-use expression of the

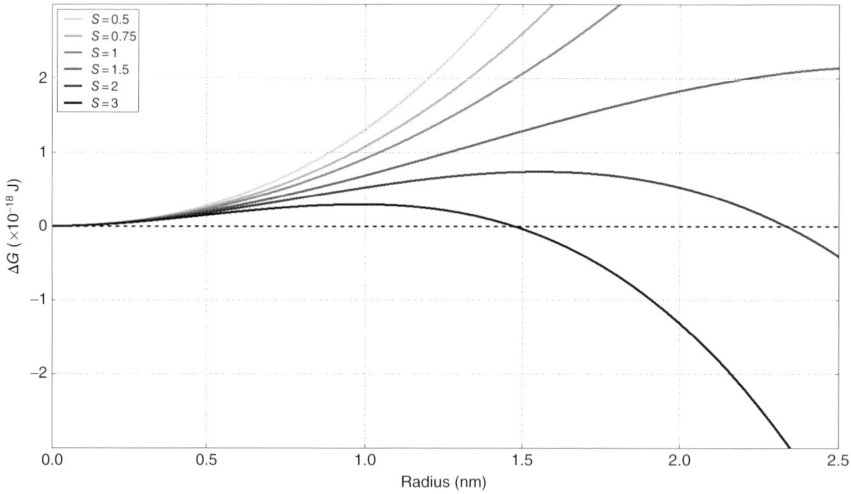

Figure 2.2 Pure water homogeneous nucleation computations for different saturation ratios ($T = 293$ K).

formation rate (Eq. (2.7)):

$$J = C \exp\left(\frac{\Delta G^*}{k_B T}\right), \tag{2.6}$$

$$J = \sqrt{\frac{2\sigma}{\pi m_{\text{part}}}} \frac{v_{\text{part}} N^2{}_1}{S} \times \exp\left[-i^* \frac{\ln(S)}{2}\right], \tag{2.7}$$

with

$$i^* = \frac{32\pi}{3} \times \frac{v_{\text{part}}^2 \sigma^3}{(k_B)^3 \ln(S)^3}. \tag{2.8}$$

These calculations show that the USM requires a high degree of supersaturation to be triggered. Moreover, the USM rates of the unary water system are still low even for 300% of saturation ($S = 3$). Those results suggest that the USM mechanism cannot be predominant in the atmosphere and might only occur in very specific conditions. For example, strong turbulence could enhance the formation of highly supersaturated zones where the USM could be triggered during a brutal cooling of an organic vapor.

2.2.1.2 The H_2SO_4–H_2O Binary System

Binary nucleation system was first studied by Flood in 1934, but its atmospheric implications were introduced by Doyle in 1961 who computed the nucleation rate of sulfate particles (see references in Kulmala and Laaksonen (1990)). This mechanism appears to be the most important binary nucleation system (BNS) in the atmosphere. Contrarily to the USM, in the BNS model none of the vapor species must be supersaturated in the gas phase but solely with respect to a mixed liquid solution particle (Kulmala and Laaksonen, 1990).

In the H_2SO_4–H_2O binary system, the variation of the Gibbs free energy during the formation of a mixed cluster of sulfuric acid (A) and water (W) is described by Eq. (2.9):

$$\Delta G = n_A(\mu_{A,\text{part}} - \mu_{A,\text{vap}}) + n_W(\mu_{W,\text{part}} - \mu_{W,\text{vap}}) + 4\pi\sigma r^2, \tag{2.9}$$

where r is the radius of the particle and $\mu_{A,\text{part}}$, $\mu_{W,\text{part}}$ and $\mu_{A,\text{vap}}$, $\mu_{W,\text{vap}}$ are the chemical potentials of each species in the mixed particle (part) and the gas phase (vap), respectively. For a given chemical species, the variation of the chemical potential from liquid to gas phase is described by Eq. (2.10):

$$\mu_{\text{part}} - \mu_{\text{vap}} = -k_B T \ln\left(\frac{p}{p_{\text{sol}}}\right), \tag{2.10}$$

where p is the partial pressure of the considered chemical species in the gas phase and p_{sol} is the vapor pressure of the considered component over a flat solution of the same composition as the particle. If we consider that (i) in the particle, water is the solvent (i.e., $[H_2O] \gg [H_2SO_4]$) and that (ii) in the gas phase, the total pressure due to all species is low compared to the pressure of the system, then chemical activities a_i of a given species i for each phase can be defined as follows:

$$a_{i,\mathrm{vap}} = \frac{p_i}{P_{\mathrm{sat},i}} \equiv S_i \quad \text{and} \quad a_{i,\mathrm{part}} = \frac{p_{\mathrm{sol},i}}{P_{\mathrm{sat},i}}.$$

Hence, from Eqs (2.9) and (2.10), ΔG can be expressed as

$$\Delta G = 4\pi\sigma r^2 - k_\mathrm{B} T \left(n_A \ln\left(\frac{S_A}{a_{A,\mathrm{drop}}}\right) + n_W \ln\left(\frac{S_W}{a_{W,\mathrm{drop}}}\right)\right). \tag{2.11}$$

In this expression of the Gibbs free energy, the radius of the particle does not appear except in the positive term due to the surface formation $4\pi\sigma r^2$. The radius r is given by the relation (2.12):

$$\frac{4}{3}\pi r^3 \rho_\mathrm{part} = n_A m_A + n_W m_W, \tag{2.12}$$

where ρ_part is the mass per unit of volume of the particle. The formation rate calculation (J) is more complicated than in the case of the USM theory and requires numerous thermodynamic data. In order to simplify the calculation, Yu (2007) has developed a simplified approach. It mainly relies on the fact that water vapor concentrations are high enough so that binary homogeneous nucleation of H_2SO_4 and H_2O can be treated as quasiunary nucleation process for H_2SO_4 in equilibrium with water vapor. Later, the same author has provided lookup tables to avoid complex calculations (Yu, 2008). An example of the formation rate variation according to the relative humidity and to sulfuric acid concentration is represented in Figure 2.3.

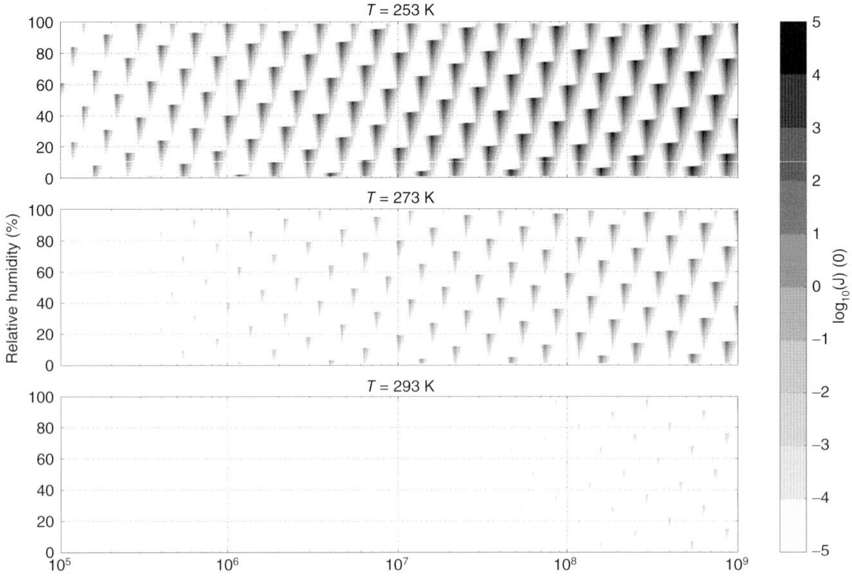

Figure 2.3 Evolution of the formation rate J as a function of the relative humidity and the sulfuric acid concentration at 250 K.

From this graphic, the BNS of H_2SO_4–H_2O predicts very low nucleation rates ($J < 1\ cm^{-3}\ s^{-1}$) for boundary layer conditions (high temperatures). However, this mechanism could occur in higher atmospheric layers where temperatures are much lower than in the low troposphere. The fact the binary nucleation scheme does not explain nucleation in the low troposphere is supported by several observations. Many studies of nucleation events report higher particle formation rates than the ones expected on the basis of binary H_2SO_4–H_2O nucleation alone (see Covert *et al.* (1992) for examples, Hoppel *et al.* (1994) for marine boundary layer, O'Dowd *et al.* (1998) for coastal environment, Kulmala *et al.* (2001) for boreal forest, or Chen *et al.* 2012 for a review). Several hypotheses were proposed to explain the enhanced observed particle production. The following sections will present other theories that have been developed to fill the gap between observations and model predictions: the participation of a third chemical compound (formation of a ternary system) or the contribution of ions to stabilize the cluster (ion-induced nucleation (IIN)).

2.2.1.3 The H_2SO_4–NH_3–H_2O Ternary System

The physics behind the ternary system nucleation (TSN) is the same than the one used for the BNS except that three species are involved in the cluster formation. The first classical ternary theory of nucleation was built by Korhonen *et al.* (1999) using the H_2SO_4–NH_3–H_2O system based on the fact that (i) ammonia, NH_3, is ubiquitous in the atmosphere and (ii) this species reduces the vapor pressure of the sulfuric acid above the solution ($p_{A,sol}$ in the term $a_{A,drop}$ of Eq. (2.11)), hence enhancing the cluster formation. As for the BNS, nucleation rate parameterizations were developed (Napari *et al.*, 2002; Merikanto *et al.*, 2007), but only few studies have confronted the model predictions to laboratory of field measurements. In 2002, during an intensive coastal environment field campaign at Mace Head, Kulmala *et al.* (2002) have shown that the measured sulfuric acid and ammonia concentrations were sufficiently high to onset TSN and the classical BNS theory could not explain the observed particle formation rate. Authors concluded that the detected particle formation events could be initiated by the homogeneous nucleation of the ternary system formed by H_2SO_4–NH_3–H_2O. Recently, state-of-the-art experiments were conducted at CERN (French acronym for "Centre for Nuclear Research"): the CLOUD experiments (see Kirkby *et al.*, 2011). Among numerous results, authors confirms that "atmospherically relevant ammonia mixing ratios of 100 parts per trillion by volume, or less, increase the nucleation rate of sulphuric acid particles more than 100–1000-fold." However, they also confirm that despite "the large enhancements in rate due to ammonia [...], atmospheric concentrations of ammonia and sulphuric acid are insufficient to account for observed boundary-layer nucleation."

2.2.1.4 The Role of Amines

As a consequence of the failure of previous theories to correctly predict either the measured nucleation rates or their functional dependence on sulfuric acid concentrations, an alternative approach has been developed to model nucleation rates as well as to propose a process-based explanation of NPF. This model, developed at the University of Minnesota (Chen *et al.*, 2012) to describe nucleation in the polluted boundary layer, is based on a sequence of acid–base reactions involving sulfuric acid, ammonia, and amines. The concept of the model is that, given the reaction constants and evaporation rates derived from measurements, it is possible to model a set of reactions from monomers to multimers:

$$A_1 \underset{E_{2,\mathrm{MV}}}{\overset{k_{11}}{\rightleftharpoons}} A_{2,\mathrm{MV}} \xrightarrow{+B} A_{2,\mathrm{LV}} \underset{E_3}{\overset{k_{21}}{\rightleftharpoons}} A_3 \xrightarrow{k_{31}} A_4 \ldots,$$

where A is for sulfuric acid, B for any base (amine or ammonia), E for evaporation of an acid unit, and the index MV and LV stand for "more volatile" and "less volatile," respectively. Such an approach is very efficient but needs many parameters to be fed with. Compared to the classical approach, this type of modeling provides to atmospheric scientist a nice alternative for the NPF study and understanding. Experiments from the CLOUD facility at CERN, associated with dynamical modeling based on quantum chemical calculations of binding energies of molecular clusters (Almeida *et al.*, 2013), confirm and measure precisely what had been pointed out by Chen *et al.*: "dimethylamine above three parts per trillion by volume can enhance particle formation rates more than 1000-fold compared with ammonia, sufficient to account for the particle formation rates observed in the atmosphere."

2.2.1.5 The Ion-Induced Nucleation

The IIN is a nucleation theory that involves the formation of charged clusters. In this particular case, the attractive potential between ions and ions and between ions and the dipole moment (induced or not) of the condensable vapor reduces the thermodynamic barrier for nucleation and hence enhances the condensational growth (e.g., Lovejoy, Curtius, and Froyd, 2004; Nadykto and Yu, 2004). This theory is based on the fact that ions are generated continuously, and ubiquitously, by galactic cosmic rays (as a function of altitude and latitude) and local sources such as radioactive decay or lightning. Mechanisms involved in this theory are the same as the ones that have been developed in the previous paragraphs. The main difference is an additional electrical term due to the charge state of each component. Nucleation mechanisms involving ions have been proposed to be important for aerosol formation in the atmosphere (e.g., Arnold, 1980; Yu and Turco, 2001). The idea of a dominant role of ions in atmospheric nucleation has been at the focus of many research efforts during the last decade. Because of their natural origin, the impact of ions on nucleation and consequent cloud formation would lower the contribution of anthropogenic emissions, if it were proven to be predominant. Ions are easier to measure at very low sizes compared to neutral clusters (see Section 2.3). Crucial experiments were again conducted at the CERN facility and provided the first experimental data to nourish the debate about the IIN importance in the atmosphere. Kirkby *et al.* (2011) found that ions increase only by an additional term between 1 and 10 the nucleation rate and that this type of nucleation event can occur in the midtroposphere but is negligible in the planetary boundary layer (PBL).

2.2.2 The Growth of New Particles

2.2.2.1 The Condensation Process

Condensation is the sticking collision of a given vapor on a condensed phase. When a vapor condenses on a particle population, the particle diameters increase while the total concentration remains equal. The condensation growth rate GR_{cond} is defined as the rate at which the diameter (or the volume) of a particle increases. For particles composed of only one species i, in a homogeneous gaseous system, $GR_{cond.i}$ is defined by Eq. (2.12):

$$GR_{cond,i} = \frac{ddp}{dt} = \frac{dV/dt}{dV/ddp}. \tag{2.13}$$

To allow growth, the particle needs to be collided by at least one vapor molecule. This collision has to be a sticking collision and should not be followed by evaporation. Before diving into the details of collision kernels, let us first define the concept of collision "regime." Historically, the aerosol science has been developed using the tools of hydrodynamics. When the studied particles are large enough, one could consider that the particle is submersed in a continuous gaseous medium. This defines the continuum regime. On the opposite, at the molecular scale, the particle embryo, that is, the molecular cluster, is affected/collided by the surrounding molecular environment. This regime is described by the kinetic theory of gas and named accordingly the kinetic regime. In between these two regimes, the transition regime collision kernel is strongly dependent on the particle size.

The collision rate K between the particle and the vapor molecule in the kinetic regime is given by

$$K = \frac{\pi}{4}(dp + dv)^2({v_{T,p}}^2 + {v_{T,v}}^2)^{1/2}C_v, \tag{2.14}$$

where dp and dv are, respectively, the particle and the vapor molecule diameter and $v_{T,p}$, $v_{T,g}$ are the mean thermal speed of the particle and the vapor molecule. In order to generalize this collision kernel to the transition and continuum regimes, it is necessary to introduce a correction factor F derived from the generalized coagulation rate (Nieminen *et al.*, 2010):

$$F = \frac{8(Dp + Dv)}{(dp + dv)({v_{T,p}}^2 + {v_{T,p}}^2)^{1/2}} \times \beta_m = \frac{4}{3}Kn\beta_m, \tag{2.15}$$

where Dp and Dv are the diffusion coefficients of the particle and the vapor molecule, respectively. The Knudsen number Kn is defined as $2\lambda/(dp + dv)$, with λ the mean free path (see Lehtinen and Kulmala, 2003). For $Kn(r) \ll 1$, the system is in the continuum regime (hydrodynamic-like regime), in free molecular regime for $Kn(r) \gg 1$, and in transition regime for $0.4 < Kn(r) < 20$. In the particular case of nucleation studies, it is commonly assumed that the system is in the free molecular regime, that is, particles are affected by the motion of individual gas molecules. Particles larger than 100 nm are treated as being submersed in a continuous gaseous medium characterized by a Knudsen number $Kn(r) > 1$. β_m is the transition regime correction factor. Many expressions exist to provide a unique equation of the collision kernel that account for the three regimes, but the most commonly used is the Fuchs' flux-matching expression (Fuchs, 1964). Gopalakrishnan and Hogan (2011) also compute such a transition regime collision kernel from mean first passage times for arbitrarily shaped particles and compare all previous expressions (Gopalakrishnan and Hogan, 2011; Gopalakrishnan, Thajudeen, and Hogan, 2011):

$$\beta_m = \frac{1 + Kn}{1 + Kn[((4/3\sigma) + 0.337) + (4/3\sigma)Kn^2]}, \tag{2.16}$$

where α is the sticking efficiency, generally assumed to be the unity. The growth rate (GR) can now be derived from those expressions:

$$\mathrm{GR}_{\mathrm{cond},i} = F\frac{dV_p}{dt}\left(\frac{ddp}{dV}\right)^{-1}, \tag{2.17}$$

$$\mathrm{GR}_{\mathrm{cond},i} = FK \times V_v\left(\frac{ddp}{dV}\right)^{-1} = FK \times \frac{dv^3}{3dp^2}, \tag{2.18}$$

where dv and dp are the condensing vapor diameter and the particle diameter. Calculation of the GR variation according to the particle size at for a given vapor concentration is represented in Figure 2.4.

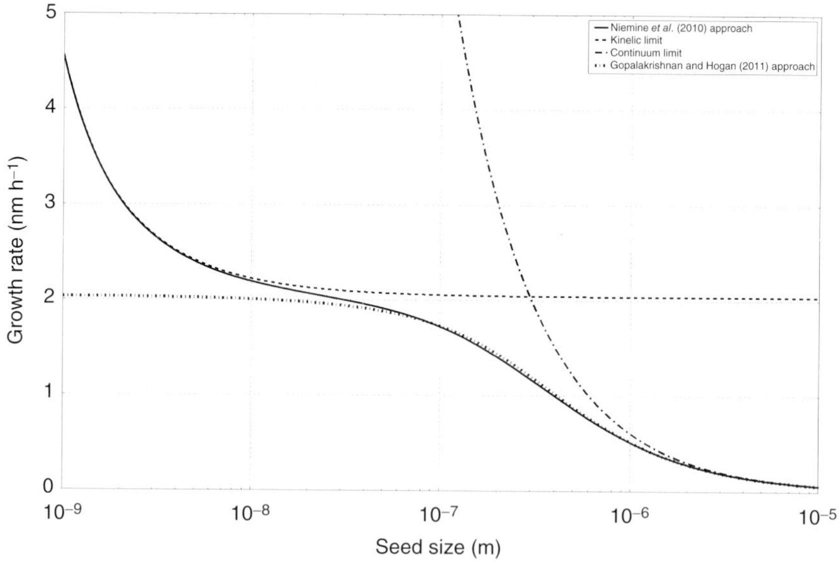

Figure 2.4 Growth rate simulation due to sulfuric acid condensation (10^{+8} molecules cm^{-3}) on a neutral seed. The difference between Nieminen and Gopalakrishnan and Hogan approaches in the kinetic regime is due to the fact Gopalakrishnan and Hogan consider the vapor as a punctual point and that the seed particle is nonmobile while the Nieminen approach accounts for the seed motion and the vapor properties.

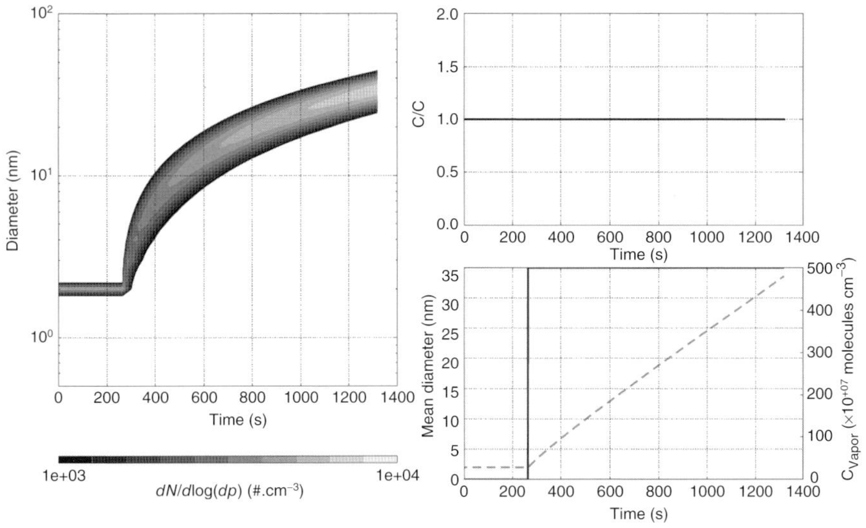

Figure 2.5 Example of growth simulation made for an aerosol population in clean atmosphere.

Also, using the expression in Eq. (2.4), it is possible to simulate the growth of an aerosol population in clean atmosphere, that is, without very large particle on which growing particles and vapor would coagulate or condense on. As an example the aerosol population dynamics of 2 nm pure sulfuric acid particles exposed to 10^8 molecules cm^{-3} of sulfuric acid vapor is plotted in Figure 2.5.

2.3
Observation and Detection Tools

Nucleation can be detected in the ambient atmosphere in the form of "events" during which a thousand to one million nanometric particles per cubic centimeter appear at the sizing lower limit of the classical instrumentation (i.e., usually 3 or 10 nm) and grow into larger sizes (i.e., 10–20 nm) in a few hours. Because nucleation itself (formation of 1 nm stable clusters) is usually not directly observed due to the sizing limitation, the events are more appropriately defined as NPF events. A typical NPF day is shown in Figure 2.5.

In Figure 2.6, one can see that during the midmorning, several thousand particles are first detected in the smallest size bins. Within a couple of hours these particles grow to sizes larger than about 10 nm. NPF itself lasts over 2 h, after which it is likely that the new particles formed represent a condensational sink for the condensable vapors that are taken up efficiently and no more new particles are nucleated.

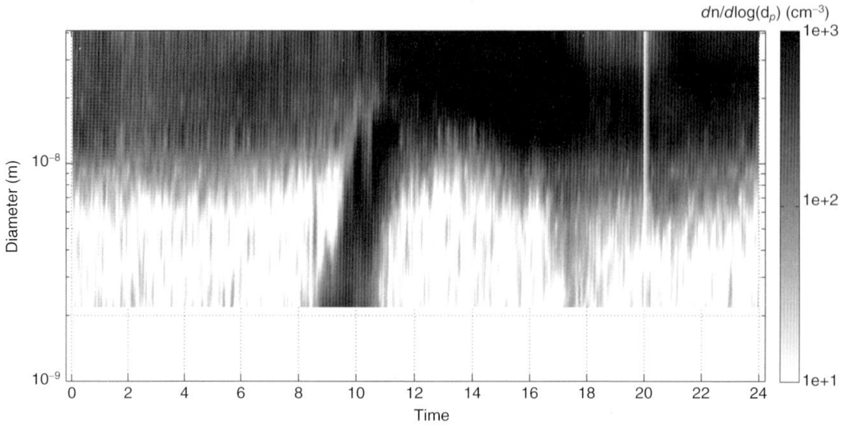

Figure 2.6 Example of a new particle formation event occurring around 10:00 (UTC) detected with a scanning mobility particle sizer (SMPS).

2.3.1 Detection Tools

2.3.1.1 Physical Characterization

The size distribution of particles larger than about 10 nm is classically obtained with a differential mobility particle sizer (DMPS) or scanning mobility particle sizer (SMPS). Both instruments consist of a neutralizing device followed by a selecting size column (differential mobility analyzer) (DMA) and finally a condensation particle counter (CPC). However, the characterization of nanoparticles is not trivial: firstly because growth by condensation/coagulation is fast, secondly because losses by diffusion and sticking to the inlets are high, and thirdly because the charging efficiency is extremely low. Consequently, instruments such as SMPS were not suited for characterization at very small particle sizes until specific columns and CPC were developed, leading to nano-DMPS or nano-SMPS in the 1990s, capable of measuring the size distribution of particles down to 3 nm. Sub-3 nm particles can be detected by recently developed CPC that combines the use of a working fluid with a lower saturation vapor pressure than *n*-butanol (used in classical CPCs) and a turbulent mixing between the dry cold particle-containing flow and a warm clean saturated air stream. In this type of CPC, activated particles still have small diameter and need to be further grown in a classical CPC in order to be optically detectable. Hence, they are rather used a "pre-CPC" or particle size magnifiers (PSM) (Vanhanen *et al.*, 2011).

Naturally charged ions and particles can be detected using an electrometer. High flow rates and electrometers detecting at several levels of the DMA column can allow to detect nanoions down to 0.4 nm with a 5 min resolution using an air ion spectrometer (AIS) (Mirme *et al.*, 2007). The neutral cluster and air ion spectrometer (NAIS) is equipped with a precharger that allows for the detection of neutral nanoclusters larger than 2 nm in the environment, the lower 2 nm size limit being dictated by the necessity to generate ions of this size to charge the clusters in the instrument.

2.3.1.2 Chemical Characterization

The determination of the chemical composition of clusters is difficult not only because of the aforementioned experimental limitations (losses during sampling, fast growth and coagulation, difficulty to charge small clusters) but also because they represent extremely low mass. The traditional approach for chemical characterization of the aerosol phase consisting in filter or impactor sampling and subsequent offline analyses does not allow a quantitative and reliable identification of particles smaller than 30 nm. An indirect approach to infer the chemical composition of nanoparticles is to measure their physical properties such as their hygroscopicity or volatility using the HTDMA or VTDMA techniques. In these instruments, nanoparticles down to 3 nm may be size-selected and humidified or volatilized after being sized again for quantifying their hygroscopic growth or volatility. While inorganic compounds have a high hygroscopicity and low volatility (except for sulfuric acid), fresh organics will show a low hygroscopicity and usually high volatility.

Another indirect approach to identify the nature of clusters and of the vapors participating to their early growth is to follow the covariation between the nanoparticle concentrations and the ones of low vapor pressure compounds present in the atmosphere using gas-phase mass spectrometry (proton transfer mass spectrometer (PTR-MS) or chemical ionization mass spectrometer (CIMS) techniques. These techniques can also be adapted to directly identify the composition of nanoparticles of the 10–20 nm size range by volatilizing them. In the thermal desorption chemical ionization mass spectrometer (TD-CIMS), aerosol particles as small as 4–7 nm can be size-selected by a DMA after being charged, collected by electrostatic deposition, and evaporated prior to analysis by a classical CIMS method. In the nanoaerosol mass spectrometer (NAMS), an aerodynamical lens and a quadrupole ion guide focus charged particles in the 7–30 nm diameter range into the center of an ion trap where they are ionized by a high-energy laser pulse.

Mass spectrometers used in atmospheric sciences to measure the composition of natural ions and nucleation clusters were initially single and triple quadrupole (Arnold *et al.*, 1999; Eisele, 1993). The low resolution of those instruments did not enable very accurate identifications of individual ions. Also very recently, atmospheric pressure interface time-of-flight (Api-TOF) mass spectrometry (Junninen *et al.*, 2010) has been applied to natural ion chemical identification. This instrument clearly identifies most ions in the negative ion mode. Yet, in the positive ion mode, most ions above 100 amu remain unidentified. A higher resolution (than the actual 3000) would be necessary in order to attribute unique stoichiometries to those peaks. With time-of-flight mass spectrometry, increased resolution is usually attained by reducing the size and therefore the original energy dispersion in the ion packet used for measuring a spectrum, which comes to the price of a lower sensitivity, as more ions are rejected.

One of the most up-to-date instrumentations for analyzing the neutral molecular cluster chemical composition are the cluster-CIMS instrument (Zhao *et al.* 2010) and the chemical ionization atmospheric pressure interface time of flight (CI-Api-TOF) (Jokinen *et al.*, 2012). The cluster-CIMS is capable of detecting masses ranging from 100 to 900 amu with a resolution close to 1 amu. However, the technique involves a chemical ionization of neutral species and hence an eventual modification of the natural clusters within the instrument.

2.3.2 Metrics for Characterizing New Particle Formation Events

An NPF event can be defined as such based on four criteria (Dal Maso *et al.*, 2005): "(i) A distinctly new mode of particles must appear in the size distribution, (ii) the mode must start in the nucleation mode, (iii) the mode must prevail over a time span of hours, and (iv) the new mode must show signs of growth." Hence, an NPF event can be characterized by the formation rate of particles of a given size i (J_i) and the GR of the newly formed particles in a given size range $i-j$ (GR_{i-j}). The growth of newly formed cluster particles is a key process that should be fast enough for these clusters to survive in the atmosphere against coagulation with larger particles. The GR calculated from the measured size distributions recorded at a fixed location are apparent, since they are based on the assumption of an Eulerian system which is spatially and temporally homogeneous. The apparent growth rate of cluster particles can be calculated from the size distribution of neutral or charged particles, which are assumed to grow at the same rate. The GR can be calculated either by following the diameter at which particles peak in the nucleation mode with time (method 1) or by searching for the maximum concentration of given size ranges (method 2):

$$\mathrm{GR}_{dp_x-dp_y} = \frac{dp_y - dp_x}{t_{(dp_y)} - t_{(dp_x)}}, \tag{2.19}$$

where dp_x is the modal diameter of the ultrafine particle population at $t(dp_x)$ (method 1) or $t(dp)$ is the time when the particle population of diameter dp is maximal (method 2). One can infer from the growth rate of newly formed particles an equivalent condensable vapor concentration responsible for this growth:

$$\mathrm{GR} = \frac{ddp}{dt} = \frac{4m_{\mathrm{vap}}\beta_m[\mathrm{vapor}]}{\rho\ dp}, \tag{2.20}$$

where m_{vap} is the mass of a vapor molecule, β_m is the transitional correction factor defined in Eq. (2.16), [Vapor] is the condensable vapor concentration, ρ is the aerosol density, and dp is the particle diameter. The reference vapor for the early growth of cluster particles has classically been sulfuric acid, but organic acids are also good candidates.

The sink of condensing vapors on the preexisting particle population is defined by the condensational sink CS:

$$CS = 4\pi D_{\text{vap}} \int_0^\infty r\beta_m N(r)dr, \tag{2.21}$$

where D_{vap} is the diffusion coefficient of the condensing vapor, r is the particle radius, and $N(r)$ is the concentration of particles of radius r.

The formation rate of particles of a given size i nm (J_i) (usually the lower size limit of the instrument measuring the number size distribution) and charged particles of size i ($J \pm i$) is determined as the number concentration increase of the particles (or charged particles) of i nm by unit of time, corrected from the loss of these particles by coagulation and growth by condensation to larger sizes:

$$J_i = \frac{dN_{i-(i+1)}}{dt} + \text{Coag}S_i \times N_{i-(i+1)} + \frac{f}{1\,\text{nm}}\text{GR}_{(i-1)-(i+1)} \times N_{i-(i+1)}, \tag{2.22}$$

$$J^{\pm}{}_i = \frac{dN^{\pm}{}_{i-(i+1)}}{dt} + \text{Coag}S_i \times N^{\pm}{}_{i-(i+1)} + \frac{f}{1\,\text{nm}}\text{GR}_{(i-1)-(i+1)} \times N^{\pm}{}_{i-(i+1)}$$
$$+ +\alpha \times N^{\pm}_{i-(i+1)} N_{<(i+1)} - \beta \times N_{i-(i+1)} N^{\pm}_{<i}, \tag{2.23}$$

where $N \pm i - (i+1)$ is the ion number concentration (positive or negative ions) [#. cm^{-3}] in diameter range from i to $i+1$ nm and $N \pm <x$ is the ion number concentration below x nm and α and β are, respectively, the ion–ion recombination coefficient and the ion-neutral attachment coefficient and are usually assumed to be equal, respectively, to 1. 6 Å ~ 10^{-6} cm^3 s^{-1} and 1 Å ~ 10^{-8} cm^3 s^{-1} (Tammet and Kulmala, 2005). The factor f represents the fraction of the aerosol population in a size range from i to $i+1$ nm which is activated for growth. When unknown, this factor is assumed to be equal to unity. CoagS_i is the coagulation sink of i nm particles (s^{-1}):

$$\text{Coag}S_i = N_i \sum_j K_{d_i,d_j} N_j, \tag{2.24}$$

where N_i is the concentration of i nm particles, N_j is the concentration of preexisting particles of diameter d_j, and K_{d_i,d_j} is the Brownian coagulation coefficient between particles of diameters d_i and d_j.

The IIN rate is a measure of the contribution of charged particles to the NPF process, and it is defined as

$$\text{IIN} = \frac{J_i^+ + J_i^-}{J_i}. \tag{2.25}$$

As it cannot be directly measured, one can derivate the nucleation rate J^* (usually assumed to be J_1) from the formation rate of particles of size i J_i :

$$J_i = J^* \exp\left[-\gamma\ dp^* \frac{\text{Coag}S_{(dp^*)}}{\text{GR}}\right], \tag{2.26}$$

where

$$\gamma = \frac{-1}{m+1}\left[\left(\frac{dp_i}{dp^*}\right)^{m+1} - 1\right] \tag{2.27}$$

and

$$m = \frac{\log(\text{Coag}S_{(dp_i)}) - \log(\text{Coag}S_{(dp^*)})}{\log(dp_i) - \log(dp^*)} \tag{2.28}$$

with assumptions that the only important sink for the clusters is their coagulation to larger preexisting particles, that they grow by condensation at a constant rate, and that the preexisting population of larger particles remains unchanged during the cluster growth.

2.3.3 Occurrence of New Particle Formation Events in the Troposphere

With the development of instruments detecting particles at the nanometer scale, NPF events have been observed in a growing number of environments, ranging from urban areas to Antarctica. In 2004, they had been reported in more than 50 locations [see Kulmala *et al.*, 2004 for a review]. With a few exceptions, NPF events were observed during daytime, in majority in the morning hours. This time of the day corresponds to the onset of photochemical reactions and also to the nocturnal stable layer breakout. The combination of both factors is in favor of NPF events. On one side condensable vapors usually associated with secondary products are created by photochemistry, and on the other side primary particles produced in the mixed boundary layer (PBL) were occasionally detected in the upper troposphere and in specific environments such as the native Eucalypt forest, in which it is believed that sulfuric acid originated from nonphotochemical reactions. AIS and NAIS measurements revealed that the charged sub-2 nm ion clusters are always present not only in the atmosphere during nucleation but also outside of the nucleation periods. Neutral clusters appear to be present at one or two orders of magnitude higher concentrations than charged clusters. The sub-2 nm neutral cluster population measured by PSM–CPC is not as ubiquitous as the charged cluster population and significantly increases during nucleation events.

2.3.3.1 Pristine and Polluted Continental Boundary Layer

Most of NPF events were reported to occur in the boundary layer of midlatitude areas due to the predominance of the measuring sites in these locations. Reported NPF event frequency varies from 25% of the time in northern Scandinavia to 60% of the time in northwestern Europe. In the continental boundary layer, NPF events often show a clear and continuous growth to larger sizes, leading to a "banana" shape of the contour-type plot shown Figure 2.5. This shape is typical of a homogeneous landscape over large distances. Measurements are conducted at a fixed location, and the fact that the development of the particle size distribution can be observed for several hours demonstrates that these NPF events are stationary over this period, happening simultaneously at the regional scale, so that the advection of air masses by the wind does not alter time-related processes.

Reported formation rates of 3 nm particles (J_3) show a very large range (from 0.01 to 100 # cm^{-3} s^{-1}), with the highest rates observed in urban areas. Cluster size (1.5–3 nm) growth rates vary from 0.3 nm h^{-1} in the rural continental background to 22 nm h^{-1} in urban environment. In the polar regions, growth rates in the 1.5–3 nm size range are among the lowest, ranging between 0.9 and 2.1 nm h^{-1}.

2.3.3.2 Coastal and Marine Boundary Layer Sites

In the marine boundary layer, the most active areas for NPF are coastal regions, where frequent and intense events were detected. NPF rates are one of the highest reported in the literature at marine coastal sites, J_3 reaching 10^4–10^5 # cm^{-3} s^{-1} (O'Dowd *et al.*, 2002). Over coastal areas, it has been shown that NPF issued from secondary processes were linearly related to the mass of seaweed mass exposed to air during low tide and sunny periods. The occurrence of nucleation and NPF over the open ocean remains an open question. Early investigations into the source of new particles in the remote marine atmosphere identify two likely sources. The first source is the marine boundary layer from the dimethyl sulfide (DMS) flux, while the second source is entrainment from the free troposphere (FT). Direct observations of NPF events in the remote marine atmosphere are scarce. The continuous growth of 15 nm particles to 40–50 nm particles over 24–48 h was detected from marine stations, indicating open ocean NPF events (O'Dowd, Monahan, and Dall'Osto, 2010). Solar radiation on seawater causes stratification favoring the dominance of high light-adapted cells, such as *Synechococcus*, whereas increase in stormy events might deepen mixed layer that would induce more opportunistic greater cell (diatoms, for instance) occurrence; both effects will likely influence marine aerosol formation.

2.3.3.3 High-Altitude Environments and Free Troposphere

The FT is much cleaner and less turbulent than the PBL. The interface between the PBL and the FT offers favorable conditions for nucleation to occur, as aerosol precursors and water vapor from lower altitudes are mixed with clean and cold air, usually under enhanced photochemical conditions. Modeling studies revealed that nucleation is expected to be a major contributor to the total particle number concentration in the upper troposphere (Merikanto *et al.*, 2009b). At high-altitude monitoring stations, the frequency of NPF event was shown to be high (from 30% to 65% of the time), up to altitudes over 5000 m a.s.l. This high frequency of the NPF events is observed to strongly impact the average aerosol size distribution daily variation, which implies that it is a significant source of particles in the upper troposphere. NPF events were shown to be significantly more frequent at high altitude than at low altitude when comparing the frequencies of NPF events simultaneously monitored at the same geographical location but different altitudes. Airborne measurements showed that nucleation occurs in the upper FT, or at the interface between the PBL and the FT, over marine areas. Clouds often form at high altitudes and may interrupt or inhibit the NPF events. On most high-altitude stations, it was shown that clouds had an inhibiting effect on NPF, both due to their role in shielding the amount of solar radiation reaching the sites and due to their role in scavenging newly formed clusters. Because the upper troposphere is naturally a condensing vapor-poor environment, polluted air masses were shown to be more favorable to NPF events, in contrast to the lower PBL. GR of the 1.5–3 nm size range are among the highest at high altitude and can reach up to 40 nm h^{-1}, even though the determination of the GR at mountainous stations can be biased by upslope winds inducing nonstationary conditions. Ions may play a more pronounced role in the upper atmosphere as IIN rates calculated for several high-altitude stations show an increase of the IIN with altitude.

2.4
Precursor Candidates for Nucleation and Early Growth from Observations

2.4.1
Continental Planetary Boundary Layer

In the continental PBL, nucleation occurs at much lower sulfuric acid concentrations than predicted by the binary homogeneous nucleation theory, and other gases may also be involved in aerosol nucleation. The detection of condensable gases suspected to be involved in the nucleation process using CIMS or PTR-MS techniques have shown that sulfuric acid was correlated to the formation rate of new clusters and organic compounds were linked to the growth of newly formed particles. The CIMS technique has been used in the past to detect organic trace gases during nucleation events. CIMS measurements first showed that organic compounds, and among them monoterpene oxidation products, actively participated to the growth of newly formed particles in forested environments (Laaksonen *et al.*, 2008). This indicates that volatile organic compound (VOC) oxidation products may have a key role in determining the spatial and temporal features of the NPF events. Using the TD-CIMS technique, 6–15 nm particles were found to be mainly composed of ammonium sulfate in several urban environments, while it was lately shown that 8 nm newly formed particles of other urban environments contained amines, which evidences that organics participate to the first step of cluster growth (Smith *et al.*, 2010). Measurements in Antarctica reveal that freshly nucleated particles have the smallest hygroscopic growth factors, indicative of a significant contribution of organics to the newly formed particles.

2.4.2
Marine Planetary Boundary Layer

Historically, sulfur species were considered the main chemical components involved in secondary marine aerosol formation via DMS emission by marine phytoplankton and its further oxidation to nonsea-salt (nss) sulfate aerosol (Charlson *et al.*, 1987). But organic compounds are also detected in secondary formed particles, and concentrations of H_2SO_4 were typically found insufficient to explain the growth of nucleated clusters to measurable sizes of 3 nm. The key species in coastal NPF events were first believed to be iodocarbons (CH_2I_2) first emitted in the gas phase and then readily photolyzed to low vapor pressure iodine compounds that produce new particles. However, an observation of molecular iodine (I_2) in the coastal boundary layer led to the proposition of new mechanisms for coastal iodine production that could explain the observed levels of iodine dioxide radicals OIO and resulting new particles. Iodine flux from I_2 is three orders of magnitude higher than that of CH_2I_2, due to a higher photolysis rate for I_2. Photochemically produced iodine atoms react with ozone to produce iodine monoxide (IO) which subsequently self-react to form OIO and I_2O_4 that will form particles and be oxidized to further species of the form I_xO_y. Species produced during coastal-type nucleation events are mainly hydrophobic. Over open marine areas, the organic fraction of the aerosol is issued from different processes than in coastal regions and was shown to be mainly water-soluble organic carbon (WSOC). A fraction of the detected WSOC is supposed to be produced via photooxidative processing of the primary organic components and/or via secondary organic aerosol (SOA) formation. Diacids and ketones have been identified in the WSOC fraction of marine Arctic aerosols; their content increased in the presence of light, suggesting a photochemical production. Water-soluble organic nitrogen (WSON) is also an important component of submicron aerosol in marine environment. Dimethyl- and diethylammonium salts (DMA+ and DEA+), formed by the reaction of dialkyl amines with acids (mainly sulfuric acid), are the second most abundant organic species, after methanesulfonic acid (MSA), in fine marine particles collected over the North Atlantic region (Facchini *et al.*, 2008). The sea surface is enriched in organic compounds which both can be reactive toward atmospheric oxidants and be light absorbers (as for polycyclic aromatic hydrocarbons (PAHs), porphyrins, and dissolved organic matter (DOM)). Volatile iodocarbons may directly be emitted from seawater upon reaction of ozone with DOM (Martino *et al.*, 2009), suggesting that emission of halogen compounds can also occur independent of the biological activity and, in this specific case, depends on the presence of the background DOM present ubiquitously at the sea surface. The importance of ozone deposition was also demonstrated upon chemical reaction with seawater components such as chlorophyll and fulvic acids. At last, light on light-absorbing species can initiate radical reactions in sea-salt water with generation of species ($X = Cl$, Br and I). Their production will increase the organic film reactivity with enhanced emission of VOCs which can potentially influence SOA formation.

2.5 Parameterizations and Chamber Experiments

As seen, none of the models described earlier are able to describe accurately the general case of atmospheric nucleation. At high altitude, the role of ions seems to increase, and also the conditions are expected to enhance nucleation of the sulfuric acid and water binary system. In order to include the nucleation into large-scale model, calculation procedures of the particle formation rate must be (i) simplified to reduce the computational burden, and (ii) if not accurate, they must provide a relevant result. As consequence to this need, different parameterizations of the nucleation process based either empirically from laboratory and field measurements or from the classical nucleation models derived from BNS or THN theories have been developed. Particle nucleation processes and in particular the H_2SO_4–H_2O system were studied in the laboratory using different techniques (fast expansion chambers, flow reactors, aerosol chambers). Nucleation studies of sulfuric acid and water are complicated because of the difficulty in producing clean and well-known concentrations of sulfuric acid; because of the necessity of having extremely homogeneous concentrations in the whole system, avoiding local supersaturations close to the inlet; and because of important losses of sulfuric acid with the reactor walls. The measurement of gaseous H_2SO_4 is challenging as well. Therefore, an exact quantification of the H_2SO_4 concentration in the nucleation zone is difficult. One of the most ambitious chamber experiments was performed at the CERN facility at the CLOUD chamber using the highest standards of cleanliness and temperature stability. Results obtained show that atmospherically relevant ammonia mixing ratios significantly increase the nucleation rate of sulfuric acid particles. APi-TOF measurements reveal that nucleation proceeds by a base stabilization mechanism in a stepwise accretion of ammonia molecules on sulfuric acid clusters. Moreover, it was shown that ions additionally increase the nucleation rate, provided that the nucleation rate lies below the limiting ion-pair production rate. Hence, it was concluded that ion-induced binary nucleation of H_2SO_4–H_2O can occur in the midtroposphere but is negligible in the boundary layer. However, even with the large enhancements in rate due to ammonia and ions, atmospheric concentrations of ammonia and sulfuric acid are insufficient to account for observed boundary layer nucleation (Kirkby *et al.*, 2011). In the real atmosphere, sulfuric acid concentrations were found to be linked to the nucleation rate J:

$$J = k_1[H_2SO_4]^{k_2}, \tag{2.29}$$

where k_2 varies between 1 and 2, which indicates that one or two molecules per cubic centimeter are necessary to reach the critical cluster size. Several studies showed that atmospheric NPF requires the presence of sulfuric acid in concentrations in excess of 10^5 molecules cm^{-3}. However, in the real atmosphere, the k_1 factor varies by several orders of magnitude, showing that sulfuric acid is not the only compound involved in nucleation and early growth and that the nucleation rate could also be written as

$$J = k_1[\text{NucSpec}][H_2SO_4]^{k_2}, \tag{2.30}$$

where organic gases are strongly suspected to also take part in the process as the nucleating species NucSpec. A parameterization has been developed from chamber experiments to account for the role of organic oxidation products in the cluster stabilization (Metzger *et al.*, 2010), but its relevance has not been confronted to atmospheric data yet. In the CLOUD experiment, amines were also shown to play a critical role in cluster formation and stabilization.

Specific chamber studies were also performed in order to investigate nucleation from marine precursors. Laboratory studies either directly used macro- or microalgae in simulation chambers or introduced precursors such as I_2 with different oxidants (lights or ozone). Realistic nucleation and growth rates could be extracted from the semicontrolled experiments involving real macroalgae. Growth rate of 3–6 nm particles varies in the 70–200 nm h^{-1} range, in agreement with field measurements (Sellegri *et al.*, 2009). The nucleation rate of 3 nm particles could be parameterized as a function of the I_2 concentration at ambient ozone levels, reaching 2800 # cm^{-3} ppt^{-1} I_2. Modeling studies examined the formation of thermodynamically stable clusters by dimer formation of IOI vapor, where the iodine precursor was assumed to be molecular I_2 emitted from seaweed. They found that a steady state was reached in less than 150 s, with predicted formation and growth rate in agreement with previous field and laboratory studies.

When compared to each other, nucleation parameterizations predict nucleation rates that differ by several orders of magnitude (Zhang *et al.*, 2010). In general best agreement was found for parameterizations derived from field measurements which were developed under specific atmospheric conditions dependent of the measurement site.

2.6 Importance of Nucleation for the Production of Aerosols and CCN at the Global Scale

Nucleated particles may reach the 50–100 nm size range by growing through the processes of condensation and intramodal coagulation. The growth rate of nucleated particles should be higher than their capture/removal by coagulation with preexisting larger particles in order for them to reach these climatic-relevant sizes. When the nucleated particles exceed 50 nm, they may become efficient light scatters and cloud CCN. Usually, the number concentration of CCN is evaluated to be comprised between the number concentration of particles larger than 100 nm (N100) and the number concentration of particles larger than 50 nm (N50). The number of nucleated particles that reach CCN sizes can be evaluated from measurements or from modeling studies. The methodology to evaluate how many particles issued from nucleation contribute to the total particle number and CCN number concentration from observations at a given location is based on the hypothesis that the location is situated in an homogeneous area. The concentrations of a given size range (N10, N50, or N100) are first evaluated prior to the nucleation event, that is, prior to the 10 nm particle concentration starting to increase. This period usually corresponds to the morning atmospheric mixing due to the development of convection, leading to a decrease of the total particle concentration and breakup of the temperature inversion. The particle concentrations are then calculated after the newly formed particles have grown to the CCN size range, but before the breakup of the nocturnal layer occurs the next day. The difference between the concentrations of particles in each size range after and before the NPF event represents the number of particles and CCN that are produced during the NPF event. The mean CCN number concentration formed by NPF event varies from 1000 # cm^{-3} for particles larger than 50 nm and 150 # cm^{-3} for particles larger than 100 nm in northern Scandinavia (Kerminen *et al.*, 2012) to 7000 #>50 cm^{-3} and 2500 #>100 cm^{-3} in northern Italy. At high altitudes, for instance, at Chacaltaya, a South American site at 5200 m a.s.l., NPF event production is corrected from preexisting particles transported to the station and represents around 2500–3500 #>50 cm^{-3} and around 400–650 #>100 cm^{-3}. The potential importance of atmospheric NPF for regional and global CCN budgets has been demonstrated by also using global models, even though uncertainties related to these studies are still large due to our lack of knowledge both in the nucleation process itself and in the species responsible for the cluster's early growth. From global models, the NPF process would account for 30–70 of the total CCN, which is relatively in agreement with measurements.

2.7 Conclusions

Nucleation is still a very active research topic in various fields (materials science, fundamental physics, chemistry, etc.). The chemical diversity of atmospheric composition is a smog screen to our understanding of atmospheric NPF. However, some simple parameterizations are able to account for observed nucleation rates, at least orders of magnitude more accurate than some process-based models. From the climatic point of view, freshly formed particles of a few nanometers (typically less than 10 nm) are not relevant but become important when they reach few tens of nanometers size at which they can act as CCN and interact with solar radiations. Reducing uncertainties on the CCN number formed from the growth of freshly formed particles depends on the aerosol population dynamics, which in turn strongly depends on the particle size and growth rate. In the future, we will need to be able to predict all steps of NPF events, from nucleation to early growth. Sulfur dioxide is considered the most important precursor gas for atmospheric nucleation particles. It is emitted into the atmosphere mostly by anthropogenic sources such as combustion of sulfur-containing fossil fuels. Therefore, aerosol nucleation in the atmosphere would be expected to be enhanced by anthropogenic activities. On the other hand, the preexisting aerosol that can take up condensing species and thereby suppress nucleation is increased as well by anthropogenic sources. At last, the role of organic substances is poorly understood, not only in the nucleation process itself but also in the cluster early growth, to which they are shown to largely contribute. Investigations on the overall anthropogenic influences on atmospheric aerosol nucleation, compared to natural NPF processes occurring in the marine atmosphere, in dust clouds transported from desert areas, and in volcanic ash plumes, are currently just emerging.

Abbreviations

AIS	air ion spectrometer
Api-ToF MS	atmospheric pressure interface mass spectrometer
BNS	binary nucleation system
CCN	cloud condensation nuclei
CERN	Center for Nuclear Research
CI-Api-TOF	chemical ionization atmospheric pressure time of flight
CIMS	chemical ionization mass spectrometer
CNT	classical nucleation theory
CPC	condensation particle counter
CS	condensational sink
DMA	differential mobility analyzer
DMPS	differential mobility particle sizer
DMS	dimethylsulfide
DOM	dissolved organic matter
FT	free troposphere
GR	growth rate
HTDMA	hygroscopicity tandem differential mobility sizer
IIN	ion-induced nucleation
MSA	methane sulfonic acid
NAIS	neutral cluster and air ion spectrometer
NPF	new particle formation
PAH	polycyclic aromatic hydrocarbons
PBL	planetary boundary layer
PSM	particle size magnifier
PTR-MS	proton transfer mass spectrometer
SOA	secondary organic aerosol
SMPS	scanning mobility particle sizer
TD-CIMS	thermal desorption chemical ionization mass spectrometer
TSN	ternary system nucleation
USM	unary system model
VOC	volatile organic compound
VTDMA	volatility tandem differential mobility analyzer
WSOC	water-soluble organic carbon
WSON	water-soluble organic nitrogen

List of Symbols

Latin

a_i	chemical activity of the specie i
d_p	particle diameter
g_{vap}	Gibbs free energy for a molecule X in the gas phase
g_{part}	Gibbs free energy for a molecule X in the particle phase
J	nucleation rate
K	collision rate between particle and vapor molecule in the kinetic regime
k_{B}	Boltzmann constant equal to $1.38 \times 10^{-23}\,\text{J K}^{-1}$
Kn	Knudsen number
$P_X(T)$	partial pressure of the species X
$P_{\text{sat},X}(T)$	saturation vapor pressure of the species X
r	particle radius
r^*	cluster critical radius
ρ_{part}	mass per unity of volume of the particle
S	vapor saturation ratio
T	air temperature

Greek

β_m	transition regime correction factor
ΔG	change in the Gibbs free energy budget
μ	chemical potential
Σ	surface tension
v_{part}	volume of a molecule in the particle

References

Almeida, J., Schobesberger, S., Kürten, A., Ortega, I.K., Kupiainen-Määttä, O., Praplan, A.P., Adamov, A., Amorim, A., Bianchi, F., Breitenlechner, M., David, A., Dommen, J., Donahue, N.M., Downard, A., Dunne, E., Duplissy, J., Ehrhart, S., Flagan, R.C., Franchin, A., Guida, R., Hakala, J., Hansel, A., Heinritzi, M., Henschel, H., Jokinen, T., Junninen, H., Kajos, M., Kangasluoma, J., Keskinen, H., Kupc, A., Kurtén, T., Kvashin, A.N., Laaksonen, A., Lehtipalo, K., Leiminger, M., Leppä, J., Loukonen, V., Makhmutov, V., Mathot, S., McGrath, M.J., Nieminen, T., Olenius, T., Onnela, A., Petäjä, T., Riccobono, F., Riipinen, I., Rissanen, M., Rondo, L., Ruuskanen, T., Santos, F.D., Sarnela, N., Schallhart, S., Schnitzhofer, R., Seinfeld, J.H., Simon, M., Sipilä, M., Stozhkov, Y., Stratmann, F., Tomé, A., Tröstl, J., Tsagkogeorgas, G., Vaattovaara, P., Viisanen, Y., Virtanen, A., Vrtala, A., Wagner, P.E., Weingartner, E., Wex, H., Williamson, C., Wimmer, D., Ye, P., Yli-Juuti, T., Carslaw, K.S., Kulmala, M., Curtius, J., Baltensperger, U., Worsnop, D.R., Vehkamaki, H., and Kirkby, J. (2013) Molecular understanding of sulphuric acid-amine particle nucleation in the atmosphere. *Nature*, **502**, 359–363. doi: 10.1038/nature12663

Arnold, F. (1980) Multi-ion complexes in the stratosphere – implications for trace gases and aerosol. *Nature*, **284**, 610–611.

Arnold, F., Curtius, J., Sierau, B., Burger, V., Busen, R., and Schumann, U. (1999) Detection of massive negative chemiions in the exhaust plume of a jet aircraft in flight. *Geophys. Res. Lett.*, **26**, 1577–1580.

Becker, R. and Döring, W. (1935) Kinetische Behandlung der Keimbildung in übersättigten Dämpfen. *Ann. Phys.*, **24**, 719–752.

Charlson, R.J., Lovelock, J.E., Andreae, M.O., and Warren, S.G. (1987) Oceanic phytoplankton, atmospheric sulfur, cloud albedo and climate. *Nature*, **326** (6114), 366–374.

Chen, M., Titcombe, M., Jiang, J., Jen, C., Kuang, C., Fischer, M.L., Eisele, F.L., Siepmann, I.J., Hanson, D.R., Zhao, J., and McMurry, P.H. (2012) Acid–base chemical reaction model for nucleation rates in the polluted atmospheric boundary layer. *Proc. Natl. Acad. Sci. U.S.A.*, **109**. doi: 10.1073/pnas.1210285109

Covert, D.S., Kapustin, V.N., Quinn, P.K., and Bates, T.S. (1992) New particle formation in the marine boundary layer. *J. Geophys. Res.*, **97**, 20581–20589.

Dal Maso, M., Kulmala, M., Riipinen, I., Wagner, R., Hussein, T., Aalto, P.P., and Lehtinen, K.E.J. (2005) Formation and growth of fresh atmospheric aerosols: eight years of aerosol size distribution data from SMEAR II, Hyytiälä, Finland. *Boreal Environ. Res.*, **10**, 5, 323–336.

Eisele, F.L., and Tanner, D.J. (1993) Measurement of the gas phase concentration of H2SO4 and methane sulfonic acid and estimates of H2SO4 production and loss in the atmosphere. *J. Geophys. Res.*, **98**, D5, 9001–9010, doi:10.1029/93JD 00031.

Facchini, M.C., Decesari, S., Rinaldi, M., Carbone, C., Finessi, E., Mircea, M., Fuzzi, S., Moretti, F., Tagliavini, E., Ceburnis, D., and O'Dowd, C.D. (2008) Important source of marine secondary organic aerosol from biogenic amines. *Environ. Sci. Technol.*, **42**, 9116–9121.

Feder, J., Russell, K.C., Lothe, J., and Pound, G.M. (1966) Homogeneous nucleation and growth of droplets in vapour. *Adv. Phys.*, **15**, 111–178. doi: 10.1080/00018736600101264

Ford, I.J. (1997) Nucleation theorems, the statistical mechanics of molecular clusters, and a revision of classical nucleation theory. *Phys. Rev. E*, **56**, 5615–5629. doi: 10.1103/PhysRevE.56.5615

Fuchs, N.A. (1964) *The Mechanics of Aerosols*, rev. edn, Pergamon Press, New York.

Gopalakrishnan, R. and Hogan, C.J. (2011) Determination of the transition regime collision kernel from mean first passage times. *Aerosol Sci. Technol.*, **45**, 1499–1509. doi: 10.1080/02786826.2011.601775

Gopalakrishnan, R., Thajudeen, T., and Hogan, C.J. (2011) Collision limited reaction rates for arbitrarily shaped particles across the entire diffusive Knudsen number range. *J. Chem. Phys.*, **135**, 054302. doi: 10.1063/1.3617251

Hoppel, W.A., Frick, G.M., and Fitzgerald, J.M. (1994) Marine boundary layer measurements of new particle formation and the effect of non precipitating clouds have on aerosol size distribution. *J. Geophys. Res.*, **99**, 14443–14495.

Jokinen, T., Sipilä, M., Junninen, H., Ehn, M., Lönn, G., Hakala, J., Petäjä, T., Mauldin, R.L., Kulmala, M., and Worsnop, D.R. (2012) Atmospheric sulphuric acid and neutral cluster measurements using CI-APi-TOF. *Atmos. Chem. Phys.*, **12**, 4117–4125.

Junninen, H., Ehn, M., Petäjä, T., Luosujärvi, L., Kotiaho, T., Kostiainen, R., Rohner, U., Gonin, M., Fuhrer, K., Kulmala, M., and Worsnop, D.R. (2010) A high-resolution mass spectrometer to measure atmospheric ion composition. *Atmos. Meas. Tech.*, **3**, 4, 1039–1053, doi:10.5194/amt-3-1039-2010.

Kerminen, V.-M., Paramonov, M., Anttila, T., Riipinen, I., Fountoukis, C., Korhonen, H., Asmi, E., Laakso, L., Lihavainen, H., Swietlicki, E., Svenningsson, B., Asmi, A., Pandis, S.N., Kulmala, M., and Petäjä, T. (2012) Cloud condensation nuclei production associated with atmospheric nucleation: a synthesis based on existing literature and new results. *Atmos. Chem. Phys.*, **12**, 12037–12059.

Kirkby, J., Curtius, J., Almeida, J., Dunne, E., Duplissy, J., Ehrhart, S., Franchin, A., Gagné, S., Ickes, L., Kürten, A., Kupc, A., Metzger, A., Riccobono, F., Rondo, L.,

Schobesberger, S., Tsagkogeorgas, G., Wimmer, D., Amorim, A., Bianchi, F., Breitenlechner, M., David, A., Dommen, J., Downard, A., Ehn, M., Flagan, R.C., Haider, S., Hansel, A., Hauser, D., Jud, W., Junninen, H., Kreissl, F., Kvashin, A., Laaksonen, A., Lehtipalo, K., Lima, J., Lovejoy, E.R., Makhmutov, V., Mathot, S., Mikkilä, J., Minginette, P., Mogo, S., Nieminen, T., Onnela, A., Pereira, P., Petäjä, T., Schnitzhofer, R., Seinfeld, J.H., Sipilä, M., Stozhkov, Y., Stratmann, F., Tomé, A., Vanhanen, J., Viisanen, Y., Vrtala, A., Wagner, P.E., Walther, H., Weingartner, E., Wex, H., Winkler, P.M., Carslaw, K.S., Worsnop, D.R., Baltensperger, U., and Kulmala, M. (2011) Role of sulphuric acid, ammonia and galactic cosmic rays in atmospheric aerosol nucleation. *Nature*, **476**, 429–433. doi: 10.1038/nature10343

Koop, T., Bookhold, J., Shiraiwa, M., and Pöschl, U. (2011) Glass transition and phase state of organic compounds: dependency on molecular properties and implications for secondary organic aerosols in the atmosphere. *Phys. Chem. Chem. Phys.*, 19238–19255. doi: 10.1039/c1cp22617g

Korhonen, P., Kulmala, M., Laaksonen, A., Viisanen, Y., McGraw, R., and Seinfeld, J.H. (1999) Ternary nucleation of the H2 SO4, NH3 and H2 O in the atmosphere. *J. Geophys. Res.*, **104**, 26349–26353.

Kulmala, M., Hämeri, K., Aalto, P.P., Mäkelä, J.M., Pirjola, L., Nilsson, E.D., Buzorius, G., Rannik, U., Dal Maso, M., Seidl, W., Hoffmann, T., Janson, R., Hansson, H.-C., Viisanen, Y., Laaksonen, A., and O'Dowd, C.D. (2001) Overview of the international project on biogenic aerosol formation in the boreal forest (BIOFOR). *Tellus B*, **54**, 324–343. doi: 10.1034/j.1600-0889.2001.530402.x

Kulmala, M., Korhonen, P., Napari, I., Karlsson, A., Berresheim, H., and O'Dowd, C.D. (2002) Aerosol formation during PARFORCE: ternary nucleation of H_2 SO_4, NH_3, and H_2O. *J. Geophys. Res.*, **107**, 8111. doi: 10.1029/2001JD000900

Kulmala, M. and Laaksonen, A. (1990) Binary nucleation of water-sulfuric acid system: Comparison of classical theories with different H_2SO_4 saturation vapor pressures. *J. Chem. Phys.*, **93**, 696–701.

Kulmala, M., Vehkamäki, H., Petäjä, T., Dal Maso, M., Lauri, A., Kerminen, V.-M., Birmili, W., and McMurry, P. (2004) Formation and growth rates of ultrafine atmospheric particles: a review of observations. *J. Aerosol Sci.*, **35**, 143–176.

Laaksonen, A., Kulmala, M., O'Dowd, C.D., Joutsensaari, J., Vaattovaara, P., Mikkonen, S., Lehtinen, K.E.J., Sogacheva, L., Dal Maso, M., Aalto, P., Petäjä, T., Sogachev, A., Yoon, Y.J., Lihavainen, H., Nilsson, D., Facchini, M.-C., Cavalli, F., Fuzzi, S., Hoffmann, T., Arnold, F., Hanke, M., Sellegri, K., Umann, B., Junkermann, W., Coe, H., Allan, J.D., Alfarra, M.R., Worsnop, D.R., Riekkola, M.-L., Hyötyläinen, T., and Viisanen, Y. (2008) The role of VOC oxidation products in continental new particle formation. *Atmos. Chem. Phys.*, **8**, 2657–2665.

Lehtinen, K.E.J. and Kulmala, M. (2003) A model for particle formation and growth in the atmosphere with molecular resolution in size. *Atmos. Chem. Phys.*, **3**, 251–257.

L'Hermite, J.-M. (2009) Growth and melting of droplets in cold vapors. *Phys. Rev. E*, **80**, 051602. doi: 10.1103/PhysRevE.80.051602

Lovejoy, E., Curtius, J., and Froyd, K. (2004) Atmospheric ion-induced nucleation of sulfuric acid and water. *J. Geophys. Res.*, **109**, D08204. doi: 10.1029/2003JD004460

Martino, M., Mills, G.P., Woeltjen, J., and Liss, P.S. (2009) A new source of volatile organoiodine compounds in surface seawater. *Geophys. Res. Lett.*, **36**, 1419–1423.

Merikanto, J., Napari, I., Vehkamäki, H., Anttila, T., and Kulmala, M. (2007) New parametrization of sulfuric acid-ammonia-water ternary nucleation rates at tropospheric conditions. *J. Geophys. Res.*, **112**, D15207. doi: 10.1029/2006JD007977

Merikanto, J., Spracklen, D.V., Mann, G.W., Pickering, S.J., and Carslaw, K.S. (2009b) Impact of nucleation on global CCN. *Atmos. Chem. Phys.*, **9**, 8601–8616.

Metzger, A., Verheggen, B., Dommen, J., Duplissy, J., Prevot, A.S., Weingartner, E., Riipinen, I., Kulmala, M., Spracklen,

D.V., Carslaw, K.S., and Baltensperger, U. (2010) Evidence for the role of organics in aerosol particle formation under atmospheric conditions. *Proc. Natl. Acad. Sci. U.S.A.*, **107**, 6646–6651. doi: 10.1073/pnas.0911330107

Mirme, A., Tamm, A., Mordas, G., Vana, M., Uin, J., Mirme, S., Bernotas, T., Laakso, L., Hirsikko, A., and Kulmala, M. (2007) A wide range multi-channel air ion spectrometer. *Boreal Environ. Res.*, **12**, 247–264.

Nadykto, A.B. and Yu, F. (2004) Formation of binary ion clusters from polar vapours: effect of the dipole-charge interaction. *Atmos. Chem. Phys.*, **4**, 385–389.

Napari, I., Noppel, M., Vehkamäki, H., and Kulmala, M. (2002) Parametrization of ternary nucléation rates for H_2SO_4–NH_3– H_2O vapors. *J. Geophys. Res.*, **107**, 4381. doi: 10.1029/2002JD002132

Nieminen, T., Paasonen, P., Manninen, H.E., Kerminen, V.-M., and Kulmala, M. (2010) Parameterization of ion-induced nucleation rates based on ambient observations. *Atmos. Chem. Phys.*, **10**, 21697–21720. doi: 10.5194/acpd-10-21697-2010

O'Dowd, C.D., Geever, M., Hill, M.K., Smith, M.H., and Jennings, S.G. (1998) New particle formation: nucleation rates and spatial scales in the clean marine coastal environment. *Geophys. Res. Lett.*, **25**, 1661–1664.

O'Dowd, C., Hämeri, K., Mäkelä, J., Väkevä, M., Aalto, P., de Leeuw, G., Kunz, G., Becker, E., Hansson, H.-C., Allen, A., Harrison, R., Berresheim, H., Kleefeld, C., Geever, M., Jennings, S., and Kulmala, M. (2002) Coastal new particle formation: environmental conditions and aerosol physico-chemical characteristics during nucleation bursts. *J. Geophys. Res.*, **107** (D19), 8107.

O'Dowd, C., Monahan, C., and Dall'Osto, M. (2010) On the occurrence of open ocean particle production and growth events. *Geophys. Res. Lett.*, **37**, L19805. doi: 10.1029/2010GL044679

Oxtoby, D.W. (1992) Homogeneous nucleation: theory and experiment. *J. Phys. Condens. Matter*, **4**, 7627–7650.

Perraud, V., Bruns, E.A., Ezell, M.J., Johnson, S.N., Yu, Y., Alexander, M.L., Zelenyuk, A., Imre, D., Chang, W.L., Dabdub, D., Pankow, J.F., and Finlayson-Pitts, B.J. (2012) Nonequilibrium atmospheric secondary organic aerosol formation and growth. *Proc. Natl. Acad. Sci. U.S.A.*, **109**, 2836–2841. doi: 10.1073/pnas.1119909109

Sellegri, K., Yoon, Y.J., Jennings, S.G., O'Dowd, C.D., Pirjola, L., Cautenet, S., Chen, H.W., and Hoffmann, T. (2009) Quantification of coastal new ultrafine particles formation from in situ and chamber measurements during the BIOFLUX campaign. *Environ. Chem.*, **36**, 260–270.

Smith, J.N., Barsanti, K.C., Friedli, H.R., Ehn, M., Kulmala, M., Collins, D.R., Scheckman, J.H., Williams, B.J., and McMurry, P.H. (2010) Observations of aminium salts in atmospheric nanoparticles and possible climatic implications. *Proc. Natl. Acad. Sci. U.S.A.*, **107**, 6634–6639. doi: 10.1073/pnas.0912127107

Tammet, H. and Kulmala, M. (2005) Simulation tool for atmospheric aerosol nucleation bursts. *J. Aerosol Sci.*, **36**, 173–196.

Vanhanen, J., Mikkilä, J., Lehtipalo, K.M., Sipilä, K., Manninen, H.E., Siivola, E., Petäjä, T., and Kulmala, M. (2011) Particle size magnifier for nano-CN detection. *Aerosol Sci. Technol.*, **45**, 533–542.

Virtanen, A., Joutsensaari, J., Koop, T., Kannosto, J., Yli-Pirilä, P., Leskinen, J., Mäkelä, J.M., Holopainen, J.K., Pöschl, U., Kulmala, M., Worsnop, D.R., and Laaksonen, A. (2010) An amorphous solid state of biogenic secondary organic aerosol particles. *Nature*, **467**, 824–827. doi: 10.1038/nature09455

Yu, F. (2007) Improved quasi-unary nucleation model for binary H_2SO_4– H_2O homogeneous nucleation. *J. Chem. Phys.*, **127**, 054301.

Yu, F. (2008) Updated H_2SO_4– H_2O binary homogeneous nucleation look-up tables. *J. Geophys. Res.*, **113**, D24201.

Yu, F. and Turco, R.P. (2001) From molecular clusters to nanoparticles: role of ambient ionization in tropospheric aerosol formation. *J. Geophys. Res.*, **106**, 4797–4814.

Zhang, Y., McMurry, P.H., Fangqun, Y., and Jacobson, M.Z. (2010) A comparative study of nucleation parametrizations: 1. Examination and evaluation of the formulations. *J. Geophys. Res.*, **115**, D20212.

Zhao, J., Eisele, F.L., Titcombe, M., Kuang, C., and McMurry, P.H. (2010) Chemical ionization mass spectrometric measurements of atmospheric neutral clusters using the Cluster–CIMS. *J. Geophys. Res.*, **115**, D8, D08205, doi: 10.1029/2009JD012606.

3
Coagulation, Condensation, Dry and Wet Deposition, and Cloud Droplet Formation in the Atmospheric Aerosol Life Cycle

Claudio Tomasi and Angelo Lupi

3.1
Introduction

Atmospheric aerosol particles are mainly generated by natural sources (windborne dust, sea spray, and volcanic particles) and in part by anthropogenic activities (such as industrial and urban aerosols from fuel combustion and vehicular traffic). They have in general low terminal fall speeds and exhibit sizes ranging from $\sim 10^{-4}$ μm (molecular clusters and small ions) to no more than 10^2 μm (fog droplets). Aerosols are formed through various primary and secondary formation processes: (i) biological solid and liquid particles are released into the atmosphere from plants (as seeds, pollen, bacteria, spores, algae, fungi) and animals (bacteria, protozoa, microorganisms, and skin fragments), contributing to yield an overall global emission flux Φ_e of biogenic organic particulate mass ranging from 15 to 70 Tg per year (Andreae and Rosenfeld, 2008); (ii) sea-salt particles of various sizes originate from the breakage of the bubble films (generated by strong wind forcing on the sea surface) and from the subsequent jet drops ejected into the atmosphere, constituting one of the major components of natural aerosols with Φ_e ranging from 3×10^3 to 2×10^4 Tg per year; (iii) smoke is generated by forest fires primarily, mainly consisting of organic compounds and elemental carbon (EC), and provides an annual particulate mass flux Φ_e ranging from 26 to 70 Tg per year; (iv) airborne mineral dust is mobilized by wind-forced saltation of soil particles in desert and semiarid regions and provides a value of Φ_e varying from 10^3 to 2.15×10^3 Tg per year; (v) volcanic particles grow in the low stratosphere after violent volcanic eruptions, through gas-to-particle (g-to-p) conversion of sulfur dioxide into solution droplets consisting of sulfuric acid (75%) and liquid water (25%), which are estimated to yield an overall emission flux Φ_e equal on average to 9 Tg per year over long periods (Tsigaridis *et al.*, 2006); and (vi) anthropogenic aerosols provide an overall flux Φ_e ranging from 40 to 130 Tg per year (Andreae and Rosenfeld, 2008).

In situ condensation of gases contributes efficiently to form new aerosol particles through various g-to-p conversion processes, in which gases may condense onto existing particles. Three major families of chemical species are involved

Atmospheric Aerosols: Life Cycles and Effects on Air Quality and Climate, First Edition.
Edited by Claudio Tomasi, Sandro Fuzzi, and Alexander Kokhanovsky.

in these processes, containing sulfur, nitrogen, and organic/carbonaceous substances:

1. Various natural and anthropogenic sulfur gases (H_2S, CS_2, COS, and dimethyl sulfide (DMS)) can be oxidized to SO_2 and then converted to sulfates. Flux Φ_e due to sulfates from natural primary and secondary sources was estimated to vary from $\sim 10^2$ to nearly 4×10^2 Tg per year (Andreae and Rosenfeld, 2008), while an additional flux of anthropogenic sulfates was estimated by Tsigaridis *et al.* (2006) to be of ~ 73 Tg per year.
2. Nitric acid can form from N_2O_5 in cloud water, and the subsequent evaporation of cloud water releases nitrate particles into the air, yielding a partial flux of nitrates from natural primary and secondary sources ranging from 12 to 27 Tg per year (Andreae and Rosenfeld, 2008) and a flux of anthropogenic nitrates equal to ~ 44 Tg per year (Tsigaridis *et al.*, 2006).
3. Organic and carbonaceous aerosols are usually produced by g-to-p conversion of gases released from the biosphere and volatile compounds, such as crude oil leaking to the Earth's surface, and biomass burning and fossil fuel combustion. They were estimated by the IPCC (2001) to yield an overall value of $\Phi_e \approx 10^2$ Tg per year. These combustion products were in large part attributed to natural carbonaceous particles emitted directly by biomass burning fires, which were estimated by Andreae and Rosenfeld (2008) to provide an annual flux varying from 26 to 70 Tg per year.

An aerosol particle is technically defined as a suspension of solid or liquid particles in a gas. It may consist of (i) a single continuous unit of solid or liquid containing many molecules held together by intermolecular forces and primarily having sizes greater than those of a molecule (with $a > 10^{-3}$ μm) or (ii) two or more such unit structures, which are held together by interparticle adhesive forces and behave as a single unit in suspension or upon deposit. Emitted directly as particles (primary aerosols) or formed in the atmosphere by g-to-p conversion processes (secondary aerosols), airborne aerosol particles exhibit sizes ranging from a few nanometer (molecular clusters) to tens of micrometer (giant particles). Therefore, three aerosol classes can be distinguished in the real cases, presenting size distribution curves characterized by the following basic particle modes:

1. Fine particle modes consisting in general of nuclei cover the diameter range $a < 5 \times 10^{-2}$ μm and exhibit number density concentrations ranging from ~ 10 cm^{-3} in the unpolluted remote regions to more than 5×10^3 cm^{-3} in the most polluted urban areas. High number concentrations may be typical of the nuclei modes including clean combustion aerosol, forest fire smoke, urban smog and polluted aerosols, fly ash, and background continental aerosol particles. In all these size distribution curves, a first nuclei mode consists of very small particles with diameters $a < 10^{-2}$ μm, while a second nuclei mode frequently spans over the 10^{-2} to 5×10^{-2} μm diameter range. Due to their small sizes, the particles belonging to these two modes account for the predominant fraction of particles by number density, but contribute by a few

percent only to the total airborne particulate mass. These particles are mainly formed through nucleation of atmospheric species generating fresh particles, condensation of hot vapors during combustion processes, and evaporation of cloud droplets and are principally lost by coagulation with larger particles or subsequently captured by cloud particles. Therefore, the composition features of these nuclei may often be characterized by prevailing fractions of soot substances, with relatively high concentrations of both EC and black carbon (BC). Due to the intense removal processes occurring in clouds and the coagulation mechanisms in cloud-free air, the residence time (or lifetime) Δt_L of the nuclei is generally shorter than 1 h in both clouds and cloud-free air masses. Because of their formation and removal processes, these small-size particles may exhibit different morphological characteristics and composition features. The smaller ones consist mainly of soot substances, and some of them are agglomerations of carbon impregnated with "tar balls," formed as a result of the incomplete combustion of carbonaceous materials. Other small gas-borne particles often originate because of incomplete combustion and consist predominantly of carbon and pieces of not well-burned matter, having often sizes $a < 10^{-2}$ μm.

2. A marked mode of fine particles with sizes greater than those of nuclei, generally called *accumulation particle mode*, extends over the 0.2–2.5 μm diameter range. These particles constitute the larger part of atmospheric aerosol load, including often significant mass fractions of continental aerosols over the agricultural land regions and sea-salt aerosol over the oceanic regions. The main sources of accumulation particles are coagulation of nuclei and condensation of vapors onto existing particles. The coagulation and condensation processes involve different aerosol types (biomass burning aerosol, volcanic fly ash, pollens of plants, and sea-salt particles), since the removal mechanisms are more efficient within the extreme size ranges and are less efficient over the central range occupied by accumulation particles. Being principally removed by precipitation scavenging and dry fallout, the accumulation particles exhibit values of lifetime Δt_L considerably higher than those typical of nuclei or coarse particles, which vary in general from less than 1 day to no more than 2 weeks (Wallace and Hobbs, 2001). For this reason, the accumulation mode is named in this way, since such particles "accumulate" there.
3. The so-called coarse mode particles have sizes ranging from 2.5 to more than 50 μm. They are mainly formed by mechanical processes, such as the mobilization of soil dust particles and/or the formation of sea spray by strong winds over the oceans, and consist most frequently of mineral dust, sea salts and not completely burned matter in forest fire smokes, and industrial and agricultural particles. Due to the numerous contributions of natural and anthropogenic substances, atmospheric coarse aerosols exhibit a wide variety of composition characteristics. The main sinks of coarse particles are precipitation scavenging and dry fallout. Therefore, the residence time Δt_L of coarse particles can vary from a few minutes to hours in clouds (due

to their rapid removal velocities) and from a few hours to more than 1 day in the cloud-free atmosphere. Because of their efficient removal processes, coarse particles exhibit atmospheric concentrations lower than 1 cm^{-3} in the remote areas of our planet, which are far from the more intense dust and maritime aerosol sources. To summarize the earlier concepts and complete the aerosol size characterization, the classification scheme proposed by Hinds (1999) is reported in Table 3.1.

The earlier remarks clearly indicate that fine and coarse particles originate separately, grow through different physical and chemical processes, and are

Table 3.1 Typical size ranges for various aerosol classes (Hinds, 1999).

Aerosol type	**Size range (μm)**
Part (a): Aerosol size characterization	
Free molecules	$10^{-3} \le a \le 2 \times 10^{-2}$
Ultrafine particles	$10^{-3} \le a \le 10^{-1}$
Fine particles	$4 \times 10^{-3} \le a \le 2.5$
Nuclei	$4 \times 10^{-3} \le a \le 5 \times 10^{-2}$
Accumulation particles	$5 \times 10^{-2} \le a \le 2.5$
Coarse particles	$2.5 \le a \le 10^{2}$
Part (b): Aerosol exterior characteristics	
Fume particles	$10^{-3} \le a \le 1$
Dust particles	$5 \times 10^{-1} \le a \le 3 \times 10^{2}$
Fog and mist droplets	$10^{-1} \le a \le 5 \times 10^{1}$
Cloud droplets	$2 \le a \le 70$
Smog particles	$5 \times 10^{-2} \le a \le 2$
Sea spray	$4 \le a \le 3 \times 10^{2}$
Smoke particles	$2 \times 10^{-3} \le a \le 1$
Part (c): Aerosol composition	
Metal fumes	$4 \times 10^{-3} \le a \le 1$
Cement dust	$2.5 \le a \le 10^{2}$
Coal dust and fly ash	$1 \le a \le 10^{2}$
Oil smoke particles	$2 \times 10^{-2} \le a \le 1$
Diesel smoke particles	$3 \times 10^{-2} \le a \le 1$
Tobacco smoke particles	$10^{-1} \le a \le 2$
Sea-salt condensation nuclei	$3 \times 10^{-2} \le a \le 5 \times 10^{-1}$
Viruses	$10^{-2} \le a \le 3 \times 10^{-1}$
Bacteria	$3 \times 10^{-1} \le a \le 15$
Pollens	$15 \le a \le 10^{2}$
Fungal spores	$6 \times 10^{-1} \le a \le 10^{2}$

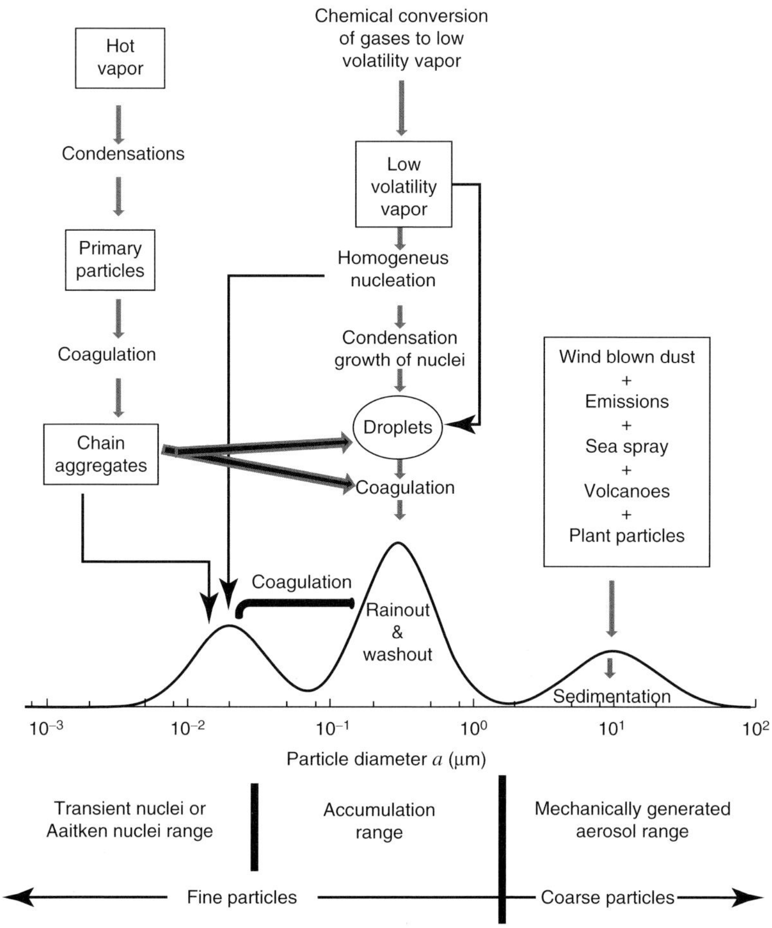

Figure 3.1 Idealized schematic of the distribution of particle surface area of an atmospheric aerosol as a function of particle diameter, showing the principal modes, sources, particle formation processes, and removal mechanisms discussed in the present study. (Adapted from a graph of Whitby and Cantrell (1976).)

finally removed from the atmosphere by different mechanisms, whose efficiency is closely related to the aerosol microphysical properties. Fine and coarse particles differ considerably also in the chemical composition and optical characteristics. Therefore, the distinction between fine and coarse particles is of basic importance when evaluating the aerosol effects on the Earth's radiation budget and human health. This concept is well illustrated in the scheme of Whitby and Cantrell (1976) shown in Figure 3.1, which describes the complex sequences and interactions of the physical and chemical processes through which the nuclei, accumulation aerosols, and coarse particles grow and evolve gradually until reaching their final removal stages.

Fine particles with $a < 0.1$ μm mainly consist of nuclei and form from hot vapor condensation, subsequently growing by coagulation and chain aggregations, until forming particles with intermediate sizes. Accumulation particles form through chemical gas-to-vapor conversions or homogeneous nucleation processes, followed by condensation onto nuclei that leads to the generation of mist and fog droplets or by coagulation in relatively dry air, which also favors the particle growth. Coarse particles are mainly generated by wind forcing on arid terrains (dust), sea spray formation in oceanic regions, plant debris, volcanic eruptions, and direct emissions from natural and/or anthropogenic sources. Accumulation particles are most efficiently removed by rainout and washout, while the removal of coarse particles occurs mainly through sedimentation, due to their supermicron sizes. The strong emissions of anthropogenic aerosols in urban and industrial areas often deteriorate the air quality and reduce the visual range (VR), posing serious threats to the human health in all cases where the emission rates largely exceed the rates at which such polluted aerosols are dispersed by winds or removed from the atmosphere through dry and wet deposition processes.

3.2
Physical Growth Processes

The atmospheric life of aerosol particles emitted by the primary sources or formed from the precursor gases is regulated by nucleation, coagulation, and condensation processes, as well as by the cloud particle formation processes leading to both scavenging and evaporation effects. Subsequently, aerosol removal may occur through dry and wet deposition or scavenging in cloud and precipitations. The sequence of these processes is well illustrated in Figure 3.2, showing that airborne aerosol particles can change their size and composition through coagulation with other particles, condensation of vapor species, and evaporation of cloud droplets. In the nucleation processes, ions, atoms, or molecules arrange themselves to form small crystals, on which additional particles deposit by coagulation and condensation causing the crystal growth and the formation of cloud condensation nuclei (hereinafter referred to as CCNs) and then liquid droplets.

Coagulation and condensation processes are described in Section 3.2, including the hygroscopic growth of particles as a function of air relative humidity (RH) and the description of the Kelvin effect. Section 3.3 describes the main aerosol sinks due to dry and wet deposition (i.e., through in-cloud scavenging and precipitation scavenging (rainout)), taking into account that the removal processes are strongly influenced by the size distribution characteristics and the chemical composition of particles. Section 3.4 deals with the formation of cloud particles, paying particular attention to the water vapor condensation processes, the basic concepts of the Köhler theory, and the role of CCNs. Finally, the concluding remarks are presented in Section 3.5, in which the most realistic evaluations of global mass burden

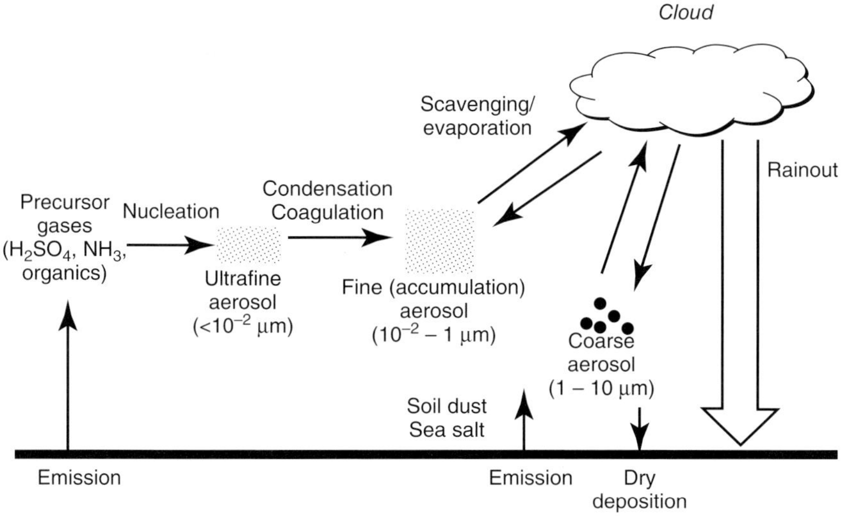

Figure 3.2 Schematic representation of the sequence of processes involving the atmospheric life of aerosol particles after the emission of the precursor gases, until their removal from the atmosphere through dry deposition and wet (rainout) mechanisms.

ranges from the various aerosol types are provided together with the estimates of atmospheric lifetime Δt_L.

The dynamic behavior of aerosols is the result of the changes affecting the airborne aerosols, which mutually interact and may behave in different ways when a vapor compound condenses onto particles and the particles collide and adhere to each other. This approach can lead to realistic evaluations of the changes caused in the size distribution features by the mass transfer of material from or to the gas phase or by the variations in the rate at which two spherical particles in Brownian motion collide and then coagulate. In the latter case, the mobility of particles decreases rapidly as their sizes increase, so that coagulation is essentially confined within the $a < 0.2\,\mu m$ diameter range. Coagulation does not remove the aerosol particles from the atmosphere, but modifies the aerosol size spectra and causes the shift of small particles toward the greater size range, where the removal mechanisms are more efficient.

3.2.1
Brownian Coagulation

Two aerosol particles suspended in a fluid may come into contact and coagulate because of their Brownian (thermal) motion or as a result of their motions produced by hydrodynamic, electrical, gravitational, or turbulent forces. To give an idea of the main coagulation features, the theory of Brownian coagulation is considered here to illustrate the following two particular cases: (i) the case where the collisions occur between particles belonging to two different monodispersed populations, of which the first consists of N_1 spherical particles having diameters a_1 and the second of N_2 spherical particles with diameters a_2, and (ii) the case in which coagulation takes place between monodispersed particles.

1. In the first case, the coagulation involves a population of particles, which have all the same diameter a and initial concentration N_0. The coagulation rate is calculated for a pair of these particles colliding to form a new stationary absorbing spherical particle, assuming that the second colliding particle is represented by a point particle located on the surface of the first one. The distribution of aerosols around the fixed particle in a continuum regime can be described in terms of the number density size distribution $N(a,t)$ of particles having diameters a at a certain time t, which satisfies the following equation:
$$\frac{\partial N(a,t)}{\partial t} = 4\,D_{\mathrm{br}}\left(\frac{\partial^2 N\,(a,t)}{\partial a^2} + \frac{2}{a}\frac{\partial N(a,t)}{\partial a}\right), \tag{3.1}$$
where N is the particle number concentration and D_{br} is the Brownian diffusion coefficient of these particles. The initial and boundary conditions valid for Eq. (3.1) are (i) $N(a, 0) = N_0$, which assumes that the particles are homogeneously distributed in space at the initial time, having particle number concentration N_0; (ii) $N(\infty, t) = N_0$, which states that the number concentration of particles infinitely far from the particle absorbing sphere is not influenced by it; and (iii) $N(a, t) = 0$, which establishes that the fixed particle is a perfect absorber, which adheres at every collision. For this triplet of boundary conditions and assuming that the sticking probability of two colliding aerosol particles is unit, the solution of Eq. (3.1) indicates that the collision rate is initially extremely fast and approaches a steady-state value of collision rate $J_{\mathrm{col}} = 4\,\pi a D_{\mathrm{br}} N_0$ for $t \gg a^2/\pi D_{\mathrm{br}}$. On this matter, Seinfeld and Pandis (2006) evaluated that such a system reaches steady-state conditions within 10^{-4} s for particles with $a = 10^{-1}$ μm and within about 0.1 s for particles with $a = 1$ μm.
2. In the second more realistic case, the particles undergo Brownian diffusion and have all diameters a_1, while the others moving in the fluid have diameters all equal to a_2. The two spherical particles coming into contact have diameters a_1 and a_2, respectively, and the stationary particle is an absorbing sphere with a diameter equal to the sum $(a_1 + a_2)$, while the colliding particle of diameter a_2 is simply represented by a point. The diffusion equation given in Eq. (3.1) governs in this case $N_2(a_2, t)$, which yields the concentration of particles with diameters a_2 at time t. Thus, the system is regulated by
$$\frac{\partial N_2(a,t)}{\partial t} = 4\,D_{12}\left(\frac{\partial^2 N_2\,(a,t)}{\partial a^2} + \frac{2}{a}\frac{\partial N_2(a,t)}{\partial a}\right), \tag{3.2}$$
where the diffusion coefficient D_{12} is equal to the sum of the two Brownian diffusion coefficients D_1 and D_2 pertaining to the particles of diameters a_1 and a_2, respectively. Equation (3.2) satisfies the following conditions: (i) N_2 $(a, 0) = N_{20}$, (ii) N_2 $(\infty, t) = N_{20}$, and (iii) N_2 $(a_1 + a_2, t) = 0$, for which the boundary condition at the distance $d = a$ (as assumed in the first case between

the two coagulation centers) is now applied at the distance $d = a_1 + a_2$. The final solution is in this case given by

$$N_2(a,t) = N_{20}\left[1 - \frac{a_1 + a_2}{a} erfc\left(\frac{a - (a_1 + a_2)}{(\pi D_{12} t)^{1/2}}\right)\right], \tag{3.3}$$

and the steady-state coagulation rate *J12* (measured in per cubic centimeter per second) between the particles having diameters a_1 and a_2 is given by

$$J_{12} = K_{12} N_1 N_2 = \frac{2k_B T_a}{3\eta} \frac{(a_1 + a_2)^2}{a_1 a_2} N_1 N_2 \tag{3.4}$$

where (i) k_B is the Boltzmann constant equal to $1.38 \times 10^{-23}\,\mathrm{J\,K^{-1}}$, (ii) η is the viscosity coefficient equal to $1.83 \times 10^{-4}\,\mathrm{g\,cm^{-1}\,s^{-1}}$ for standard conditions of air, and (iii) N_1 and N_2 are the total number concentrations of the particles having diameters a_1 and a_2, respectively.

Equation (3.4) indicates that the steady-state coagulation rate is proportional to coagulation coefficient K_{12} and the product of the two number concentrations of the colliding particles. It can be applied to Brownian motion only over time intervals considerably longer than the relaxation time of such particles and for distances appreciably greater than the aerosol mean free path ℓ. In all cases where ℓ is comparable to the half of particle diameter, the boundary condition at the absorbing particle surface needs to be corrected to account for the nature of the diffusion processes occurring near the surface. This implies that the diffusion equations can be applied to the Brownian motion only for (i) long time intervals compared to the relaxation time and (ii) distances $d > \ell$. On this matter, Fuchs (1964) suggested that the coagulation coefficient K_{12} used in Eq. (3.4) needs to be corrected by using the following equation:

$$K_{12} = 2\pi(a_1 + a_2)\, D_{12} \psi_c \tag{3.5}$$

in which the coagulation correction factor ψ_c is assumed to increase from 1.4×10^{-2} for particles with diameter $a = 2 \times 10^{-3}\,\mu\mathrm{m}$ to about unit for particles with $a = 2\,\mu\mathrm{m}$. The values of coefficient K_{12} obtained using the Fuchs (1964) formula are shown in Figure 3.3 as a function of particle diameter a_1 over the range from 10^{-3} to $10\,\mu\mathrm{m}$ for some selected values of diameter a_2 increasing from 10^{-2} to $10\,\mu\mathrm{m}$. Such calculations give a measure of the variations of this parameter as a function of particle diameter for (i) absolute temperature $T = 298.16\,\mathrm{K}$, (ii) particle mass density $\rho = 1\,\mathrm{g\,cm^{-3}}$, (iii) mean free path $\ell = 6.86 \times 10^{-2}\,\mu\mathrm{m}$, and (iv) viscosity coefficient of air $\eta = 1.83 \times 10^{-4}\,\mathrm{g\,cm^{-1}\,s^{-1}}$. It can be seen that the smallest value of K_{12} occurs when both particles have the same sizes and follow the lowest curve with $a_1 = a_2$, while K_{12} rises rapidly when ratio a_1/a_2 decreases, due to the synergism between the two particles. A large particle may appear to be slow in the movement from the Brownian motion perspective, but, because of its large surface area, it constitutes a large target for the small faster particles. Thus, collisions between large particles appear to be not frequent because such particles move slowly, and therefore the mutual collisions between them occur with a

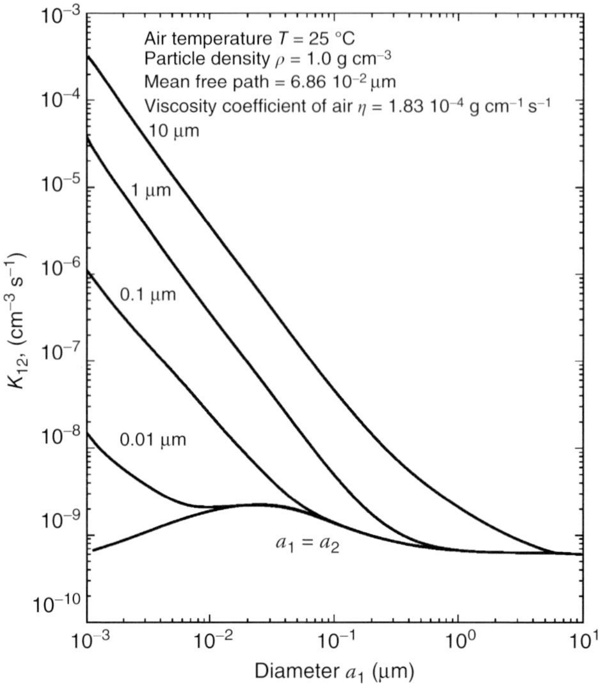

Figure 3.3 Dependence curves of Brownian coagulation coefficient K_{12} on particle diameter in a coagulation process occurring in air at a temperature of 25 °C and involving pairs of spherical aerosol particles, the former having diameter a_1 and the latter having diameter a_2 equal to 10^{-2} μm, 10^{-1} μm, 1 μm, and 10 μm, and diameter $a_2 = a_1$ over the whole diameter range from 10^{-3} to 10 μm (gray curve), providing in the last case the patterns of coagulation rate K_{11}. All the five curves have been calculated according to the Fuchs (1964) theory. To use this graph, find the smaller of the two particles as the abscissa and then locate the line corresponding to the larger particle. (Adapted from a graph of Seinfeld and Pandis (2006).)

rather low frequency with respect to the pairs of small particles. Consequently, both particles tend to miss each other because their cross-sectional areas for collision are both small. Conversely, very small particles exhibit relatively higher velocities and are subject more easily to mutual collisions. Figure 3.5 shows that the target area is proportional to the square of a_2, and the Brownian diffusion coefficient decreases only as diameter a_1 increases, while the first particles grow in sizes. The highest value of coagulation coefficient for collisions among equal-sized particles is obtained for $a_1 \approx 2 \times 10^{-2}$ μm. In the continuum regime is the Knudsen number $Kn \ll 1$ for $a_1 = a_2$ in Eq. (3.4), both particles being assumed to have diameters $a > 1$ μm. The coagulation coefficient K_{11} pertaining to the case with $a_1 = a_2$ results to be (i) equal to the ratio $(8k_B T)/(3\eta)$ in the continuum regime, therefore being independent of the particle size, and (ii) equal to 4 $(6k_B T/\rho)^{1/2} a_1^{1/2}$ in a free molecular coagulation regime with $a_1 = a_2$, therefore increasing as diameter a_1 increases. This suggests that the maximum in the coagulation coefficient for equal-sized particles reflects a balance between particle mobility and cross-sectional area for collision.

For $a_2 \gg a_1$, coefficient K_{12} in the continuum regime approaches the limit equal to the ratio between terms $2k_B Ta_2$ and $3\eta a_1$ and tends to reach an asymptotic value equal to $\sqrt{(3k_B T/\rho)}(4a_2^{\,2}/a_1^{\,3/2})$ in the free molecular regime. Comparing these limits and assuming a fixed value of a_1, it can be found that K_{12} increases more rapidly as a_2 increases in the free molecular regime than in the continuum regime. In fact, the coagulation rate is proportional to the collision efficiency under the free molecular regime, while coagulation is comparatively insensitive under the continuum regime to the changes in correction factor ψ_c adopted in Eq. (3.5). For example, for $a_1 = 2\ \mu m$, the coagulation rate K_{11} decreases by 5% only as ψ_c decreases from 1.0 to 0.25, and inversely a 5% decrease in the coagulation rate implies a value of $\psi_c = 0.60$ for $a_1 = 0.2\ \mu m$ (Fuchs, 1964).

New particles can be formed through g-to-p conversions or combustion processes over the size range $a < 10$ nm. Since the diffusion coefficient is proportional to particle diameter a_1, the coagulation due to Brownian diffusion becomes very effective for these small particles. Neglecting electrical effects and considering the case in which two spherical particles having diameters a_1 and a_2 adhere within a distance equal to $(a_1 + a_2)/2$, the rate dN/dt of coagulation can be evaluated as equal to the double integral of the product of coagulation coefficient K_{12} by the number concentrations N_1 and N_2 of the two particles having diameters a_1 and a_2, respectively, where K_{12} is evaluated according to Eq. (3.5). Besides their sizes, the particles are characterized not only by their diffusion coefficients but also by their mean thermal velocity, while the carrier gas is characterized by its absolute temperature T and viscosity coefficient η. Using these assumptions, the aging of an atmospheric aerosol population due to coagulation can be modeled over appropriate time and size intervals.

The earlier results indicate that coagulation strongly depends on the product of the two interacting particle concentrations and is most effective for very small particles colliding with very large aerosols, as it occurs, for instance, during the scavenging of ultrafine particles by cloud droplets. This is clearly shown in Figure 3.4, presenting the size dependence curves of coagulation coefficient K_{12} defined by Fuchs (1964) for unequal spherical particles over the $10^{-3} \leq a \leq 10\ \mu m$ range. It can be noted that $K_{12}(a_1, a_2)$ describes a wide minimum for nearly equal sizes of the two colliding particles and reaches the highest values for the largest difference between the two particle diameters. Because of the coagulation between pairs of aerosol particles, the temporal evolutionary patterns of the size distribution curve $N(a, t)$ given in Eq. (3.3) can be described in terms of the coagulation coefficient K_{12} defined in Eq. (3.5), which is in general given by a double integral of function J_{12} represented in Eq. (3.4) over the whole intervals of a_1 and a_2, for the Fuchs (1964) assumptions of K_{12} represented in Figure 3.4, and correction factor ψ_c adopted in Eq. (3.5).

An example of the variations in the size distribution curve $N(a, t)$ obtained by applying the Fuchs (1964) theory to represent the effects of Brownian coagulation

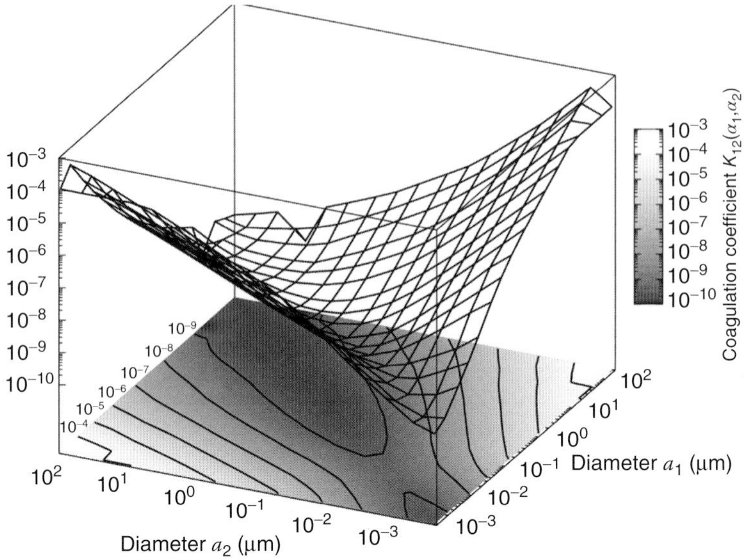

Figure 3.4 Three-dimensional representation of the coagulation coefficient $K_{12}(a_1, a_2)$ of unequal spherical particles (measured in cubic centimeter per second) defined in Eq. (3.5) according to the Fuchs (1964) theory applied to a pair of spherical particles having diameters a_1 and a_2, respectively, both varying over the $10^{-3} \leq a \leq 10\,\mu m$ range.

is shown in Figure 3.5a, obtained at initial time $t_0 = 0$ and after 12 h over the $10^{-2} \leq a \leq 10\,\mu m$ diameter range. The example shows that the strongest changes in the curve of $N(a, t)$ occur over the nuclei size range, causing in practice a nearly total shift of the particles with $a < 10^{-2}\,\mu m$ toward the $0.1 \leq a \leq 0.2\,\mu m$ range and therefore leading to a marked decrease in the number concentration of nuclei. However, due to the relatively small sizes of these fresh coagulating particles, the earlier variations do not provide a marked change in the overall aerosol mass concentration. In the example shown in Figure 3.5a, the particles were assumed to be all spherical. In real cases, chain-like aggregates and irregular-shape particles are frequently observed during the coagulation process, behaving in a way that cannot be realistically predicted using a spherical particle model. The enlargement of cloud drops and crystals by coalescence or adhesion occurs in general when

Figure 3.5 (a) Example of variation in the number density size distribution curve of coagulated aerosol particles plotted versus the particle diameter a, as evaluated at time $t = 0$ (open circles, before coagulation) and at time $t = 12$ h (solid squares, after coagulation) in an exercise made by applying the Fuchs (1964) theory to the initial size distribution curve. Note that such a variation in the particle number density size distribution curve is characterized by more marked changes within the small particle size range $a < 5 \times 10^{-2}\,\mu m$ and gradually weaker changes as particle diameter increases. (b) Schematic representation of the differential particle volume balance for the derivation of Eq. (3.7) used to represent the particle growth by condensation.

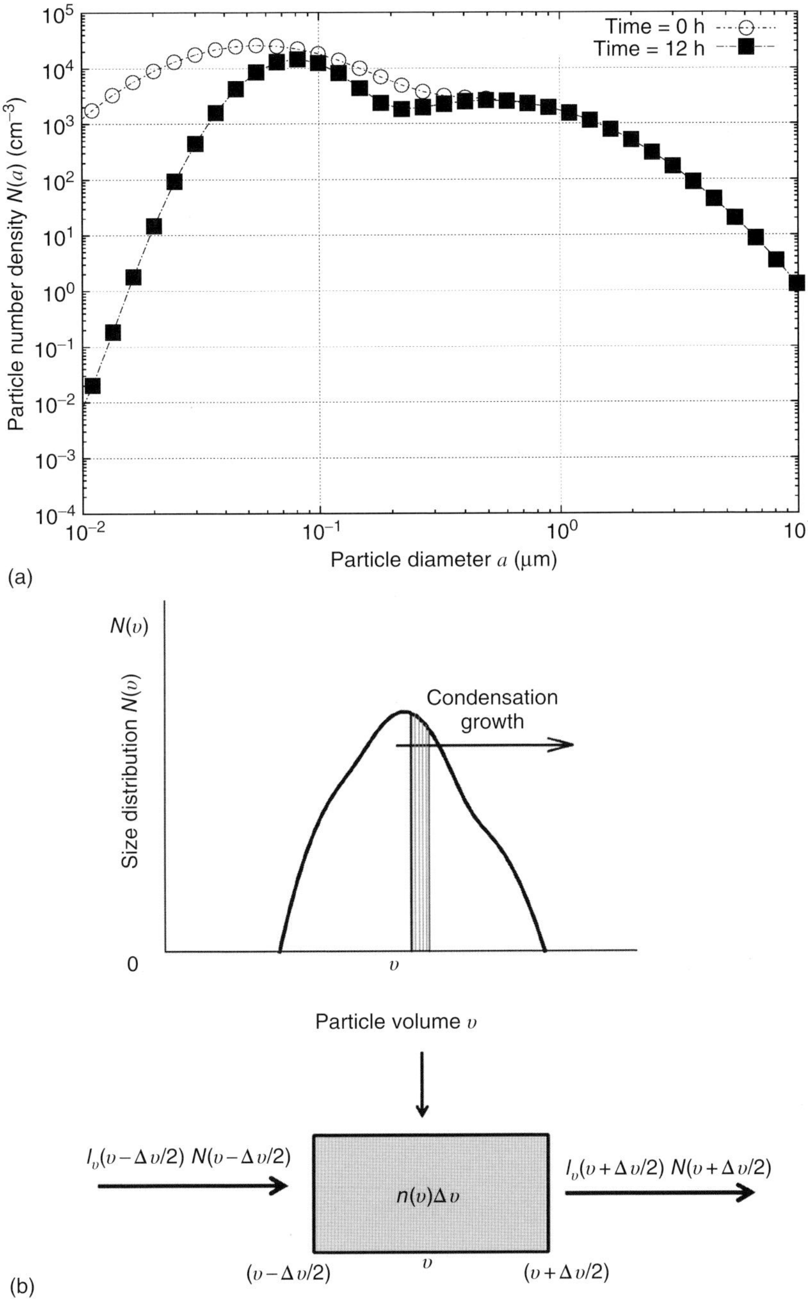
Time = 0 h
Time = 12 h
Particle number density $N(a)$ (cm^{-3})
Particle diameter a (μm)
(a)
N(υ)
Size distribution N(υ)
Condensation growth
0
υ
Particle volume υ
$I_\upsilon(\upsilon-\Delta\upsilon/2)\ N(\upsilon-\Delta\upsilon/2)$
$n(\upsilon)\Delta\upsilon$
$I_\upsilon(\upsilon+\Delta\upsilon/2)\ N(\upsilon+\Delta\upsilon/2)$
$(\upsilon-\Delta\upsilon/2)$
υ
$(\upsilon+\Delta\upsilon/2)$
(b)

they collide with one another, presenting features similar to those of coagulation between spherical aerosol particles. The coagulation of cloud elements is due to numerous physical processes causing particles to collide, such as (i) Brownian coagulation, which results from the chaotic thermal motions of particles; (ii) turbulent coagulation, which results from turbulent movements of individual particles; (iii) electrical coagulation, produced by opposite electric charges on the particles; and (iv) gravitational coagulation, caused by the differences between the fall rates of particles having different sizes and shapes, which can considerably contribute to enlarge cloud particles during precipitations. In nature, such different coagulation processes often act simultaneously, and this may greatly complicate the process and its theoretical description.

3.2.2
Growth by Condensation of Gases onto Preexisting Particles

As shown in Figure 3.5a, function $N(a, t)$ defines the particle size distribution curve at time t. However, it is often more useful to use the particle volume size distribution function $N(v, t)$ in all cases where the evolutionary patterns of the mass particle content are evaluated in place of number concentration $N(a, t)$. This is the case when a vapor condenses onto particles or a material evaporates from the aerosol phase to pass to the gaseous phase, since the variable $v = (\pi/6)a^3$ is a parameter capable to better represent the aerosol size distribution features, being proportional to the particle mass, and therefore more suitable for particulate mass content calculations. The differential $N(v, t)dv$ can be used in all the mass or volume calculations to obtain the number concentration of particles having volumes ranging from v to $v + dv$. In addition, the total aerosol mass concentration can be appropriately calculated by integrating $N(v, t)$ over the whole particle volume range by simply multiplying such an overall volume by the average particulate mass density.

Assuming that the particles are not in equilibrium with the gas phase (the vapor pressure of the particles being not equal to the partial pressure of the same particles in the gas phase), the change rate of mass m of a certain spherical particle of diameter a induced by the g-to-p conversion of a certain ith molecular species results to be given by

$$\frac{dm}{dt} = \frac{2\pi a D_i \mu_i}{RT_a} f(K_n, k_{\text{ma}})(p_i - p_{\text{eq},i}), \tag{3.6}$$

where D_i is the diffusion coefficient of the i-th species suspended in air, μ_i is its molecular weight, and $f(K_n, k_{\text{ma}})$ is the correction factor due to noncontinuum effects and imperfect surface accommodation, expressed as a function of Knudsen number K_n and molecular accommodation coefficient k_{ma}. The difference between the vapor pressure p_i of the i-th species far from the particle and the corresponding equilibrium vapor pressure $p_{\text{eq},\,i}$ constitutes the driving force for the transport of the molecules toward the particle. The condensation growth rate $I_v(v)$ can be defined as the rate at which a certain particle volume v changes as a function of time. Thus, assuming that the aerosol particle contains only one component presenting constant density ρ, the growth rate $I_\upsilon(\upsilon)$ can be expressed in terms of the following equation:

$$I_\upsilon(\upsilon) = \frac{d\upsilon}{dt} = \frac{2\pi a D_i \mu_i}{\rho R T_a} f(K_n, k_{\mathrm{ma}})(p_i - p_{\mathrm{eq},i}). \tag{3.7}$$

Figure 3.5b shows that the size distribution curve of a particle polydispersion growing by condensation moves to the right. Within the infinitesimal slice $\Delta\upsilon$ of the distribution centered at υ and extending from $\upsilon - (\Delta\upsilon/2)$ to $\upsilon + (\Delta\upsilon/2)$, the particle number at time t is equal to $N(\upsilon, t)\Delta\upsilon$. As a result of both condensation and subsequent growth, the particles enter the slice from the left and exit from the right. The instantaneous rate of entry from the left is proportional to $N(\upsilon - (\Delta\upsilon/2), t)$ at the left boundary of the slice, and their volume growth rate is equal to $I_\upsilon(\upsilon - (\Delta\upsilon/2), t)$. Therefore, particles enter the slice from the left with a rate equal to $I_\upsilon(\upsilon - (\Delta\upsilon/2), t)\, N(\upsilon - (\Delta\upsilon/2), t)$. Similarly, particles grow out of the distribution slice represented in Figure 3.5b, emerging from the right boundary with a rate equal to $I_\upsilon(\upsilon + (\Delta\upsilon/2), t)\, N(\upsilon + (\Delta\upsilon/2), t)$. After a certain period Δt, the number of particles inside the fixed slice varies from $N(\upsilon, t)\Delta\upsilon$ to $N(\upsilon, t + \Delta t)\Delta\upsilon$, such a number change being equal to the net flux into the slice, that is, to the difference $N(\upsilon, t + \Delta t)\,\Delta\upsilon - N(\upsilon, t)\Delta\upsilon$. Going to the limit of the earlier difference, it can be found that

$$\frac{\partial N(\upsilon, t)}{\partial t} = -\frac{\partial}{\partial \upsilon}[I_\upsilon(\upsilon, t) N(\upsilon, t)], \tag{3.8}$$

where $I_\upsilon(\upsilon, t)$ is represented in terms of Eq. (3.7). Equation (3.8) is called *condensation equation* and provides the mathematical expression of the rate followed by a particle size distribution $N(\upsilon, t)$ varying because of the condensation flux I_υ (υ, t), in which all the other processes capable of influencing the distribution curve, such as sources, removal, coagulation, and nucleation, are totally neglected. A particular analytic version of the condensation equation can be obtained assuming that the size distribution curve $N(a)$ is log-normal, that is,

$$N(a) = \frac{N_0}{\sqrt{2\pi} a s} \exp\left(-\frac{\ln^2(a/a_c)}{2s^2}\right), \tag{3.9}$$

where (i) shape parameter s is given by the natural logarithm of the standard deviation σ of the size distribution curve, a_c is the mode diameter measured in micrometer, and N_0 is the overall number density particle concentration.

Equation (3.9) represents the distribution curve at the initial time, while the size distribution curve at time t is represented by the following time-dependent function:

$$N(a, t) = \frac{2\,a}{(a^2 - 2\varphi_g\ t)} \frac{N_0}{\sqrt{2\pi}\ s} \exp\left(-\frac{\ln^2\left[2\left(\frac{a^2 - 2\varphi_g\ t}{2}\right)^{1/2} / a_c\right]}{2s^2}\right), \tag{3.10}$$

in which φ_g is equal to $a\ (da/dt)$. For an aerosol population undergoing pure growth without particle sources or sinks, the total particle number is preserved.

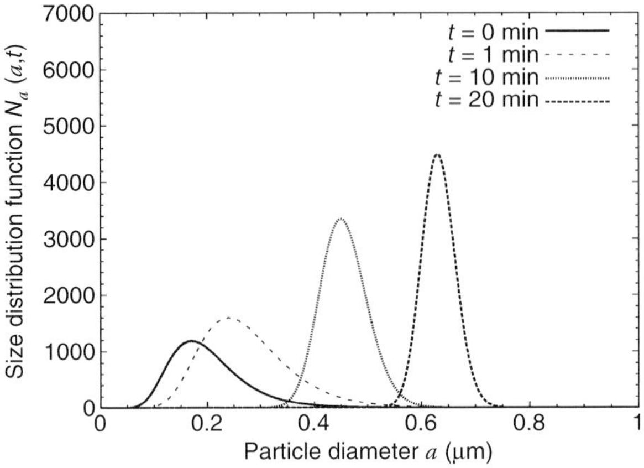

Figure 3.6 Evolution of a log-normal size distribution function of $N_a(a, t)$ as a function of aerosol particle diameter a at different times of its coagulation growth: (i) $t = 0$ min (solid curve, with mode diameter $a_{max} \approx 0.2$ μm, and standard deviation $\sigma = 1.50$) assuming in Eq. (3.7), diffusion coefficient $D_i = 10^{-1}$ cm^2 s^{-1}, molecular weight $\mu_i = 10^2$ g mol^{-1}, difference $p_i - p_{eq,i} = 1$ ppb, absolute temperature $T = 298$ K, and particulate density $\rho = 1.0$ g cm^{-3}; (ii) $t = 1$ min (dashed curve); (iii) $t = 10$ min (dashed and dotted curve); and (iv) $t = 20$ min (dotted curve).

To be valid, Eq. (3.10) must satisfy the following integral relation at all times:

$$N_0 = \int_{2\sqrt{2\varphi_g}}^{\infty} N(a, t) da. \tag{3.11}$$

The lower limit of the integral in Eq. (3.11) is determined by the characteristic curve passing through the origin, since the solution is valid only for diameter $a > \sqrt{2\varphi_g t}$. The integrand function in Eq. (3.11) is often represented by using a log-normal distribution curve, as in the example shown in Figure 3.6, which describes the gradual growth of an aerosol log-normal distribution curve as a function of time, by representing a sequence of gradually more marked peaks characterized by increasing values of a_c corresponding to different times of the coagulation growth phase.

3.2.3
The Kelvin Effect

In situ condensation of gases exerts a strong influence on the airborne aerosol life, because gases are converted into particles through two basic processes in which mass transfer occurs from the gaseous phase to the particulate one: (i) gases may condense onto existing particles, thereby increasing the mass but not the number of aerosols, or (ii) new particles may form through homogeneous nucleation. The former process is favored when the surface area S of preexisting particles is markedly large and supersaturation conditions exist, while the latter contributes to form new particles having sizes in general smaller than 10^{-2} μm. These processes are very efficient in the atmosphere, since the global mean mass concentration of aerosol particles generated by g-to-p conversion is estimated to exceed that of the aerosols directly emitted from anthropogenic sources and to be comparable to that of natural aerosols. Two main parameters regulate the aforementioned nucleation process, which are the curved particle/gas interface and supersaturation ratio:

$$S_{S-S} = p_{\mathrm{eq},i,r} / P_{\mathrm{eq},i,f}, \tag{3.12}$$

between the equilibrium partial pressure $p_{\mathrm{eq},i,r}$ of the ith vapor species over a particle having diameter a and the corresponding equilibrium vapor pressure $p_{\mathrm{eq},i,f}$ over the flat surface. For $S_{S-S} > 1$, the Gibbs free energy reduction ΔG due to condensation reaches a maximum in a droplet having diameter a^*, where the equilibrium is metastable. From the condition $\Delta G =$ maximum, an expression relating the diameter a^* to the supersaturation rate can be derived, called Kelvin equation, which assumes the following form:

$$\ln S_{S-S} = \zeta v_{\mathrm{mol}} (k_{\mathrm{B}} T a^*)^{-1}, \tag{3.13}$$

where ζ is the bulk surface tension corrected for particle diameter a, and v_{mol} is the volume occupied by the molecules of the ith species within the droplet, both pressure and temperature being assumed to be constant. Following the classical nucleation theory, an approximate expression can be derived to represent the nucleation rate J_{nucl} of droplets with diameters a^*, which can be kept constant if the gaseous vapor is continuously reinforced. Numerous physical processes can concur to create supersaturation conditions, such as adiabatic expansion, isothermal compression, and isobaric cooling. Gas-phase chemical reactions can also lead to the formation of supersaturated condensable species such as sulfuric acid. We assume here that the aqueous solution aerosol has a flat surface. Therefore, the effects of curvature on the vapor pressure of the ith species over the particle surface need to be related to both the vapor pressure of this species over the particle surface and the vapor pressure over a flat surface. The change of the Gibbs free energy accompanying the formation of a single drop of the pure ith species (having diameter a and containing n molecules of the gaseous substance) is given by the difference $\Delta G = G_{\mathrm{drop}} - G_{\mathrm{pv}}$ between the Gibbs free energies for the droplet and pure vapor, respectively. If N_{tot} is the initial total number of vapor molecules, the number of vapor molecules remaining after the drop formation is equal to $N_{\mathrm{res}} = N_{\mathrm{tot}} - n$. Therefore, if G_{vap} and G_{liq} are the Gibbs free energies of a molecule in the vapor and liquid phases, respectively, we can write that as

$$\Delta G = N_{\mathrm{res}} G_{\mathrm{vap}} + n G_{\mathrm{liq}} + \pi a^2 \gamma_S - N_{\mathrm{tot}} G_{\mathrm{vap}}, \tag{3.14}$$

where $\pi a^2 \gamma_S$ is the free energy associated with an interface with curvature determined for diameter a and surface tension γ_S, the latter parameter being the energy amount required to yield a unit increase in the droplet surface area. Equation (3.14) can be therefore written as

$$\Delta G = n(G_{\mathrm{liq}} - G_{\mathrm{vap}}) + \pi a^2 \gamma_S, \tag{3.15}$$

where the number n of molecules in the drop is related to the droplet diameter a through the following relationship:

$$n = \frac{\pi a^3}{6 v_{\text{liq}}}, \tag{3.16}$$

where v_{liq} is the volume occupied by a molecule in the liquid phase. Combining Eqs. (3.15) and (3.16), the following equation is obtained:

$$\Delta G = \frac{\pi a^3}{6 v_{\text{liq}}} (G_{\text{liq}} - G_{\text{vap}}) + \pi a^2 \gamma_S. \tag{3.17}$$

with some approximations, the difference $G_{\text{liq}} - G_{\text{vap}}$ for saturation conditions can be evaluated as

$$G_{\text{liq}} - G_{\text{vap}} = -k_B T \ln S_{\text{sat}}, \tag{3.18}$$

where saturation ratio S_{sat} is equal to the ratio $p_j,f/p_{\text{eq},j}$ between the vapor pressure $p_{j,f}$ of the pure jth species over a flat surface and the equilibrium partial pressure $p_{\text{eq},j}$ of the same species over the liquid. Substituting Eq. (3.18) into Eq. (3.17), the following expression is finally obtained for the Gibbs free energy change:

$$\Delta G = -\frac{\pi a^3 k_B T}{6 v_{\text{liq}}} \ln S_{\text{sat}} + \pi a^2 \gamma_S. \tag{3.19}$$

Figure 3.7 shows the general representation of the dependence features of ΔG on diameter a. If we have $S_{\text{sat}} < 1$ in Eq. (3.18), both terms in Eq. (3.19) are positive and ΔG increases monotonically as a increases. Conversely, for $S_{\text{sat}} > 1$ in Eq. (3.18), the Gibbs free energy reduction ΔG defined in Eq. (3.19) results to be given by the sum of a negative term and a positive term. The first right term including the surface tension dominates on the second right term in all cases presenting small

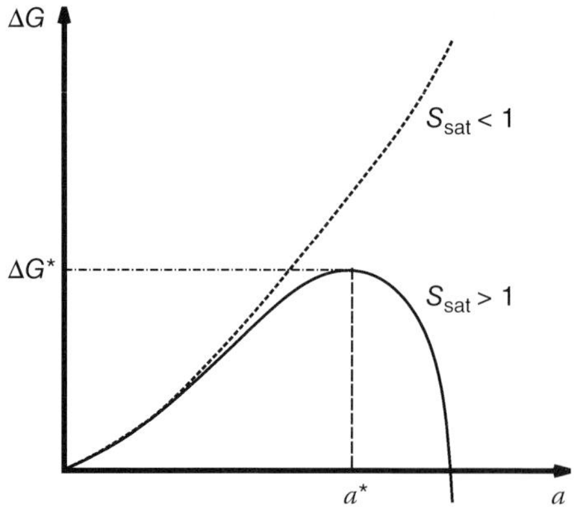

Figure 3.7 Gibbs free energy change for the formation of a droplet having a diameter a from a vapor with saturation ratio equal to S_{sat}. In the case represented for $S_{\text{sat}} < 1$ (dashed curve), ΔG is given by Eq. (3.19) reported in the text, in which both terms are positive. In the case represented for $S_{\text{sat}} > 1$ (solid curve), ΔG results to be negative, the first term of Eq. (3.19) being negative and the second term positive.

values of a. Thus, the behavior of ΔG for these small particles is similar to that considered in the previous case with $S_{sat} < 1$. The first right term of Eq. (3.19) becomes gradually more important as a increases, until reaching its maximum ΔG^* at $a = a^*$, and subsequently it decreases for increasing values of a. The critical droplet diameter a^* is therefore equal to the ratio between the terms $2\gamma_S v_{liq}$ and $k_B T \ln S_{sat}$, which relates in practice the equilibrium diameter of a droplet consisting of a pure substance to (i) the physical properties of the substance (represented by γ_S and v_{liq}) and (ii) the saturation ratio S_{sat} of its environment (given by ratio $p_{eq,j}/p_{j,f}$). This relationship can be rearranged obtaining the so-called Kelvin equation:

$$S_{sat} = \frac{p_{eq,j}}{p_{j,f}} = \exp\left(\frac{4\gamma_S \mu_{liq}}{RT\rho_{liq} a}\right) \tag{3.20}$$

in which (i) γ_S is the surface tension of the liquid substance; (ii) μ_{liq} is the molecular weight of the pure liquid substance; (iii) R is the universal constant of gases, equal to 8.31446 J mol^{-1} K^{-1}; and (iv) ρ_{liq} is the liquid mass density of the droplet.

Equation (3.20) indicates that the vapor pressure over a curved interface always exceeds that of the same substance over a flat surface. A rough physical interpretation of the Kelvin equation states that the vapor pressure of a liquid is determined by the energy necessary to (a) separate a molecule from the attractive forces exerted by its neighbors and (b) bring it to the gas phase. When a curved interface exists, as in the case of a small liquid droplet, there are fewer molecules that are more closely adjacent to a molecule on the surface than those on the flat surface. Consequently, it is easier for the molecules on the surface of a small droplet to escape into the vapor phase. The vapor pressure over a curved interface is always greater than that exerted over a plane surface. Seinfeld and Pandis (2006) estimated that the saturation ratio S_{sat} for a pure water droplet increases over a flat interface by 2.1% for a droplet with diameter $a = 10^{-1}$ μm and by 23% for a droplet with $a = 10^{-2}$ μm, finding that the Kelvin effect begins to become important for aqueous particles when a is equal to ~50 nm.

3.2.4
Hygroscopic Growth of Particles by Water Vapor Condensation

During their atmospheric life, aerosol particles adjust continuously to the variations in the RH conditions of the surrounding air by uptaking or releasing water vapor molecules. Water vapor condensation can lead to the formation of particles at different stages of their life: (i) aerosols grow by condensation as a function of RH over its range <100%, through features depending closely on the particulate matter composition, (ii) haze particles can appreciably grow within the ground layer as RH increases beyond 65–70% and cause often marked reductions of VR to less than 1 km as RH exceeds 80–90%, (iii) cloud particles form at various atmospheric levels for RH > 100%, and (iv) fog particles frequently form within the ground layer, constituting a polydispersed size distribution curve of water droplets over the $a > 1$ μm range. In some cases, suspensions of the so-called mists are observed for high RH conditions, which are liquid or solid dispersed particles grown by water vapor condensation near the ground after volatilization from melted substances. They often contain pollutants and dust particles with sizes mainly ranging from 0.2 to 2 μm, typical of accumulation particles. These small water droplets are floating or falling, often similar to a drizzling rain. They provide in general good visibility conditions and present narrow unimodal size distribution curves with a_c mainly ranging from 10^{-1} to 1.4 μm. Fog droplets grow in general by condensation of water vapor for RH > 100% on small particles containing a large fraction of water-soluble substances. Near the ground, they frequently exhibit sizes <1 μm, while mature fog droplets formed by radiative cooling or transported from sea by advection of moist air masses exhibit in general sizes ranging from 4 to 40 μm (Tampieri and Tomasi, 1976).

The earlier remarks indicate that it is important to distinguish between the two regimes governing the hygroscopic growth of aerosols for RH < 100% and the growth of fog and cloud droplets for RH ≥ 100%. Before showing some experimental data on the water vapor condensation growth of natural aerosols, it is useful to define the concept of the deliquescence point (DP) of a water-soluble substance. The term *deliquescence* is used to indicate the process by which a substance absorbs water molecules from the atmosphere until it dissolves in the absorbed water forming a solution. Therefore, DP is the value of RH at which a specific pure salt starts to change passing from the crystalline (or solid) phase to the solute phase. Deliquescence occurs when the partial vapor pressure of the solution is lower than the partial pressure e_w of water vapor in the surrounding air. The salt mixture is initially solid for very low RH and its solution starts when the increasing ambient RH reaches the lowest DP among those of the salts present in the mixture. A substance that absorbs moisture from the air but not necessarily for RH = DP is called *hygroscopic*, which means that it is capable to attract and hold water molecules from the surrounding air, through either absorption or adsorption processes. Below the DP value, a particle acquires only a little water mass through surface effects. A diluted salt solution forms when RH exceeds its DP value, having a concentration which remains in equilibrium with the water vapor partial pressure e_w during the whole condensation process.

Examining a numerous set of experimental data, Winkler (1988) developed a general concept of the condensation growth of mixed aerosol particles and derived simple formulas in order to parameterize the RH dependence of aerosol particles. An aerosol particle suspended in the dry air consists in general of both insoluble matter fragments and various soluble substances. The soluble components are for the major part in a crystalline state at very low RH, except unbuffered strong acids (such as sulfuric and nitric acids), and take up very small amounts of water for poorly humid air conditions. Since the soluble components have different DP values, the first still undissolved fraction of crystalline matter present in the mixture starts to dissolve as RH increases until reaching the lowest of such DP values, while the other salts remain substantially undissolved since RH is lower than their DPs. As RH continues to increase, the various soluble components gradually dissolve until reaching their complete solution when RH becomes equal to the highest DP value among those of the soluble substances contained in the initial aerosol particle. For such a high value of RH, all the water-soluble materials are dissolved, and the aerosol particle becomes a solution droplet containing only the pieces of insoluble matter, which may be suspended inside the droplet or located at its surface. A further increase in RH has the effect of diluting the solution inside the droplet more and more. These hygroscopic growth processes play a fundamental role in the formation of hazes and fog particles and hence of most accumulation and coarse particles commonly observed in the atmosphere.

In the present study, it is important to take into account that a certain soluble substance does not have a unique DP during the life of an aerosol particle, since it changes with time because of the hysteresis processes that can alter considerably both the chemical and structural characteristics of the particle during its life, caused by the alternation of high-RH and relatively low-RH conditions during the long sequence of nocturnal cooling and diurnal heating of air by solar radiation. In these daily cycles, a particle exhibits the same size as RH rises or sinks for RH > DP, since it continues to be in water vapor equilibrium. When RH decreases below DP, the salt solution does not recrystallize at the same RH at which it was previously dissolved but remains in a supersaturation state until the probability of a crystallization nucleus formation within the solution remains high enough.

Calculations of the water vapor condensation growth factor ϕ_g equal to the ratio between the equivalent diameter a(RH) of an aerosol particle growing in wet air and the initial diameter a_o of the same particle for dry air (RH = 0%) conditions were carried out by Hänel (1976) and by Hänel and Bullrich (1978) for different aerosol samples, collected in the middle of the Atlantic Ocean or at urban and rural/mountain sites in Germany. The various aerosol particle types were found to present considerably different hygroscopic characteristics and size distribution features. The analysis of these particulate samples indicated that the water mass condenses upon the aerosol particles also for low values of water activity A_w and assumes significant values for A_w ranging from 0.50 to 0.70. This implies that the dependence of the mean particulate density on RH is very small for low air moisture conditions and rather high for RH > 60%. Evaluations of the mean linear mass increase coefficient k_μ were obtained by Hänel (1976) for the earlier samples, represented as a function of (i) water uptake m_w/m_o per unit mass of dry matter and (ii) increasing or decreasing values of water activity A_w. For these data, Hänel and Bullrich (1978) calculated the values of growth factor $\phi_g = a(\text{RH})/a_0$ defined in the preceding text, for the following aerosol models: (i) model A, for pure maritime aerosols; (ii) model B, for maritime aerosols mixed with Saharan dust; (iii) model C, for urban aerosol; and (iv) model D, for background continental aerosol sampled at a mountain site. The four aerosol types were assumed to be characterized by different chemical composition features and hygroscopic properties, therefore presenting different patterns during the growth phase (for increasing RH) and the subsequent metastable phase (for decreasing RH), until reaching low RH conditions at the end of the daily 24-h cycle. The values of the growth factor ϕ_g are given in Table 3.2 for the four aerosol models mentioned earlier and seven values of the initial equivalent volume diameter a_o

Table 3.2 Values of the growth factor $\phi_g = a(\mathrm{RH})/a_o$ of a spherical aerosol particle suspended in humid air as a function of increasing and decreasing air RH (%) for different initial equivalent volume diameters a_o of the dry particle (RH = 0%) defined by Hänel and Bullrich (1978) for (i) model A, consisting of maritime aerosol sampled on 13–16 April 1969 over the central Atlantic Ocean (with dry particulate matter mass density $\rho_o = 2.45\ \mathrm{g\,cm^{-3}}$); (ii) model B, consisting of maritime aerosol sampled on 16–25 April 1969 over the Atlantic Ocean, containing Saharan dust (with dry particulate matter mass density $\rho_o = 2.64\ \mathrm{g\,cm^{-3}}$); (iii) model C, consisting of urban aerosol sampled in January 1970 at Mainz (Germany) (with dry particulate matter mass density $\rho_o = 2.85\ \mathrm{g\,cm^{-3}}$); and (iv) model D, consisting of background continental aerosol sampled in summer 1970 on the top of the Hohenpeissenberg elevation (1000 m a.m.s.l.) (Germany) (with dry particulate matter mass density $\rho_o = 1.85\ \mathrm{g\,cm^{-3}}$).

	Values of initial equivalent volume diameter a_o (μm)						
	2×10^{-2}	6×10^{-2}	2×10^{-1}	6×10^{-1}	2	6	20
Model A							
Incr. RH (%)							
20	1.007	1.007	1.007	1.007	1.007	1.007	1.007
40	1.015	1.015	1.015	1.015	1.015	1.015	1.015
60	1.048	1.048	1.048	1.048	1.048	1.048	1.048
65	1.102	1.102	1.102	1.102	1.102	1.102	1.102
70	1.200	1.214	1.214	1.214	1.214	1.214	1.214
75	1.299	1.364	1.415	1.457	1.495	1.503	1.503
80	1.414	1.603	1.632	1.635	1.638	1.645	1.645
85	1.606	1.705	1.740	1.751	1.754	1.754	1.754
90	1.731	1.855	1.906	1.921	1.927	1.928	1.929
95	1.905	2.147	2.257	2.291	2.303	2.307	2.309
97.5	2.066	2.475	2.707	2.784	2.811	2.819	2.822
99.0	2.216	2.924	3.423	3.606	3.676	3.697	3.704
99.5	2.281	3.219	4.051	4.439	4.595	4.642	4.658
Decr. RH (%)							
70	1.226	1.382	1.466	1.493	1.502	1.510	1.510
60	1.096	1.096	1.096	1.096	1.096	1.096	1.096
50	1.049	1.049	1.049	1.049	1.049	1.049	1.049
40	1.030	1.030	1.030	1.030	1.030	1.030	1.030
20	1.011	1.011	1.011	1.011	1.011	1.011	1.011
Model B							
Incr. RH (%)							
20	1.007	1.007	1.007	1.007	1.007	1.007	1.007
40	1.014	1.014	1.014	1.014	1.014	1.014	1.014
60	1.030	1.030	1.030	1.030	1.030	1.030	1.030
65	1.053	1.053	1.053	1.053	1.053	1.053	1.053
70	1.112	1.156	1.156	1.156	1.156	1.156	1.156
75	1.179	1.218	1.245	1.257	1.262	1.264	1.264
80	1.229	1.312	1.334	1.339	1.340	1.341	1.341
85	1.312	1.371	1.393	1.401	1.404	1.404	1.404
90	1.371	1.458	1.499	1.512	1.516	1.517	1.518
95	1.462	1.608	1.675	1.697	1.704	1.707	1.707

Table 3.2 (*continued*)

	Values of initial equivalent volume diameter a_o (μm)						
	2×10^{-2}	6×10^{-2}	2×10^{-1}	6×10^{-1}	2	6	20
97.5	1.534	1.733	1.922	1.991	2.017	2.025	2.027
99.0	1.587	1.962	2.359	2.525	2.591	2.610	2.617
99.5	1.607	2.098	2.713	3.058	3.207	3.252	3.268
Decr. RH (%)							
80	1.297	1.341	1.341	1.341	1.341	1.341	1.341
75	1.228	1.295	1.307	1.311	1.312	1.314	1.314
70	1.153	1.209	1.250	1.269	1.271	1.275	1.275
65	1.137	1.137	1.137	1.137	1.137	1.137	1.137
60	1.069	1.069	1.069	1.069	1.069	1.069	1.069
40	1.027	1.027	1.027	1.027	1.027	1.027	1.027
20	1.010	1.010	1.010	1.010	1.010	1.010	1.010
Model C							
Incr. RH (%)							
20	1.008	1.008	1.008	1.008	1.008	1.008	1.008
40	1.016	1.016	1.016	1.016	1.016	1.016	1.016
60	1.086	1.086	1.086	1.086	1.086	1.086	1.086
65	1.123	1.138	1.138	1.138	1.138	1.138	1.138
70	1.158	1.185	1.205	1.205	1.205	1.205	1.205
75	1.185	1.263	1.299	1.309	1.313	1.341	1.341
80	1.247	1.404	1.439	1.449	1.453	1.463	1.463
85	1.400	1.525	1.567	1.579	1.583	1.585	1.587
90	1.533	1.681	1.744	1.763	1.770	1.772	1.773
95	1.711	1.983	2.112	2.153	2.168	2.172	2.173
97.5	1.867	2.330	2.604	2.696	2.730	2.740	2.744
99.0	2.017	2.841	3.453	3.679	3.768	3.794	3.803
99.5	2.085	3.191	4.201	4.623	4.800	4.844	4.863
Decr. RH (%)							
85	1.452	1.573	1.603	1.612	1.615	1.617	1.617
80	1.320	1.448	1.502	1.518	1.523	1.526	1.526
75	1.249	1.330	1.365	1.376	1.379	1.382	1.382
70	1.202	1.243	1.271	1.282	1.286	1.286	1.286
65	1.168	1.183	1.192	1.196	1.197	1.197	1.197
60	1.143	1.143	1.143	1.143	1.143	1.143	1.143
40	1.048	1.048	1.048	1.048	1.048	1.048	1.048
20	1.014	1.014	1.014	1.014	1.014	1.014	1.014
Model D							
Incr. RH (%)							
20	1.004	1.004	1.004	1.004	1.004	1.004	1.004
40	1.011	1.011	1.011	1.011	1.011	1.011	1.011
60	1.036	1.036	1.036	1.036	1.036	1.036	1.036
65	1.047	1.047	1.047	1.047	1.047	1.047	1.047
70	1.061	1.067	1.067	1.067	1.067	1.067	1.067
75	1.081	1.114	1.129	1.132	1.135	1.135	1.135

(*continued overleaf*)

Table 3.2 (*continued*)

	Values of initial equivalent volume diameter a_o (μm)						
	2×10^{-2}	6×10^{-2}	2×10^{-1}	6×10^{-1}	2	6	20
80	1.114	1.210	1.246	1.257	1.261	1.264	1.264
85	1.201	1.337	1.369	1.378	1.381	1.384	1.384
90	1.332	1.441	1.482	1.493	1.497	1.500	1.500
95	1.444	1.582	1.644	1.664	1.671	1.674	1.674
97.5	1.511	1.716	1.934	2.014	2.044	2.053	2.056
99.0	1.560	1.985	2.480	2.695	2.781	2.807	2.816
99.5	1.578	2.157	2.968	3.356	3.521	3.571	3.589
Decr. RH (%)							
85	1.262	1.422	1.430	1.434	1.436	1.436	1.436
80	1.181	1.248	1.281	1.293	1.298	1.301	1.301
75	1.136	1.244	1.244	1.244	1.244	1.244	1.244
70	1.098	1.109	1.109	1.109	1.109	1.109	1.109
65	1.075	1.075	1.075	1.075	1.075	1.075	1.075
60	1.058	1.058	1.058	1.058	1.058	1.058	1.058
40	1.026	1.026	1.026	1.026	1.026	1.026	1.026
20	1.012	1.012	1.012	1.012	1.012	1.012	1.012

of the dry particle (for RH = 0%), ranging from 2×10^{-2} to 20 μm. Examples of growth curves of ϕ_g as a function of increasing and decreasing RH is shown in Figure 3.8 for model A of pure maritime aerosols and three different values of initial dry air diameter a_o pertaining to a typical nucleus having $a_o = 2\times10^{-2}$ μm, an accumulation particle with a_o = v.6 μm, and a coarse particle with a_o = 20 μm.

The results given in Table 3.2 and the example shown in Figure 3.8 clearly indicate that both accumulation and coarse particles are subject to largely increase as a result of water vapor condensation over the RH > 65% range.

In particular, it can be noted that growth factor ϕ_g increases considerably as RH exceeds 90% and increases even more as RH approaches 99% and the saturation conditions, exceeding the value of 3 for all the four considered aerosol types. Therefore, pronounced variations in the size distribution curves of the four aerosol populations are expected to occur as RH assumes values higher than 90%, while less marked changes are usually observed during the decreasing phase of RH. To give a measure of the most common multimodal characteristics of an aerosol polydispersion, an example of a multimodal size distribution curve $N(a)$ is shown in Figure 3.9 for a general case of an aerosol population consisting of (i) a first mode of nuclei arising from homogeneous nucleation processes; (ii) an Aitken nuclei mode consisting of particles with diameters $a < 0.10$ μm, mainly formed through secondary nucleation/coagulation and coagulation/condensation processes; (iii) an accumulation mode over the $0.2 \leq a \leq 1$ μm size range, mainly formed through chemical reactions occurring in cloud droplets and by condensation in moist air; (iv) a coarse particle mode over

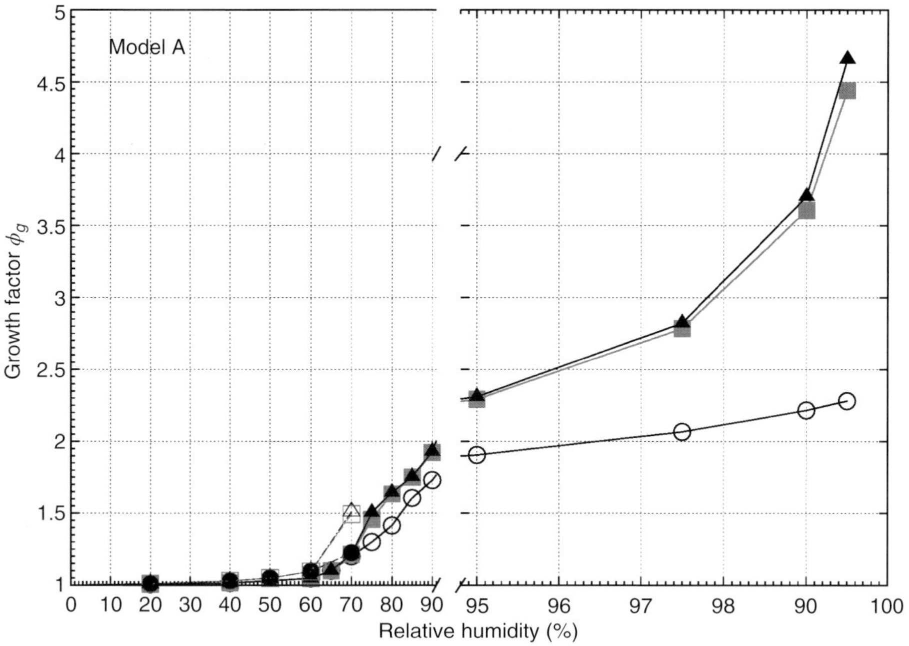

Figure 3.8 Variations of growth factor $\phi_g = a(\text{RH})/a_o$ given by the ratio between the equivalent diameter $a(\text{RH})$ of an aerosol particle suspended in wet air and the equivalent diameter a_o of the initial dry air particle (for RH = 0) as a function of increasing RH (lower dashed curve) and decreasing RH (upper dashed and dotted curve), as determined by Hänel and Bullrich (1978) for (a) aerosol model A, consisting of maritime aerosols sampled on 13–16 April 1969 over the central Atlantic Ocean (with dry particulate matter mass density $\rho_o = 2.45\ \text{g cm}^{-3}$) and (b) three representative values of initial diameter a_o equal to (i) 2×10^{-2} μm (indicative of a nucleus and represented by open circles), (ii) 6×10^{-1} μm (indicative of an accumulation particle and represented by gray squares), and (iii) 20 μm (indicative of a coarse particle and represented by solid triangles).

the $2 \leq a \leq 10$ μm range, mainly consisting of mixed maritime (wind-borne) and mineral dust particles; and (v) a giant particle mode, usually centered at diameters $a > 20$ μm, which principally consist of particles produced by fragmentation of soil matter mobilized by winds.

3.3
Aerosol Removal Processes

Deposition is the process by which aerosols collect or deposit themselves on solid surfaces, causing a decrease in their atmospheric concentration. It includes two main classes of removal processes that are commonly called dry and wet deposition:

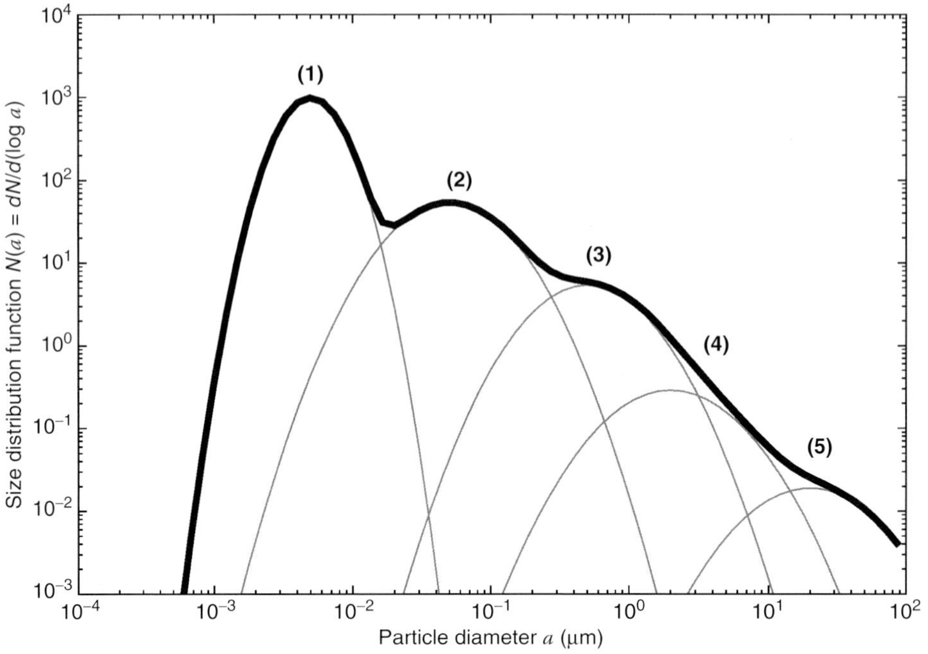

Figure 3.9 Example of a multimodal size distribution curve of particle number density $N(a)$ over the overall diameter range from 10^{-4} to 10^{2} μm, consisting of (i) a nucleation mode formed by homogeneous nucleation of new particles; (ii) an Aitken nuclei mode consisting of particles having diameters a mainly ranging from 10^{-2} to 10^{-1} μm and formed for the most part through secondary processes, such as nucleation/condensation of low vapor pressure substances and coagulation/condensation; (iii) an accumulation mode mainly consisting of particles with diameters a ranging from 0.2 to 2 μm and mainly formed through the chemical reactions occurring in cloud droplets; (iv) a coarse particle mode mainly consisting of primary particles having diameters ranging from 2 to 10 μm and principally formed by wind-borne mechanical processes on oceanic surfaces (forming sea-salt droplets) and land surfaces (mobilizing mineral dust particles); and (v) a giant particle mode produced for the most part by fragmentation of terrain particles by winds.

1. Dry deposition may be caused by (i) gravitational sedimentation (settling) of particles; (ii) interception, which mainly involves the small particles following the streamlines flowing in proximity of an obstacle and colliding with the obstacle itself when they flow too close to it; (iii) impaction, occurring when small particles interfacing a bigger particle are not able to follow the curved streamlines of the flow due to their inertia and hit or impact this particle; (iv) diffusion (Brownian motion), by which aerosol particles move randomly due to collisions with gas molecules, such collisions leading to further collisions with either obstacles or to coagulation between colliding particles; (v) turbulence, where turbulent eddies in the air favor the transfer of colliding particles; and (vi) other processes, such as thermophoresis, turbophoresis,

diffusiophoresis, and electrophoresis. The rate of deposition is lowest for particles having intermediate sizes, since the most effective deposition processes involve mainly (a) very large particles settling out quickly through sedimentation or impaction processes and (b) small particles, which are mainly removed through Brownian diffusion and subsequent coagulation within a few hours, until they achieve a diameter $a \approx 0.30\ \mu m$, beyond which they don't coagulate any more.

2. Wet deposition occurs when atmospheric hydrometeors (such as raindrops, snowflakes, or ice crystals) scavenge aerosol particles. This process may include (i) below-cloud scavenging, which occurs when falling rain droplets or snow particles collide with aerosol particles through Brownian diffusion, interception, impaction, and turbulent diffusion, and (ii) in-cloud scavenging, taking place when aerosol particles get into cloud droplets or ice crystals, acting as cloud nuclei or being captured by them through collision. These processes are described in the following subsections.

3.3.1 Dry Deposition of Aerosol Particles

Dry deposition is one of the most important processes occurring in the atmosphere, by which trace gases and particles are removed from the atmosphere in the absence of precipitations. The relative importance of the removal effects by dry deposition depends on various factors, such as (i) the atmospheric turbulence level, (ii) the chemical and water solubility characteristics of the particles, and (iii) the nature of the surface and terrain characteristics. In fact, the level of turbulence governs the rate at which aerosols are delivered down to the surface, especially within the layer nearest to the ground, such a rate varying as a function of aerosol size, density, and morphological characteristics. The surface roughness characteristics constitute also a very important factor in dry deposition, since a smooth surface may lead to particle bounce-off, and canopies generally promote dry deposition. To simulate the microphysical pathways by which particles travel from the bulk atmosphere to individual surface elements where they adhere, it can be assumed that the dry deposition flux F_d is directly proportional to the local aerosol concentration C_a at a reference height of 10 m or less above the surface, that is,

$$F_d = -u_d C_a, \tag{3.21}$$

where F_d is the amount of particulate matter depositing to a unit surface area per unit time, C_a is the aerosol particle number per unit volume, and u_d is the vertical deposition velocity of aerosols, measured in units of length per unit time. Considering that particle concentration C_a is a function of height z above the surface, deposition velocity u_d is also a function of z and must be related to a reference height at which C_a is measured. By convention, a downward flux is negative, so that u_d is positive for the depositing particles. The advantage of the earlier representation of deposition velocity is that all the numerous complexities of the dry deposition process are expressed by means of parameter u_d only. The disadvantage is that u_d is given by a variety of physical and chemical processes, which are difficult to be properly specified in real cases. Flux F_d is assumed to be constant up to the reference height at which C_a is determined. Equation (3.21) can be readily adapted in atmospheric models to account for dry deposition and is used as a surface boundary condition in the atmospheric diffusion equation.

Dry deposition includes in general the following three processes: (i) aerodynamic transport by turbulent diffusion down through the atmospheric surface layer until to a very thin layer of stagnant air just adjacent to the surface; (ii) Brownian transport across this thin stagnant layer of air until reaching the surface, called the *quasi-laminar sublayer*; and (iii) uptake at the surface. All the three mechanisms contribute to determine the deposition velocity u_d, affected by eddy transport of aerosols in the surface layer as well as by sedimentation of the particles having large sizes. As a consequence of the no-slip boundary condition for airflow at a surface, the air at an infinitesimal distance above a stationary surface is also stationary. The surface consists generally of a large number of irregularly shaped obstacles, such as leaves and buildings. The quasi-laminar sublayer has a depth Δz of the order of millimeters adjacent to each surface, in which the air is more or less stationary, and the transport occurs by both diffusion and sedimentation. Particle removal can actually take place within such a quasi-laminar sublayer by (i) interception, when the particles move sufficiently close to an obstacle and collide with it, and (ii) impaction, when particles cannot follow the abrupt changes of direction of the mean flow and their inertia carries them across the sublayer to the surface. The final step in the dry deposition process is the uptake of particles, which adhere to the surface, whose moisture and stickiness are important factors. Solid particles may bounce off a smooth surface, while liquid particles are more likely to adhere upon contact.

The deposition process can be usefully illustrated in terms of an electrical resistance analogy, in which the transport of particulate matter to the surface is assumed to be governed by the three following resistances in series: (i) aerodynamic resistance Ω_{aer}, (ii) quasi-laminar layer resistance Ω_{qll}, and (iii) surface or vegetation canopy resistance Ω_{can}. The total resistance Ω_{tot} to deposition of aerosols is thus given by the inverse of the deposition velocity:

$$u_d^{-1} = \Omega_{\mathrm{tot}} = \Omega_{\mathrm{aer}} + \Omega_{\mathrm{qll}} + \Omega_{\mathrm{can}}. \tag{3.22}$$

The overall resistance Ω_{tot} includes also the sedimentation effect. Therefore, it is realistic to assume that the three terms Ω_{aer}, Ω_{qll}, and Ω_{can} are located within the three distinct atmospheric layers near the surface, leading to obtain well-distinct values of particle concentrations at the top level of the surface layer and the top level of the quasi-laminar layer, while the surface-level concentration is null because of the perfect removal occurred within the surface and quasi-laminar layers. For these assumptions, Eq. (3.22) can be expressed in the following dependence form of particle dry deposition on resistances Ω_{aer} and Ω_{qll}:

$$u_d = \Omega_{\mathrm{tot}}^{-1} = [1/(\Omega_{\mathrm{aer}} + \Omega_{\mathrm{qll}} + \Omega_{\mathrm{aer}}\Omega_{\mathrm{qll}}u_s)] + u_s \tag{3.23}$$

where the particle settling velocity u_s is used to define the sedimentation flux of particles and a virtual resistance in parallel was associated with u_s.

The aerodynamic and quasi-laminar resistances Ω_{aer} and Ω_{qll} are affected by wind speed, vegetation height, leaf size, and atmospheric stability. In general, the sum $(\Omega_{aer} + \Omega_{qll})$ decreases as wind speed and vegetation height increase. Therefore, smaller resistances and higher deposition rates are expected to occur over tall forests but not over short grass. Typical aerodynamic layer resistance values calculated for a wind speed of $4\ m\ s^{-1}$ are equal to $\sim 60\ s\ m^{-1}$ over grass for $\Delta z = 10$ cm, $20\ s\ m^{-1}$ over crop for $\Delta z = 10$ cm, and $10\ s\ m^{-1}$ over conifer forests for $\Delta z = 10$ m. In particular, the following remarks can be made to better illustrate the dependence features of aerodynamic and quasi-laminar resistances Ω_{aer} and Ω_{qll} on the ambient conditions:

1. Turbulent transport is the mechanism bringing aerosols from the bulk atmosphere down to the surface. Therefore, it determines the aerodynamic resistance Ω_{aer}. The turbulence intensity depends mainly on the stability conditions of the lower atmosphere and the surface roughness and can be determined on the basis of microphysical measurements of the surface canopy characteristics and meteorological parameters, such as wind speed, temperature, and incoming solar radiation flux. During the diurnal hours of a day, the turbulence intensity is typically high over a reasonably thick well-mixed layer, thus exposing a considerably large reservoir of particulate matter to potential surface deposition. During the night, stable stratification of the atmosphere near the surface often reduces the intensity and vertical extent of the turbulence, in this way diminishing efficiently the overall dry deposition flux. Resistance Ω_{aer} is independent of the aerosol types involved, although gravitational settling must be taken into account for both accumulation and coarse particles.
2. The quasi-laminar resistance model for dry deposition postulates that adjacent to the surface exists a quasi-laminar layer, across which the resistance to transfer depends on molecular properties of the particulate substance and structural characteristics of the surface. This layer does not usually correspond to a laminar boundary layer in the classical sense, but rather results from the combination of many viscous layers adjacent to the obstacles constituting the overall effective structure of the surface. The depth of this layer constantly changes in response to turbulent shear stresses adjacent to the surface or surface elements. Consequently, this layer may only exist intermittently on plant leaves, which are often in continuous motion. On a theoretical basis, the existence of a quasi-laminar layer depends physically on the smoothness and the shape of the surface or surface elements and to some extent on the variability of the near-surface turbulence. Across the quasi-laminar sublayer, particles are transported analogously by Brownian impaction. In addition, because of their finite sizes, particles are also removed by interception on elements of the surface. The inverse of quasi-laminar resistance Ω_{qll} is a function of particle diameter a over the range from 10^{-3} to 10^{2} μm, presenting a minimum centered at $a \approx 0.8$ μm for particle density varying from 1 to

$2\,\mathrm{g\,cm^{-3}}$ for friction velocity $u^* = 10\,\mathrm{m\,s^{-1}}$ and close to $1\,\mu\mathrm{m}$ for $u^* = 1\,\mathrm{m\,s^{-1}}$. If Ω_{aer} can be neglected, the overall particle dry deposition velocity is given by the sum $(1/\Omega_{qll}) + u_s$. The term $1/\Omega_{qll}$ defines the component of the overall dry deposition velocity accounting for the processes occurring in the surface layer other than settling. It assumes large values for very small particles because of the better efficiency of Brownian diffusion to transport such particles across the surface layer, while impaction and interception lead to effective removal for very large particles. The result is that $1/\Omega_{qll}$ achieves a minimum over the $0.10 \leq a \leq 1.00\,\mu\mathrm{m}$ range where none of the removal process is particularly effective.

3. The surface resistance Ω_{can} for aerosol particles suspended over a vegetation canopy is the most difficult to be evaluated because of the complexity of the vegetation cover types. Numerous pathways are available from the quasi-laminar layer to the vegetation canopy or ground, including uptake by plant tissue inside leaf pores (stomata), the waxy skin of some leaves (cuticle), and mesophyll of leaves, deposition to the soil, and reactions with wetted surfaces. Each of these pathways can be represented by an appropriate resistance term to aerosol transport. The total canopy resistance Ω_{can} can be thus determined as the sum of the various resistance terms in parallel, chosen among those due to uptake by water surfaces (Ω_{cw}), uptake by the ground covered by soil or snow (Ω_{cg}), and foliar resistance (Ω_{cf}). If a water surface is present, Ω_{can} can be expressed in terms of Ω_{cf} and Ω_{cw} only, these resistances acting in parallel since the material can take only one deposition pathway, either to vegetation or to the ground surface. The foliar resistance Ω_{cf} can itself be divided into resistance to uptake at the cuticle (epidermal) surfaces (Ω_{cut}) and at the plant's stomata (Ω_{sto}), being both weighted using the leaf area index (LAI) (equal to the ratio between the leaf area and the overall area) as weight. Therefore, the earlier approach allows one to define the specific contributions given by terms Ω_{cg} and/or Ω_{cw} to foliar resistance Ω_{cf}.

3.3.2
Wet Deposition of Aerosol Particles

Wet deposition includes various natural processes by which atmospheric gases and aerosols are scavenged by atmospheric hydrometeors (cloud and fog drops, rain, snow) and delivered to the ground. Wet deposition of gases and aerosols may occur by (i) precipitation scavenging (in raining clouds); (ii) cloud-water interception, in which cloud droplets impact on the terrain, occurring most commonly at the top of tall mountains; (iii) fog deposition, through removal of gases and particulate matter by settling fog droplets; and (iv) snow deposition, through removal of material during snowstorms. Three steps are necessary in all these wet removal processes, since gases and aerosols are first brought into the presence of condensed water, subsequently scavenged by the hydrometeors, and finally delivered to the oceanic and land surfaces. These atmospheric compounds may undergo chemical transformations during each one of the earlier steps. The first wet deposition step involving aerosols occurs principally through nucleation scavenging during the cloud formation and collisions of particles with cloud droplets (through activation and subsequent growth of a part of aerosols to become cloud droplets); in the opposite direction, evaporation of cloud droplets contributes to produce new particles. The second step is given the below-cloud scavenging of aerosols occurring during rain- or snowfall, since the water drops and snowflakes collide with airborne particles and collect them; the reversible step is given by the evaporation of raindrops to form new aerosols. In the third step, the raindrops and snowflakes are deposited to the surface, while a further wet deposition process may occur through direct interception of aerosols by cloud particles.

These remarks indicate that the wet removal pathways depend on multiple and complex processes and involve numerous physical phases, which are influenced by the earlier processes on a variety of length scales, varying from those of the molecular transport and chemical reactions between molecules ($10^4 - 10^{-2}$ μm) to those of aerosol particles and cloud droplets ($10^{-3} - 10^2$ μm) and then from those of ice and snow particles and raindrops ($10^{-6} - 10^{-2}$ m) to those of thunderstorms, squall lines, and perturbation fronts ($10^1 - 10^6$ m). The rates of the earlier aerosol removal processes are determined in the following subsections, to better illustrate the removal fluxes using the concepts of scavenging coefficients, varying as a function of location, time, rainstorm characteristics, and aerosol size distribution shape parameters. The study of these processes involve the three vapor, solid, and aqueous phases, the aqueous phase being present in several forms (cloud water, rain, snow, ice crystals, sleet, hail), each of them having its particular size resolution.

3.3.2.1 In-Cloud Scavenging (Rainout)

The in-cloud scavenging of aerosol particles consists of two combined processes: the first is the so-called nucleation scavenging occurring through the growth of CCNs until becoming cloud droplets, and the second is the collection of a part of the remaining aerosols by cloud or rain droplets. Nucleation scavenging is a very efficient process, often capable of incorporating most aerosol mass in cloud drops. Conversely, interstitial aerosol collection by cloud droplets is rather slow and in general scavenges only a negligible fraction of the interstitial aerosol mass. Aerosol particles entering the base of a cloud serve as CCNs into which water vapor condenses to form cloud droplets. If a CCN is partially or completely soluble in water, it can dissolve in the new droplet to produce a solution droplet. Therefore nucleation scavenging of aerosols in clouds refers to activation and subsequent growth of a part of the aerosol particles to become cloud droplets. If $C_{i,o}$ is the mass concentration of the ith aerosol species suspended in air at the cloud base before the cloud formation, and $C_{i,\text{cloud}}$ and $C_{i,\text{int}}$ are its concentrations given in mass per volume of air in the aqueous and interstitial aerosol phases, respectively, the cloud mass scavenging ratio F_i can be defined for the ith aerosol species as

$$F_i = (C_{i,o} - C_{i,int})/C_{i,o} = C_{i,\text{cloud}}/C_{i,o}, \tag{3.24}$$

Since there is no production or removal of the ith aerosol species in the cloud, parameter $C_{i,o}$ is equal to the sum $C_{i,\text{int}} + C_{i,\text{cloud}}$. Correspondingly, the number density scavenging ratio F_N can be defined as

$$F_N = (N_{o,\text{cloud}} - N_{\text{int}})/N_{o,\text{cloud}}, \tag{3.25}$$

where $N_{o,\text{cloud}}$ is the aerosol number concentration before cloud formation and N_{int} is the number concentration of interstitial aerosols. Since particles with diameters $a > \sim 0.5\,\mu\text{m}$ become cloud droplets in a typical cloud, and provide the predominant fraction of the aerosol mass, one would expect a nearly unit value of their mass activation efficiency. In some cases, aerosol scavenging efficiency was estimated to largely exceed 0.7 in various cloud environments. Conversely, low number scavenging efficiencies are expected to characterize the clouds influenced by anthropogenic sources, because polluted aerosols exhibit a prevailing number concentration of small-size particles.

Aerosol species can be incorporated into cloud particles and raindrops inside the raining cloud. This process exerts a strong influence on the initial raindrop concentration $C_{\text{aq},0}$, before raindrops start to fall below the cloud base. The local removal rate of an irreversibly soluble gas (for instance, HNO_3) having concentration C_g is given for a droplet number density size distribution $N(a)$ by

$$W_{ic} = C_g \int_0^\infty K_c(\pi/16)a^2 N(a)da = \Lambda C_g, \tag{3.26}$$

where Λ is the scavenging coefficient, which assumes a value of approximately 5 s for the scavenging of HNO_3 in a typical cloud. In this case, the uptake process in a cloud is very rapid with respect to the air parcel motions inside the cloud. For less soluble gases (e.g., SO_2), the uptake process is very complicated depending not only on the cloud droplet size distribution curve but also on the properties of the other species participating in the aqueous-phase reactions (e.g., H_2O_2) or exerting a control on the cloud droplet pH (e.g., NH_3). Once emitted into the atmosphere, SO_2 and NO_x are oxidized to sulfate and nitrate through both gas- and aqueous-phase processes, while organic acids are simultaneously produced during the oxidation of emitted organic compounds. The result of these reactions is the existence in the atmosphere of acids in the gas phase, such as HNO_3, HCl, HCOOH, and CH_3COOH, in both the aerosol phase (sulfate, nitrate, chloride, organic acids, etc.) and the aqueous phase. These acidic species are removed from the atmosphere by both wet and dry deposition, and the overall process is commonly named *acid deposition*, while the removal of acidic material by rain and wet deposition is called *acid rain*. Therefore, the acid deposition processes include acid rain, dry deposition of both acid vapor and acid particles, and other wet removal phenomena such as acid fogs and cloud interception. The effects attributable to acid rain actually result from the combination of wet and dry deposition occurrences.

Materials scavenged by cloud droplets (in-cloud scavenging) can be removed by wet deposition without rain formation if the cloud comes in contact with the ground surface. This wet deposition of intercepted cloud water can be locally important, especially along the slopes of mountains which are high enough as to regularly intercept clouds. In general, the concentrations of SO_4^{2-}, NO_3^-, H^+, and NH_4^+ ions in cloud water were found to be 5–10 times higher than those measured inside the precipitations at the same sites. Most cloud droplets have sizes $a > 5\,\mu m$, and their main removal mechanism is the gravitational settling for flat terrain and low wind speeds.

The most susceptible sites to wet deposition by cloud droplet interception and impaction are the mountain summits, in cases where they are located above the average cloud base. The deposition of cloud water depends closely on the height of such sites in all the areas where the surface is covered by ecosystems such as forest canopies or complex vegetation-covered terrain and relatively high turbulence conditions occur. In these cases, the wet removal rate depends not only on the cloud droplet size distribution but also on sizes, shapes, and distribution of forest and vegetated canopy elements protruding into the airflow, where turbulent transport occurs over and among such inhomogeneous surfaces. The driving force for cloud-water interception is not well known: droplet deposition velocities were estimated to be equal to 2–5 cm s^{-1} on grass and 10–20 cm s^{-1} in forests under orographic clouds. These values are comparable with those of settling velocity and equal to 0.3 cm s^{-1} for a 10-μm diameter droplet, 1.2 cm s^{-1} for a 20-μm droplet, and 7.5 cm s^{-1} for a 50-μm droplet. The volume mean diameters are equal to around 10 μm, indicating that turbulence can significantly increase the deposition velocity of the cloud drops even over a flat surface located in mountain tops for high wind speeds and marked surface roughness characteristics. Cloud droplet removal rate depends critically on characteristics of the forest canopy, since the cloud-water deposition at the edge of a forest is evaluated to be 4–5 times greater than in a closed forest. The liquid water content (LWC) of a cloud and the shape features of the cloud droplet size distribution curves (especially those over the large droplet size range) can also exert a significant influence on the overall wet deposition rate. Field data recorded in the Eastern United States and Europe indicate that cloud-water deposition yields a substantial contribution to the wet deposition inputs at high elevation sites. Cloud droplets are typically more acidic than precipitation droplets collected at the ground. They are generally small and not subject to the dilution associated with the growth of raindrops and snowflakes nor to the neutralization associated with the capture of surface-derived NH_3 and alkaline particles suspended in layers at lower altitudes. Therefore the interception of these droplets provides a route by which concentrated solutions of sulfate and nitrate can be transferred to the foliage in high-elevation areas exposed to clouds.

3.3.2.2 **Interstitial Aerosol Scavenging by Cloud Droplets**

Interstitial aerosol particles collide with cloud droplets and are removed from cloud interstitial air. The coagulation theory described in Section 3.2.1 can be correctly used to quantify the rate and effects of such a removal process. If $n_{ia}(a, t)$ is the aerosol number density size distribution of interstitial aerosol particles in a cloud at time t, and $n_d(a, t)$ is the cloud droplet number density size distribution at time t, the loss rate of aerosol particles per unit volume of air due to scavenging by cloud drops is governed by the following equation:

$$\frac{\partial n(a,t)}{\partial t} = -n_{ia}(a,t)\int_0^{\infty} k_{col}(a,a_d)n_d(a_d,t)da_d, \tag{3.27}$$

where $k_{col}(a, a_d)$ is the collection coefficient for collisions between an interstitial aerosol particle of diameter a and a cloud droplet of diameter a_d. The scavenging coefficient $\Lambda_d(a)$ for interstitial aerosols and the cloud droplet polydispersion can be therefore defined as

$$\Lambda_d(a,t) = -\frac{1}{n_{ia}(a,t)}\frac{\partial n(a,t)}{\partial t} = \int_0^{\infty} k_{col}(a,a_d)n_d(a_d,t)da_d. \tag{3.28}$$

If the scavenging coefficient $\Lambda_d(a)$ does not vary with time, the time-dependent function of the aerosol number density size distribution of interstitial aerosol particles in a cloud is given by

$$n_{ia}(a,t) = n_{ia}(a,0)\exp[-\Lambda_d(a)t]. \tag{3.29}$$

Assuming that the cloud droplets are stationary, one can state that the particles are captured by Brownian diffusion. It was pointed out in Section 3.2.1 that coefficient K_{12} in the continuum regime of Brownian coagulation for $a_2 \gg a_1$ approaches the limit equal to $\frac{2k_B T a_2}{3\eta a_1}$ and tends to assume an asymptotic value equal to $\sqrt{\frac{3k_B T}{\rho}}(a_2{}^2/\sqrt{a_1{}^3})$ in the free molecular regime. Thus the collection coefficient $k_{col}(a, a_d)$ considered in Eq. (3.28) can be estimated on the basis of the earlier evaluations made for a case in which the particles are collected by a falling drop. Assuming that this cloud has an LWC = 0.5 g m^{-3} and that all drops have diameters $a_d = 10\,\mu m$, resulting in an overall drop number concentration $N_d = 955\,cm^{-3}$, Eq. (3.29) can be simplified by inserting the following evaluation of

$$\Lambda_d(a) = N_d k_{col}(a, 10\,\mu m), \tag{3.30}$$

where $k_{col}(a, 10\,\mu m)$ can be calculated on the basis of the coagulation theory described in Section 3.2.1. The calculations of the corresponding particle collision efficiencies and lifetimes are given in Table 3.3, indicating that nuclei with sizes $a < 10$ nm are scavenged within a few minutes in the cloud, whereas aerosol particles with $a > 10^{-1}$ μm are in practice not collected by droplets during the cloud lifetime. The earlier results indicate that for average residence times of air parcels in clouds equal to about 1 h only the nuclei can be efficiently collected by cloud droplets. These particles often represent a significant fraction of the aerosol load, and therefore their total number concentration may be abundantly reduced, while the shape of the aerosol number size distribution curve changes dramatically. However, these particles contain only a very low mass fraction. Therefore, the mass size distribution curve is expected to change only slightly because of such a collection process.

Table 3.3 Estimates of collection coefficient $k_{col}(a, a_d = 10\,\mu m)$, scavenging coefficient $\Lambda_d(a)$ for interstitial aerosol particles, and aerosol lifetimes $\Delta t_L = [\Lambda_d(a)]^{-1}$ in a nonraining cloud (having number density concentration $N_d = 955\,cm^{-3}$, cloud droplet diameter $a_d = 10\,\mu m$, and LWC = $0.5\,g\,m^{-3}$) at standard atmospheric conditions for various values of aerosol particle diameter a.

Aerosol diameter a (μm)	$k_{col}(a,a_d = 10\,\mu m)$ ($cm^{-3}\,s^{-1}$)	$\Lambda_d(a)$ (s^{-1})	Lifetime Δt_L
0.002	4×10^{-5}	3.8×10^{-2}	0.4 min
0.010	1.6×10^{-6}	1.5×10^{-3}	11 min
0.005	1.1×10^{-7}	1.0×10^{-4}	2.8 h
0.10	2.2×10^{-8}	2.1×10^{-5}	13 h
1.0	1.03×10^{-9}	9.8×10^{-7}	11.8 days
10	3×10^{-10}	2.9×10^{-7}	40 days

3.3.2.3 Precipitation Scavenging

The term *precipitation scavenging* refers to the removal of gases and particles by clouds and precipitations. This process is of crucial importance for cleansing the atmosphere of pollutants and lowering the frequency of acid rain episodes at the ground. Aerosol particles are incorporated into cloud droplets through nucleation scavenging, as explained in the previous subsection. Additional processes by which particles may be captured by hydrometeors are the so-called diffusional collection and inertial collection: (i) diffusional collection refers to the migration of particles through the air by diffusion to reach the hydrometeors, this process being most important for the submicron particles, which diffuse through the air more readily than the larger particles, and (ii) inertial collection refers to the collision of particles with hydrometeors occurring because of their differential fall speeds, this process being similar to the collision and coalescence of droplets. Since very small particles follow closely the streamlines around a falling hydrometeor, they often avoid to be captured. Consequently, inertial collection is important only for particles having sizes greater than a few micrometer.

The relative contributions of nucleation scavenging, aqueous-phase chemical reactions, and precipitation scavenging to the amount of any chemical substance contained in the hydrometeors reaching the ground depend on the ambient air conditions and the nature of the cloud. In particular, the incorporation of sulfate occurring inside to the hydrometeors originating in warm clouds can provide an important acid rain contribution. For a warm cloud situated in heavily polluted urban air, the approximate contributions to the sulfate content of rain at the ground level are aqueous-phase chemical reactions (61%), nucleation scavenging (37%), and precipitation scavenging below the cloud base (2%). The corresponding approximate percentages for a warm cloud situated in clean marine air are 14%, 75%, and 11%, respectively. The principal reason of so markedly different percentages for polluted and clean clouds is that polluted air contains much greater concentrations of SO_2 than clean air. Therefore, the sulfate production by the aqueous-phase chemical reactions is much more intense in polluted air than in clean air, since these reactions are considerably faster in polluted air. To better illustrate the basic principles involved in the process, it is useful to mention that the conversion of SO_2 to H_2SO_4 in cloud water leads to the dissolution of gaseous SO_2 in cloud water and the subsequent formation of bisulfite and sulfite ions.

After SO_2H_2O (aq.), HSO_3^{2-} (aq.), and SO_3^{2-} (aq.) are formed within a cloud droplet, they are oxidized very quickly to sulfate. Their oxidation rates depend on the oxidant and, in general, on the pH of the droplet. The fastest oxidant in the atmosphere, over a wide range of pH values, is hydrogen peroxide (H_2O_2), while ozone (O_3) also serves as a fast oxidant for pH values higher than ~5.5. With regard to this, it is useful to mention that the pH of pure water in contact only with its own vapor is 7, and the pH of rainwater in contact with very clean air is 5.6. The lowering of pH in clean air is due to the absorption of CO_2 into the rainwater and the formation of carbonic acid. The pH of rainwater in polluted air can be significantly lower than 5.6, and this gives rise to acid rain. Such a high acidity is due to the incorporation of gaseous and particulate pollutants into the rain by various mechanisms such as nucleation scavenging, dissolution of gases in cloud droplets, aqueous-phase chemical reactions, precipitation scavenging, and sulfate formation in precipitations. In addition to sulfate, many other chemical species contribute to enhance the acidity of rain, such as nitrate and sulfuric acid and nitric acid emitted from coal-fired electric power plants.

The aerosol particles released from evaporating clouds may be larger and more soluble than an original CCN on which the cloud droplet forms. Therefore, also the cloud-processed particles can serve as CCNs at lower supersaturation levels together with the original CCNs involved in cloud formation. In a similar way to that of the formation of small water droplets by combination of water molecules in highly supersaturated air (homogeneous nucleation), the molecules of two gases can combine under appropriate conditions to form aerosol particles through a process usually named *homogeneous-bimolecular nucleation*. The conditions favoring such a formation process of new particles are (i) the high concentrations of the two gases; (ii) the low ambient concentrations of preexisting particles, which would otherwise provide a large surface area onto which the gases could condense; and (iii) the low temperatures favoring the condensed phase. These conditions can be satisfied in the outflow regions of clouds, since some of the particles carried upward in a cloud are removed by clouds and precipitations. Consequently, the air detrained from a cloud may contain relatively low concentrations of particles, but the RH of the detrained air is in general high, due to cloud droplet evaporation. The air is also detrained in deep convective clouds, near cloud tops where the temperature is lower. All these conditions favor the production of new particles in the detrained air. For example, O_3 can be subject to photolysis forming the OH radical: this can oxidize SO_2 to form gaseous sulfuric acid, which can subsequently combine with water vapor through homogeneous-bimolecular nucleation to form solution droplets of H_2SO_4. The formation of particles in the outflow regions of convective clouds may act on a large scale to supply high number concentrations of particles to the upper troposphere in the tropics and subtropics as well as to the subtropical marine boundary layer.

As a raindrop falls through the air, it collides with airborne particles and collects them. Having a diameter a_f, it sweeps the volume of a cylinder equal to $(\pi/4)a_f^2U_f(a_f)$, in a unit time, where U_f is the raindrop falling velocity. As a first approximation, the droplet would collect all the particles that are in this volume. In reality, if all the other aerosol particles have sizes equal to a, a collision will occur if the center of the particle is inside the cylinder with an overall radius equal to $(a+a_f)/2$. The particles are themselves settling with velocity $u_s(a)$. Therefore, the collision volume per unit time is actually equal to $(\pi/4)\ (a+a_f)^2\ [U_f(a_f)-u_s(a)]$. The major complication arises from the fact that the falling drop perturbs the air around it, creating a flow field in which the flow streamlines diverge around the drop. Thus, as the raindrop approaches the particle, it exerts a force on it and modifies its trajectory.

Whether a collision will in fact occur depends on the sizes of the drop and the particle and their relative locations. Prediction of the actual trajectory of the particle is a complicated problem. Solutions are expressed in terms of collision efficiency $E_{\text{col}}(a, a_f)$, which is defined in the same manner as made for the drop-to-drop collision efficiency. Therefore, $E_{\text{col}}(a, a_f)$ provides in practice the fraction of particles having diameters a which are contained within the collision volume of a drop of diameter a_f. In this approach, $E_{\text{col}}(a, a_f)$ can be used as a correction factor which accounts for the interaction between the falling raindrop and the aerosol particles. If the aerosol number density size distribution below the cloud is given by $n_{\text{bc}}(a)$, the number of collisions between particles having diameters belonging to the range from a to $a+da$ and a drop of diameter a_f is given by

$$(\pi/4)(a_f+a)^2\left[U_f\left(a_f\right)-u_s(a)\right]E_{\text{col}}(a,a_f)n_{bc}(a)da. \tag{3.31}$$

The mass accumulation rate of these particles by a single falling drop can be calculated by substituting the number density size distribution $n_{\text{bc}}(a)$ with the mass concentration size distribution $M(a)$ to obtain

$$(\pi/4)(a_f+a)^2\left[U_f\left(a_f\right)-u_s(a)\right]E_{\text{col}}(a,a_f)M(a)da. \tag{3.32}$$

The total collection rate for the mass of the particles having diameters a is thus obtained by integrating the term given in Eq. (3.32) over the whole size distribution of collector drops

$$W_{\text{bc}}=M(a)da\int_0^{\infty}(\pi/4)(a_f+a)^2\left[U_f\left(a_f\right)-u_s(a)\right]E_{\text{col}}(a,a_f)N(a_f)da_f, \tag{3.33}$$

where $N(a_f)$ is the falling raindrop number density size distribution function.

The two following approximations can generally be made in Eq. (3.33): (i) $U_f(a_f)\gg u_s(a)$ and (ii) $(a_f+a)^2\cong a_f$. For these approximations, Eq. (3.33) becomes

$$W_{\text{bc}}=M(a)da\int_0^{\infty}(\pi/4){a_f}^2U_f(a_f)E_{\text{col}}(a,a_f)N(a_f)da_f. \tag{3.34}$$

The below-cloud scavenging (rainout) rate of aerosol particles having a diameter a can be thus written as

$$dM(a)/dt = -\Lambda(a)M(a), \tag{3.35}$$

where the scavenging coefficient $\Lambda(a)$ is given by

$$\Lambda(a) = \int_0^\infty (\pi/4)a_f{}^2 U_f(a_f)E_{\text{col}}(a, a_f)N(a_f)da_f. \tag{3.36}$$

Therefore, to calculate the aerosol scavenging rate for a given aerosol diameter a, one needs to know the droplet number density size distribution $N(a)$ and the collision efficiency $E_{\text{col}}(a, a_f)$ characterizing the precipitation scavenging process. The total aerosol mass scavenging rate can be calculated by integrating Eq. (3.36) over the whole aerosol size distribution as follows:

$$dM_{\text{bc}}/dt = \frac{d}{dt}\int_0^\infty M(a)da = -\int_0^\infty \Lambda(a)M(a)da \tag{3.37}$$

Finally, the rainfall rate $R_{r,0}$ (usually measured in mm per hour) is related to the raindrop number density size distribution $N_{\text{rd}}(a_f)$ through the following analytic form:

$$R_{r,0} = \int_0^\infty \left(\frac{\pi}{6}\right) a_f{}^3 U_f(a_f)N_{rd}(a_f)da_f. \tag{3.38}$$

Routine measurements of $R_{r,0}$ are available from ground-based meteorological measurements. The evaluation of reliable raindrop size distribution curves presents serious problems in real cases because of their wide variability during both a single rainstorm and from one event to another.

On the Raindrop–Aerosol Collision Efficiency The collision efficiency $E_{\text{col}}(a, a_f)$ is equal to the ratio between the total number of collisions occurring between rain droplets and particles and the total number of particles in an area equal to the effective cross-sectional area of droplets. The unit value of E_{col} implies that all particles in the geometric volume swept out by a falling drop are collected. Usually it is $E_{\text{col}} \ll 1$, although E_{col} can exceed the unit value under particular conditions, for charged particles, for instance. It is very difficult to find a theoretical solution of the Navier–Stokes equation for prediction of $E_{\text{col}}(a, a_f)$ in the general raindrop–aerosol interaction case. Such difficulties arise mainly from the fact that the aerosol sizes vary by several orders of magnitude, and the large raindrop sizes cause complicated flow patterns (drop oscillations, wake creation, eddy shedding, etc.). Aerosol particles can be collected by a falling drop as a result of their Brownian diffusion. Such a random motion of the particles can bring some of them in contact with the drop and tend to increase the collection efficiency E_{col}. Because the Brownian diffusion of particles decreases rapidly as a increases, the removal mechanism is expected to be the most important for aerosol particles having diameters $a < 2 \times 10^{-1}$ μm. Inertial impaction occurs when a particle is not able to follow the rapidly curving streamlines around the falling spherical drop

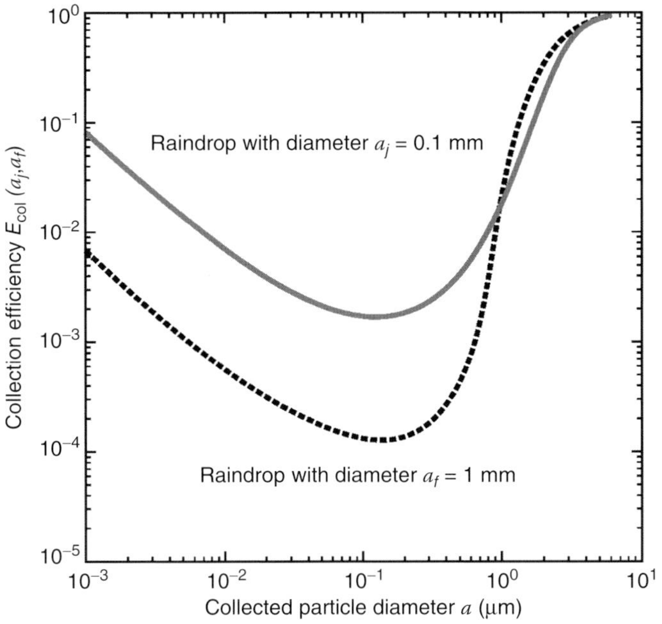

Figure 3.10 Semiempirical correlation curves of collection efficiency $E_{col}(a, a_f)$ of two drops (having diameters a and a_f, respectively) as a function of the collected particle diameter a (measured in micrometer) for particles assumed to have unit density ρ, as evaluated by Slinn (1983) for raindrop diameters $a_f = 0.1$ mm (gray curve) and $a_f = 1$ mm (dashed curve). (Adapted from a graph of Seinfeld and Pandis (2006).)

and, because of its inertia, continues to move toward the drop until being eventually captured. Inertial impaction increases in importance as the aerosol size a increases and accelerates the scavenging of particles with $a > 1$ μm. These remarks indicate that, whereas the scavenging of small and large particles is expected to be efficient, scavenging of accumulation particles having diameters ranging from 10^{-1} to 2 μm is expected to be relatively slow. Finally, interception takes place when a particle follows the streamlines of flow around an obstacle and comes into contact with the raindrop. If the streamline on which the particle center lies is within a distance equal to the particle radius or less from the drop surface, interception is expected to occur being closely related to the inertial impaction.

The dependence curves of collection efficiency E_{col} on the collected particle diameter a are shown in Figure 3.10 for two values of raindrop diameter a_f equal to 2×10^{-1} and 2 mm, as calculated by Slinn (1983) on the basis of a semiempirical expression fitting a set of experimental data. As expected, Brownian diffusion dominates for $a < 0.10$ μm, and impaction and interception control the removal for larger values of a. The characteristic minimum in E_{col} occurs in the regime when the particles are too large for being subject to an appreciable Brownian diffusivity but are at the same time too small for being effectively collected by either impaction or interception. For very large aerosol particles (with $a > 20$ μm) and for the extremely small nuclei (with $a < 2 \times 10^{-3}$ μm), parameter E_{col} approaches the unit value. However, for particles close to the minimum (with a ranging from about 0.5 to 1.5 μm), the drops collect only the particles which are very close to the center of the volume swept by the raindrop.

On the Scavenging Rates Using the Slinn (1983) semiempirical curves determined in Figure 3.10 for the collision efficiency $E_{col}(a, a_f)$, one can estimate the scavenging coefficient and the scavenging rate for a rain event. The calculation requires the knowledge of the size distribution curves of raindrops and below-cloud aerosols. The scavenging coefficient $\Lambda(a)$ determined in Eq. (3.36) describes the rate of removal by rain for particles having diameters a and raindrop number density size distribution $N(a_f)$. If one assumes that all the raindrops have the same diameter a_f and the overall number concentration is equal to N_f, Eq. (3.36) can be simplified as follows:

$$\Lambda(a) = (\pi/4)a_f{}^2 U_f(a_f) E_{col}(a, a_f) N_f, \tag{3.39}$$

where the collection efficiency $E_{col}(a, a_f)$ depends on aerosol diameter a only, and the number concentration of drops can be estimated using Eq. (3.38) as a function of the rainfall rate,

$$R_{r,0} = (\pi/6)a_f{}^3 U_f(a_f) N_f. \tag{3.40}$$

For monodispersed aerosols and raindrops, Eq. (3.39) provides the scavenging coefficient $\Lambda(a)$ as equal to three times the ratio $E_{col}(a, a_f)R_{r,0}/2a_f$. The scavenging coefficient for such a simple case is shown in Figure 3.11 for $R_{r,0} = 1\ \text{mm h}^{-1}$ and two raindrops having diameters equal to 0.2 and 2 mm. The graph provides evidence of the sensitivity of the scavenging coefficient to the aerosol and raindrop sizes, suggesting that realistic size distributions of aerosol particles and raindrops are needed in order to correctly estimate the ambient scavenging rates.

The total aerosol mass scavenging rate is given by Eq. (3.37). Replacing the aerosol mass distribution with the number distribution, the overall mass scavenging rate is given by

$$dM_{bc}/dt = -\int_0^{\infty} (\pi/6)a^3 n_{bc}(a)\rho\Lambda(a)da. \tag{3.41}$$

The present calculations can be simplified by defining a mean mass scavenging coefficient Λ_m, for which the following relationship,

$$(\pi\rho/6)\int_0^{\infty} a^3 n_{bc}(a)\Lambda(a)da = \Lambda_m(a)(\pi\rho/6)\int_0^{\infty} a^3 n_{bc}(a)da, \tag{3.42}$$

was obtained with

$$\Lambda_m = \int_0^{\infty} \Lambda(a)a^3 n_{bc}(a)da \Big/ \int_0^{\infty} a^3 n_{bc}(a)da. \tag{3.43}$$

For a certain particle density ρ, the following equation is obtained from Eqs. (3.41)–(3.43):

$$dM_{bc}/dt = -\Lambda_m M_{bc}, \tag{3.44}$$

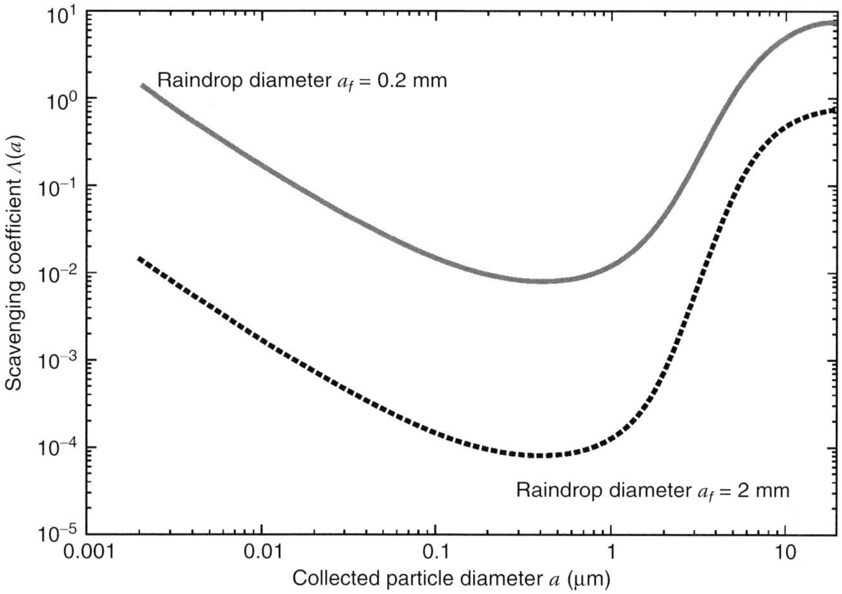

Figure 3.11 Dependence curves of scavenging coefficient $\Lambda(a)$ as a function of collected particle diameter a for monodispersed particles collected by monodispersed raindrops with diameters a_f equal to 0.2 mm (gray solid curve) and 2 mm (dashed curve), assuming a rainfall intensity of 1 mm h^{-1}. (Adapted from a graph of Seinfeld and Pandis (2006).)

in which all the scavenging information is effectively incorporated into one parameter only.

On the Below-Cloud Scavenging (Washout) The complexity of the wet removal process led early investigators to attempt to quantify the relationship between airborne species concentrations, meteorological conditions, and wet deposition rates by lumping the effects of these processes into a few parameters, with the purpose of describing the processes occurring inside and below the clouds, such as rain formation, aerosol–cloud interactions, gas–cloud interactions, aerosol–gas interactions, and interactions between rain and gaseous atmospheric compounds.

The rate of transfer of an aerosol particle into rain droplets below a cloud can be approximated by the first-order relationship

$$W_{\text{part/rain}} = \Lambda_p / C_p, \tag{3.45}$$

where Λ_p is the scavenging coefficient for particulate matter. It is in general a function of location, time, rainstorm characteristics, and aerosol size distribution. Equation (45) can be correctly used in the real cases only if the scavenging is independent of the amount of previously scavenged particulate matter. If $C_p(z, t)$ is the vertical profile of the aerosol concentration of a species at time t in a horizontally homogeneous atmosphere "washed out" by rain, the below-cloud scavenging rate $J_{\text{bc}}(t)$ of aerosols, measured per unit mass area and unit time, is given as a function of time by

$$J_{bc}(t) = \int_0^h \Lambda_p(z,t)/C_p(z,t)dz, \tag{3.46}$$

where h is the cloud base height and $\Lambda_p(z, t)$ is the height-dependent scavenging coefficient for aerosols measured per unit time. The removal rate of aerosols from the cloud is often referred to as the "rainout rate" and the rate of below-cloud scavenging of aerosols as the "washout rate." If the atmospheric profile of C_p (z, t) located below the cloud is homogeneous, the concept of average scavenging coefficient Λ_g^* can be adopted and inserted into Eq. (3.46), thus obtaining

$$J_{bc}(t) = \Lambda_g{}^* h C_p(t). \tag{3.47}$$

Another empirical parameter often used in wet removal studies is the scavenging ratio w_s, defined as the ratio between the concentration of aerosol particles collected by precipitations and the overall concentration of aerosol particles in air:

$$w_s = C_{p,\text{rain}}/C_{p,\text{air}}, \tag{3.48}$$

where (i) $C_{p,\text{rain}}$ is the wet aerosol particle concentration in precipitation falling at the surface and (ii) $C_{p,\text{air}}$ is the aerosol particle concentration in the air mass located aloft the point where the storm enters. However, both of these concentration parameters are very difficult to be experimentally measured inside the space where aerosols enter the storm.

Another useful parameter is the washout ratio w_r, given by the ratio between the particle concentration in surface-level precipitation and the aerosol concentration in surface-level air. The corresponding net wet deposition flux F_w can be expressed in terms of $C_{p,\text{rain}}$, evaluated as a function of the two ground-level space coordinates and time, that is, by

$$F_w = C_{p,\text{rain}}(x, y, 0, t) R_{r,0}, \tag{3.49}$$

where $R_{r,0}$ is the precipitation rate, defined in Eq. (3.38) and usually measured in millimeter per hour. Typical values of $R_{r,0}$ are 0.5 mm h^{-1} for drizzle and 25 mm h^{-1} for heavy rain.

On the basis of the definition of flux F_w given in Eq. (3.49), the wet deposition velocity u_w of aerosols can be determined as the ratio between wet deposition flux F_w and aerosol concentration $C_{p,\text{air}}$ $(x, y, 0, t)$ measured at the ground level, that is,

$$u_w = F_w/C_{p,\text{air}}(x, y, 0, t). \tag{3.50}$$

It is interesting to note that on the basis of Eqs. (3.49) and (3.50), the relationship between wet deposition velocity u_w and washout ratio w_r is given by

$$u_w = F_w/C_{p,\text{air}}(x, y, 0, t) = C_{p,\text{rain}}(x, y, 0, t) R_{r,0}/C_{p,\text{air}}(x, y, 0, t) = w_r R_{r,0}. \tag{3.51}$$

3.3.2.4 Wet Deposition in Fogs

Fogs can be viewed as clouds that are in contact with the ground surface. Therefore, the removal mechanisms described in the preceding text to occur in clouds can be also used to explain the wet deposition processes by fogs. They are created during cooling of air next to the ground, either by radiation to space (radiation fogs) or by contact with a surface (advection fogs), and in general are distinguished from the clouds intercepting a mountain slope. Both radiation and advection fogs can be created in heavily polluted air masses characterized in general by very strong thermal inversions within the ground layer: for these conditions, urban fogs scavenge actively the acidic particles and gases by the same mechanisms as clouds and are often characterized by low values of pH. The composition and pH of fogs vary as a function of time, the concentrations of its major constituents being rather high at the beginning of a fog. As the fog develops, LWC rises, droplets are diluted, and acidity drops, and then evaporation takes place as the air is heated, while RH decreases, and the pH is lowered again. Fogs have LWC values usually varying between 0.1 and $0.2\,g\,m^{-3}$ and often consist of droplets having sizes of $\sim 10\,\mu m$ and presenting much higher number concentrations than clouds. Consequently, the removal processes of the major aerosol species result to be 5–20 times faster than those observed in the atmosphere without fogs.

3.3.2.5 Nucleation of Ice Particles

Empirical observations clearly indicate that water may readily supercool and water clouds are often found at air temperatures $T < 0\,°C$. If a cloud extends above the $0\,°C$ thermal level, it is called a *cold cloud*, and the water droplets suspended in these clouds for $T < 0\,°C$ are usually called *supercooled droplets*. Cold clouds may also contain ice particles. In cases where a cold cloud contains contemporaneously ice particles and supercooled droplets, the cloud is said to be "mixed," while a cloud consisting entirely of ice particles is said to be "glaciated." A supercooled droplet is in an unstable state. In freezing episodes, it needs enough water molecules to come together within the droplet to form an embryo of ice large enough to survive and grow, through a process which is analogous to that leading to the formation of a water droplet from the vapor phase (see Section 3.2.4). If an ice embryo within a droplet exceeds a certain critical size, its growth is expected to cause a decrease in the energy of the system, while any increase in the size of an ice embryo smaller than the critical size causes an increase in total energy. In the latter case, from an energetic point of view, it is preferable for the embryo to break up. If a water droplet contains no foreign particles, it can freeze only by homogeneous nucleation. Because the numbers and sizes of the ice embryos forming by chance aggregations both increase as the air temperature decreases below a certain temperature depending on the volume of available water mass, the occurrence of freezing by homogeneous nucleation is virtually sure.

The results of laboratory experiments on the freezing of very pure water droplets, which were probably nucleated homogeneously, evidenced that the median freezing temperature increases on average from around $-41\,°C$ for droplets having equivalent diameters of $\sim 1\,\mu m$ to around $-35\,°C$ for droplets having diameters close to 10^2 μm (Wallace and Hobbs, 2001). Therefore, homogeneous nucleation of freezing generally occurs in the atmosphere only within high-altitude clouds. If a droplet contains a special particle type acting as "freezing nucleus," it may freeze by heterogeneous nucleation. In this process, water molecules are collected onto the surface of the particle to form an icelike structure, which may increase in size causing the droplet to freeze. Because the formation of the ice structure is aided by the freezing nucleus, and the ice embryo also starts off with the sizes of the freezing nucleus, heterogeneous nucleation can occur at much higher temperatures than those observed in the homogeneous nucleation process.

Laboratory experiments conducted on the heterogeneous freezing of droplets consisting of distilled water showed that most (but not all) of the foreign particles were removed. A large number of droplets having diameters ranging from less than 60 μm to more than 2 cm were cooled, for temperatures at which half of the droplets had frozen. The results indicate that the median freezing temperature increases as the size of the droplet increases, showing regular features while passing from a median freezing temperature of about −33 °C measured for a drop diameter equal to 46 μm to a temperature of −17 °C for drop diameters close to 1 cm. It was assumed in the preceding text that the particle which initiates the freezing is contained within the droplet. However, cloud droplets may also be frozen if a suitable particle in the air comes into contact with the droplet. In this case, freezing is said to occur by contact nucleation, and the particle is referred to as a *contact nucleus*. Laboratory experiments suggest that some particles can cause a drop to freeze by contact nucleation at temperatures higher by several degrees more easily than in cases where they are embedded in the drop. Certain particles in the air also serve as centers upon which ice can form directly from the vapor phase. In these cases, they are referred to as *deposition nuclei*. Ice can form by deposition provided that the air is supersaturated with respect to ice and the temperature is low enough. If the air is supersaturated with respect to water, a suitable particle may serve either as a freezing nucleus or a deposition nucleus: in the first case, liquid water first condenses onto the particle and subsequently freezes, while in the latter case there is no intermediate liquid phase, at least on the macroscopic scale.

An ice nucleating particle is in general called *ice nucleus*, without specifying its particular icing effects. The temperature at which a particle can cause ice formation generally depends on the mechanism by which the particle nucleates the ice as well as on the specific structural characteristics of particulate matter. Particles with molecular spacings and crystallographic arrangements similar to those of ice (which has a hexagonal structure) tend to be very effective as ice nuclei, although this is neither a necessary nor a sufficient condition for having a good ice nucleus. Most effective ice nuclei are virtually insoluble in water. Some inorganic soil particles (mainly clays) can nucleate ice at temperatures higher than −15 °C and probably play an important role in nucleating ice in clouds. For instance, 87% of the snow crystals collected on the ground have been found to contain clay mineral particles at their centers, and more than half of these contained kaolinite. Many

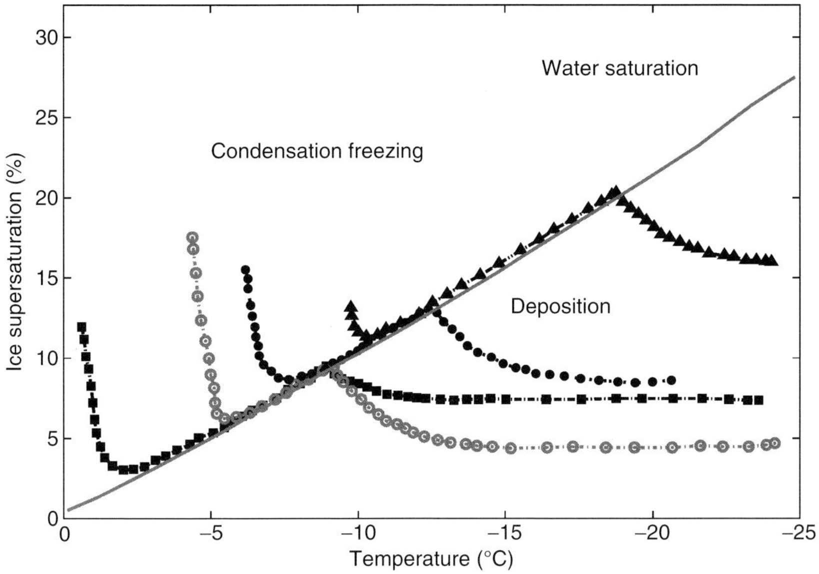

Figure 3.12 Ice supersaturation as a function of air temperature for water saturation (gray thick curve) and for condensation-freezing and ice deposition conditions. Ice nucleation starts above the indicated lines, obtained for methaldehyde (solid squares), silver iodide (gray circles), lead iodide (solid circles), and kaolinite (solid triangles). (Adapted from graphs of Schaller and Fukuta (1979) and Wallace and Hobbs (2001).)

organic materials are effective ice nucleators. Ice nuclei active at −4 °C were found in seawater rich in plankton. The results of laboratory measurements on condensation freezing and deposition shown in Figure 3.12 for different materials (such as silver iodide, lead iodide, methaldehyde, and kaolinite) indicate that the onset of ice nucleation occurs at higher temperatures under water-supersaturated conditions (for which condensation freezing is possible) than under water-subsaturated conditions (for which only ice deposition occurs). For instance, kaolinite serves as an ice nucleus at −10.5 °C for water saturation, but at 17% supersaturation with respect to ice and at subsaturation with respect to water: a kaolinite particle needs to have a temperature equal to about −20 °C in order to act as an ice nucleus. In some cases, a particle first served as an ice nucleus, all of the visible ice being evaporated from it, for a temperature no higher than −5 °C and RH < 35% with respect to ice, and subsequently, it served as an ice nucleus at a temperature that is a few degrees higher than that recorded during the initial phase. This two-stage process is referred to as *preactivation*. Thus, ice crystals from upper-level clouds which evaporate before reaching the ground may leave behind preactivated ice nuclei.

Worldwide measurements of ice nucleus concentration C_{in} (giving the number of ice nuclei per liter) made as a function of temperature indicate that this quantity tends to be higher in the Northern than in the Southern Hemisphere. Wallace and Hobbs (2001) showed that C_{in} determined by means of millipore filter measurements increases on average from a median value of around 10 m^{-3} at temperature $T = -10\,°C$ to values ranging from 5×10^2 for $T = -15\,°C$ to more than $8 \times 10^2\,m^{-3}$ for $T = -21\,°C$. Mixing chamber measurements provided values of $C_{in} > 10^4\,m^{-3}$ at $T = -20\,°C$ in the Northern Hemisphere and increasing on average from $\sim 10^3\,m^{-3}$ for $T = -21\,°C$ to around $10^4\,m^{-3}$ in the Southern Hemisphere for T nearly equal to $-25\,°C$. In the Antarctic atmosphere, C_{in} was found to generally increase from $\sim 3 \times 10^2\,m^{-3}$ at $T = -17\,°C$ to more than $5 \times 10^4\,m^{-3}$ at $T = -25\,°C$. Expansion chamber measurements yielded values of C_{in} increasing from $\sim 2 \times 10^3\,m^{-3}$ at $T = -20\,°C$ to about $10^4\,m^{-3}$ at $T = -23\,°C$ in the Northern Hemisphere and increasing from $\sim 6 \times 10^2\,m^{-3}$ at $T = -20\,°C$ to $\sim 4 \times 10^3\,m^{-3}$ at $T = -23\,°C$ in the Southern Hemisphere. However, concentration C_{in} can sometimes vary by several orders of magnitude within a few hours. On average, C_{in} at temperature T tends to follow an empirical relationship defining the natural logarithm of C_{in} as proportional to the difference between a temperature T_{in} (typically close to $-20\,°C$), at which one ice nucleus per liter is active, and air temperature T, presenting a proportionality coefficient k_{in} varying in general from about 0.3 to 0.8. For $k_{in} = 0.6$, C_{in} is estimated to increase by about 10 times for every decrease of $4\,°C$ in temperature T. In urban air, the total concentration of aerosol is equal to around $10^8\,l^{-1}$, and, therefore, only about one particle every 10^8 aerosols acts as an ice nucleus at $T = -20\,°C$.

Atmospheric ice crystals exhibit a variety of shapes, in which hexagonal prismatic features are most frequently observed (Pruppacher and Klett, 1997). Considering the Kelvin effect theory for water droplets, the case of a spherical ice particle with diameter a can be examined, finding that the vapor pressure p_i of water over the ice particle surface is given by

$$p_i = p_{sat,i} \exp(4\mu_{wat}\gamma_{ice}/RT\rho_i a) \tag{3.52}$$

where $p_{sat,i}$ is the vapor pressure of water over a flat ice surface, γ_{ice} the surface energy of ice in air, and ρ_i the ice density. On this matter, Pruppacher and Klett (1997) reviewed theoretical and experimental values of surface tension γ_{ice} finding values of this parameter ranging most frequently from 0.10 to $0.11\,N\,m^{-1}$. Being higher than the corresponding water/air value, the surface tension γ_{ice} results in a relative vapor increase (p/p_{sat}) greater than that determined for a water droplet having the same size. The freezing temperature decreases as the ice particle sizes decrease. This behavior was observed to be particularly pronounced for crystals having sizes $a < 20\,nm$, being further enhanced by the solute effect for solute concentrations higher than 0.1 per mole of solute.

Ice particles can be formed through different mechanisms requiring the presence of a particle acting as ice nucleus: (i) water vapor adsorption onto the ice nucleus surface and transformation to ice (deposition mode), (ii) transformation of a supercooled droplet to become an ice particle (freezing mode), and (iii) collision of a supercooled droplet with an ice nucleus to form an ice particle (contact mode). The formation of ice particles without the disposability of ice nuclei is possible only at very low temperatures $T < -40\,°C$ (Hobbs, 1995). The presence of ice nuclei allows ice formation at higher temperatures. Aerosols serving as ice nuclei are rather different from those acting as CCNs. Ice-forming nuclei are usually insoluble in water and have chemical bonding and crystallographic structures similar to those of ice. Dust particles (especially clay particles such as kaolinite) and combustion particles (containing metal oxides) were often found to serve as ice nuclei.

Wallace and Hobbs (2001) pointed out that concentration C_{in} is quite variable in the atmosphere. An empirical relation proposed by Pruppacher and Klett (1997) to define the quantity C_{in} per liter of air as a function of air temperature T states that on average,

$$C_{in} = \exp[0.6(253 - T)], \tag{3.53}$$

where T is measured in degree Celsius. This relationship suggests that ice nucleation in the atmosphere is a very selective process. For example, at temperature $T = -20\,°C$, the atmosphere typically contains one ice nucleus per liter of air and 10^6 particles per liter, such a ratio indicating that only one out of a million particles can serve as ice nucleus on average. Measurements of ice particle concentrations for cloud top temperatures below $-10\,°C$ were found to provide concentrations varying from 0.1 to 2×10^2 per liter, while the concentrations vary from 10 to 300 per liter for $T < -20\,°C$. These ice crystal number concentrations observed in real cases often exceed by several orders of magnitude the usual value of C_{in}. Proposed explanations for such an enhancement of ice particle concentration include breakup of primary ice particles, ice sprinter production during droplet freezing, and unusually high supersaturation conditions.

3.4 Formation of Cloud Particles

Clouds form by condensation of water vapor onto airborne aerosols when RH exceeds the saturation level. Air becomes commonly supersaturated through the uplift of moist air accompanied by adiabatic expansion and cooling, as it usually occurs for local turbulent convention in frontal systems, in large cyclonic areas, or along the slopes of a mountain. Condensation of liquid droplets from water vapor molecules in the absence of other external nuclei is called *homogeneous nucleation*. This process requires supersaturation levels of several hundred percent that are not attained in the real atmosphere. Most frequently, aerosol particles act as CCNs to enable cloud formation at the supersaturation levels commonly observed in the atmosphere for values of RH differing by less than 1% from the saturation level. The process of cloud droplet formation onto CCNs is called *heterogeneous nucleation*. The water vapor condensation on aerosol particles is described by the Köhler (1936) theory, which takes into account the effects due to the curved surface of the cloud droplet (Kelvin effect) as well as the partial vapor pressure characteristics of each chemical component present in an ideal solution. These aspects are discussed in the following three subsections, the first dealing with the nucleation of water vapor condensation in clouds, the second presenting the main concepts of the Köhler (1936) theory and its applications to real cases, and the third describing the role of CCNs in the cloud formation process.

3.4.1 Water Vapor Condensation

In the homogeneous nucleation, a pure water droplet is formed by condensation from a supersaturated vapor in perfectly clean air and therefore without the aid of airborne particles. Its first stage in water vapor nucleation is given by the chance collisions of a number of water molecules in the vapor phase to form small embryonic water droplets having sizes that are large enough to remain intact (without effects due to other particles). Considering the case of a small water droplet of volume $V = \pi a^3/6$ and surface area $S = \pi a^2/4$, which originates from condensation of pure supersaturated water vapor at constant temperature and pressure, we can perform the same analysis made in the previous Section 3.2.2 in terms of Eqs. (3.14)–(3.19) also in the present case regarding water molecules only. We obtain here that the net increase ΔG_w in the Gibbs free energy reduction of the water droplet system with diameter a is given by

$$\Delta G_w = \pi a^2 \gamma_S - n\frac{\pi a^3}{6}(G_{\text{vap}} - G_{\text{liq}}) = \pi a^2 \gamma_S - \frac{\pi a^3}{6} n k_B T \ln(e_w/e_s), \qquad (3.54)$$

where (i) γ_S is the surface tension of the liquid water (i.e., the work required to create a unit area of vapor–liquid interface, also called *interfacial energy* between water vapor and liquid water); (ii) n is the number of water molecules per unit volume of liquid water; (iii) G_{vap} and G_{liq} are the Gibbs free energies per molecule in the vapor and liquid phases, respectively; (iv) e_w and e_s are the partial pressure of water vapor and the saturation vapor pressure over a plane surface of liquid water at absolute temperature T, respectively; and (v) k_B is the Boltzmann constant.

For subsaturated conditions, we have $e_w < e_s$ in Eq. (3.54). Therefore, $\ln(e_w/e_s)$ is negative, and ΔG_w is always positive, since it increases as a increases and exhibits features similar to those of the dashed curve shown in Figure 3.7 for the case with saturation ratio $S_{\text{sat}} < 1$. This means that the greater are the sizes of the embryonic water droplet forming in a subsaturated vapor and the greater is the increase in the energy ΔG_w of the system. Because this system approaches an equilibrium state by reducing its energy, the formation of water droplets is clearly not favored under subsaturated conditions. Due to random collisions of water molecules, very small embryonic droplets continually form (and evaporate) in such a subsaturated vapor.

For supersaturated conditions, we have $e_w > e_s$ in Eq. (3.54), and therefore $\ln(e_w/e_s)$ is positive. In this case, ΔE assumes either positive or negative values, depending on diameter a. The variation of ΔE as a function of a for $e_w > e_s$ and, hence, for $S_{\text{sat}} > 1$ is also shown in the example of Figure 3.7, which indicates that ΔG_w initially increases as a increases, reaches a maximum value $\Delta G_w = \Delta G_w^*$ at diameter a^*, and then decreases as a continues to increase. This means that embryonic water droplets with $a < a^*$ tend to evaporate under supersaturated conditions, causing a consequent decrease in ΔG_w. However, droplets growing by chance collisions are predicted to continue to grow spontaneously by condensation from the vapor phase, since this will produce a decrease in ΔG_w. When $a = a^*$ is reached, a droplet can grow or evaporate infinitesimally without any change in ΔG_w. The following expression of diameter a^* can be obtained as a function of e_w by assuming a null value of the derivative $d(\Delta G_w)/dr$ at $a = a^*$:

$$a^* = \frac{4\gamma_S}{nk_B T \ln(e_w/e_s)}. \tag{3.55}$$

Equation (3.55) is generally referred to as "Kelvin equation for water vapor condensation." It can be used for two specific purposes: (i) to calculate the diameter a^* of a droplet which is in unstable equilibrium with a given water vapor pressure e_w, in the sense that if the droplet begins to grow by condensation, it will continue to do so and if the droplet begins to evaporate, it will continue to evaporate, and (ii) to determine the saturation vapor pressure e_w over a droplet of diameter a^*. The RH value at which a droplet of diameter a^* is in unstable equilibrium is equal to $100\times$ (e_w/e_s), where ratio e_w/e_s can be easily obtained by inversion of Eq. (3.55). The gradual decrease of supersaturation RH from about 112% to 100% was shown by Wallace and Hobbs (2001) as a function of droplet diameter over its range from 2×10^{-2} to 2 μm, estimating that (i) a pure water droplet of diameter $a = 2 \times 10^{-2}$ μm requires a value of RH equal to ~112% (i.e., a supersaturation of ~12%) to be in unstable equilibrium with its environment, (ii) RH rapidly decreases as a increases until reaching a value of around 101% for $a = 0.2$ μm, and (iii) a droplet with $a = 2$ μm requires an RH = 100.12% only (i.e., a supersaturation of 0.12%).

3.4.2
The Köhler Theory

The supersaturation conditions observed in natural clouds and formed as a result of the adiabatic ascent of air rarely exceed a few percent. From the earlier remarks, it can be argued that embryonic droplets of pure water having sizes $a \approx 2 \times 10^{-2}$ μm and formed by chance collisions of water molecules would be well below the critical diameter required for their survival in air. Consequently, we can state that droplets do not form in natural clouds by homogeneous nucleation of pure water, but instead they form on atmospheric aerosols through heterogeneous nucleation. Considering that the atmosphere contains aerosol particles ranging in size from less than 1 μm to several tens of micrometer, it is realistic to assume that (i) only the hygroscopic particles among them can serve as centers upon which water vapor condenses and (ii) droplets can form and grow on these particles at much lower supersaturations than those required for homogeneous nucleation. For example, if sufficient water condenses onto a completely wettable particle with $a = 0.6$ μm to form a thin film of water over the particle surface, the water film will be in unstable equilibrium with air having a supersaturation of 0.4%. In cases where the supersaturation conditions should be slightly greater than 0.4%, water would condense onto the water film and the droplet would increase in size. Since some of the particles suspended in air are soluble in water, they dissolve wholly or in part when water condenses onto them, so that a solution droplet turns out to be formed (but not a pure water droplet).

The saturation vapor pressure of water adjacent to a solution droplet (for instance, a water droplet containing some dissolved material, such as sodium chloride or ammonium sulfate) is lower than that adjacent to a pure water droplet having the same sizes. The fractional reduction in the water vapor pressure is given by Raoult's law:

$$e'/e = F_{\mathrm{mol}}, \tag{3.56}$$

where (i) e' is the saturation vapor pressure of water adjacent to a solution droplet containing a mole fraction F_{mol} of pure water, (ii) e is the saturation vapor pressure of water adjacent to a pure water droplet of the same size and at the same temperature, and (iii) the mole fraction F_{mol} of pure water is defined as the number of moles of pure water in the solution divided by the total number of moles in the solution of pure water and dissolved material.

If we consider a solution droplet of diameter a, containing a mass m of a dissolved material of molecular weight μ_{sol}, and take into account that each molecule of the material dissociates into n_i ions in water, the effective number of moles of the material in the droplet is equal to the ratio between $1000 n_i m$ and μ_{sol}. If the density of the solution is ρ_{sol} and the molecular weight of water is μ_{wat}, the number of moles of pure water in the droplet is equal to $10^3 \left(\frac{\pi a^3}{6}\rho_{\mathrm{sol}} - m\right)/\mu_{\mathrm{wat}}$. Therefore, the mole fraction of pure water in the droplet is given by

$$F_{\mathrm{mol}} = \frac{\left(\frac{\pi a^3}{6}\rho_{\mathrm{sol}} - m\right)/\mu_{\mathrm{wat}}}{\left[\left(\frac{\pi a^3}{6}\rho_{\mathrm{sol}} - m\right)/\mu_{\mathrm{wat}}\right] + n_i(m/\mu_{\mathrm{sol}})}. \tag{3.57}$$

Combining Eqs. (55)–(57) and replacing γ_S and n used in Eq. (3.55) by $\gamma_S{}'$ and n' to indicate the surface energy and number concentration of water molecules, respectively, the following expression for the saturation vapor pressure e' adjacent to a solution droplet of diameter a can be obtained from Eq. (3.57):

$$e'/e_s = \frac{\exp\left(\frac{4\gamma_S{}'}{n' k_B T a}\right)}{\left[1 + \frac{n_i m \mu_{\mathrm{wat}}}{\mu_{\mathrm{sol}}\left(\frac{\pi a^3}{6}\rho_{\mathrm{sol}} - m\right)}\right]}. \tag{3.58}$$

This equation has been derived following the Köhler (1936) theory and can be correctly used to describe the process in which water vapor condenses to form liquid cloud droplets. It is based on the combination of two expressions: (i) the first defines the Kelvin effect, which regulates the increase in saturation vapor pressure over a curved surface, and (ii) the second is based on the thermodynamic concepts of Raoult's law, which relates the mole fraction of water molecules to the water vapor pressure. The latter describes the water equilibrium over a flat

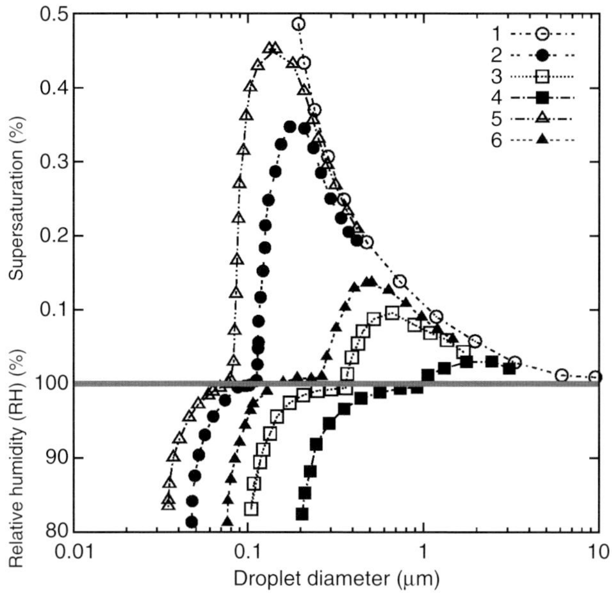

Figure 3.13 Variations of RH (%) and supersaturation (%) as a function of droplet diameter *a*, as measured in the adjacent surroundings of droplets consisting of (i) pure water and adjacent to solution droplets containing the following fixed masses of salt (solid curve with open circles), (ii) 10^{-19} kg of NaCl (dashed curve with solid circles), (iii) 10^{-18} kg of NaCl (dotted curve with open squares), (iv) 10^{-17} kg of NaCl (dotted curve with solid squares), (v) 10^{-19} kg of $(NH_4)_2SO_4$ (dashed and dotted curve with open triangles), and (vi) 10^{-18} kg of $(NH_4)_2SO_4$ (dashed and dotted curve with solid triangles). Note the discontinuity in the ordinate at RH = 100%. (Adapted from a graph of Pruppacher (1973).)

solution, establishing that the total vapor pressure of the ideal solution depends on the vapor pressure of each chemical component and the mole fraction of the component present in the solution. Therefore, Eq. (3.58) regulates the growth of cloud droplets under supersaturation conditions and can be reliably used to calculate the saturation vapor pressure e' (for RH $= 100 \times (e'/e_s)$ or the supersaturation equal to $100 \times [(e'/e_s) - 1]$), adjacent to a solution droplet with a specified diameter a. The corresponding Köhler curve can be obtained by plotting the variation in RH for supersaturation conditions adjacent to a solution droplet as a function of its diameter, according to Eq. (3.58), as made in the six examples shown in Figure 3.13.

Below a certain droplet diameter, the value of RH adjacent to a solution droplet is lower than that evaluated for the equilibrium conditions over a plane surface of pure water at the same temperature (i.e., 100%). As the droplet increases, the solution becomes weaker, the Kelvin curvature effect becomes gradually more intense, and the value of RH adjacent to the droplet becomes in practice the same measured adjacent to a pure water droplet of the same size. To better understand the physical meaning of the Köhler curves, it is useful to examine more carefully

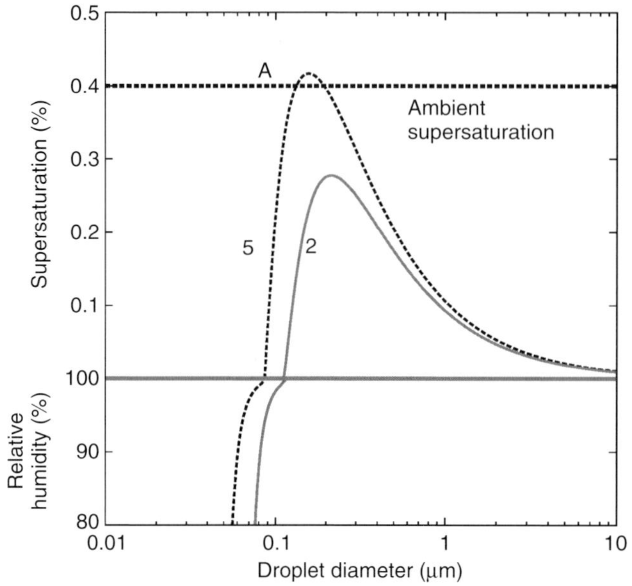

Figure 3.14 Köhler curves describing the variations obtained for two of the cases shown in Figure 13: curve 2 (gray solid curve), which refers to a solution droplet containing 10^{-19} kg of NaCl, and curve 5 (dashed black curve), which refers to a solution droplet containing 10^{-19} kg of $(NH_4)_2SO_4$. The horizontal gray line refers to RH = 100%. The horizontal dotted line refers to the ambient supersaturation value of case A discussed in the text. (Adapted from a graph of Wallace and Hobbs (2001).)

the two Köhler curves represented in Figure 3.13, for solution droplets containing 10^{-19} kg of NaCl (curve 2) and 10^{-19} kg of $(NH_4)_2SO_4$ (curve 5), respectively:

1. In the former case, if one supposes that a particle of NaCl with mass 10^{-19} kg is placed in air presenting a water supersaturation of 0.4%, it can be seen that the solution droplet grows when condensation occurs on this particle and follows the growth patterns of curve 2 shown in Figure 3.14. Thus, the supersaturation adjacent to the surface of this solution droplet will initially increase, but the supersaturation adjacent to the droplet remains below the ambient supersaturation level even at the peak of its Köhler curve, which is lower than ambient supersaturation level as shown in Figure 3.14. Consequently, the droplet will grow over the peak in its Köhler curve and down over the right-hand side of this curve to form a fog or cloud droplet. A droplet which has passed over the peak in its Köhler curve and continues to grow is said to be "activated."
2. In the latter case, a particle of $(NH_4)_2SO_4$ with mass 10^{-19} kg is considered and placed in the same ambient supersaturation of 0.4%. Condensation will occur on the particle also in this case, and the particle will grow as a solution droplet along the Köhler curve 5 until reaching point A, in which the supersaturation adjacent to the droplet is equal to the ambient supersaturation. If

the droplet located at point A should grow slightly, the supersaturation adjacent to it would increase above the ambient supersaturation, and therefore the droplet would evaporate back to point A. If the droplet at point A should evaporate slightly, the supersaturation adjacent to it would decrease below the ambient supersaturation, and the droplet would grow by condensation back to point A. Therefore, the solution droplet at point A is in stable equilibrium with the ambient supersaturation: should the ambient supersaturation change by a very small increment, point A would correspondingly shift and the equilibrium size of the droplet would change accordingly. Droplets in this state are said to be "inactivated" and commonly called "haze droplets," as mentioned in the preceding text.

To better describe the different behaviors of the solution droplets of NaCl and $(NH_4)_2SO_4$ for some discrete values of dry diameter a_0, a set of Köhler curves is shown in Figure 3.15 over the wet diameter range from 0.1 to 10^2 μm. It can be seen in Figure 3.15 that all the six curves pass through a maximum, and all the maxima are located at the critical droplet diameter a_{max}, which corresponds to a critical saturation value S_c reached by each solution droplet. The saturation value

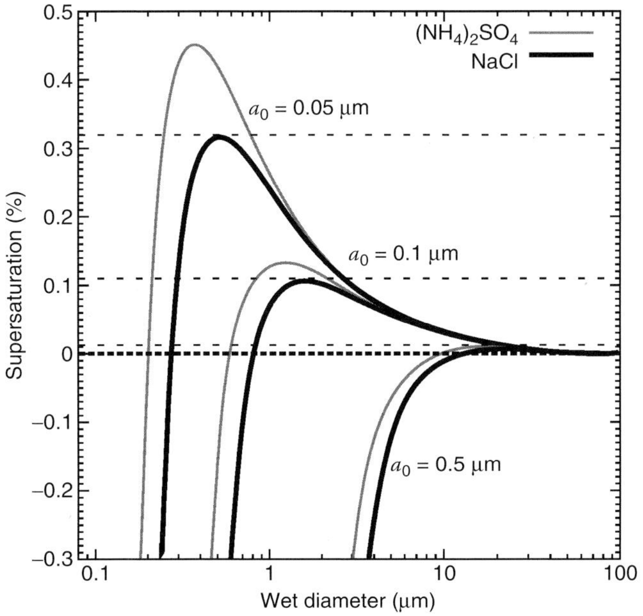

Figure 3.15 Köhler curves drawn for the $(NH_4)_2SO_4$ solution (gray curves) and NaCl solution (solid curves) and particles having three different values of the dry initial diameter $a_o = 5 \times 10^{-2}$ μm, $a_o = 10^{-1}$ μm, and $a_o = 5 \times 10^{-1}$ μm at air absolute temperature $T = 293$ K. The supersaturation is defined in ordinate as the saturation minus 1: for instance, supersaturation = 1% corresponds to an RH = 101%. The corresponding short-dashed horizontal lines indicate the three values of critical saturation S_c for NaCl. (Adapted from a graph of Seinfeld and Pandis (2006).)

then slowly decreases as diameter increases beyond a_{max} until assuming sizes typical of cloud droplets. Each Köhler curve is a visual representation of the Köhler equation. It individuates the supersaturation at which the cloud drop is in equilibrium with the environment over a certain droplet size range, the shapes of these curves depending closely on the amount and composition of the solution droplets.

It was shown in the previous section that a pure water droplet cannot remain in stable equilibrium with its surroundings during a long time interval, since a small perturbation of either the droplet itself or its surroundings causes a spontaneous droplet growth or a shrinkage. When considering a droplet lying on the portion of the Köhler curve for which $a < a_{max}$, it can be assumed that the atmospheric ambient saturation is equal to S_{amb}. Thus, a droplet will constantly experience small perturbations caused by the gain or loss of a few molecules of water. If the droplet grows slightly with the addition of a few molecules of water, its equilibrium vapor pressure reached for a momentary larger size is greater than the fixed ambient value, and the droplet will lose water by evaporation, eventually returning to its original equilibrium state. The same evolutionary patterns occur if the droplet loses a few molecules of water, since its equilibrium vapor pressure decreases becoming lower than the ambient pressure, and water will condense on the droplet which returns to its original size. This indicates that droplets located in the rising part of the Köhler curve are in stable equilibrium with their environment.

If a droplet is considered to be located on the portion of the curve for which $a > a_{max}$, a slight perturbation occurs, associated with its growth by a few molecules of water, and a slightly larger size is reached by it for equilibrium vapor pressure is lower than the ambient one. Thus water molecules will continue to condense on the droplet, which becomes even larger in size. Conversely, a slight shrinkage leads to having a droplet with a higher equilibrium vapor pressure than the ambient, which therefore continues to evaporate: (i) in cases where the droplet consists of pure water, it will evaporate completely, and (ii) in cases where the droplet contains a solute, it will diminish in size until intersecting the ascending branch of the Köhler curve corresponding to stable equilibrium conditions. The descending branches of the curves describe the sequences of unstable equilibrium states.

If the ambient saturation ratio S_{amb} is lower than the critical saturation S_{cri} for a given particle, the particle will be in equilibrium following the ascending part of the Köhler curve. If the condition $1 < S_{amb} < S_{cri}$ is observed, there are two equilibrium states and therefore two different diameters that correspond to S_{amb}, of which one pertains to a stable state and the other to an unstable state. The particle can reach stable equilibrium only at the state corresponding to the smaller diameter. If the ambient saturation ratio S_{amb} exceeds the particle critical saturation S_{cri}, there is no feasible equilibrium size for the particle. For any particle diameter, ambient saturation S_{amb} will exceed the equilibrium saturation at the particle surface, and the particle will grow indefinitely. In such a way, a droplet can grow to a size much larger than the original size of the dry particle. Through this process, particles having small diameters $\approx 10^{-2}$ μm can grow one billion times in mass to become cloud or fog droplets having diameters $a > 10\,\mu m$, those of the cloud droplets being all greater than their critical value a_{max}.

3.4.3
The Cloud Condensation Nuclei

As clearly shown in Figure 3.2, atmospheric aerosols have sizes ranging from a few tens of angstroms ($1\,Å = 10^{-8}$ cm) to more than $10^2\,\mu m$ (i.e., $>10^{-2}$ cm). Aitken nuclei and transient nuclei have sizes smaller than $5 \times 10^{-2}\,\mu m$, while accumulation particle sizes vary from less than $10^{-1}\,\mu m$ to $2.5\,\mu m$, and coarse particle sizes are greater than $2.5\,\mu m$. Atmospheric aerosols belonging to the first subset serve as particles upon which water vapor condenses to form droplets, which are activated and grow further by condensation to form cloud droplets at supersaturations of about 0.1 – 1% within clouds. These particles are called CCNs. As stated earlier, a particle is more readily wetted by water when it has large sizes. The supersaturation at which the particle can serve as a CCN is lower when its solubility is higher. For example, to serve as a CCN at 1% supersaturation, completely wettable water-insoluble particles need to have diameters no greater than $0.2\,\mu m$, and soluble particles can serve as CCNs at 1% supersaturation even if they have diameters $a < 2 \times 10^{-2}\,\mu m$. The CCNs consist usually of a mixture of soluble and insoluble components and constitute in this case the so-called mixed nuclei. The capability of a particle to serve as a nucleus for a cloud droplet formation does not only depend on its size but also on its chemical composition and local supersaturation conditions. Therefore, the number of particles acting as CCNs inside a certain aerosol population varies as a function of water supersaturation. For a given mass of soluble material in the particle, there is a particular value S_{amb} of the ambient water vapor supersaturation, below which the particle exists in a stable state and above which it spontaneously grows to become a cloud droplet having a diameter equal to at least $10\,\mu m$. For marine stratiform clouds, for which supersaturations are in the range of 0.1 – 0.5%, the minimum diameter of a CCN ranges usually from 5×10^{-2} to $1.4 \times 10^{-1}\,\mu m$. On average, an air parcel will spend a few hours in a cloud followed by a few days outside clouds. Thus, the average lifetime of a CCN is about 1 week, during which a CCN is subject to experience on average 5 – 10 cloud activation/evaporation cycles before being removed from the atmosphere by precipitations.

The CCN number concentration varies from less than $10^2\,cm^{-3}$ in remote marine regions to many thousands per cubic centimeter of air in polluted urban areas, as can be seen in Table 3.4, showing that higher concentrations of CCNs are measured near the Earth's surface more frequently in continental air masses than in marine environments. The ratio between the CCN concentration at 1% supersaturation and the total particle number concentration is on average equal to around 0.4 ± 0.2 in marine air and lower than 0.01 in continental air, although sometimes rising to ~0.10. Such very low values of this ratio measured in continental air can be plausibly attributed to the very high number density of very small particles, which are not activated at low supersaturations. Concentration

Table 3.4 Average measurements of CCN number concentration in various aerosol air masses above different regions of the Earth.

Sampling area	CCN concentration (cm^{-3})
Florida (the United States) (offshore marine air mass)	10^2 to 2×10^2
Azores (Portugal) (marine air mass)	3×10^2 to 10^3
Arctic (clean air mass)	3×10^1 to 4×10^1
North Atlantic Ocean (marine air mass)	10^2 to 4×10^2
Equatorial Pacific Ocean (marine air mass)	5×10^1 to 2×10^2
Remote North Pacific Ocean (marine air mass)	2.5×10^1 to 1.3×10^2
Polluted North Pacific Ocean (marine air mass)	3×10^2 to 4×10^2
North America (continental air mass)	10^2 to 3.5×10^3
Australia (continental air mass)	5×10^1 to 2×10^3
Global scale (average)	6×10^2

of CCNs over land declines by about five times on average when passing from the planetary boundary layer to the upper free troposphere, while those over the oceanic regions remain fairly constant or sometimes increase with height until reaching the maximum concentration just above the mean cloud level.

A considerably higher production rate of CCNs is given by burning smoke particles originated from the combustion of vegetation canopies, which is estimated over a global scale to vary from 10^{12} to 10^{15} particles per kilogram of burned biomass material. Conversely, oceanic particles do not provide important concentrations of CCNs; the most part of CCNs consisting of marine sulfates mainly originated from gaseous DMS and methane sulfonic acid (MSA). For this reason, clouds formed over the oceans can release small sulfate particles. An intense source of CCNs is the g-to-p conversion, producing particles with sizes up to 0.1–0.2 μm, which can act as CCNs when they are soluble or wettable. Gas-to-particle conversion mechanisms associated with solar radiation reaching the ground are estimated to produce CCNs during the day, giving rise to a maximum in CCN concentrations in the late afternoon, about 1–3 h before sunset.

3.5 Concluding Remarks

The present study highlights that the aerosol growth processes mainly occur through coagulation and/or condensation in the cloudless atmosphere, leading to cloud droplet formation within clouds. They concur to create the most favorable conditions for the subsequent removal processes caused by "dry deposition" of aerosols by gravitational settling or "wet deposition" through incorporation of particles into cloud droplets during the precipitation formation phase. These processes establish in practice the limits of the aerosol life in the atmosphere. It is well known that the aerosol size distribution exhibits in general multimodal features over the size range from less than 10^{-3} μm to more than 50 μm, with typical size ranges occupied by the various particle modes, as shown in Figure 3.9. Since the geographic distribution of particle sources is highly irregular, the tropospheric aerosols vary widely in concentration and composition from one region to another of our planet, often being mixtures of particles from different natural and anthropogenic sources. Whereas atmospheric trace gases exhibit values of atmospheric lifetime Δt_L ranging from less than 1 s to more than 10^2 s, the lifetime of tropospheric particles varies on average from a few days to a pair of weeks. In general, particles with sizes $<10^{-2}$ μm have lifetimes $\Delta t_L < 1$ day, since the major removal mechanisms for aerosol particles in this size range are diffusion toward cloud particles and coagulation. Particles with sizes $a > 20$ μm have often lifetimes shorter than 1 day, since they are efficiently removed by sedimentation, impaction onto surfaces, and precipitation scavenging. In contrast, particles having sizes ranging from about 0.1 to 2 μm are removed from the atmosphere through very weak sink processes. Actually, these accumulation particles are originated by very effective mechanisms, such as coagulation of Aitken nuclei and condensation, or are left behind by the evaporation of cloud droplets. Consequently, they have relatively long lifetimes, reaching several hundred days in the upper troposphere but not exceeding 10–15 days in the middle and lower troposphere, due to the strong efficiencies of precipitation scavenging and impaction effects. For this reason, the accumulation particles exhibit in general values of lifetime Δt_L varying from a few days to 1–2 weeks, while coarse particles have lifetimes Δt_L ranging from a few hours to a few days. The average values of Δt_L are given in Table 3.5 for the most important aerosol types, together with the ranges of the global mass burden for natural, anthropogenic, and overall aerosols, and the average estimates of global mass burden determined by Pósfai and Buseck (2010).

The sizes, number density, and chemical composition of airborne aerosol particles are strongly altered by several physical mechanisms and solution processes until they are ultimately removed by natural scavenging and dry and wet deposition. The most important of such removal processes are gravitational settling and dry deposition on surfaces in the lower atmosphere, while precipitation scavenging acts more efficiently at altitudes higher than a few hundred meters. Some of the physical and chemical processes causing the aging of atmospheric particles are more effective in one regime of particle size than in another. As air rises through a cloud and becomes slightly supersaturated with water vapor (i.e., with RH exceeding 100%), cloud droplets form on CCNs (mainly constituted by soluble airborne aerosol particles having number density concentrations varying from 10^2 to 3×10^3 cm^{-3}) and grow by condensation of water vapor. As the droplets grow and collide with each other, they become raindrops assuming sizes which increase during their fall by accreting small cloud particles. Due to these mechanisms, the atmospheric content of each particle polydispersion can be seen as the result of a steady-state equilibrium between emissions and removal mechanisms. For instance, considering all the sulfur-containing compounds in the troposphere, it can be stated that the total mass of these substances in the troposphere is equal

Table 3.5 Present evaluations of the mean atmospheric lifetime (measured in days) of aerosols originated from different sources and corresponding estimates of the ranges and average values of global mass burden (Tg) determined by Pósfai and Buseck (2010) for the various aerosol types.

Aerosol types	Mean atmospheric lifetime Δt_L (days)	Global mass burden ranges (Tg)			Average global mass burden (Tg) (Pósfai and Buseck, 2010)
		Natural	Anthropogenic	Overall	
Sea salt	0.5 ± 0.4	4.1–27.4	-	4.1–27.4	15
Mineral dust	4 ± 2	11.0–23.6	0.5–2.7	11.5–26.7	18 ± 5[a)]
Biogenic primary organic	2 ± 1	0.09–0.38	—	0.09–0.38	0.2
Biomass burning organic	3 ± 2	0.21–0.58	0.10–2.22	0.31–2.8	—
Tropospheric volcanic	3 ± 2	0.03–0.74	—	0.03–0.74	—
Stratospheric volcanic	910 ± 180	10–25000	—	10–25000	—
Sulfates	4 ± 2	1.2–4.1	0.5–1.3	1.7–5.4	2.8
Nitrates	9 ± 3	0.30–0.67	—	0.30–0.70	0.49
Industrial dust	4 ± 2	—	0.44–1.42	0.44–1.42	1.1
Biogenic secondary organic	10 ± 5	0.08–2.27	—	0.08–2.3	0.8
Black carbon (soot)	3 ± 2	—	0.07–0.12	0.07–0.12	0.1
Overall biological from isoprene, monoterpenes, and VOCs	3 ± 2	6.9–8.2	—	6.9–8.2	—
Carbonaceous from hydrocarbons and VOCs	3 ± 2	—	0.12–0.74	0.12–0.74	—

a) Against the 8–36 Tg range determined by Zender, Miller, and Tegen (2004).

to about 4×10^9 Tg, while the total mass $M_{\text{tot-}S}$ of sulfur-containing compounds in the troposphere is equal to 4 Tg. If natural and anthropogenic sources of sulfur would be estimated to contribute with an entering global annual flux Φ_e of sulfate aerosol particles equal to ~200 Tg per year, the mean lifetime of these tropospheric particles should be given by the ratio $M_{\text{tot-}S}/\Phi_e$ yielding a value of $\Delta t_L = 7.3$ days. In the case of mineral dust, Δt_L depends closely on the particle sizes, because submicron particles can have lifetimes of several weeks and supermicron particles are quickly removed from the atmosphere due to gravitational settling. Similarly, sea-salt particles have sizes covering the range from $\sim 5 \times 10^{-2}$

Table 3.6 Evaluations of the average lifetimes for various aerosol types, as derived from the estimates obtained through different studies conducted over the past 10 years.

Aerosol types	Size range, composition, and origins	Lifetime Δt_L (days)	References
Sea-salt particles	Overall diameter range of coarse particles	0.7	Present study
	Overall diameter range of coarse particles	~0.8	Stier *et al.* (2006)
	Diameter range $a < 1$ µm	7.1	Stier *et al.* (2006)
	Diameter range from 1 to 16 µm	0.44	Stier *et al.* (2006)
	Overall diameter range	0.54	Andreae and Rosenfeld (2008)
	Overall diameter range	0.5 ± 0.4	Pósfai and Buseck (2010)
Mineral dust	Overall size spectrum	4.6	Present study
	Insoluble fraction	4.7	Stier *et al.* (2006)
	Diameter range $a < 1$ µm	10.4	Andreae and Rosenfeld (2008)
	Diameter range from 1 to 2.5 µm	9.2	Andreae and Rosenfeld (2008)
	Diameter range from 2.5 to 10 µm	2.2	Andreae and Rosenfeld (2008)
	Overall diameter range	~4	Andreae and Rosenfeld (2008)
	Overall diameter range	4.2 ± 2.7	Pósfai and Buseck (2010)
	Overall diameter range of Asian dust in the Arctic	8–16	Ménégoz *et al.* (2012)
Biogenic primary organic aerosol	Overall diameter range	2	Present study
	Diameter range $a \leq 2$ µm	4.6	Andreae and Rosenfeld (2008)
	Overall diameter range	1.7 ± 1.1	Pósfai and Buseck (2010)
Tropospheric volcanic aerosol	Overall diameter range	3	Present study
		3.5	Andreae and Rosenfeld (2008)
Mt. Pinatubo sulfuric acid droplets in the low stratosphere	Overall diameter range	2.5 ± 1.2	Tomasi and Lupi (see Chapter 1)
Sulfate tropospheric aerosols	Overall diameter range	4	Present study
	Overall diameter range	3.8–4.3	Stier *et al.* (2006)
	Overall diameter range	5.1	Stier *et al.* (2006)
	Overall diameter range of biogenic sulfates	7.7	Stier *et al.* (2006)

(*continued overleaf*)

Table 3.6 (continued)

Aerosol types	Size range, composition, and origins	Lifetime Δt_L (days)	References
	Overall diameter range of anthropogenic sulfates	4.2	Andreae and Rosenfeld (2008)
	Overall diameter range	4.2 ± 2.3	Pósfai and Buseck (2010)
	Overall diameter range of sulfates in the Arctic	4–10	Ménégoz *et al.* (2012)
Nitrate tropospheric aerosols	Overall diameter range	9	Present study
		9.9	Andreae and Rosenfeld (2008)
		9.2 ± 3.5	Pósfai and Buseck (2010)
Industrial dust	Overall diameter range	4	Present study
		4.0	Andreae and Rosenfeld (2008)
		4.7 ± 2.5	Pósfai and Buseck (2010)
Biogenic secondary organic aerosol	Overall diameter range	10	Present study
		10.4	Andreae and Rosenfeld (2008)
		6.8 ± 6.3	Pósfai and Buseck (2010)
Organic aerosols mainly consisting of isoprene, monoterpenes, and VOCs	Overall diameter range	3	Present study
		0.5–2	Stier *et al.* (2006)
		3.9	Liousse *et al.* (1996)
		4.5	Cooke *et al.* (1999)
		3.4	Cooke, Ramaswamy, and Kasibhatla (2002)
Biomass burning carbonaceous aerosols	Diameter range $a \leq 0.1$ μm	3	Present study
Fossil fuel combustion aerosols	Overall diameter range	7.9	Cooke and Wilson (1996)
		3.9	Liousse *et al.* (1996)
		5.3	Cooke *et al.* (1999)
		4.3	Cooke, Ramaswamy, and Kasibhatla (2002)
Black carbon particles	Overall diameter range	5.8–6.9	Stier *et al.* (2006)
	Diameter range $a \leq 2$ μm	3.7	Andreae and Rosenfeld (2008)
	Overall diameter range	3.3 ± 0.9.	Pósfai and Buseck (2010)
	Overall diameter range in the Arctic	8–16	Ménégoz *et al.* (2012)

to more than 10 μm and correspondingly exhibit a large range of Δt_L. Volcanic primary dust particles suspended in the troposphere have also sizes varying over a wide range, with Δt_L varying from a few hours to 1–2 weeks. Calculations of Δt_L can be useful in estimating how far from its source a species is likely to remain airborne before its removal from the atmosphere. On the basis of the Seinfeld and Pandis (2006) remarks and evaluations, it can be estimated that the atmospheric aerosol lifetime Δt_L varies from about 1 week for sulfur-containing particles to ~2 weeks for mineral dust insoluble particles. Considering the evaluations of the average annual global emission flux Φ_e available in the literature of the past 10 years and taking into account the evaluations of the prevailing mass fractions of fine and coarse particles in the particulate mass loading M_L, an overall picture of the average evaluations of Δt_L was obtained in Table 3.6. The results indicate that the aerosol lifetime can vary from 0.5 days for sea-salt particles (Pósfai and Buseck, 2010) to more than 10 days for mineral dust accumulation particles and biogenic secondary aerosols (Andreae and Rosenfeld, 2008).

Abbreviations

DMS	dimethylsulfide
LAI	leaf area index, given by the ratio of the total leaf area to the overall ground area
LWC	liquid water content of clouds and fogs
MSA	methane sulfonic acid

List of Symbols

Latin

ℓ	mean free path of a diffusing aerosol particle
a	diameter of a spherical aerosol particle (measured in μm)
a_c	mode diameter of the aerosol size-distribution curve
A_w	water activity
C_0	particle concentration at the ground level
C_1	particle concentration at the top level of the canopy layer
C_2	particle concentration at the top level of the quasi-laminar layer
C_3	particle concentration at the top level of the surface layer
C_a	local concentration of aerosols at a reference height of 10 m or less above the surface
$C_{\mathrm{aq},0}$	initial mass concentration of the raindrops before rain
C_g	mass concentration of a soluble gas (like HNO_3) in the cloud before rain
$C_{i,\mathrm{cloud}}$	mass concentration of aerosol species i in the aqueous phase
$C_{i,\mathrm{int}}$	mass concentration of aerosol species i in the interstitial aerosol

$C_{i,o}$	mass concentration of aerosol species i before the cloud formation
C_{in}	concentration of ice nuclei in the atmosphere per liter of air
$C_p(z,t)$	vertical profile of the aerosol concentration of a certain type at time t in a horizontally homogeneous atmosphere "washed out" by rain
$C_{p,air}$	aerosol particle concentration in air entering the storm
$C_{p,rain}$	wet aerosol particle concentration in the precipitation measured at the surface
d	distance between two coagulation centers
D_1	Brownian diffusion coefficient of an aerosol particle having radius r_1
$D_{12} = D_1 + D_2$	sum of the Brownian diffusion coefficients of the aerosol particles having radii equal to r_1 and r_2, respectively
D_2	Brownian diffusion coefficient of an aerosol particle having radius r_2
D_{br}	Brownian diffusion coefficient for aerosol particles
dM_{bc}/dt	total aerosol mass scavenging rate
dN/dt	coagulation rate of aerosol particle number concentration
DP	deliquescence point of a water-soluble component of an aerosol particle
e	saturation vapor pressure of water adjacent to a pure water droplet of the same size and at the same temperature T considered in evaluating e'
e'	saturation vapor pressure of water adjacent to a solution droplet containing a mole fraction F_{mol} of pure water at a certain air temperature T
$E_{col}(r, r_f)$	collision efficiency between a drop of radius r_f and particles of radius r
e_s	saturation vapor pressure of water vapor over a plane surface of water at air temperature T
e_w	partial pressure of water vapor over a plane surface of water in air at temperature T
F_d	vertical dry deposition flux of aerosols
F_i	cloud mass scavenging ratio of aerosol species i
F_{mol}	mole fraction of pure water, giving the number of moles of pure water in the solution divided by the total number of moles (pure water plus dissolved material) in the solution
F_N	number scavenging ratio of aerosol particles
F_w	net wet deposition flux
G_{drop}	Gibbs free energy for the droplet of radius r
G_{liq}	Gibbs free energies of a molecule in the liquid phase
G_{pv}	Gibbs free energy for pure vapor
G_{vap}	Gibbs free energy of a molecule in the vapor phase
$I_v(v)$	condensation growth rate of aerosol particle volume

J_{12}	steady-state coagulation rate between a pair of spherical particles having radii r_1 and r_2 (measured in $cm^{-3}\,s^{-1}$)
$J_{bc}(t)$	below-cloud scavenging rate of aerosols as a function of time
J_{col}	collision rate between aerosol particles
J_{nucl}	nucleation rate
k_μ	mean linear mass increase coefficient of an aerosol particle during the condensation phase
K_{11}	coagulation rate coefficient between equal spherical aerosol particles, both having radius r_1
K_{12}	coagulation rate coefficient between spherical aerosol particles of radii r_1 and r_2
k_B	Boltzmann constant equal to $1.38 \times 10^{-23}\,J\,K^{-1}$
$k_{col}(r, r_d)$	collection coefficient for collisions between an interstitial aerosol particle of radius r and a cloud droplet of radius r_d
k_{in}	proportionality coefficient between the natural logarithm of concentration C_{in} of ice nuclei and the difference between temperature T_{in} (at which 1 ice nucleus per liter is active, typically of about −20 °C) and air temperature T
Kn	Knudsen number, defined as the ratio of molecular mean free path ℓ to a representative physical length scale
m	mass of a dissolved material of molecular weight μ_S
$M(r)$	mass concentration size-distribution function
M_{bc}	total aerosol mass in the below-cloud column during rain
M_L	aerosol mass loading in the atmospheric column (measured in $g\,cm^{-2}$)
M_p	mass concentration of aerosol particles (measured in $g\,cm^{-3}$)
$M_{tot\text{-}S}$	total atmospheric mass content of sulfur-containing compounds
m_w/m_o	water uptake per unit mass of a dry-air aerosol particle during the condensation phase
N	number density concentration of aerosol particles
n	number of water molecules per unit volume of liquid solution
$N(v, t)$	size-distribution curve of particle volume v at time t
$N(r)$	lognormal size distribution of particle number density concentration
n'	number concentration of water molecules in the solution droplet
N_0	overall number concentration of aerosol particles
N_1 (or N_2)	number density concentration of aerosol particles with radius r_1 (or with radius r_2)
$N_a(r, t)$	size-distribution curve of aerosol particle number density concentration at time t
$n_{bc}(r)$	number density size distribution of aerosol particles below the precipitation cloud
$n_d(r, t)$	cloud droplet number density size distribution at time t
N_f	overall number concentration of falling rain drops of radius r_f

n_i	number of ions dissociated in water for each molecule of the dissolved material in a solution droplet
N_{ia}	number concentration of interstitial aerosol particles
$n_{\text{ia}}(r,t)$	aerosol number density size distribution of interstitial aerosol particles in a cloud at time t
$N_{o,\text{cloud}}$	aerosol number concentration before cloud formation
$N_{\text{rd}}(r_f)$	number density size distribution of falling raindrops during a precipitation scavenging event
N_{res}	number of vapor molecules remaining after the drop formation
N_{tot}	initial total number of vapor molecules before drop formation
$p_{\text{eq},i,f}$	equilibrium partial vapor pressure of species i over the flat surface
$p_{\text{eq},i,r}$	equilibrium partial vapor pressure of species i over a spherical particle with radius r
$p_{\text{eq},j}$	equilibrium partial vapor pressure of pure species j over the liquid
pH	parameter used in chemistry to measure the acidity or basicity of an aqueous solution
p_i	vapor pressure of water over the ice particle surface
$p_{j,f}$	equilibrium partial vapor pressure of pure species j over a flat surface
$p_{\text{sat},i}$	vapor pressure of water over a flat ice surface
r	radius of a spherical aerosol particle (measured in μm)
R	universal constant of gases (=8.31446 J mol^{-1} K^{-1})
r^*	critical radius of the spherical particle, at which ΔG reaches its maximum value
r_c	mode radius of an aerosol size-distribution curve
r_d	radius of a cloud droplet colliding with an aerosol particle of radius r
r_f	radius of the falling raindrop during a precipitation scavenging event
$R_{r,0}$	rainfall rate measured at the ground level in mm h^{-1}
$R_{r,0}$	rainfall rate at ground level
$s = \ln(\sigma)$	parameter of the lognormal distribution
$S = \pi a^2$	surface area of a spherical aerosol particle having diameter a
$S(a)$	lognormal size-distribution function of aerosol particle surface area
S_{amb}	ambient saturation ratio of a cloud droplet
S_{cri}	critical saturation of a cloud droplet
S_{S-S}	supersaturation rate of a vapor species
T	absolute air temperature (measured in K)
t	time
T_{in}	air temperature at which 1 ice nucleus per liter is active (≈−20 °C)
u^*	friction velocity (related to wind speed WS)

u_d	dry deposition velocity of aerosol particles
$U_f(r_f)$	falling velocity of the raindrop having radius r_f in a precipitation scavenging event
$u_s(r)$	settling velocity of falling aerosol particles of radius r in a precipitation scavenging event
u_w	wet deposition velocity of an aerosol particle
$V(r)$	lognormal size distribution of aerosol particle volume
pH	numeric scale used to specify the acidity or basicity of an aqueous solution (values of pH less than 7 are said to be acidic and those greater than 7 are said to be basic or alkaline)
V_{bc}	aerosol particle volume concentration below cloud
VR	standard visual range (measured in km)
$W_{bc}(r)$	total rate of collection of mass of all particles of radius r
W_{ic}	local rate of removal of an irreversibly soluble gas (for instance, HNO_3)
w_r	washout ratio between the concentration of particulate matter in surface-level precipitation and the concentration of aerosol particles in surface-level air
z	altitude above mean sea level (measured in m a.m.s.l.)

Greek

$\upsilon = (4/3)\pi r^3$	volume of a spherical aerosol particle
$\varphi_g = 4r(dr/dt) = a(da/dt)$	parameter used to describe the particle growth by condensation of gases
ρ	aerosol particle mass density (measured in g cm^{-3}, for RH $\neq$ 0%)
ρ_o	dry aerosol particle mass density (measured in g cm^{-3}, for RH = 0%)
ρ_w	mass density of wet aerosol particles collected by a falling rain drop
ζ	bulk surface tension for a spherical particle having radius r
σ	shape parameter of lognormal size distribution
η	viscosity coefficient of air (equal to 1.83×10^{-4} g cm^{-1} s^{-1} for standard conditions of air)
ψ_c	Fuchs (1964) correction factor for coagulation coefficient K_{12}
Ω_{aer}	aerodynamic resistance to aerosol deposition
Ω_{can}	surface (or vegetation canopy) resistance to aerosol deposition
Ω_{cf}	foliar resistance, given by the sum of Ω_{cut} and Ω_{sto}
Ω_{cg}	resistance to uptake by the ground surfaces covered by soil or snow
Ω_{cut}	resistance to aerosol uptake at the cuticle (epidermal) surface
Ω_{cw}	resistance to uptake by water surfaces
Ω_{qll}	quasi-laminar layer resistance to aerosol deposition
Ω_{sto}	resistance to aerosol uptake at the plant's stomata
Ω_{tot}	total resistance to aerosol deposition

$\Lambda_d(r)$	scavenging coefficient for interstitial aerosols and the full cloud droplet population
$\Lambda(a)$	scavenging coefficient for an aerosol particle having diameter a
Λ_g^*	average scavenging coefficient of aerosol particles for a homogeneous vertical profile of $C_p(z, t)$ in the atmospheric region below the cloud
$\Lambda_p(z, t)$	scavenging coefficient for aerosol particles at height z and time t
Φ_e	annual global emission flux of aerosol particulate mass (measured in Tg per year)
ΔG	Gibbs free energy reduction
$\phi_g = a(RH)/a_o$	growth factor by water vapor condensation of an aerosol particle having initial dry-air diameter a_o
ΔG_w	net increase in the Gibbs free energy reduction of the water droplet system
$\rho_{\rm i}$	ice density (measured in $\rm g\,cm^{-3}$)
$\gamma_{\rm ia}$	surface energy of ice in air
$\rho_{\rm liq}$	liquid droplet mass density
$\upsilon_{\rm liq}$	volume occupied by a molecule in the liquid phase
$\upsilon_{\rm mol}$	volume occupied by a molecule in a spherical droplet
$\rho_{\rm o}$	particulate matter mass density (measured in $\rm g\,cm^{-3}$, for dry-air conditions (RH = 0%))
$\gamma_{\rm S}$	surface tension of a liquid droplet, giving the amount of energy required to yield a unit increase in the surface area of the droplet
γ'_S	surface energy of water molecules
Δt_L	lifetime of aerosol particles
$\Delta t_{\rm rel}$	relaxation time of a collected aerosol particle
Δz	depth of the quasi-laminar sublayer adjacent to the surface (measured in millimeter)
$\mu_{\rm i}$	molecular weight of gaseous species i
$\mu_{\rm liq}$	molecular weight of a pure liquid substance
$\mu_{\rm sol}$	molecular weight of dissolved material in a solution droplet
$\mu_{\rm wat}$	molecular weight of water

References

Andreae, M.O. and Rosenfeld, D. (2008) Aerosol-cloud-precipitation interactions. Part 1. The nature and sources of cloud-active aerosols. *Earth Sci. Rev.*, **89**, 13–41. doi: 10.1016/j.earscirev.2008.03.001

Cooke, W.F., Liousse, C., Cachier, H., and Feichter, J. (1999) Construction of a 1° x 1° fossil fuel emission data set of carbonaceous aerosol and implementation and radiative impact in the ECHAM4 model. *J. Geophys. Res.*, **104** (D18), 22137–22162. doi: 10.1029/1999JD900187

Cooke, W.F., Ramaswamy, V., and Kasibhatla, P. (2002) A general circulation model study of the global carbonaceous distribution. *J. Geophys. Res.*, **107** (D16), 4279. doi: 10.1029/2001JD001274

Cooke, W.F. and Wilson, J.J.N. (1996) A global black carbon aerosol model. *J. Geophys. Res.*, **101** (D14), 19395–19409. doi: 10.1029/96JD00671

Fuchs, N.A. (1964) *The Mechanics of Aerosols* (translated from the Russian by Daisley, R.E. and Fuchs, M., translation edited by C.N. Davies), Pergamon Press, Oxford, 422 pp.

Hänel, G. (1976) The properties of atmospheric aerosol particles as functions of the relative humidity at thermodynamic equilibrium with the surrounding moist air. *Adv. Geophys.*, **19**, 73–188. doi: 10.1016/S0065-2687(08)60142-9

Hänel, G. and Bullrich, K. (1978) Physico-chemical property models of tropospheric aerosol particles. *Beitr. Phys. Atmos.*, **51**, 129–138.

Hinds, W.C. (1999) *Aerosol Technology: Properties, Behavior, and Measurement of Airborne Particles*, 2nd edn, John Wiley & Sons, Inc., New York, 504 pp.

Hobbs, P.V. (1995) in *Aerosol-Cloud-Climate Interactions* (ed. P.V. Hobbs), Academic Press, San Diego, CA, pp. 33–73.

IPCC (2001) *Climate Change 2001. The Scientific Basis* (eds J.T. Houghton *et al.*), Cambridge University Press, New York, 881 pp.

Köhler, H. (1936) The nucleus in and the growth of hygroscopic droplets. *Trans. Faraday Soc.*, **32**, 1152–1161. doi: 10.1039/TF9363201152

Liousse, C., Penner, J.E., Chuang, C., Walton, J.J., Eddleman, H., and Cachier, H. (1996) A global three-dimensional model study of carbonaceous aerosols. *J. Geophys. Res.*, **101** (D14), 19411–19432. doi: 10.1029/95JD03426

Ménégoz, M., Voldoire, A., Teyssèdre, H., Salas y Mélia, D., Peuch, V.-H., and Gouttevin, I. (2012) How does the atmospheric variability drive the aerosol redisence time in the Arctic region? *Tellus Ser. B*, **64**, 11596. doi: 10.3402/tellusb.v6410.11596

Pósfai, M. and Buseck, P.R. (2010) Nature and climate effects of individual tropospheric aerosol particles. *Annu. Rev. Earth Planet. Sci.*, **38**, 17–43. doi: 10.1146/annurev.earth.031208.100032

Pruppacher, H.R. (1973) The role of natural and anthropogenic pollutants in cloud and precipitation formation, in *Chemistry of the Lower Atmosphere* (ed. S.I. Rasool), Plenum Press, New York, 335 pp.

Pruppacher, H.R. and Klett, J.D. (1997) *Microphysics of Clouds and Precipitation*, 2nd edn, Kluwer Academic Publishers, Dordrecht.

Schaller, R.C. and Fukuta, N. (1979) Ice nucleation by aerosol particles: experimental studies using a wedge-shaped ice thermal diffusion chamber. *J. Atmos. Sci.*, **36**, 1788–1802. doi: 10.1175/1520-0469(1979)036<1788:INBAPE>2.0.CO;2

Seinfeld, J.H. and Pandis, S.N. (2006) *Atmospheric Chemistry and Physics, from Air Pollution to Climate Change*, 2nd edn, John Wiley & Sons, Inc., Hoboken, NJ, 1225 pp.

Slinn, W.G.N. (1983) A potpourri of deposition and resuspension questions, in *Precipitation Scavenging, Dry Deposition, and Resuspension*, vol. **2**, Proceedings of the 4th International Conference, Santa Monica, CA, November 29–December 3, 1982 (Pruppacher, H.R., Semonin, R.G., and Slinn, W.G.N. coordinators), Elsevier, New York.

Stier, P., Feichter, J., Kloster, S., Vignati, E., and Wilson, J. (2006) Emission-induced nonlinearities in the global aerosol system: results from the ECHAM5-HAM aerosol-climate model. *J. Clim.*, **19** (16), 3845–3862. doi: 10.1175/JCLI3772.1

Tampieri, F. and Tomasi, C. (1976) Size distribution models of fog and cloud droplets in terms of the modified gamma function. *Tellus*, **28** (4), 333–347. doi: 10.1111/j.2153-3490.1976.tb00682.x

Tsigaridis, K., Krol, M., Dentener, F.J., Balkanski, Y., Lathière, J., Metzger, S., Hauglustaine, D.A., and Kanakidou, M. (2006) Change in global aerosol composition since preindustrial times. *Atmos. Chem. Phys.*, **6**, 5143–5162. doi: 10.5194/acp-6-5143-2006

Wallace, J.M. and Hobbs, P.V. (2001) *Atmospheric Science: An Introductory Survey*, International Geophysics, 2nd edn, vol. **92**, Academic Press, Burlington, MA, USA, 484 pp.

Whitby, K.T. and Cantrell, B.K. (1976) Atmospheric aerosols: characteristics and measurement. International Conference on Environmental Sensing and Assessment (ICESA), Institute of Electrical and Electronic Engineers (IEEE), Las Vegas,

NV. September 14–19, 1975, Paper 29–1, 6 pp.

Winkler, P. (1988) The growth of atmospheric aerosol particles with relative humidity. *Phys. Scr.*, **37** (2), 223–230. doi: 10.1088/0031-8949/37/2/008

Zender, C.S., Miller, R.L., and Tegen, I. (2004) Quantifying mineral dust mass budgets: terminology, constraints, and current estimates. *Eos Trans. Am. Geophys. Union*, **85** (48), 509–512. doi: 10.1029/2004EO480002

4
Chemical Composition of Aerosols of Different Origin

Stefania Gilardoni and Sandro Fuzzi

4.1
Introduction

Atmospheric aerosol particles exhibit a wide range of sizes, from nanometers to micrometers, and a wide range of shapes. Their chemical composition usually differs among the different size ranges and even among particles within a given size range.

These properties distinguish aerosols from trace gases, for which knowledge of the mixing ratio univocally defines their abundance and allows the definition of their properties, reactions, and effects. On the contrary, defining the aerosol particles suspended in the atmosphere involves a multidimensional approach which takes into account the aerosol size distribution and the chemical composition of the myriad of particles that, in principle, are all different from each other.

The chemical composition affects aerosol properties like radiative effects, hygroscopic growth, reactivity, ability to form cloud droplets, and, consequently, their effects on the environment, from visibility impairment to effects on human health and from effects on the biosphere to material deterioration and to climate change.

Atmospheric particles have a multitude of sources, primary and secondary or natural and anthropogenic, and their chemical composition in any given region of the world is the result of not only the type and strength of the sources affecting the area but also of the physical and chemical atmospheric processes undergone by the aerosol and its precursors. In fact, aerosol particles are strictly coupled with the atmospheric gas phase and the liquid phase of clouds, and understanding the properties and effects of the atmospheric aerosol requires that gas, aerosol, and cloud phases are treated as a single system.

The aim of this chapter is to provide a description of the size-dependent aerosol chemical composition in different areas of the world, characterized by different sources and environmental conditions, and of the processes affecting the chemical composition itself. The final paragraph then highlights some still open issues in atmospheric aerosol science related to aerosol chemistry.

Atmospheric Aerosols: Life Cycles and Effects on Air Quality and Climate, First Edition.
Edited by Claudio Tomasi, Sandro Fuzzi, and Alexander Kokhanovsky.

4.2
Global Distribution and Climatology of the Main Aerosol Chemical Constituents

As a first approximation, atmospheric aerosol can be defined as composed by inorganic and organic constituents and by elemental carbon (EC). The mass of inorganic aerosol in fine particles is generally dominated by ammonium sulfate and ammonium nitrate.

Figure 4.1 shows, as an example, the total mass of PM_1 and the percent distribution of these main classes of constituents in various environments characterized by different level of pollution, from a remote marine area to an area strongly affected by forest fires.

Coarse inorganic components include carbonate, metal oxides, aluminosilicates, and silicates. Organic constituents and EC can be found both within fine and coarse particles.

4.2.1
Definition of Primary and Secondary Inorganic and Organic Aerosol Compounds

Atmospheric aerosol components are traditionally classified into primary and secondary, where the term primary refers to those pollutants emitted in the atmosphere directly in the particulate phase, while secondary aerosol identifies particulate matter that forms in the atmosphere through chemical and physical transformation, heterogeneous phase chemical reaction, gas-phase particle conversion, gas-phase oxidation, and partitioning on preexisting particles. This kind of classification is useful to distinguish those species whose concentration is controlled by emissions, transport, and dilution from others whose concentration

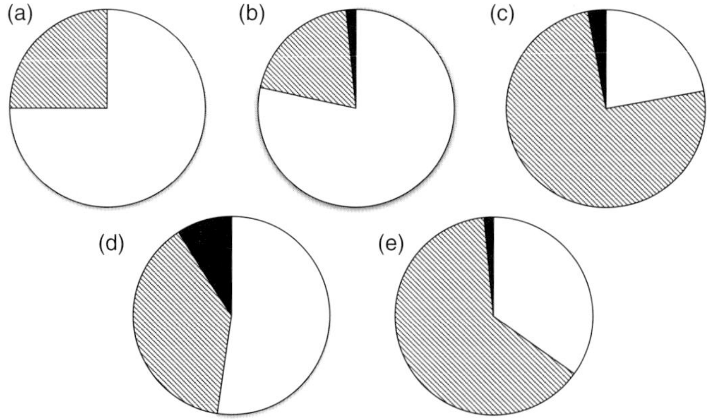

Figure 4.1 Mass of PM_1 and percent distribution of inorganic compounds (white), organic compounds (dash), and elemental carbon (black) in various environments: (a) marine clean environment, (b) marine polluted, (c) biomass burning aerosol, (d) continental polluted regions, and (e) remote forested area.

is governed by more complex mechanisms and the simultaneous presence of different chemical species, for example, oxidants and radical species.

A detailed description of traditional classification into primary and secondary pollutants is presented in Chapter 1. Recently, a better understanding of the chemical and physical mechanisms controlling aerosol lifetime in the atmosphere shows that the threshold between primary aerosol particles and precursors of secondary aerosol is not as sharp as one may think. This is especially evident for the organic aerosol (OA) component. Some organic species immediately emitted in the particulate phase (traditionally described as primary pollutants) could partition into the gas phase after dilution in ambient conditions. These species are referred to as intermediate volatile organic compounds (IVOCs). Once in the gas phase, these species could then be oxidized or undergo oligomerization and fragmentation reactions. As a consequence their volatility is modified. When the IVOC volatility is reduced, as in the case of oxidation or oligomerization processes, the reaction products can partition back into the particulate phase, contributing to the formation of what would be traditionally called secondary organic aerosol (SOA; Robinson *et al.*, 2007). Another example of primary organic aerosol (POA) contributing to SOA formation is through heterogeneous phase chemistry. Some chemical species emitted in the aerosol phase and hydrophilic (water soluble) in certain atmospheric conditions can undergo solubilization in atmospheric water, like cloud, fog droplets, or aerosol water. In the aqueous phase these species undergo oxidation or oligomerization reactions. These reactions lead to the formation of organic molecules with lower volatility compared to the original chemical species (or precursors), which stay in the particulate phase after water evaporation, contributing to SOA formation (Figure 4.2).

Taking into account this kind of mechanisms, the static representation of POA is replaced by a more dynamic description. A new general framework to describe

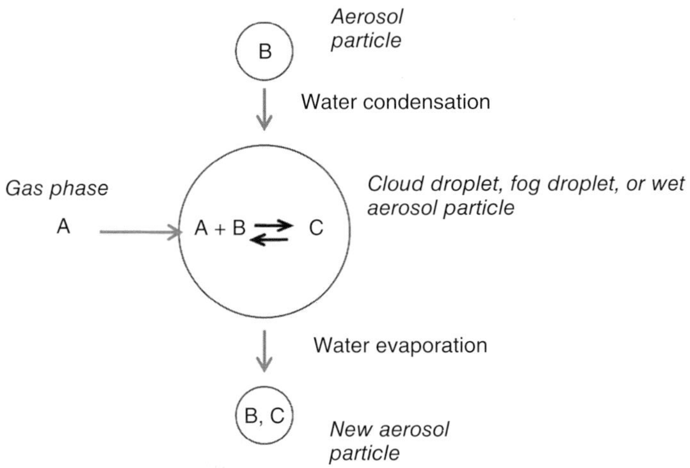

Figure 4.2 Schematic representation of secondary aerosol formation through heterogeneous phase chemistry.

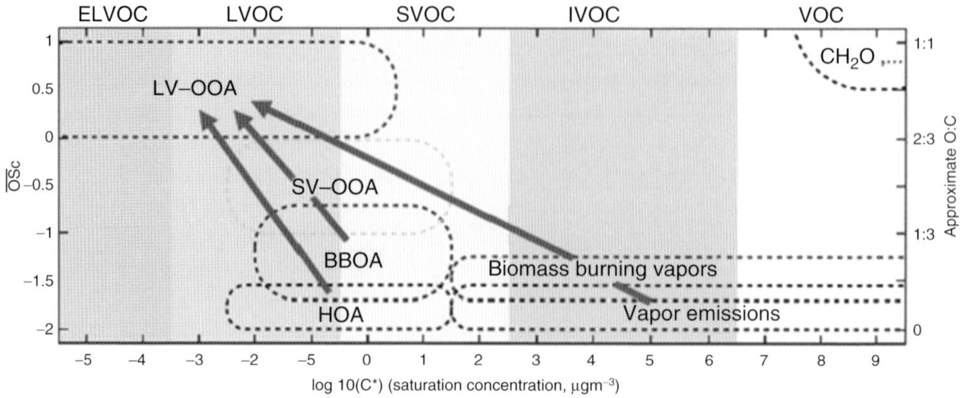

Figure 4.3 Evolution of organic aerosol in the atmosphere. (Donahue *et al.*, 2012 http://www.atmos-chem-phys.net/12/615/2012/. Used under CC-BY 3.0 license https://creativecommons.org/licenses/by/3.0/.)

evolution of OA in the atmosphere is proposed by Jimenez *et al.* (2009) and described by the scheme reported in Figure 4.3. In this representation the fate of organic compounds, both in the gas phase and in the aerosol phase, depends on their volatility. In Figure 4.3 the *x*-axis ranges over 14 orders of magnitude in volatility, spanning from extremely low-volatile organic compounds (ELVOCs), low-volatile organic compounds (LVOCs), semivolatile organic compounds (SVOCs), IVOCs, and volatile organic compounds (VOCs). At atmospheric conditions, VOCs are gaseous species, while SVOCs and IVOCs can partition between the gas and the aerosol phase. LVOCs and ELVOCs are found exclusively in the aerosol phase. The *y*-axis shows the carbon oxidation state (OS_c). The volatility of aerosol organic compounds varies depending on the chemical transformation taking place in the atmosphere, mainly oxidation, oligomerization, and fragmentation. Oxidation and oligomerization lead to an increase in the carbon oxidation state, a decrease in volatility, and preferential partitioning in the particulate phase, while fragmentation reactions lead to an increase in volatility and eventually preferential partitioning into the gas phase.

4.2.2 Aerosol Global Budgets

4.2.2.1 Organic Aerosol

OA is composed by thousands of different molecular species. OA is emitted both by anthropogenic and natural sources. Anthropogenic sources include fossil fuel burning (industrial activities, energy production, transport), burning of biofuels (wood, agricultural waste, animal waste, urban solid waste, and charcoal combustion), and industry emissions (metallurgic industry and cement industry). Natural sources are open biomass burning (wild fires), biogenic aerosol (primary biogenic aerosol particles (PBAP) and SOA from biogenic precursors), and marine aerosol. All these sources contribute both to primary and secondary OA. A detailed description of processes responsible for secondary OA formation is reported in paragraph 4.b.

Estimate of OA budget at the global scale is traditionally performed using a bottom-up approach. This approach estimates SOA budget combining emission data of SOA precursors, which include biogenic VOCs (mainly isoprene and terpenes) and anthropogenic VOCs, with laboratory data that describe reaction yields characteristic of SOA formation from the different precursors. Models estimate that the SOA budget ranges between 9 and 50 TgC per year from biogenic precursors and 1–9 TgC per year from anthropogenic precursors (Kanakidou *et al.*, 2005; Henze *et al.*, 2008). Emissions of POA are about 34 TgC per year of which 9 TgC per year are from anthropogenic sources and 25 TgC per year are from open biomass burning, like forest fires (Bond *et al.*, 2004). The resulting total budget of OA from a bottom-up approach is equal to 50–90 TgC per year.

More recently, OA budget estimates have been performed using a top-down approach, that is, constraining calculations with ambient observations in a sort of inverse modelling. The top-down approach assumes that OA is internally mixed with sulfate, and OA budget can be derived from the sulfate budget. Ambient observations show that the OA to sulfate ratio varies between 1 : 2 to 2 : 1 (on average 1 : 1). Considering that the OA to organic carbon (OC) mass ratio is 2 : 1 and the sulfur to sulfate mass ratio is 1 : 3, the carbon to sulfur ratio in ambient aerosol is about 3 : 2. The sulfur global budget is relatively well known and is equal to 50–110 TgS per year. Based on the previous assumptions, the best estimate of the OC global budget is 150 TgC per year. Considering that 34 TgC per year correspond to primary OA, this approach indicates that SOA budget is equal to 115 TgC per year. This value is significantly higher than the values calculated using the bottom-up approach. The discrepancy between the top-down and the bottom-up results clearly indicate that the models based on oxidation chamber experiment results underestimate SOA formation, likely due to the underestimation of VOC emission inventories and the lack of detailed knowledge of the formation mechanisms and precursors important for SOA formation (Hallquist *et al.*, 2009).

Recently, satellite observations have been combined with chemical transport models to constrain the global OA budget. Based on aerosol optical depth (AOD) measurements, it is estimated that the OA budget is 150 TgC per year. This value agrees with the top-down calculations (Heald *et al.*, 2010) and emphasizes the need for a better knowledge of the sources, mechanisms, and yields responsible for SOA formation.

4.2.2.2 **Black Carbon Aerosol**

Black carbon (BC) is emitted exclusively by combustion processes, due to the incomplete combustion of fuels, including biofuels like wood, animal waste, and agricultural waste, as well as fossil fuels such as coal, kerosene, oil, and natural gas. Anthropogenic activities/sectors responsible for BC emissions include transport, power generation, industry, residential heating, and agricultural activities. Forest fires are the main natural sources of BC.

The estimate of BC emission inventory ranges between 4 and 22 TgC per year. The large uncertainty range is mainly due to the variability of carbon emission factors, that is, the amount of BC emitted per unit of burned fuel mass. The emission factors depend strongly on the fuel type and on the combustion conditions.

The relative contributions of fossil fuel, biofuel, and open burning (i.e., forest fires and agricultural burning) to the global BC budget are 38%, 20%, and 42%, respectively (Bond *et al.*, 2004).

4.2.2.3 Sulfur Aerosol

In the atmospheric aerosol sulfur is present mainly as sulfate. Globally, a small fraction of sulfate is emitted as primary particles associated with sea salt, while the majority of sulfate is produced from the oxidation of sulfur dioxide through gas-phase and aqueous-phase oxidation with a time constant of a few days.

In the atmosphere the oxidation reaction of sulfur dioxide (or sulfur IV) to sulfur trioxide (or sulfur VI) is thermodynamically favored but is slow, thus its contribution to aerosol sulfate can be neglected. Instead, in the gas phase, sulfur oxide oxidation takes place through a radical mechanism:

$$SO_2 + OH\bullet + M = HOSO_2\bullet + M$$
$$HOSO_2\bullet + O_2 = HO_2\bullet + SO_3$$
$$SO_3 + H_2O + M = H_2SO_4 + M$$

In the aqueous phase, sulfur(IV) can be converted into sulfur(VI) through:

1. Oxidation by dissolved ozone:

$$SO_2 + O_3 = SO_3 + O_2$$

2. Oxidation by hydrogen peroxide:

$$SO_2 + H_2O = HSO_3^- + H^+$$
$$HSO_3^- + H_2O_2 = SO_2OOH^- + H_2O$$
$$SO_2OOH^- + H^+ = H_2SO_4$$

3. Oxidation by dissolved oxygen catalyzed by iron and manganese ions:

$$2\,SO_2 + O_2 = 2\,SO_3$$

4. Radical oxidation:

$$HSO_3^- + OH\cdot = SO_3^- + H_2O$$
$$SO_3^- + O_2 = SO_5^-$$
$$SO_5^- + SO_5^- = 2SO_4^- + O_2$$
$$SO_4^- + HSO_3^- = SO_3^- + H^+ + SO_4^{2-}$$

The rate of mechanisms (1), (2), and (3) depends on the pH of the water solution. At pH lower than 4–5, the dominant pathway is the oxidation by hydrogen peroxide. At pH higher than 5, ozone oxidation is the fastest

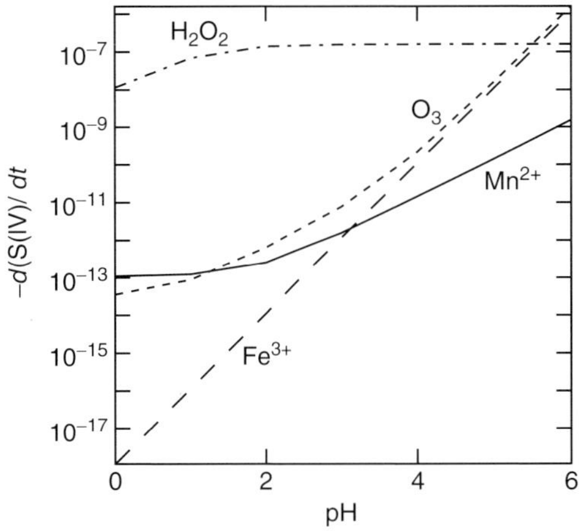

Figure 4.4 Dependence of sulfur(IV) oxidation rate on water solution acidity. Rate is calculated for the following conditions: SO_2 (g) 3 ppb, O_3 (g) 30 ppb, H_2O_2 (g) 1 ppb, Fe^{3+} 0.3 μM, Mn^{2+} 0.03 μM.

mechanism. The catalyzed reaction by oxygen may be important at higher pH as well. The dependence of oxidation rate as a function of solution acidity is reported in Figure 4.4.

Sulfate can also be produced by the oxidation of reduced sulfur species, like hydrogen sulfide (H_2S) and DMS, emitted by biogenic sources. Both H_2S and DMS are oxidized by the OH radical in the gas phase.

Anthropogenic sources of sulfur (energy production or processing, biomass burning, and shipping) accounts for about 80% of the global sulfur budget. Table 4.1 reports the global budget of main sulfur species from the analysis of 20 modelling studies (Faloona, 2009).

In the late 1980s many developed countries have introduce environmental control policies to reduce continental sulfur emissions, especially to stop acid rains and their damages to vegetation. Emission reduction led to reduction in sulfate ambient concentration, especially in Europe and North America. So far, in Asian countries, sulfur emissions and sulfate ambient concentrations are still increasing.

More recently, attempts are made to reduce sulfur emissions also over the ocean, with the introduction of the Sulfur Emission Control Area (SECA), which so far include the North Sea and the Baltic Sea, together with most of the United State and Canadian coasts. The protocol that defines the SECA imposes more stringent limit to the sulfur content of ship fuel in these areas compared to the other sea regions.

4.2.2.4 **Nitrogen Aerosol Species**

Ammonium and nitrate account for the largest fraction of nitrogen in the atmospheric aerosol. Nitrate ion is formed by the reaction of NO_2 with the OH radical and by the hydrolysis of N_2O_5 in aqueous phase:

a. $NO_2 + OH^* + M = HNO_3 + M$
$NO_2 + O_3 = NO_3 + O_2$
b. $NO_3 + NO_2 = N_2O_5$
$N_2O_5 + H_2O = 2HNO_3$

Note that mechanism (a) is about 10 times faster than the oxidation of sulfur dioxide in the gas phase. During daytime, the conversion rate of NO_2 is about 10–50% per hour. It follows that the mechanism 2 is active only during nighttime, that is, when photochemical activity is low.

The ammonium ion is formed by the reaction of ammonia with acidic species, like nitric acid or sulfuric acid. Atmospheric aerosol nitrogen budget is strongly bound to nitrogen oxides (NO_x) and ammonia (NH_3) budget.

Table 4.2 reports the global emission inventory data for NO_x and NH_3 referred to the year 2005, as calculated by the EDGAR model. Global emission of NO_x and NH_3 are about 120 and 50 Tg per year, respectively. The largest emissions of NO_x are due to power generation and transportation, including public electricity and heat production (29%), road transportation (22%), and international shipping (11%). Agricultural related processes, that is, manure management, soil emissions, and manure in pasture/range/paddock accounts for the largest emissions of ammonia globally (about 80%).

Table 4.1 Global budget of main sulfur species in TgS per year.

		Models' median	Standard deviation
DMS			
	Sources	19.4	4.4
SO_2			
	Sources	95.7	10.9
Anthropogenic		67.2	10
Ship		—	0.6
Biomass		2.3	
Volcanic		7.8	2.9
DMS		18.5	4.5
	Sinks	96.4	—
Dry deposition		34.6	9.9
Wet deposition		7.3	5
Oxidation		51.6	—
NSS-sulfate			
	Sources	53.3	13
Direct		2	0.6
Heterogeneous oxidation		42	10.3
Homogeneous oxidation		11	2.9
	Sinks	57.6	—
Dry deposition		6.4	3.5
Wet deposition		44.6	12.6

Table 4.2 Global NO_x and NH_3 emissions (in teragram) based on the EDGAR data for the year 2005.

Sectors	NO_x	NH_3
Power generation	37.5	0.1
Transport	49.6	0.5
Industry	14.1	3.4
Agriculture	5.5	39.4
Open fires	11.3	4.6

http://edgar.jrc.ec.europa.eu/overview.php?v=41.

4.2.2.5 Dust Aerosol

Mineral dust dominates ambient aerosol in several large regions of the planet. The largest fraction of mineral dust has natural sources (resuspension of soil dust particles), while anthropogenic activities (i.e., agricultural activities or transportation) account for a smaller percentage of the dust burden (<25%). Natural dust budget is dominated by emissions from large desert areas. Table 4.3 reports the annual mean dust fluxes from different global models.

Global dust emissions are dominated by emissions from the Saharan desert, which accounts for 51–69% of total emissions. The Australian desert is the largest source of dust in the Southern Hemisphere, with a contribution that varies between 2.5% and 15% of the total.

Dust particles are removed from the atmosphere through dry and wet deposition, as a consequence particle lifetime is about 2–3 days. Lifetime varies with regions, depending on the effects of transport and precipitation on deposition rate. Size distribution of dust emissions peaks at about 7 μm, while the size of ambient dust concentrations shows a maximum around 2–3 μm. Fast dry deposition of larger particles is responsible for the different size distribution between emissions and ambient data.

Table 4.3 Dust emissions in Tg per year from six different studies.

Africa		Asia			America		Australia	Total
North	South	Arabia	Central	East	North	South		
1087 (58%)	63(3%)	221 (12%)	140 (8%)	214 (11%)	2 (0.1%)	44 (2%)	106 (6%)	1877
693 (65%)		101 (10%)	96 (9%)				52 (5%)	1060
1114 (67%)		119 (7.2%)		54 (3.2%)	132 (8%)			1654
980 (66%)		415 (28%)			8 (0.5%)	35 (2%)	37 (3%)	1490
1430 (69%)		496 (24%)			9 (0.4%)	55 (3%)	61 (3%)	2073
517 (51%)		43 (4%)	163 (16%)	50 (5%)	53 (5%)		148 (15%)	1019

Tanaka and Chiba, 2006 and reference therein.

4.2.3
Main Regional Differences and Seasonal Variations of Aerosol Chemical Composition

4.2.3.1 Urban Aerosol

Fine particle concentration (PM_1 and $PM_{2.5}$) in urban areas of developed countries varies between 10 and 40 $\mu g\,m^{-3}$, while PM_{10} ranges from 30 to 50 $\mu g\,m^{-3}$, with higher values during the cold season. Polluted urban areas in East Asia typically exhibit higher concentrations, with values exceeding 300 $\mu g\,m^{-3}$. The higher concentrations observed during the cold season are generally due to the higher emissions also affected by residential heating and accumulation of pollutants in the shallow boundary layer and consequent lack of dispersion (see Figure 4.5).

As said in section 4.2, fine particles are mainly composed by OA and ammonium nitrate and sulfate. OA accounts for 25% to 50% of fine particle mass (PM_1 and $PM_{2.5}$), while the remaining fraction is dominated by ammonium nitrate or ammonium sulfate, depending on the source of precursor gases in the surrounding areas. In the central and eastern part of the United States, where a large fraction of electrical energy is produced by coal combustion, a significant source of SO_2, the inorganic aerosol is dominated by ammonium sulfate. In those areas characterized by intense agricultural activities, like the Central Valley of California and the Po Valley in Italy, the secondary inorganic aerosol is mainly composed by ammonium nitrate (see Table 4.4 for mass and chemical composition variability of aerosol in Europe). Due to the larger volatility of ammonium nitrate relative to sulfate, the warm season is generally associated with an increase of the mass fraction of ammonium sulfate relative to that of ammonium nitrate. BC accounts for 5–10% of fine particle mass with concentrations ranging between 5 and 10 $\mu g\,m^{-3}$.

In the urban areas, the highest values of fine particle concentration are found at curbside locations, where $PM_{2.5}$ levels can be as high as twice the urban background. Curbsides are characterized by higher concentrations of BC, OA,

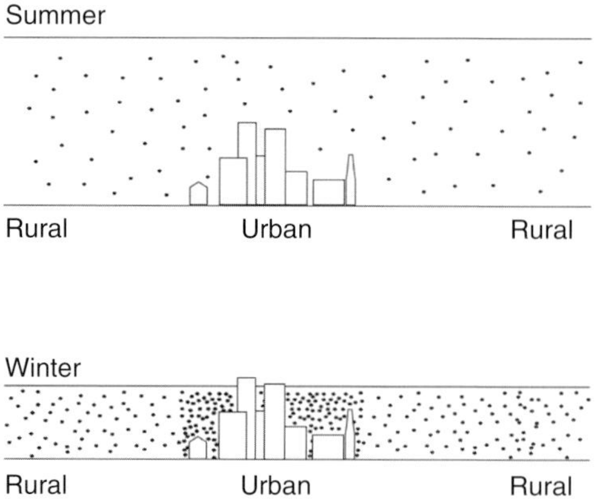

Figure 4.5 Seasonal variation of pollutant dispersion in urban and rural environments.

Table 4.4 Average mass and chemical composition of $PM_{2.5}$ in Europe in microgram per cubic meter.

Category	Site	Min. dust	Sea dust	nss-sulfate	Ammonium	Nitrate	Organic	Black carbon	$PM_{2.5}$
Natural	Sevettijarvi (FI)	0.15	0.46	1.29	0.15	0.02	1.12	NM	4.5
Natural	Skreadalen (N)	0.18	0.29	1.28	NM	NM	2.33	NM	8.01
Natural	Birkenes (N)	0.25	0.34	1.61	NM	NM	1.59	NM	7.94
Rural	Chaumont (CH)	0.77	0.11	2.41	0.84	0.29	1.71	1.7	10.52
Rural	Monagrega (E)								
Rural	Illmitz (A)	0.26	0.11	3.97	2.04	2.28	4.12	5.49	23.92
Near city	Waasmunster (B)	0.98	0.13	5.04	NM	NM	15.12	NM	39.55
Near city	Melpitz (D)	1.8	NM	2.69	1.99	3.43	10.83	NM	24.17
Near city	Ispra (IT)								
Urban	Zuerich (CH)	1.18	0.17	3.27	1.84	2.68	3.6	4.44	24.5
Urban	Basel (CH)	1.4	0.25	3.8	1.83	2.47	2.57	4.21	25.77
Urban	Gent (B)	1.05	1.31	6.17	NM	NM	13.7	NM	39.8
Urban	Bologna (IT)	2.89	0.84	4.62	4.12	8.54	7.55	8.24	44.23
Curbside	Barcelona (E)	3.79	0.81	5.6	3.31	4.27	1.96	11.89	52.85
Curbside	Bern (CH)	1.35	0.21	2.67	1.4	2.32	4.07	6.68	40.22
Curbside	Wien (A)	2.37	0.41	4.44	2.18	3.57	7.84	8.62	54.58

Putaud *et al.* (2004).

and mineral dust. BC and OA are emitted by vehicle as engine exhaust, while mineral dust is produced by resuspension of soil dust from road pavement abrasion.

Coarse particle concentration in urban background and curbside locations are about 5–20 $\mu g\,m^{-3}$. These particles are mainly composed by mineral dust, produced by nonexhaust vehicle emissions, that is, particulate matter from the abrasion of tire wear, break wear, road wear, and road dust suspension.

4.2.3.2 **Rural Aerosol**

PM_1 and $PM_{2.5}$ concentrations in rural background areas range from a few micrograms per cubic meter to 20 $\mu g\,m^{-3}$. Chemical composition resembles that of urban background aerosol with sometimes a larger mass fraction of OA and a smaller contribution from BC. Despite the similarity in composition, it is worth pointing out that OA in rural regions exhibits different properties with respect to the urban OA. OA can be represented as composed by a less oxygenated fraction, also identified as hydrocarbon-like organic aerosol (HOA), and a more oxygenated factor, defined as oxidized organic aerosol (OOA). The OOA fraction is mainly representative of secondary or aged OA. Rural OA is more oxidized than urban OA. On average OOA accounts for 95% of rural OA and for only 60% of urban OA (Zhang *et al.*, 2007), indicating a larger contribution of secondary or aged species away from emission sources, mainly located in urban environments.

Rural coarse aerosol mass varies between a few micrograms per cubic meter to 5 μg m^{-3} and is mainly composed by OA. Coarse OA in rural areas derives mainly from PBAP. PBAP are particles of biological origin emitted directly in the aerosol phase, like fungi, spores, bacteria, viruses, and biological debris. Some PBAP, like viruses, have sizes in the nanometer range, but the majority of these particles are emitted in the supermicron size range. Typical size ranges are 0.1–4 μm for bacteria, 0.5–15 μm for fungal spores, and 10–30 μm for pollens. Large PBAP are usually removed by dry deposition close to the emission area due to their large size. Nevertheless some PBAP can be transported to high altitude where they are incorporated into clouds and transported over long distances. An example is the transport of PBAP associated with desert dust transport from North Africa to Southern Europe.

4.2.3.3 Continental Regional Background Aerosol

Regional background aerosol concentration varies from a few micrograms per cubic meter to 10–15 μg m^{-3}. Regional or natural background aerosol concentration and composition vary significantly across the world. Table 4.5 reports the composition of background aerosol in different regions. Generally, sulfate and OA account of the largest fraction of fine background aerosol. BC is a minor component, but its concentration is roughly constant (<1 μg m^{-3}) in

Table 4.5 Concentration and composition of regional background aerosol in different areas of the world.

Region	$PM_{2.5}$	Sulfate	Nitrate	Black carbon	Organic aerosol
Arctic	1–1.5	0.3–2	0.04–0.15	0.1	0.7
North America					
Canada	1–3	0.7–2	0.08–1	0.6	4
U.S. marine	2–6	0.5–1.6	0.16–0.3	0.14–0.3	1.2–2
U.S. continental	1.7–5	0.6–2	0.15–0.6	0.12–0.2	0.8–1.5
Europe					
Coastal	5	1–4	0.1–0.5	0.1–0.5	0.5–3
Continental	5–30	1–4	0.1–0.5	0.3–1	0.5–10
Mediterranean	15–30	2–7	1–3	<1	7–10
Asia					
Continental	8–20	0.2–6	0.08–2	0.15–1.5	3–40
Coastal	—	0.3–2	0.04–0.6	0.3	4–9
North Pacific	1	0.1–1	0.005–0.8	0.02–1	0.3–4
North Atlantic	6–8	0.15–3	0.005–0.5	0.01–0.4	0.3–1.5
South America	3–50	0.07–3.6	0.08–0.3	0.2–5	0.2–3
South Atlantic Ocean	—	0.5–2	0.001	<0.01–0.01	0.16
Africa	10	0.9–4	0.04–3	1–4	7–20
South Pacific	—	0.3–5	0.01–0.3	1b	4
Antarctica	—	0.1–0.65	0.016–0.17	0.001–0.13	0.2

From Hidy and Blanchard (2005) and Minguillón *et al.* (2015).

many remote sites. Particle mass and sulfate mass concentration are higher in continental regions than in remote marine areas. Nitrate concentration increases from marine to continental regions, with higher levels in sites occasionally influenced by anthropogenic emissions (Hidy and Blanchard, 2005).

In Europe, fine regional background aerosol is composed by OA (40–60%), nitrate (5–20%), sulfate (5–20%), ammonium (10%), and BC (5%). OA and sulfate mass fractions increase during the warm season, while nitrate mass fraction increases during the cold season (Putaud *et al.,* 2004). In Southern Europe the transport of Saharan dust from Africa can increase the mineral dust mass fraction up to 8–14% of $PM_{2.5}$ mass (Querol *et al.,* 2009).

In North America the fine regional background aerosol is dominated by OA (25–50%) and sulfate (25–40%). Soil-derived aerosol account for less than 10% of $PM_{2.5}$ mass, although larger mass fractions are observed on the east coast (20%) due to transport of dust from Africa. BC (5%), nitrate (5–10%), and ammonium (5–10%) are minor components (Hidy and Blanchard, 2005).

The coarse regional background aerosol is mainly composed by sea salt and mineral dust.

4.2.3.4 Marine Background Aerosol

Marine background aerosol is composed by sea salt, sulfate, and OA. Sea salt is produced by bubble bursting from braking waves and capillary actions on the ocean surface due to the wind. Sea salt particles are present both in the fine and coarse modes. Fine marine particle mass is dominated by primary and SOA. POA is formed through the transfer of organic matter suspended in the ocean surface water into the atmosphere through bubble bursting. SOA is generated by the oxidation of VOCs, such as DMS, aliphatic amines, and isoprene. Figure 4.6 reports

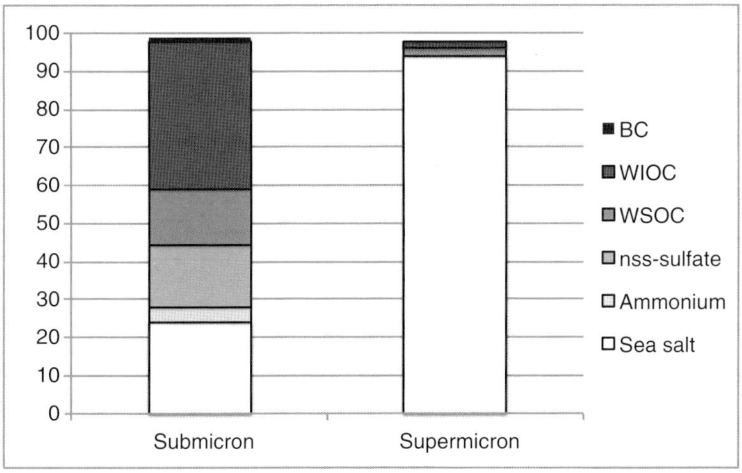

Figure 4.6 Chemical composition of submicron and supermicron marine aerosol; WSOC is water-soluble organic carbon, WIOC is water-insoluble organic carbon, and BC is black carbon.

the size-segregated chemical composition of marine aerosol measured at Mace Head, Ireland.

4.3
Size Distributions of Aerosol Chemical Compounds

Within a given environment the chemical composition of submicron particles is usually different from that of coarse mode particles, reflecting the different sources and formation mechanisms. Likewise, anthropogenically affected areas exhibit a very different aerosol chemical composition with respect to unperturbed regions.

4.3.1
Aerosol Size-Resolved Chemical Composition in Polluted Areas

Aerosol number size distribution in environments affected by anthropogenic emissions shows a maximum below 100 nm, while the mass size distribution is generally bimodal with a maximum in the submicron range and a second maximum in the supermicron range. The aerosol size distribution in urban areas is extremely variable, depending on the proximity to emission sources. For example, close to high traffic roads, the mass size distribution is characterized by a significant contribution from particles smaller than 100 nm, but their concentration decreases moving toward urban background areas. In rural environments number and mass size distributions are quite similar to the urban area distributions, although coarse particles accounts for a larger fraction of aerosol mass.

Figure 4.7 shows the typical mass size distribution of aerosol chemical constituents in an area affected by anthropogenic emissions. Dust from soil suspension, road traffic suspension, or transport are found in the supermicron region. Ammonium nitrate, ammonium sulfate, and organic compose the accumulation

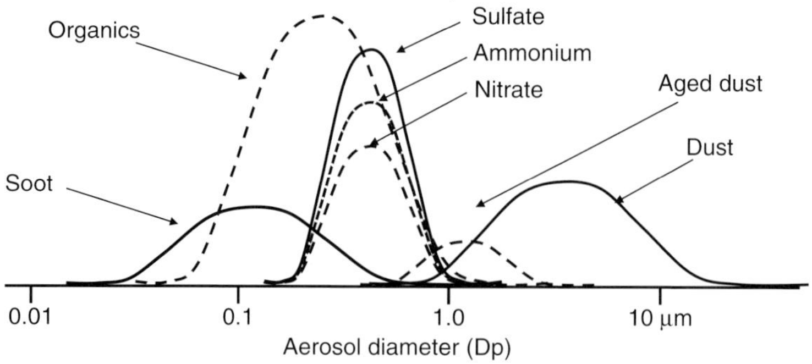

Figure 4.7 Chemical component mass size distribution in areas affected by anthropogenic emissions.

mode particles, while particles smaller than 100 nm consist mainly of OA and soot. Close to the coast, the sea salt contribution can be seen in the accumulation mode region.

4.3.1.1 **Secondary Inorganic Aerosol (Ammonium Sulfate and Nitrate)**

Often ammonium sulfate and nitrate have two modes in the accumulation region, one around 200 nm and a second one around 700 nm. The smaller mode results from the condensation of secondary inorganic species from the gas phase. The mode around 700 nm, also known as droplet mode, is due to the heterogeneous aqueous-phase reactions that lead to the formation of the inorganic salts.

In conditions of water vapor supersaturation, or even at high relative humidity, aerosol particles uptake water forming cloud and fog droplet or simply wet aerosol particles. Soluble gases, like sulfur dioxide, nitric acid, and ammonia dissolve into the aqueous medium, where several reactions can take place. As already explained in section 4.2.2.3, in the aqueous phase sulfur dioxide is oxidized to sulfate by reaction with ozone, oxygen peroxide, or oxygen under the catalytic effect of manganese and iron ions. The relative importance of the oxidation mechanism depends on the water-phase acidity. Within the aqueous medium, the acid–base reaction of sulfate, nitrate, and ammonia leads to the formation of ammonium sulfate and nitrate. These species are nonvolatile, or less volatile than water, thus water evaporation (or cloud and fog dissipation) leaves behind an aerosol enriched in secondary inorganic aerosol whose particle diameter is larger than the initial particles, typically in the range 400–700 nm. To understand the importance of the aqueous-phase reaction for the formation of secondary inorganic aerosol, it is worth noting that more than 50% aerosol sulfate is produced from the oxidation in the aqueous phase of SO_2 and subsequent reaction with the ammonium ion.

4.3.1.2 **Organic Aerosol**

Size distribution of OA is strongly related to its sources. As an example, Figure 4.8 reports the submicron size distribution of OA, together with the inorganic components, in a few urban and rural environments.

Urban OA generally exhibits a bimodal distribution with a maximum around 100 nm and a second maximum around 400–600 nm. Rural OA has a broader distribution centered around 400–600 nm, with sometimes a shoulder around 200–300 nm. These differences can be explained by the different OA sources and processing.

Processed and SOA exhibits a maximum around 400–600 nm. When condensation of fresh secondary species is observed, the size distribution might show a tail toward smaller diameters, even below 100 nm. Primary emissions from fossil combustion contribute to the formation of an additional mode between 30 and 100 nm, typical of fresh traffic emissions. Biomass burning, instead, contributes to the OA distribution with two modes at 200–300 nm and at 500 nm, due to fresh and aged emissions, respectively.

In rural environments, especially during spring and summer, OA contributes also to the coarse mode through the PBAP.

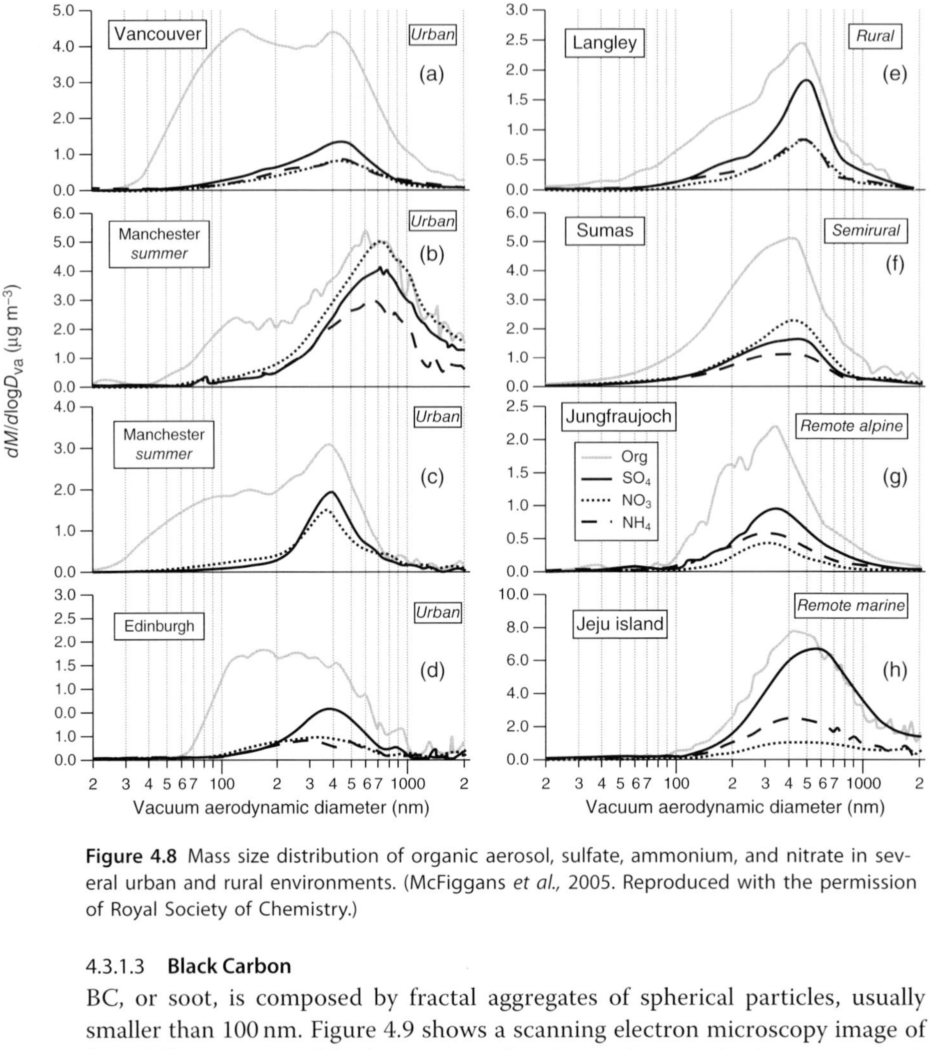

Figure 4.8 Mass size distribution of organic aerosol, sulfate, ammonium, and nitrate in several urban and rural environments. (McFiggans *et al.*, 2005. Reproduced with the permission of Royal Society of Chemistry.)

4.3.1.3 Black Carbon

BC, or soot, is composed by fractal aggregates of spherical particles, usually smaller than 100 nm. Figure 4.9 shows a scanning electron microscopy image of four different types of fresh soot particles. Depending on the interaction with other species, during and immediately after combustion, fresh soot particles are classified as embedded, partly coated, bare, and with inclusion.

Figure 4.10 shows the BC ambient size distribution, together with the size distribution of HOA and OOA, proxies of POA and SOA, respectively. The size distribution of BC close to emission sources (as shown in Figure 4.10) peaks around 100 nm D_{va} (vacuum aerodynamic diameter). Below 100 nm BC accounts for about 50% of particle mass, with the remaining mass mainly composed by OA. A second BC mode is found in particles larger than 200 nm, usually internally mixed with OA and secondary inorganic aerosol. These particles are formed during aging of fresh primary emissions. Thus, the relative contribution of the two peaks varies depending on the proximity to emission sources.

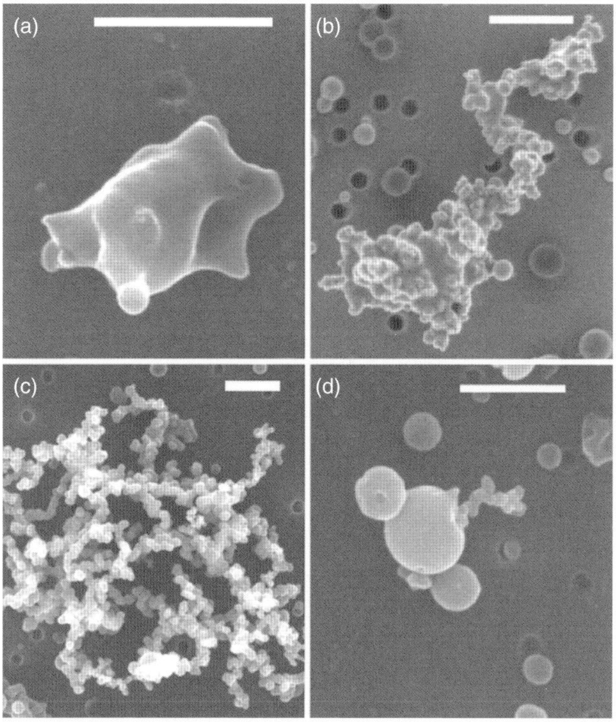

Figure 4.9 Scanning electron microscopy of soot particles: embedded (a), partly coated (b), bare (c), and with inclusion (d). (China *et al.*, 2013. Reproduced with the permission of Nature Publishing Group.)

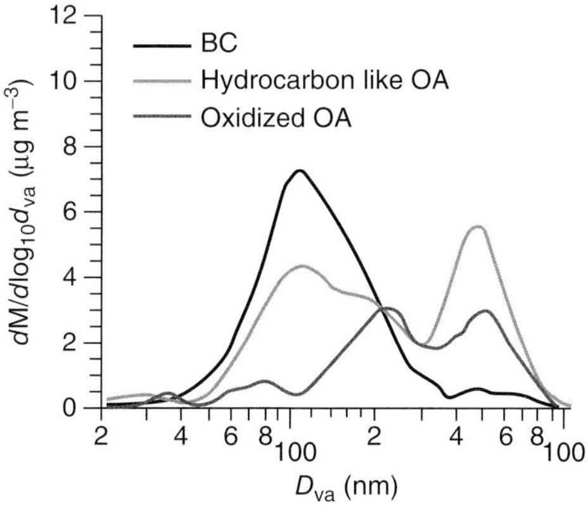

Figure 4.10 Size distribution of black carbon and organic aerosol downwind of traffic emission sources. (Massoli *et al.*, 2012. Reproduced with permission of Taylor and Francis.)

4.3.1.4 Dust

Dust size distribution at a certain location depends on the distance from emission sources. At emission, dust size distribution does not depend on wind speed and shows a maximum between 5 and 10 μm. Dust size distribution far from sources presents a maximum that ranges between 1 and 10 μm, depending on the efficiency of removal by wet and dry deposition of larger particles during atmospheric transport.

4.3.2 Aerosol Size-Resolved Chemical Composition in Unperturbed Environments

Aerosol size distributions of four different types of unperturbed, or less anthropogenically impacted regions, are presented here: the rain forest, the desert, the high altitude mountain regions, and the polar regions.

4.3.2.1 Rain Forest

Figure 4.11 reports the size distribution of major elements and ions measured during dry and wet periods in the Amazon forest.

The dry period is characterized by higher aerosol concentration due to high emission from wood burning, while the wet season shows lower concentration due to the reduced number of fires as well as to higher aerosol removal efficiency by wet deposition. The size distribution of crustal elements, such as calcium, silicon, aluminum, and iron presents a single mode in the supermicron range. Ammonium exhibits a unimodal distribution with a maximum in the submicron range, where ammonium nitrate and ammonium sulfate are formed by condensation or cloud processing. Unlike ammonium, sulfate and nitrate exhibit a second mode in the coarse region, due to the condensation of sulfur dioxide and nitric acid onto coarse particles containing alkaline species. The nitrate supermicron mode is observed also in the other forest areas, like the boreal forest, where the number concentration of fine particles is limited, and the presence of alkaline species in the coarse mode (sea salt) favors the condensation of the volatile nitric acid on supermicron particles (Cavalli *et al.*, 2006). In the dry season OA exhibits a maximum in the submicron region, due to primary biomass burning emissions and SOA, as well as to biogenic emissions. In the wet period, a significant fraction of OA is observed also within coarse particles. Large part of this coarse OA is due to the PBAP, although a contribution from condensation of SOA is also present (Fuzzi *et al.*, 2007).

4.3.2.2 High Altitude Mountain Regions

High altitude mountain regions are usually affected by free-tropospheric air masses, so the aerosol concentration is usually lower during winter and tend to increase during summer, especially during daytime, due to the increase of boundary layer height and the influence of pollutants transported from the lower altitude surrounding regions.

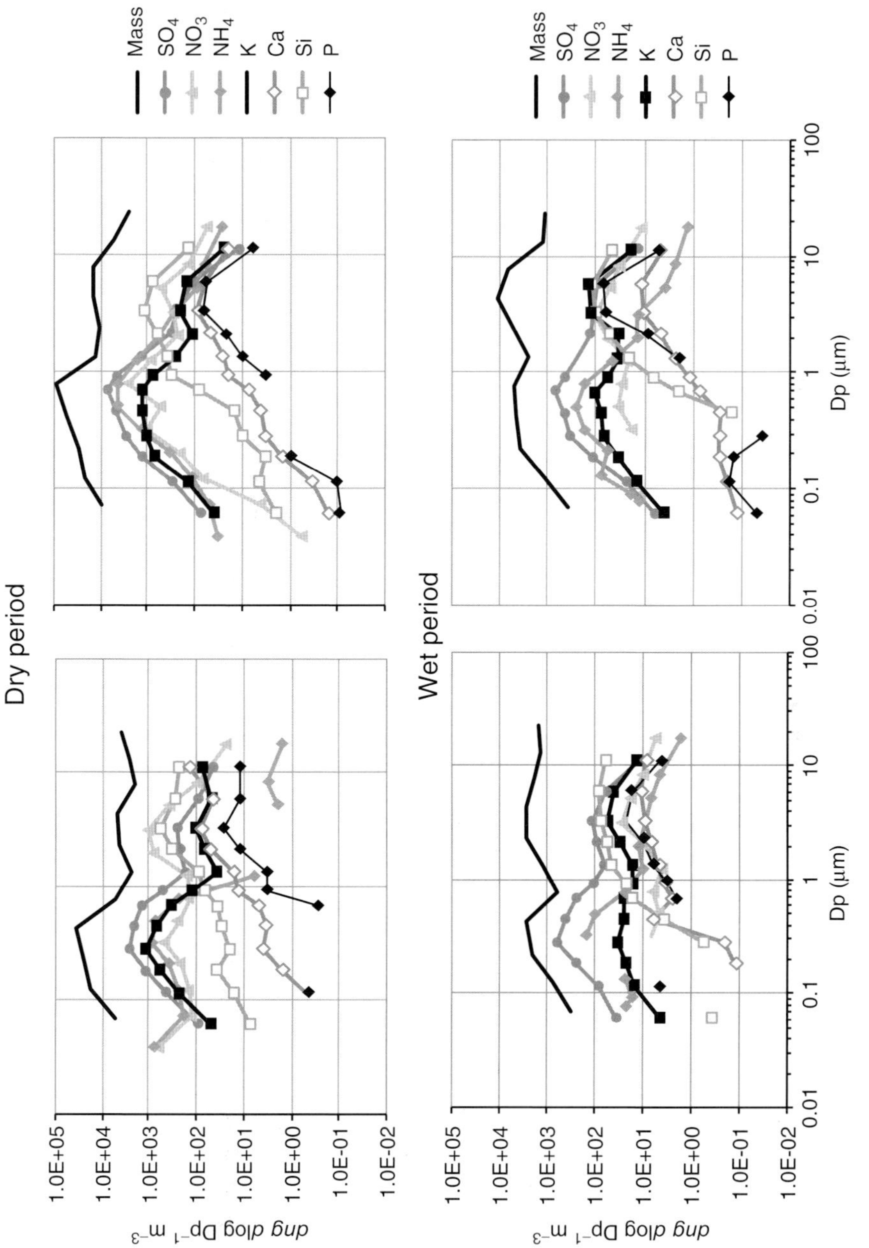

Figure 4.11 Size distribution of main inorganic species during the wet and the dry season in the Amazon forest. (Fuzzi *et al.*, 2007. Reproduced with permission of Wiley.)

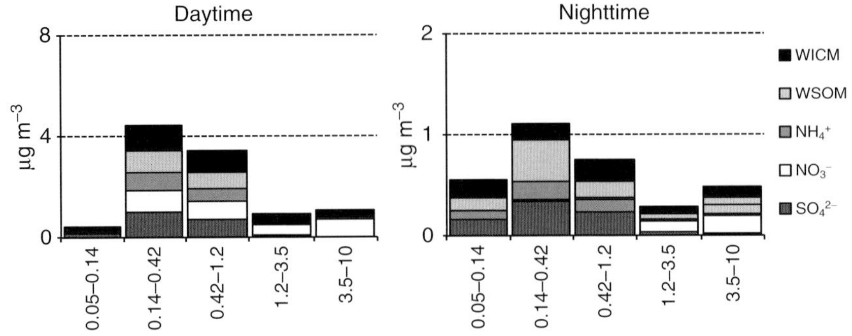

Figure 4.12 Size-segregated aerosol chemical composition at the Mt. Cimone station during summer and winter. (Carbone *et al.*, 2010. Reproduced with permission of Elsevier.)

Figure 4.12 reports the aerosol size-segregated composition at Mt. Cimone, a high altitude site (2165 m a.s.l.) of the Italian Apennines. During summer, ammonium sulfate and carbonaceous aerosol are characterized by a unimodal distribution with a maximum in the submicron range, specifically in the condensation mode between 140 and 420 nm. During daytime, nitrates are uniformly distributed over the coarse and fine particles, while during nighttime, they are observed only in the coarse particle range, associated with calcium, a tracer of mineral dust (Carbone *et al.*, 2010). The absence of nitrate in the fine particles during nighttime is likely due to the fact that the site resides in the free troposphere at night, and a lack of ammonia, typically emitted at lower altitudes by agricultural practices, does not allow the formation of ammonium nitrate in the condensation mode.

Nitrate exhibit a similar behavior at the high altitude site of Jungfraujoch (Switzerland), where their concentration increases in the coarse particles during summer compared to winter (Cozic *et al.*, 2008). Under the influence of dust transport events, high altitude sites show an increase in the coarse particle concentration of mineral elements, such as silicon, calcium, aluminum, and iron.

4.3.2.3 **Polar Regions**

Sea salt is the major component of aerosol mass in the polar regions throughout the year. During winter chloride is depleted relative to sodium due to the reaction of particles with acidic gaseous species that leads to the formation of hydrogen chloride, a gaseous species that moves chloride into the gas phase. Sea salt size distribution shows a maximum in the coarse mode in particles larger than 1–2 μm. Nitrate size distribution follows the same distribution of sodium, indicating that the main formation mechanism of aerosol nitrate is the adsorption of nitric acid on sea salt particle surface. Nevertheless, submicron particles, especially in the range 250 nm–1 μm, are enriched in nitrate relative to sodium, because of the larger surface area of the submicron mode particles relative to the coarse particles. Non-sea salt sulfate and methanesulfonic acid (MSA) are formed from the oxidation of sulfur organic species, mainly released by marine

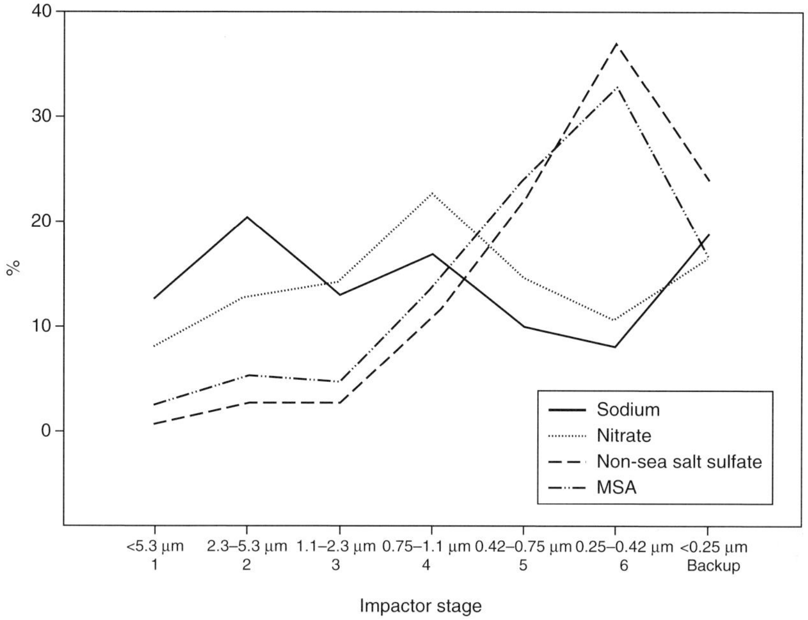

Figure 4.13 Annual average size-segregated aerosol chemical composition at Halley, Antarctica. (Rankin and Wolff, 2003 Reproduced with the permission of Wiley.)

biota, like DMS. Their concentrations are higher in spring and summer, and their size distributions peak in the condensation mode (250–420 nm). Figure 4.13 shows the size distribution of main polar aerosol constituents observed in Halley, Antarctica (Rankin and Wolff, 2003).

4.3.3 Long-Term Changes of Aerosol Chemical Components

Long-term measurements of aerosol chemical composition are not very common, and most of the sites where this kind of information is available are located in Europe, North America, and the Arctic.

In North America, sulfate, BC, and OC measurements have been performed for over 20 years at 44 sites of the IMPROVE network (1995–present). Sulfate, BC, and OC decreased by 30–40% both in urban and rural locations, reproducing the same decreasing trend observed for emissions in the United States over the same time period (Chin *et al.*, 2014). In California long-term measurements of $PM_{2.5}$ chemical composition show a decreasing trend of sulfate and nitrate concentrations from the 1980s, similar to the trend of their gas-phase precursor concentrations (NO_x and SO_2). The downward trend is due the decrease of automotive emissions and to the reduction of fuel sulfur content.

However, over the last 30 years, the largest reduction in particulate matter was observed over Europe. In Europe, long-term measurements of aerosol chemical composition started in the late 1970s to study the effect of emissions on acid

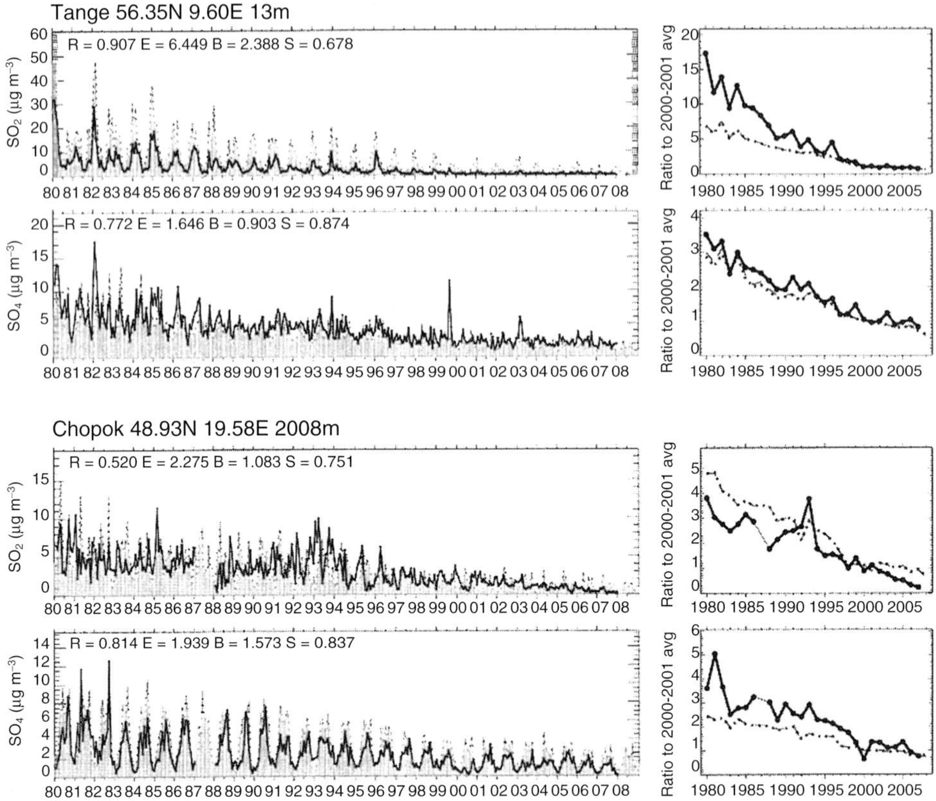

Figure 4.14 Average surface SO_2 and sulfate concentrations from 1980 to 2008 at two European sites, Tange in Denmark and Chopok in Slovakia (data from EMEP network). Observations are shown in black lines and model results on gray bars. (Chin *et al.*, 2014 http://ntrs.nasa.gov/archive/nasa/casi.ntrs.nasa.gov/20140013029.pdf created under creative commons license 3.0)

depositions. Sulfate measurements are available in 64 sites of the EMEP network starting from that period, while BC and OC measurements started only later, after the year 2000. Sulfate concentration decreased by 3–10 times, depending on the location, over the last 30 years in all Europe, although SO_2 decreased by almost 20 times (Chin *et al.*, 2014) (Figure 4.14).

The concentration of anthropogenic pollutants in the Arctic region is strongly affected by transport from the lower latitudes, especially from Europe. Similarly to Europe, sulfate in the Arctic region decreased significantly during the last 20 years, with concentrations in the 2000s two to three times lower than in the 1980s (Chin *et al.*, 2014). Sulfate and BC decreased from the mid 1980s till the late 1990s and remained relatively steady since, similarly to the emissions in Eurasia (Gong *et al.*, 2010).

Ground-based measurements in East and South Asia do not allow any discussion of long-term trends. However, AOD from satellite measurements indicates an increase of combustion aerosols over the last 30 years, but the high concentration of dust in Asia makes it difficult to correlate the increasing aerosol trend to increasing anthropogenic emissions. Long-term measurements of Asian outflow at Gosan, in Korea, show that sulfate concentration in total suspended particles decreased by 60% from 1993 to 2003 and then started to increase until present. This time trend is strictly related to the changes of SO_2 emissions in China, which exhibit a decreasing trend before 2003 followed by a steep increase afterward (Kim, Kim, and Kang, 2011).

4.4 Issues Related to Aerosol Chemical Composition

The enormous advances in aerosol measurement technologies over the last decade has allowed for an increasingly accurate understanding of the chemical composition and physical properties of atmospheric particles and of their processes in the atmosphere. Nevertheless, many open issues still remain in the atmospheric aerosol science, mainly connected to OC/EC speciation and sources and to aerosol–cloud interaction, which are addressed in this paragraph.

4.4.1 Characterization of the Aerosol Carbonaceous Fraction

4.4.1.1 Soot: BC or EC

Soot is emitted together with a large variety of other organic and inorganic chemical species, thus its determination is performed indirectly based on the aerosol optical properties, aerosol thermo-optical properties, or on the response of aerosol to laser-induced incandescence. Depending on the empirical method used for its determination, soot is defined as BC (optical method and light incandescence) or EC (thermal–optical method).

Thermo-optical analysis allows the simultaneous determination of EC and OC. Aerosol samples collected on quartz filter substrate are heated following a specific thermal protocol, first in helium atmosphere (stage 1) and then in a helium/oxygen atmosphere (stage 2). Ideally, OC would evolve during the first stage, while EC would combust in the oxidizing conditions of stage 2. In real samples, part of the OC converts into pyrolyzed carbon in stage 1, which combust during stage 2, leading to an overestimation of the EC. To correct for the pyrolytic carbon (PC), the optical properties of aerosol are continuously monitored during the different heating steps: when the PC forms the transmission of a laser beam passing through the sample decreases, when the PC is burnt under the oxidizing conditions, the laser transmission increases again. The PC is thus quantified as the amount of carbon that evolves during the first minutes of heating under oxidizing conditions before the laser transmission reaches the prepyrolysis value. The thermal protocols most commonly employed by international monitoring networks

Table 4.6 Temperature protocol commonly used for measurements by thermal–optical analysis.

Step	NIOSH 5040 *T*, duration (°C, s)	IMPROVE *T*, duration (°C, s)	EUSAAR *T*, duration (°C, s)
He1	250, 60	120, 150–580	200, 120
He2	500, 60	250, 150–580	300, 150
He3	650, 60	450, 150–580	450, 180
He4	850, 90	550, 150–580	650, 180
He/O21a	650, 30	550, 150–580	500, 120
He/O22	750, 30	700, 150–580	550, 120
He/O23	850, 30	800, 150–580	
He/O24	940, 120		
He/O25			

Cavalli *et al.* (2010).

are NIOSH5040, IMPROVE, and EUSAAR protocol. Temperature and duration of each heating step is reported in Table 4.6.

The accuracy of EC determination by thermal–optical analysis requires that the following assumptions are verified: (i) the optical properties of PC and EC are identical to guarantee that the pyrolysis correction is adequate, (ii) OC is completely removed during the first heating stage in helium atmosphere, and (iii) EC combustion does not begin in stage 1. The three assumptions rarely hold, but the three protocols have been optimized to minimize potential biases.

BC mass concentration is estimated from light attenuation measurements. Light attenuation is measured comparing the intensity of light passing through a reference path and the light passing through the aerosol deposited on a filter. In this case the attenuation data need to be corrected for:

1. Backscattering (a portion of the incident light will be backscattered by the aerosol deposited on the collection substrate, leading to an overestimation of light attenuation).
2. Multiple scattering (filter fibers promotes the scattering of light that passes through the filter, increasing its optical path and leading to an overestimation of BC mass).
3. Shadowing effect (high particle loading can prevent the interaction of light with BC particles hidden into the collection substrate, underestimating the BC mass).

A schematic representation of these effects is reported in Figure 4.15.

To calculate BC mass concentration, light attenuation is then converted into equivalent black carbon (EBC) concentration based on the mass absorption cross-section (MAC) at the same wavelength (λ) of the light attenuation measurements:

$$\mathrm{EBC} = s(\lambda)/\mathrm{MAC}(\lambda)[\mu\mathrm{g\,m^{-3}}] = [\mathrm{Mm^{-1}}]/[\mathrm{m^2\,g^{-1}}]$$

Determination of BC from light attenuation implies three assumptions that are rarely verified:

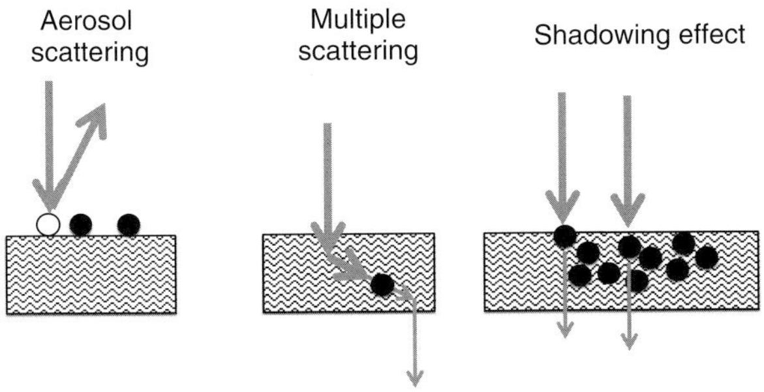

Figure 4.15 Schematic representation of potential artefacts affecting the accurate determination of BC light attenuation.

1. Carbonaceous species are the only absorbing species at the measurement wavelength.
2. Organic species do not contribute to light absorption.
3. The relationship between light attenuation and BC concentration is well known and predictable.

Determination of BC based on incandescence measurements is increasingly widespread. Soot particles, illuminated by a laser source, absorb the light and are heated up to very high temperature (about 4000 K). As a consequence, soot emits an incandescence signal, which is proportional to soot particle volume. With a proper calibration, BC mass concentration can then be determined.

4.4.1.2 **Organic Aerosol**

OA is a mixture of thousands of different chemical species, which includes aliphatic and aromatic hydrocarbons, polycyclic aromatic compounds, aldehydes, ketones, carboxylic acids and polyacids, ammines, alkylnitrates, polyols, and sugars. Volatility and polarity of the organic species in particulate matter ranges over orders of magnitude, as reported in section 4.2.1. In literature the organic fraction of particulate matter is also indicated with the terms OC and organic mass (OM). OC refers to the amount of carbon that composes the OA and is usually quantified with thermo-optical methods, together with the EC. OM is the mass of OA estimated from the OC and assuming a constant and known OM to OC ratio so that

$$\mathrm{OM} = \mathrm{OC} \times \mathrm{OM/OC}$$

The OM to OC ratio is measured by mass spectrometric techniques (aerosol mass spectrometers) or is determined from the organic functional group analysis by infrared spectrometry. OM to OC ratios reported in literature range between 1.2 for fresh OA and 2.4 for highly processed OA (Aiken *et al.*, 2008).

OA can be measured online and offline. Online techniques provide a real-time measurement of OA content and properties at the resolution of minutes, removing sampling artefacts, risks of sample contamination, and risks of losses during sample preparation. The scheme in Figure 4.16 illustrates the different available

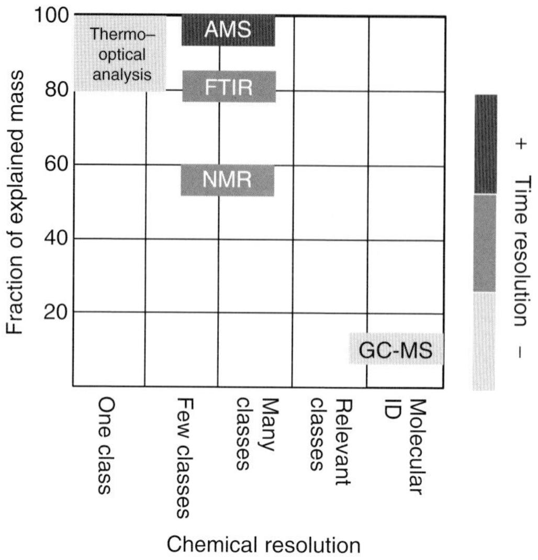

Figure 4.16 Organic aerosol measurement techniques.

techniques for OA measurements that differ for different degrees of explained mass and for the degree of chemical speciation.

OA is quantified as OC with thermo-optical analysis described in the previous paragraph. This technique allows quantification of OC, thus no speciation is performed, but the quantification is complete (100% of OA is described). Molecular species in the aerosol can be identified and quantified using chromatographic techniques coupled with mass spectrometry, like gas chromatography–mass spectrometry (GC-MS) and high-performance liquid chromatography–mass spectrometry (HPLC-MS). These techniques are based on the separation of organic molecules and their subsequent identification with a mass spectrometer. This approach requires first the extraction of organic molecules from the aerosol matrix and eventually their chemical derivatization prior to analysis. This approach allows the identification of individual molecular species. Although extremely useful to study specific organic tracers or toxic compounds, this approach does not produce a complete representation of the OA since only a small fraction (10–40%) of its mass is speciated.

Organic functional group analysis is performed with spectrometric techniques, like Fourier transform infrared spectrometry (FT-IR) and nuclear magnetic resonance (NMR) spectrometry. The functional groups of OA absorb infrared radiation at different wavelengths. From the absorption intensity it is possible to quantify the different organic functional groups, of which the most common are carbon–oxygen single and double bonds, hydrogen–carbon, hydrogen–oxygen, and hydrogen–nitrogen single bonds. From the organic functional group composition, it is then possible to reconstruct completely the OA mass. Similarly to FT-IR, the proton NMR is employed to quantify specific organic functional groups, like the alkyl and aryl moieties, and the carbon–oxygen single and double bonds. This technique is usually employed for the analysis of the water-soluble OA fraction.

Recently, mass spectrometry has been applied for the online characterization of OA. The time-of-flight aerosol mass spectrometer (TOF-AMS) allows the online quantification of OA, together with inorganic aerosol main components (sulfate, nitrate, ammonium, and chloride). The aerosol is introduced into the aerosol-sampling chamber at a pressure of 1.8 hPa through a critical orifice of 100 μm and an aerodynamic lens that focuses particles between 50 and 600 nm into a narrow beam. Particles are then accelerated at a terminal velocity that depends on their size, so that particle size can be retrieved from the time that particles need to cross the sizing chamber. After the sizing chamber, particles impact a tungsten vaporizer at 600 °C, and the molecules are then ionized by electron impact at 70 eV. The ion beam is then focused by electrostatic lenses into the mass spectrometer where mass spectra of organic and inorganic molecules are recorded (Figure 4.17). The analysis of the mass spectra allows the quantification of OA, together with the determination of its elemental composition.

4.4.2 Sources of BC and OA

4.4.2.1 Black Carbon

BC sources can be grouped into four sectors: transportation, industry, residential heating, and open burning. Transportation emissions are mainly due to diesel

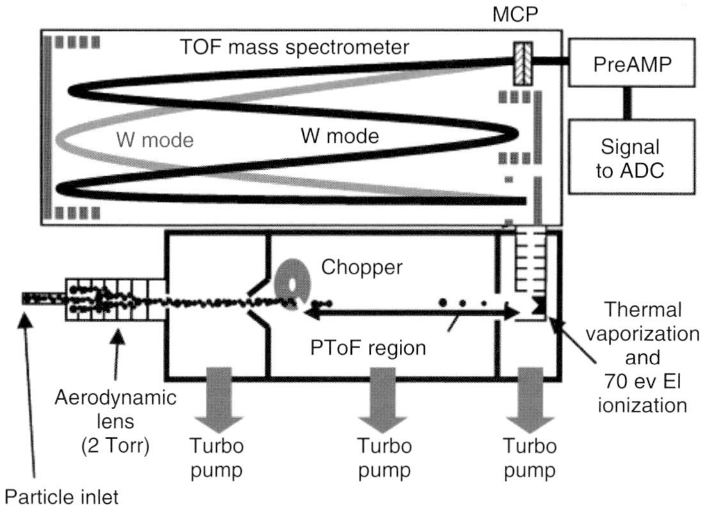

Figure 4.17 Schematic description of the time-of-flight aerosol mass spectrometer. (DeCarlo *et al.*, 2006. Reproduced with the permission of American Chemical Society.)

Table 4.7 Best estimates of bottom-up BC emissions.

BC	BC	BC range
All sources		
Energy	4770	1220
Open	2760	800
Total	7530	2020
1750 background		
Energy related	390	
Open burning	1020	
Total background	1410	
Industrial era		
Energy related	4380	
Open burning	1740	
Total background	6120	

fuel combustions. At global scale, the largest source of BC is open burning. Open burning emissions are mainly due to savannah and forest fires, which accounts for 45% and 40% of open burning BC emissions, respectively (Table 4.7). The relative contribution of the other sectors depends on the location and the time of the year. Coal and biomass burning for residential heating accounts for 60% to 80% of BC emissions in Asia and Africa, while diesel fuel combustion accounts for about 70% of BC emissions in Europe, North America, and South America.

The largest BC emissions are observed in South America, Africa, and Asia, with open burning dominating South American and African emissions and industrial coal combustion and biofuel combustion for cooking dominating the Asian emissions. The highest BC emissions from open burning in Africa take place from May to August and in November–December, while in South America the highest emissions are observed between August and October. At higher latitudes open burning emissions are not as relevant.

Global BC emissions from a bottom-up approach are equivalent to 7500 Gg per year with an uncertainty ranging from 2000 to 29 000 Gg per year. The contribution from energy-related sectors (transportation, industry, and residential heating) is 4800 Gg per year with an uncertainty range of 1200–15 000 Gg per year. The main source of uncertainty in the energy production sectors are the emission factors measured in the laboratory which do not exactly represent ambient conditions, the uncertainty in BC and OC speciation, the difficulty to quantify the number of emitting sources (vehicles, furnaces, etc.), and to the limited information on the fuel consumption in some sectors like industrial and residential solid fuel combustion (Table 4.8).

The open burning estimate of global BC emission ranges between 740 and 12 800 Gg per year. The uncertainty of open burning BC emissions is dominated by the uncertainty in the emission factors. BC measurements are subject to biases depending in the measurement techniques employed. These biases are especially large for the open burning BC due to the influence of coemitted species on

Table 4.8 Grades of refinement of BC emission inventories.

Region	Energy use		
	Urban	Rural	Open burning
North America	High	High	Medium+
Latin America	Medium+	Low	Low
Europe	High	High	Medium+
EECCA	Low	Low	Medium
Middle East	Low	Low	Low
Pacific	Low	Low	Low
Africa	Low	Low	Medium
East Asia	High	Medium+	Medium
South Asia	Medium+	Low	Low
Southeast Asia	Low	Low	Low

Bond *et al.* (2013); EECCA stays for Eastern Europe, Caucasus and Central Asia.

thermal and optical aerosol properties, which are used for the BC determination. In addition, BC measurements based on thermal and optical properties can differ significantly, further increasing the BC measure uncertainty. Uncertainty in fuel load and combustion completeness is about a factor of two, while data on combusted areas are uncertain within 10%.

Models are used to predict the trend in BC emissions in the future. Different models agree in predicting a reduction by 50% of 2010 emissions by the end of the century at the global scale. These predictions rely on the assumption that cleaner fuels will be used and that emission factors will decrease with the increase of population income. On the contrary, at regional scale, several scenarios indicate an increase in emissions in Asia, Middle East, and Africa during the first decades of the century.

4.4.2.2 **Organic Aerosol**

The OA global budget has been reported in paragraph 2.2. Global primary OC budget is about 34 Tg per year. The sources of primary OC are different from those of BC. In fact, contained combustion accounts for a small fraction of OC, mainly due to the larger OC emission factors with respect to open burning. Globally, open burning accounts for 74% of OC emissions, while fossil fuel combustion and biofuel combustion emit 7% and 19% of total OC, respectively. The uncertainty range of the primary OC budget is 17–77 Tg per year. This uncertainty is mainly due to the uncertainty of the emission factors from open burning and the uncertainty associated to the low-technology combustion emissions.

The largest OA uncertainty affects the SOA budget, mainly due to the still limited knowledge of SOA formation mechanisms, their yields under ambient conditions, and the knowledge on SOA precursors. SOA indicates OA particles formed in the atmosphere following physical or chemical transformation of precursors in gas or aerosol phase, including nucleation, gas-phase chemistry followed by condensation, condensed-phase chemistry, and aging.

Nucleation forms a critical nucleus that then grows to a detectable size. Organic molecules are essential for nucleation since they stabilize critical nucleus formed by sulfuric acid molecules, decreasing the nucleation barrier energy and favoring the nucleus growth (Zhang, 2010).

The traditional description of SOA formation implies the oxidation of gas-phase precursors in the atmosphere to form low-volatile products that then condense on preexisting particles. The condensation equilibrium is described by the Pankow equation:

$$K = S_p/(S_g M),$$

where K is the partitioning constant, M is the mass concentration of particulate matter into which the product can dissolve (i.e., OA), and S_p and S_g are the concentrations of the oxidation products in the particle and the gas phase, respectively. It follows that the fraction of condensed matter F is

$$F = S_p/(S_p + S_g) = 1/(1 + c^*/M)$$

So the condensation equilibrium is shifted toward the condensed phase at higher OA mass concentration and at lower volatility (c^*) of the oxidation products. Oxidation and oligomerization reactions form compounds with lower volatility compared to their precursors, so they contribute substantially to the formation of SOA. Oxidation reactions add moieties like carboxylic acid, carbonyl, alcohol, and nitrate groups to the OC chain, while oligomerization increases the carbon chain length.

SOA formation takes place also in the condensed phase. OA precursors condense in the liquid water associated to aerosol particles, where oxidation and oligomerization reactions form low-volatility products that stay in the aerosol phase even after water evaporation. These reactions are likely responsible for the formation of highly oxidized and water-soluble SOA.

OA formation through aging is schematically represented in Figure 4.18. The volatile species R is transformed into S1 that can partition into the aerosol phase. S1 can be converted, both in the gas or aerosol phase, into second-generation products (S2 in the gas phase and P1 in the aerosol phase). S2 partitions again between the gas and the aerosol phase. Third-generation products can then be formed, increasing the mass of SOA and its complexity. Aging explains the

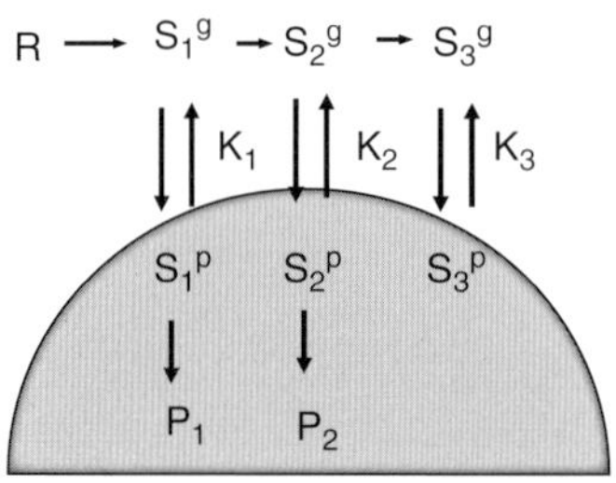

Figure 4.18 Schematic representation of OA aging.

presence in the aerosol phase of species with large molecular weight, large number of functional groups, and high degree of oxidation.

A comprehensive description of oxidation, oligomerization, and fragmentation reactions involving OA formation and aging is reported in Kroll and Seinfeld (2008).

4.4.3 Effect of Organic and Inorganic Chemical Composition on Aerosol Activity as Cloud Condensation Nuclei and Ice Nuclei

4.4.3.1 Cloud Condensation Nuclei

The cloud condensation nuclei (CCN) efficiency of an aerosol population is defined by the ratio between the number of particles that activate at a certain supersaturation and the total number of particles (CCN/CN). CCN ability is determined by particle size and the content of hydrophilic chemical species. Size-resolved CCN measurements shows that the cutoff diameter, that is, the diameter at which CCN efficiency is equal to 50%, varies only slightly with aerosol types, indicating that CCN concentration is mainly determined by the aerosol population size distribution (Dusek *et al.*, 2006). At constant size distribution, CCN efficiency is affected by the aerosol composition. Figure 4.19 shows the CCN efficiency for different particle types at $S = 0.4\%$. At sizes of 50 nm, only hygroscopic particles, like pure sulfate and sea salt particles, are completely activated, while regional background aerosol, which likely contains a fraction of carbonaceous and less hygroscopic species, reaches 50% CCN efficiency only at 70 nm. The CCN efficiency of pyrogenic particles, that is, particles from combustion processes, is larger than 50% only for diameters above 120 nm.

At a given supersaturation, the Kohler theory can be used to predict the critical diameter based on the physical and chemical properties of the aerosol

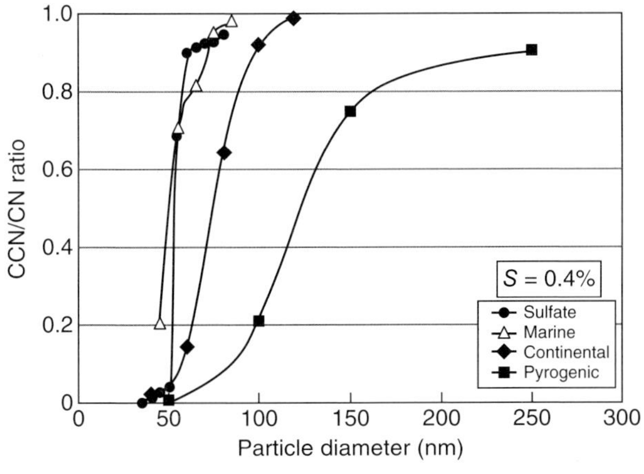

Figure 4.19 CCN efficiency at $S = 0.4\%$ as a function of aerosol size for different aerosol types. (Andreae and Rosenfeld, 2008, Reproduced with the permission of Elsevier.)

population, including solute mass, molecular weight, bulk density, dissociable ions, and activity coefficient. Ambient aerosol is generally composed by internally mixed particles, that is, particles where a variety of different chemical species coexist, including organic and inorganic components. The exact chemical composition of the organic fraction, which often dominates the aerosol mass, is still unknown, thus the application of the Kohler theory is difficult.

The k-Köhler theory is a simplified parameterization of the Köhler theory that describes the CCN efficiency of an aerosol population based on the hygroscopicity parameter k. The supersaturation ratio over a particle is defined as

$$S(D) = D^3 - D_d^{\ 3}/(D^3 - D_d^{\ 3}(1-k))\exp(4\sigma_{s/a}M_w/RT\rho_w D),$$

where D is the particle diameter, D_d is the dry particle diameter, $\sigma_{s/a}$ is the surface tension, M_w is the water molecular weight, R is the universal gas constant, T is the air temperature, ρ_w is the water density, and k is the hygroscopicity parameter defined as

$$k = \Sigma\varepsilon_i k_i.$$

k ranges between 0, for nonhygroscopic particles, and 1.4 for very hygroscopic salts like sodium chloride. The k-Köhler theory applies both at supersaturated and subsaturated conditions. k can be empirically determined below saturation (RH < 100%) from growth factor (GF) measurements and above saturation (RH > 100%) from CCN measurements.

A list of k values for different chemical species and derived from GF and CCN measurements is reported in Table 4.9. Generally, ambient aerosol exhibits k values between 0.1 and 0.9 (Petters and Kredenweis, 2007). Although k for inorganic salts observed in ambient aerosol are relatively well known, OA k are characterized by a larger uncertainty. k-Org of ambient and smog chamber generated OA varies between 0.04 and 0.22, with lower values for fresh/less aged OA and higher values for more aged/processed OA (Chang *et al.*, 2010).

Laboratory and field experiments have shown that the hygroscopicity of OAs increases with increasing oxygen content. For this reason, a simplified parameterization has been suggested to estimate k from the oxygen to carbon ratio of the organic fraction:

$$k\text{-org} = a * O/C + b.$$

Table 4.10 reports the coefficients of this equation proposed by two different authors.

Although this simple parameterization explains a certain number of field observations, the previous equation does not apply to individual molecular species.

4.4.3.2 **Ice Nuclei**

Homogeneous ice nucleation is the formation of ice crystals from freezing of liquid water without the presence of solid particles and takes place at supercooled conditions, that is, below −38 °C. At ambient conditions, heterogeneous ice nucleation is more common and requires the presence of ice nuclei (IN). IN are

Table 4.9 *k* values for organic and inorganic species determined experimentally with GF and CCN measurements.

Compound	GF derived			CCN derived		
	k low	*k* mean	*k* up	*k* low	*k* mean	*k* up
$(NH_4)_2SO_4$	0.33	0.53	0.72	N/A	0.61	N/A
NH_4NO_3	N/A	N/A	N/A	0.577	0.67	0.753
NaCl	0.91(6)	1.12	1.33	N/A	1.28	N/A
H_2SO_4	N/A	1.19	N/A	N/A	0.9	N/A
$NaNO_3$	N/A	0.8	N/A	N/A	0.88	N/A
$NaHSO_4$	N/A	1.01	N/A	N/A	0.91	N/A
Na_2SO_4	N/A	0.68	N/A	N/A	0.8	N/A
$(NH_4)_3H(SO_4)_2$	N/A	0.51	N/A	N/A	0.65	N/A
Malonic acid	0.28	0.44	0.6	0.199	0.227	0.255
Glutaric acid	0.12	0.2	0.28	0.054	0.088	0.016
				0.113	0.195	0.376
Glutamic acid	N/A	0.154	N/A	0.113	0.182	0.319
				0.158	0.248	0.418
Succinic acid	N/A	<0.006	N/A	0.166	0.231	0.295
Adipic acid	N/A	<0.006	N/A	N/A	0.096	N/A
Levoglucosan	0.15	0.165	0.18	0.193	0.208	0.223
Phthalic acid	N/A	0.059	N/A	0.022	0.051	0.147
Homophthalic acid	N/A	0.081	N/A	0.048	0.094	0.212
Leucine	N/A	N/A	N/A	0.001	0.002	0.003
Pinic acid	N/A	N/A	N/A	0.158	0.248	0.418
Pinonic acid	N/A	<0.006	N/A	0.063	0.106	0.196
Norpinic acid	N/A	N/A	N/A	0.113	0.182	0.318
Poly(acrylic acid)		0.06	N/A	0.04	0.051	0.068
SOA from alpha-pinene	N/A	0.022	N/A	0.014	0.1 ± 0.04	0.091
		0.07		0.028		0.229
		0.037				
SOA from beta-pinene	N/A	0.022	N/A	0.033	0.1 ± 0.04	0.178
		0.009		0.033		0.106

Petters and Kreidenweis (2007).

Table 4.10 Parameterization of organic hygroscopicity as a function of organic oxygen content.

References	a	b
Lambe *et al.*, (2011)	0.18	0.03
Chang et al. (2010)	0.29	0

solid and insoluble particles with favorable surface properties that can initiate freezing of water at temperature well above the supercooled conditions. Typically, less than one in a million of atmospheric particles can act as IN.

Heterogeneous nucleation takes place through four main mechanisms: immersion, contact, deposition, and condensation. Immersion freezing occurs when the

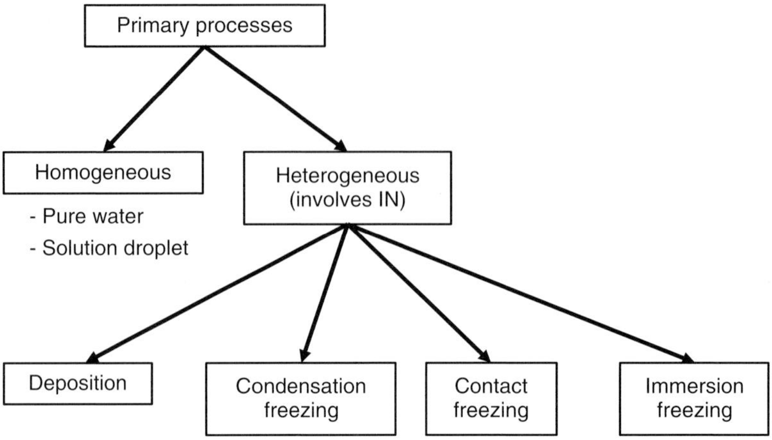

Figure 4.20 Schematic summary of ice formation processes.

insoluble particle is immersed in a water droplet and initiates freezing. Contact freezing implies the collision of the IN with a supercooled droplet. Deposition freezing is the formation of ice after condensation of gas-phase water on an ice nucleus. Condensation freezing implies the condensation of liquid water on an ice nucleus and subsequent freezing (Figure 4.20).

Mineral dust particles are known as effective IN. IN formation can occur through all four freezing mechanisms, although immersion freezing is favored when particles contain even small amount of water-soluble material. The relevance of mineral dust as IN is due to their presence throughout the atmosphere, even far from their source regions. For example, over the United States, dust from Africa or Asia can dominate fine aerosol mass in spring and summer (DeMott *et al.*, 2003).

Soot particles can act as IN as well. The ability of soot to act as IN is affected by the presence on the particle surface of functional groups able to form hydrogen bonds with water molecules (Gorbunov *et al.*, 2001), though the IN activity is reduced by the presence of organic or sulfate coating (Andreae and Rosenfeld, 2008).

Some PBAP, like some species of bacteria, pollens, and spores, can also act as IN. This is likely due the symmetry of their cell walls constituted by an amino acid structure, which act as a template for ice formation. At a global scale, PBAP account for a small fraction of IN, but their role is significant at regional scale, like in the Amazon region, where PBAP sources are particularly important (Prenni *et al.*, 2009).

Abbreviations

BC	black carbon
CCN	cloud condensation nuclei
CN	total number of particles

DMS	dimethylsulfide
EBC	equivalent black carbon
EC	elemental carbon
EECCA	Eastern Europe, Caucasus, and Central Asia
ELVOC	extremely low-volatile organic compound
EMEP	European Monitoring and Evaluation Programme
FT-IR	Fourier transform infrared (spectrometry)
GC-MS	gas chromatography–mass spectrometry
GF	growth factor
HOA	hydrocarbon like organic aerosol
HPLC-MS	high performance liquid chromatography–mass spectrometry
IN	ice nuclei
IVOC	intermediate volatile organic compound
LVOC	low volatile organic compound
MAC	mass absorption cross-section
MSA	methanesulfonic acid
NMR	nuclear magnetic resonance (spectrometry)
OA	organic aerosol
OC	organic carbon
OM	organic mass
OOA	oxidized organic aerosol
PBAP	primary biogenic aerosol particles
PC	pyrolytic carbon
SECA	sulfur Emission Control Area
SOA	secondary organic aerosol
SVOC	semivolatile organic compound
TOF-AMS	time-of-flight aerosol mass spectrometer
VOC	volatile organic compound
WICM	water-insoluble carbonaceous matter
WIOC	water-insoluble organic carbon
WSOC	water-soluble organic carbon
WSOM	water-soluble organic matter

List of Symbols

Latin

c^*	volatility of an oxidation product
D	particle diameter
D_d	dry particle diameter
D_{va}	vacuum aerodynamic diameter
F	fraction of condensed matter
k	hygroscopicity parameter
K	partitioning constant

M	particle mass concentration
M_w	water molecular weight
OS_c	carbon oxidation state
pH	numeric scale used to specify the acidity or basicity of an aqueous solution, approximately equal to the negative of the logarithm to base 10 of concentration
PM_1	24-hour average mass concentration of particles having aerodynamic diameter <1.0 μm
$PM_{2.5}$	24-hour average mass concentration of particles having aerodynamic diameter <2.5 μm
PM_{10}	24-hour average mass concentration of particles having aerodynamic diameter <10 μm
R	universal gas constant, equal to 8.31446 J K^{-1} mol^{-1}
S_g	concentration of the oxidation products in the gas phase
S_p	concentration of the oxidation products in the particle phase
T	air temperature

Greek

ρ_w	water density
$\sigma_{s/a}$	surface tension

References

Aiken, A.C., Decarlo, P.F., Kroll, J.H., Worsnop, D.R., Huffman, J.A., Docherty, K., Ulbrich, I.M., Mohr, C., Kimmel, J.R., Sueper, D., Sun, Y., Zhang, Q., Trimborn, A., Northway, M., Ziemann, P.J., Canagaratna, M.R., Onasch, T.B., Alfarra, R., Prevot, A.S.H., Dommen, J., Duplissy, J., Metzger, A., Baltensperger, U., and Jimenez, J.L. (2008) O/C and OM/OC ratios of primary, secondary, and ambient organic aerosols with high-resolution time-of-flight aerosol mass spectrometry. *Environ. Sci. Technol.*, **42**, 4478–4485. doi: 10.1021/es703009q.

Andreae, M.O. and Rosenfeld, D. (2008) Aerosol-cloud-precipitation interactions. Part 1. The nature and sources of cloud-active aerosols. *Earth Sci. Rev.*, **89**, 13–41. doi: 10.1016/j.earscirev.2008.03.001

Bond, T.C., Doherty, S.J., Fahey, D.W., Forster, P.M., Berntsen, T., DeAngelo, B.J., Flanner, M.G., Ghan, S., Kärcher, B., Koch, D., Kinne, S., Kondo, Y., Quinn, P.K., Sarofim, M.C., Schultz, M.G., Schulz, M., Venkataraman, C., Zhang, H., Zhang, S., Bellouin, N., Guttikunda, S.K., Hopke, P.K., Jacobson, M.Z., Kaiser, J.W., Klimont, Z., Lohmann, U., Schwarz, J.P., Shindell, D., Storelvmo, T., Warren, S.G., and Zender, C.S. (2013) Bounding the role of black carbon in the climate system: a scientific assessment. *J. Geophys. Res.*, **118** (11), 5380–5552. doi: 10.1002/jgrd.50171.

Bond, T.C., Streets, D.G., Yarber, K.F., Nelson, S.M., Woo, J.-H., and Klimont, Z. (2004) A technology-based global inventory of black and organic carbon emissions from combustion. *J. Geophys. Res.*, **109**, D14. doi: 10.1029/2003JD003697

Carbone, C., Decesari, S., Mircea, M., Giulianelli, L., Finessi, E., Rinaldi, M., Fuzzi, S., Marinoni, A., Duchi, R., Perrino, C., Sargolini, T., Vardè, M., Sprovieri, F., Gobbi, G.P., Angelini, F., and Facchini, M.C. (2010) Size-resolved aerosol chemical composition over the Italian Peninsula during typical summer and winter conditions. *Atmos. Environ.*, **44** (39), 5269–5278. doi: 10.1016/j.atmosenv.2010.08.008

Cavalli, F., Facchini, M.C., Decesari, S., Emblico, L., Mircea, M., Jensen, N.R., and Fuzzi, S. (2006) Size-segregated aerosol chemical composition at a boreal site in southern Finland, during the QUEST project. *Atmos. Chem. Phys.*, **6**, 993–1002. doi: 10.5194/acp-6-993-2006

Cavalli, F., Viana, M., Yttri, K.E., Genberg, J., and Putaud, J.-P. (2010) Toward a standardised thermal-optical protocol for measuring atmospheric organic and elemental carbon: the EUSAAR protocol. *Atmos. Meas. Tech.*, **3**, 79–89. doi: 10.5194/amt-3-79-2010

Chang, R.Y.-W., Slowik, J.G., Shantz, N.C., Vlasenko, A., Liggio, J., Sjostedt, S.J., Leaitch, W.R., and Abbatt, J.P.D. (2010) The hygroscopicity parameter (kappa) of ambient organic aerosol at a field site subject to biogenic and anthropogenic influences: relationship to degree of aerosol oxidation. *Atmos. Chem. Phys.*, **10**, 5047–5064. doi: 10.5194/acp-10-5047-2010

Chin, M., Diehl, T., Tan, Q., Prospero, J.M., Kahn, R.A., Remer, L.A., Yu, H., Sayer, A.M., Bian, H., Geogdzhayev, I.V., Holben, B.N., Howell, S.G., Huebert, B.J., Hsu, N.C., Kim, D., Kucsera, T.L., Levy, R.C., Mishchenko, M.I., Pan, X., Quinn, P.K., Schuster, G.L., Streets, D.G., Strode, S.A., Torres, O., and Zhao, X.-P. (2014) Multidecadal aerosol variations from 1980 to 2009: a perspective from observations and a global model. *Atmos. Chem. Phys.*, **14**, 3657–3690. doi: 10.5194/acp-14-3657-2014

China, S., Mazzoleni, C., Gorkowski, K., Aiken, A.C., and Dubey, M.K. (2013) Morphology and mixing state of individual freshly emitted wildfire carbonaceous particles. *Nat. Commun.*, **4**, 2122. doi: 10.1038/ncomms3122

Cozic, J., Mertes, S., Verheggen, B., Cziczo, D.J., Gallavardin, S.J., Walter, S., Baltensperger, U., and Weingartner, E. (2008) Black carbon enrichment in atmospheric ice particle residuals observed in lower tropospheric mixed phase clouds. *J. Geophys. Res.*, **113**, D15. doi: 10.1029/2007JD009266

DeCarlo, P.F., Kimmel, J.R., Trimborn, A., Northway, M.J., Jayne, J.T., Aiken, A.C., Gonin, M., Fuhrer, K., Horvath, T., Docherty, K.S., Worsnop, D.R., and Jimenez, J.L. (2006) Field-deployable, high-resolution, time-of-flight aerosol mass spectrometer. *Anal. Chem.*, **78** (24), 8281–8289. doi: 10.1021/ac061249n

DeMott, P.J., Sassen, K., Poellot, M.R., Baumgardner, D., Rogers, D.C., Brooks, S.D., Prenni, A.J., and Kreidenweis, S.M. (2003) African dust aerosols as atmospheric ice nuclei. *Geophys. Res. Lett.*, **30**, 14. doi: 10.1029/2003GL017410

Donahue, N.M., Kroll, J.H., Pandis, S.N., and Robinson, A.L. (2012) A two-dimensional volatility basis set – Part 2: diagnostics of organic-aerosol evolution. *Atmos. Chem. Phys.*, **12**, 615–634. doi: 10.5194/acp-12-615-2012

Dusek, U., Frank, G.P., Hildebrandt, L., Curtius, J., Schneider, J., Walter, S., Chand, D., Drewnick, F., Hings, S., Jung, D., Borrmann, S., and Andreae, M.O. (2006) Size matters more than chemistry for cloud-nucleating ability of aerosol particles. *Science*, **312** (5778), 1375–1378. doi: 10.1126/science.1125261

Faloona, I. (2009) Sulfur processing in the marine atmospheric boundary layer: a review and critical assessment of modeling uncertainties. *Atmos. Environ.*, **43** (18), 2841–2854. doi: 10.1016/j.atmosenv.2009.02.043

Fuzzi, S., Decesari, S., Facchini, M.C., Cavalli, F., Emblico, L., Mircea, M., Andreae, M.O., Trebs, I., Hoffer, A., Guyon, P., Artaxo, P., Rizzo, L.V., Lara, L.L., Pauliquevis, T., Maenhaut, W., Raes, N., Chi, X., Mayol-Bracero, O.L., Soto-Garcia, L.L., Claeys, M., Kourtchev, I., Rissler, J., Swietlicki, E., Tagliavini, E., Schkolnik, G., Falkovich, A.H., Rudich, Y., Fisch, G., and Gatti, L.V. (2007) Overview of the inorganic and organic composition of size-segregated aerosol in Rondonia, Brazil, from the biomass-burning period to the onset of the wet season. *J. Geophys. Res.*, **112**, D1. doi: 10.1029/2005jd006741

Gong, S.L., Zhao, T.L., Sharma, S., Toom-Sauntry, D., Lavoué, D., Zhang, X.B., Leaitch, W.R., and Barrie, L.A. (2010) Identification of trends and interannual variability of sulfate and black carbon in the Canadian High Arctic: 1981–2007. *J. Geophys. Res.*, **115**, D7. doi: 10.1029/2009JD012943

Gorbunov, B., Baklanov, A., Kakutkina, N., Windsor, H.L., and Toumi, R. (2001) Ice nucleation on soot particles. *J. Aerosol Sci.*, **32** (2), 199–215. doi: 10.1016/S0021-8502(00)00077-X

Hallquist, M., Wenger, J.C., Baltensperger, U., Rudich, Y., Simpson, D., Claeys, M., Dommen, J., Donahue, N.M., George, C., Goldstein, A.H., Hamilton, J.F., Herrmann, H., Hoffmann, T., Iinuma, Y., Jang, M., Jenkin, M.E., Jimenez, J.L., Kiendler-Scharr, A., Maenhaut, W., McFiggans, G., Mentel, T.F., Monod, A., Prévot, A.S.H., Seinfeld, J.H., Surratt, J.D., Szmigielski, R., and Wildt, J. (2009) The formation, properties and impact of secondary organic aerosol: current and emerging issues. *Atmos. Chem. Phys.*, **9**, 5155–5236. doi: 10.5194/acp-9-5155-2009

Heald, C.L., Ridley, D.A., Kreidenweis, S.M., and Drury, E.E. (2010) Satellite observations cap the atmospheric organic aerosol budget. *Geophys. Res. Lett.*, **37**, L24808. doi: 10.1029/2010GL045095

Henze, D.K., Seinfeld, J.H., Ng, N.L., Kroll, J.H., Fu, T.-M., Jacob, D.J., and Heald, C.L. (2008) Global modeling of secondary organic aerosol formation from aromatic hydrocarbons: high- vs. low-yield pathways. *Atmos. Chem. Phys.*, **8**, 2405–2420. doi: 10.5194/acp-8-2405-2008

Hidy, G.M. and Blanchard, C.L. (2005) The midlatitude North American background aerosol and global aerosol variation. *J. Air Waste Manage. Assoc.*, **55** (11), 1585–1599. doi: 10.1080/10473289.2005.10464761

Jimenez, J.L., Canagaratna, M.R., Donahue, N.M., Prévot, A.S.H., Zhang, Q., Kroll, J.H., DeCarlo, P.F., Allan, J.D., Coe, H., Ng, N.L., Aiken, A.C., Docherty, K.S., Ulbrich, I.M., Grieshop, A.P., Robinson, A.L., Duplissy, J., Smith, J.D., Wilson, K.R., Lanz, V.A., Hueglin, C., Sun, Y.L., Tian, J., Laaksonen, A., Raatikainen, T., Rautiainen, J., Vaattovaara, P., Ehn, M., Kulmala, M., Tomlinson, J.M., Collins, D.R., Cubison, M.J., Dunlea, E., Huffman, J.A., Onasch, T.B., Alfarra, M.R., Williams, P.I., Bower, K., Kondo, Y., Schneider, J., Drewnick, F., Borrmann, S., Weimer, S., Demerjian, K., Salcedo, D., Cottrell, L., Griffin, R., Takami, A., Miyoshi, T., Hatakeyama, S., Shimono, A., Sun, J.Y., Zhang, Y.M., Dzepina, K., Kimmel, J.R., Sueper, D., Jayne, J.T., Herndon, S.C., Trimborn, A.M., Williams, L.R., Wood, E.C., Middlebrook, A.M., Kolb, C.E., Baltensperger, U., and Worsnop, D.R. (2009) Evolution of organic aerosols in the atmosphere. *Science*, **326** (5959), 1525–1529. doi: 10.1126/science.1180353

Kanakidou, M., Seinfeld, J.H., Pandis, S.N., Barnes, I., Dentener, F.J., Facchini, M.C., Van Dingenen, R., Ervens, B., Nenes, A., Nielsen, C.J., Swietlicki, E., Putaud, J.P., Balkanski, Y., Fuzzi, S., Horth, J., Moortgat, G.K., Winterhalter, R., Myhre, C.E.L., Tsigaridis, K., Vignati, E., Stephanou, E.G., and Wilson, J. (2005) Organic aerosol and global climate modelling: a review. *Atmos. Chem. Phys.*, **5**, 1053–1123. doi: 10.5194/acp-5-1053-2005

Kim, N.K., Kim, Y.P., and Kang, C.-H. (2011) Long-term trend of aerosol composition and direct radiative forcing due to aerosols over Gosan: TSP, PM10, and PM2.5 data between 1992 and 2008. *Atmos. Environ.*, **45** (34), 6107–6115. doi: 10.1016/j.atmosenv.2011.08.051

Kroll, J.H. and Seinfeld, J.H. (2008) Chemistry of secondary organic aerosol: formation and evolution of low-volatility organics in the atmosphere. *Atmos. Environ.*, **42** (16), 3593–3624. doi: 10.1016/j.atmosenv.2008.01.003

Lambe, A.T., Onasch, T.B., Massoli, P., Croasdale, D.R., Wright, J.P., Ahern, A.T., Williams, L.R., Worsnop, D.R., Brune, W.H., and Davidovits, P. (2011) Laboratory studies of the chemical composition and cloud condensation nuclei (CCN) activity of secondary organic aerosol (SOA) and oxidized primary organic aerosol (OPOA). *Atmos. Chem. Phys.*, **11**, 8913–8928. doi: 10.5194/acp-11-8913-2011

Massoli, P., Fortner, E.C., Canagaratna, M.R., Williams, L.R., Zhang, Q., Sun, Y., Schwab, J.J., Trimborn, A., Onasch, T.B., Demerjian, K.L., Kolb, C.E., Worsnop, D.R., and Jayne, J.T. (2012) Pollution gradients and chemical characterization of particulate matter from vehicular traffic near major roadways: results from the 2009 Queens College Air Quality Study in NYC. *Aerosol*

Sci. Technol., **46** (11), 1201–1218. doi: 10.1080/02786826.2012.701784

McFiggans, G., Alfarra, M.R., Allan, J., Bower, K., Coe, H., Cubison, M., Topping, D., Williams, P., Decesari, S., Facchini, C., and Fuzzi, S. (2005) Simplification of the representation of the organic component of atmospheric particulates. *Faraday Discuss.*, **130**, 341–362. doi: 10.1039/B419435G

Minguillón, M.C., Ripoll, A., Pérez, N., Prévôt, A.S.H., Canonaco, F., Querol, X., and Alastuey, A. (2015) Chemical characterization of submicron regional background aerosols in the Western Mediterranean using an Aerosol Chemical Speciation Monitor. *Atmos. Chem. Phys. Discuss.*, **15**, 36. doi: 10.5194/acpd-15-965-2015

Petters, M.D., and Kreidenweis, S.M. (2007) A single parameter representation of hygroscopic growth and cloud condensation nucleus activity. *Atmos. Chem. Phys.*, **7**, 1961–1971. doi: 10.5194/acp-7-1961-2007

Prenni, A.J., Petters, M.D., Kreidenweis, S.M., Heald, C.L., Martin, S.T., Artaxo, P., Garland, R.M., Wollny, A.G., and Pöschl, U. (2009) Relative roles of biogenic emissions and Saharan dust as ice nuclei in the Amazon basin. *Nat. Geosci.*, **2**, 401–404. doi: 10.1038/NGEO517

Putaud, J.-P., Raes, F., Van Dingenen, R., Brüggemann, E., Facchini, M.-C., Decesari, S., Fuzzi, S., Gehrig, R., Hüglin, C., Laj, P., Lorbeer, G., Maenhaut, W., Mihalopoulos, N., Müller, K., Querol, X., Rodriguez, S., Schneider, J., Spindler, G., ten Brink, H., Tørseth, K., and Wiedensohler, A. (2004) European aerosol phenomenology-2: chemical characteristics of particulate matter at kerbside, urban, rural and background sites in Europe. *Atmos. Environ.*, **38** (16), 2579–2595. doi: 10.1016/j.atmosenv.2004.01.041

Querol, X., Alastuey, A., Pey, J., Cusack, M., Pérez, N., Mihalopoulos, N., Theodosi, C., Gerasopoulos, E., Kubilay, N., and Koçak, M. (2009) Variability in regional background aerosols within the Mediterranean. *Atmos. Chem. Phys.*, **9**, 4575–4591. doi: 10.5194/acp-9-4575-2009

Rankin, A.M. and Wolff, E.W. (2003) A year-long record of size-segregated aerosol composition at Halley, Antarctica. *J. Geophys. Res.*, **108** (D24), 4775. doi: 10.1029/2003JD003993

Robinson, A.L., Donahue, N.M., Shrivastava, M.K., Weitkamp, E.A., Sage, A.M., Grieshop, A.P., Lane, T.E., Pierce, J.R., and Pandis, S.N. (2007) Rethinking organic aerosols: semivolatile emissions and photochemical aging. *Science*, **315** (5818), 1259–1262. doi: 10.1126/science.1133061

Tanaka, T.Y. and Chiba, M. (2006) A numerical study of the contributions of dust source regions to the global dust budget. *Global Planet. Change*, **52** (1-4), 88–104. doi: 10.1016/j.gloplacha.2006.02.002

Zhang, R. (2010) Getting to the critical nucleus of aerosol formation. *Science*, **328** (5984), 1366–1367. doi: 10.1126/science.1189732

Zhang, Q., Jimenez, J.L., Canagaratna, M.R., Allan, J.D., Coe, H., Ulbrich, I., Alfarra, M.R., Takami, A., Middlebrook, A.M., Sun, Y.L., Dzepina, K., Dunlea, E., Docherty, K., DeCarlo, P.F., Salcedo, D., Onasch, T., Jayne, J.T., Miyoshi, T., Shimono, A., Hatakeyama, S., Takegawa, N., Kondo, Y., Schneider, J., Drewnick, F., Borrmann, S., Weimer, S., Demerjian, K., Williams, P., Bower, K., Bahreini, R., Cottrell, L., Griffin, R.J., Rautiainen, J., Sun, J.Y., Zhang, Y.M., and Worsnop, D.R. (2007) Ubiquity and dominance of oxygenated species in organic aerosols in anthropogenically-influenced Northern Hemisphere midlatitudes. *Geophys. Res. Lett.*, **34** (13), L13801. doi: 10.1029/2007GL029979.

5 Aerosol Optics

Alexander A. Kokhanovsky

5.1 Introduction

The average size of most aerosol particles is close to the wavelength of the visible light. Therefore, optical methods are suitable for studies of aerosol properties using various instruments on ground, aircraft, ship, and from space. The electromagnetic Maxwell theory is used to model the optical signals (with account for light absorption, emission, and multiple scattering effects) at the input of various optical devices such as LIDARs, spectrometers, polarimeters, sunphotometers, and radiometers. The main task of the aerosol optics is to understand the processes, which govern the interaction of electromagnetic waves with solid and liquid aerosol particles in the optical range of electromagnetic spectrum. The problem is usually solved via a three-stage process. At the first step, the scattering and absorption characteristics of a single particle of a given shape, internal structure, and chemical composition are calculated using Maxwell theory. Then the derived characteristics are averaged taking into account that the size, shape, and refractive index of particles vary in any given volume of atmospheric air. The derived averaged characteristics are used as parameters in the vector integrodifferential radiative transfer equation with correspondent boundary conditions. This enables the calculation of transmission and reflection functions of aerosol layers and also respective polarization characteristics such as the degree of polarization, ellipticity (if any), and the orientation of the polarization ellipse both for reflected and transmitted light. The optical properties of internal light field can be calculated as well. It should be pointed out that the vector radiative transfer equation can be derived from the Maxwell theory under several assumptions (Mishchenko, 2014) although this equation has been initially derived using simple conservation energy arguments.

The aerosol optics concepts are of great importance both for the analysis of the field measurements and also for the estimation of radiative forcing and remote sensing signals. They can be used to explain a number of atmospheric optics phenomena as observed by a naked eye.

Atmospheric Aerosols: Life Cycles and Effects on Air Quality and Climate, First Edition.
Edited by Claudio Tomasi, Sandro Fuzzi, and Alexander Kokhanovsky.

In this chapter, we introduce main concepts of aerosol optics including radiative transfer equation and also local scattering and absorption characteristics used in studies of optical properties of aerosol media.

5.2 Absorption

Liquid and solid particles suspended in atmosphere absorb, emit, and scatter electromagnetic radiation. These processes are characterized by corresponding cross sections, which are defined as the ratios of the total energy absorbed, scattered, or attenuated (scattered and absorbed) to the incident radiation intensity. In particular, it follows for the absorption cross section (Stratton, 1941):

$$C_{\text{abs}} = -k \int_V \text{Im}(\varepsilon) \frac{\vec{E}\vec{E}^* dV}{\vec{E}_0 \vec{E}_0^*}. \tag{5.1}$$

Here $k = 2\pi/\lambda$, λ is the wavelength, ε is the dielectric permittivity inside an aerosol particle, V is its volume, $\vec{E}$ is the electric vector inside the particle, and $\vec{E}_0$ is the electric vector of incident electromagnetic wave. The absorption cross section is measured in square micrometer.

The vector $\vec{E}$ can be calculated solving the wave equation

$$\Delta\vec{E} + k^2 m^2 \vec{E} = 0 \tag{5.2}$$

with the corresponding boundary conditions depending on the internal structure of the aerosol particle and also on its shape. Here

$$m = \sqrt{\varepsilon} \equiv n - i\chi \tag{5.3}$$

is the complex refractive index of an aerosol particle. The analytical solution of Eq. (5.2) is known for various shapes of aerosol particles including spheres, cylinders, and spheroids. In some particular cases the results can be obtained directly from Eq. (5.1). For instance, if the refractive index of particles is close to that of the surrounding medium, one may assume that $\vec{E} \approx \vec{E}_0$. Then it follows from Eq. (5.1)

$$C_{\text{abs}} = \alpha V \tag{5.4}$$

for any shape of a homogeneous aerosol particle, where $\alpha = (4\pi/\lambda)\chi$ is the bulk absorption coefficient. We have used the relationship: $\text{Im}(\varepsilon) = -2n\chi$, which follows from Eq. (5.3), and also the fact that the value of n is close to 1.

One can introduce the volumetric absorption coefficient

$$k_{\text{abs}} = NC_{\text{abs}}, \tag{5.5}$$

where N is the number concentration of aerosol particles. This coefficient has the dimension of the inverse length (m^{-1}). It follows from Eqs (5.4) and (5.5):

$$k_{\text{abs}} = N\alpha V. \tag{5.6}$$

Introducing the volumetric concentration $c_v = NV$, one derives from Eq. (5.6)

$$k_{\text{abs}} = \alpha c_v. \tag{5.7}$$

Equation 5.7 is accurate only in the case, if the refractive index of a scatter is close to that of air. However, the refractive index of aerosol particles in the visible ($n \approx 1.3-1.7$) differs considerably from that of air. Therefore, Eq. (5.7) is not very accurate. More accurate equations can be obtained in the assumption of small ($x \ll \lambda$) and large ($x \gg \lambda$) aerosol particles. Here, $x = 2\pi r/\lambda$ is the dimensionless size parameter and r is the radius of a particle.

In particular, it follows for the internal field of an aerosol particle much smaller than the wavelength of the incident light (Stratton, 1941):

$$\vec{E} = \frac{3}{2+\varepsilon}\vec{E}_0. \tag{5.8}$$

One derives from Eq. (5.1) under this assumption:

$$C_{\text{abs}} = A\alpha V, \tag{5.9}$$

where taking into account that $\chi << n$ for aerosol particles

$$A = \frac{9n}{(2+n^2)^2}. \tag{5.10}$$

Also it follows:

$$k_{\text{abs}} = A\alpha c_v. \tag{5.11}$$

Note that the value of $A = 1$ at $n = 1$ as it should be (see Eq. 5.4). It follows from Eq. (5.10) that the value of $A \approx 0.75$ at $n = 1.5$ in the case of particles much smaller than the wavelength of the incident light.

In the case of large spherical particles with the radius $r \gg \lambda$, the calculation of the absorption cross section is reduced to the evaluation of the following integral (Kokhanovsky, 2004):

$$C_{\text{abs}} = \frac{\pi r^2}{2}\sum_{j=1}^{2}\int_0^1 \frac{(1-R_j)(1-\exp(-c\xi))}{1-R_j\exp(-c\xi)}d\sigma. \tag{5.12}$$

Here $c = \alpha d$ is the absorption path length on the diameter d of the particle (or the attenuation parameter),

$$R_1 = \left[\frac{\upsilon - n\xi}{\upsilon + n\xi}\right]^2, \qquad R_2 = \left[\frac{n\upsilon - \xi}{n\upsilon + \xi}\right]^2 \tag{5.13}$$

are Fresnel reflectance coefficients, $\upsilon = \sqrt{1-\sigma^2}$, $\xi = \sqrt{1-\sigma/n^2}$ and we assumed that $\chi/n \ll 1$, which is a valid assumption for most of particles suspended in atmosphere in the optical spectral range.

It follows from Eq. (5.12) that the absorption efficiency factor

$$Q_{\text{abs}} = \frac{C_{\text{abs}}}{\pi r^2} \tag{5.14}$$

of optically large scatterers depends only on two parameters: the attenuation parameter c and the refractive index n. Therefore, different aerosol particles having similar values of c and n will have very close values of Q_{abs}. The refractive index is a very important parameter because it determines the intensity of light reflected from the surface of an aerosol particle via Fresnel reflectance coefficients (in the framework of the geometrical optics approximation valid for large particles). Also for the larger refractive indices, there is an effect of strong deviation of incident light to the center of an aerosol particle. Therefore, the rays penetrating into the aerosol particle are absorbing more strongly for larger refractive indices as compared to the smaller ones. This result can be also derived from Eq. (5.12) assuming that $c \to 0$. Then it follows that $R_j \exp(-c\xi) \approx R_j$ and integral 5.12 can be evaluated analytically. The result is

$$C_{abs} = \pi r^2 \left[1 - \frac{2n^2\,(\varsigma\,(b) - \varsigma(1))}{c^2}\right], \tag{5.15}$$

where $\varsigma(b) = (1 + cb)\exp(-cb)$, $b = \sqrt{1 - n^{-2}}$. It appears that this equation is accurate at any value of c, if the refractive index of particles is close to 1. Otherwise, Eq. (5.12) must be used. It follows from Eq. (5.15) that$C_{abs} \to \pi r^2$ as $c \to \infty$ instead of the correct limit (see Eq. 5.12) $C_{abs} \to (1 - \mathfrak{R})\pi r^2$. The integral

$$\mathfrak{R} = \frac{1}{2}\sum_{j=1}^{2}\int_0^1 R_j(\sigma)d\sigma \tag{5.16}$$

can be evaluated analytically (Gershun, 1937):

$$\mathfrak{R} = w_1 \ln n + w_2 \ln\left\{\frac{n-1}{n+1}\right\} + w_3, \tag{5.17}$$

where

$$w_1 = \frac{8n^4(1+n^4)}{(n^4-1)^2(n^2+1)}, \qquad w_2 = \frac{n^2(n^2-1)^2}{(n^2+1)^3}, \qquad w_3 = \frac{p(n)}{3(n^4-1)(n^2+1)(n+1)}, \tag{5.18}$$

$$p(n) = \sum_{l=0}^{7} \zeta_l n^l, \quad \zeta_l \equiv (-1, -1, -3, 7, -9, 13, -7, 3). \tag{5.19}$$

Therefore, it follows in the limit of strongly absorbing aerosol particles:

$$Q_{abs} = 1 - \mathfrak{R}. \tag{5.20}$$

The dependence of R on the refractive index n is given in Table 5.1. Clearly, the equality $Q_{abs} + \mathfrak{R} = 1$ is just the manifestation of the energy conservation law. The value of $\mathfrak{R}$ is proportional the amount of reflected energy and Q_{abs} is proportional to the amount of the absorbed energy. It is assumed that the absorption is so strong that any ray penetrating inside the particle is absorbed there (therefore, there is no transmission). As a matter of fact a particle is similar to a semiinfinite absorbing plane – parallel plate in this case. For such plates, there is also no transmission

Table 5.1 The dependence of the reflectivity $\mathfrak{R}$ on the refractive index n.

n	1.34	1.4	1.5	1.6	1.7
$\mathfrak{R}$	0.0675	0.0768	0.0918	0.1063	0.1203

term and the total absorbed light energy is equal to the difference of the energy of the incident light and reflected from the plate light.

It follows from Eq. (5.15) as $c \to 0$:

$$C_{\text{abs}} = \Phi(n)\alpha V, \tag{5.21}$$

where $\Phi(n) = n^2(1 - (1 - n^{-2})^{3/2})$.

Therefore, one drives for the volumetric absorption coefficient:

$$k_{\text{abs}} = \Phi(n)\alpha c_v. \tag{5.22}$$

This equation is similar to Eq. (5.7) except it is more accurate for the case of large particles. One derives from Eq. (5.22) that $\Phi \to 1$ as $n \to 1$. For aerosol particles, it follows from Eq. (5.22): $\Phi > 1$ and Φ increases with n. This means that the particles with larger values of n also absorb more energy as compared to particles with smaller values of n, if they have the same imaginary part of refractive index and concentration. The same is true for nonspherical particles (Kokhanovsky and Macke, 1997). However, the value of Φ is generally larger for nonspherical particles as compared to spheres (Kokhanovsky and Nauss, 2005; Räisänen *et al.*, 2015). The dependence $Q_{\text{abs}}(c)$ for the typical values of the aerosol refractive index calculated using Eq. (5.12) is shown in Figure 5.1. It follows from Figure 5.1 that the asymptotic values as shown in Eq. (5.2) are reached at $c = 10$. However, the results at $c = 5$ are already close to the asymptotic values. The decrease in refractive index leads to the increase of Q_{abs} at $c = \text{const}$ and large values of c.

The emission of radiation by a particle is directly proportional to the absorption cross section and the Planck function $B(\lambda, T)$. In particular, the total flux F of thermal radiation emitted by a particle per unit time can be easily found, if the absorption cross section is known (Shifrin, 1968):

$$F(\lambda) = 4\pi C_{\text{abs}}(\lambda)B(\lambda, T). \tag{5.23}$$

The emission is strong for highly absorbing particles. This leads to warming of surrounding air.

It follows from Eq. (5.1) in the general case of a spherical particles having radius r after substitution of correspondent equation for the electric field inside an aerosol particle and subsequent analytical integration (Babenko, Astafyeva, and Kuzmin, 2003):

$$C_{\text{abs}} = -\frac{\lambda^2}{2\pi}\sum_{l=1}^{\infty}\frac{2l+1}{|\xi_l^2|}\left[\frac{\text{Im}\left(m^*D_l\right)}{NG_l - D_l} + \frac{\text{Im}(mD_l)}{G_l - ND_l}\right], \tag{5.24}$$

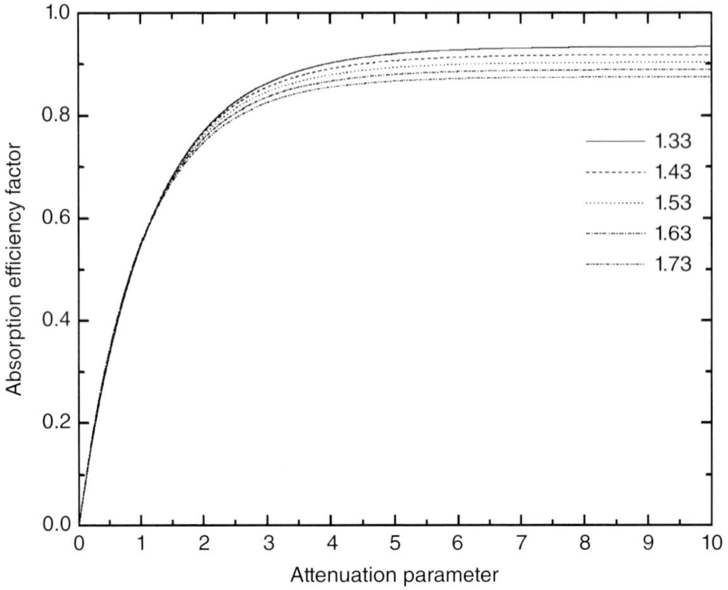

Figure 5.1 The dependence of the absorption efficiency factor on the attenuation parameter at various values of the refractive index.

where $D_l = \psi_l'(mx)/\psi_l(mx)$, $G_l = \xi_l'(x)/\xi_l(x)$, $\psi_l(mx)$, and $\xi_l(x)$ are well-known Riccati–Bessel functions (Babenko, Astafyeva, and Kuzmin, 2003). This equation gives exact result for the absorption cross section of a spherical particle in the framework of the electromagnetic theory developed by Maxwell. Similar equations can be obtained for multilayered spheres (Babenko, Astafyeva, and Kuzmin, 2003) and also for nonspherical particles (Mishchenko, 2014). The example of calculations using Eq. (5.24) is shown in Figure 5.2. It follows from Figure 5.2 that the increase in absorption damps high-frequency oscillations on the curves $Q_{\text{abs}}(x)$. Similar effects occur due to account for the aerosol polydispersity.

Usually aerosol particles have different sizes in a unit volume of aerosol medium (say, radii from r_1 to r_2 distributed with the particle size distribution (PSD) $f(r)$). Therefore, the volumetric absorption coefficient can be presented as follows:

$$k_{\text{abs}} = N\langle C_{\text{abs}}\rangle. \tag{5.25}$$

The angular brackets mean the averaging with respect to the PSD:

$$\langle C_{\text{abs}}\rangle = \int_{r_1}^{r_2} C_{\text{abs}}(r) f(r)\, dr. \tag{5.26}$$

Instead of number concentration N, one can use the volumetric concentration $c_v = N/\langle V\rangle$ or the mass concentration $c_m = N\langle m\rangle \equiv \rho c_v$, where $\langle m\rangle = \rho\langle V\rangle$ is the average mass of an aerosol particle and ρ is the density of particles. For instance,

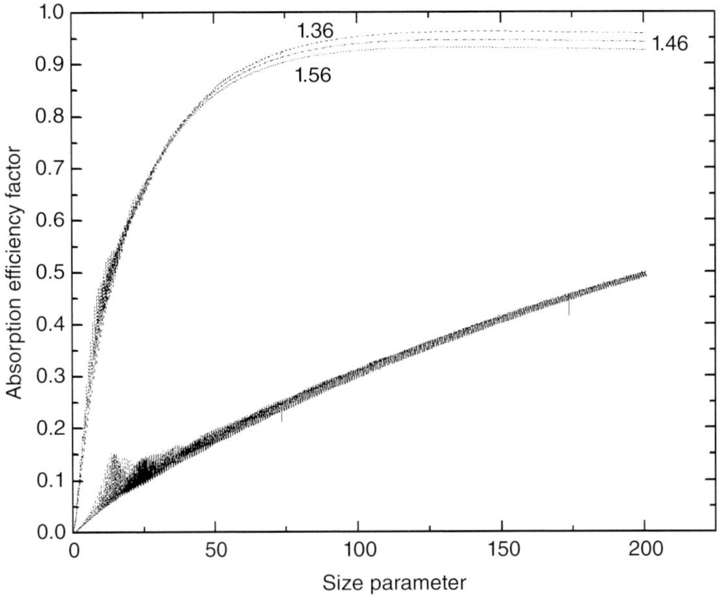

Figure 5.2 The dependence of the absorption efficiency factor on the size parameter at various values of the refractive index ($n = 1.36$, 1.46, 1.56). The imaginary part of the complex refractive index is equal to 0.01 for the upper curves. It is equal to 0.001 for the lower curves.

the following relationship holds:

$$k_{\text{abs}} = \frac{c_m}{\langle m \rangle} \int_{r_1}^{r_2} C_{\text{abs}}(r) f(r)\, dr. \tag{5.27}$$

The calculations of the absorption coefficient according Eq. (5.27) are usually done for the lognormal size distribution:

$$f(r) = \frac{1}{\sqrt{2\pi} rs} \exp\left[-\frac{\ln^2(r/r_0)}{2s^2}\right] \tag{5.28}$$

and $r_1 = 0.05\,\mu\text{m}$ and $r_2 = 20\,\mu\text{m}$. The parameters of the lognormal size distribution for main aerosol types are given in Table 5.2. The spectral dependences of the complex refractive index are presented in Tables 5.3 and 5.4.

5.3 Scattering

Aerosol particles not only absorb but also they scatter incident radiation. If particles are large, they act as small imperfect lenses focusing light in the forward scattering hemisphere. Then the backward scattered light flux is small. As a matter of fact only particles much smaller than the wavelength of incident light

Table 5.2 The parameters of the lognormal size distribution for the typical aerosol types.

N	Aerosol type	r_0(μm)	*s*
1	Continental (fine)	0.0804	0.43
2	Oceanic (coarse)	0.5469	0.72
3	Clean continental (coarse)	0.3400	0.72
4	Polluted continental (coarse)	0.9181	0.63
5	Smoke (coarse)	0.4624	0.81
6	Dust (coarse)	0.7879	0.60
7	Volcanic (coarse)	0.59	0.56

The dust aerosol is assumed to be nonspherical and described by the model introduced by Dubovik *et al.* (2006). The parameters for the dust case correspond to the case of the equivolume spheres.

Table 5.3 Real part of the refractive index of various types of particulate matter at several wavelengths.

λ(nm)	410	443	490	555	670	763	765	865	910	1370	1650	2130
Oceanic	1.36	1.36	1.36	1.35	1.35	1.35	1.35	1.345	1.345	1.340	1.333	1.307
Dust	1.56	1.56	1.56	1.56	1.56	1.56	1.56	1.56	1.56	1.56	1.56	1.56
Continental	1.42	1.42	1.42	1.42	1.43	1.43	1.43	1.44	1.44	1.44	1.45	1.40
Smoke	1.53	1.53	1.53	1.54	1.55	1.56	1.56	1.585	1.58	1.50	1.50	1.40
Volcanic	1.50	1.50	1.50	1.50	1.50	1.50	1.50	1.50	1.50	1.50	1.48	1.46
Continental (fine)	1.418	1.417	1.414	1.411	1.408	1.406	1.406	1.4	1.4	1.382	1.371	1.321

It is assumed that the real part of the refractive index is the same for clean and polluted continental aerosols.

Table 5.4 Imaginary part of the refractive index of various types of particulate matter.

λ(nm)	Oceanic	Dust	Fine mode
410	5×10^{-9}	3.222×10^{-3}	0.0023
443	4×10^{-9}	2.900×10^{-3}	0.0023
490	3×10^{-9}	2.437×10^{-3}	0.0023
555	3×10^{-9}	1.908×10^{-3}	0.0026
670	3×10^{-8}	1.300×10^{-3}	0.0027
763	3×10^{-7}	1.155×10^{-3}	0.0035
765	3×10^{-7}	1.155×10^{-3}	0.0035
865	5×10^{-6}	10^{-3}	0.0043
910	6×10^{-6}	$10–^{3}$	0.0047
1370	10^{-4}	10^{-3}	0.0066
1650	2×10^{-4}	10^{-3}	0.0070
2130	10^{-3}	10^{-3}	0.0037

It is assumed that the imaginary part of aerosol refractive index for smoke, continental clean/polluted aerosol, and volcanic aerosol are spectrally neutral and equal to 0.01, 0.01, 0.001, and 0.008, respectively.

produce symmetric scattering in both hemispheres (dipole scattering). In general, aerosol particles can be considered as an array of interacting dipoles and the final scattering pattern is determined by the interference of waves scattered by the system of dipoles.

The scattered light energy in a given direction specified by the angles (θ, φ)inside solid angle $d\Omega = \sin\theta d\theta d\varphi$ can be presented as

$$dW(\Omega) = C^d_{\rm sca}(\Omega)\, I_0 d\Omega, \tag{5.29}$$

where $C^d_{\rm sca}(\Omega)$ is the so-called differential scattering cross section. It follows from Eq. (5.29) for the total scattered energy:

$$W = \int_{4\pi} C^d_{\rm sca}(\Omega)\, I_0 d\Omega \tag{5.30}$$

or introducing the total (integrated on all angles) scattering cross section:

$$W = C_{\rm sca} I_0, \tag{5.31}$$

where

$$C_{\rm sca} = \int_{4\pi} C^d_{\rm sca}(\Omega)\, d\Omega. \tag{5.32}$$

The normalized differential cross section is called the phase function

$$p(\Omega) = \frac{4\pi}{C_{\rm sca}} C^d_{\rm sca}(\Omega) \tag{5.33}$$

with the normalization condition:

$$\int_{4\pi} p(\Omega)\, d\Omega = 4\pi. \tag{5.34}$$

Due to this normalization condition, isotropic (not dependent on the direction) scattering has the phase function equal to unity. For spherical particles and also for randomly oriented ensembles of nonspherical particles we can perform the integration with respect to the azimuthal angle. Then Eq. (5.34) is reduced to

$$\frac{1}{2}\int_0^\pi p(\theta)\sin\theta d\theta = 1. \tag{5.35}$$

The angle θ between incident and scattered light beams is called the scattering angle. The average cosine of scattering angle or the asymmetry parameter

$$g = \frac{1}{2}\int_0^\pi p(\theta)\sin\theta\cos\theta d\theta \tag{5.36}$$

is often used to characterize the asymmetry of light scattering pattern. This parameter is equal to zero for the symmetric light scattering patterns (say, for the dipole scattering with $p(\theta) = 0.75(1 + \cos^2\theta)$) and close to 1.0 for the highly extended in the forward direction phase functions. Usually it is in the range 0.6–0.8 for atmospheric aerosol in the visible.

The differential scattering cross section for an aerosol particle of radius a can be found solving the wave equation with correspondent boundary conditions. It follows in the case of spheres:

$$C^{d}_{\text{sca}}(\theta) = \frac{2\pi I(\theta)}{k^2}, \tag{5.37}$$

where $k = 2\pi/\lambda$, $I(\theta) = \left[I_1(\theta) + I_2(\theta)\right]/2$, $I_j(\theta) = \left|S_j(\theta)\right|^2$ with amplitude functions determined by the following series:

$$S_1(\theta) = \sum_{j=1}^{\infty} \frac{2j+1}{j(j+1)}[a_j\pi_j(\theta) + b_j\tau_j(\theta)], \tag{5.38}$$

$$S_2(\theta) = \sum_{j=1}^{\infty} \frac{2j+1}{j(j+1)}[a_j\tau_j(\theta) + b_j\pi_j(\theta)], \tag{5.39}$$

where $\pi_j(\theta) = P_j^1(\cos\theta)/\sin\theta$, $\tau_j(\theta) = dP_j^1(\cos\theta)/d\theta$, $P_j^1(\cos\theta)$ is the associate Legendre polynomial and

$$a_j = \frac{\psi_j'(mx)\psi_j(x) - m\psi_j(mx)\psi_j'(x)}{\psi_j'(mx)\xi_j(x) - m\psi_j(mx)\xi_j'(x)}, \tag{5.40}$$

$$b_j = \frac{m\psi_j'(mx)\psi_j(x) - \psi_j(mx)\psi_j'(x)}{m\psi_j'(mx)\xi_j(x) - \psi_j(mx)\xi_j'(x)}. \tag{5.41}$$

It follows that the differential scattering cross section depends on the size parameter $x = ka$. Therefore, particles having different radii but the same values of the size parameter have the same scattering patterns (at the same value of the refractive index). Knowing the differential scattering cross section, one can also evaluate the scattering cross section and the asymmetry parameter calculating the corresponding angular integrals. The result is (van de Hulst, 1981; Kokhanovsky, 2006):

$$C_{\text{sca}} = \frac{2\pi}{k^2}\sum_{j=1}^{\infty}(2j+1)\left[\left|a_j\right|^2 + \left|b_j\right|^2\right], \tag{5.42}$$

$$g = \frac{4\pi}{k^2 C_{\text{sca}}}\text{Re}\sum_{j=1}^{\infty}\left[\frac{2j+1}{j(j+1)}a_j b_j^* + \frac{j(j+2)}{j+1}(a_j a_{j+1}^* + b_j b_{j+1}^*)\right]. \tag{5.43}$$

The results of calculations of the scattering efficiency factor $Q_{\text{sca}} = C_{\text{sca}}/\pi a^2$ are presented in Figure 5.3 at the refractive index equal to 1.36, 1.46, and 1.56, $\chi = 0.01$. It follows from Figure 5.3 that particles with the diameter smaller than the wavelength of the incident light beam are characterized by larger values of Q_{sca} for larger refractive indices at a given value of the size parameter. The curve $Q_{\text{sca}}(x)$ has oscillations due to the interference effects and approaches the asymptotic value of $1 + \mathfrak{R}$ (see Table 5.1) for absorbing aerosol particles (Kokhanovsky, 2004). The limiting value is equal to 2.0 in case of nonabsorbing aerosol (Shifrin, 1968).

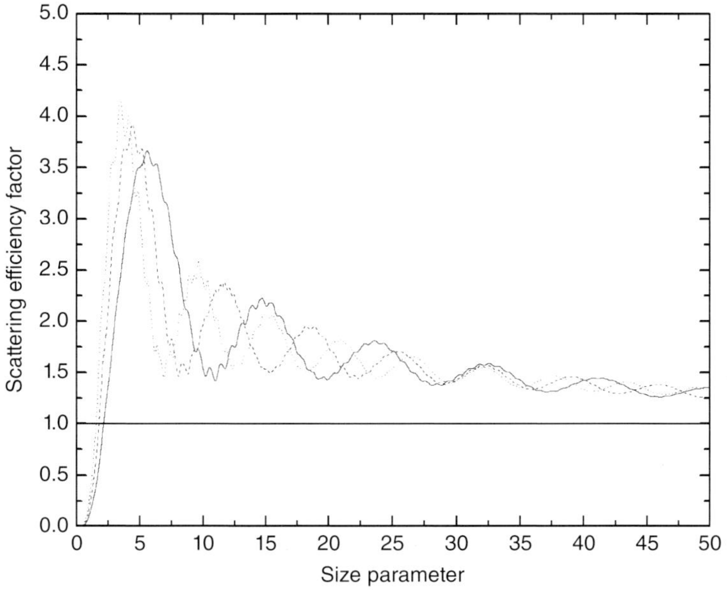

Figure 5.3 The dependence of the scattering efficiency factor on the size parameter at various values of the refractive index ($n = 1.36$, 1.46, 1.56). The imaginary part of the complex refractive index is equal to 0.01. The maximal values of the efficiency factor decrease with n.

In case of polydispersed aerosol media, one should carry out averaging with respect to the size distribution. In particular, it follows for the scattering coefficient:

$$k_{\text{sca}} = N\langle C_{\text{sca}}\rangle, \tag{5.44}$$

where

$$\langle C_{\text{sca}}\rangle = \int_{r_1}^{r_2} C_{\text{sca}}(r) f(r)\, dr. \tag{5.45}$$

It follows for the phase function

$$\langle p(\theta)\rangle = \frac{\displaystyle\int_{a_1}^{a_2} p(\theta, r)\, C_{\text{sca}}(r) f(r)\, dr}{\langle C_{\text{sca}}\rangle}, \tag{5.46}$$

where

$$p(\theta, a) = \frac{4\pi^2 (I_1(\theta, a) + I_2(\theta, a))}{k^2 C_{\text{sca}}(a)}. \tag{5.47}$$

The asymmetry of the phase function is characterized by the asymmetry parameter:

$$\langle g\rangle = \int_{r_1}^{r_2} \frac{C_{\text{sca}}(r) g(r) f(r)\, dr}{\langle C_{\text{sca}}\rangle}. \tag{5.48}$$

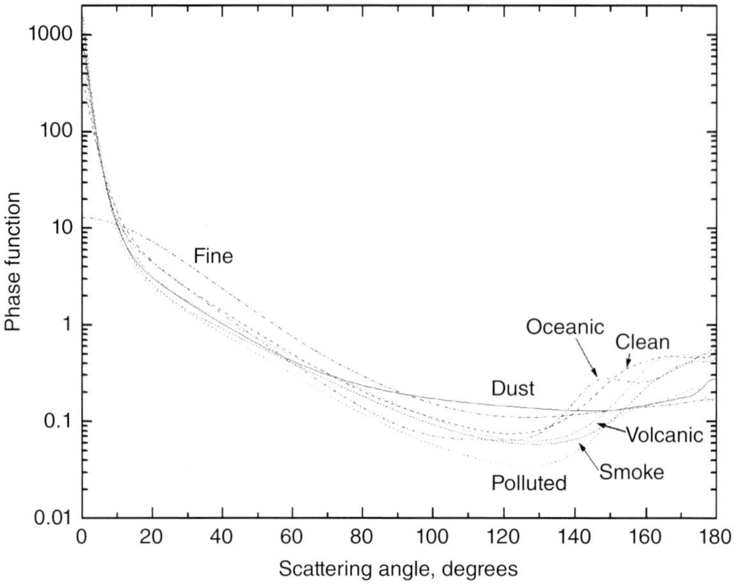

Figure 5.4 The phase functions of typical aerosol media (see Table 5.2).

We show the phase functions of typical aerosol media in Figure 5.4. It follows that the behavior of phase functions is quite different in the angular range 140–180°. This can be used for aerosol typing using optical measurements.

5.4
Polarization

Aerosol particles not only alter the incident light field intensity. The polarization characteristics of the incident light beam are also changed.

Let us take the scattering plane containing the incident and scattered beam as the reference plane. Then it follows for the components of the electric vector of scattered light field (van de Hulst, 1981):

$$\begin{pmatrix} E_l^s \\ E_r^s \end{pmatrix} = \frac{\widehat{S}(\theta)\exp(ik(z-z_0))}{ikz_0}\begin{pmatrix} E_l^0 \\ E_r^0 \end{pmatrix}. \tag{5.49}$$

Here z_0 is the distance to the observation point, z is the coordinate along the direction of light propagation, the subscript $l(r)$ denotes the component of the electric vector parallel (perpendicular) to the scattering plane,

$$\widehat{S}(\theta) = \begin{pmatrix} S_2(\theta) & 0 \\ 0 & S_1(\theta) \end{pmatrix}. \tag{5.50}$$

Equation 5.48 can be written in a short notation:

$$\vec{E}^{s} = \frac{\alpha}{kz_0}\widehat{S}(\theta)\vec{E}^{0}, \tag{5.51}$$

where $\alpha = -i\exp(ik(z - z_0))$. It follows for the density matrix:

$$\widehat{\rho}^{s} \equiv \vec{E} \otimes \vec{E}^{+} = \frac{1}{k^2 z_0{}^2}\widehat{S}\widehat{\rho}^{i}\widehat{S}^{+}. \tag{5.52}$$

The density matrix is related to the Stokes vector via the following relationship:

$$\widehat{\rho} = \frac{1}{2}\{I\widehat{\sigma}_1 + Q\widehat{\sigma}_2 + U\widehat{\sigma}_3 + V\widehat{\sigma}_4\}, \tag{5.53}$$

where

$$\widehat{\sigma}_1 = \begin{pmatrix} 1 & 0 \\ 0 & 1 \end{pmatrix}, \quad \widehat{\sigma}_2 = \begin{pmatrix} 1 & 0 \\ 0 & -1 \end{pmatrix}, \quad \widehat{\sigma}_3 = \begin{pmatrix} 0 & 1 \\ 1 & 0 \end{pmatrix}, \quad \widehat{\sigma}_4 = \begin{pmatrix} 0 & -i \\ i & 0 \end{pmatrix}. \tag{5.54}$$

Therefore, it follows:

$$I_l = Tr(\sigma_l \rho), \tag{5.55}$$

where I_l are the components of the Stokes vector (I,Q,U,V) and we used the property: $Tr(\sigma_p \sigma_q) = 2\delta_{pq}$. Here Tr means the trace operation.

Therefore, one can derive

$$I_l = \frac{1}{k^2 z_0{}^2} Tr(\widehat{\sigma}_j \widehat{S}\widehat{\rho}^{i}\widehat{S}^{+}). \tag{5.56}$$

Taking into account that

$$\widehat{\rho}^{i} = \frac{1}{2}\sum_{l=1}^{4} I_{0l}\sigma_l \tag{5.57}$$

we derive

$$I_l = \frac{P_{jk} I_{0k}}{k^2 z_0{}^2}, \tag{5.58}$$

where

$$P_{jk} = \frac{1}{2} Tr(\widehat{\sigma}_j \widehat{S}\widehat{\rho}^{i}\widehat{S}^{+}). \tag{5.59}$$

The nonzero elements of this matrix have the following form:

$$P_{11} = P_{22} = \frac{I_1 + I_2}{2}, \quad P_{12} = P_{21} = \frac{I_1 - I_2}{2}, \tag{5.60}$$

$$P_{33} = P_{44} = \mathrm{Re}(S_1 S_2^*), \quad P_{34} = -P_{43} = \mathrm{Im}(S_1 S_2^*). \tag{5.61}$$

The matrix elements P_{jk} can be derived from the first principles using the electromagnetic Maxwell theory (van de Hulst, 1981). Let us establish their physical sense. For this, let us imagine that we illuminate the elementary volume of a an aerosol medium composed of spherical particles by light beam with the Stokes vector with the components I_0, Q_0, U_0, V_0. Then the Stokes vector of scattered light beam has the following components:

$$I = P_{11}I_0 + P_{12}Q_0, \qquad Q = P_{12}I_0 + P_{11}Q_0,$$
$$U = P_{44}U_0 + P_{34}V_0, \qquad V = -P_{34}U_0 + P_{44}V_0. \tag{5.62}$$

The physical sense of the ratios

$$m_{12} = \frac{P_{12}}{P_{11}}, \qquad m_{34} = \frac{P_{34}}{P_{11}}, \qquad m_{44} = \frac{P_{44}}{P_{11}}, \tag{5.63}$$

can be derived using various assumptions on the polarization state of the incident light beam. For instance assuming that the incident light is unpolarized (parameters Q_0, U_0, and V_0 vanish), one derives from Eq. (5.62) for the nonzero Stokes vector components of scattered light:

$$Q = P_{12}I_0, \quad I = P_{11}I_0. \tag{5.64}$$

Taking into account that the degree of polarization of linearly polarized scattered light is defined as $P = -Q/I$, one obtains that $-m_{12}$ is equal to the degree of polarization of linearly polarized scattered light under illumination of the aerosol medium by the unpolarized light beam (see Figure 5.5).

Let us assume now that the incident light is linearly polarized with the azimuth $-45°$ ($I_0 = -U_0$ and Q_0, V_0 vanish). Then it follows from Eq. (5.62):

$$I = P_{11}I_0, \qquad Q = P_{12}I_0, \qquad U = P_{44}U_0, \qquad V = -P_{34}U_0. \tag{5.65}$$

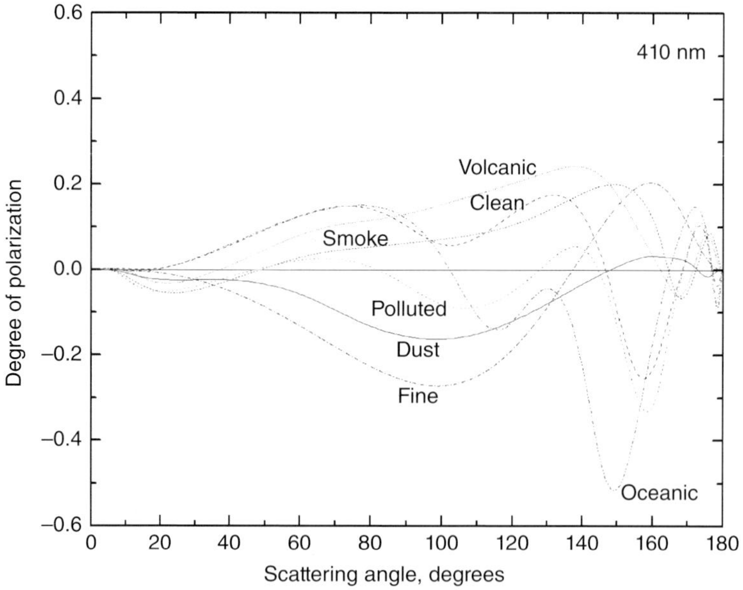

Figure 5.5 The same as in Figure 5.4 except for the degree of polarization.

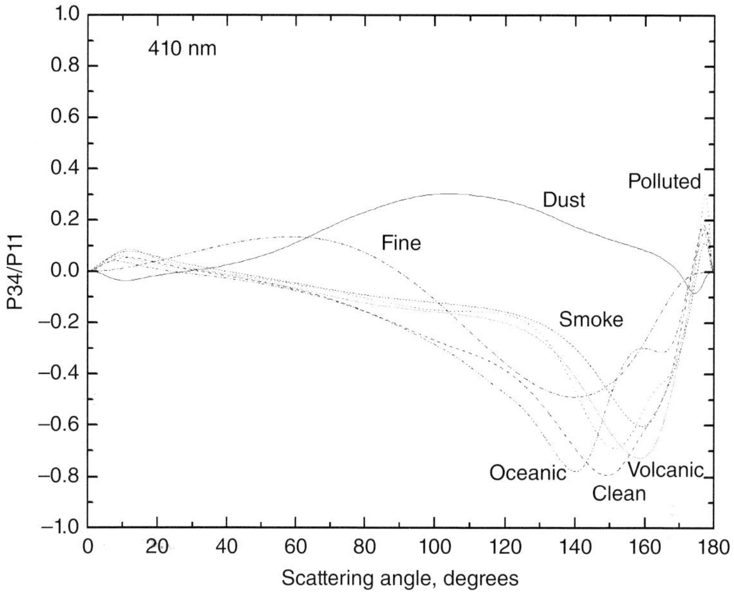

Figure 5.6 The same as in Figure 5.4 except for the ratio P34/P11.

One concludes that light becomes the circular polarized with the degree of circular polarization (V/I) equal to $-m_{34}$. Therefore, this element describes the process of linear to elliptic light transformation process (see Figure 5.6). The value of m_{44} is equal to the degree of circular polarization of scattered light but under illumination of the aerosol medium by the right-hand circular polarized light beam (see Eq. 5.62 at $I_0 = V_0$ with other Stokes vector elements equal to zero). Therefore, this element shows the distraction of circular polarization by a light scattering process. This distraction depends strongly on the scattering angle being larger at side scattering angles (see Figure 5.7). The degree of circular polarization is equal to 100% at the exact backscattering geometry. However, light becomes left-hand polarized (for incident 100% right-hand polarized light). Therefore, scattering medium can serve as an optical medium for the right-to-left-hand circular polarized light converter.

5.5 Extinction

It is known that light attenuates (on the length L) in homogeneous absorbing media according to the following law:

$$I = I_0 \exp(-k_{abs} L), \tag{5.66}$$

where I_0 is the incident light intensity, I is the intensity of the transmitted light, and k_{abs} is the absorption coefficient. In case, if also scattering occurs, Eq. (5.66)

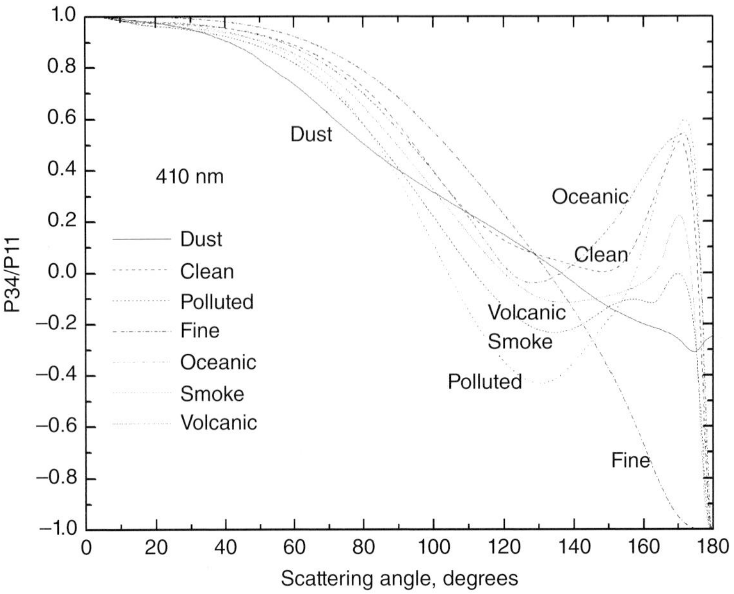

Figure 5.7 The same as in Figure 5.4 except for the ratio P44/P11.

still holds, but one must use the extinction coefficient

$$k_{\text{ext}} = k_{\text{abs}} + k_{\text{sca}} \tag{5.67}$$

instead of just k_{abs} in Eq. (5.66). So the correct equation is

$$I = I_0 \exp(-k_{\text{ext}} L). \tag{5.68}$$

The value of

$$V = \frac{3.92}{k_{\text{ext}}} \tag{5.69}$$

is equal to the meteorological range. The meteorological range gives a distance from which the absolutely black screen becomes invisible against the sky at the horizon. Clearly, the meteorological range is smaller in case of dense aerosol plumes positioned between the observer and a black screen. Let us assume that $k_{\text{ext}} = 10^{-4}\,\text{m}^{-1}$, then it follows: $V = 39.2$ km. One can easily observe large objects (say, buildings) on horizontal paths on the distances 30–40 km depending on the atmospheric turbidity. Therefore, the notion of the meteorological range is a useful one. The extinction efficiency factor ($Q_{\text{ext}} = Q_{\text{abs}} + Q_{\text{sca}}$) is shown in Figure 5.8.

It follows that this factor approaches value of 2 for very large particles (Shifrin, 1968).

Otherwise it depends on the size and chemical composition of aerosol particles. We may also conclude that the meteorological range is influenced not only by the concentration of aerosol particles but also by their dimension, shape, and

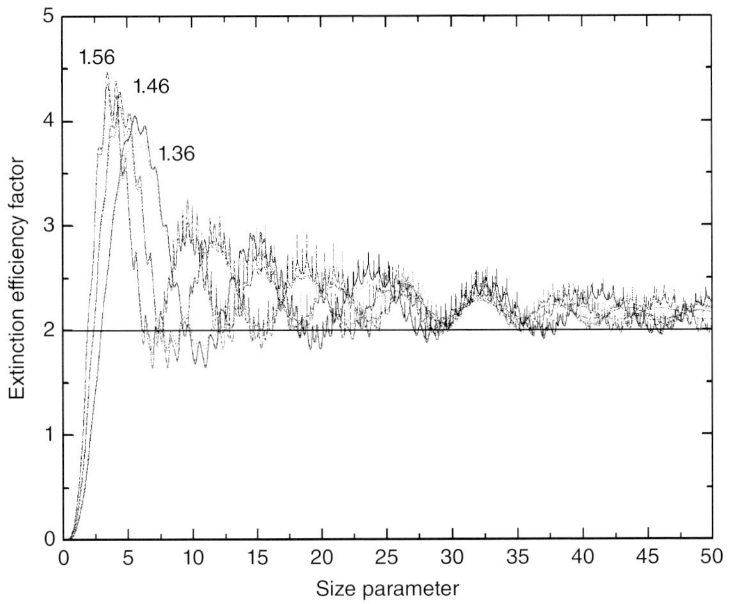

Figure 5.8 The dependence of the extinction efficiency factor on the size parameter at various values of the refractive index ($n = 1.36$, 1.46, 1.56). The imaginary part of the complex refractive index is equal to 0, 0.001, and 0.01. The maximal values of the efficiency factor decrease with n.

chemical composition. It should be pointed out that the extinction coefficient is a function of altitude with generally smaller values at higher levels in atmosphere (with some exclusions like dust outbreaks to the pristine atmosphere). The aerosol optical thickness is defined as the integral of the extinction coefficient with respect to the vertical coordinate z:

$$\tau_a = \int_0^{H_{\text{toa}}} k_{\text{ext}}(z)\,dz. \tag{5.70}$$

5.6 Radiative Transfer

Multiple scattering plays an important role in atmospheric radiative transfer processes. The light intensity at the wavelength λ in the direction specified by the angles (ϑ, φ) can be described by the following scalar radiative transfer equation:

$$\cos\vartheta \frac{dI_\lambda(\vartheta_0, \vartheta, \varphi)}{d\tau} + I_\lambda(\vartheta_0, \vartheta, \varphi) - \frac{\omega_0}{4\pi}\int_0^{\pi}\int_0^{2\pi} I_\lambda(\vartheta_0, \vartheta, \varphi)p(\theta)\cos\vartheta\, d\vartheta d\varphi = 0 \tag{5.71}$$

with corresponding boundary conditions. Here $\tau = k_{\text{ext}}z$ is the optical vertical coordinate (optical depth), z is the geometrical vertical coordinate, ϑ_0 is the zenith incidence angle (equal to zero for the overhead sun), $\omega_0 = \sigma_{sca}/\sigma_{ext}$ is the single-scattering albedo. Equation 5.71 does not account for the electromagnetic nature of light and, therefore, it provides useful approximation being less accurate for smaller particles. The errors of Eq. (5.71) could reach 10% for molecular atmosphere (Rozanov and Kokhanovsky, 2006).

As a matter of fact it is more appropriate to use the system of four coupled radiative transfer equations (instead of Eq. 5.71) with corresponding boundary conditions for finding the intensity of light propagated in aerosol atmosphere:

$$\cos\vartheta \frac{d\vec{I}_\lambda(\vartheta_0, \vartheta, \varphi)}{d\tau} + \vec{I}_\lambda(\vartheta_0, \vartheta, \varphi) - \frac{\omega_0}{4\pi}\int_0^\pi \int_0^{2\pi} \widehat{\mathfrak{R}}\vec{I}_\lambda(\vartheta_0, \vartheta, \varphi)\cos\vartheta d\vartheta d\varphi = 0. \tag{5.72}$$

Here $\vec{I}_\lambda(\vartheta_0, \vartheta, \varphi)$ is the Stokes vector with the components (I, Q, U, V) as described above. The matrix $\widehat{\mathfrak{R}}$ is realted to the phase matrix described above and can be calculated via following equation:

$$\widehat{\mathfrak{R}} = \widehat{\Lambda}(\pi - \gamma_s)\widehat{P}(\theta)\widehat{\Lambda}(-\gamma_i), \tag{5.73}$$

where $\widehat{P}(\theta)$ is the scattering matrix, θ is the scattering angle and the rotation matrix $\widehat{\Lambda}$ is defined as

$$\widehat{\Lambda}(\alpha) = \begin{pmatrix} 1 & 0 & 0 & 0 \\ 0 & c & -s & 0 \\ 0 & s & c & 0 \\ 0 & 0 & 0 & 0 \end{pmatrix}, \tag{5.74}$$

where $c = \cos 2\alpha$ and $s = \sin 2\alpha$. The angles γ_i and γ_s are the angles between the vertical planes containing the incident (i) and scattered (s) light beams, respectively.

The description of various codes to solve the radiative transfer equation is given by Kokhanovsky *et al.* (2010). There one can also find the tables of Stokes parameters for various geometries of observation both for transmitted and reflected solar light.

Integrodifferential Eqs (5.71), (5.72) (unlike wave Eq. 5.2) cannot be solved analytically. However, a lot of various approximate techniques have been developed to derive approximate solutions of these equations. Let us derive now the solution of Eq. (5.71) for the case of optically thin vertically homogeneous aerosol layer. We assume that the intensity $I_\lambda(\mu_0, \mu, \varphi)$ I can be represented in the following form:

$$I(\mu_0, \mu, \varphi) = I(\mu_0, \mu, \varphi) + F_0\delta(\mu - \mu_0)\delta(\phi - \phi_0), \tag{5.75}$$

where $\mu_0 = \cos\vartheta_0$, $\mu = \cos\vartheta$, $\varphi = \phi - \phi_0$ is the relative azimuthal angle, ϕ is the observation direction azimuthal angle, and ϕ_0 is the incidence direction azimuthal angle, F_0 is the incident light flux and we omitted the subscript λ. The substitution of Eq. (5.75) into Eq. (5.71) gives taking into account the properties of the Dirac delta function:

$$\mu\frac{dI(\vartheta_0,\vartheta,\varphi)}{d\tau}+I(\vartheta_0,\vartheta,\varphi) - \frac{\omega_0}{4\pi}\int_0^{\pi}\int_0^{2\pi} I(\vartheta_0,\vartheta,\varphi)p(\theta)\cos\vartheta d\vartheta d\varphi - \frac{\omega_0 p(\theta_s)}{4\pi}F_0\exp(-\tau/\mu_0)=0, \tag{5.76}$$

$$\cos\theta_s = -\mu\mu_0 + ss_0\cos(\phi-\phi_0), \tag{5.77}$$

$s_0 = \sin\vartheta_0$, $s = \sin\vartheta$. The boundary conditions for this equation in the case of black underlying surface can be formulated as follows: there is no diffuse light intensity $I(\mu_0,\mu,\varphi)$ coming to the scattering layer from below or above. The integral in Eq. (5.76) can be neglected under assumption that the multiple light scattering effects are weak. This is usually the case for pristine atmospheric aerosol over remote oceanic areas in the near infrared part of electromagnetic spectrum. Then it follows from Eq. (5.76):

$$\mu\frac{dI(\vartheta_0,\vartheta,\varphi)}{d\tau}+I(\vartheta_0,\vartheta,\varphi)-\frac{\omega_0}{4\pi}p(\theta_s)F_0\exp(-\tau/\mu_0)=0. \tag{5.78}$$

This equation can be solved analytically. Let us introduce a new variable $x=\tau/\mu$. Then it follows from Eq. (5.78) after multiplying it by e^x:

$$\frac{d(I(x)e^x)}{dx}-\frac{\omega_0}{4\pi}p(\theta_s)F_0\exp(-(s-1)x)=0, \tag{5.79}$$

where $s=\mu/\mu_0$. Therefore, one derives for a vertically homogeneous light scattering layer:

$$I(x)=\frac{\omega_0 e^{-x}}{4\pi}p(\theta_s)F_0\int_{\varsigma}^{x}\exp(-(s-1)x')dx', \tag{5.80}$$

or taking into account boundary conditions

$$I_\downarrow(x)=\frac{\omega_0 e^{-x}}{4\pi}p(\theta_s)F_0\int_0^{x}\exp(-(s-1)x')dx', \tag{5.81}$$

$$I_\uparrow(x)=\frac{\omega_0 e^{-x}}{4\pi}p(\theta_s)F_0\int_{x_0}^{x}\exp(-(s-1)x')dx', \tag{5.82}$$

where $I_\downarrow(x)$ is the intensity of diffuse light propagating downwards and $I_\downarrow(x)$ is the intensity of diffuse light propagating downwards, $x_0=\tau_a/\mu$ is the vertical coordinate at the bottom of the layer, and $x=0$ is the value of the coordinate at the top of the layer. These equations can be easily generalized to account for the vertical inhomogeneity of aerosol medium. It follows from Eqs (5.80) and (5.81):

$$I_\downarrow(x)=\frac{\omega_0}{4\pi(s-1)}p(\theta_s)F_0\left\{e^{-x}-e^{-sx}\right\}, \tag{5.83}$$

$$I_\uparrow(x)=\frac{\omega_0}{4\pi(s-1)}p(\theta_s)F_0\left\{e^{-x-(s-1)x_0}-e^{-sx}\right\}. \tag{5.84}$$

These equations enable the calculation of light intensity propagated downwards and upwards at any level inside the aerosol layer. It follows at the lower ($x = x_0$) and upper ($x = 0$) boundaries of a scattering layer:

$$I_{\downarrow}(x) = \frac{\omega_0}{4\pi(s-1)} p(\theta_s) F_0 \{e^{-x_0} - e^{-sx_0}\}, \tag{5.85}$$

$$I_{\uparrow}(x) = \frac{\omega_0}{4\pi(s-1)} p(\theta_s) F_0 \left\{e^{-(s-1)x_0} - 1\right\}. \tag{5.86}$$

Let us introduce the transmission T and reflection R functions as

$$T = \frac{\pi I_{\downarrow}(x)}{\mu_0 F_0}, \qquad R = \frac{\pi I_{\downarrow}(x)}{\mu_0 F_0}. \tag{5.87}$$

Then it follows from Eqs (5.84) and 5.85:

$$T = \frac{\omega_0}{4\pi(\mu_0 - \mu)} p(\theta_s) F_0 \left\{e^{-\tau_a/\mu_0} - e^{-\tau_a/\mu}\right\}, \tag{5.88}$$

$$R = \frac{\omega_0}{4\pi(\mu_0 + \eta)} p(\theta_s) F_0 \left\{1 - e^{-\tau_a(1/\mu_0 + 1/\eta)}\right\}, \tag{5.89}$$

where $\eta = |\mu|$.

5.7 Image Transfer

Atmospheric aerosol influences visibility of various objects in atmosphere. The simplest way to estimate the atmospheric turbidity is via calculation of the meteorological visibility range. However, there are more advanced approaches based on the notion of the optical transfer function (OTF) and its changes due to the presence of various types of atmospheric aerosol particles. The OTF is defined in the way described below.

Any diffused light source can be considered as a superposition of point light sources. The image created by the optical system of any object with the irradiance $F_0(\vec{r}\,')$ ($\vec{r}\,'$ is the radius vector of an arbitrary point in the object plane) is a linear superposition of images of point sources:

$$F(\vec{r}) = \int_{-\infty}^{\infty}\int_{-\infty}^{\infty} S(\vec{r}, \vec{r}\,')\, F_0(\vec{r}\,')\, d\vec{r}\,', \tag{5.90}$$

where $\vec{r}$ is the radius vector of an arbitrary point in the image plane. The function $S(\vec{r}, \vec{r}\,')$ plays a central role in the image transfer theory. It is called the point spread function (PSF). This function describes the process of the transformation of the object irradiance $F_0(\vec{r})$ in the initial plane to the image irradiance in the image plane. Equation 5.90 takes the following form in the frequency domain:

$$F(\vec{v}) = S(\vec{v}) F_0(\vec{v}). \tag{5.91}$$

Therefore, the complex operation of integration is substituted by the ordinary multiplication. The functions in Eq. (5.91) are defined as follows:

$$F(\vec{v}) = \int_{-\infty}^{\infty}\int_{-\infty}^{\infty} F(\vec{r}\,')\exp(-i\vec{v}\,\vec{r}\,')d\vec{r}\,',$$

$$F_0(\vec{v}) = \int_{-\infty}^{\infty}\int_{-\infty}^{\infty} F_0(\vec{r}\,')\exp(-i\vec{v}\,\vec{r}\,')d\vec{r}\,', \tag{5.92}$$

$$S(\vec{v}) = \int_{-\infty}^{\infty}\int_{-\infty}^{\infty} S(\vec{r}\,')\exp(-i\vec{v}\,\vec{r}\,')d\vec{r}\,'. \tag{5.93}$$

The function $S(\vec{v})$ is called the optical transfer function. It depends on the space frequency $\vec{v}$. The absolute value of OTF normalized to its value at zero spatial frequency is called the modulation transfer function (MTF). The MTF depends on the properties of media between am object and an image. If this function is known for a given medium, then one can determine the irradiance distribution in the image plane for a given irradiance distribution in the object plane:

$$F(\vec{r}) = \frac{1}{4\pi^2}\int_{-\infty}^{\infty}\int_{-\infty}^{\infty} F(\vec{v})\exp(i\vec{v}\,\vec{r})d\vec{v}. \tag{5.94}$$

Usually, the image transfer is studied along axis OZ perpendicular to the aerosol layer. In this case the OTF is a real function and does not depend on the azimuthal angle. In the case of aerosol media with large particles it can be approximated by the following expression (Kokhanovsky, 2004):

$$S(v) = \exp(-\tau(1 - \omega_0\gamma(v))), \tag{5.95}$$

where

$$\gamma(v) = \frac{\alpha\sqrt{\pi}}{vz} erf\left\{\frac{vz}{2\alpha}\right\}, \tag{5.96}$$

z is the vertical coordinate, α is the shape parameter of the phase function $p(\theta)$, approximated as $4\alpha^2\exp(-\alpha^2\theta)$, and

$$erf\{x\} = \frac{2}{\sqrt{\pi}}\int_0^x \exp(-y^2)dy. \tag{5.97}$$

Taking into account that

$$erf\left\{\frac{vz}{2\alpha}\right\} = \frac{vz}{\alpha\sqrt{\pi}}\left\{1 - \frac{v^2z^2}{12\alpha^2}\right\} \tag{5.98}$$

at small dimensionless frequencies $\upsilon = vz$, one concludes that

$$\gamma(v) = 1 - \frac{\upsilon^2}{12\alpha^2}. \tag{5.99}$$

It follows from Eqs (5.95) and (5.99) that OTF takes larger values for aerosol media with larger particles (larger values of the shape parameter α). Therefore, the image quality is better for objects observed through aerosol media with larger particles (at the same values of AOT and single scattering albedo). It follows from Eqs (5.95) and (5.99) that

$$S(0) = \exp(-\tau_{\text{abs}}), \tag{5.100}$$

where $\tau_{\text{abs}} = (1 - \omega_0)\tau_a$ is the absorbing optical depth. Therefore, only absorbing processes influence the OTF at the zero spatial frequency. More details on the image transfer in aerosol media can be found in Zege, Ivanov, and Katsev (1991).

Abbreviations

AOT	aerosol optical thickness
BC	black carbon
LUTs	lookup tables
NIR	near-infrared
UV	ultra-violet
OTF	optical transfer function
MTF	modulation transfer function

List of Symbols

Latin

C_{abs}	absorption cross section
C_{sca}	scattering cross section
C_{ext}	extinction cross section
k_{abs}	absorption coefficient
k_{sca}	scattering coefficient
k_{ext}	extinction coefficient
p	phase function
d	particle diameter
B	Planck function
T	temperature
R	Fresnel reflectance coefficient
$\vec{E}$	electric vector
m	refractive index
n	real part of refractive index
V	volume, meteorological range
P	phase matrix
c_v	volumetric concentration
c_m	mass concentration
$f(r)$	size distribution
N	number of particles in unit volume
r	particle radius
r_0	median radius

x	size parameter
W	scattered light energy
I	intensity
S	optical transfer function

Greek

α	bulk absorption coefficient
ε	dielectric permittivity
λ	wavelength
ρ	density
k	imaginary part of refractive index
θ	scattering angle
ω_0	single-scattering albedo
υ	spatial frequency
τ_a	aerosol optical thickness (AOT)
τ_{abs}	absorbing optical depth
Ω	solid angle

References

Babenko, V.A., Astafyeva, L.G., and Kuzmin, V.N. (2003) *Electromagnetic Scattering in Disperse Media: Inhomogeneous and Anisotropic Particles*, 1st edn, Springer-Praxis Books/Environmental Sciences, Chichester, 434 pp.

Dubovik, O., Sinyuk, A., Lapyonok, T., Holben, B.N., Mishchenko, M., Yang, P., Eck, T.F., Volten, H., Muñoz, O., Veihelmann, B., van der Zande, W.J., Leon, J.-F., Sorokin, M., and Slutsker, I. (2006) Application of spheroid models to account for aerosol particle non-sphericity in remote sensing of desert dust. *J. Geophys. Res.*, **111**, D11208. doi: 10.1029/2005JD006619.

Gershun, A.A. (1937) On the problem of diffuse light transmission. *GOI Proc.* (Proceed. State Optical Institute, St. Petersburg, Russia), **4**, 12–40.

van de Hulst, H.C. (1981) *Light Scattering by Small Particles*, Dover Publications Inc., New York, 470 pp.

Zege, E.P., Ivanov, A.P., and Katsev, J.L. (1991) *Image Transfer Though Light Scattering Medium*, Springer, Berlin, 349 pp.

Kokhanovsky, A.A. (2004) *Light Scattering Media Optics*, Springer Science & Business Media, Chichester, 299 pp.

Kokhanovsky, A.A. (2006) *Cloud Optics*, Springer, Dordrecht, 181 pp.

Kokhanovsky, A.A., Budak, V.P., Cornet, C., Duan, M., Emde, C., Katsev, I.L., Klyukov, D.A., Korkin, S.V., Labonnote, L.C., Mayer, B., Min, Q., Nakajima, T., Ota, Y., Prikhach, A.S., Rozanov, V.V., Yokota, T., and Zege, E.P. (2010) Benchmark results in vector atmospheric radiative transfer. *J. Quant. Spectrosc. Radiat. Transfer*, **111** (12-13), 1931–1946. doi: 10.1016/j.jqsrt.2010.03.005

Kokhanovsky, A.A. and Macke, A. (1997) Integral light-scattering and absorption characteristics of large, nonspherical particles. *Appl. Opt.*, **36**, 8785–8790. doi: 10.1364/AO.36.008785

Kokhanovsky, A.A. and Nauss, T. (2005) Satellite-based retrieval of ice cloud properties using a semianalytical algorithm. *J. Geophys. Res.*, **110**, D19206. doi: 10.1029/2004JD005744

Mishchenko, M.I. (2014) *Electromagnetic Scattering by Particles and Particle Groups: An Introduction*, Cambridge

University Press, Cambridge. doi: 10.1017/CBO9781139019064, 435 pp.

Räisänen, P., Kokhanovsky, A., Guyot, G., Jourdan, O., and Nousiainen, T. (2015) Parameterization of single-scattering properties of snow. *Cryosphere*, **9**, 1277–1301. doi: 10.5194/tc-9-1277-2015

Rozanov, V.V. and Kokhanovsky, A.A. (2006) The solution of the vector radiative transfer equation using the discrete ordinates technique: selected applications. *Atmos. Res.*, **79** (3-4), 241–265. doi: 10.1016/j.atmosres.2005.06.006

Shifrin, K.S. (1951) *Light Scattering in Turbid Media*, Gostekhteorizdat, Moscow-Leningrad [English translation: Shifrin, K.S. (1968) *Scattering of Light in Turbid Media*, NASA Tech. Trans. TT F-447, 1968, NASA, Washington, DC)].

Stratton, J.A. (1941) *Electromagnetic Theory*, John Wiley & Sons, Inc., New York, 640 pp.

6
Aerosol Models

Claudio Tomasi, Mauro Mazzola, Christian Lanconelli, and Angelo Lupi

6.1
Introduction

Airborne aerosol particles consist of fragments of solid particulate matter (PM) mixed with solute substances for humid air conditions. Their number concentration usually ranges from a few hundreds per cubic centimeter of air in the Earth's remote regions to more than 10^4 cm^{-3} in the most polluted urban areas, having sizes varying from 5×10^{-4} to more than 50 μm. Aerosols originate mainly from natural sources, such as wind-borne dust, sea spray generated by wind forcing on the ocean surface, and volcanic emissions, and in a lower part from anthropogenic activities (by combustion of fossil fuels and emission of pollutants). Formed directly as particles (primary aerosol) or through gas-to-particle (g-to-p) conversion processes (secondary aerosol), the wet-air aerosol particles can be technically defined as a suspensions of solid matter and liquid water (LW). During their atmospheric life, aerosols can grow and change their sizes and composition by condensation of vapor species (mainly water vapor) and coagulation with other particles, often becoming fog or cloud droplets. Finally, aerosols are removed from the atmosphere through (i) gravitational deposition at the surface (dry deposition) or (ii) incorporation into cloud droplets during the formation of precipitations (wet deposition). These processes concur to shorten the tropospheric residence times of aerosols, which can vary from a few days to a few weeks, and therefore are considerably longer than those of atmospheric trace gases usually ranging from less than 1 s to a few minutes.

Primary natural aerosols mainly consist of sea-salt particles originated by the wind forcing on the oceanic surface, mineral dust mobilized by winds in desert and semiarid regions, biogenic aerosols released by plants and animals, smokes from forest fires, stratospheric dust formed from debris and gases emitted by violent volcanic eruptions, and particles formed through reactions between natural

Atmospheric Aerosols: Life Cycles and Effects on Air Quality and Climate, First Edition.
Edited by Claudio Tomasi, Sandro Fuzzi, and Alexander Kokhanovsky.

gases. Primary anthropogenic aerosols are mainly produced by fossil fuel combustion, industrial processes, vehicular and airplane transport, nonindustrial fugitive sources (roadway dust from paved and unpaved roads, wind erosion of cropland, construction, etc.), and burning of urban and agricultural biomass waste products. The secondary aerosols of natural origins consist mainly of sulfate particles from SO_2 and sulfur compounds emitted by hypothermic and volcanic sources or produced by marine organisms, nitrate particles formed from nitrogen oxides, and organic aerosols from biogenic volatile organic compounds (VOCs), while significant amounts of sulfuric acid (SA) and water droplets can form at stratospheric levels from the volcanic SO_2 or the biological SO_2 molecules of marine origin. Secondary anthropogenic aerosols are formed through the g-to-p conversion processes of SO_2, NO_x, and organic substances emitted into the atmosphere by man-made activities, in which condensable vapors lead either to the nucleation of new particles or to the growth of existing particles by condensation. In these cases, both physical and chemical processes contribute to the accretion of precursors (most frequently molecular clusters), which do not have by themselves sizes large enough for being easily counted as particles during the initial stage of their atmospheric life. Therefore, aerosols have sizes ranging from a few hundreds of nanometer (molecular clusters), such as those produced by biomass combustion, to very large particles having sizes greater than 10–20 μm, as in the cases of mineral dust, sea salt, and volcanic debris.

The aforementioned processes lead to the formation of four basic size classes of airborne aerosols. The so-called ultrafine particles constitute the first class, mainly consisting of small nuclei with sizes up to about 10^{-2} μm. The particles of the second class are the so-called Aitken nuclei, presenting sizes ranging mainly from 5×10^{-3} to 5×10^{-2} μm, and mainly generated by combustion and g-to-p conversion and principally removed by coagulation with larger particles. The particles of the third class mainly form through coagulation and condensational growth of nuclei, having sizes ranging from about 0.1 to 2.5 μm. They are commonly named "accumulation" particles since they tend to accumulate over such a size range, being subsequently activated as cloud condensation nuclei (CCN) and/or removed by the rainout and washout mechanisms occurring in clouds and precipitations. The particles of the fourth class constitute the so-called coarse particle mode, which exhibit sizes $a \geq 2.5$ μm and are strongly removed from the atmosphere after reasonably short times by both gravitational settling and wet removal.

An aerosol population may include a variety of particles belonging to the four aforementioned size classes. The ultrafine particle class includes the tiny particles formed through nucleation and the small combustion particles, which are often agglomerations of soot substances and carbon particles impregnated with "tar," formed because of the incomplete combustion of carbonaceous material. The nuclei and accumulation modes include particles having sizes a ranging from 5×10^{-1} to 2.5 μm, which form through secondary processes and chemical reactions occurring in cloud droplets. They also include smog and fume particles generated by condensation from the vapor state, generally after volatilization from melted substances. Haze particles belong usually to the accumulation particle class and generally consist of a combination of water droplets, pollutants, and industrial dust with sizes $a < 1\ \mu m$. The coarse particle class may include primary sea-salt aerosols generated by wind-borne mechanical processes on oceanic surfaces, mineral dust produced by wind erosion and fragmentation of the soil particles, and mist and fog droplets.

The previous remarks indicate that both coagulation and condensation processes can efficiently modify the size-distribution curves of the Aitken nuclei and accumulation mode particles, causing a shift of particle concentration from the accumulation particle range to the coarse particle range. These processes are accomplished by marked variations in the shape of the aerosol size-distribution curves and in the chemical composition characteristics of atmospheric aerosols, which cause important changes in the optical parameters of both accumulation and coarse particles.

6.2 Modeling of the Optical and Microphysical Characteristics of Atmospheric Aerosol

The sizes of accumulation and coarse particles are comparable with the wavelengths of incoming solar radiation, mainly ranging from 0.3 to 4.0 μm. Therefore, these airborne aerosols produce very intense scattering (and absorption) effects on the shortwave radiation and weaker interactions with the terrestrial radiation emitted by the surface toward space. Such scattering and absorption effects on the incoming shortwave radiation cause marked changes in the radiation budget of the surface–atmosphere system, when aerosol optical thickness (AOT) is higher than 0.10 at the visible and near-infrared wavelengths, inducing pronounced direct aerosol radiative forcing (DARF) effects at both surface and top levels of the atmosphere and, therefore, within the atmosphere. The present study provides a picture of the main shape parameters of the number density size-distribution curves characterizing the aerosol types most frequently observed in the atmosphere and describes the most relevant optical characteristics of aerosols. Particular attention is paid in the present study to evaluate the multimodal features of the size-distribution curves of particle number concentration, surface area, and volume, and since the number concentration provides the measure of the contributions given by various aerosol classes to the overall particle polydispersion, the surface area gives the measure of the relevance of the extinction effects produced by aerosols, and the volume size-distribution indicates the particle size range within which the most important contribution is given to the PM mass concentration.

The volume extinction coefficient $\beta_{ext}(\lambda)$ of a certain aerosol particle polydispersion at wavelength λ is given by the sum of volume scattering coefficient $\beta_{sca}(\lambda)$ and volume absorption coefficient $\beta_{abs}(\lambda)$. Each of these three volume coefficients can be calculated for a spherical aerosol particle of diameter a as the product of the geometrical cross section $\pi a^2/4$ by the corresponding efficiency factor chosen among factors $K_{ext}(a, \lambda)$, $K_{sca}(a, \lambda)$, and $K_{abs}(a, \lambda)$, all varying as a function of

a, λ and complex particulate refractive index $m(\lambda) = n(\lambda) - i\chi(\lambda)$. The scattering efficiency factor $K_{sca}(a, \lambda)$ varies as a function of the so-called Mie size parameter $x = \pi a/\lambda$, increasing rapidly with x until reaching a first maximum at a value of x varying from about 4 (for LW) to about 6 (for dry-air continental aerosols) and describing a series of damped oscillations as a function of x, which become gradually less pronounced tending to reach an asymptotic value close to 2 for values of $x > 50$ and exhibit extinction features well described by classical optics theory. The Mie scattering and absorption effects produced by aerosol particles on the shortwave radiation depend strongly on particle size a and refractive index parts $n(\lambda)$ and $\chi(\lambda)$. Therefore, aerosol particle polydispersions of different origins can often present largely different spectral patterns of the three efficiency factors, varying with the shape parameters of the size-distribution curves and PM refractive index.

On the basis of these remarks, the present study is aimed at characterizing the microphysical and optical parameters of the airborne aerosol particles depending on the chemical composition characteristics of PM:

a. The number density size-distribution curve $N(a)$ of spherical aerosol particles as a function of diameter a can be in most cases realistically represented using the following lognormal function to represent each particle mode:

$$N(a) = dN/da = \frac{N_{tot}}{\sqrt{2\pi}a(\ln \sigma_g)} \exp\left(-\frac{(\ln a - \ln a_m)^2}{2(\ln \sigma_g)^2}\right), \quad (6.1)$$

where (i) diameter a usually varies over the range from 2×10^{-4} to more than 50 μm; (ii) N_{tot} is the total aerosol number concentration (measured in per cubic centimeter) of the single-particle mode, which is labeled with symbol N_i when it pertains to ith mode of a multimodal size-distribution curve; (iii) σ_g is the so-called geometric standard deviation; and (iv) a_m is the so-called geometrical mean diameter. The shape parameters σ_g and a_m define in practice the shape of each unimodal lognormal aerosol size-distribution curve, the diameter a_m being given by the following ratio:

$$a_m = \int_0^\infty N(a)da/N_{tot} \quad (6.2)$$

which is in general equal to the diameter at which the lognormal curve exhibits its maximum number concentration value. For all the most commonly used size-distribution curves of airborne aerosols, the mean diameter a_m in practice equals to the mode diameter a_c of the unimodal size-distribution curve.

b. The complex refractive index $m(\lambda) = n(\lambda) - i\chi(\lambda)$ of PM, which varies largely with particulate origins and the composition of PM. The real part $n(\lambda)$ at visible wavelengths ranges usually from 1.33 (LW) to 1.65 (dust particles, mainly containing quartz and silicates), while the imaginary part $\chi(\lambda)$ varies commonly from less than 10^{-8} (LW) to more than 0.4 (soot substances), as shown in Figure 6.1, where the spectral curves of $n(\lambda)$ and $\chi(\lambda)$ are presented for various aerosol types defined by the World Climate Programme (WCP) (WCP-112, 1986) and for LW by Hale and Querry (1973).

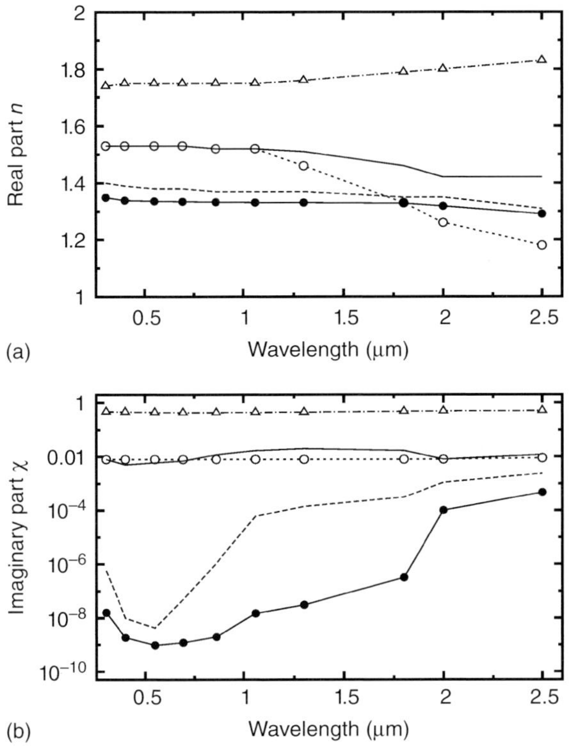

Figure 6.1 (a) Spectral curves of the real part $n(\lambda)$ of particulate matter refractive index over the 0.30–2.50 μm wavelength range for (i) oceanic aerosol (dashed curve), mainly containing sea salts; (ii) water-insoluble substances (dotted curve with open circles), mainly consisting of soil dust; (iii) water-soluble substances (solid curve), mainly containing sulfate and nitrate components; (iv) soot substances (dotted and dashed curve with open triangles), mainly produced by natural and anthropogenic combustion processes (World Climate Programme, WCP-112, 1986); and (v) liquid water (solid curve with solid circles), according to Hale and Querry (1973). (b) As in the (a) for the imaginary part $\chi(\lambda)$ of the same five substances.

c. The volume extinction, scattering, and absorption coefficients $\beta_{ext}(\lambda)$, $\beta_{sca}(\lambda)$, and $\beta_{abs}(\lambda)$ calculated per unit path by applying the Mie theory to the size-distribution curves defined at point (a) for the values of $n(\lambda)$ and $\chi(\lambda)$ determined at point (b).
d. The single scattering albedo $\omega(\lambda)$ of an aerosol particle population, defined as the ratio between $\beta_{sca}(\lambda)$ and $\beta_{ext}(\lambda)$ (with $\beta_{ext}(\lambda) = \beta_{sca}(\lambda) + \beta_{abs}(\lambda)$), which provides in practice the percentage of incoming radiation subject to extinction which is scattered by aerosols. Parameter $\omega(\lambda)$ assumes values ranging from 1.0 for nonabsorbing particles to less than 0.5 for strongly absorbing particles.
e. The phase function $p(\cos\theta)$, defined in radiative transfer studies as the energy scattered per unit solid angle in a given direction and normalized to the average energy scattered in all directions. It is evaluated as a function of the cosine

of scattering angle θ between the incident radiation and the scattered radiation and depends on the shape parameters of the aerosol size-distribution curve and complex refractive index $m(\lambda)$, varying as a function of wavelength λ. This parameter shows different spectral features passing from one aerosol type to another and yields the percentages of the forward, lateral, and backward scattering components produced by aerosols. Some examples of the spectral curves of $p(\cos\theta)$ are shown in Figure 6.2 for continental, maritime, urban, desert dust, biomass burning smoke (BBS), and stratospheric volcanic aerosol

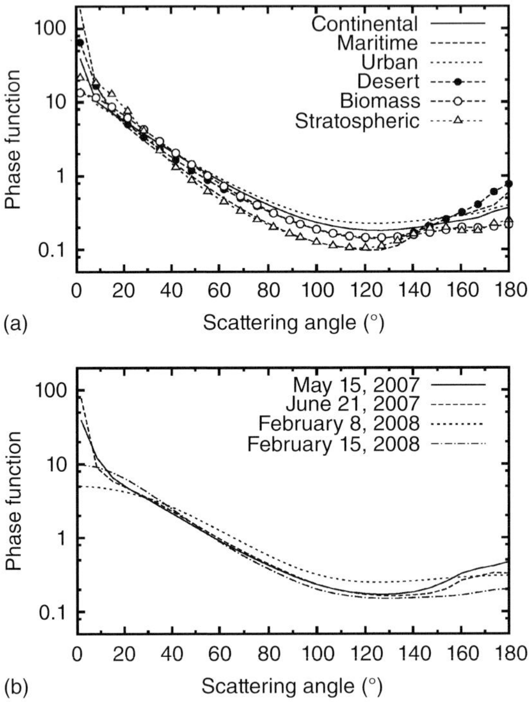

Figure 6.2 (a) Angular dependence curves of phase function $p(\cos\theta)$ calculated at wavelength $\lambda = 0.55\ \mu m$ by Vermote *et al.* (1997b) for the six following aerosol types: (i) continental (solid curve), (ii) maritime (dashed curve), (iii) urban (dotted curve), (iv) desert dust (dashed curve with small solid circles), (v) biomass burning (dashed curve with small open circles), and (vi) stratospheric (dashed and dotted curve with small open triangles). (b) As in the (a) for the aerosol particle contents analyzed by means of sky-brightness measurements performed in almucantar using the Prede radiometer (POM-02L model) at the San Pietro Capofiume (Italy) station on four measurement days during the AEROCLOUDS field campaigns: (i) 15 May 2007 (solid curve), for maritime–continental aerosol transported from Atlantic Ocean, southern France, and Ligurian Sea; (ii) 21 June 2007 (dashed curve), for continental-polluted aerosol from central and southern Italy, with marine aerosols from southern Mediterranean Sea and desert dust from the Libyan coasts; (iii) 8 February 2008 (dotted curve), for continental–anthropogenic aerosol from northwestern Europe; and (iv) 15 February 2008 (dotted and dashed curve), for continental aerosol from eastern Europe and northern Balkans.

types, calculated using the Second Simulation of the Satellite Signal in the Solar Spectrum (6S) code (Vermote *et al.*, 1997a,b). Four additional examples are also shown in Figure 6.2 for some aerosol cases, as derived from the sky-brightness measurements performed in the almucantar using a Prede POM-02L sun/sky radiometer of the SKYNET network at the ISAC-CNR San Pietro Capofiume station (hereinafter referred to as SPC), about 25 km northeast from Bologna (Italy), on some spring and winter days of 2007–2008 during the AEROCLOUDS field campaign.

f. The asymmetry factor $g(\lambda)$, defined in terms of the following integral:

$$g(\lambda) = \frac{1}{2}\int_{-1}^{+1} p(\cos\theta)\cos\theta \; d(\cos\theta) \tag{6.3}$$

provides a measure of the preferred scattering direction of the radiation beam encountering an aerosol particle polydispersion: (i) a null value of $g(\lambda)$ indicates that the scattering directions are evenly distributed between forward and backward directions, as in the case of isotropic scattering, which is typical of small particles; (ii) positive values of $g(\lambda)$ indicate that scattering is strongly peaked in the forward direction (i.e., with $\theta < 90°$), as it is typical of large Mie particles, yielding in general values of $g(\lambda)$ close to +1 at visible wavelengths; and (iii) negative values of $g(\lambda)$ indicate that aerosol scattering is predominantly given in the backward direction (i.e., with $\theta > 90°$), thus peaking at $\theta = 180°$. Typical values of $g(\lambda)$ at visible wavelengths were estimated to range from ∼0.60 for urban aerosols to ∼0.80 for ice crystals and ∼0.85 for cloud water droplets.

A number of aerosol extinction models of different origins are described in the present study to characterize their scattering and absorption properties. They were in part found in the literature and in part determined by us for different particle populations to provide the shape parameters of the multimodal number density size-distribution curves of $N(a)$, the spectral series of $n(\lambda)$ and $\chi(\lambda)$, and the spectral values of coefficients $\beta_{ext}(\lambda)$ and $\beta_{sca}(\lambda)$ and optical parameters $\omega(\lambda)$ and $g(\lambda)$, obtained for the following 43 aerosol extinction models: (i) the set of three models defined by Vermote *et al.* (1997a,b) to represent the optical characteristics of continental, maritime, and urban aerosol particles for dry-air conditions; (ii) the four aerosol extinction models for background desert dust (BDD) (Shettle, 1984), El Chichón volcanic stratospheric (EVS) (King, Harshvardhan, and Arking, 1984), BBS (Remer *et al.*, 1998), and continental-polluted aerosol determined by us on the basis of the size-distribution and composition parameters measured in the Po Valley (Italy) by Carbone *et al.* (2010); (iii) eight aerosol models determined by Tomasi *et al.* (2013) by differently combining the 6S basic components of maritime and continental aerosols in wet air for relative humidity RH = 50%; (iv) the ten optical properties of aerosols and clouds (OPAC) aerosol models defined by Hess, Koepke, and Schult (1998) for RH = 50%; (v) the four classical aerosol models defined by Shettle and Fenn (1979) for rural, maritime, tropospheric, and urban aerosols; (vi) seven additional aerosol models defined by Tomasi *et al.* (2013) for a pair of Saharan dust models, a pair of

BBS particle polydispersions and three pre- and post-Pinatubo volcanic aerosol models; and (vii) seven polar aerosol models determined by Tomasi *et al.* (2015a) to represent the radiative properties of Arctic and Antarctic aerosols. Detailed descriptions of the microphysical and optical characteristics of the 43 aerosol extinction models are presented in the following subsections.

6.2.1
The 6S Code Aerosol Extinction Models

The 6S radiative transfer code was prepared by Vermote *et al.* (1997a,b) to (i) evaluate the effects of the radiative transfer processes occurring in the surface – atmosphere system, (ii) simulate the radiative effects occurring along the sun path and the atmospheric path from the Earth's surface to the satellite-borne sensor, (iii) analyze the remote sensing observations conducted from satellite platforms over sea and land surfaces, and (iv) estimate the aerosol extinction effects on the outgoing flux of solar radiation. Particular attention was paid in the 6S code to the aerosol scattering and absorption effects, defining the four 6S basic aerosol components consisting of dustlike (DL), water-soluble (WS), oceanic (OC), and soot (SO) PM, each component having a well-defined unimodal size-distribution curve and specific radiative characteristics at the selected visible and near-infrared wavelengths.

6.2.1.1 The Four 6S Basic Aerosol Components

The DS, WS, OC, and SO 6S aerosol components were assumed by Vermote *et al.* (1997b) to have the spectral and microphysical characteristics of the four basic components defined by the International Radiation Commission (WMO, 1983), which established in particular that the SO component consists of both soluble and insoluble organic substances. The spectral and geometric patterns of the previously defined parameters $N(a)$, $n(\lambda)$, $\chi(\lambda)$, $\beta_{ext}(\lambda)$, $\beta_{sca}(\lambda)$, $\beta_{abs}(\lambda)$, $p(\cos\theta)$, $\omega(\lambda)$, and $g(\lambda)$ were determined with great accuracy for the four 6S components at 11 selected wavelengths from 0.30 to 3.75 μm for (i) the three scattering angles of 0°, 90°, and 180° and (ii) 80 supplementary Gaussian angles ranging between 0° and 180°. The lognormal number density size-distribution curves of the four 6S components are shown in Figure 6.3 for dry-air (RH = 0%) conditions over the $10^{-3} \leq a \leq 10^{2}$ μm diameter range. The lognormal size-distribution curve $N(a)$ was assumed to have the analytical form given in Eq. (6.1) for all the four 6S components, with the shape parameters and volume and mass fractions of the DL, OC, WS, and SO unimodal size-distribution curves reported in Table 6.1, together with the parameters defining the dry-air and wet-air particulate mass density, the particle growth by condensation, and the average single scattering albedo. To determine the most reliable values of the average growth factor φ_r for the four 6S basic components due to the RH increase from 0% to 50%, the particle growth simulation models defined by Hänel (1976) and Hänel and Bullrich (1978) were taken into account: those of the sea spray and desert dust particles for the OC and DL components, respectively; those of urban aerosols

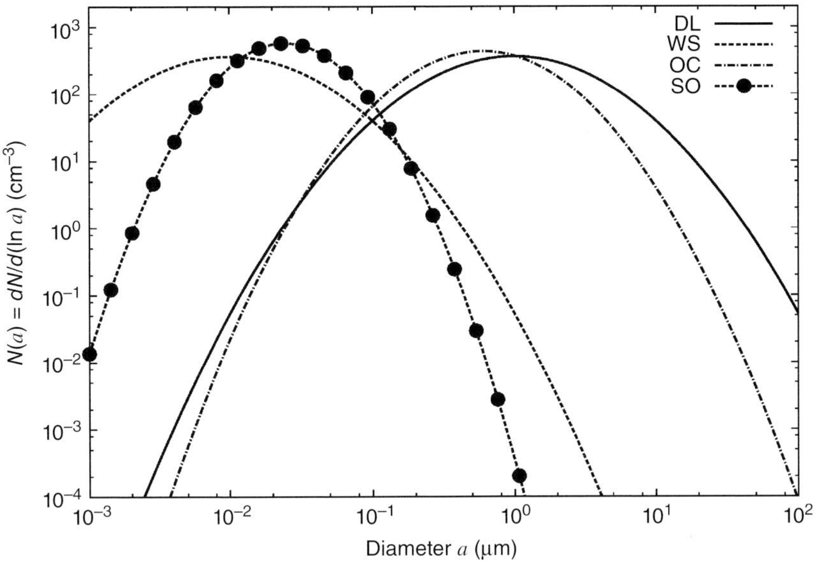

Figure 6.3 Size-distribution curves of particle number concentration $N(a)$ over the $10^{-3} \leq a \leq 10^2$ μm diameter range determined by Vermote *et al.* (1997b) to represent the four 6S basic aerosol components – dustlike (DL) (solid curve), water soluble (WS) (dashed curve), oceanic (OC) (dotted and dashed curve), and soot (SO) (dashed curve solid circles) – for dry-air conditions. Each lognormal size-distribution curve is normalized to give the overall particle number density $N_T = 10^3$ cm^{-3}.

for the SO component; and those of background (BG) continental (mountain) aerosols for the WS component. Using the estimates of growth factor φ_r given in Table 6.1, the particle size-distribution curves of the four 6S basic components were modified with respect to those defined for dry-air conditions by assuming the wet-air values of mode diameter $a_c = a_m \varphi_r$ and the same values of standard deviation σ_g (as reported in Table 6.1). The overall particle volumes V of the original 6S lognormal size-distribution curves of dry-air particles and the overall particle volumes V_g $(= V + \Delta V)$ of the new lognormal size-distribution curves after the particle growth were also calculated, obtaining the percentage volume ratios $\Delta V/V$ reported in Table 6.1 for the four 6S basic components and the values of wet-air (RH = 50%) mass density ρ_w and the mass fractions Γ_p of dry PM and Γ_w of LW calculated for water condensation, according to Hänel (1976) and Hänel and Bullrich (1978) for the sea spray, maritime, continental, and urban particles and to Volz (1972a,b, 1973) for the WS and SO components and for Saharan dust and coal-fired dust.

Figure 6.3 clearly shows that the WS and SO dry-air particles exhibit predominant number concentrations of nuclei and accumulation particles, with values of mode diameter a_c equal to 10^{-2} and 0.236 μm, respectively, while the OC and DL components exhibit rather low contents of nuclei and mainly consist of accumulation/coarse particles, their values of a_c being equal to 0.60 and 1.00 μm,

Table 6.1 Shape parameters and particulate mass density and mass fractions of the four 6S basic components dustlike (DL), water soluble (WS), oceanic (OC), and soot (SO), used for both dry-air (RH = 0%) and wet-air (RH = 50%) conditions to determine the physicochemical and radiative properties of the four 6S basic aerosol components used by Vermote *et al.* (1997a,b) to define the continental (6S-C), maritime (6S-M), and urban (6S-U) aerosol extinction models.

6S basic component	Dry particle mode diameter a_m (μm)	Geometric standard deviation σ_g	Average growth radius factor φ_r	Dry particle volume increase ratio $\Delta V/V$	Dry-air mass density ρ (g cm^{-3})	Wet-air (RH = 50%) mass density ρ_w (g cm^{-3})	Mass fraction of dry particulate matter Γ_p	Mass fraction of liquid water Γ_w	Average single scattering albedo ω
Dustlike (DL)	1.00	2.99	1.008	0.023	2.36	2.329	0.990	0.0098	0.702
Water soluble (WS)	10^{-2}	2.99	1.018	0.055	1.86	1.815	0.971	0.0287	0.923
Oceanic (OC)	6.00×10^{-1}	2.51	1.069	0.223	2.25	2.022	0.910	0.0901	0.996
Soot (SO)	2.36×10^{-2}	2.00	1.028	0.087	1.62	1.570	0.949	0.0511	0.149

The last column provides the average values of the single scattering albedo ω calculated over the 0.30–3.75 μm wavelength range.

respectively. The optical parameters of the four 6S basic components were calculated over the 0.30–3.75 μm wavelength range, using the four unimodal particle size-distribution curves shown in Figure 6.3 for dry-air conditions, obtaining the spectral values of parameters $n(\lambda), \chi(\lambda)$, $\beta_{ext}(\lambda)$, $\beta_{sca}(\lambda)$, $\omega(\lambda)$, and $g(\lambda)$ reported in Table 6.2 at 11 selected wavelengths.

These calculations show that (i) the highest spectral values of $n(\lambda)$ in the visible are provided by the SO component and the lowest ones by the OC component, (ii) the SO component exhibits the highest values of $\chi(\lambda)$ over the entire spectral range, and (iii) the spectral patterns of $\beta_{ext}(\lambda)$ vary only slightly with wavelength for the DL and OC components and decrease slowly for the WS and SO components. The spectral values of $\omega(\lambda)$ calculated for the OC component were found to be close to unit at all wavelengths, while those of the WS component

Table 6.2 Spectral values of real part $n(\lambda)$ and imaginary part $\chi(\lambda)$ of particulate matter refractive index, volume extinction coefficient $\beta_{ext}(\lambda)$ and volume scattering coefficient $\beta_{sca}(\lambda)$ (both measured in per kilometer and calculated for overall number density concentration $N_{tot} = 10^3$ cm^{-3}), single scattering albedo $\omega(\lambda)$, and asymmetry factor $g(\lambda)$, defined by Vermote *et al.* (1997a,1997b) at 11 selected wavelengths from 0.30 to 3.75 μm for the four 6S basic aerosol components DL, WS, OC, and SO assumed for dry-air conditions.

λ (μm)	$n(\lambda)$	$\chi(\lambda)$	$\beta_{ext}(\lambda)$ (km^{-1})	$\beta_{sca}(\lambda)$ (km^{-1})	$\omega(\lambda)$	$g(\lambda)$
Dustlike (DL)						
0.300	1.530	8.0×10^{-3}	1.821×10^{1}	1.122×10^{1}	0.616	0.861
0.400	1.530	8.0×10^{-3}	1.839×10^{1}	1.158×10^{1}	0.630	0.854
0.488	1.530	8.0×10^{-3}	1.857×10^{1}	1.199×10^{1}	0.646	0.850
0.515	1.530	8.0×10^{-3}	1.862×10^{1}	1.211×10^{1}	0.650	0.848
0.550	1.530	8.0×10^{-3}	1.869×10^{1}	1.226×10^{1}	0.656	0.846
0.633	1.530	8.0×10^{-3}	1.884×10^{1}	1.261×10^{1}	0.669	0.841
0.694	1.530	8.0×10^{-3}	1.895×10^{1}	1.285×10^{1}	0.678	0.837
0.860	1.520	8.0×10^{-3}	1.923×10^{1}	1.346×10^{1}	0.700	0.828
1.536	1.400	8.0×10^{-3}	2.016×10^{1}	1.544×10^{1}	0.766	0.830
2.250	1.220	9.0×10^{-3}	1.952×10^{1}	1.556×10^{1}	0.797	0.904
3.750	1.270	1.1×10^{-2}	1.917×10^{1}	1.571×10^{1}	0.820	0.871
Water soluble (WS)						
0.300	1.530	5.0×10^{-3}	8.663×10^{-4}	8.402×10^{-4}	0.970	0.653
0.400	1.530	5.0×10^{-3}	7.790×10^{-4}	7.505×10^{-4}	0.963	0.642
0.488	1.530	5.0×10^{-3}	6.276×10^{-4}	6.050×10^{-4}	0.964	0.633
0.515	1.530	5.0×10^{-3}	5.895×10^{-4}	5.683×10^{-4}	0.964	0.631
0.550	1.530	6.0×10^{-3}	5.456×10^{-4}	5.223×10^{-4}	0.957	0.628
0.633	1.530	6.0×10^{-3}	4.571×10^{-4}	4.375×10^{-4}	0.957	0.621
0.694	1.530	7.0×10^{-3}	4.052×10^{-4}	3.849×10^{-4}	0.950	0.616
0.860	1.520	1.2×10^{-2}	2.960×10^{-4}	2.703×10^{-4}	0.913	0.610
1.536	1.510	2.3×10^{-2}	1.180×10^{-4}	9.504×10^{-5}	0.805	0.571
2.250	1.420	1.0×10^{-2}	3.891×10^{-5}	3.263×10^{-5}	0.839	0.549
3.750	1.452	4.0×10^{-3}	1.137×10^{-5}	1.001×10^{-5}	0.881	0.432

(*continued overleaf*)

Table 6.2 (Continued)

λ (µm)	$n(\lambda)$	$\chi(\lambda)$	$\beta_{ext}(\lambda)$ (km^{-1})	$\beta_{sca}(\lambda)$ (km^{-1})	$\omega(\lambda)$	$g(\lambda)$
Oceanic (OC)						
0.300	1.388	1.00×10^{-8}	3.530	3.530	1.000	0.793
0.400	1.385	9.90×10^{-9}	3.560	3.560	1.000	0.790
0.488	1.382	6.41×10^{-9}	3.638	3.638	1.000	0.785
0.515	1.381	3.70×10^{-9}	3.657	3.657	1.000	0.785
0.550	1.381	4.26×10^{-9}	3.685	3.685	1.000	0.783
0.633	1.377	1.62×10^{-8}	3.739	3.739	1.000	0.781
0.694	1.376	5.04×10^{-8}	3.769	3.769	1.000	0.780
0.860	1.372	1.09×10^{-6}	3.828	3.828	1.000	0.778
1.536	1.359	2.43×10^{-4}	3.745	3.728	0.996	0.782
2.250	1.334	8.50×10^{-4}	3.315	3.276	0.988	0.797
3.750	1.398	2.90×10^{-3}	2.837	2.758	0.972	0.749
Soot (SO)						
0.300	1.750	4.7×10^{-1}	8.899×10^{-4}	2.319×10^{-4}	0.261	0.417
0.400	1.750	4.6×10^{-1}	8.602×10^{-4}	2.295×10^{-4}	0.267	0.395
0.488	1.750	4.5×10^{-1}	6.582×10^{-4}	1.515×10^{-4}	0.230	0.357
0.515	1.750	4.5×10^{-1}	6.130×10^{-4}	1.344×10^{-4}	0.219	0.346
0.550	1.750	4.4×10^{-1}	5.529×10^{-4}	1.151×10^{-4}	0.208	0.334
0.633	1.750	4.3×10^{-1}	4.491×10^{-4}	8.142×10^{-5}	0.181	0.308
0.694	1.750	4.3×10^{-1}	3.958×10^{-4}	6.424×10^{-5}	0.162	0.289
0.860	1.750	4.3×10^{-1}	2.954×10^{-4}	3.554×10^{-5}	0.120	0.244
1.536	1.770	4.6×10^{-1}	1.498×10^{-4}	5.571×10^{-6}	0.0372	0.119
2.250	1.810	5.0×10^{-1}	1.048×10^{-4}	1.383×10^{-6}	0.0145	0.0576
3.750	1.900	5.7×10^{-1}	6.895×10^{-5}	1.983×10^{-7}	0.0029	0.0205

decrease gradually from 0.97 to ~0.80 over the $0.30 \leq \lambda \leq 1.54$ µm range, those of the DL component present relatively low values slowly increasing from nearly 0.6 in the visible to ~0.8 at infrared wavelengths, and those of the SO component are very low at all wavelengths, decreasing from less than 0.27 in the visible to less than 0.05 for $\lambda > 1.0$ µm, due to the strong absorption characteristics of the soot substances. The values of $g(\lambda)$ were estimated over the 0.30–3.75 wavelength range to vary between 0.62 and 0.82 for the DL component, decrease from 0.65 to 0.43 for the WS component and from 0.42 to 0.02 for the SO component, and increase from 0.75 to 0.80 for the OC component.

6.2.1.2 The Three 6S Aerosol Models

Using the set of parameters provided in Table 6.1 for the four 6S basic components, the following three tropospheric aerosol models were determined: (i) the continental (6S-C) aerosol (trimodal) model, consisting of volume percentages equal to 70% DL, 29% WS, and 1% SO; (ii) the maritime (6S-M) aerosol (bimodal) model, consisting of volume percentages equal to 95% OC and 5% WS; and (iii) the urban (6S-U) aerosol (trimodal) model, consisting of volume percentages equal to

61% WS, 22% SO, and 17% DL. The shape parameters of the various modes giving form to the size-distribution curves of the three aerosol models are given in Table 6.3 for values of number density concentration N_i assumed for each mode, yielding a total multimodal concentration $N_T = 10^3\ cm^{-3}$. The mean dry-air particulate mass density ρ was calculated for each of the three 6S models as a linear combination of the dry-air mass density values of the 6S basic components, weighted by their mass percentage contents reported in Table 6.4, together with the corresponding number density and volume percentages. For these assumptions, the mass percentages of the 6S components are equal to (i) 74.8% DL, 24.4% WS, and 0.7% SO for the 6S-C aerosol model; (ii) 4.2% WS and 95.8% OC for the 6S-M aerosol model; and (iii) 21.2% DL, 60.0% WS, and 18.8% SO for the 6S-U aerosol model. Using these mass percentages as weights of the radiative parameters defined in Table 6.2 for the four 6S basic components, the values of monochromatic coefficients $\beta_{sca}(0.55\ \mu m)$, $\beta_{abs}(0.55\ \mu m)$, and $\beta_{ext}(0.55\ \mu m)$, monochromatic single scattering albedo $\omega(0.55\ \mu m)$, average single scattering albedo ω over the $0.30-3.75\ \mu m$ spectral range, and best-fit Ångström (1964) exponent α calculated over the $0.40-0.86\ \mu m$ wavelength range are given in Table 6.3 for the three dry-air 6S-C, 6S-M, and 6S-U aerosol models.

The number density size-distribution curves of $N(a)$ for the 6S-C, 6S-U, and 6S-M dry-air aerosol models are shown in Figure 6.4 over the $10^{-3}-10^2\ \mu m$ diameter range, providing evidence of the significant differences between the number concentrations of accumulation and coarse particles over the $5 \times 10^{-1} \le a \le 5\ \mu m$ size range, where the number concentration $N(a)$ of 6S-M particles is higher by more than one order of magnitude than those of the 6S-U model and by nearly two orders of magnitude than those of the 6S-C model.

Using as weights the mass concentration percentages reported in Table 6.4 for the four 6S basic components, the weighted average spectral values of dry-air PM refractive index parts $n(\lambda)$ and $\chi(\lambda)$ have been calculated for the three 6S aerosol models at the 11 selected wavelengths, obtaining the results given in Table 6.5. Real part $n(\lambda)$ varies in the visible from less than 1.40 (6S-M model) to more than 1.55 (6S-U model) and decreases at longer wavelengths for all the three models, presenting the most marked decrease for the 6S-C model until assuming a value lower than 1.32 for $\lambda > 2\ \mu m$. Imaginary part $\chi(\lambda)$ was found to vary between 2×10^{-4} (6S-M model) and 10^{-1} (6S-U model) at the visible wavelengths. The spectral values of $\beta_{ext}(\lambda)$ and $\beta_{sca}(\lambda)$ are given in Table 6.5, presenting marked variations over the $0.30-3.75\ \mu m$ wavelength range. Correspondingly, the spectral values of $\omega(\lambda)$ given in Table 6.5 for the 6S-M model are very close to unit at all wavelengths, while those of model 6S-C decrease slowly from ~0.90 in the visible to less than 0.80 in the infrared, and those of model 6S-U decrease from ~0.70 in the visible to less than 0.40 at $\lambda > 2\ \mu m$. The values of $g(\lambda)$ determined for model 6S-C model were estimated to be rather stable and close to 0.63 at visible and near-infrared wavelengths and then appreciably increase to 0.77 at 3.75 μm, while those of model 6S-M were found to slowly increase from 0.74 to 0.79 over the whole wavelength range and those of model 6S-U model to range between 0.58 and 0.60.

Table 6.3 Values of the shape parameters N_i (for $i = 1, 2, 3$ for the first, second, and third mode, respectively), geometric standard deviation σ_g, and mode diameter a_c used to define the unimodal size distributions giving form to the 6S dry-air multimodal aerosol models and calculate the mean particle mass density ρ (g cm^{-3}) and the following main radiative parameters: (i) volume extinction, scattering, and absorption coefficients at the 0.55 μm wavelength, (ii) single scattering albedo ω at the 0.55 μm wavelength, (iii) average single scattering albedo ω, and (iv) mean Ångström's exponent α calculated over the 0.40–0.86 μm wavelength range, as obtained by Vermote *et al.* (1997b) for the dry-air aerosol models 6S-C (continental, trimodal), 6S-M (maritime, bimodal), and 6S-U (urban, trimodal).

6S dry-air aerosol model	First mode			Second mode			Third mode			ρ (g cm^{-3})	β_{ext} (0.55 μm) (km^{-1})	β_{sca} (0.55 μm) (km^{-1})	β_{abs} (0.55 μm) (km^{-1})	ω (0.55 μm)	ω	α
	$N_{i=1}$ (cm^{-3})	σ_g	a_c (μm)	$N_{i=2}$ (cm^{-3})	σ_g	a_c (μm)	$N_{i=3}$ (cm^{-3})	σ_g	a_c (μm)							
Continental (6S-C)	741.396	2.990	10^{-2}	258.598	2.000	0.0236	0.006	2.990	1.000	2.208	5.77×10^{-4}	5.15×10^{-4}	6.26×10^{-5}	0.892	0.850	1.249
Maritime (6S-M)	997.201	2.990	10^{-2}	2.799	2.500	0.600	—	—	—	2.231	2.06×10^{-3}	2.04×10^{-3}	2.20×10^{-5}	0.989	0.986	0.229
Urban (6S-U)	967.683	2.990	10^{-2}	32.249	2.000	0.0236	0.068	2.990	1.000	1.892	5.46×10^{-4}	3.53×10^{-4}	1.93×10^{-4}	0.646	0.553	1.459

The unimodal values of N_i are calculated to give a value of the overall particle number density $N_T = 10^3$ cm^{-3} for all the three 6S dry-air multimodal aerosol models.

Table 6.4 Number percentages N_j (%), volume percentages V_j (%), and mass percentages M_j (%) of the four 6S basic components – dustlike (DL), water soluble (WS), oceanic (OC), and soot (SO) – assumed by Vermote *et al.* (1997a,b) to give form to the continental (6S-C), maritime (6S-M), and urban (6S-U) aerosol models for the dry-air (RH = 0%) particulate mass density values of ρ equal to 2.36 g cm^{-3} for the DL component, 1.86 g cm^{-3} for the WS component, 2.25 g cm^{-3} for the OC component, and 1.62 g cm^{-3} for the SO component.

Aerosol model	Number percentages N_j (%)				Volume percentage V_j (%)				Mass percentages M_j (%)			
	DL	WS	OC	SO	DL	WS	OC	SO	DL	WS	OC	SO
Continental (6S-C)	2.265×10^{-4}	93.830	—	6.110	70.0	29.0	—	1.0	74.832	24.434	—	0.7338
Maritime (6S-M)	—	99.958	4.208×10^{-2}	—	—	5.0	95.0	—	—	4.169	95.830	—
Urban (6S-U)	1.651×10^{-5}	59.251		40.749	17.0	61.0	—	22.0	21.203	59.962	—	18.835

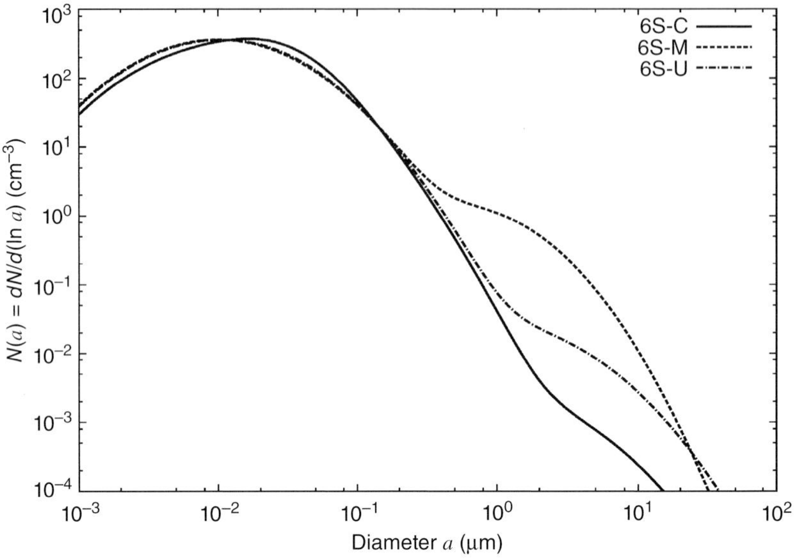

Figure 6.4 Size-distribution curves of particle number density $N(a)$ over the $10^{-3} \leq a \leq 10^2$ μm diameter range determined by Vermote *et al.* (1997b) to represent the three dry-air aerosol models of continental (6S-C) particles (solid curve), maritime (6S-M) particles (dashed curve), and urban (6S-U) particles (dashed and dotted curve). All the three multimodal size-distribution curves are normalized to give the overall particle number concentration $N_T = 10^3$ cm^{-3}.

6.2.2
The 6S Additional Aerosol Models

In presenting the 6S code, Vermote *et al.* (1997a) suggested the opportunity of determining some supplementary aerosol extinction models, which cannot be simply represented by varying the contents of the four 6S basic aerosol components. Two aerosol models were defined by Tomasi *et al.* (2013) to represent the particle radiative characteristics of (i) the BDD model based on the Shettle (1984) measurements and (ii) the EVS model based on the measurements analyzed by King, Harshvardhan, and Arking (1984). The third model is here defined by us for BBS aerosol particles on the basis of the measurements conducted by Remer *et al.* (1998) during the burning season in the Amazon Basin and Cerrado region of Brazil. The fourth aerosol extinction model, called Winter Po Valley (WPV) aerosol model, is here proposed to represent the optical characteristics of the aerosol particles suspended within the planetary boundary layer (PBL) of the Po Valley (Italy) on winter cloudless-sky days. It was defined on the basis of the *in situ* particulate composition measurements conducted by Carbone *et al.* (2010) at the SPC station within various size ranges in February 2008 and taking into account the spectral measurements of the aerosol extinction and scattering parameters performed on winter cloud-free days using the Prede POM-02L sun/sky

Table 6.5 Spectral values of real part $n(\lambda)$ and imaginary part $\chi(\lambda)$ of particulate matter refractive index, volume extinction coefficient $\beta_{ext}(\lambda)$ and volume scattering coefficient $\beta_{sca}(\lambda)$ (both measured in per kilometer and calculated for overall multimodal number density concentration $N_T = 10^3$ cm^{-3}), single scattering albedo $\omega(\lambda)$, and asymmetry factor $g(\lambda)$, defined by Vermote *et al.* (1997a,b) at 11 selected wavelengths from 0.30 to 3.75 μm for the continental (6S-C) model, maritime (6S-M) model, and urban (6S-U) model assumed for dry-air conditions.

λ (μm)	$n(\lambda)$	$\chi(\lambda)$	$\beta_{ext}(\lambda)$ (km^{-1})	$\beta_{sca}(\lambda)$ (km^{-1})	$\omega(\lambda)$	$g(\lambda)$
6S continental (6S-C) model						
0.300	1.532	1.22×10^{-2}	9.27×10^{-4}	8.46×10^{-4}	0.914	0.642
0.400	1.532	1.22×10^{-2}	8.12×10^{-4}	7.32×10^{-4}	0.902	0.643
0.488	1.532	1.22×10^{-3}	6.62×10^{-4}	5.92×10^{-4}	0.894	0.636
0.515	1.532	1.22×10^{-3}	6.25×10^{-4}	5.51×10^{-4}	0.881	0.635
0.550	1.532	1.47×10^{-3}	5.82×10^{-4}	5.15×10^{-4}	0.885	0.6339
0.633	1.532	1.47×10^{-3}	4.94×10^{-4}	4.31×10^{-4}	0.873	0.629
0.694	1.532	1.71×10^{-3}	4.44×10^{-4}	3.91×10^{-4}	0.881	0.627
0.860	1.522	2.93×10^{-3}	3.32×10^{-4}	2.78×10^{-4}	0.837	0.628
1.536	1.430	5.62×10^{-3}	1.62×10^{-4}	1.22×10^{-4}	0.752	0.638
2.250	1.273	2.44×10^{-3}	8.71×10^{-5}	6.62×10^{-5}	0.759	0.732
3.750	1.319	9.80×10^{-4}	5.92×10^{-5}	4.65×10^{-5}	0.785	0.771
6S maritime (6S-M) model						
0.300	1.392	2.085×10^{-4}	2.26×10^{-3}	2.22×10^{-3}	0.982	0.736
0.400	1.391	2.085×10^{-4}	2.25×10^{-3}	2.21×10^{-3}	0.985	0.737
0.488	1.388	2.085×10^{-4}	2.12×10^{-3}	2.11×10^{-3}	0.992	0.739
0.515	1.387	2.085×10^{-4}	2.10×10^{-3}	2.08×10^{-3}	0.991	0.744
0.550	1.387	2.502×10^{-4}	2.07×10^{-3}	2.05×10^{-3}	0.989	0.742
0.633	1.383	2.502×10^{-4}	2.01×10^{-3}	1.95×10^{-3}	0.973	0.748
0.694	1.382	2.919×10^{-4}	1.96×10^{-3}	1.95×10^{-3}	0.992	0.743
0.860	1.378	5.014×10^{-4}	1.88×10^{-3}	1.85×10^{-3}	0.985	0.750
1.536	1.365	1.192×10^{-3}	1.68×10^{-3}	1.65×10^{-3}	0.982	0.772
2.250	1.338	1.232×10^{-3}	1.42×10^{-3}	1.40×10^{-3}	0.985	0.791
3.750	1.400	2.946×10^{-3}	1.19×10^{-3}	1.17×10^{-3}	0.979	0.746
6S urban (6S-U) model						
0.300	1.571	9.134×10^{-2}	8.90×10^{-4}	6.05×10^{-4}	0.680	0.603
0.400	1.571	9.134×10^{-2}	8.12×10^{-4}	5.36×10^{-4}	0.661	0.602
0.488	1.571	8.945×10^{-2}	6.42×10^{-4}	4.28×10^{-4}	0.666	0.596
0.515	1.571	8.945×10^{-2}	6.01×10^{-4}	3.97×10^{-4}	0.659	0.595
0.550	1.571	8.817×10^{-2}	5.46×10^{-4}	3.53×10^{-4}	0.647	0.594
0.633	1.571	8.629×10^{-2}	4.52×10^{-4}	2.98×10^{-4}	0.659	0.590
0.694	1.571	8.688×10^{-2}	4.02×10^{-4}	2.57×10^{-4}	0.641	0.587
0.860	1.563	8.988×10^{-2}	3.02×10^{-4}	1.82×10^{-4}	0.602	0.585
1.536	1.536	1.021×10^{-1}	1.39×10^{-4}	5.96×10^{-5}	0.428	0.565
2.250	1.451	1.021×10^{-1}	6.83×10^{-5}	2.35×10^{-5}	0.344	0.581
3.750	1.498	1.121×10^{-1}	3.67×10^{-5}	1.04×10^{-5}	0.282	0.577

radiometer (Tomasi *et al.*, 2015a) together with ground-based measurements conducted with a nephelometer and a particle soot absorption photometer (PSAP).

The four aerosol models have the following size-distribution characteristics:

a. The BDD aerosol model represents a multimodal population of desert dust particles mobilized in the Saharan region and transported over the Po Valley by intercontinental winds across the Mediterranean Sea. Its number concentration size-distribution curve was assumed to consist of three lognormal curves, each having the analytical form given by Eq. (6.1) for a triplet of unimodal values of N_i yielding the overall multimodal particle number density $N_T = 10^3$ cm^{-3} and unimodal values of standard deviation σ_g and mode diameter a_m calculated according to the shape parameters proposed by Shettle (1984) and given in Table 6.6. The trimodal BDD size-distribution curve $N(a)$ is shown in Figure 6.5 over the $10^{-3} \leq a \leq 10^2$ μm range, presenting peak number concentrations higher than 10^2 cm^{-3} for the nuclei and accumulation modes and lower than 10^{-3} cm^{-3} for the coarse particle mode, as can be clearly seen in Figure 6.5.
The dry-air particulate mass density ρ was assumed to be equal to 2.50 g cm^{-3}, according to Volz (1973). The spectral values of $n(\lambda)$ and $\chi(\lambda)$ determined for the BDD model are given in Table 6.7 by assuming (i) the values of $n(\lambda)$ according to Carlson and Benjamin (1980) at wavelengths $\lambda < 0.6$ μm, those of the 6S DL basic component over the $0.6 \leq \lambda \leq 2.5$ μm range and those of Volz (1973) at longer wavelengths, and (ii) the values of $\chi(\lambda)$ according to Volz (1973), Carlson and Caverly (1977), Patterson, Gillette, and Stockton (1977), Patterson (1977, 1981), and Carlson and Benjamin (1980). Using the size-distribution shape parameters given in Table 6.6 and the refractive index data of Table 6.7, the spectral values of parameters $\beta_{ext}(\lambda)$, $\beta_{sca}(\lambda)$, $\omega(\lambda)$, and $g(\lambda)$ were calculated for the BDD model obtaining the evaluations reported in Table 6.7 at the 11 previously selected wavelengths. It can be noted that the BDD values of $\beta_{ext}(\lambda)$ calculated for $N_T = 10^3$ cm^{-3} are greater by more than two orders of magnitude than those of the three 6S original models, since the BDD model has considerably higher number concentrations of both accumulation and coarse particles than those assumed in the 6S models. The BDD values of $\omega(\lambda)$ increase gradually with wavelength, from around 0.80 at $\lambda = 0.30 - 0.98$ at the 1.54 and 2.25 μm wavelengths. The values of $g(\lambda)$ were estimated to slowly decrease from 0.73 to 0.66 over the wavelength range from 0.30 to 1.54 μm.

b. The El Chichón Stratospheric Volcanic (ESV) aerosol model was determined by Tomasi *et al.* (2013) through analyzing the King, Harshvardhan, and Arking (1984) measurements conducted at the Mauna Loa Observatory (Hawaii, United States) to analyze the optical and microphysical parameters of the El Chichón volcanic particles. The size-distribution curve was assumed by King, Harshvardhan, and Arking (1984) to consist of three modes: (i) the first represented in terms of the modified gamma function,

$$dN(r)/d(\log r) = C_1 r^2 \exp(-1.98r/r_c) \tag{6.4}$$

Table 6.6 Values of the shape parameters N_i, σ_g, and a_c used to define the multimodal number size distributions, particle mass density ρ (g cm^{-3}), and main radiative parameters (volume extinction, scattering, and absorption coefficients at the 0.55 μm wavelength; single scattering albedo ω at the 0.55 μm wavelength, mean single scattering albedo ω, weighted average single scattering albedo ω^*, and mean Ångström's exponent α calculated over the 0.40–0.86 μm wavelength range), as obtained for the dry-air aerosol models: background desert dust (BDD) (trimodal), El Chichón stratospheric volcanic (ESV) (trimodal), biomass burning smoke (BBS) (bimodal), and winter polluted Po Valley (WPP) (trimodal).

Dry-air aerosol models	First mode			Second mode			Third mode			ρ (g cm^{-3})	β_{ext} (0.55 μm) (km^{-1})	β_{sca} (0.55 μm) (km^{-1})	β_{abs} (0.55 μm) (km^{-1})	ω (0.55 μm)	ω	α
	N_1 (cm^{-3})	σ_g	a_c (μm)	N_2 (cm^{-3})	σ_g	a_c (μm)	N_3 (cm^{-3})	σ_g	a_c (μm)							
BDD	542.142	2.104	10^{-3}	457.857	3.120	2.18×10^{-2}	1.0×10^{-3}	1.860	6.24	2.50[a]	2.02×10^{-2}	1.81×10^{-2}	2.05×10^{-3}	0.898	0.913	0.316
ESV	592.000	2.500	0.011	400.000	1.500	0.270	8.000	1.100	1.00	1.65[b]	4.71×10^{-1}	4.71×10^{-1}	—	1.000	0.997	0.321
BBS	$10^3 - N_2$	3.981	0.130	7.5×10^{-8}	18.20	11.5	—	—	—	2.10	5.17	3.88	1.30	0.749	0.789	−0.051
WPV	945.294	2.990	5×10^{-3}	54.69	2.00	1.18×10^{-2}	1.6×10^{-2}	2.99	0.50	1.97	6.01×10^{-4}	5.18×10^{-4}	8.31×10^{-5}	0.862	0.803	1.158

The unimodal values of N_i are calculated to give a value of the overall particle number density $N_T = 10^3$ cm^{-3} for all the four multimodal aerosol models.

a) Volz (1973).

b) King, Harshvardhan, and Arking (1984).

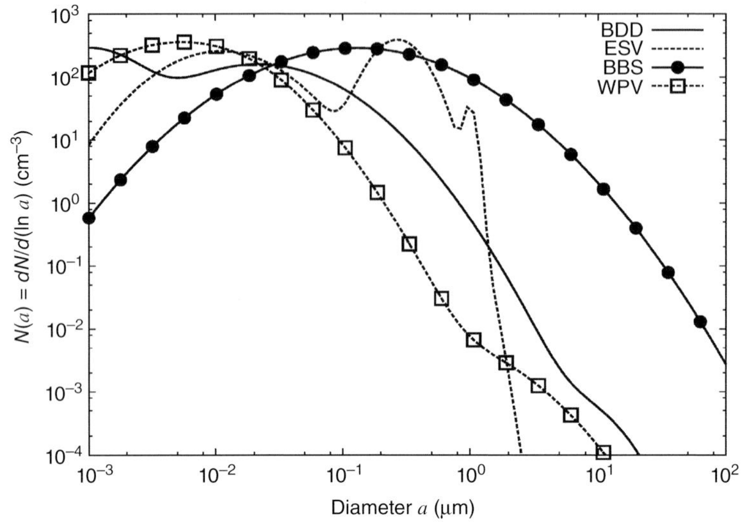

Figure 6.5 Size-distribution curves of particle number density $N(a)$ over the $10^{-3} \leq a \leq 10^{2}$ μm diameter range determined to represent the following four additional aerosol models: (i) background desert dust (BDD) model (solid curve) of Shettle (1984); (ii) El Chichón stratospheric volcanic (ESV) aerosol model (dashed curve) of King, Harshvardhan, and Arking (1984); (iii) BBS particle model (solid curve with solid circles), based on the Remer *et al.* (1998) measurements; and (iv) winter Po Valley (WPV) aerosol model (dashed curve with open squares), defined on the basis of the AOT measurements conducted by Tomasi *et al.* (2015a) and the chemical composition data of Carbone *et al.* (2010). All the four multimodal size-distribution curves are normalized to give the overall particle number concentration $N_T = 10^3$ cm^{-3}.

Table 6.7 Spectral values of real part $n(\lambda)$ and imaginary part $\chi(\lambda)$ of particulate matter refractive index, volume extinction coefficient $\beta_{ext}(\lambda)$ and volume scattering coefficient $\beta_{sca}(\lambda)$ (both measured in per kilometer and calculated for overall number density concentration $N_T = 10^3$ cm^{-3}), single scattering albedo $\omega(\lambda)$, and asymmetry factor $g(\lambda)$, defined by Vermote *et al.* (1997a,b) at 11 selected wavelengths from 0.30 to 3.75 μm for the four BDD, ESV, BBS, and WPV multimodal models of dry-air aerosol particles.

λ (μm)	$n(\lambda)$	$\chi(\lambda)$	$\beta_{ext}(\lambda)$ (km^{-1})	$\beta_{sca}(\lambda)$ (km^{-1})	$\omega(\lambda)$	$g(\lambda)$
Background desert dust (BDD) trimodal model						
0.300	1.540	2.00×10^{-2}	2.26×10^{-2}	1.81×10^{-2}	0.799	0.730
0.400	1.540	1.80×10^{-2}	2.15×10^{-2}	1.72×10^{-2}	0.801	0.726
0.488	1.540	9.85×10^{-3}	2.07×10^{-2}	1.70×10^{-2}	0.825	0.714
0.515	1.540	8.45×10^{-3}	2.05×10^{-2}	1.81×10^{-2}	0.884	0.696
0.550	1.540	5.90×10^{-3}	2.02×10^{-2}	1.81×10^{-2}	0.898	0.691
0.633	1.540	3.40×10^{-3}	1.93×10^{-2}	1.78×10^{-2}	0.926	0.681
0.694	1.540	3.35×10^{-3}	1.86×10^{-2}	1.77×10^{-2}	0.953	0.674
0.860	1.530	1.10×10^{-3}	1.69×10^{-2}	1.62×10^{-2}	0.957	0.668
1.536	1.409	1.65×10^{-3}	1.17×10^{-2}	1.15×10^{-2}	0.986	0.656
2.250	1.228	3.40×10^{-3}	6.71×10^{-3}	6.55×10^{-3}	0.976	0.712
3.750	1.278	1.05×10^{-2}	3.13×10^{-3}	1.97×10^{-3}	0.629	0.805

Table 6.7 (*continued*)

λ (μm)	$n(\lambda)$	$\chi(\lambda)$	$\beta_{ext}(\lambda)$ (km^{-1})	$\beta_{sca}(\lambda)$ (km^{-1})	$\omega(\lambda)$	$g(\lambda)$
El Chichón stratospheric volcanic (ESV) trimodal model						
0.300	1.452	6.00×10^{-8}	5.09×10^{-1}	5.09×10^{-1}	1.000	0.747
0.400	1.440	3.70×10^{-9}	4.46×10^{-1}	4.46×10^{-1}	1.000	0.723
0.488	1.432	3.80×10^{-9}	4.76×10^{-1}	4.76×10^{-1}	1.000	0.747
0.515	1.431	4.50×10^{-9}	4.77×10^{-1}	4.77×10^{-1}	1.000	0.752
0.550	1.430	1.07×10^{-8}	4.71×10^{-1}	4.71×10^{-1}	1.000	0.751
0.633	1.430	1.60×10^{-8}	4.39×10^{-1}	4.39×10^{-1}	1.000	0.741
0.694	1.429	2.12×10^{-8}	4.13×10^{-1}	4.13×10^{-1}	1.000	0.734
0.860	1.426	1.80×10^{-7}	3.61×10^{-1}	3.61×10^{-1}	1.000	0.722
1.536	1.403	1.37×10^{-4}	2.11×10^{-1}	2.11×10^{-1}	1.000	0.707
2.250	1.368	1.77×10^{-3}	9.80×10^{-2}	9.79×10^{-2}	0.999	0.649
3.750	1.396	1.31×10^{-1}	2.29×10^{-2}	2.22×10^{-2}	0.970	0.488
Biomass burning smoke (BBS) bimodal model						
0.300	1.430	3.62×10^{-3}	5.06	3.56	0.7051	0.8330
0.400	1.430	3.57×10^{-3}	5.10	3.65	0.7169	0.8300
0.488	1.430	3.54×10^{-3}	5.15	3.79	0.7372	0.8269
0.515	1.430	3.54×10^{-3}	5.15	3.82	0.7429	0.8238
0.550	1.430	3.50×10^{-3}	5.17	3.88	0.7492	0.8216
0.633	1.430	3.46×10^{-3}	5.20	3.98	0.7640	0.8186
0.694	1.430	3.46×10^{-3}	5.24	4.06	0.7743	0.8170
0.860	1.425	3.46×10^{-3}	5.30	4.22	0.7954	0.8120
1.536	1.377	3.58×10^{-3}	5.45	4.61	0.8458	0.7938
2.250	1.310	3.96×10^{-3}	5.43	4.74	0.8721	0.8051
3.750	1.370	5.28×10^{-3}	5.09	4.56	0.8968	0.8325
Winter Po Valley (WPV) trimodal model						
0.300	1.537	1.97×10^{-2}	8.86×10^{-4}	7.73×10^{-4}	0.873	0.658
0.400	1.537	1.94×10^{-2}	8.38×10^{-4}	7.26×10^{-4}	0.866	0.656
0.488	1.537	1.91×10^{-2}	6.85×10^{-4}	5.93×10^{-4}	0.866	0.647
0.515	1.537	1.91×10^{-2}	6.46×10^{-4}	5.59×10^{-4}	0.865	0.644
0.550	1.537	1.95×10^{-2}	6.01×10^{-4}	5.18×10^{-4}	0.862	0.642
0.633	1.537	1.92×10^{-2}	5.11×10^{-4}	4.39×10^{-4}	0.860	0.635
0.694	1.537	1.99×10^{-2}	4.58×10^{-4}	3.91×10^{-4}	0.853	0.631
0.860	1.527	2.36×10^{-2}	3.46×10^{-4}	2.84×10^{-4}	0.820	0.628
1.536	1.491	3.25×10^{-2}	1.54×10^{-4}	1.08×10^{-4}	0.705	0.629
2.250	1.384	2.45×10^{-2}	7.72×10^{-5}	5.08×10^{-5}	0.658	0.694
3.750	1.422	2.27×10^{-2}	5.17×10^{-5}	3.48×10^{-5}	0.674	0.719

$n(\lambda)$ – Carlson and Benjamin (1980).
$\chi(\lambda)$ – Patterson, Gillette, and Stockton (1977) and Patterson (1977).
$\chi(\lambda)$ – Remer *et al.* (1998).

expressed as a function of particle radius r (measured in micrometer), with columnar particle number content $C_1 = 1.674 \times 10^{11}$ cm^{-2} μm^{-2} and mode radius $r_c = 0.11$ μm, as defined by McClatchey, Bolle, and Kondratyev (1980) to represent the BG unperturbed conditions of the standard radiation atmosphere; and (ii) the second and third modes represented using the lognormal curve defined in terms of Eq. (6.1) with values of σ_g equal to 1.5 and 1.1 and values of a_c equal to 0.27 and 1.00 μm, providing for the second mode a value of columnar particle number content equal to 50 times that of the third mode. The values of these shape parameters were derived by King, Harshvardhan, and Arking (1984) examining the balloon-borne data recorded by Hofmann and Rosen (1983a,b) at altitudes of 21.5–24.5 km. above the Mauna Loa Observatory, a few months after the El Chichón (Mexico) eruption occurred in spring 1982. In the ESV model, we have substituted the original unimodal modified gamma curve (represented in Eq. (6.4)) with a lognormal curve providing similar size-distribution characteristics, therefore obtaining a more homogeneous representation of the three particle modes. The values of the lognormal shape parameters were calculated for the three modes by applying a best-fit procedure to the original concentration data measured by McClatchey, Bolle, and Kondratyev (1980) and obtaining the values given in Table 6.6 yielding an overall value of multimodal number density $N_T = 10^3$ cm^{-3}. The corresponding trimodal size-distribution curve of $N(a)$ calculated for dry-air conditions is shown in Figure 6.5, where it is compared with the trimodal size-distribution curve of the BDD aerosol model. The comparison shows that the number concentration $N(a)$ of ESV particles is considerably lower than that of the BDD model for $a < 2 \times 10^{-2}$ μm range and becomes higher by more than one order of magnitude over the $2 \times 10^{-1} \leq a \leq 2$ μm range, where the ESV curve exhibits a complex accumulation mode given by the linear combination of a first accumulation mode centered at $a_c = 0.27$ μm and a weaker peak of volcanic particles centered at $a_c = 1.00$ μm, as given in Table 6.6.

The "long-lived" sulfate aerosols forming in the stratosphere after the El Chichón eruption were primarily generated through oxidation processes (Turco, Whitten, and Toon, 1982). These particles were nearly spherical LW droplets consisting of a 75% mass concentration of SA (Hofmann and Rosen, 1983a,b) and a residual 25% of LW. The spectral values of $n(\lambda)$ were assumed in the ESV aerosol model according to the measurements performed by Palmer and Williams (1975) for a 75% (by weight) aqueous solution of H_2SO_4 and those of $\chi(\lambda)$ taken according to (i) the evaluations made by Burley and Johnston (1992) over the $0.25 \leq \lambda \leq 0.34$ μm wavelength range, (ii) those of Hummel, Shettle, and Longtin (1988) at wavelengths from 0.35 to 0.70 μm, and (iii) those made by Palmer and Williams (1975) at wavelengths $\lambda > 0.70$ μm. The spectral values of $n(\lambda)$ and $\chi(\lambda)$ are reported in Table 6.7 at the 11 previously selected wavelengths, being typical of H_2SO_4 and H_2O solution droplets. The spectral values of parameters $\beta_{ext}(\lambda)$, $\beta_{sca}(\lambda)$, $\omega(\lambda)$, and $g(\lambda)$ are also given in Table 6.7, indicating that considerably greater values of $\beta_{ext}(\lambda)$ were obtained for the ESV particle model than those of the 6S-M aerosol models, due to the higher content of coarse particles. The ESV spectral values of $\omega(\lambda)$ are very close to unit over the 0.30–2.5 μm wavelength range, due to the very poor absorption properties of these volcanic particles, while the spectral values of $g(\lambda)$ are mostly close to 0.75 at visible wavelengths and slowly decrease at the higher infrared wavelengths until reaching a value of 0.49 at the 3.75 μm wavelength.

c. The BBS model represents a bimodal aerosol extinction model based on the remote sensing measurements performed by Remer *et al.* (1998), who retrieved more than 800 volume size-distribution curves from the AERONET measurements conducted in Brazil over a 3-year period to represent the aerosol optical characteristics during the burning season in the Amazon Basin and Cerrado region. The average number size-distribution curve of this model was assumed to consist of two lognormal modes, both having an analytical form described by Eq. (6.1) with $a_c = 0.26$ μm and $\sigma_g = 3.981$ for the accumulation particle mode and $a_c = 23$ μm and $\sigma_g = 18.20$ for the coarse particle mode, as given in Table 6.6. The refractive index was assumed by Remer *et al.* (1998) to be given by $n(\lambda) = 1.43$ and $\chi(\lambda) = 3.5 \times 10^{-3}$ at the visible wavelengths, for which a value of $\omega(\lambda)$ close to 0.90 was obtained by assuming that the particles are covered by an external mixed layer of black carbon (BC). The calculations of the number concentrations of accumulation and coarse-mode particles in the Cerrado and forest regions have been made by us on the basis of the field data of Remer *et al.* (1998), obtaining very low values of the average coarse particle concentration N_2 equal to 2.4×10^{-8} cm^{-3} over the Cerrado and 7.5×10^{-8} cm^{-3} over the forest region for AOT equal to 0.20 at wavelength $\lambda = 0.67$ μm. To determine the BBS model, we have considered the second scenario for the forest area and assumed the average values of the unimodal accumulation and coarse particle concentrations N_1 and N_2, with the size-distribution shape parameters reported in Table 6.6 for the two aerosol modes, yielding the overall concentration $N_T = 10^3$ cm^{-3} and the refractive index data reported in Table 6.7. The bimodal size-distribution curve $N(a)$ obtained for these assumptions is shown in Figure 6.5, presenting in practice a continuous and regular lognormal size-distribution curve over the whole size range from 10^{-3} to 10^2 μm. The volume of each mode was found by Remer *et al.* (1998) to increase with AOT, thus causing a gradual variation in the ratio between accumulation and coarse particle unimodal volumes, accompanied by less pronounced changes in the optical parameters of the two modes. The dry-air particulate mass density ρ of the BBS model resulting from the aforementioned combustion processes was assumed to be intermediate between those of the DL (with a mass fraction $M_{DL} = 50\%$), WS (mass fraction $M_{WS} = 46\%$), and soot substances (mass fraction $M_{SO} = 4\%$), obtaining an average value of $\rho = 2.10$ g cm^{-3}. The optical characteristics of BBS aerosols in the visible were found to be markedly dominated by the accumulation particle mode, which was found to exhibit spectral features that do not appreciably vary passing from the forest region to the Cerrado area. The spectral values of parameters $\beta_{ext}(\lambda)$, $\beta_{sca}(\lambda)$, $\omega(\lambda)$, and $g(\lambda)$ are also given in Table 6.7. The spectral values of $n(\lambda)$ and $\chi(\lambda)$ have been calculated at the selected wavelengths by assuming that their spectral variations are proportional to those of the SO and DL basic components. The values of single scattering albedo parameters $\omega(0.55\,\mu m)$ and ω and those of exponent α were calculated according to the assumptions made earlier for the dry-air BBS aerosol model observed over the Amazon Basin forests and provided in Table 6.6. The values of $\omega(\lambda)$ were estimated to increase from 0.70 at $\lambda = 0.30\,\mu m$ to 0.80 at $\lambda = 0.86\,\mu m$ and to subsequently increase even more until reaching a value of ~0.90 at $\lambda = 3.75\,\mu m$, while the values of $g(\lambda)$ varied slowly from 0.83 at $\lambda = 0.30\,\mu m$ to around 0.80 at $\lambda = 2.25\,\mu m$.

d. The WPV aerosol model is here defined by us to represent the microphysical and optical characteristics of the PM suspended within the PBL over the Po Valley (Italy) for winter anticyclonic and cloud-free sky conditions. The particles were sampled using a five-stage Berner impactor with 50% size cut at 0.05, 0.14, 0.42, 1.2, 3.5, and 10 μm aerodynamic diameter, obtaining the chemical composition data for various size classes and evaluating the mass percentages of water-soluble organic matter (WSOM), water-insoluble organic matter (WINOM), and water-soluble inorganic aerosol components (WSIM) consisting mainly of NH_4^+, Na^+, K^+, Ca_2^+, Mg_2^+, Cl^-, NO_3^-, and SO_4^{2-} ions (Carbone *et al.*, 2010). The daytime measurements provided mass fractions of 24% WINOM, 19% WSOM, 1% unknown substances, and 56% WSIM (given by 8% SO_4^{2-}, 34% NO_3^-, and 14% NH_4^+) over the $5 \times 10^{-2} \leq a \leq 10\,\mu m$ range, with a predominant content of accumulation particles. On the basis of these results, we represented the WPV aerosol model for the diurnal evaluations of the mass percentages equal to 73% water-soluble substances (represented by the 6S WS basic component), 24% DL particles (represented by the 6S DL component), and 3% soot substances (represented by the 6S SO component). For these assumptions, the overall WPV size-distribution curve was assumed to consist of the following three particle modes: (i) a nuclei mode of WS substances, having $a_c = 10^{-2}\,\mu m$, $\sigma_g = 2.99$, and dry-air mass density $\rho = 1.86\,g\,cm^{-3}$; (ii) a second nuclei mode of soot substances, having $a_c = 2.36 \times 10^{-2}\,\mu m$, $\sigma_g = 2.00$, and $\rho = 1.62\,g\,cm^{-3}$; and (iii) a third accumulation/coarse mode of dust particles, having $a_c = 1.00\,\mu m$, $\sigma_g = 2.99$, and $\rho = 2.36\,g\,cm^{-3}$. The multimodal size-distribution curve of number density concentration was found to exhibit appreciably lower values than those deemed for the BDD and BBS aerosol models over the size range from $2 \times 10^{-2}\,\mu m$ to about 10 μm.

For the previous assumptions, an overall dry-air mass density $\rho = 1.97\,g\,cm^{-3}$ was obtained for a predominant content of very small water-soluble particles, a moderate content of soot particles, and a relatively low content of dust particles. For these composition features, the spectral values of parameters $n(\lambda)$, $\chi(\lambda)$, $\beta_{ext}(\lambda)$, $\beta_{sca}(\lambda)$, $\omega(\lambda)$, and $g(\lambda)$ were calculated at the 11 selected wavelengths obtaining the values given in Table 6.7. The results show that (i) $n(\lambda)$ assumes stable values equal to 1.537 in the visible and decreasing values in the infrared to around 1.400 at the 2.25 and 3.70 μm wavelengths, (ii) $\chi(\lambda)$ varies between 1.97×10^{-2} and 3.3×10^{-2} over the 0.30–1.54 μm spectral range, (iii) $\beta_{ext}(\lambda)$ decreases from 8.9×10^{-4} to 5.1×10^{-4} km^{-1} in visible wavelengths, (iv) $\beta_{sca}(\lambda)$ decreases from 7.7×10^{-4} to less than 4×10^{-4} km^{-1} at visible wavelengths, (v) $\omega(\lambda)$ decreases from 0.87 at $\lambda = 0.30$ μm to less than 0.70 at $\lambda = 2.25$ μm and $\lambda = 3.75$ μm, and (vi) $g(\lambda)$ decreases from 0.66 to 0.63 in the visible. The value of $\omega(0.55\ \mu m) = 0.862$ given in Table 6.7 for the WPV model agrees closely with the value of $\omega_0(0.55\ \mu m) = 0.88$ derived by Tomasi *et al.* (2015a) from the ground-based nephelometer and PSAP measurements conducted at the SPC station in February 2008.

6.2.3
The 6S Modified (M-Type) Aerosol Models

The 6S-C, 6S-M, and 6S-U aerosol models and the BDD, ESV, BBS, and WPV aerosol models described in the previous two subsections were calculated for dry-air PM. In the real cases, the air RH conditions observed in the troposphere for cloudless conditions vary most frequently from more than 30% to around 70% during the central hours of the day. In order to define a set of mixed aerosol models for moist air conditions with RH ≈ 50%, we have calculated the multimodal size-distribution curves and optical parameters of aerosol models obtained as linear combinations of the 6S WS, OC, DL, and SO components, which are moderately grown by condensation with limited uptakes of LW (Hänel, 1976). In these wet-air cases, the spectral values of $\beta_{ext}(\lambda)$ and $\omega(\lambda)$ are expected to be only slightly higher than those given in Table 6.5 for dry-air continental, maritime, and urban particles.

Using the growth parameters of the four 6S basic components determined in Table 6.1, we calculated a set of eight mixed (M-type) wet-air aerosol models by modifying step by step the mass fractions of the four WS, OC, DL, and SO basic components and the mass fraction of LW. The volume percentages of the five components adopted for the eight M-type models are given in Table 6.8, together with the values of (i) mean Ångström (1964) exponent $\alpha(0.40-0.86\ \mu m)$ calculated for the values of coefficient $\beta_{ext}(\lambda)$, (ii) average single scattering albedo ω, and (iii) density ρ_w of wet-air (RH = 50%) particulate mass. The mass fractions of the eight M-type aerosol models were then calculated by using the volume percentages of the WS, OC, DL, SO, and LW components given in Table 6.8 and the dry-air mass density values of the WS, OC, DL, and SO components of Vermote *et al.* (1997b) given in Table 6.1, for unit value of LW mass density, determining the percentages given in Table 6.9.

In particular, we assumed that (i) model M-1 consists of sea-salt particles only, therefore containing a very small fraction of absorbing coarse particles; (ii) model M-2 consists predominantly of sea-salt particles, containing a few mass percents of water-soluble substances; (iii) model M-8 consists of continental aerosols only; (iv) model M-14 presents a soot particulate mass concentration very close to that chosen by Vermote *et al.* (1997b) for the 6S-U model, characterizing a heavy

Table 6.8 Volume percentages of the four 6S basic wet-air (RH = 50%) aerosol components such as water soluble (WS), oceanic (OC), dustlike (DL), and soot (SO) and of liquid water (LW) used to give form to the eight M-type modified wet aerosol models (labeled with letter M and increasing numbers from 1 to 14) having different composition, each model being normalized to the overall particle number density $N_T = 10^3\ cm^{-3}$.

6S wet-air aerosol components	Volume percentages for the 14 M-type modified wet aerosol models for RH = 50%							
	M-1 (pure oceanic)	M-2 (maritime)	M-3 (mixed maritime–continental)	M-5 (mixed maritime–continental)	M-8 (pure continental)	M-10 (continental polluted)	M-12 (anthropogenic–continental)	M-14 (heavy polluted)
WS	—	0.0408	0.0940	0.2128	0.2788	0.4546	0.5451	0.5560
OC	0.8177	0.7761	0.6151	0.2314	—	—	—	—
DL	—	—	0.1452	0.4813	0.6729	0.4267	0.2481	0.1160
SO	—	—	—	—	0.0096	0.0464	0.0927	0.1800
LW	0.1823	0.1831	0.1457	0.0745	0.0387	0.0723	0.1141	0.1480
$\alpha(0.40-0.86\ \mu m)$	−0.092	0.243	0.544	0.967	1.149	1.239	1.277	1.301
ω	0.995	0.986	0.966	0.925	0.852	0.771	0.687	0.615
$\rho_w\ (g\ cm^{-3})$	2.022	2.005	2.023	2.127	2.161	2.000	1.864	1.748

The corresponding values of the mean Ångström's exponent $\alpha(0.40-0.86\ \mu m)$ (calculated for the values of volume extinction coefficient $\beta_{ext}(\lambda)$ determined at the 0.400, 0.488, 0.515, 0.550, 0.633, 0.694, and 0.860 μm wavelengths), mean single scattering albedo ω, weighted average single scattering albedo ω^*, and wet particulate mass density ρ_w measured in gram per cubic centimeter are given in the last lines.

Table 6.9 Mass percentages of the dry-air 6S basic components WS, OC, DL, and SO and liquid water (LW) assumed for determining the eight M-type wet-air (RH = 50%) aerosol models.

M-type aerosol model	Mass percentages of the 6S dry-air basic components and LW adopted to give form to the radiative parameters of the 14 M-type aerosol models as linear combinations of the various components				
	WS	OC	DL	SO	LW
M-1	—	90.98	—	—	9.02
M-2	3.79	87.08	—	—	9.13
M-3	8.54	67.60	16.74	—	7.12
M-5	18.61	24.48	53.41	—	3.50
M-8	24.00	—	73.49	0.72	1.79
M-10	42.28	—	50.35	3.76	3.61
M-12	53.41	—	30.84	9.74	6.01
M-14	59.30	—	15.70	16.51	8.49

polluted aerosol case, very similar to that determined by Bush and Valero (2002) for aerosol samples collected in the equatorial region of the Indian Ocean during the INDOEX experiment for heavy anthropogenic pollution conditions; (v) the intermediate models M-3 and M-5 represent mixed maritime–continental aerosol populations, characterized by decreasing mass fractions of sea salt, water-soluble substances, and LW and increasing mass fractions of DL particles, with soot mass fractions $M_{SO} \leq 0.50\%$; and (vi) the intermediate models M-10 and M-12 represent the continental-polluted and mixed anthropogenic–continental aerosols, respectively, which contain increasing mass fractions of WS, SO, and LW components and decreasing concentrations of DL particles and do not contain OC sea-salt particles, while the mass fraction of LW was calculated by taking into account the hygroscopic properties of the water-soluble substances.

The number density size-distribution curves of the eight M-type wet-air aerosol models are shown in Figure 6.6 to provide evidence of their multimodal features. The most pronounced variations from one model to another appear to occur over the accumulation/coarse particle size range from 0.8 to 20 μm, mainly associated with the variations in the number concentration percentages of the OC and DL components, leading to significant changes in the accumulation and coarse particle contents. The calculations of the spectral values of $n(\lambda)$, $\chi(\lambda)$, $\beta_{ext}(\lambda)$, $\beta_{sca}(\lambda)$, $\omega(\lambda)$, and $g(\lambda)$ were made at the 11 selected wavelengths from 0.30 to 3.75 μm for the eight M-type wet-air aerosol models shown in Figure 6.6, obtaining the data given in Table 6.10 for these aerosol models normalized to give $N_T = 10^3$ cm^{-3}. The calculations provided (i) spectral values of $n(\lambda)$ decreasing with wavelength and in general increasing as the maritime particle and LW mass contents gradually decrease passing from M-1 to M-14; (ii) spectral values of $\chi(\lambda)$ gradually increasing with the wavelength for all models and increasing as one passes from M-1 to M-14; and (iii) spectral values of $\beta_{ext}(\lambda)$ calculated for the M-1 model

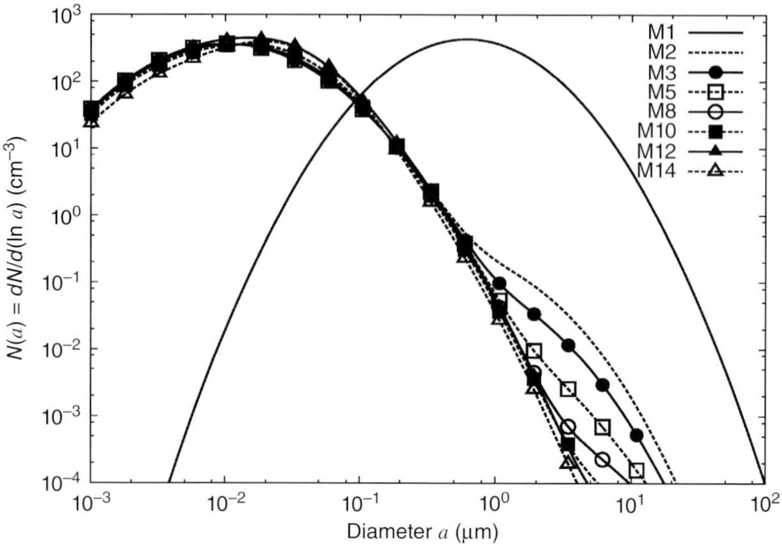

Figure 6.6 Size-distribution curves of particle number density $N(a)$ over the $10^{-3} \leq a \leq 10^2$ μm diameter range for (i) model M-1 (pure oceanic aerosol) (solid curve), (ii) model M-2 (maritime aerosol) (dashed curve), (iii) model M-3 (mixed maritime–continental aerosol) (solid curve with solid circles), (iv) model M-5 (mixed maritime–continental aerosol) (dashed curve with open squares) models, (v) model M-8 (pure continental aerosol) (solid curve with open circles) model, (vi) model M-10 (continental-polluted aerosol) (dashed curve with solid squares), (vii) model M-12 (continental-polluted aerosol) (dashed curve with solid triangles) models, and (viii) model M-14 (heavy polluted aerosol) (dashed curves with open triangles) model. All the eight aerosol models were defined by Tomasi *et al.* (2013) as linear combinations of the 6S basic aerosol components and are normalized to give the total particle number concentration $N_T = 10^3$ cm^{-3}.

which are greater by about three orders of magnitude than those determined for the other seven M-type models at all wavelengths, due to the considerably higher number concentration of coarse particles (~50 cm^{-3}) compared to the partial concentration of fine particles (~950 cm^{-3}). It is worth noting that the spectral values of $\beta_{ext}(\lambda)$ calculated for concentration $N_T = 10^3$ cm^{-3} are produced by gradually higher number concentrations of fine particles as one passes from the M-2 to the M-14 model and gradually lower number concentrations of coarse particles, assuming values lower than 2×10^{-3} km^{-1} for model M-2, equal to $\sim 3 \times 10^{-4}$ km^{-1} for model M-3 model, and gradually lower for the subsequent M-type models. Therefore, $\beta_{ext}(\lambda)$ was found to decrease with wavelength, assuming smaller values by more than 70% for the M-14 model with respect to M-2 model. The spectral values of $\beta_{sca}(\lambda)$ were estimated to decrease as λ increases for all the eight M-type models, becoming gradually lower as one passes from the M-2 model to the M-14 model, with an overall decrease of about 80%. The spectral values of $\omega(\lambda)$ slowly decrease with increasing λ, passing from nearly unit values of the M-1 model to a

Table 6.10 Spectral values of real part $n(\lambda)$ and imaginary part $\chi(\lambda)$ of particulate matter refractive index, volume extinction coefficient $\beta_{ext}(\lambda)$ and volume scattering coefficient $\beta_{sca}(\lambda)$ (both measured in per kilometer and calculated for overall number density concentration $N_T = 10^3$ cm^{-3}), single scattering albedo, $\omega(\lambda)$ and asymmetry factor $g(\lambda)$, defined by Tomasi *et al.* (2013) at 11 selected wavelengths from 0.30 to 3.75 μm for the 14 M-type aerosol models proposed by Tomasi *et al.* (2013) for wet-air (RH = 50%) conditions.

λ (μm)	$n(\lambda)$	$\chi(\lambda)$	$\beta_{ext}(\lambda)$ (km^{-1})	$\beta_{sca}(\lambda)$ (km^{-1})	$\omega(\lambda)$	$g(\lambda)$
M-1 model (pure oceanic)						
0.300	1.379	8.00×10^{-9}	3.94×10^{0}	3.94×10^{0}	0.998	0.792
0.400	1.382	9.30×10^{-9}	4.05×10^{0}	4.05×10^{0}	1.000	0.796
0.488	1.379	5.90×10^{-9}	4.13×10^{0}	4.13×10^{0}	1.000	0.787
0.515	1.377	3.50×10^{-9}	4.19×10^{0}	4.19×10^{0}	1.000	0.782
0.550	1.377	4.00×10^{-9}	4.20×10^{0}	4.20×10^{0}	1.000	0.786
0.633	1.374	1.60×10^{-8}	4.27×10^{0}	4.27×10^{0}	1.000	0.780
0.694	1.372	4.90×10^{-8}	4.27×10^{0}	4.27×10^{0}	1.000	0.785
0.860	1.368	3.00×10^{-8}	4.35×10^{0}	4.35×10^{0}	1.000	0.783
1.536	1.355	2.30×10^{-4}	4.33×10^{0}	4.31×10^{0}	0.995	0.783
2.250	1.330	8.00×10^{-4}	3.87×10^{0}	3.83×10^{0}	0.988	0.801
3.750	1.397	2.96×10^{-3}	3.36×10^{0}	3.26×10^{0}	0.971	0.753
M-2 model (maritime)						
0.300	1.385	2.00×10^{-4}	2.52×10^{-3}	2.46×10^{-3}	0.975	0.736
0.400	1.388	2.20×10^{-4}	2.30×10^{-3}	2.27×10^{-3}	0.987	0.743
0.488	1.385	2.20×10^{-4}	2.19×10^{-3}	2.17×10^{-3}	0.989	0.743
0.515	1.385	2.20×10^{-4}	2.17×10^{-3}	2.15×10^{-3}	0.990	0.741
0.550	1.384	2.60×10^{-4}	2.13×10^{-3}	2.11×10^{-3}	0.989	0.746
0.633	1.381	2.60×10^{-4}	2.06×10^{-3}	2.04×10^{-3}	0.990	0.745
0.694	1.379	3.10×10^{-4}	2.01×10^{-3}	1.99×10^{-3}	0.990	0.751
0.860	1.375	5.30×10^{-4}	1.93×10^{-3}	1.90×10^{-3}	0.986	0.757
1.536	1.363	1.23×10^{-3}	1.73×10^{-3}	1.708×10^{-3}	0.982	0.770
2.250	1.335	1.20×10^{-3}	1.48×10^{-3}	1.46×10^{-3}	0.984	0.795
3.750	1.399	3.01×10^{-3}	1.26×10^{-3}	1.22×10^{-3}	0.970	0.750
M-3 model (mixed maritime–continental)						
0.300	1.415	1.63×10^{-3}	1.611×10^{-3}	1.536×10^{-3}	0.953	0.697
0.400	1.424	2.01×10^{-3}	1.374×10^{-3}	1.332×10^{-3}	0.969	0.706
0.488	1.422	2.01×10^{-3}	1.224×10^{-3}	1.189×10^{-3}	0.971	0.707
0.515	1.420	2.01×10^{-3}	1.191×10^{-3}	1.158×10^{-3}	0.972	0.706
0.550	1.420	2.11×10^{-3}	1.137×10^{-3}	1.102×10^{-3}	0.969	0.710
0.633	1.419	2.10×10^{-3}	1.059×10^{-3}	1.028×10^{-3}	0.971	0.711
0.694	1.418	2.20×10^{-3}	9.948×10^{-4}	9.635×10^{-4}	0.969	0.717
0.860	1.412	2.67×10^{-3}	8.993×10^{-4}	8.625×10^{-4}	0.959	0.728
1.536	1.378	3.88×10^{-3}	7.121×10^{-4}	6.774×10^{-4}	0.951	0.754
2.250	1.317	3.25×10^{-3}	5.716×10^{-4}	5.525×10^{-4}	0.967	0.790
3.750	1.378	4.61×10^{-3}	4.848×10^{-4}	4.645×10^{-4}	0.958	0.753
M-5 model (mixed maritime–continental)						
0.300	1.484	4.92×10^{-3}	1.20×10^{-3}	1.12×10^{-3}	0.933	0.659
0.400	1.494	5.44×10^{-3}	9.50×10^{-4}	9.03×10^{-4}	0.951	0.664
0.488	1.493	5.44×10^{-3}	7.91×10^{-4}	7.51×10^{-4}	0.949	0.661
0.515	1.491	5.44×10^{-3}	7.51×10^{-4}	7.13×10^{-4}	0.949	0.660

(continued overleaf)

Table 6.10 (*Continued*)

λ (μm)	$n(\lambda)$	$\chi(\lambda)$	$\beta_{ext}(\lambda)$ (km^{-1})	$\beta_{sca}(\lambda)$ (km^{-1})	$\omega(\lambda)$	$g(\lambda)$
0.550	1.491	5.64×10^{-3}	7.09×10^{-4}	6.69×10^{-4}	0.943	0.661
0.633	1.491	5.64×10^{-3}	6.19×10^{-4}	5.83×10^{-4}	0.942	0.659
0.694	1.490	5.83×10^{-3}	5.59×10^{-4}	5.23×10^{-4}	0.935	0.660
0.860	1.483	6.78×10^{-3}	4.51×10^{-4}	4.10×10^{-4}	0.908	0.669
1.536	1.410	8.93×10^{-3}	2.69×10^{-4}	2.32×10^{-4}	0.864	0.700
2.250	1.285	7.15×10^{-3}	1.79×10^{-4}	1.61×10^{-4}	0.900	0.771
3.750	1.336	7.68×10^{-3}	1.31×10^{-4}	1.19×10^{-4}	0.909	0.763
M-8 model (pure continental)						
0.300	1.526	1.13×10^{-2}	1.13×10^{-3}	1.00×10^{-3}	0.885	0.646
0.400	1.530	1.03×10^{-2}	8.68×10^{-4}	7.83×10^{-4}	0.902	0.646
0.488	1.529	1.02×10^{-2}	7.10×10^{-4}	6.38×10^{-4}	0.900	0.641
0.515	1.528	1.02×10^{-2}	6.71×10^{-4}	6.03×10^{-4}	0.899	0.639
0.550	1.528	1.05×10^{-2}	6.19×10^{-4}	5.52×10^{-4}	0.892	0.638
0.633	1.528	1.04×10^{-2}	5.30×10^{-4}	4.71×10^{-4}	0.889	0.634
0.694	1.528	1.06×10^{-2}	4.69×10^{-4}	4.13×10^{-4}	0.881	0.632
0.860	1.519	1.18×10^{-2}	3.58×10^{-4}	3.02×10^{-4}	0.843	0.634
1.536	1.427	1.46×10^{-2}	1.77×10^{-4}	1.34×10^{-4}	0.755	0.647
2.250	1.272	1.25×10^{-2}	9.69×10^{-5}	7.42×10^{-5}	0.766	0.743
3.750	1.319	1.31×10^{-2}	6.61×10^{-5}	5.22×10^{-5}	0.789	0.781
M-10 model (continental polluted)						
0.300	1.528	2.71×10^{-2}	1.12×10^{-3}	9.22×10^{-4}	0.823	0.635
0.400	1.535	2.31×10^{-2}	8.49×10^{-4}	7.12×10^{-4}	0.838	0.634
0.488	1.535	2.28×10^{-2}	6.82×10^{-4}	5.71×10^{-4}	0.837	0.628
0.515	1.535	2.28×10^{-2}	6.37×10^{-4}	5.32×10^{-4}	0.836	0.625
0.550	1.535	2.29×10^{-2}	5.91×10^{-4}	4.91×10^{-4}	0.831	0.624
0.633	1.535	2.24×10^{-2}	4.98×10^{-4}	4.13×10^{-4}	0.829	0.619
0.694	1.535	2.29×10^{-2}	4.39×10^{-4}	3.61×10^{-4}	0.821	0.616
0.860	1.525	2.50×10^{-2}	3.31×10^{-4}	2.59×10^{-4}	0.782	0.615
1.536	1.458	3.07×10^{-2}	1.52×10^{-4}	1.02×10^{-4}	0.673	0.605
2.250	1.328	2.73×10^{-2}	7.11×10^{-5}	4.51×10^{-5}	0.635	0.664
3.750	1.372	2.84×10^{-2}	3.98×10^{-5}	2.52×10^{-5}	0.632	0.702
M-12 model (anthropogenic–continental)						
0.300	1.529	4.75×10^{-2}	1.13×10^{-3}	8.60×10^{-4}	0.761	0.625
0.400	1.545	4.45×10^{-2}	8.45×10^{-4}	6.54×10^{-4}	0.773	0.623
0.488	1.545	4.36×10^{-2}	6.67×10^{-4}	5.14×10^{-4}	0.771	0.617
0.515	1.545	4.37×10^{-2}	6.29×10^{-4}	4.84×10^{-4}	0.769	0.615
0.550	1.545	4.33×10^{-2}	5.77×10^{-4}	4.42×10^{-4}	0.765	0.613
0.633	1.544	4.25×10^{-2}	4.89×10^{-4}	3.73×10^{-4}	0.763	0.608
0.694	1.544	4.30×10^{-2}	4.29×10^{-4}	3.23×10^{-4}	0.753	0.606
0.860	1.534	4.59×10^{-2}	3.19×10^{-4}	2.24×10^{-4}	0.703	0.603
1.536	1.490	5.46×10^{-2}	1.40×10^{-4}	8.23×10^{-5}	0.597	0.585
2.250	1.384	5.11×10^{-2}	6.69×10^{-5}	3.39×10^{-5}	0.507	0.617
3.750	1.428	5.44×10^{-2}	3.37×10^{-5}	1.55×10^{-5}	0.460	0.633

Table 6.10 (*Continued*)

λ (μm)	$n(\lambda)$	$\chi(\lambda)$	$\beta_{ext}(\lambda)$ (km^{-1})	$\beta_{sca}(\lambda)$ (km^{-1})	$\omega(\lambda)$	$g(\lambda)$
M-14 model (heavy polluted)						
0.300	1.542	8.66×10^{-2}	1.14×10^{-3}	8.08×10^{-4}	0.709	0.615
0.400	1.557	7.17×10^{-2}	8.49×10^{-4}	6.09×10^{-4}	0.717	0.614
0.488	1.557	7.02×10^{-2}	6.66×10^{-4}	4.75×10^{-4}	0.713	0.607
0.515	1.557	7.01×10^{-2}	6.26×10^{-4}	4.45×10^{-4}	0.711	0.605
0.550	1.557	6.93×10^{-2}	5.80×10^{-4}	4.10×10^{-4}	0.707	0.604
0.633	1.557	6.79×10^{-2}	4.76×10^{-4}	3.35×10^{-4}	0.703	0.599
0.694	1.557	6.85×10^{-2}	4.25×10^{-4}	2.95×10^{-4}	0.693	0.596
0.860	1.545	7.19×10^{-2}	3.16×10^{-4}	2.06×10^{-4}	0.652	0.594
1.536	1.523	8.34×10^{-2}	1.44×10^{-4}	7.25×10^{-5}	0.505	0.570
2.250	1.439	8.09×10^{-2}	6.70×10^{-5}	2.73×10^{-5}	0.408	0.581
3.750	1.483	8.77×10^{-2}	3.40×10^{-5}	1.12×10^{-5}	0.328	0.565

value close to 0.70 for the M-14 model, such a behavior being mainly due to the increasing mass fraction of soot substances. The spectral values of $g(\lambda)$ were found to decrease from ~0.78 at visible wavelengths for the M-1 pure OC aerosol model to ~0.60 for the M-14 heavy polluted aerosol model. The average single scattering albedo ω calculated over the 0.30–3.75 μm wavelength range was estimated to decrease slowly from ~1.00 for the M-1 pure OC aerosol model for to ~0.62 for the M-14 heavy polluted aerosol model, which mainly consist of WS substances (predominantly sulfate), slowly increasing fractions of soot substances and gradually smaller fractions of DL substances, and LW; correspondingly, the wet-air particulate mass density ρ_w was estimated to slowly increase from 2.02 to 2.16 g cm^{-3} passing from the M-1 to the M-8 model and to decrease from 2.00 to 1.75 g cm^{-3} passing from the M-10 to the M-14 model, due to the variable contributions of the wet-air 6S basic components and LW.

6.2.4
The OPAC Aerosol Models

OPAC is a software package containing a set of aerosol models representing the optical characteristics of various aerosol types, water droplets, and ice crystals at 61 wavelengths chosen over the 0.25–40 μm wavelength range covering the whole short- and longwave transmission spectrum of the atmosphere for eight values of RH from 0% to 99% (Hess, Koepke, and Schult, 1998). The aerosol optical parameters were first calculated for 10 basic aerosol components, each having the individual lognormal particle size-distribution curve determined using the shape parameters given in Table 6.11 and presenting a specific chemical particulate composition. They are:

(1) The water-insoluble (INS) component, mainly consisting of soil particles with a very limited content of organic matter (OM).

Table 6.11 Shape parameters and microphysical properties of the lognormal size-distribution curves defined for the 10 OPAC wet-air (RH = 50%) aerosol components (Hess, Koepke, and Schult, 1998).

OPAC aerosol component	Geometric standard deviation σ_g	Mode diameter a_c (μm) of the number density size distribution	Mode diameter a_c (μm) of the volume size distribution	Lower limit a_{min} (μm) of the diameter range	Upper limit a_{max} (μm) of the diameter range	Wet particle mass density ρ_w (g cm^{-3})
Insoluble (INS)	2.51	9.42×10^{-1}	12.00	0.01	40.0	2.0
Water soluble (WAS)	2.24	4.24×10^{-2}	0.30	0.01	40.0	1.8
Soot (SOO)	2.00	2.36×10^{-2}	0.10	0.01	40.0	1.0
Sea salt (accumulation mode) (SAM)	2.03	4.18×10^{-1}	1.88	0.01	40.0	2.2
Sea salt (coarse mode) (SCM)	2.03	3.50	15.80	0.01	120.0	2.2
Mineral dust (nucleation mode) (MNM)	1.95	1.40×10^{-1}	0.54	0.01	40.0	2.6
Mineral dust (accumulation mode) (MAM)	2.00	7.80×10^{-1}	3.20	0.01	40.0	2.6
Mineral dust (coarse mode) (MCM)	2.15	3.80	22.00	0.01	120.0	2.6
Mineral transported (MTR)	2.20	5.00×10^{-1}	3.00	0.04	10.0	2.5
Sulfate droplets (SDR)	2.03	6.95×10^{-2}	0.31	0.01	40.0	1.7

(2) The water-soluble (WAS) component, assumed to be mainly originated through g-to-p conversion processes and containing mass percentages of anthropogenic sulfate, nitrate, organic substances, and a small fraction of dimethyl sulfide of OC origin.

(3) The soot (SOO) component, mainly containing insoluble BC (strongly absorbing the solar radiation) and consisting of (i) small particles, which do not grow with increasing RH and have mass density equal to 1 g cm^{-3}, being in general fluffy with space inside, and (ii) particles presenting chain-like features, which exhibit size distributions containing a marked fraction of very small particles.

(4, 5) The two sea-salt particle components, both present in seawater, of which the first gives form to the sea-salt accumulation particle mode (SAM), and the second presents a sea-salt coarse particle mode (SCM), both originated by wind forcing for different surface-level wind speed conditions over the various size ranges (Koepke *et al.*, 1997).

(6, 7, 8) The three mineral aerosol components, consisting of mixtures of quartz and clay minerals and based on three distinct unimodal curves representing the nucleation (MNM), accumulation (MAM), and coarse (MCM) particle components, in which the relative amounts of large particles vary as a function of the atmospheric turbidity conditions.

(9) The mineral particle (MTR) component of desert origin, consisting mainly of mineral dust mobilized in desert regions and transported over long distances and assumed to contain a relatively small amount of large particles whose sizes do not increase as RH increases.

(10) The sulfate (SDR) component, consisting of 75% H_2SO_4 and a residual 25% H_2O to describe the sulfate particles often sampled in the Antarctic atmosphere and in the low stratosphere, where BG aerosol layers of relatively small particles are present after strong volcanic eruptions, consisting mainly of SA droplets.

The upper and lower limits of the aforementioned unimodal particle size-distribution curves are given in Table 6.11, as defined by Hess, Koepke, and Schult (1998) to calculate the Mie aerosol radiative parameters. The overall mass concentration M^* of each size-distribution curve was measured in microgram per cubic centimeter per unit number concentration and calculated with a cutoff diameter equal to 15 μm. The number and volume mode diameters of the 10 basic aerosol components subject to condensation growth and the two diameter range limits given in Table 6.11 were assumed to increase with increasing RH according to Hänel and Zankl (1979). Each lognormal size-distribution was determined for RH = 50% by calculating the mode diameter increase and assuming that the values of σ_g given in Table 6.11 remain unchanged as RH increases from 0% to 50%. The values of wet-air mass density ρ_w were calculated using the growth factors of Hänel and Zankl (1979) obtaining the values given in Table 6.11.

The unimodal size-distribution curves of particle number density $N(a)$ were calculated for the values of a_c and σ_g given in Table 6.11, assuming an overall number concentration $N_T = 10^3$ cm^{-3} for each of the 10 components and using the values of a_c varying from 2.36×10^{-2} μm (SOO component) to 3.80 μm (MCM mineral dust coarse component).

The calculations of the optical parameters of the 10 OPAC basic components were made for wet-air (RH = 50%) conditions at the 61 wavelengths from 0.30 to 3.75 μm and for the spectral values of $n(\lambda)$ and $\chi(\lambda)$ estimated by Shettle and Fenn (1979), Deepak and Gerber (1983), d'Almeida, Koepke, and Shetlle (1991), and Koepke *et al.* (1997). The following results were obtained: (i) spectral values of $n(\lambda)$ ranging from 1.33 to more than 1.70 at visible wavelengths and $\chi(\lambda)$ characterized by the most marked variations over the 0.30–1.50 μm spectral range, (ii) spectral values of $\beta_{ext}(\lambda)$ varying by several orders of magnitude passing from 0.30 to 3.75 μm for each component, and (iii) spectral values of $\omega(\lambda)$ varying at visible wavelengths from nearly unit values for the WAS, SCM, and SDR components to a value lower than 0.3 for the SOO component.

The 10 OPAC aerosol models were determined as linear combinations of the 10 OPAC basic aerosol components described earlier by varying the number concentrations of the various components to calculate the optical and microphysical characteristics of the 10 following OPAC aerosol models observed in the various areas:

1. The *continental clean* (CC) aerosol model represents the aerosol particle population monitored in remote continental areas, therefore only slightly affected by anthropogenic pollution, and presents a totally negligible soot particulate mass concentration M_{soo}, as assumed in Table 6.12.
2. The *continental average* (CA) aerosol model represents a mixed continental and anthropogenic aerosol particle population, containing a relatively low mass concentration of soot insoluble substances.
3. The *continental-polluted* (CP) aerosol model represents the particle population usually sampled in highly polluted areas because of anthropogenic and industrial activities, with a value of soot mass concentration $M_{soo} = 2$ μg m^{-3} and a mass concentration M_{wss} of water-soluble substances which is more than double that usually found in the CA aerosol model, as reported in Table 6.12.
4. The *urban* (UR) aerosol model represents a strong pollution case monitored in urban areas, with a relatively high value of $M_{soo} \approx 7.8$ μg m^{-3} and values of M_{wss} and M_{ins} yielding mass concentrations of water-soluble and water-insoluble substances, considerably higher than those assumed for the CP aerosols.
5. The *desert* (DE) aerosol model represents the dust load transported from desert areas, consisting of various mineral aerosol components combined with the WAS basic component containing an important fraction of water-soluble substances mixed with a mineral dust component containing nuclei, accumulation, and coarse particles.
6. The *maritime clean* (MC) aerosol model represents a particle population sampled in undisturbed remote maritime areas, not affected by appreciable

Table 6.12 Values of the particle number concentration N_j and mass mixing ratios Γ_j of the basic components giving form to the 10 OPAC wet (RH = 50%) aerosol models defined by Hess, Koepke, and Schult (1998) for the values of particle number and mass concentration originally adopted by them.

OPAC models with their acronyms (given in brackets)	Components	Particle number concentrations N_j (cm^{-3}) of the basic components	Mass mixing ratios Γ_j of the basic components	Particle number concentration (cm^{-3})	Particle mass concentration (μg m^{-3})
Continental clean (CC)	Water soluble	999.94	0.591	2 600	5.2
	Insoluble	6×10^{-2}	0.409	0.15	3.6
Continental average (CA)	Soot	542.47	0.021	8 300	5×10^{-1}
	Water soluble	457.50	0.583	7 000	14.0
	Insoluble	2.6×10^{-2}	0.396	0.40	9.5
Continental polluted (CP)	Soot	686.00	0.044	34 300	2.1
	Water soluble	313.99	0.658	15 700	31.4
	Insoluble	10^{-3}	0.298	0.60	14.2
Urban (UR)	Soot	822.79	0.079	130 000	7.8
	Water soluble	177.21	0.563	28 000	56.0
	Insoluble	10^{-3}	0.358	1.50	35.6
Desert (DE)	Water soluble	869.51	0.018	2 000	4.0
	Mineral dust (nucleation)	117.17	0.033	269.5	7.5
	Mineral dust (accumulation)	13.26	0.747	30.5	168.7
	Mineral dust (coarse)	6.2×10^{-2}	0.202	0.142	45.6
Maritime clean (MC)	Water soluble	986.84	0.071	1 500	3.0
	Sea salt (accumulation)	13.16	0.908	20	38.6
	Sea salt (coarse)	2.0×10^{-3}	0.021	3.2×10^{-3}	9×10^{-1}

(*continued overleaf*)

Table 6.12 (*Continued*)

OPAC models with their acronyms (given in brackets)	Components	Particle number concentrations N_j (cm^{-3}) of the basic components	Mass mixing ratios Γ_j of the basic components	Particle number concentration (cm^{-3})	Particle mass concentration (μg m^{-3})
Maritime polluted (MP)	Soot	575.56	0.006	5 180	3×10^{-1}
	Water soluble	422.22	0.160	3 800	7.6
	Sea salt (accumulation)	2.22	0.814	20	38.6
	Sea salt (coarse)	4.0×10^{-4}	0.019	3.2×10^{-3}	9×10^{-1}
Maritime tropical (MT)	Water soluble	983.33	0.058	590	1.2
	Sea salt (accumulation)	16.6668	0.928	10	19.3
	Sea salt (coarse)	2.0×10^{-4}	0.014	1.3×10^{-3}	3×10^{-1}
Arctic (AR)	Soot	802.80	0.044	5 300	3×10^{-1}
	Water soluble	196.91	0.382	1 300	2.6
	Sea salt (accumulation)	2.88×10^{-1}	0.544	1.9	3.7
	Insoluble	1.50×10^{-3}	0.029	10^{-2}	2×10^{-1}
Antarctic (AN)	Sulfate	998.78	0.910	42.9	2.0
	Sea salt (accumulation)	1.09	0.045	4.70×10^{-2}	10^{-1}
	Mineral transported	1.23×10^{-1}	0.045	5.30×10^{-3}	10^{-1}

anthropogenic influences, and hence without soot particles, presenting a limited mass concentration of WAS particles used to represent the nonsea-salt (nss) sulfate particles. The sea-salt particle concentration was assumed to present a value of 20 cm^{-3} for wind speed WS = 8.9 m s^{-1}.

7. The *maritime-polluted* (MP) aerosol model represents an aerosol type usually sampled in the maritime environment subject to strong anthropogenic impact, with highly variable mass concentrations of SOO and WAS basic components, yielding average values of $M_{soo} = 0.3\ \mu g\ m^{-3}$ and $M_{wss} = 7.6\ \mu g\ m^{-3}$, respectively, and supplemental concentrations of the SAM and SCM sea-salt components approximately equal to those adopted in the MC aerosol model.
8. The *maritime tropical* (MT) aerosol model represents a case characterized by a very low mass concentration M_{wss} and WS = 5 m s^{-1}, which is lower than those assumed for the MC and MP models, therefore presenting a lower number concentration of sea-salt particles.
9. The *Arctic* (AR) aerosol model represents the polar airborne particles present on average in the Arctic region at latitudes higher than 70°N, therefore with a relatively high mass concentration of soot particles transported from the midlatitude continental areas: therefore, this model seems to be particularly suitable to represent the BG Arctic aerosol optical characteristics observed by Tomasi *et al.* (2012) during the late winter and spring months.
10. The *Antarctic* (AN) aerosol model represents the airborne particles sampled at the inner Antarctic sites during the austral summer, which contains relatively low mass concentrations of mineral dust and sea-salt particles transported from coastal sites and rather high number concentrations of nss sulfate aerosols (Tomasi *et al.*, 2012).

Table 6.12 provides a summarized picture of the values of particle number concentration N_j and mass mixing ratios Γ_j of the OPAC basic components giving form to the OPAC aerosol models determined by Hess, Koepke, and Schult (1998) for wet-air (RH = 50%) conditions. The multimodal size-distribution curves of $N(a)$ determined for these 10 aerosol models are shown in Figure 6.7, all normalized to the overall particle number density $N_T = 10^3\ cm^{-3}$. All curves (except that of the AN model) exhibit similar features over the range $a < 5 \times 10^{-2}$ μm and differ appreciably one from another over the diameter range $a \geq 10^{-1}$ μm, where the highest content of coarse particles is given by the DE model and the lowest by the UR and CP models.

The spectral values of optical parameters $n(\lambda)$, $\chi(\lambda)$, $\beta_{ext}(\lambda)$, $\beta_{sca}(\lambda)$, $\omega(\lambda)$, and $g(\lambda)$ were calculated using the OPAC code for all the 10 aerosol extinction models listed in Table 6.12. They are given in Table 6.13 for wet-air (RH = 50%) conditions, showing that (i) $n(\lambda)$ ranges from 1.38 to 1.50 in the visible; (ii) $\chi(\lambda)$ varies from about 10^{-3} to 5×10^{-1} at visible wavelengths; (iii) $\beta_{ext}(\lambda)$ decreases more or less rapidly as a function of wavelength λ, closely depending on the mass fractions of fine and coarse particle modes; (iv) $\beta_{sca}(\lambda)$ decreases as λ increases and assumes values differing from those of $\beta_{ext}(\lambda)$ by percentages proportional to the

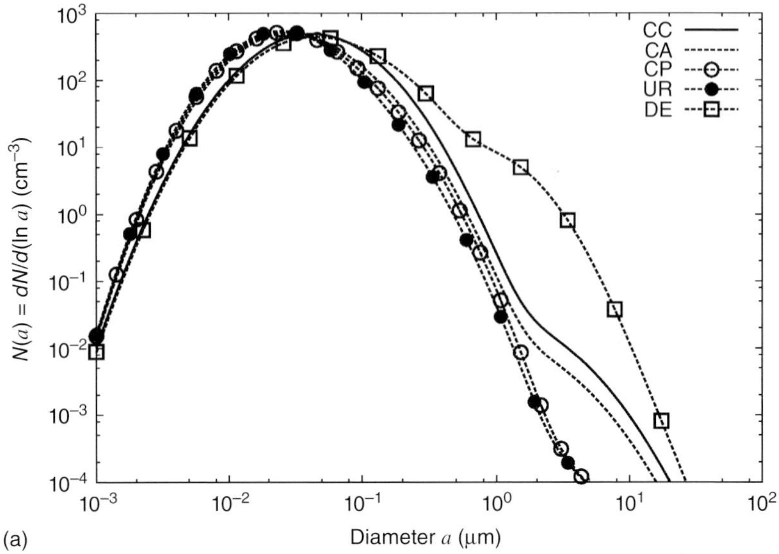

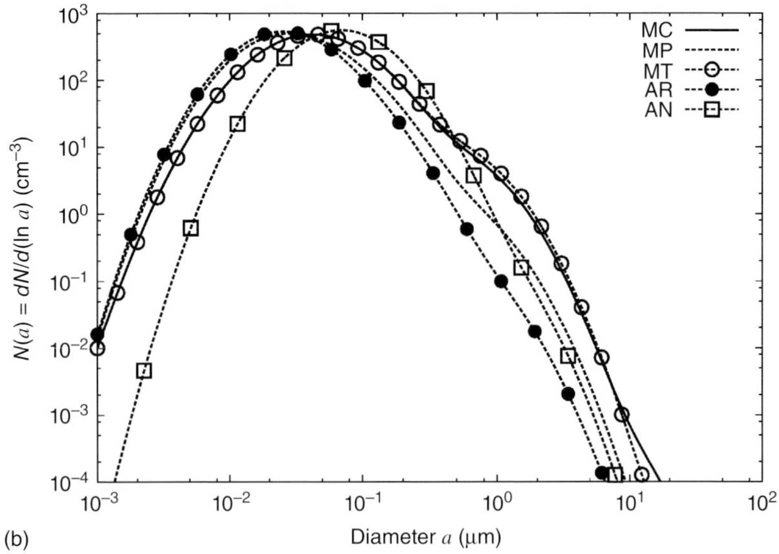

Figure 6.7 (a) Size-distribution curves of particle number density $N(a)$ over the $10^{-3} \leq a \leq 10^{2}$ μm diameter range for the five OPAC wet-air (RH = 50%) aerosol models labeled by Hess, Koepke, and Schult (1998) with the acronyms CC (solid curve), CA (dashed curve), CP (dashed curve with open circles), UR (dashed curve with solid circles), and DE (dashed curve with open squares) listed in Table 6.18. (b) As in the (a) for the five OPAC wet-air (RH = 50%) aerosol models labeled with acronyms MC (solid curve), MP (dashed curve), MT (dashed curve with open circles), AR (dashed curve with solid circles), and AN (dashed curve with open squares) listed in Table 6.18. All the 10 aerosol models are calculated for RH = 50% and normalized to give the overall particle number concentration $N_T = 10^{3}$ cm^{-3}.

Table 6.13 Spectral values of real part $n(\lambda)$ and imaginary part $\chi(\lambda)$ of particulate matter refractive index, volume extinction coefficient $\beta_{ext}(\lambda)$ and volume scattering coefficient $\beta_{sca}(\lambda)$ (both measured in per kilometer and calculated for overall number density concentration $N_T = 10^3$ cm^{-3}), single scattering albedo $\omega(\lambda)$, and asymmetry factor $g(\lambda)$, defined by Hess, Koepke, and Schult (1998) at 11 selected wavelengths from 0.30 to 3.75 μm for the 10 OPAC wet-air (RH = 50%) aerosol models.

λ (μm)	$n(\lambda)$	$\chi(\lambda)$	$\beta_{ext}(\lambda)$ (km^{-1})	$\beta_{sca}(\lambda)$ (km^{-1})	$\omega(\lambda)$	$g(\lambda)$
Continental clean (CC) model						
0.300	1.474	1.35×10^{-2}	3.24×10^{-2}	3.14×10^{-2}	0.971	0.715
0.400	1.467	1.01×10^{-2}	2.63×10^{-2}	2.55×10^{-2}	0.967	0.701
0.500	1.465	7.77×10^{-3}	2.03×10^{-2}	1.95×10^{-2}	0.964	0.687
0.550	1.464	6.86×10^{-3}	1.78×10^{-2}	1.71×10^{-2}	0.959	0.681
0.600	1.464	6.09×10^{-3}	1.60×10^{-2}	1.52×10^{-2}	0.956	0.674
0.650	1.463	5.44×10^{-3}	1.42×10^{-2}	1.35×10^{-2}	0.952	0.667
0.700	1.463	4.89×10^{-3}	1.27×10^{-2}	1.21×10^{-2}	0.948	0.661
0.850	1.456	3.62×10^{-3}	9.47×10^{-3}	8.77×10^{-3}	0.924	0.647
1.550	1.377	1.46×10^{-3}	3.83×10^{-3}	3.26×10^{-3}	0.852	0.635
2.250	1.258	7.32×10^{-4}	1.90×10^{-3}	1.64×10^{-3}	0.859	0.736
3.750	1.314	2.64×10^{-2}	1.27×10^{-3}	1.10×10^{-3}	0.872	0.770
Continental average (CA) model						
0.300	1.454	6.82×10^{-3}	9.63×10^{-2}	8.72×10^{-2}	0.908	0.705
0.400	1.444	5.10×10^{-3}	7.80×10^{-2}	7.05×10^{-2}	0.903	0.692
0.500	1.441	3.90×10^{-3}	5.98×10^{-2}	5.37×10^{-2}	0.898	0.679
0.550	1.440	3.44×10^{-3}	5.26×10^{-2}	4.70×10^{-2}	0.893	0.674
0.600	1.440	3.05×10^{-3}	4.70×10^{-2}	4.18×10^{-2}	0.889	0.667
0.650	1.439	2.72×10^{-3}	4.18×10^{-2}	3.70×10^{-2}	0.885	0.661
0.700	1.439	2.45×10^{-3}	3.75×10^{-2}	3.30×10^{-2}	0.880	0.655
0.850	1.434	1.82×10^{-3}	2.80×10^{-2}	2.39×10^{-2}	0.852	0.641
1.550	1.376	7.46×10^{-4}	1.15×10^{-3}	8.79×10^{-3}	0.764	0.632
2.250	1.286	3.88×10^{-4}	5.94×10^{-3}	4.41×10^{-3}	0.737	0.732
3.750	1.347	4.47×10^{-2}	3.89×10^{-3}	2.95×10^{-3}	0.759	0.769
Continental-polluted (CP) model						
0.300	1.448	5.01×10^{-3}	2.31×10^{-1}	1.99×10^{-1}	0.866	0.697
0.400	1.437	3.72×10^{-3}	1.86×10^{-1}	1.60×10^{-1}	0.860	0.684
0.500	1.434	2.83×10^{-3}	1.42×10^{-1}	1.21×10^{-1}	0.854	0.671
0.550	1.433	2.48×10^{-3}	1.24×10^{-1}	1.05×10^{-1}	0.849	0.665
0.600	1.432	2.19×10^{-3}	1.11×10^{-1}	9.34×10^{-2}	0.844	0.658
0.650	1.432	1.96×10^{-3}	9.82×10^{-2}	8.23×10^{-2}	0.839	0.652
0.700	1.432	1.75×10^{-3}	8.78×10^{-2}	7.32×10^{-2}	0.833	0.645
0.850	1.427	1.29×10^{-3}	6.50×10^{-2}	5.21×10^{-2}	0.801	0.631
1.550	1.381	5.03×10^{-4}	2.54×10^{-2}	1.75×10^{-2}	0.686	0.606
2.250	1.306	2.53×10^{-4}	1.27×10^{-2}	7.92×10^{-3}	0.620	0.689
3.750	1.369	6.01×10^{-2}	7.61×10^{-3}	4.79×10^{-3}	0.629	0.737

(*continued overleaf*)

Table 6.13 (*Continued*)

λ (μm)	$n(\lambda)$	$\chi(\lambda)$	$\beta_{ext}(\lambda)$ (km^{-1})	$\beta_{sca}(\lambda)$ (km^{-1})	$\omega(\lambda)$	$g(\lambda)$
Urban (UR) model						
0.300	1.469	3.38×10^{-3}	4.89×10^{-1}	3.79×10^{-1}	0.778	0.682
0.400	1.460	2.50×10^{-3}	3.95×10^{-1}	3.04×10^{-1}	0.769	0.670
0.500	1.457	1.90×10^{-3}	3.00×10^{-1}	2.29×10^{-1}	0.760	0.659
0.550	1.456	1.67×10^{-3}	2.63×10^{-1}	1.98×10^{-1}	0.754	0.654
0.600	1.455	1.48×10^{-3}	2.35×10^{-1}	1.76×10^{-1}	0.748	0.649
0.650	1.455	1.32×10^{-3}	2.09×10^{-1}	1.55×10^{-1}	0.741	0.643
0.700	1.455	1.18×10^{-3}	1.87×10^{-1}	1.38×10^{-1}	0.734	0.637
0.850	1.450	8.86×10^{-4}	1.41×10^{-1}	9.86×10^{-2}	0.699	0.626
1.550	1.399	3.74×10^{-4}	5.96×10^{-2}	3.50×10^{-2}	0.586	0.620
2.250	1.318	2.08×10^{-4}	3.29×10^{-2}	1.72×10^{-2}	0.517	0.719
3.750	1.381	7.62×10^{-2}	2.08×10^{-2}	1.12×10^{-2}	0.540	0.763
Desert (DE) model						
0.300	1.528	6.58×10^{-2}	1.50×10^{-1}	1.09×10^{-1}	0.719	0.800
0.400	1.527	6.31×10^{-2}	1.45×10^{-1}	1.14×10^{-1}	0.785	0.770
0.500	1.527	6.09×10^{-2}	1.40×10^{-1}	1.19×10^{-1}	0.851	0.740
0.550	1.527	6.01×10^{-2}	1.38×10^{-1}	1.22×10^{-1}	0.883	0.727
0.600	1.527	5.94×10^{-2}	1.37×10^{-1}	1.22×10^{-1}	0.896	0.719
0.650	1.527	5.88×10^{-2}	1.35×10^{-1}	1.23×10^{-1}	0.907	0.712
0.700	1.527	5.83×10^{-2}	1.34×10^{-1}	1.23×10^{-1}	0.915	0.707
0.850	1.527	5.73×10^{-2}	1.32×10^{-1}	1.22×10^{-1}	0.925	0.699
1.550	1.527	5.31×10^{-2}	1.22×10^{-1}	1.13×10^{-1}	0.929	0.691
2.250	1.521	4.66×10^{-2}	1.07×10^{-1}	9.72×10^{-2}	0.907	0.696
3.750	1.498	1.27×10^{-2}	7.36×10^{-2}	6.67×10^{-2}	0.907	0.688
Maritime clean (MC) model						
0.300	1.388	4.32×10^{-2}	6.45×10^{-2}	6.41×10^{-2}	0.995	0.759
0.400	1.378	4.13×10^{-2}	6.27×10^{-2}	6.25×10^{-2}	0.996	0.757
0.500	1.375	4.01×10^{-2}	6.10×10^{-2}	6.08×10^{-2}	0.996	0.755
0.550	1.373	3.97×10^{-2}	6.03×10^{-2}	6.01×10^{-2}	0.996	0.756
0.600	1.370	3.93×10^{-2}	5.97×10^{-2}	5.95×10^{-2}	0.996	0.758
0.650	1.369	3.89×10^{-2}	5.92×10^{-2}	5.90×10^{-2}	0.997	0.759
0.700	1.369	3.86×10^{-2}	5.86×10^{-2}	5.84×10^{-2}	0.997	0.760
0.850	1.365	3.74×10^{-2}	5.67×10^{-2}	5.65×10^{-2}	0.996	0.766
1.550	1.352	2.94×10^{-2}	4.48×10^{-2}	4.44×10^{-2}	0.992	0.778
2.250	1.321	2.02×10^{-2}	3.06×10^{-2}	3.01×10^{-2}	0.982	0.787
3.750	1.392	6.24×10^{-3}	2.08×10^{-2}	2.03×10^{-2}	0.975	0.712
Maritime-polluted (MP) model						
0.300	1.391	1.13×10^{-2}	9.77×10^{-2}	9.29×10^{-2}	0.950	0.737
0.400	1.380	9.93×10^{-3}	8.94×10^{-2}	8.54×10^{-2}	0.956	0.736
0.500	1.376	9.00×10^{-3}	8.11×10^{-2}	7.80×10^{-2}	0.962	0.736
0.550	1.375	8.64×10^{-3}	7.78×10^{-2}	7.49×10^{-2}	0.964	0.737
0.600	1.372	8.35×10^{-3}	7.53×10^{-2}	7.27×10^{-2}	0.965	0.738
0.650	1.371	8.09×10^{-3}	7.29×10^{-2}	7.05×10^{-2}	0.967	0.740
0.700	1.371	7.86×10^{-3}	7.08×10^{-2}	6.86×10^{-2}	0.969	0.742
0.850	1.367	7.28×10^{-3}	6.55×10^{-2}	6.35×10^{-2}	0.970	0.749
1.550	1.354	5.29×10^{-3}	4.77×10^{-2}	4.63×10^{-2}	0.970	0.768

Table 6.13 (*Continued*)

λ (μm)	$n(\lambda)$	$\chi(\lambda)$	$\beta_{ext}(\lambda)$ (km^{-1})	$\beta_{sca}(\lambda)$ (km^{-1})	$\omega(\lambda)$	$g(\lambda)$
2.250	1.323	3.54×10^{-3}	3.18×10^{-2}	3.07×10^{-2}	0.963	0.781
3.750	1.394	1.41×10^{-2}	2.14×10^{-2}	2.05×10^{-2}	0.959	0.709
Maritime tropical (MT) model						
0.300	1.388	5.11×10^{-2}	3.03×10^{-2}	3.01×10^{-2}	0.996	0.763
0.400	1.378	4.96×10^{-2}	2.98×10^{-2}	2.97×10^{-2}	0.996	0.760
0.500	1.375	4.88×10^{-2}	2.93×10^{-2}	2.92×10^{-2}	0.997	0.758
0.550	1.373	4.85×10^{-2}	2.91×10^{-2}	2.90×10^{-2}	0.997	0.759
0.600	1.370	4.83×10^{-2}	2.89×10^{-2}	2.88×10^{-2}	0.997	0.760
0.650	1.369	4.79×10^{-2}	2.88×10^{-2}	2.87×10^{-2}	0.997	0.761
0.700	1.369	4.76×10^{-2}	2.86×10^{-2}	2.85×10^{-2}	0.997	0.763
0.850	1.365	4.64×10^{-2}	2.78×10^{-2}	2.77×10^{-2}	0.997	0.768
1.550	1.352	3.70×10^{-2}	2.22×10^{-2}	2.20×10^{-2}	0.993	0.779
2.250	1.321	2.54×10^{-2}	1.52×10^{-2}	1.50×10^{-2}	0.983	0.788
3.750	1.392	5.59×10^{-3}	1.04×10^{-2}	1.01×10^{-2}	0.976	0.712
Arctic (AR) model						
0.300	1.411	4.21×10^{-3}	2.58×10^{-2}	2.14×10^{-2}	0.828	0.699
0.400	1.399	3.29×10^{-3}	2.17×10^{-2}	1.81×10^{-2}	0.834	0.695
0.500	1.396	2.67×10^{-3}	1.77×10^{-2}	1.49×10^{-2}	0.840	0.692
0.550	1.394	2.44×10^{-3}	1.61×10^{-2}	1.35×10^{-2}	0.842	0.692
0.600	1.392	2.24×10^{-3}	1.49×10^{-2}	1.25×10^{-2}	0.844	0.693
0.650	1.391	2.08×10^{-3}	1.37×10^{-2}	1.16×10^{-2}	0.846	0.693
0.700	1.391	1.93×10^{-3}	1.28×10^{-2}	1.09×10^{-2}	0.848	0.694
0.850	1.388	1.61×10^{-3}	1.07×10^{-2}	9.03×10^{-3}	0.846	0.702
1.550	1.372	9.32×10^{-4}	6.18×10^{-3}	5.21×10^{-3}	0.843	0.732
2.250	1.337	5.78×10^{-4}	3.81×10^{-3}	3.18×10^{-3}	0.829	0.759
3.750	1.407	4.71×10^{-2}	2.42×10^{-3}	2.03×10^{-3}	0.837	0.700
Antarctic (AN) model						
0.300	1.446	2.22×10^{-1}	9.71×10^{-3}	9.70×10^{-3}	0.999	0.772
0.400	1.441	2.03×10^{-1}	8.73×10^{-3}	8.72×10^{-3}	0.999	0.771
0.500	1.439	1.80×10^{-1}	7.74×10^{-3}	7.73×10^{-3}	0.999	0.771
0.550	1.438	1.69×10^{-1}	7.25×10^{-3}	7.24×10^{-3}	0.999	0.769
0.600	1.438	1.57×10^{-1}	6.78×10^{-3}	6.77×10^{-3}	0.999	0.766
0.650	1.437	1.47×10^{-1}	6.33×10^{-3}	6.32×10^{-3}	0.999	0.763
0.700	1.437	1.37×10^{-1}	5.90×10^{-3}	5.90×10^{-3}	0.999	0.759
0.850	1.431	1.11×10^{-1}	4.79×10^{-3}	4.79×10^{-3}	0.999	0.748
1.550	1.418	4.42×10^{-2}	1.92×10^{-3}	1.91×10^{-3}	0.996	0.690
2.250	1.361	1.96×10^{-2}	8.41×10^{-4}	8.18×10^{-4}	0.968	0.644
3.750	1.416	4.05×10^{-3}	5.92×10^{-4}	3.23×10^{-4}	0.546	0.536

relative content of absorbing substances; (v) $\omega(\lambda)$ assumes values varying between 0.74 (CP model) and 1.00 (MC and AN models) in the visible; and (iv) $g(\lambda)$ slowly decreases as λ increases, assuming the two lowest values for the CP and UR models and presenting rather high values for the DE and AN models and the three maritime models.

A general picture of the main radiative properties of the 10 OPAC wet-air (RH = 50%) aerosol models defined by Hess, Koepke, and Schult (1998) is given in Table 6.14, in which the monochromatic values of $\beta_{ext}(0.55\,\mu m)$, $\beta_{sca}(0.55\,\mu m)$, and $\beta_{abs}(0.55\,\mu m)$ are provided for the 10 OPAC aerosol models. The Ångström exponent $\alpha(0.40-0.85\,\mu m)$ calculated for the values of $\beta_{ext}(\lambda)$ determined at the seven OPAC wavelengths ranging from 0.40 to 0.85 μm was found to range in Table 6.13 from ~0.09 (MT model) to more than 1.4 (CP model). Parameter $\omega(0.55\,\mu m)$ estimated for RH = 50% was estimated to vary from 0.742 (UR) to 0.997 (MT), while the single scattering albedo ω averaged over the 0.40–3.75 μm wavelength range was evaluated to vary from 0.686 (UR model) to 0.993 (MC and MT models), and asymmetry factor $g(0.55\,\mu m)$ was estimated to range from 0.665 (CP) to 0.764 (AN). Correspondingly, the wet-air particulate mass density varied from 1.30 (MT) to 2.56 g cm^{-3} (UR).

6.2.5 The Aerosol Models of Shettle and Fenn (1979)

Four classical aerosol extinction models (hereinafter referred to as SF) were determined by Shettle and Fenn (1979) to represent the optical characteristics of rural, maritime, tropospheric, and urban particle populations for seven values of RH ranging from 0% to 99%. The size-distribution curves of the four aerosol models were represented by means of unimodal or bimodal lognormal size-distribution functions $N(a)$, each mode having the analytical form defined by Eq. (6.1) for the values of particle number density N_i (cm^{-3}), mode diameter a_c (μm), and standard deviation σ_g given in Table 6.15. Only one lognormal curve was adopted by Shettle and Fenn (1979) to give form to the SF-M (maritime) and SF-T (tropospheric) models, while pairs of lognormal curves were used to give form to the SF-R (rural) and SF-U (urban) models. In particular:

1. The rural (SF-R) aerosol model represents an aerosol population not influenced by urban and/or industrial sources, whose particles are composed of a mixture of 70% of water-soluble substances (mainly ammonium and calcium sulfates, with a low content of organic compounds) and 30% of DL aerosols. The size-distribution curve of $N(a)$ for the SF-R model is shown in Figure 6.8 for decreasing RH = 50%, presenting the bimodal features associated with the nuclei and accumulation/coarse rural particle modes. The spectral values of $n(\lambda)$ and $\chi(\lambda)$ were determined at 11 selected wavelengths from 0.25 to 3.75 μm, according to Volz (1972a, 1973). They are given in Table 6.16, presenting values of $n(\lambda)$ gradually decreasing from ~1.52 at visible wavelengths to less than 1.40 for $\lambda > 2\,\mu m$ and values of $\chi(\lambda)$ ranging from 8×10^{-3} to less than 7×10^{-3} at visible wavelengths and subsequently decreasing to 6×10^{-3} at $\lambda = 3.75\,\mu m$.
   The calculations of parameters $\beta_{ext}(\lambda)$, $\beta_{sca}(\lambda)$, $\omega(\lambda)$, and $g(\lambda)$ were made at the same wavelengths and given in Table 6.16, indicating that $\beta_{ext}(\lambda)$ is greater than 10^{-2} km^{-1} in the visible and slowly decreases with λ until reaching a value

Table 6.14 Values of the volume extinction coefficient $\beta_{ext}(0.55\ \mu m)$, volume scattering coefficient $\beta_{sca}(0.55\ \mu m)$, volume absorption coefficient $\beta_{abs}(0.55\ \mu m)$, Ångström's exponent $\alpha(0.40-0.85\ \mu m)$, monochromatic single scattering albedo $\omega(0.55\ \mu m)$, average single scattering albedo ω determined over the 0.40–3.75 μm spectral range, monochromatic asymmetry factor $g(0.55\ \mu m)$, and particulate mass density ρ_w ($g\ cm^{-3}$) calculated by Hess, Koepke, and Schult (1998) for the 10 OPAC wet-air (RH = 50%) aerosol models, all normalized to give the overall particle number density $N_T = 10^3\ cm^{-3}$.

Radiative and physical parameters	OPAC aerosol models (RH = 50%)									
	CC	CA	CP	UR	DE	MC	MP	MT	AR	AN
$\beta_{ext}(0.55\ \mu m)$	1.783×10^{-2}	5.256×10^{-2}	1.240×10^{-1}	2.630×10^{-1}	1.382×10^{-1}	6.027×10^{-2}	7.778×10^{-2}	2.907×10^{-2}	1.607×10^{-2}	7.245×10^{-3}
$\beta_{sca}(0.55\ \mu m)$	1.709×10^{-2}	4.696×10^{-2}	1.053×10^{-1}	1.984×10^{-1}	1.220×10^{-1}	6.005×10^{-2}	7.494×10^{-2}	2.899×10^{-2}	1.353×10^{-2}	7.240×10^{-3}
$\beta_{abs}(0.55\ \mu m)$	7.400×10^{-4}	5.600×10^{-3}	1.870×10^{-2}	6.460×10^{-2}	1.620×10^{-2}	2.200×10^{-4}	2.840×10^{-3}	8.000×10^{-5}	2.540×10^{-3}	5.000×10^{-6}
$\alpha(0.40-0.85\ \mu m)$	1.386	1.389	1.427	1.397	0.127	0.131	0.410	0.088	0.952	0.999
$\omega(0.55\ \mu m)$	0.959	0.893	0.849	0.754	0.883	0.996	0.964	0.997	0.842	0.979
ω	0.926	0.847	0.782	0.686	0.886	0.993	0.964	0.993	0.840	0.950
$g(0.55\ \mu m)$	0.681	0.674	0.665	0.654	0.727	0.756	0.737	0.759	0.692	0.769
$\rho_w\ (g\ cm^{-3})$	1.61	1.59	1.52	2.56	1.97	1.32	1.31	1.30	1.33	1.77

Table 6.15 Shape parameters and characteristics of the unimodal lognormal size-distribution curves used by Shettle and Fenn (1979) in the calculations of the shape parameters defining the four SF aerosol models determined for RH = 50%, all normalized to the overall particle number density concentration $N_T = 10^3$ cm^{-3}.

Aerosol model	Modes	Aerosol type of the unimodal size-distribution curve	Unimodal lognormal size-distribution shape parameters		
			Particle number concentrations N_1 (1st mode) and N_2 (2nd mode)	Mode diameter a_c (μm)	Geometric standard deviation σ_g
Rural (SF-R)	1st mode (water soluble) 2nd mode (dustlike)	Mixture of water-soluble and dustlike aerosols	999.875 0.125	5.500×10^{-2} 8.754×10^{-1}	2.239 2.512
Urban (SF-U)	1st mode (rural) 2nd mode (sootlike)	Rural aerosol mixture with sootlike aerosols	999.875 0.125	5.126×10^{-2} 8.226×10^{-1}	2.239 2.512
Maritime (SF-M)	1st mode (sea salt)	Sea-salt solution in water of oceanic origin	1000.00	3.422×10^{-1}	2.512
Tropospheric (SF-T)	1st mode (rural fine particles)	Rural aerosol mixture	1000.00	5.500×10^{-2}	2.239

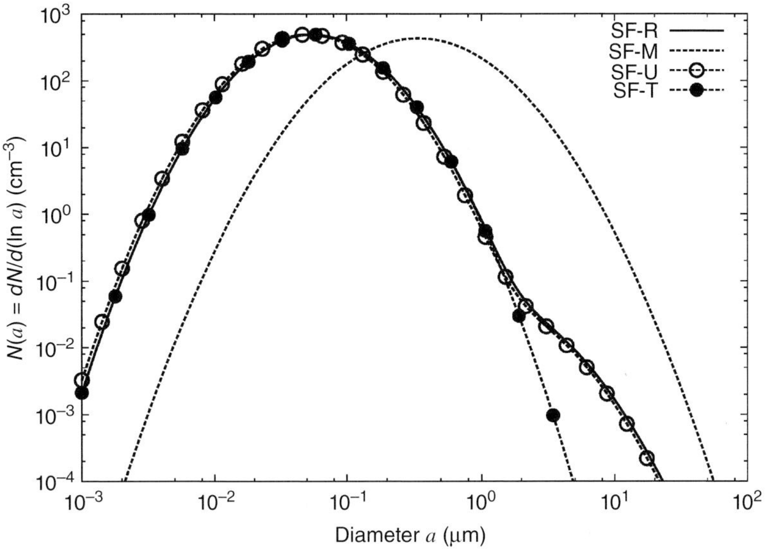

Figure 6.8 Size-distribution curves of particle number density $N(r)$ over the $10^{-3} \leq a \leq 10^{2}$ μm diameter range for (i) rural aerosol model (SF-R) (solid curve), (ii) maritime aerosol model (SF-M) (dashed curve), (iii) urban aerosol model (SF-U) (dashed curve with open circles), and (iv) tropospheric aerosol model (SF-T) (dashed curve with solid circles), defined by Shettle and Fenn (1979) for RH = 50%. All the four models are normalized to give the overall particle number concentration $N_T = 10^3$ cm^{-3}.

Table 6.16 Spectral values of real part $n(\lambda)$ and imaginary part $\chi(\lambda)$ of particulate matter refractive index, volume extinction coefficient $\beta_{ext}(\lambda)$ and volume scattering coefficient $\beta_{sca}(\lambda)$ (both measured in km^{-1} and calculated for overall number density concentration $N_T = 10^3$ cm^{-3}), single scattering albedo $\omega(\lambda)$, and asymmetry factor $g(\lambda)$, defined by Shettle and Fenn (1979) at 11 selected wavelengths from 0.30 to 3.75 μm for the SF-R, SF-U, SF-M, and SF-T models of wet-air (RH = 50%) aerosol particles.

λ (μm)	$n(\lambda)$	$\chi(\lambda)$	$\beta_{ext}(\lambda)$ (km^{-1})	$\beta_{sca}(\lambda)$ (km^{-1})	$\omega(\lambda)$	$g(\lambda)$
Rural (SF-R) bimodal model						
0.300	1.521	8.00×10^{-3}	1.76×10^{-2}	1.67×10^{-2}	0.934	0.684
0.337	1.521	5.59×10^{-3}	1.62×10^{-2}	1.54×10^{-2}	0.949	0.676
0.550	1.520	6.26×10^{-3}	1.01×10^{-2}	9.52×10^{-3}	0.943	0.653
0.694	1.520	6.92×10^{-3}	7.59×10^{-3}	7.25×10^{-3}	0.934	0.639
1.060	1.510	1.36×10^{-2}	4.24×10^{-3}	3.67×10^{-3}	0.865	0.622
1.536	1.496	1.76×10^{-2}	2.44×10^{-3}	2.00×10^{-3}	0.798	0.637
2.000	1.369	7.64×10^{-3}	1.49×10^{-3}	1.30×10^{-3}	0.870	0.707
2.250	1.356	9.22×10^{-3}	1.35×10^{-3}	1.15×10^{-3}	0.849	0.728
2.700	1.326	4.12×10^{-2}	1.33×10^{-3}	7.67×10^{-4}	0.576	0.782
3.392	1.381	9.40×10^{-3}	1.07×10^{-3}	9.27×10^{-4}	0.866	0.741
3.750	1.396	6.00×10^{-3}	1.01×10^{-3}	8.78×10^{-4}	0.914	0.731

(continued overleaf)

Table 6.16 (*Continued*).

λ (μm)	$n(\lambda)$	$\chi(\lambda)$	$\beta_{ext}(\lambda)$ (km^{-1})	$\beta_{sca}(\lambda)$ (km^{-1})	$\omega(\lambda)$	$g(\lambda)$
Urban (SF-U) bimodal model						
0.300	1.554	9.24×10^{-2}	1.51×10^{-2}	9.82×10^{-3}	0.649	0.733
0.337	1.555	9.08×10^{-2}	1.41×10^{-2}	9.18×10^{-3}	0.653	0.713
0.550	1.555	8.58×10^{-2}	9.24×10^{-3}	6.00×10^{-3}	0.648	0.609
0.694	1.555	8.45×10^{-2}	7.19×10^{-3}	4.56×10^{-3}	0.635	0.648
1.060	1.547	9.15×10^{-2}	4.34×10^{-3}	2.46×10^{-3}	0.567	0.623
1.536	1.518	9.83×10^{-2}	2.76×10^{-3}	1.35×10^{-3}	0.489	0.634
2.000	1.445	9.61×10^{-2}	1.97×10^{-3}	8.59×10^{-4}	0.436	0.692
2.250	1.437	0.92×10^{-2}	1.79×10^{-3}	7.54×10^{-4}	0.422	0.711
2.700	1.414	1.28×10^{-1}	1.66×10^{-3}	5.91×10^{-4}	0.356	0.744
3.392	1.473	1.09×10^{-1}	1.37×10^{-3}	5.78×10^{-4}	0.420	0.739
3.750	1.488	1.10×10^{-1}	1.29×10^{-3}	5.55×10^{-4}	0.429	0.739
Maritime (SF-M) unimodal model						
0.300	1.481	1.64×10^{-6}	2.84×10^{-2}	2.76×10^{-2}	0.970	0.707
0.337	1.480	3.29×10^{-7}	2.73×10^{-2}	2.67×10^{-2}	0.979	0.701
0.550	1.470	8.54×10^{-9}	2.17×10^{-2}	2.14×10^{-2}	0.984	0.692
0.694	1.461	8.74×10^{-8}	1.86×10^{-2}	1.91×10^{-2}	0.985	0.691
1.060	1.444	1.64×10^{-4}	1.57×10^{-2}	1.54×10^{-2}	0.978	0.698
1.536	1.434	5.09×10^{-4}	1.27×10^{-2}	1.24×10^{-2}	0.975	0.711
2.000	1.424	1.02×10^{-3}	1.05×10^{-2}	1.03×10^{-2}	0.983	0.723
2.250	1.413	1.71×10^{-3}	9.55×10^{-3}	9.32×10^{-3}	0.976	0.729
2.700	1.362	9.18×10^{-3}	7.75×10^{-3}	6.89×10^{-3}	0.889	0.764
3.392	1.469	5.34×10^{-3}	7.64×10^{-3}	7.29×10^{-3}	0.954	0.691
3.750	1.452	1.78×10^{-3}	6.74×10^{-3}	6.60×10^{-3}	0.982	0.692
Tropospheric (SF-T) unimodal model						
0.300	1.521	8.00×10^{-3}	1.67×10^{-2}	1.59×10^{-2}	0.948	0.677
0.337	1.520	5.60×10^{-3}	1.53×10^{-2}	1.47×10^{-2}	0.962	0.669
0.550	1.520	6.26×10^{-3}	9.16×10^{-3}	8.80×10^{-3}	0.961	0.640
0.694	1.520	6.92×10^{-3}	6.64×10^{-3}	6.34×10^{-3}	0.955	0.621
1.060	1.510	1.36×10^{-2}	3.24×10^{-3}	2.92×10^{-3}	0.901	0.579
1.536	1.469	1.76×10^{-2}	1.40×10^{-3}	1.15×10^{-3}	0.822	0.533
2.000	1.369	7.65×10^{-3}	4.82×10^{-4}	4.04×10^{-4}	0.838	0.499
2.250	1.357	9.22×10^{-3}	3.59×10^{-4}	2.77×10^{-4}	0.772	0.474
2.700	1.328	4.12×10^{-2}	4.26×10^{-4}	1.34×10^{-4}	0.315	0.425
3.392	1.361	9.40×10^{-3}	1.51×10^{-4}	9.95×10^{-5}	0.660	0.382
3.750	1.396	6.00×10^{-3}	1.09×10^{-4}	7.98×10^{-5}	0.735	0.360

of ~10^{-3} km^{-1} at 3.75 μm. Correspondingly, $\beta_{sca}(\lambda)$ exhibit lower values by 5% on average at the visible wavelengths, yielding values of $\omega(\lambda)$ equal to ~0.94 and decreasing to less than 0.80 at $\lambda = 2.70$ μm, while $g(\lambda)$ was estimated to decrease from 0.68 at $\lambda = 0.30$ μm to less than 0.63 in the near-infrared, and to slowly increase with λ until reaching a value of ~0.73 at $\lambda = 3.75$ μm.

2. The urban (SF-U) aerosol model was represented by combining the unimodal size distribution of BG rural aerosols, consisting of small urban

particles composed by water-soluble substances only, with the secondary aerosol unimodal size distribution consisting of accumulation/coarse urban aerosols, mainly containing combustion products and industrial smokes, and represented by Shettle and Fenn (1979) as a mixture of rural aerosol (80%) and carbonaceous sootlike particles (20%). This bimodal size distribution of urban aerosols were assumed to have the same values of $N(a)$ and σ_g adopted in the previous case to give form to the SF-R model and the unimodal values of a_c very similar to those of the SF-R model. Therefore, the SF-U bimodal size distribution obtained for RH = 50% and shown in Figure 6.8 exhibits very similar characteristics to those of the SF-R aerosol model. The spectral values of refractive index assumed for sootlike PM were calculated according to the evaluations made by Twitty and Weinman (1971) for soot substances, while their variations as a function of RH were determined using the growth factors proposed by Hänel (1976) for his urban aerosol model (model 5). The spectral values of $n(\lambda)$ and $\chi(\lambda)$ are given in Table 6.16 at the 11 selected wavelengths for decreasing RH = 50%, together with those of parameters $\beta_{ext}(\lambda)$, $\beta_{sca}(\lambda)$, $\omega(\lambda)$, and $g(\lambda)$. The spectral values of $n(\lambda)$ are appreciably higher than those of the SF-R model at all wavelengths, decreasing from ~1.55 at visible and near-infrared wavelengths to less than 1.50 at $\lambda > 2\,\mu m$, and the values of $\chi(\lambda)$ vary from 9.2×10^{-2} to ~8.6×10^{-2} at visible wavelengths and from 8.4×10^{-2} to around 10^{-1} in the infrared. The values of $\beta_{ext}(\lambda)$ range from 1.5×10^{-2} to $9.2 \times 10^{-3}\,km^{-1}$ at visible wavelengths and then decrease as λ increases until becoming smaller than $2 \times 10^{-3}\,km^{-1}$ for $\lambda > 2\,\mu m$. Correspondingly, $\beta_{sca}(\lambda)$ assume values smaller than $\beta_{ext}(\lambda)$ by about 35% at visible wavelengths, providing values of $\omega(\lambda)$ equal to ~0.65 at visible wavelengths and ~0.42 at $\lambda > 2\,\mu m$. Correspondingly, $g(\lambda)$ was estimated to decrease from 0.73 at $\lambda = 0.30\,\mu m$ to 0.61 at $\lambda = 0.55\,\mu m$, subsequently increasing from 0.62 to 0.74 in the infrared.

3. The maritime (SF-M) aerosol model was represented using a lognormal size-distribution curve of sea-salt particles formed through the evaporation of sea spray droplets. These particles were assumed to mainly consist of OC aerosols, mixed with BG continental aerosols to constitute a fairly uniform maritime aerosol model, which is representative of the marine aerosol usually sampled within the 2–3 km depth PBL over the ocean surface. The number density size-distribution curve of the SF-M model is shown in Figure 6.8, as obtained for the number concentration and size-distribution shape parameters reported in Table 6.15. The spectral values of parameters $\chi(\lambda)$, $\beta_{ext}(\lambda)$, $\beta_{sca}(\lambda)$, $\omega(\lambda)$, and $g(\lambda)$ are given in Table 6.16. The dry-air values of $n(\lambda)$ and $\chi(\lambda)$ were defined according to Volz (1972b), as well as those assumed for a decreasing value of RH = 50% by using the growth factor estimated by Hänel (1976) for sea spray aerosols. The wet-air values of $n(\lambda)$ were found to decrease from 1.48 to 1.46 in the visible and decrease until reaching values varying between 1.36 and 1.47 at wavelengths $\lambda > 2\,\mu m$, while those of $\chi(\lambda)$ were assumed to be all very small at visible wavelengths, ranging from 1.6×10^{-6} to less than 10^{-6}, and to vary from 8.7×10^{-8} to ~10^{-2} at infrared

wavelengths. The corresponding values of $\beta_{ext}(\lambda)$ were estimated to vary from 2.8×10^{-2} to 2.0×10^{-2} km^{-1} at visible wavelengths, being mainly due to the poorly absorbing coarse particles of marine origin. For this reason, the values of $\beta_{sca}(\lambda)$ result to be only slightly lower than those of $\beta_{ext}(\lambda)$, yielding values of $\omega(\lambda)$ ranging from 0.95 to 0.98 over the whole wavelength range, while $g(\lambda)$ was estimated to assume values close to 0.70 in the visible.

4. The tropospheric (SF-T) aerosol model represents a unimodal particle population suspended in the troposphere above the PBL. These BG tropospheric particles were therefore assumed to have the same composition of the SF-R PM, giving form to a size distribution of dry particles consisting of 70% water-soluble mass and 30% DL particulate mass. The size distribution of the SF-T model was obtained by removing the second mode of large particles from the SF-R model and using the same shape parameters given in Table 6.15 for the first SF-R mode consisting mainly of water-soluble substances. This choice was made by Shettle and Fenn (1979) considering that the fine particles suspended above the PBL have a longer residence time than the coarse particles concentrated within the PBL, implying a differential loss of coarse particles and a large predominance of accumulation particle concentration, which leads to have only one lognormal size distribution of aerosols, such as that shown in Figure 6.8. The spectral values of $n(\lambda)$ and $\chi(\lambda)$ given in Table 6.16 for the SF-T model are similar to those determined for the SF-R model, while the spectral values of $\beta_{ext}(\lambda)$, $\beta_{sca}(\lambda)$, $\omega(\lambda)$, and $g(\lambda)$ given in Table 6.16 indicate that $\beta_{ext}(\lambda)$ assumes slightly lower values than those of SF-R aerosols in the visible and appreciably lower values for $\lambda > 2$ μm and $\beta_{sca}(\lambda)$ is lower than $\beta_{ext}(\lambda)$ by a few percents at visible wavelengths and by around 20–30% in the infrared. Correspondingly, $\omega(\lambda)$ was estimated be equal to ~0.96 at visible wavelengths and to decrease from 0.95 to 0.73 as λ increases from 0.70 to 3.75 μm, while $g(\lambda)$ is equal to ~0.64 in the visible and decreases from 0.62 to 0.36 at infrared wavelengths.

The monochromatic values of coefficients $\beta_{ext}(0.55\ \mu m)$, $\beta_{sca}(0.55\ \mu m)$, and $\beta_{abs}(0.55\ \mu m)$ are given in Table 6.17 for the four SF aerosol models, giving values of Ångström's (1964) exponent $\alpha(0.40 - 0.86\ \mu m)$ ranging from 0.028 (SF-M) to 1.355 (SF-T).

The monochromatic value of $\omega(0.55\ \mu m)$ was found to vary from 0.648 (SF-U) to 0.984 (SF-M), giving an average value of ω ranging from 0.585 (SF-U) to 0.996 (SF-M), while $g(0.55\ \mu m)$ was evaluated to range from 0.609 (SF-U) to 0.692 (SF-M). Table 6.17 provides the values of wet-air (decreasing RH = 50%) particulate mass density obtained for (i) the SF-R and SF-T models, using the particle growth estimates made by Hänel (1976) for model 6 based on aerosol samplings made at Hohenpeissenberg (Germany) in summer 1970; (ii) the SF-U model, using the evaluations made by Hänel (1976) for model 5 pertaining to an urban aerosol sample collected at Mainz (Germany) in January 1970; and (iii) the SF-M model, defined according to the estimates made by Hänel (1976) for model 2 of sea spray aerosol.

Table 6.17 Values of the volume extinction coefficient $\beta_{ext}(0.55\ \mu m)$, volume scattering coefficient $\beta_{sca}(0.55\ \mu m)$, volume absorptionR coefficient $\beta_{abs}(0.55\ \mu m)$, Ångström's exponent $\alpha(0.40-0.86\ \mu m)$, monochromatic single scattering albedo $\omega(0.55\ \mu m)$, average single scattering albedo determined over the 0.40–3.75 μm spectral range, monochromatic asymmetry factor $g(0.55\ \mu m)$, and particulate mass density ρ_w (g cm^{-3}) of the SF-R, SF-U, SF-M, and SF-T aerosol models calculated by Shettle and Fenn (1979) for decreasing relative humidity RH = 50% and overall particle number density $N_T = 10^3$ cm^{-3}.

Optical and physical parameters	SF-R (rural)	SF-U (urban)	SF-M (maritime)	SF-T (tropospheric)
$\beta_{ext}(0.55\ \mu m)$ (km^{-1})	1.009×10^{-2}	9.240×10^{-3}	2.171×10^{-2}	9.160×10^{-3}
$\beta_{sca}(0.55\ \mu m)$ (km^{-1})	9.520×10^{-3}	5.995×10^{-3}	2.136×10^{-2}	8.798×10^{-3}
$\beta_{abs}(0.55\ \mu m)$ (km^{-1})	5.787×10^{-4}	3.245×10^{-3}	3.575×10^{-4}	3.616×10^{-4}
$\alpha(0.40-0.86\ \mu m)$	1.193	1.050	0.028	1.355
$\omega(0.55\ \mu m)$	0.943	0.648	0.984	0.961
ω	0.914	0.585	0.996	0.904
$g(0.55\ \mu m)$	0.653	0.609	0.692	0.640
ρ_w (g cm^{-3})	1.778	2.677	1.878	1.408

6.2.6
The Seven Additional Aerosol Models of Tomasi *et al.* (2013)

Seven aerosol models were defined by Tomasi *et al.* (2013) to represent (i) a pair of multimodal aerosol polydispersions of Saharan desert dust, which were defined by examining some dust samples collected by Tomasi, Prodi, and Tampieri (1979) in Northern Italy; (ii) two polydispersions of BBS particles sampled by Carr (2005) at Jabiru (Australia), the first in the free troposphere (FT) and the second within the atmospheric boundary layer (BL); (iii) a particle polydispersion present in the low stratosphere during a volcanic quiescence period before the Mt Pinatubo (Luzon, Philippines) volcanic eruption occurred on 15 June 1991; and (iv) two stratospheric volcanic particle polydispersions sampled during two post-Pinatubo airborne campaigns conducted in two different regions and at different altitudes (Pueschel *et al.*, 1993; Tomasi, Vitale, and Tarozzi, 1997):

1. The Saharan dust (SD-1) model is based on the first of two size spectra measured by Tomasi, Prodi, and Tampieri (1979) at Sestola (Italy) on 18–19 May 1977, by examining a particle sample deposited by rain during a desert dust transport episode from north Africa. The size-distribution curve was found to consist of three modes, which were represented in terms of the Deirmendjian (1969) modified gamma function and recently best-fitted by a triplet of lognormal curves by Tomasi *et al.* (2013) giving the trimodal overall particle number concentration $N_T = 10^3$ cm^{-3}. The best-fit values of the shape parameters of the three modes are given in Table 6.18, and the SD-1 trimodal size-distribution curve $N(a)$ is shown in Figure 6.9, consisting of a first mode of nuclei/accumulation particles with $a_c \approx 0.10\ \mu m$, a second mode

Table 6.18 Values of the shape parameters N_i, σ_g, and a_c of the unimodal size distributions and of the main physical and radiative parameters giving form to the seven additional aerosol models proposed in the present study: particle mass density ρ (g cm^{-3}), volume extinction coefficient β_{ext}(0.55 μm), volume scattering coefficient β_{sca}(0.55 μm), and volume absorption coefficient β_{abs}(0.55 μm), all measured in per kilometer and obtained for overall number concentration $N_o = 10^3$ cm^{-3}; average single scattering albedo ω, and mean Ångström's exponent α determined over the 0.40–0.86 μm wavelength range.

Aerosol models	Shape parameters of the lognormal unimodal distributions									ρ (g cm^{-3})	β_{ext} (0.55 μm)	β_{sca} (0.55 μm)	β_{abs} (0.55 μm)	ω (0.55 μm)	ω	α
	First mode			Second mode			Third mode									
	N_1 (cm^{-3})	σ_g	a_c (μm)	N_2 (cm^{-3})	σ_g	a_c (μm)	N_3 (cm^{-3})	σ_g	a_c (μm)							
SD-1	940	2.00	0.102	59	1.50	1.108	1	1.40	3.070	2.60[a]	4.129×10^{-1}	3.438×10^{-1}	6.910×10^{-2}	0.833	0.881	−0.090
SD-2	733	1.78	0.240	261.5	2.12	0.898	5.5	1.40	4.036	2.60[a]	1.554	1.192	3.620×10^{-1}	0.767	0.819	−0.050
FT	1000	1.546	0.044	—	—	—	—	—	—	1.480[b]	6.357×10^{-5}	4.292×10^{-5}	2.065×10^{-5}	0.675	0.444	2.425
BL	404.218	1.533	0.101	594.886	1.591	0.131	0.896	1.644	1.210	1.485[b]	1.562×10^{-2}	1.484×10^{-2}	7.800×10^{-4}	0.950	0.942	1.546
PV-1	1000	1.80	0.140	—	—	—	—	—	—	1.658[c]	3.590×10^{-2}	3.584×10^{-2}	6.00×10^{-5}	0.998	0.895	1.688
PV-2	586.207	1.50	0.180	413.83	1.50	0.620	—	—	—	1.65[d]	5.724×10^{-1}	5.724×10^{-1}	—	1.00	0.923	0.186
PV-3	757.785	1.40	0.120	214.533	1.50	0.360	27.682	1.20	1.500	1.65[d]	1.923×10^{-1}	1.923×10^{-1}	—	1.00	0.913	0.054

a) Tomasi, Prodi, and Tampieri (1979).
b) Carr (2005).
c) Pueschel *et al.* (1989).
d) Hofmann and Rosen (1983a).

of accumulation/coarse particles with $a_c \approx 1.11\,\mu m$, and the third mode of coarse particles having $a_c \approx 3\,\mu m$, which gives form to a pronounced peak and a relevant right wing. The SD-1 dust model provides spectral values of $n(\lambda)$ calculated by Tomasi *et al.* (2013) at 23 selected wavelengths chosen over the 0.25–3.70 μm range on the basis of the data provided by Hänel (1968, 1972) for his aerosol model sampled over the Atlantic in April 1969 and containing Saharan dust, those given by Volz (1973) for Saharan dust samples collected over the Caribbean region, and those defined by Vermote *et al.* (1997b) for the 6S DL component. The spectral values of $\chi(\lambda)$ were determined at the same 23 wavelengths according to the estimates of Hänel (1968, 1972), Volz (1973), Patterson (1977), and Patterson, Gillette, and Stockton (1977) for Saharan dust particles. The spectral values of parameters $n(\lambda)$, $\chi(\lambda)$, $\beta_{ext}(\lambda)$, $\beta_{sca}(\lambda)$, $\omega(\lambda)$, and $g(\lambda)$ obtained at the 11 previously selected wavelengths for the SD-1 model are given in Table 6.19, showing that $n(\lambda)$ decreases from 1.62 to 1.46 over the $0.30 \leq \lambda \leq 3.75\,\mu m$ range, while $\chi(\lambda)$ varies between 1.5×10^{-2} at $\lambda = 0.30\,\mu m$ and 5.3×10^{-3} at $\lambda = 3.75\,\mu m$. Coefficients $\beta_{ext}(\lambda)$ and $\beta_{sca}(\lambda)$ were estimated to slowly increase as λ increases from 0.30 to 0.86 μm, giving a negative value of exponent $\alpha = -0.09$, due to the predominant extinction effects by accumulation and coarse particles. Parameter $\omega(\lambda)$ is equal to 0.74 at $\lambda = 0.30\,\mu m$ and increases with wavelength until reaching a value higher than 0.96 at $\lambda = 3.75\,\mu m$, while $g(\lambda)$ decreases from 0.78 to 0.66 as λ increases from 0.30 to 3.75 μm. The corresponding values of coefficients $\beta_{ext}(\lambda)$, $\beta_{sca}(\lambda)$, and $\beta_{abs}(\lambda)$ at the 0.55 μm wavelength are given in Table 6.18, together with the average single scattering albedo $\omega = 0.88$ and the value of mass density $\rho = 2.60\,g\,cm^{-3}$, obtained according to Hänel (1968, 1972) and in good agreement with the Volz (1973) estimate of $2.50\,g\,cm^{-3}$.

2. The Saharan dust (SD-2) model is based on the size-distribution curve determined by Tomasi, Prodi, and Tampieri (1979) analyzing the second particle sample deposited by rain at Sestola on 18–19 May 1977. These experimental data were best-fitted by Tomasi *et al.* (2013) obtaining a triplet of lognormal curves having the shape parameters reported in Table 6.18, which give form to the SD-2 size-distribution curve normalized to $N_T = 10^3\,cm^{-3}$ and shown in Figure 6.9. The comparison between the SD-1 and SD-2 multimodal curves clearly indicates that the SD-2 model exhibits more limited contents of nuclei and small-size accumulation particles and considerably more marked contents of large-size accumulation and coarse particles, presenting three modes centered at diameters a_c close to 0.24, 0.90, and 4 μm, the third mode yielding a right wing with considerably higher concentration values than those of the SD-1 model. The spectral values of $n(\lambda)$, $\chi(\lambda)$, $\beta_{ext}(\lambda)$, $\beta_{sca}(\lambda)$, $\omega(\lambda)$, and $g(\lambda)$ obtained for the SD-2 model are given in Table 6.19 at the 11 selected wavelengths. The values of $n(\lambda)$ and $\chi(\lambda)$ were assumed to be equal to those defined for the SD-1 aerosol model. Coefficients $\beta_{ext}(\lambda)$ and $\beta_{sca}(\lambda)$ were estimated to slowly increase as λ increases from 0.30 to 1.536 μm, providing a slightly negative best-fit value of $\alpha = -0.05$. Correspondingly, $\omega(\lambda)$ was estimated to increase from 0.67 to 0.96 over the $0.30 \leq \lambda \leq 3.75\,\mu m$

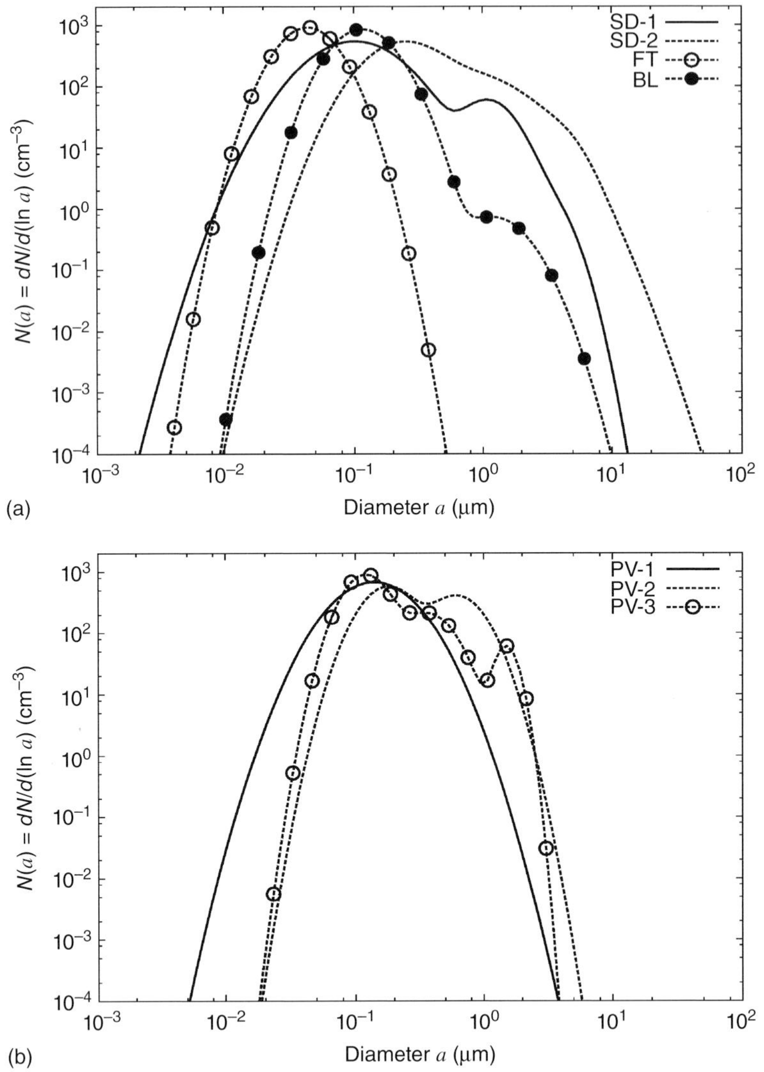

Figure 6.9 (a) Size-distribution curves of particle number density $N(r)$ over the $10^{-3} \leq a \leq 10^{2}$ μm diameter range for (i) Saharan dust model SD-1 (solid curve), (ii) Saharan dust model SD-2 (dashed curve), (iii) BBS model for particles suspended in the free troposphere (FT) (dashed curve with open circles), and (iv) BBS model for particles suspended in the boundary layer (BL) (dashed curve with solid circles). (b) As in the (a), for (i) model PV-1 representing the background stratospheric aerosol during a volcanic quiescence period in Antarctica (solid curve), (ii) model PV-2 representing the 2-month aged stratospheric volcanic aerosol after the Pinatubo volcanic eruption (dashed curve), and (iii) model PV-3 representing the 9-month aged stratospheric volcanic aerosol after the Pinatubo volcanic eruption (dashed curve with open circles). All the seven aerosol models are normalized to give the overall particle number density $N_T = 10^{3}$ cm^{-3}.

Table 6.19 Spectral values of real part $n(\lambda)$, imaginary part $\chi(\lambda)$, volume extinction coefficient $\beta_{ext}(\lambda)$, volume scattering coefficient $\beta_{sca}(\lambda)$ (both measured in per kilometer and calculated for overall number density concentration $N_T = 10^3$ cm^{-3}), single scattering albedo $\omega(\lambda)$, and asymmetry factor $g(\lambda)$, as defined by us at 11 selected wavelengths from 0.30 to 3.75 μm for the seven additional models of wet-air (RH = 50%) aerosol particles determined by Tomasi *et al.* (2013).

λ (μm)	$n(\lambda)$	$\chi(\lambda)$	$\beta_{ext}(\lambda)$ (km^{-1})	$\beta_{sca}(\lambda)$ (km^{-1})	$\omega(\lambda)$	$g(\lambda)$
First Saharan dust (SD-1) trimodal model						
0.300	1.620	1.50×10^{-2}	4.01×10^{-1}	2.96×10^{-1}	0.740	0.783
0.400	1.580	1.32×10^{-2}	4.06×10^{-1}	3.15×10^{-1}	0.777	0.755
0.488	1.572	1.21×10^{-2}	4.10×10^{-1}	3.32×10^{-1}	0.810	0.730
0.515	1.567	1.12×10^{-2}	4.11×10^{-1}	3.39×10^{-1}	0.825	0.722
0.550	1.560	1.12×10^{-2}	4.13×10^{-1}	3.44×10^{-1}	0.833	0.717
0.633	1.556	1.07×10^{-2}	4.18×10^{-1}	3.56×10^{-1}	0.853	0.703
0.694	1.551	1.10×10^{-2}	4.23×10^{-1}	3.64×10^{-1}	0.860	0.699
0.860	1.550	1.25×10^{-2}	4.35×10^{-1}	3.79×10^{-1}	0.870	0.695
1.536	1.545	6.10×10^{-3}	4.18×10^{-1}	3.99×10^{-1}	0.955	0.681
2.250	1.524	5.80×10^{-3}	3.18×10^{-1}	3.06×10^{-1}	0.963	0.671
3.750	1.460	5.30×10^{-3}	1.50×10^{-1}	1.44×10^{-1}	0.963	0.656
Second Saharan dust (SD-2) trimodal model						
0.300	1.620	1.50×10^{-2}	1.504	1.009	0.672	0.829
0.400	1.580	1.32×10^{-2}	1.526	1.083	0.710	0.807
0.488	1.572	1.21×10^{-2}	1.545	1.148	0.743	0.788
0.515	1.567	1.12×10^{-2}	1.549	1.176	0.759	0.781
0.550	1.560	1.12×10^{-2}	1.554	1.192	0.767	0.777
0.633	1.556	1.07×10^{-2}	1.564	1.232	0.788	0.764
0.694	1.551	1.10×10^{-2}	1.571	1.249	0.795	0.758
0.860	1.550	1.25×10^{-2}	1.587	1.272	0.802	0.745
1.536	1.545	6.10×10^{-3}	1.619	1.491	0.921	0.689
2.250	1.524	5.80×10^{-3}	1.557	1.469	0.943	0.685
3.750	1.460	5.30×10^{-3}	1.234	1.188	0.963	0.717
Biomass burning smoke (FT) unimodal model						
0.300	1.524	9.45×10^{-3}	2.49×10^{-4}	2.12×10^{-4}	0.905	0.322
0.400	1.524	9.45×10^{-3}	1.64×10^{-4}	1.35×10^{-4}	0.824	0.250
0.488	1.524	9.35×10^{-3}	8.87×10^{-5}	6.67×10^{-5}	0.752	0.186
0.515	1.524	9.35×10^{-3}	7.53×10^{-5}	5.48×10^{-5}	0.727	0.171
0.550	1.524	1.02×10^{-2}	6.36×10^{-5}	4.29×10^{-5}	0.675	0.154
0.633	1.524	1.00×10^{-2}	4.24×10^{-5}	2.53×10^{-5}	0.597	0.122
0.694	1.524	1.10×10^{-2}	3.46×10^{-5}	1.78×10^{-5}	0.513	0.103
0.860	1.523	1.58×10^{-2}	2.66×10^{-5}	7.67×10^{-6}	0.288	0.0700
1.536	1.499	2.55×10^{-2}	1.74×10^{-5}	7.03×10^{-7}	0.0404	0.0226
2.250	1.409	1.44×10^{-2}	6.89×10^{-6}	1.08×10^{-7}	0.0156	0.0101
3.750	1.445	1.00×10^{-2}	2.75×10^{-6}	1.62×10^{-8}	0.0059	0.0037
Biomass burning smoke (BL) trimodal model						
0.300	1.563	4.75×10^{-3}	3.40×10^{-2}	3.27×10^{-2}	0.964	0.663
0.400	1.563	4.75×10^{-3}	2.61×10^{-2}	2.50×10^{-2}	0.958	0.631

(*continued overleaf*)

Table 6.19 (*Continued*).

λ (μm)	$n(\lambda)$	$\chi(\lambda)$	$\beta_{ext}(\lambda)$ (km^{-1})	$\beta_{sca}(\lambda)$ (km^{-1})	$\omega(\lambda)$	$g(\lambda)$
0.488	1.566	4.78×10^{-3}	1.90×10^{-2}	1.81×10^{-2}	0.953	0.603
0.515	1.566	4.75×10^{-3}	1.74×10^{-2}	1.66×10^{-2}	0.952	0.595
0.550	1.566	4.72×10^{-3}	1.56×10^{-2}	1.48×10^{-2}	0.950	0.586
0.633	1.565	4.80×10^{-3}	1.24×10^{-2}	1.17×10^{-2}	0.945	0.568
0.694	1.565	4.80×10^{-3}	1.08×10^{-2}	1.01×10^{-2}	0.942	0.560
0.860	1.556	4.99×10^{-3}	8.09×10^{-3}	7.57×10^{-3}	0.936	0.561
1.536	1.494	5.55×10^{-3}	5.56×10^{-3}	5.26×10^{-3}	0.946	0.687
2.250	1.371	6.40×10^{-3}	3.46×10^{-3}	3.25×10^{-3}	0.939	0.760
3.750	1.420	1.09×10^{-2}	2.05×10^{-3}	1.85×10^{-3}	0.901	0.676
Background stratospheric (PV-1) aerosol unimodal model						
0.300	1.455	2.00×10^{-4}	7.15×10^{-2}	7.14×10^{-2}	0.999	0.726
0.400	1.444	2.00×10^{-4}	5.62×10^{-2}	5.61×10^{-2}	0.999	0.710
0.488	1.434	2.00×10^{-4}	4.27×10^{-2}	4.26×10^{-2}	0.999	0.696
0.515	1.435	2.20×10^{-4}	3.96×10^{-2}	3.96×10^{-2}	0.999	0.690
0.550	1.435	2.40×10^{-4}	3.59×10^{-2}	3.58×10^{-2}	0.998	0.681
0.633	1.433	2.60×10^{-4}	2.83×10^{-2}	2.82×10^{-2}	0.998	0.661
0.694	1.433	2.80×10^{-4}	2.40×10^{-2}	2.39×10^{-2}	0.998	0.646
0.860	1.430	4.80×10^{-4}	1.54×10^{-2}	1.54×10^{-2}	0.996	0.604
1.536	1.407	1.10×10^{-3}	3.35×10^{-3}	3.29×10^{-2}	0.981	0.450
2.250	1.370	2.10×10^{-3}	9.48×10^{-4}	8.72×10^{-4}	0.919	0.328
3.750	1.398	1.30×10^{-1}	2.73×10^{-3}	1.84×10^{-4}	0.069	0.177
Stratospheric post-Pinatubo 2-month volcanic (PV-2) aerosol bimodal model						
0.300	1.452	1.07×10^{-8}	5.07×10^{-1}	5.07×10^{-1}	1.000	0.688
0.400	1.440	1.07×10^{-8}	5.40×10^{-1}	5.40×10^{-1}	1.000	0.713
0.488	1.432	1.07×10^{-8}	5.68×10^{-1}	5.68×10^{-1}	1.000	0.734
0.515	1.431	1.07×10^{-8}	5.71×10^{-1}	5.71×10^{-1}	1.000	0.739
0.550	1.431	1.07×10^{-8}	5.72×10^{-1}	5.72×10^{-1}	1.000	0.743
0.633	1.430	1.56×10^{-8}	5.61×10^{-1}	5.61×10^{-1}	1.000	0.750
0.694	1.429	2.21×10^{-8}	5.42×10^{-1}	5.42×10^{-1}	1.000	0.753
0.860	1.426	1.80×10^{-7}	4.67×10^{-1}	4.67×10^{-1}	1.000	0.749
1.536	1.403	1.37×10^{-5}	1.86×10^{-1}	1.85×10^{-1}	0.999	0.672
2.250	1.368	1.77×10^{-3}	6.60×10^{-2}	6.46×10^{-2}	0.979	0.554
3.750	1.396	1.31×10^{-1}	6.37×10^{-2}	1.64×10^{-2}	0.257	0.325
Stratospheric post-Pinatubo 9-month volcanic (PV-3) aerosol trimodal model						
0.300	1.452	1.07×10^{-8}	2.53×10^{-1}	2.53×10^{-1}	1.000	0.783
0.400	1.440	1.07×10^{-8}	2.28×10^{-1}	2.28×10^{-1}	1.000	0.742
0.488	1.432	1.07×10^{-8}	2.06×10^{-1}	2.06×10^{-1}	1.000	0.705
0.515	1.431	1.07×10^{-8}	1.99×10^{-1}	1.99×10^{-1}	1.000	0.692
0.550	1.431	1.07×10^{-8}	1.92×10^{-1}	1.92×10^{-1}	1.000	0.678
0.633	1.430	1.56×10^{-8}	1.89×10^{-1}	1.89×10^{-1}	1.000	0.671
0.694	1.429	2.21×10^{-8}	1.96×10^{-1}	1.96×10^{-1}	1.000	0.685
0.860	1.426	1.80×10^{-7}	2.20×10^{-1}	2.20×10^{-1}	1.000	0.736
1.536	1.403	1.37×10^{-4}	1.56×10^{-1}	1.56×10^{-1}	0.999	0.760
2.250	1.368	1.77×10^{-3}	6.79×10^{-2}	6.70×10^{-2}	0.986	0.685
3.750	1.396	1.31×10^{-1}	4.91×10^{-2}	1.84×10^{-2}	0.375	0.453

range, and $g(\lambda)$ to decrease from 0.83 to less than 0.70 over the same spectral range. The values of $\beta_{ext}(\lambda)$, $\beta_{sca}(\lambda)$, and $\beta_{abs}(\lambda)$ at $\lambda = 0.55$ μm are given in Table 6.18, together with the values of $\omega(0.55\ \mu m) = 0.77$ and $\omega = 0.82$, while the mass density was assumed to be that of the SD-1 model.

3. The BBS (FT) aerosol model was determined by Tomasi *et al.* (2013) examining the field data obtained by Carr (2005) analyzing some aerosol samples collected at Jabiru (Australia) in September 2003 within the FT for nearly dry-air conditions. Model FT was found to consist of fine particles only, including only one mode consisting of both nuclei and accumulation particles, giving form to the lognormal size-distribution curve shown in Figure 6.9 obtained in terms of Eq. (6.1) for $N_T = 10^3\ cm^{-3}$, $a_c = 4.4 \times 10^{-2}$ μm, and $\sigma_g = 1.546$, reported in Table 6.18. An average fine particle density of 1.48 g cm^{-3} was estimated by Carr (2005) for the FT particles. The spectral values of $n(\lambda)$ and $k(\lambda)$ were determined by Tomasi *et al.* (2013) assuming the mass fractions $M_{ws} = 92\%$ for water-soluble substances, $M_{md} = 5\%$ for mineral dust (containing mainly quartz and silicates), $M_{ss} = 1.9\%$ for sea salt, and $M_{so} = 1.1\%$ for soot substances, obtaining the values of $n(\lambda)$ and $\chi(\lambda)$ given in Table 6.19. Parameter $n(\lambda)$ was estimated to be equal to 1.524 over the visible and near-infrared range and $\chi(\lambda)$ to slightly increase from 9.45×10^{-3} at $\lambda = 0.30$ μm to 1.00×10^{-2} at $\lambda = 3.75$ μm. These values agree very well with the average values of $n(0.55\ \mu m) = 1.558$ and $\chi(0.55\ \mu m) = 1.06 \times 10^{-2}$ measured by Carr (2005). For the aforementioned data given in Tables 6.18 and 6.19, the spectral values of $\beta_{ext}(\lambda)$, $\beta_{sca}(\lambda)$, $\omega(\lambda)$, and $g(\lambda)$ were calculated obtaining the results reported in Table 6.19, showing that $\beta_{ext}(\lambda)$ decreases by about one order of magnitude over the 0.30–0.86 μm wavelength range providing a very high best-fit value of $\alpha = 2.4$, and $\beta_{sca}(\lambda)$ decreases by nearly two orders of magnitude over the 0.30–3.75 μm range, for which $\omega(\lambda)$ was found to decrease from ~0.90 to less than 0.30 as λ increases from 0.30 to 0.86 μm. Correspondingly, parameter $g(\lambda)$ was estimated to decrease from 0.32 to less than 0.10 over the whole spectral range.

4. The BBS (BL) aerosol model was determined by Tomasi *et al.* (2013) using the data collected by Carr (2005) at Jabiru (Australia) in September 2003 within the atmospheric BL, presenting nearly dry-air conditions. The BL model was assumed to consist of particles having an average particle density equal to 1.49 g cm^{-3}, due to the predominant contribution of coarse particles. It exhibits a trimodal size-distribution curve consisting of the nuclei, accumulation, and coarse particle modes presenting the three lognormal curves shown in Figure 6.9, each mode being represented in terms of Eq. (6.1) for the values of N_i, a_c, and σ_g given in Table 6.18. The spectral values of $n(\lambda)$ were determined by assuming an average chemical composition with mass fractions $M_{md} = 46\%$, $M_{ws} = 35\%$, $M_{ss} = 15\%$, and $M_{so} = 4\%$, for which the average value of $n(0.55\ \mu m)$ was found to be only slightly higher than 1.546, as measured by Carr (2005). The corresponding spectral values of $\chi(\lambda)$ were derived according to the previous mass fraction estimates, finding an average value of 4.75×10^{-3} in the visible. For the previous assumptions, $\beta_{ext}(\lambda)$ was

found to decrease from 3.4×10^{-2} km^{-1} at $\lambda=0.30$ μm to about 8×10^{-3} km^{-1} at $\lambda=0.86$ μm, giving a best-fit value of $\alpha=1.55$, and $\beta_{sca}(\lambda)$ to correspondingly decrease from ~3.3×10^{-2} km^{-1} at $\lambda=0.30$ μm to 1.8×10^{-3} km^{-1} at $\lambda=3.75$ μm. These calculations provided values of $\omega(\lambda)$ decreasing from 0.96 at $\lambda=0.30$ μm to about 0.90 at $\lambda=3.75$ μm and values of $g(\lambda)$ slowly decreasing from 0.66 to 0.56 over the 0.30–0.86 μm wavelength range and subsequently increasing at longer wavelengths until reaching a value of ~0.68 at $\lambda=3.75$ μm. The comparison made in Figure 6.9 between the FT and BL size-distribution curves gives evidence of the fact that model BL includes important number fractions of both accumulation and coarse particles, while model FT consists mainly of accumulation particles.

5. The BG stratospheric (PV-1) aerosol model was represented using a single unimodal lognormal size-distribution curve, which was measured by Pueschel *et al.* (1989) over the Antarctic continent in 1987 during a long volcanic quiescence period. Therefore, model PV-1 represents realistically the atmospheric turbidity conditions observed at low stratospheric levels before the Mt Pinatubo eruption in June 1991. The BG stratospheric aerosol was estimated to be composed of a 60–80% aqueous SA solution at temperatures ranging from about 190 to 230 K (Hofmann and Rosen, 1983a,b) and to be globally distributed, being mainly formed through oxidation of carbonyl sulfide (OCS), which is rapidly dissociated by UV solar radiation to form SA. OCS is the primary component of the natural stratospheric aerosol, since other sulfur-containing species emitted at the surface, such as SO_2, DMS, and CS_2, do not persist long enough in the upper troposphere and are not efficiently transported to the low stratosphere.[1)]

   The size distribution of model PV-1 based on the Pueschel *et al.* (1989) data was defined using the analytical form of Eq. (6.1) for the shape parameters defined in Table 6.18, obtaining the unimodal curve shown in Figure 6.9. The optical characteristics were determined by assuming a particulate chemical composition given by mass fractions of 72% SA, 24% LW, and 4% water-soluble (nitrate) substances, as estimated by Pueschel *et al.* (1989). Therefore, the spectral values of $n(\lambda)$ and $\chi(\lambda)$ were first calculated over the

1) A state of unperturbed background stratospheric aerosol is relatively rare. Only a few major volcanic eruptions were so violent in the past 50 years as to inject significant quantities of SO_2 directly into the lower and midstratosphere, as observed during the eruptions of Mt Agung (Bali, Indonesia) in February and May 1963, El Chichón (Mexico) in March and April 1982, and Mt Pinatubo (Luzon, Philippines) in June 1991. In particular, the Mt Pinatubo eruption was estimated to have furnished a stratospheric aerosol mass contribution of 30 Tg, which enhanced the stratospheric aerosol levels for at least 2 years. After this event, the sulfuric acid aerosol particles were gradually distributed over the global scale within a few months, causing optical density characteristics of the lower stratosphere totally hiding those of the natural background aerosol. The subsequent stratosphere's relaxation to background conditions was estimated to occur totally within a time of 2–3 years. Thus, assuming an average frequency of strong volcanic eruptions equal to 10–20 years, it can be realistically assume that the low stratospheric aerosol is seldom in a totally unperturbed state by volcanic emissions.

0.36–3.70 μm range for a 75% solution of SA using the values of $n(\lambda)$ given by Palmer and Williams (1975) over the entire spectral range and the values of $\chi(\lambda)$ given by Hummel, Shettle, and Longtin (1988) over the 0.36–0.70 μm range and Palmer and Williams (1975) over the 0.70–3.70 μm range. All the values were then reduced by Tomasi *et al.* (2013) to the 72% SA mass fraction and complementary mass fractions of 24% LW (using the Hale and Querry (1973) data) and 4% WS (using the 6S WS data of Vermote *et al.*, 1997b). Parameter $n(\lambda)$ was estimated to decrease from 1.455 at $\lambda = 0.30$ μm to 1.370 at $\lambda = 2.25$ μm, and $\chi(\lambda)$ to assume rather low values (equal to $\sim 2 \times 10^{-4}$) at all the visible and near-infrared wavelengths, which gradually increase in the infrared. For the size-distribution shape parameters given in Table 6.18 and the complex refractive index data given in Table 6.19, parameters $\beta_{ext}(\lambda)$, $\beta_{sca}(\lambda)$, $\omega(\lambda)$, and $g(\lambda)$ were calculated finding that $\beta_{ext}(\lambda)$ decreases from $\sim 7 \times 10^{-2}$ km^{-1} to less than 2×10^{-2} km^{-1} over the 0.30–0.86 μm wavelength range, yielding a very high best-fit value of α close to 1.70, and $\beta_{sca}(\lambda)$ decreases presenting spectral values which are only slightly lower than those of $\beta_{ext}(\lambda)$, due to the very weak absorption properties of these stratospheric particles. Consequently, $\omega(\lambda)$ was found to assume values very close to unit over the $0.30 \leq \lambda \leq 0.86$ μm range and equal to 0.98 and 0.92 at the 1.536 and 2.25 μm wavelengths, respectively. The spectral values of $g(\lambda)$ were found to decrease from 0.73 at $\lambda = 0.30$ μm to 0.60 at $\lambda = 0.86$ μm and to assume more steeply decreasing values at the longer infrared wavelengths.

6. The volcanic stratospheric (PV-2) aerosol model was assumed to have the bimodal size-distribution curve defined by Pueschel *et al.* (1993) from *in situ* sampling measurements performed at a stratospheric altitude of 16.5 km to represent the post-Pinatubo volcanic particle load formed 2 months after the Mt Pinatubo eruption. The shape parameters of the bimodal PV-2 size-distribution curve are given in Table 6.18, giving form to the overall number concentration size-distribution curve shown in Figure 6.9, which consists of two modes having analytical forms given by Eq. (6.1) for values of $a_c = 0.18$ μm and $a_c = 0.62$ μm, respectively, both modes consisting mostly of accumulation particles. The complex refractive index of these stratospheric particles was determined according to Pueschel *et al.* (1993) for particulate mass fractions of 75% SA and 25% LW. The spectral values of $n(\lambda)$ were calculated over the 0.30–3.75 μm wavelength range for an aqueous solution of 75% SA, according to Palmer and Williams (1975), and those of $\chi(\lambda)$ according to Hummel, Shettle, and Longtin (1988) for $\lambda \leq 0.70$ μm and Palmer and Williams (1975) for $\lambda > 0.70$ μm. For such assumptions, $n(\lambda)$ was estimated to decrease from 1.452 at $\lambda = 0.30$ μm to 1.368 at $\lambda = 2.25$ μm, and $\chi(\lambda)$ to assume very low values of $\sim 10^{-8}$ at all the visible and wavelengths and to slowly increase throughout the near-infrared spectral range. The spectral values of $\beta_{ext}(\lambda)$, $\beta_{sca}(\lambda)$, $\omega(\lambda)$, and $g(\lambda)$ are given in Table 6.19, showing that $\beta_{ext}(\lambda)$ and $\beta_{sca}(\lambda)$ are very similar from 0.30 to 1.54 μm, because of the pure scattering properties of these stratospheric droplets, and differ only slightly at the 2.25 and 3.75 μm wavelengths. Due to the fact that $\beta_{ext}(\lambda)$

slowly increases with wavelength, the best-fit value of $\alpha = 0.186$ was obtained, indicating that aerosol extinction is predominantly due to large accumulation particles. Correspondingly, $\omega(\lambda)$ assumes unit values at all the visible and infrared wavelengths and slowly decreases in the infrared until becoming equal to 0.26 at $\lambda = 3.75\,\mu m$. Asymmetry factor $g(\lambda)$ was estimated to increase from 0.69 to 0.75 over the 0.30–0.86 μm wavelength range and to decrease considerably at the longer wavelengths.

7. The volcanic stratospheric (PV-3) aerosol model was determined by Tomasi *et al.* (2013) for a linear combination of three modes represented by the lognormal size-distribution curves derived by Pueschel *et al.* (1993) from *in situ* airborne sampling measurements conducted at stratospheric altitudes of around 12.5 km, about 9 months after the Mt Pinatubo eruption. The shape parameters of the three lognormal modes are given in Table 6.18, and the trimodal size-distribution curve of model PV-3 is shown in Figure 6.9, consisting of three modes centered at $a_c = 0.12\,\mu m$, $a_c = 0.36\,\mu m$, and $a_c = 1.50\,\mu m$. It can be noted that PV-3 model contains a predominant content of accumulation particles together with a consistent number fraction of accumulation/coarse particles causing a narrow and marked peak between 1 and 2 μm diameters. The complex refractive index of model PV-3 was defined over the 0.30–3.75 μm range using the Palmer and Williams (1975) and Hummel, Shettle, and Longtin (1988) data sets. Therefore, the spectral values of $n(\lambda)$ and $\chi(\lambda)$ were assumed to be equal to those of model PV-2. The spectral values of $\beta_{ext}(\lambda)$ and $\beta_{sca}(\lambda)$ are given in Table 6.19. They are very similar over the whole 0.30–1.54 μm wavelength range and present only small discrepancies at the 2.25 and 3.75 μm wavelengths. Coefficient $\beta_{ext}(\lambda)$ was estimated to slowly decrease with wavelength over the $0.30 \leq \lambda \leq 0.63\,\mu m$ range, providing a best-fit value of $\alpha = 0.05$. Unit values of $\omega(\lambda)$ were obtained at all the visible and infrared wavelengths, which subsequently decrease to less than 0.38 at $\lambda = 3.75\,\mu m$. Parameter $g(\lambda)$ was estimated to decrease from 0.78 to 0.67 over the 0.30–0.63 μm wavelength range and to increase at the longer wavelengths until assuming a value of 0.76 at $\lambda = 1.54\,\mu m$.

The comparison of the PV-1, PV-2, and PV-3 size-distribution curves made in Figure 6.9 clearly shows that the predominant contents of the PV-2 and PV-3 stratospheric aerosols are given by accumulation and small-size coarse particles, associated with the formation of volcanic solution droplets, while the PV-1 size-distribution curve of BG stratospheric particles mostly consists of nuclei and accumulation particles having sizes mainly ranging from 10^{-2} to $2 \times 10^{-1}\,\mu m$.

6.2.7 The Polar Aerosol Models

A large set of ground-based and shipborne sun-photometer measurements was examined by Tomasi *et al.* (2015a) to characterize the spectral features of AOT measured over the visible and near-infrared wavelength range at numerous

Arctic and Antarctic stations. In addition, the main shape characteristics of the fine, accumulation, and coarse particle modes and their chemical composition derived from *in situ* measurements were taken into account to define the average multimodal characteristics of the aerosol size-distribution curves and the average values of parameters β_{ext}, β_{sca}, $\omega(\lambda)$, and $g(\lambda)$. On the basis of numerous *in situ* measurements conducted in the most remote polar regions over the last decades, the following seven polar aerosol models were defined by Tomasi *et al.* (2015b) by combining the following pairs of lognormal curves both having the analytical form given in Eq. (6.1):

1. The winter–spring Arctic haze (AH) aerosol bimodal model, consisting of a total fine particle content $N_f = 9.25 \times 10^9$ cm^{-2} and an accumulation/coarse particle content $N_{a/c} = 1.75 \times 10^3$ cm^{-2} in the vertical atmospheric column of unit cross section, giving average values of AOT = 0.12 at wavelength $\lambda = 0.50$ μm and $\alpha = 1.48$, which closely agree with those obtained from the ground-based sun-photometer measurements conducted in winter and spring at various Arctic sites.
2. The Arctic summer background (ASB) aerosol bimodal model, with $N_f = 1.58 \times 10^9$ cm^{-2} and $N_{a/c} = 3.52 \times 10^0$ cm^{-2}, obtained for AOT(0.50 μm) = 0.08 and $\alpha = 1.60$.
3. The Asian dust (AD) bimodal model with $N_f = 1.25 \times 10^7$ cm^{-2} and $N_{a/c} = 1.13 \times 10^2$ cm^{-2} obtained for AOT(0.50 μm) = 0.22 and $\alpha = 0.26$, as measured at Barrow during some AD transport episodes occurred in spring 2002 (Stone *et al.*, 2007).
4. The boreal forest fire smoke (BFFS) particle bimodal model with $N_f = 1.24 \times 10^9$ cm^{-2} and $N_{a/c} = 1.03 \times 10^5$ cm^{-2}, giving AOT(0.50 μm) = 0.22 and $\alpha = 1.50$ and presenting spectral features similar to those observed by Mielonen *et al.* (2013) in 2010 at some Northern Finland sites during wildfire particle transport episodes from Russian regions, characterized by values of $\omega \approx 0.93$.
5. The Antarctic austral summer coastal (AASC) aerosol bimodal model with $N_f = 6.15 \times 10^6$ cm^{-2} and $N_{a/c} = 4.37 \times 10^0$ cm^{-2}, measured at Neumayer and found to give AOT(0.50 μm) = 0.045 and $\alpha = 0.78$.
6. The austral summer Antarctic Plateau (ASAP) aerosol bimodal model with $N_f = 1.27 \times 10^9$ cm^{-2} and $N_{a/c} = 3.83 \times 10^{-1}$ cm^{-2}, measured at South Pole with average values of AOT(0.50 μm) = 0.018 and $\alpha = 1.49$.
7. The Antarctic austral winter coastal (AAWC) aerosol bimodal model with $N_f = 3.73 \times 10^6$ cm^{-2} and $N_{a/c} = 2.62 \times 10^2$ cm^{-2}, measured at Neumayer on austral winter days for AOT(0.50 μm) = 0.035 and $\alpha = 0.65$.

The previous values of particle number contents N_f and $N_{a/c}$ were normalized to give an overall particle number concentration $N_T = 10^3$ cm^{-3} for all the seven aerosol models, obtaining the pairs of first mode concentration N_1 of nuclei and second mode concentration N_2 of accumulation/coarse particles given in

Table 6.20 for all the models. The values of σ_g and a_c are given in Table 6.20 for the various modes of the seven bimodal size-distribution curves shown in Figure 6.10.

Taking into account the field measurements of particulate chemical composition made at the various Arctic and Antarctic sites for the submicron and supermicron aerosols (Tomasi *et al.*, 2012), the optical characteristics of the nuclei and accumulation/coarse particles were defined for the unimodal mass percentages of the four 6S basic components DL, WS, OC, and SO given in Table 6.20 for dry-air (RH = 0%) conditions. The unimodal mass densities of the nuclei and accumulation/coarse particle modes were calculated for the unimodal mass percentages and the unimodal particulate mass concentrations Γ_1 and Γ_2 of the two modes given in Table 6.20, obtaining the values of average mass density ρ_1 for nuclei and average mass density ρ_2 for accumulation/coarse particles provided in Table 6.20. The values of ρ_1 and ρ_2 differ appreciably from one mode to the other by less than 1% for the BFFS modes to nearly 17% for the AAWC modes. The average mass percentages of the 6S basic components determined for the pairs of modes were used as weights for each bimodal model to calculate the spectral values of $n(\lambda)$, $\chi(\lambda)$, $\beta_{ext}(\lambda)$, $\beta_{sca}(\lambda)$, $\omega(\lambda)$, and $g(\lambda)$ at the 11 previously selected wavelengths. The results are given in Table 6.20 for the AH, ASB, AD, and BFFS models of dry-air Arctic aerosol particles and for the AASC, ASAP, and AAWC dry-air bimodal models of Antarctic aerosols. The results indicate that the real part $n(\lambda)$ of refractive index for the Arctic aerosols were found to decrease as a function of wavelength, ranging from 1.487 (AH model) to 1.533 (BFFS model) at $\lambda = 0.55\ \mu m$, and those of $\chi(\lambda)$ varied between 7×10^{-3} (AD model) and 2.2×10^{-2} (AH model). With regard to the Antarctic aerosols, the values of $n(\lambda)$ were estimated to decrease slowly with λ, and vary from 1.453 (AASC model) to 1.526 (AAWC model) at $\lambda = 0.55\ \mu m$, and those of $\chi(\lambda)$ from 5×10^{-3} (AASC model) to more than 6×10^{-3} (ASAP model). Coefficient $\beta_{ext}(\lambda)$ was found to decrease more or less rapidly with λ, giving a set of best-fit values of exponent α ranging between ~1.00 (AASC model) and 1.87 (ASB model). The values of $\beta_{sca}(\lambda)$ were estimated to be appreciably lower than those of $\beta_{ext}(\lambda)$, providing values of $\omega(\lambda) > 0.90$ at all the visible and near-infrared wavelengths for the ASB and AD Arctic aerosol models, while lower values of $\omega(\lambda)$ were in general found, decreasing from 0.88 to 0.77 for the AH model over the $0.30 \leq \lambda \leq 1.54\ \mu m$ range and from 0.91 to 0.73 for the BFFS model, due to the higher mass contents of soot substances. Parameter $g(\lambda)$ was found to slowly decrease with wavelength for all the seven polar aerosol models, varying on average between 0.61 (ASB) and 0.77 (AAWC) at $\lambda = 0.55\ \mu m$.

6.3
General Remarks on the Aerosol Particle Number, Surface, and Volume Size-Distribution Functions

Airborne aerosol particles contain a variety of substances (sulfate, nitrate, ammonium, OM, crustal matter, sea salt, metal oxides, hydrogen ions, and

Table 6.20 Values of the shape parameters N_j, σ_g, and a_m used to define the bimodal number size-distribution curves; mass percentages of the four basic 6S dry-air dustlike (DL), water-soluble (WS), oceanic (OC), and soot (SO) components adopted to define the complex refractive index of the lognormal curves of the first fine-mode and the second accumulation/coarse-mode particles, giving form to the four dry-air (RH = 0%) Arctic aerosol models and the three dry-air (RH = 0%) Antarctic aerosol models; unimodal mass densities ρ_1 and ρ_2 of the two modes; unimodal particle volumes V_1 and V_2 of the two modes; and unimodal particulate mass contents Γ_1 and Γ_2 of the two modes for the polar aerosol models AH (winter–spring Arctic haze aerosol), ASB (Arctic summer background aerosol), AD (Asian dust aerosol), BFFS (boreal forest fire smoke aerosol), AASC (Antarctic austral summer coastal aerosol), ASAP (austral summer Antarctic Plateau aerosol), and AAWC (Antarctic austral winter coastal aerosol).

Aerosol models	First mode			Second mode			6S basic component mass percentages of the first particle mode				6S basic component mass percentages of the second particle mode				ρ_1 (g cm^{-3})	ρ_2 (g cm^{-3})	V_1 (μm^{3}cm^{-3})	V_2 (μm^{3}cm^{-3})	Γ_1 (g cm^{-3})	Γ_2 (g cm^{-3})
	N_1 (cm^{-3})	σ_g	a_m (μm)	N_2 (cm^{-3})	σ_g	a_m (μm)	OC	WS	DL	SO	OC	WS	DL	SO						
Arctic haze (AH)	$10^3 - N_2$	2.24	4.2×10^{-2}	1.89×10^{-4}	2.03	6.0×10^{-1}	35	39	22	4	58	4	38	—	2.10	2.276	6.99×10^{6}	4.42×10^{1}	1.47×10^{-5}	1.01×10^{-10}
Arctic summer background aerosol (ASB)	$10^3 - N_2$	1.95	7.0×10^{-2}	2.23×10^{-6}	2.03	3.50	18	35	45.8	1.2	71	1	28	—	2.16	2.277	1.36×10^{6}	8.71	2.93×10^{-6}	1.98×10^{-11}
Asian dust (AD)	$10^3 - N_2$	1.95	1.4×10^{-1}	9.04×10^{-3}	2.15	2.60	—	24	76	—	4	6	90	—	2.24	2.326	6.74×10^{6}	3.55×10^{4}	1.51×10^{-5}	8.25×10^{-8}
Boreal forest fire smoke (BFFS)	$10^3 - N_2$	2.00	7.8×10^{-2}	8.31×10^{-2}	2.00	2.4×10^{-1}	—	—	97.9	2.1	—	—	98.4	1.6	2.34	2.348	2.92×10^{6}	2.73×10^{3}	6.84×10^{-6}	6.41×10^{-9}

(continued overleaf)

Table 6.20 (*Continued*)

Aerosol models	First mode			Second mode			6S basic component mass percentages of the first particle mode				6S basic component mass percentages of the second particle mode				ρ_1 (g cm^{-3})	ρ_2 (g cm^{-3})	V_1 (μm^{3}cm^{-3})	V_2 (μm^{3}cm^{-3})	Γ_1 (g cm^{-3})	Γ_2 (g cm^{-3})
	N_1 (cm^{-3})	σ_g	a_m (μm)	N_2 (cm^{-3})	σ_g	a_m (μm)	OC	WS	DL	SO	OC	WS	DL	SO						
Antarctic summer coastal aerosol (AASC)	$10^3 - N_2$	2.03	1.4×10^{-1}	7.11×10^{-4}	2.10	3.00	52	40	7.5	0.5	61	16.5	22.5	—	2.10	2.210	1.38×10^7	2.86×10^3	2.89×10^{-5}	6.32×10^{-9}
Austral summer Antarctic Plateau aerosol (ASAP)	$10^3 - N_2$	2.24	4.2×10^{-2}	3.02×10^{-7}	2.03	3.50	11.2	86.4	2.2	0.2	86	6	8	—	1.91	2.235	6.99×10^6	1.179	1.34×10^{-5}	2.64×10^{-12}
Antarctic austral winter coastal aerosol (AAWC)	$10^3 - N_2$	2.03	1.4×10^{-1}	7.02×10^{-2}	2.03	1.00	—	90	10	—	95	—	5	—	1.91	2.256	1.38×10^7	3.98×10^4	2.63×10^{-5}	8.98×10^{-8}

The unimodal values of N_j are calculated for overall particle number concentration $N_T = 10^3$ cm^{-3} for all the seven aerosol models.

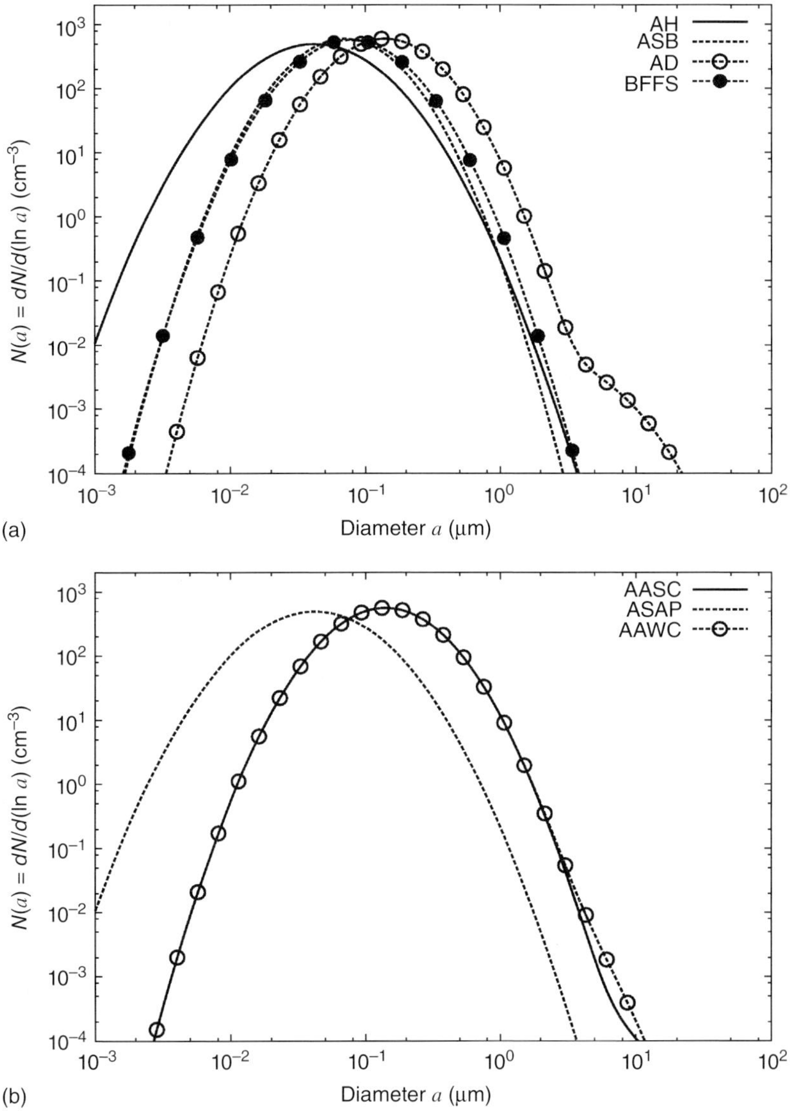

Figure 6.10 (a) Size-distribution curves of particle number density $N(r)$ over the $10^{-3} \leq a \leq 10^2$ μm diameter range for the following four Arctic aerosol models: (i) model AH, representing the winter–spring Arctic haze aerosol (solid curve), (ii) model ASB, representing the Arctic summer background aerosol (dashed curve), (iii) model AD, representing the Asian dust transported over the Arctic (dashed curve with open circles), and (iv) model BFFS, representing the boreal forest fire smoke particles (dashed curve with solid circles), determined by Tomasi *et al.* (2015a,b). (b) As in the (a), for the following three Antarctic aerosol models: (v) model AASC, representing the Antarctic austral summer coastal aerosol (solid curve), (vi) model ASAP, representing the austral summer aerosol on the Antarctic Plateau (dashed curve), and (vii) model AAWC, representing the Antarctic austral winter coastal aerosol (dashed curve with open circles), determined by Tomasi *et al.* (2015a,b). All the aerosol models are normalized to give the overall particle number density $N_T = 10^3$ cm^{-3}.

water), of which (i) sulfates, ammonium, organic and elemental carbon (EC), and transition metals predominantly constitute fine particles ($a < 2\,\mu m$); (ii) crustal materials, including Si, Ca, Mg, Al, and Fe, and biogenic organic particles (pollen, spores, plant fragments) are the usual constituents of the coarse particle mode (for $a > 2.5\,\mu m$); and (iii) nitrate can be found in both accumulation and coarse particle components, as a result of reactions between nitric acid and ammonia, producing ammonium nitrate, or between nitric acid and already existing coarse particles. The formation pathways of fine particles are chemical reactions, nucleation, condensation, coagulation, and cloud/fog formation processes, their main sources being the combustion of cool, oil, gasoline, diesel, and wood, the g-to-p conversion of NO_x, SO_3, and VOCs, and smelters and mills. The main sources of coarse particles are the mechanical disruption and suspension processes of dust particles mobilized by strong winds, as well as wind forcing over the OC surface generating sea spray, which dominate in the marine regions. Additional sources of coarse particles are (i) resuspension of industrial dust; (ii) suspension of soil, due to farming, mining, and unpaved roads; (iii) biological processes; and (iv) construction/demolition activities of buildings (Seinfeld and Pandis, 2006). Therefore, fine particles are in general largely soluble and hygroscopic, being subject to significant growth by condensation, while coarse particles are differently composed as they are transported from one area or another, since such particles may include OC sea salt, resuspended dust, coal and oil fly ash, crustal elements (Si, Al, Ti, and Fe oxides and $CaCO_3$), plant pollens, spores, animal debris, and tire wear debris, being mainly insoluble.

Aerosol extinction effects depend closely on the Mie parameter $x = \pi a / \lambda$. This implies that accumulation and coarse particles having sizes ranging from 10^{-1} to $10\,\mu m$ contribute more efficiently to create the atmospheric turbidity conditions, whereas condensation nuclei produce relatively weak optical effects despite their relatively high number concentration. The larger part of the total aerosol mass consists of aerosols having sizes ranging from 10^{-1} to more than $50\,\mu m$, and only a fraction of 10–20% is given by condensation nuclei. Due to their large sizes, the atmospheric residence time of coarse particles varies usually from minutes to days, and their travel distance does not exceed some tens of kilometers. Conversely, because of their different sizes and mass density characteristics, the travel distance of fine particles may range from hundreds to thousands of kilometers, having consequently an atmospheric lifetime varying from a few days to weeks, which closely depends on their morphological characteristics.

6.3.1 The Aerosol Particle Number Size-Distribution Function

The various emission and growth processes provide different contributions to the overall particle concentration, giving form to multimodal size-distribution curves of particles often characterized by at least two fine particle modes and in general only one coarse particle mode. An example of the multimodal characteristics of an aerosol size-distribution curve consisting of particles originated from

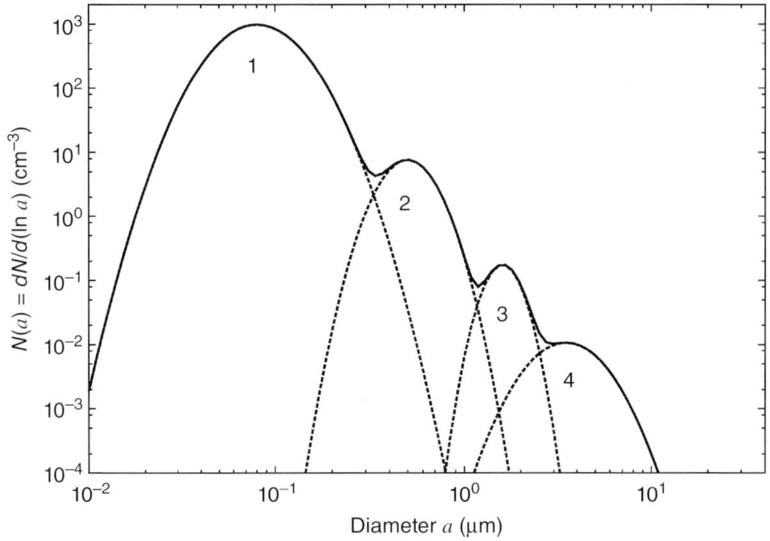

Figure 6.11 Example of a four-modal best-fit aerosol number density lognormal size-distribution curves of background continental aerosols, determined from the multimodal size-distribution curve represented by Tomasi and Tampieri (1977) over the $10^{-2} \leq a \leq 50$ µm diameter range as a combination of four modes of particles sampled at altitudes of 6–7 km over continental Russia.

different sources is shown in Figure 6.11, presenting the size-distribution curve of continental aerosols obtained by Tomasi and Tampieri (1977) for an airborne aerosol sample collected at altitudes of 6–7 km over the central part of Russia.

It can be seen that the overall size-distribution curve consists of four modes: (i) the first given by fine particles with diameters a ranging from 2×10^{-2} to 0.4 µm and peak number concentration higher than 10^3 cm^{-3} µm^{-1}; (ii) the second mode of accumulation particles occupies the $0.4 \leq a \leq 1.4$ µm range and has a peak concentration of $\sim 10^1$ cm^{-3} µm^{-1}; (ii) the third mode of accumulation/coarse particles exhibits sizes ranging from 1.4 to 2.4 µm and a peak concentration of $\sim 2 \times 10^{-1}$ cm^{-3} µm^{-1}; and (iv) the fourth mode consists of coarse particles with $a > 2.4$ µm and peak number concentration close to 3×10^{-2} cm^{-3} µm^{-1}. This multimodal curve is given by contributions from different sources, like BBS generated by forest fires, mobilization and transport of dust from remote desert regions, and anthropogenic emissions from urban and industrial activities.

Aerosol particle number concentration is often equal to $10^7 - 10^8$ cm^{-3} in urban areas, with equivalent diameters varying by more than four orders of magnitude, from a few nm to nearly 10^2 µm. Thus, the mass of a particle with diameter $a = 10$ µm is equivalent to the mass of 1 billion of particles with $a = 10$ nm, and a particle with $a = 10^{-2}$ µm contains $\sim 10^4$ molecules, while a particle with $a = 1$ µm contains about 10^{10} molecules. Airborne aerosols produced by combustion processes (for instance, emissions from automobiles, power generation, and wood burning) have usually sizes of a few nm, while windblown dust, pollens, plant fragments, and sea-salt particles have generally sizes greater than 1 μm. To describe their complex multimodal features, it is very useful to develop mathematical functions suitable for representing realistically the aerosol size-distribution curves. In the example of multimodal size-distribution curves shown in Figure 6.11 for a BG continental aerosol population, each number density of a particle mode per cubic centimeter of air is represented by

$$N(a) = dN/da, \tag{6.5}$$

where $N(a)$ is a function of particle diameter a (measured in μm) and usually expressed in $\mu m^{-1}\,cm^{-3}$. It is often shown on a logarithmic scale in order to achieve a better graphical representation of the particle concentration for various aerosol size classes. In most cases, the aerosols have sizes $a < 0.1$ μm. Consequently, the function $N(a)$ exhibits in general a narrow spike near the origin, which does not appear if one uses the logarithmic scale in abscissa. The use of function $N(a)$ expressed in terms of Eq. (6.1) allows us to determine the value of $N(a)da$, yielding the number of particles per cm^3 of air, which have diameters ranging from a to $(a + da)$, and to calculate the total particle number N_T per cm^3, given by the following equation:

$$N_T = \int_0^\infty N(a)da \tag{6.6}$$

in which the $N(a)$ is implicitly assumed to exhibit continuous features over the whole diameter range.

The results shown in Figure 6.11 clearly indicate that the number concentration of aerosol particles consists of a certain number of unimodal contributions formed as a result of various emission, *in situ* formation, and condensation growth processes. The volume particle size distribution is in general dominated by two modes: (i) the overall accumulation particle mode, with a ranging from ~0.1 to ~2 μm, and (ii) the coarse particle mode, with a ranging from ~2 to more than 50 μm. As pointed out previously, the accumulation mode particles are mainly produced by primary emissions; condensation of secondary sulfates, nitrates, and organics from the gaseous phase; and coagulation of smaller particles. In numerous cases, the accumulation mode consists of two overlapping submodes, which consist of small particles grown by condensation and haze droplets of greater sizes, respectively. The condensation submode is the result of primary particle emissions and growth of smaller particles by coagulation and vapor condensation, while the droplet submode is most often formed during the cloud formation processes involving a part of the accumulation particles. Being usually produced by mechanical processes such as wind or erosion (dust, sea salt, pollens, etc.), the coarse particles mostly consist of primary material with fractions of secondary sulfates and nitrates. An example of average multimodal size-distribution curve of particle number concentration is shown in Figure 6.12 over wide size range from 10^{-3} to 10^2 μm. It includes usually an ultrafine particle mode formed by nucleation and the Aitken nuclei, accumulation, and coarse particle modes. It is worth noting that nuclei start their atmospheric life as primary particles and secondary material condenses on them as they are transported through the atmosphere. Therefore, they

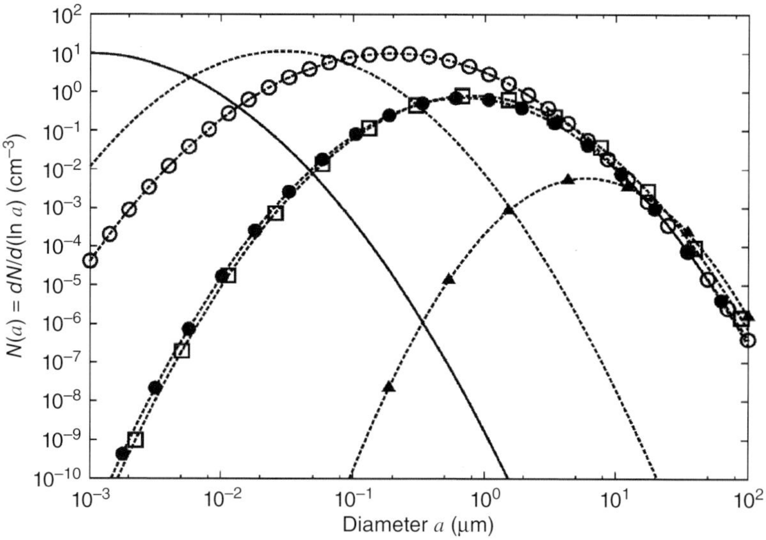

Figure 6.12 Typical curves of number concentration size distribution $N(a) = dN/d(\ln a)$ of atmospheric aerosols presenting different modes over the $10^{-3} \leq a \leq 10^{2}$ μm diameter range: (i) nucleation mode (solid curve), (ii) Aitken nuclei mode (dashed curve), (iii) condensation submode (dashed curve with open circles), (iv) accumulation mode (dashed curve with solid circles), (v) droplet submode (dashed curve with open squares), and (vi) coarse mode (dashed curve with solid triangles). Note that accumulation mode and droplet submode are very similar.

contribute to form at least in part an evident accumulation mode in the volume and mass distribution curves. The accumulation particles with sizes $a > 0.1$ μm are negligible in number compared to the particles with $a < 0.1$ μm, but actually they provide the predominant contribution to the overall aerosol volume.

The coarse particles have sizes $a < 2.5$ μm and are most frequently emitted from primary sources or grown by water vapor condensation. These modes have number concentrations varying by several orders of magnitude throughout the diameter range. The number concentration modes, such as those shown in Figure 6.12, can be represented by means of simple mathematical functions, among which the lognormal distribution function is the most commonly used in the literature, since it provides in general a good fit of the experimental data and can be easily employed in atmospheric optics applications using only a few shape parameters. When represented in terms of Eq. (6.1), the lognormal size-distribution function is defined by the pair of shape parameters σ_g and a_c, from which all the principal characteristics of the lognormal distribution curve can be easily calculated, since (i) the product $\sigma_g a_c$ determines the diameter below which the particle number is equal to 84.1% N_T, (ii) a value of $\sigma_g = 1$ can be used to represent the monodisperse aerosol distribution function, and (iii) any other value of $\sigma_g \neq 1$ indicates that a fraction of 67% N_T particles lie in the range from a_c/σ_g to $a_c\sigma_g$ and a number of 95% N_T lies in the range from a_c/σ_g^2 to $a_c\sigma_g^2$.

In particular, the examples of number density size-distribution curve given in Figure 6.12 refer to the following modes: (i) a nucleation mode, with $\sigma_g = 2.80$ and $a_c = 10^{-3}$ μm, usually consisting of fresh aerosols created *in situ* from the gas phase by nucleation, which dominate the ultrafine aerosol number size-distribution curve in urban and rural environments; (ii) an Aitken nuclei mode, with $\sigma_g = 2.50$ and $a_c = 3 \times 10^{-2}$ μm, mainly composed of nuclei with sizes ranging from ~10 to 100 nm and presenting number concentrations of $\sim 10^5$ cm^{-3} and mass concentrations smaller than 0.05 μg m^{-3} in most cases; (iii) a condensation submode, with $\sigma_g = 2.90$ and $a_c = 2 \times 10^{-1}$ μm; (iv) an accumulation mode, with $\sigma_g = 2.50$ and $a_c = 7 \times 10^{-1}$ μm; (v) a droplet submode, with $\sigma_g = 2.50$ and $a_c = 8 \times 10^{-1}$ μm; and (vi) a coarse mode, with $\sigma_g = 2.00$ and $a_c = 6$ μm.

6.3.2
The Aerosol Surface, Volume, and Mass Size Distributions

Several optical and composition parameters of aerosol populations depend on the shapes of the size-distribution curves of particle surface area and particle volume. Assuming that all the particles are spherical, the aerosol surface area size distribution $S(a)$ can be defined as the surface area per cubic centimeter of air given by particles having diameters ranging from a to $(a + da)$. Thus, all the particles belonging to this infinitesimal size range have effectively the same diameter a and surface area equal to πa^2. This means that there are $N(a)da$ particles in this size range, whose surface area is given by the product

$$S(a) = \pi a^2 N(a), \tag{6.7}$$

measured in square micrometer per cubic centimeter. The total surface area S_T of the aerosol per cubic centimeter of air is given by

$$S_T = \pi \int_0^\infty a^2 N(a)da = \int_0^\infty S(a)da, \tag{6.8}$$

being S_T equal to the area which lies below the $S(a)$ curve drawn in the example shown in Figure 6.13, for the continental aerosol case observed by Tomasi *et al.* (2015a) at the SPC station on 19 December 2007 (08:22 UTC), examining the AEROCLOUDS winter field campaign measurements performed during an episode of aerosol transport from eastern Europe over Northern Italy. The

Figure 6.13 Example of trimodal size-distribution curves of (a) particle number density $N(a) = dN/d(\ln a)$, (b) particle surface area $S(a) = dS/d(\ln a)$, and (c) particle volume $V(a) = dV/d(\ln a)$, obtained over the $10^{-3} \leq a \leq 10^2$ μm diameter range by Tomasi *et al.* (2015a) from AOT measurements performed during an episode of continental aerosol transport from eastern Europe observed at SPC (Po Valley, Northern Italy) on 19 December 2007 (08:22 UTC), during the AEROCLOUDS winter field campaign. The trimodal number concentration curve is normalized to $N_T = 10^3$ cm^{-3}. The areas below the three curves provide the total aerosol number N_T, total surface area S_T, and total volume V_T, respectively.

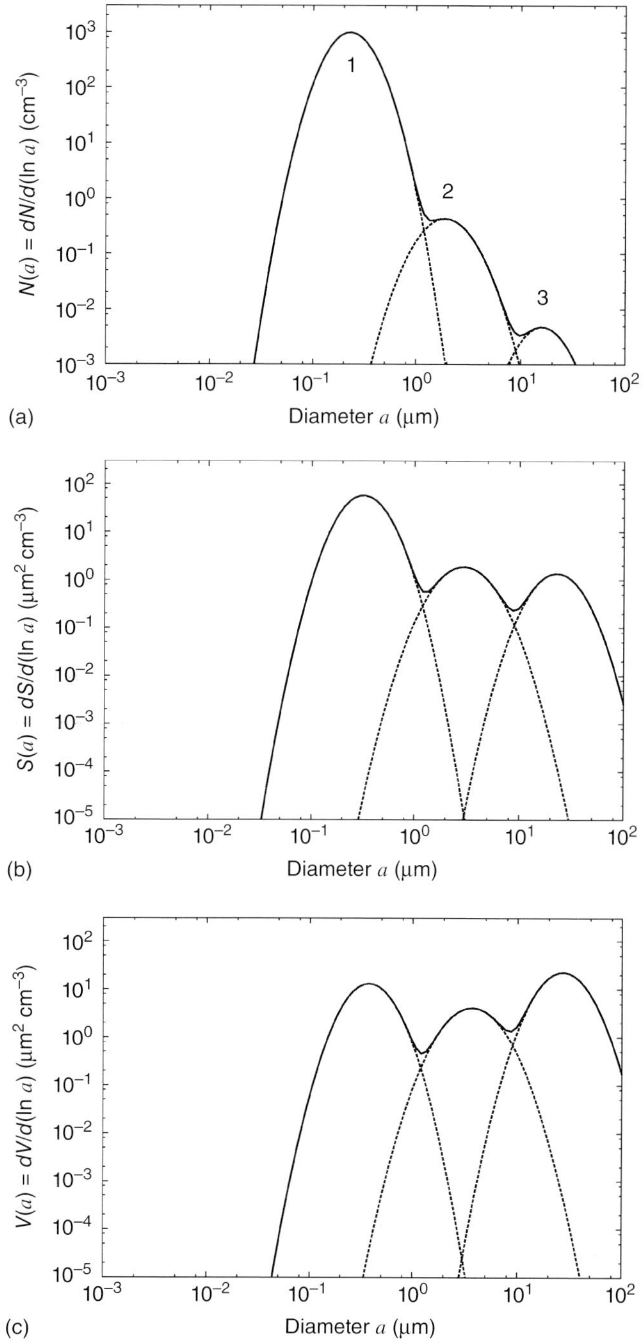
$N(a) = dN/d(\ln a)$ (cm^{-3})
10^{3}
10^{2}
10^{1}
10^{0}
10^{-1}
10^{-2}
10^{-3}
1
2
3
10^{-3} 10^{-2} 10^{-1} 10^{0} 10^{1} 10^{2}
(a)
Diameter a (μm)
$S(a) = dS/d(\ln a)$ (μm^2 cm^{-3})
10^{2}
10^{1}
10^{0}
10^{-1}
10^{-2}
10^{-3}
10^{-4}
10^{-5}
10^{-3} 10^{-2} 10^{-1} 10^{0} 10^{1} 10^{2}
(b)
Diameter a (μm)
$V(a) = dV/d(\ln a)$ (μm^2 cm^{-3})
10^{2}
10^{1}
10^{0}
10^{-1}
10^{-2}
10^{-3}
10^{-4}
10^{-5}
10^{-3} 10^{-2} 10^{-1} 10^{0} 10^{1} 10^{2}
(c)
Diameter a (μm)

size-distribution curves of particle number density $N(a) = dN/d(\ln a)$, particle surface area $S(a) = dS/d(\ln a)$, and particle volume $V(a) = dV/d(\ln a)$ exhibit three modes over the $10^{-3} \leq a \leq 10^2$ μm diameter range, of which the first mode of nuclei/accumulation particles is characterized by $\sigma_g = 1.50$ and $a_c = 0.23$ μm, the second mode of mixed accumulation/coarse particles by $\sigma_g = 1.60$ and $a_c = 1.90$ μm, and the third mode of coarse particles by $\sigma_g = 1.52$ and $a_c = 16.0$ μm.

The aerosol volume distribution $V(a)$ can be defined as the volume of particles per cubic centimeter of air, having diameters ranging from a to $(a + da)$. Therefore, it is represented by

$$V(a) = \frac{\pi}{6} a^3 N(a), \tag{6.9}$$

measured in cubic micrometer per cubic centimeter, while the total aerosol volume V_T per cubic centimeter of air is given by

$$V_T = \frac{\pi}{6} \int_0^\infty a^3 N(a) da = \int_0^\infty V(a) da, \tag{6.10}$$

which is equal to the area below the $V(a)$ curve drawn in the example of Figure 6.13. If all the particles have density ρ_p (g cm^{-3}), the size-distribution curve of particulate mass is given by

$$M(a) = (\rho_p/10^6) V(a) = (\rho_p/10^6) \left(\frac{\pi}{6}\right) a^3 N(a), \tag{6.11}$$

measured in micrograms per cubic centimeter of air, where the factor 10^6 is used to convert the units of density ρ_p from gram per cubic centimeter to microgram per cubic centimeter and to maintain the units of $M(a)$ expressed in micrograms per cubic centimeter of air. The representation of the mass concentration size-distribution function $M(a)$ is very useful to individuate the particulate size components providing the larger fractions of aerosol mass. On this matter, it is worth to mention that BG urban aerosols are in general characterized by overall mass concentration M_p in air ranging from 10^{-2} to 1 mg m^{-3}, while fogs provide values of M_p ranging from 5 to 50 mg m^{-3}, and dust storms yield values of M_p varying from 5×10^2 to 10 g m^{-3}.

Considering that aerosol sizes vary over several orders of magnitude, the use of the size-distribution functions $N(a)$, $S(a)$, $V(a)$, and $M(a)$ over the linear scale of diameter is often inconvenient, since all the unimodal size-distribution functions of number density usually occupy only the size range from a few nanometer to no more than 10^2 μm, which constitutes a very limited part of the overall particle diameter range. In addition, functions $N(a)$, $S(a)$, $V(a)$, and $M(a)$ may vary by several orders of magnitude passing from the ultrafine particle mode to the coarse particle mode, as shown in Figure 6.12 for parameter $N(a)$. To circumvent this scale problem, the horizontal axis can be better scaled in logarithmic intervals, in this way including several orders of magnitude of diameter a in the graph, as shown in Figure 6.13. It is true that plotting the function $N(a)$ on semilogarithmic axes can yield a somewhat distorted picture of the size-distribution curve of aerosol number concentration, since the area below the curve no longer corresponds to the aerosol number concentration. However, the example of Figure 6.13

clearly shows that more than 60% of the particles belong to the first mode with $a_c \approx 2 \times 10^{-2}$ µm and that the volume distribution is dominated by the coarse particle mode having a varying from ~2 to ~50 µm and to a lesser extent by the accumulation particle mode having a ranging from ~0.1 to ~2 µm. For this reason, it seems more appropriate in general to represent the particle number concentration, surface area, and aerosol volume using the logarithmic scales on both the abscissa and ordinate axes.

6.4 Size-Distribution Characteristics of Various Aerosol Types

A different representation of the ambient aerosol size-distribution characteristics is obtained if one focuses on the number of particles instead of their surface area of their volume or mass contents. The overall particle number concentration varies most commonly from 50 to 10^4 cm^{-3} in remote continental areas, 2×10^3 to 10^4 cm^{-3} in rural areas, 10^2 to 4×10^2 cm^{-3} in oceanic regions, and 10^5 to 4×10^6 cm^{-3} in urban polluted areas. Correspondingly, values of PM_1 equal to 0.5–2.5 µg m^{-3} are measured in remote continental areas, 2.5–8.0 µg m^{-3} in rural areas, 1.0–4.0 µg m^{-3} in marine areas, and 30–150 µg m^{-3} in urban polluted areas, against values of PM_{10} equal to 2–10 µg m^{-3} in remote continental areas, 10–40 µg m^{-3} in rural areas, ~10 µg m^{-3} in marine areas, and 100–300 µg m^{-3} in urban polluted areas (Seinfeld and Pandis, 2006). Due to their various origins, the aforementioned aerosol types exhibit markedly different characteristics of the multimodal size-distribution curves, as highlighted in the following paragraphs.

6.4.1 Remote Continental Aerosols

Studies of the multimodal characteristics of remote continental aerosols were conducted by Whitby (1978) over the $5 \times 10^{-5} \leq a \leq 50$ µm size range, finding that these particle size-distribution curves consist in general of a first mode of ultrafine particles occupying the size range $a < 10^{-2}$ µm, with number concentration varying between 10 and 10^2 cm^{-3}; a second pronounced mode of accumulation particles over the 10^{-2}–2.5 µm size range; and a third coarse particle mode with $a > 2.5$ µm. Similar results were obtained by Jaenicke (1993), finding that primary particles consisting of dust, pollens, plant waxes, and secondary oxidation products are the main components of remote continental aerosols, for which size-distribution curve $N(a)$ is characterized by two modes, the first given by ultrafine and fine particles and having mode diameter $a_c = 2 \times 10^{-2}$ µm and the second of accumulation particles with $a_c = 10^{-1}$ µm. To give an idea of the bimodal size-distribution characteristics of these aerosols, two examples of the size-distribution curves $N(a)$, $S(a)$, and $V(a)$ of remote continental aerosols are shown in the left-hand side part of Figure 6.14: (i) the bimodal aerosol model of Jaenicke (1993), presenting a first mode with $a_c = 2 \times 10^{-2}$ µm and $\sigma_g = 2.0$, and a second mode with $a_c = 1.5 \times 10^{-1}$ µm and $\sigma_g = 2.2$, and (ii) the bimodal OPAC

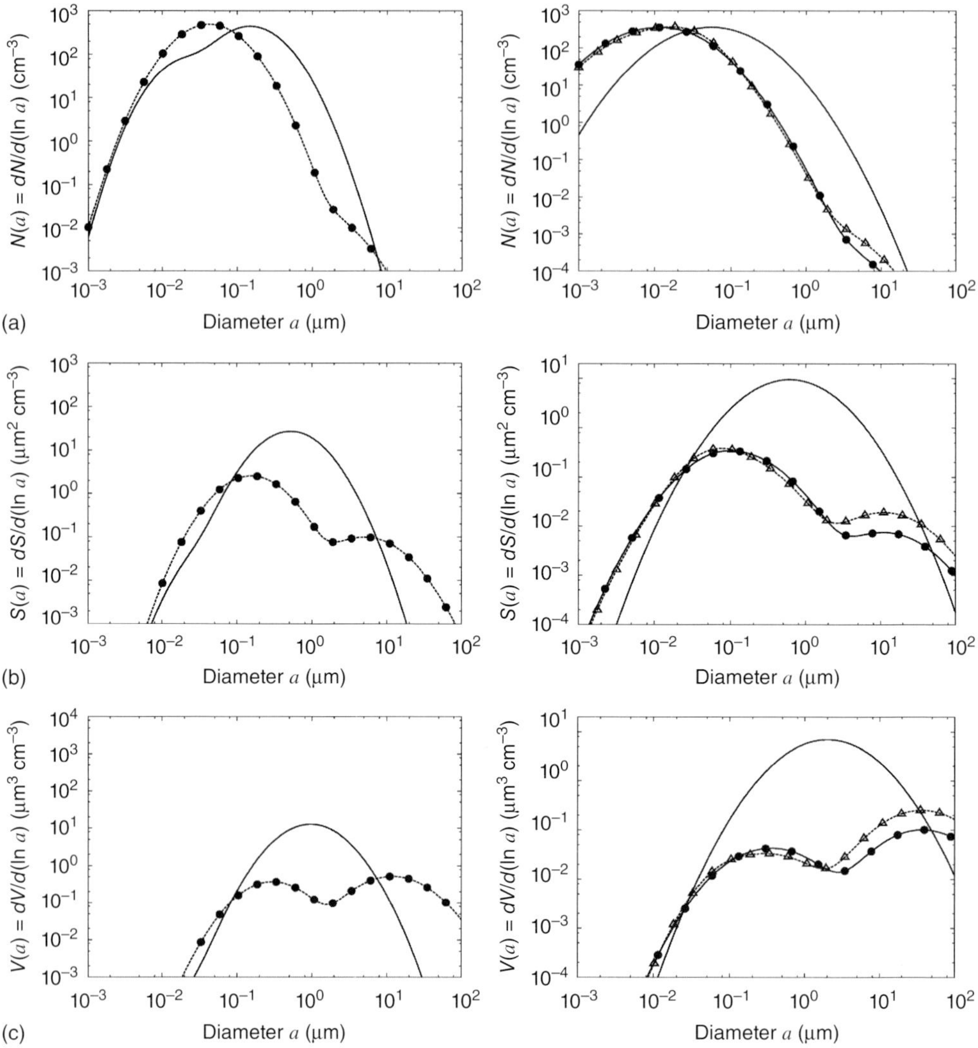

Figure 6.14 Left-hand side: Multimodal size-distribution curves of (a) particle number concentration $N(a) = dN/d(\ln a)$, (b) particle surface area $S(a) = dS/d(\ln a)$, and (c) particle volume $V(a) = dV/d(\ln a)$ defined over the $10^{-3} \le a \le 10^{2}$ μm diameter range, determined for the following remote continental aerosol models: (i) the model determined by Jaenicke (1993) (solid curves) and (ii) the OPAC continental clean aerosol model CC (dashed curves with solid circles) defined by Hess, Koepke, and Schult (1998). Both number concentration size-distribution curves are normalized to $N_T = 10^3$ cm^{-3}. Right-hand side: As in the left-hand side part, for the following free tropospheric aerosol models: (i) model SF-T (solid curves) of Shettle and Fenn (1979), (ii) continental aerosol model 6S-C (dashed curves with open triangles) of Vermote *et al.* (1997a), and (iii) pure continental aerosol model M-8 (solid curves with solid circles) of Tomasi *et al.* (2013). The three number concentration size-distribution curves are normalized to $N_T = 10^3$ cm^{-3}.

CC aerosol model of Hess, Koepke, and Schult (1998), consisting of two modes with $a_c = 4.24 \times 10^{-2}$ μm and $\sigma_g = 2.24$, and $a_c = 9.42 \times 10^{-1}$ μm and $\sigma_g = 2.51$, respectively. The graph clearly shows that large-width maxima characterize the curves of $S(a)$ and $V(a)$ obtained for the two models, in which the left wing of the Jaenicke (1993) model made by nuclei agrees closely with that of the CC model, while its accumulation particle mode causes more marked effects on the curves of $N(a)$, $S(a)$, and $V(a)$ than those produced by the CC model, which appears to yield more equally distributed contributions to the aerosol total surface area and total volume.

6.4.2
Free Tropospheric Aerosols

BG aerosols are present in the FT above the clouds and exhibit in general bimodal size-distribution curves, with a nuclei mode peaked at $a_c = 6 \times 10^{-3}$ μm and an accumulation mode with $a_c = 2.6 \times 10^{-1}$ μm, respectively (Jaenicke, 1993). The free tropospheric aerosol size spectra typically provide higher number concentrations of accumulation particles than in the lower part of the troposphere, due to the more efficient precipitation scavenging and deposition processes acting on nuclei and coarse particles. A nucleation mode often characterizes the number size-distribution curve of these aerosol particles, presenting very high number aerosol concentrations in the upper FT at the tropical latitudes, where high numbers of small particles form from SA by nucleation in the convective regions and near the cloud edges, until becoming effective CCNs. Lower aerosol number concentrations are measured in the midlatitude FT, but with larger sizes, presenting multimodal features similar to those shown in Figure 6.11. The size-distribution curves of $N(a)$, $S(a)$, and $V(a)$ determined for the SF-T (tropospheric) aerosol model of Shettle and Fenn (1979) are compared in the right-hand side part of Figure 6.14 with those obtained for the 6S-C (continental) model of Vermote *et al.* (1997b) and the M-8 (pure continental) model of Tomasi *et al.* (2013). It can be noted that the SF-T accumulation and coarse particle number concentrations are considerably higher than those of the 6S-C and M-8 models, which exhibit very similar curves over the whole diameter range. Therefore, the SF-T model produces a more marked contribution to the total particle surface area S_T, which is considerably higher than those of the 6S-C and M-8 models by nearly two orders of magnitude. Similarly, the SF-T model presents a more peaked size-distribution curve of $V(a)$, providing a larger contribution to the total volume V_T than those given together by the 6S-C and M-8 accumulation and coarse modes by three and two orders of magnitude, respectively.

6.4.3
Rural-Continental Aerosols

These aerosols mainly originate from natural sources, although they are sometimes considerably influenced by anthropogenic emissions (Hobbs, Bowdle, and

Radke, 1985). Their number size-distribution curves are in general characterized by the combination of a nuclei mode with a mixed nuclei/accumulation mode and a coarse particle mode, which contributes to yield a value of PM_{10} concentration equal to around 20 μg m^{-3} for moderate pollution conditions (Jaenicke, 1993). Two aerosol models can be reliably used among those determined earlier to represent the a rural-continental aerosol population: (i) the SF-R (rural) aerosol model of Shettle and Fenn (1979) and (ii) the OPAC CA aerosol model of Hess, Koepke, and Schult (1998). Their size-distribution curves are shown in the left-hand side part of Figure 6.15, giving evidence of the predominant contributions provided by the nuclei/accumulation mode to both total number and surface and of the more limited contribution to the total volume with respect to the coarse particle mode.

To complete the examination of the size-distribution characteristics of these aerosols, the following three multimodal size-distribution curves of $N(a)$ for rural-continental aerosols are also shown in the graph, as retrieved by Tomasi *et al.* (2015a) from field spectral measurements of AOT: (i) the bimodal size-distribution curve derived from a set of AERONET measurements conducted during the DOE/ARM/AIOP field campaign on 8 and 9 May 2003, in north-central Oklahoma (United States) (Ferrare *et al.*, 2006), found to consist of an accumulation mode with $a_c \approx 0.2$ μm and $\sigma_g = 1.60$ and a mixed accumulation/coarse particle mode with $a_c = 1.76$ μm and $\sigma_g = 1.73$; (ii) the bimodal size-distribution curve retrieved from the AERONET Level 2.0 AOT measurements conducted at Lecce (Italy) on 10 June 2003 (10:03 UTC), found to consist of a nuclei/accumulation mode with $a_c = 0.22$ μm and $\sigma_g = 1.60$ and a coarse particle mode with $a_c = 2.40$ μm and $\sigma_g = 1.75$; and (iii) the size-distribution curve retrieved from the AEROCLOUDS measurements performed at the SPC station in the Po Valley (Italy) using a Prede POM-02L sun/sky radiometer on 15 October 2007 (11:45 UTC), found to consist of a mixed nuclei/accumulation

Figure 6.15 Left-hand side: As in Figure 6.14, for the following rural-continental aerosol models: (i) model SF-R (solid curves) of Shettle and Fenn (1979), (ii) OPAC continental average aerosol model CA (dashed curves with open diamonds) of Hess, Koepke, and Schult (1998), (iii) continental clean aerosol model (solid curves with solid circles) of Tomasi *et al.* (2015a) (9 May 2003 (14:08 UTC), DOE/ARM/AIOP field campaign), (iv) continental aerosol model (dashed curves with open squares) of Tomasi *et al.* (2015a) (10 June 2003 (10:03 UTC), PRIN-2004 field campaign, Lecce (Italy)), and (v) continental aerosol model (solid curves with solid triangles) of Tomasi *et al.* (2015a) (15 October 2007 (11:45 UTC), summer AEROCLOUDS field campaign at San Pietro Capofiume (Italy)). The five multimodal aerosol number concentration size-distribution curves are normalized to $N_T = 10^3$ cm^{-3}. Right-hand side: As in the left-hand side part, for the following continental-polluted aerosol models: (i) OPAC model CP (solid curves) of Hess, Koepke, and Schult (1998), (ii) model M-10 (dashed curves) of Tomasi *et al.* (2013), (iii) anthropogenic–continental aerosol model M-12 (solid curves with solid circles) of Tomasi *et al.* (2013), and (iv) heavy polluted aerosol model M-14 (dashed curves with open squares) of Tomasi *et al.* (2013). The four aerosol number size-distribution curves are normalized to $N_T = 10^3$ cm^{-3}.

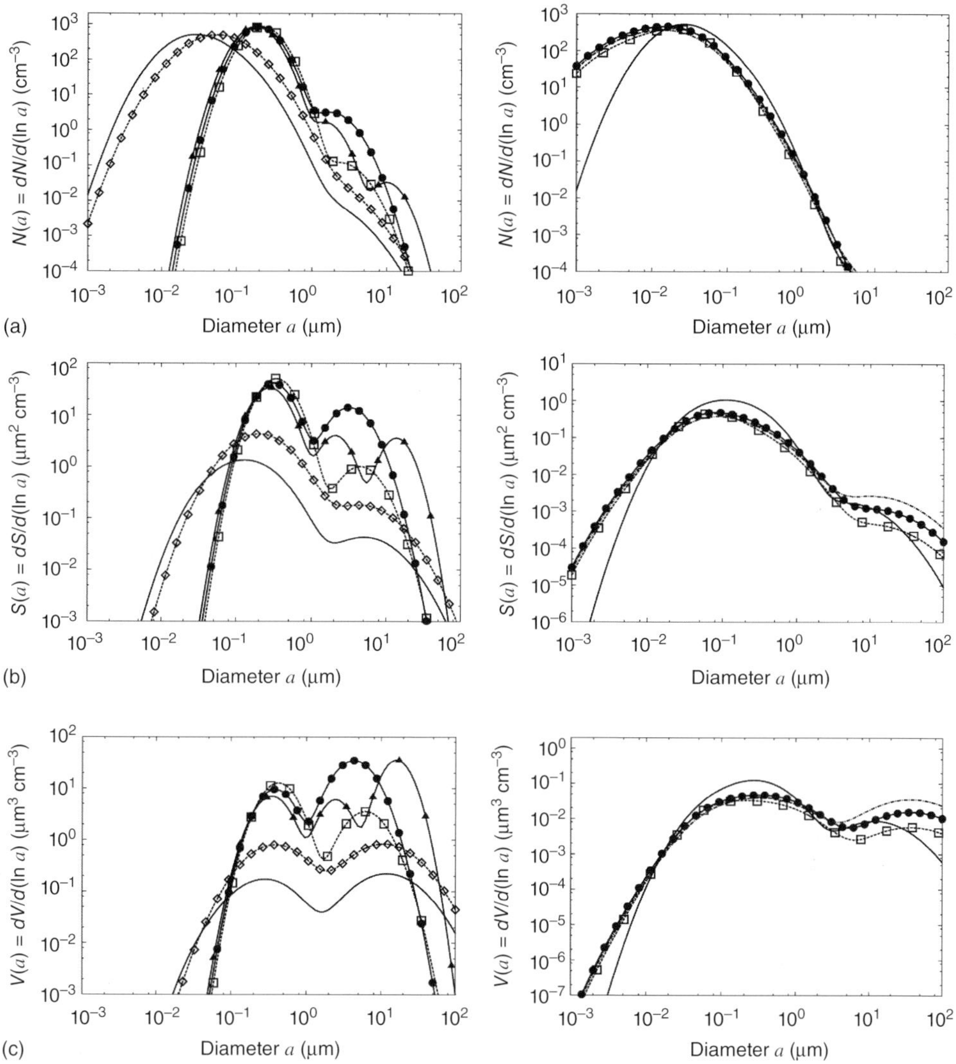

mode with $a_c = 0.18\ \mu m$ and $\sigma_g = 1.60$, an accumulation mode with $a_c = 1.50\ \mu m$ and $\sigma_g = 1.50$, and a coarse particle mode with $a_c = 10\ \mu m$ and $\sigma_g = 1.49$. The size-distribution curves of $N(a)$, $S(a)$, and $V(a)$ determined by Tomasi *et al.* (2015a) for these three experimental measurements are compared in the left-hand side part of Figure 6.15 with those of the SF-R model and OPAC CA aerosol model of Hess, Koepke, and Schult (1998), showing that the curves obtained for the three experimental data sets agree satisfactorily with those of the SF-R and the CA models, although presenting more peaked characteristics of the accumulation and coarse particle modes, which contribute more efficiently to enhance the total contents N_T, S_T, and V_T within narrower size ranges.

6.4.4 Continental-Polluted Aerosols

Aerosols of this type are usually measured over large areas of North America, Europe, and Asia and the surrounding oceanic areas, as shown by Takemura *et al.* (2005) using the SPRINTARS aerosol chemical transport model simulations of AOT made over a global scale for the four major aerosol types (soil dust, BC and OM, sulfate, and sea salt). The size-distribution curves of $N(a)$, $S(a)$, and $V(a)$ relative to four CP aerosol models are shown in the right-hand side part of Figure 6.15, where those of the OPAC CP aerosol model of Hess, Koepke, and Schult (1998) are compared with those of the M-10 CP aerosol model, those of the M-12 anthropogenic–continental aerosol model, and those of the M-14 heavy polluted aerosol model, defined by (Tomasi *et al.*, 2013). The comparison shows a close agreement of the four size-distribution curves of $N(a)$ given by the four models, with a more peaked contribution provided by the CP accumulation particles. A predominant contribution is given by the CP accumulation particles to the total surface area S_T and total volume V_T, while the M-10 model yields appreciably larger surface and volume contributions than the other models over the size range $a > 5\,\mu m$.

6.4.5 Maritime Clean Aerosols

Oceans constitute the main sources of sea-salt aerosols, yielding a global annual mean production rate estimated to vary from 10^3 to 3×10^3 Tg per year (IPCC, 2001). A pair of number and volume size distributions of marine aerosols in clean maritime air were shown by Seinfeld and Pandis (2006), consisting

Figure 6.16 Left-hand side: As in Figure 6.14, for the following maritime aerosol models: (i) model 6S-M (solid curves) of Vermote *et al.* (1997b), (ii) pure oceanic aerosol model M-1 (dashed curves with solid squares) of Tomasi *et al.* (2013), (iii) model M-2 (solid curves with solid diamonds) of Tomasi *et al.* (2013), (iv) OPAC maritime clean aerosol model MC (dashed curves with open circles) of Hess, Koepke, and Schult (1998), (v) OPAC maritime tropical aerosol model MT (solid curves with solid circles) of Hess, Koepke, and Schult (1998), (vi) model SF-M (dashed curves with open squares) of Shettle and Fenn (1979), and (vii) the bimodal aerosol model (dashed curves with open triangles) of Tomasi *et al.* (2015a) determined for pure maritime aerosol observed during the Aerosols99 cruise. The seven aerosol number size-distribution curves are normalized to $N_T = 10^3\ cm^{-3}$. Right-hand side: As in the left-hand side part, for the following maritime-polluted aerosol models: (i) OPAC model MP (solid curves) of Hess, Koepke, and Schult (1998), (ii) mixed maritime–continental aerosol model M-3 (dashed curves with open circles) of Tomasi *et al.* (2013), (iii) mixed maritime–continental aerosol model M-5 (solid curves with solid circles) of Tomasi *et al.* (2013), and (iv) trimodal mixed marine/continental aerosol model (dashed curves with open squares) retrieved by Tomasi *et al.* (2013) from the AEROCLOUDS measurements conducted at SPC (Po Valley, Italy) on 1 July 2007 (07:45 UTC), during an aerosol transport episode from the Atlantic Ocean, southern France, and Ligurian Sea. The four number size-distribution curves are normalized to $N_T = 10^3\ cm^{-3}$.

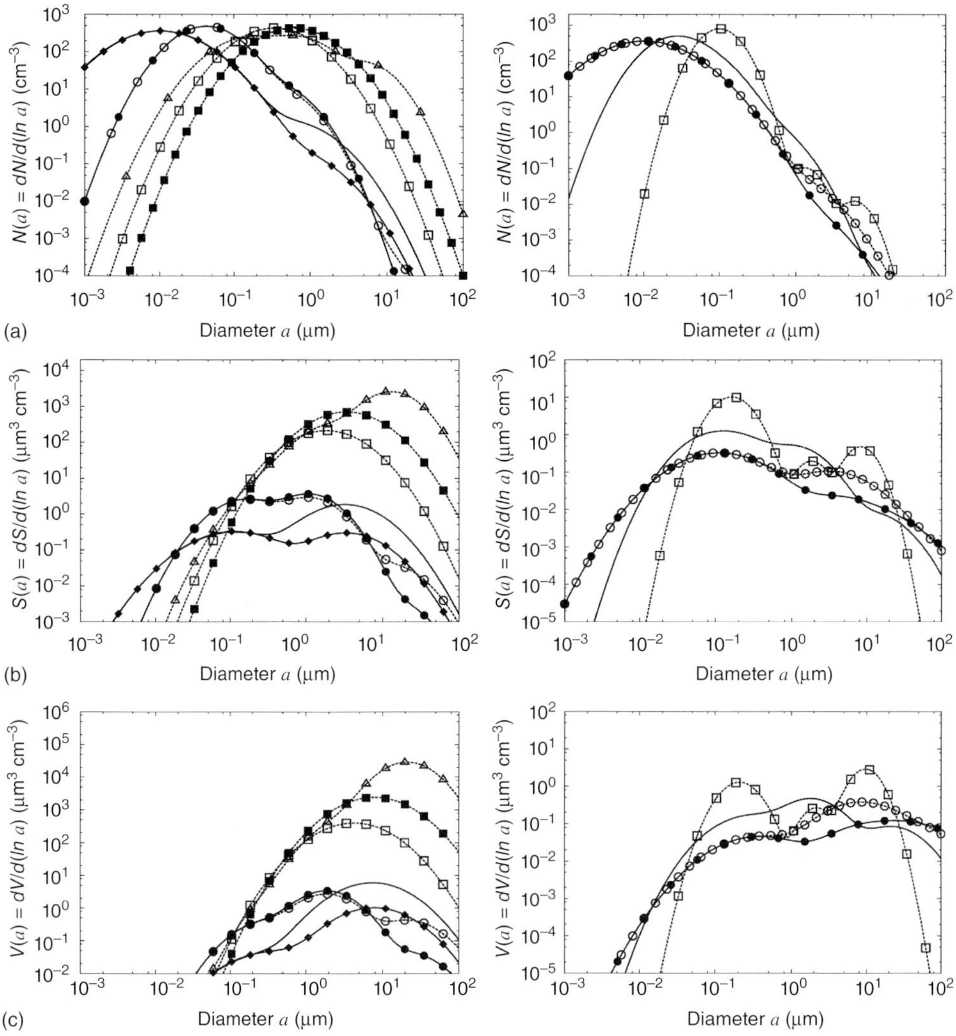

of a trimodal size-distribution curve with an Aitken nuclei mode centered at $a_c \approx 0.05\,\mu m$, an accumulation particle mode with $a_c \approx 0.25\,\mu m$, and a coarse particle mode covering the size range from 3 to around 50 μm, with $a_c \approx 6.0\,\mu m$. The size-distribution curves of $N(a)$, $S(a)$, and $V(a)$ obtained for the 6S-M (maritime) aerosol model of Vermote *et al.* (1997b) are shown in the left-hand side part of Figure 6.16 for a comparison with those of (i) the M-1 (pure oceanic) and M-2 (maritime) models of Tomasi *et al.* (2013), (ii) the OPAC MC and MT aerosol models of Hess, Koepke, and Schult (1998), (iii) the SF-M (maritime) aerosol model of Shettle and Fenn (1979), and (iv) the bimodal size-distribution curve determined by Tomasi *et al.* (2015a) for pure maritime aerosol from the AOT spectral series measured during the Aerosols99 cruise. The size-distribution curves determined for the previous seven aerosol models exhibit a large variety of combinations of accumulation and coarse particle modes, with prevailing contributions given by the accumulation particles to the total content N_T and by coarse particles to the total contents S_T and V_T. Particularly marked volume contributions are given by the coarse particle modes of the Aerosols99, M-1 (pure oceanic), and SF-M aerosol models, since they consist of oceanic sea-salt particles presenting higher percent number concentrations than those of the 6S-M, M-2, MC, and MT aerosol models.

6.4.6
Maritime-Polluted Aerosols

The trimodal size-distribution curves of $N(a)$, $S(a)$, and $V(a)$ determined for the OPAC MP aerosol model of Hess, Koepke, and Schult (1998) give an idea of the multimodal features of these mixed particles, consisting of four modes centered at around 2.4×10^{-2}, 4.2×10^{-2}, 4.2×10^{-1}, and 3.80 μm. The curves of MP model are compared in the right-hand side part of Figure 6.16 with those of the M-3 and M-5 (mixed maritime–continental) aerosol models defined by Tomasi *et al.* (2013) and those retrieved by Tomasi *et al.* (2015a) from the AEROCLOUDS measurements performed on 1 July 2007 (07:45 UTC), during an aerosol transport episode from the Atlantic Ocean, southern France, and Ligurian Sea over the Po Valley (Italy). The comparison indicates that the size-distribution curves of $N(a)$, $S(a)$, and $V(a)$ obtained for the MP model agree satisfactorily with those of the M-3 and M-5 (maritime–continental) aerosol models. Conversely, quite different patterns are shown by the AEROCLOUDS size-distribution curve, which is characterized by a more peaked nuclei/accumulation particle mode centered at $a_c \approx 10^{-1}$ μm, a pair of less pronounced modes produced by the accumulation/coarse particles with $a_c \approx 1.5$ μm, and coarse particles with $a_c \approx 6$ μm. However, the second of these coarse particle modes causes marked variations in the curves of $S(a)$ and $V(a)$ within its relatively narrow range of a, which provide appreciably higher maxima by at least one order of magnitude than those given by the M-3 and M-5 aerosol models The multimodal characteristics of the $S(a)$ and $V(a)$ obtained by Tomasi *et al.* (2015a) from the CLEARCOLUMN measurements conducted during a mixed maritime–continental aerosol transport case on 5 July 1997 (17:14 UTC) (Tomasi *et al.*, 2003).

6.4.7
Desert Dust

It is often transported after its mobilization toward the adjacent oceanic regions (Jaenicke and Schutz, 1978). The shape parameters of the desert dust size-distribution curves are similar to those of remote continental aerosol, varying strongly with the wind speed. The trimodal size-distribution curve defined by Jaenicke (1993) for desert dust presents a triplet of modes centered at values of a_c close to 10^{-2}, 5×10^{-2}, and 10 μm, respectively. The size-distribution curves of

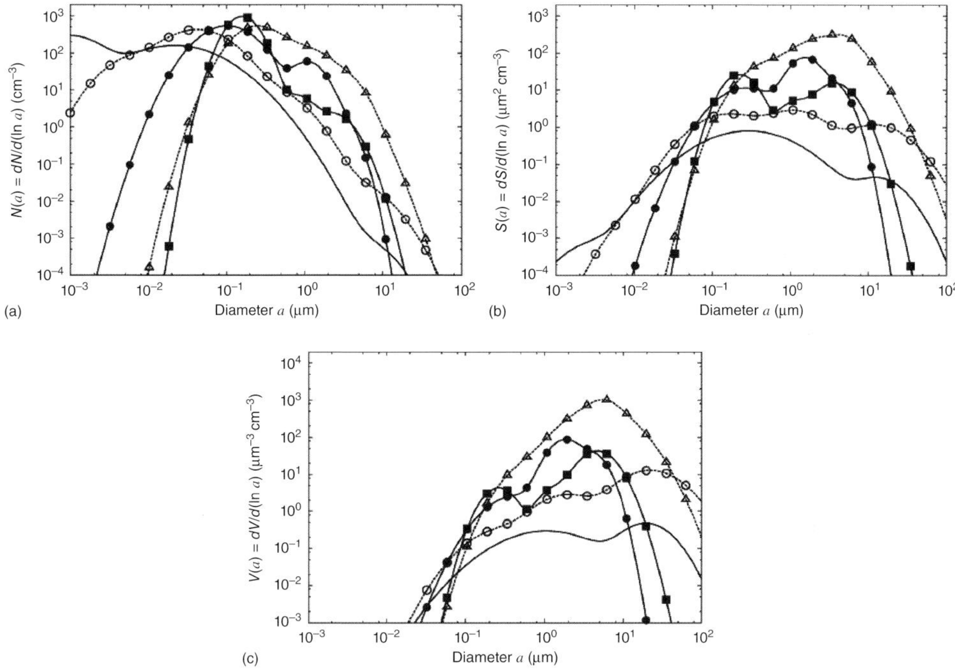

Figure 6.17 As in Figure 6.14, for the following aerosol models: (i) OPAC desert dust model DE (dashed curves with open circles) of Hess, Koepke, and Schult (1998), (ii) trimodal background desert dust model BDD (solid curves) of Shettle (1984), (iii) Saharan dust model SD-1 (solid curves with solid circles) of Tomasi *et al.* (2013), (iv) Saharan dust model SD-2 (dashed curves with open triangles) of Tomasi *et al.* (2013), and (v) mixed anthropogenic–continental–Saharan dust aerosol model (solid curves with solid squares) observed by Tomasi *et al.* (2015a) during the PRIN-2004 experiment (30 August 2003 (07:30 UTC)) conducted at Lecce (Italy). The five desert dust number size-distribution curves are normalized to $N_T = 10^3$ cm^{-3}.

$N(a)$, $S(a)$, and $V(a)$ obtained for the OPAC DE dust model of Hess, Koepke, and Schult (1998) are compared in Figure 6.17 with those of (i) the trimodal BDD dust particle model of Shettle (1984), (ii) the SD-1 and SD-2 Saharan dust models of Tomasi *et al.* (2013), and (iii) the mixed anthropogenic–continental and Saharan dust aerosol retrieved by Tomasi *et al.* (2015a) at Lecce (Italy) on 30 August 2003 (07:30 UTC), from the AERONET measurements conducted by Perrone *et al.* (2005) at Lecce, during a transport episode of mixed anthropogenic–continental aerosol from central and southern Italy at levels lower than 1 km and Saharan dust moving from north Africa at levels between 1.5 and 3.5 km. The comparison shows that the present desert dust models exhibit largely different size-distribution curves of $N(a)$, $S(a)$, and $V(a)$, characterized by strong discrepancies which presumably arise from the largely different morphological and physical characteristics of soil particles mobilized in the various geographical desert regions.

6.4.8 Biomass Burning Aerosols

Biomass burning is an important source of combustion aerosol particles, mainly located in the tropics, where uncontrolled combustion processes cause the major emissions of such absorbing particles. The intensity of this source varies not only with the vegetation type but also with the moisture content of the burned matter, the ambient temperature and humidity, and the local wind speed and may be comparable with that of urban industrialized regions, as, for instance, in the case of large-scale savannah burning in the dry season in southern Africa, leading to very high regional-scale pollution levels, or during the burning season in South America reaching its highest intensity in September over large areas, due only in part to natural events. A vegetated area equal to about half the area of Europe is estimated to burn globally each year because of natural forest fires (many initiated by lightning) or deliberate deforestation activities (e.g., in the Amazon Basin and in the savannas of southern Africa) associated with refertilization of soils and grazing and the use of wood for heating and cooking (e.g., in Africa, India, and Southeast Asia). Many of the emissions from biomass burning are carcinogens and often cause significant degradation to air quality on local and regional scales and over long distances for favorable wind conditions, exerting global effects on atmospheric chemistry and climate.

Bimodal size-distribution curves were retrieved by Tomasi *et al.* (2015a) from the AERONET Level 1.5 spectral series of AOT measured at the ARM Southern Great Plains (SGP) Climate Research Facility site, during the DOE/ARM/AIOP field campaign of May 2003 conducted in north-central Oklahoma (United States), consisting of rural aerosols mixed with Siberian forest fire smoke particles. The left-hand side part of Figure 6.18 shows the comparison of the size-distribution curves of $N(a)$, $S(a)$, and $V(a)$ with those obtained for (i) bimodal model BBS derived by Remer *et al.* (1998) from AOT measurements conducted during the burning season in the Amazon Basin, (ii) aerosol model FT of Tomasi *et al.* (2013), and (iii) aerosol model BL of Tomasi *et al.* (2013), and (iv) the aforementioned results of Tomasi *et al.* (2015a). The comparison indicates that important contributions to $N(a)$, $S(a)$, and $V(a)$ are given by the coarse particle mode in spite of the lower concentration of these particles by two orders of magnitude with respect to those of the nuclei and accumulation modes.

6.4.9 Urban Aerosols

Aerosols of this kind are mixtures of primary particulate emissions from industries, vehicular traffic, power generation, and natural sources with secondary particles formed by g-to-p conversion processes. The aerosol number concentration size-distribution curve is quite variable in urban areas, since it exhibits extremely high concentrations of small particles close to the sources (such as highways and industrial plants) and rapidly decreasing concentrations with the distance from

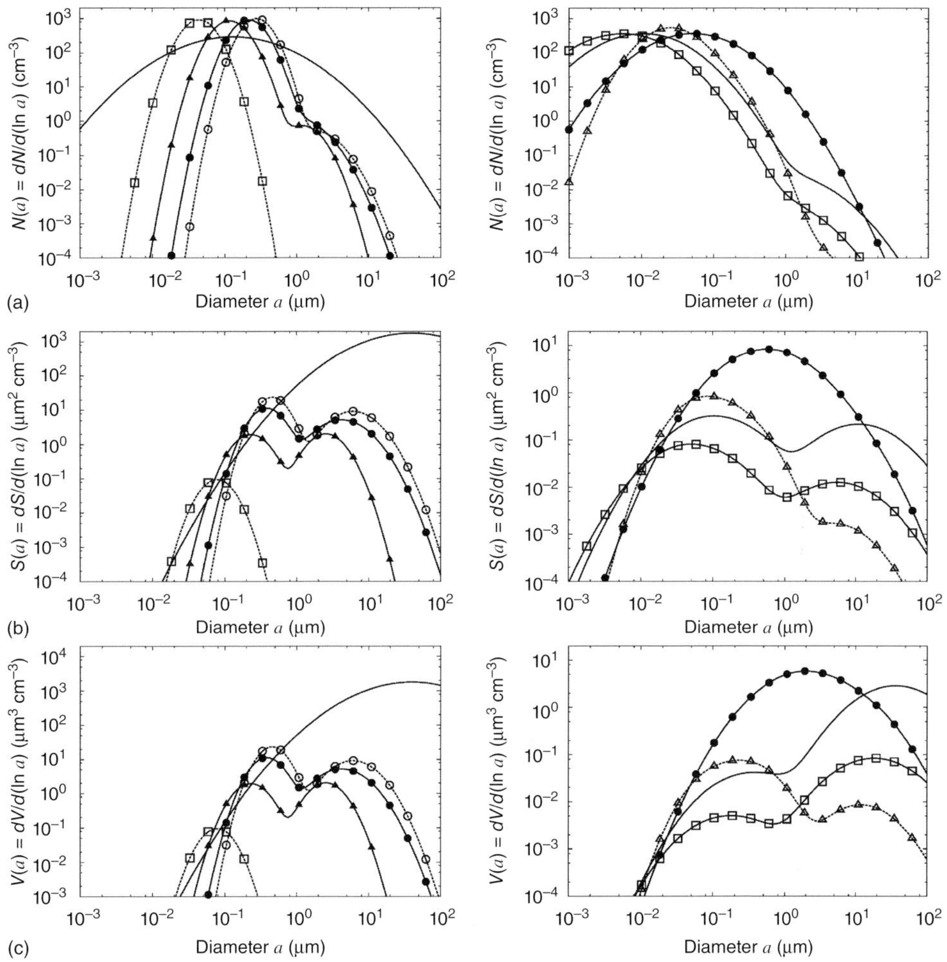

Figure 6.18 Left-hand side: As in Figure 6.14, for the following aerosol models: (i) continental-polluted aerosol model, containing soot and Siberian forest fire (SFF) smoke particles (solid curves with solid circles), retrieved by Tomasi *et al.* (2015a) from the DOE/ARM/AIOP field measurements conducted on 25 May 2003 (22:58 UTC); (ii) continental-polluted aerosol model, containing soot and SFF smoke particles (dashed curves with open circles), retrieved by Tomasi *et al.* (2015a) from the DOE/ARM/AIOP field measurements conducted on 27 May 2003 (16:28 UTC); (iii) bimodal model BBS (solid curves), derived from the measurements conducted by Remer *et al.* (1998) during the burning season in the Amazon Basin and Cerrado region of Brazil; (iv) aerosol model FT (dashed curves with open squares) of Tomasi *et al.* (2013); and (v) aerosol model BL (solid curves with solid triangles) of Tomasi *et al.* (2013). The five aerosol number size-distribution curves are normalized to $N_T = 10^3$ cm^{-3}. Right-hand side: As in the left-hand side part for the following urban aerosol models: (i) model 6S-U (solid curves) of Vermote *et al.* (1997b), (ii) OPAC model UR (dashed curves with open triangles) of Hess, Koepke, and Schult (1998), (iii) model SF-U (solid curves with solid circles) of Shettle and Fenn (1979), and (iv) winter-polluted Po Valley aerosol model WPV (solid curves with open squares), determined by using the field data of Carbone *et al.* (2010). The four urban aerosol number size-distribution curves are normalized to $N_T = 10^3$ cm^{-3}.

the strong emission sites. The sources of nuclei, accumulation, and coarse urban particles are quite different: Aitken and accumulation mode particles contain both primary aerosol from combustion sources and secondary particulate matter composed of sulfate, nitrate, ammonium, and organic substances, while coarse particles are generated by mechanical natural and anthropogenic processes, mainly consisting of dust, sea salt, fly ash, and tire wear particles. Sampling measurements of the typical urban aerosol size composition parameters indicate that sulfate, nitrate, and ammonium exhibit usually a pair of submodes containing both condensation particles and droplets, centered at mode diameters of around 0.2 and 0.7 μm, besides a pronounced coarse-mode nitrate and sodium and chloride ions (Seinfeld and Pandis, 2006).

The number size distribution is in general dominated by nuclei, and the surface area size distribution by particles with sizes ranging from 0.1 to 0.5 μm, and the volume size distribution usually exhibits a distinct accumulation particle mode together with a marked coarse particle mode, as shown in the examples given in the right-hand side part of Figure 6.18, where the following multimodal size-distribution curves of urban-polluted aerosols are shown: (i) the urban aerosol model 6S-U of Vermote *et al.* (1997b), (ii) the OPAC urban aerosol model UR of Hess, Koepke, and Schult (1998), (iii) the urban aerosol model SF-U of Shettle and Fenn (1979), and (iv) the winter-polluted Po Valley aerosol model WPV defined in the present study using the field measurements of particulate concentration carried out by Carbone *et al.* (2010). The various size-distribution curves of $N(a)$, $S(a)$, and $V(a)$ indicate that the accumulation particle mode provides the most important contribution to the curves of $N(a)$, while accumulation particles mainly contribute to give form to the size-distribution curve of $S(a)$, and coarse particles provide a predominant contribution to $V(a)$.

6.4.10
Polar Arctic Aerosols

Particles sampled in summer at the Arctic sites for BG conditions exhibit in general relatively low mass concentrations, since they are not appreciably influenced by the PM transported from midlatitudes. During the winter and spring season, transport episodes of pollutant gases and aerosols from North America and Eurasia are frequent and cause often a sharp increase in the AH concentration within the ground layer. In spring, AD is often mobilized over the desert regions of Central Asia and in part transported toward the Arctic, especially over Alaska and Canada. In summer, BFFS particles are generated by frequent forest fires in northern America (Alaska, Yukon, and Northwest Territories) and in central and eastern Siberia and then transported toward the remote Arctic regions. The numerous *in situ* sampling measurements conducted at the Arctic sites showed that the size-distribution curve of $N(a)$ for BG Arctic aerosol exhibits sometimes monodisperse characteristics with $a_c \approx 0.15$ μm (Seinfeld and Pandis, 2006) or bimodal features with two distinct modes centered at 0.75 and 8 μm. During the winter and early spring (most frequently from February to April), the Arctic aerosol was found to be significantly affected by anthropogenic aerosols, yielding ground-level aerosol number concentrations $N_0 > 200\,cm^{-3}$, with a nucleation mode centered at $a_c = 5 \times 10^{-2}\,\mu m$ and an accumulation mode at $a_c \approx 0.2\,\mu m$. Polar aerosol contains sea-salt particles, mineral dust, sulfates from local and remote anthropogenic sources, and carbonaceous material from midlatitude polluted areas, presenting mass concentrations lower than $5\,\mu g\,m^{-3}$, with sulfate yielding roughly 8% of the total mass content in summer and 24% in the other seasons. Mineral dust is the main aerosol constituent of the AD, and combustion products dominate in the BFFS particles. The size-distribution curves of $N(a)$, $S(a)$, and $V(a)$ determined by Tomasi *et al.* (2015a,b) for the aerosol models ASB, AH, AD, and BFFS shown in Figure 6.10 are compared in the left-hand side part of Figure 6.19 with the OPAC aerosol model AR of Hess, Koepke, and Schult (1998), giving a measure of the contributions of the nuclei/accumulation and coarse particle modes to the total aerosol number, surface area, and volume. The graph shows that the size-distribution curve of OPAC Arctic model AR exhibits intermediate features between those of aerosol models AD and ASB, in which the coarse particle modes provide significant contributions to the total number, surface area, and volume. Conversely, both models AH and BFFS present a dominant mode of accumulation/coarse particles in the curve of $N(a)$ and significant contributions to curves $S(a)$ and $V(a)$ within the size range from 1 to 10 μm.

6.4.11
Polar Antarctic Aerosols

Particles sampled at the coastal and high-altitude Antarctic sites exhibit in general very low mass concentrations during the whole year, since the sampling stations are very far from the most intense sources of particles (dust, anthropogenic pollution, forest fires) located in the midlatitude land regions of the southern hemisphere, and the contributions of maritime wind-forced particles and crustal wind-mobilized particles from local natural sources are in general rather weak because of the large ice- and snow-covered areas throughout the whole year. At the Antarctic coastal sites, both fine and coarse particle mass contents were found to be predominantly given by sea-salt aerosols, with (i) a first mode of Aitken nuclei, centered at a_c close to $7 \times 10^{-2}\,\mu m$ and a mass percentage of ~1%; (ii) an accumulation particle mode with $a_c \approx 0.30\,\mu m$ (14%); (iii) a first coarse particle mode with $a_c \approx 2\,\mu m$ (22%); and (iv) a second coarse particle mode with $a_c \approx 6.5\,\mu m$ (63%). The nuclei/accumulation particles consisted on average of 72% sulfate and 19% sea salt, with minor fractions of ammonium and nitrate, and coarse particles of ~80% sea salt and ~15% sulfate and dust. On the Antarctic Plateau, the aerosol size-distribution curves were found to consist of a nuclei/accumulation particle mode mainly composed of nss sulfate and MSA substances together with minor fractions of water-soluble substances during the austral summer, when aerosols are predominantly deposited on the ground by strong subsidence effects from the FT and the coarse particle mode mainly contains sea salt. On the basis of the previous measurements, the following

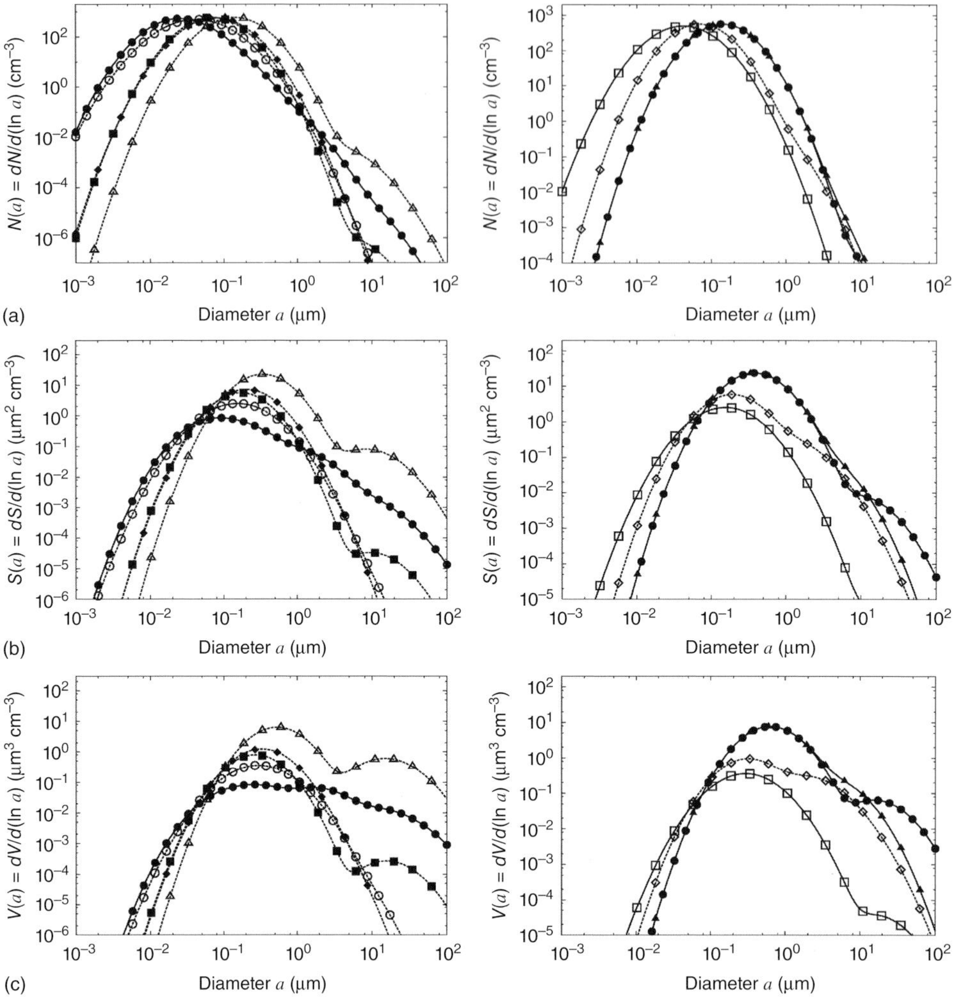

Figure 6.19 Left-hand side: As in Figure 6.14, for the following Arctic aerosol models: (i) OPAC model AR (solid curves with solid circles) of Hess, Koepke, and Schult (1998), (ii) model ASB for Arctic summer background aerosol (dashed curves with solid squares) of Tomasi *et al.* (2015a), (iii) model AH for Arctic haze (dashed curves with open circles) of Tomasi *et al.* (2015a), (iv) model AD for Asian dust (dashed curves with open triangles) of Tomasi *et al.* (2015a), and (v) model BFFS for boreal forest fire smoke particles (dashed curves with solid diamonds) of Tomasi *et al.* (2015a). The five aerosol number size-distribution curves are normalized to $N_T = 10^3$ cm^{-3}. Right-hand side: As in the left-hand side for the following Antarctic aerosol models: (i) OPAC model AN (dashed curves with open diamonds) of Hess, Koepke, and Schult (1998), (ii) model AASC for Antarctic austral summer coastal aerosols (solid curves with solid circles) of Tomasi *et al.* (2015a,b), (iii) model ASAP for austral summer particles on the Antarctic Plateau (solid curves with open squares) of Tomasi *et al.* (2015a,b), and (iv) model AAWC for Antarctic austral winter coastal particles (solid curves with solid triangles) of Tomasi *et al.* (2015a,b). The four aerosol number size-distribution curves are normalized to $N_T = 10^3$ cm^{-3}.

aerosol models were determined by Tomasi *et al.* (2015b): (i) model AASC, represented by a combination of a nuclei/accumulation particle mode with a coarse sea-salt particle mode; (ii) model ASAP, given by the combination of a fine sulfate particle mode with a coarse sea-salt particle mode; and (iii) model AAWC, consisting of a nuclei/accumulation particle mode composed of nss sulfate and MSA and a pronounced coarse mode of sea-salt aerosol mainly transported from the large ice-free oceanic areas offshore of Antarctica. The size-distribution curves of $N(a)$, $S(a)$, and $V(a)$ obtained for the three Antarctic aerosol models are compared in the right-hand side part of Figure 6.19 with the OPAC Antarctic aerosol model AN of Hess, Koepke, and Schult (1998), which presents appreciably lower values of $N(a)$ than those of coastal aerosol models AASC and AAWC within the coarse particle size range and considerably higher values of $N(a)$ than those of high-altitude model ASAP. The size-distribution curves of $S(a)$ and $V(a)$ provide important coarse particle contributions in model AN as well as in models AASC and AAWC.

6.4.12
Stratospheric Volcanic Aerosols

BG aerosols are often sampled at low stratospheric levels during the volcanic quiescence periods over the $0.2 \leq a \leq 4\,\mu m$ size range, presenting a number concentration peak of $10^{-1}\,cm^{-3}$ at altitudes of 17–20 km, rapidly decreasing beyond the 23 km level. These particles are composed of ∼75% SA and ∼25% LW and constitute the so-called stratospheric sulfate layer, which is often named "Junge layer" to honor the memory of Prof. C. E. Junge who discovered it in the late 1950s. As mentioned earlier, these stratospheric sulfate aerosols are primarily produced by the oxidation of SO_2 to SO_3 occurring in the lower stratosphere through a sequence of three chemical reactions, the first involving OH and a third molecule, the second a molecule of O_2, and the third an oxygen atom and a third molecule, after which SO_3 reacts with H_2O to form H_2SO_4. The conversion of SA vapor to liquid H_2SO_4 can occur by (a) combination of molecules of H_2SO_4 and H_2O and/or of H_2SO_4, H_2O, and HNO_3 to form new primarily SA droplets or by (b) vapor condensation of H_2SO_4, H_2O, and HNO_3 onto the surfaces of preexisting particles with sizes $a > 0.30\,\mu m$. Model calculations suggest that the second mechanism is the more likely route in the stratosphere. The tropical stratosphere is probably the major region where such a nucleation process occurs, the aerosols being then transported to higher latitudes by large-scale atmospheric motions. The size-distribution curves of $N(a)$, $S(a)$, and $V(a)$ of BG stratospheric aerosol model PV-1 are shown in Figure 6.20 to illustrate the atmospheric turbidity conditions of the low stratosphere during the volcanic quiescence periods in Antarctica (Pueschel *et al.*, 1989; Tomasi *et al.*, 2013). They are compared in Figure 6.20 with the curves obtained for the three following aerosol models: (i) trimodal model ESV defined by Tomasi *et al.* (2013) on the basis of the King, Harshvardhan, and Arking (1984) data (see Figure 6.5); (ii) aerosol model PV-2, consisting of stratospheric post-Pinatubo 2-month aged volcanic particles; and (iii) aerosol model PV-3, consisting

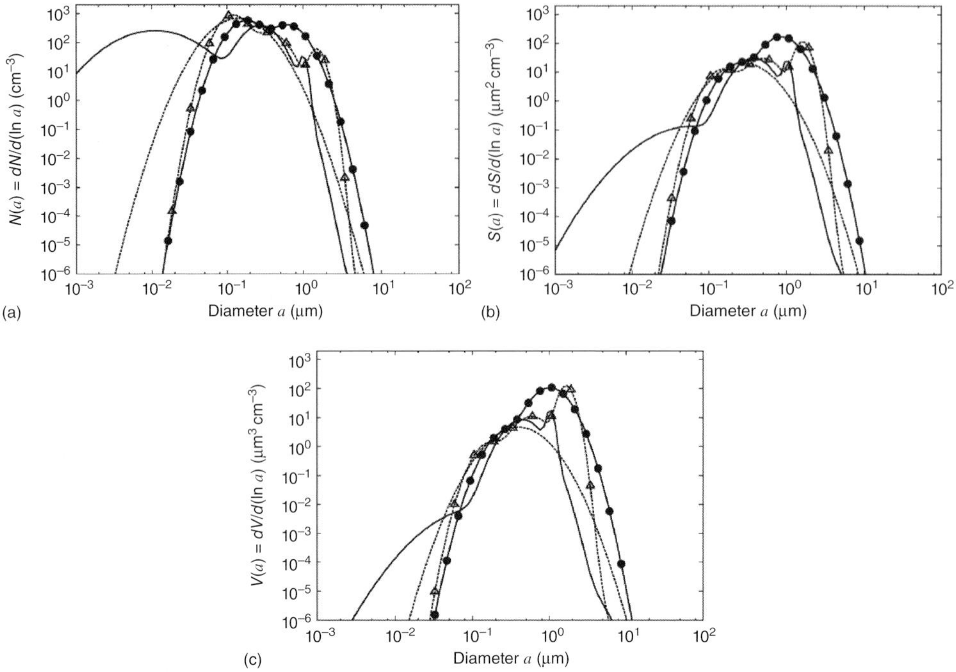

Figure 6.20 As in Figure 6.14, for the following aerosol models: (i) aerosol model ESV for the El Chichón stratospheric volcanic particles (solid curves), defined in the present study using the King, Harshvardhan, and Arking (1984) data; (ii) aerosol model PV-1 for background stratospheric particles (dashed curves) during the long volcanic quiescence period over Antarctica before the Mt Pinatubo eruption (Tomasi *et al.*, 2013); (iii) aerosol model PV-2 for the 2-month aged stratospheric Mt Pinatubo volcanic particles (solid curves with solid circles), defined by (Tomasi *et al.*, 2015a) on the basis of the airborne spectral measurements of the Mt Pinatubo volcanic aerosol extinction coefficient conducted in July–August 1991 over the Caribbean region by Pueschel *et al.* (1993); and (iv) aerosol model PV-3 for the 9-month aged stratospheric Mt Pinatubo volcanic particles (dashed curves with open triangles), defined by (Tomasi *et al.*, 2015a) from airborne measurements of the Mt Pinatubo volcanic aerosol extinction coefficient performed in March 1992 by Pueschel *et al.* (1993). The four stratospheric volcanic particle number size-distribution curves are normalized to $N_T = 10^3$ cm^{-3}.

of stratospheric post-Pinatubo 9-month aged volcanic particles. The comparison highlights that the various models closely agree one with the others, except model ESV presenting a considerably higher number concentration of the nuclei mode than those given by the three PV models of Tomasi *et al.* (2013).

6.5 Concluding Remarks

A numerous set of aerosol number, surface, and volume size-distribution curves are provided in the present study together with their main optical characteristics observable for the various aerosol types in the Earth's atmosphere. Part of these aerosol models is here defined with their main radiative parameters, such as the real and imaginary parts of refractive index, the volume extinction and scattering coefficients, the single scattering albedo, and the asymmetry factor calculated at various visible and infrared wavelengths. Because of their different sources, the aerosol size-distribution curves were estimated to exhibit pronounced multimodal characteristics over a very large size range, from less than 2×10^{-3} μm to more than 50 μm, consisting of ultrafine, nuclei, accumulation, and coarse particle modes. Among them, accumulation and coarse particles were found to yield the most important scattering and absorption effects on the incoming solar radiation, presenting spectral features which vary with their origins and the air moisture conditions. A characterization of the multimodal number, surface, and volume size-distribution curves is made in the present study for remote continental, rural-continental, free tropospheric, CP, MC, mixed maritime–continental, desert dust, biomass burning, urban/industrial, polar Arctic, polar Antarctic, and stratospheric volcanic aerosols. The perturbations directly induced by airborne aerosols on the radiation budget of the surface–atmosphere climate system are mainly caused by scattering and absorption of incoming solar radiation. These effects on the radiative transfer processes occurring in the atmosphere produce in general a marked decrease in the flux density of global solar radiation reaching the surface, due in part to aerosol absorption, and a relative increase in the solar radiation fraction reflected back to space. Within the atmospheric layers where aerosols contain significant mass fractions of soot and other organic substances, a relevant percentage of incoming solar radiation is absorbed, leading to an appreciable warming of such atmospheric layers (Kaufman, Tanré, and Boucher, 2002). The present sets of optical data calculated for 43 aerosol extinction models can be conveniently used to evaluate the scattering and absorption effects produced by the various aerosol types. In addition, the analysis of the size-distribution curves of $N(a)$, $S(a)$, and $V(a)$ shown in Figures 6.14–6.20 for numerous aerosol models subdivided into 11 aerosol classes provides a measure of the contributions given by nuclei, accumulation, and coarse particle modes to the overall number concentration, particulate extinction, and volume of the aerosols associated with various environmental conditions of the Earth's atmosphere.

Abbreviations

AD	Asian dust
AH	Arctic Haze
AOT	aerosol optical thickness
BC	black carbon
BG	background
BL	atmospheric boundary layer
CCNs	Cloud Condensation Nuclei
DARF	direct aerosol radiative forcing
DL	dustlike component of basic aerosol in the 6S code
DMS	dimethylsulfide

EC	elemental carbon
FT	free troposphere
HITRAN	high-resolution transmission database
INS	water-insoluble component of the OPAC aerosol models
LW	liquid water
MAM	accumulation particle component of the OPAC mineral aerosol models
MBL	marine boundary layer
MCM	coarse particle component of the OPAC mineral aerosol models
MNM	nucleation particle component of the OPAC mineral aerosol models
MSA	methane sulfonic acid
MTR	mineral particle component of the OPAC aerosol models
nss	non-sea-salt (used for aerosol particles)
OC	oceanic component of basic aerosol in the 6S code
OM	organic matter component of the OPAC aerosol models
OPAC	optical properties of aerosols and clouds
PBL	planetary boundary layer of the atmosphere
PM	particulate matter
PSAP	particle soot absorption photometer
RH	relative humidity of air
6S	Second Simulation of the Satellite Signal in the Solar Spectrum code
SA	sulfuric acid H_2SO_4
SAM	sea-salt accumulation particle mode
SCM	sea-salt coarse particle mode
SDR	sulfate component of the OPAC aerosol models
SO	soot component of basic aerosol in the 6S code
SOO	soot component of the OPAC aerosol models
SPC	San Pietro Capofiume
SSA	single-scattering albedo
VOCs	volatile organic compounds
WAS	water-soluble component of the OPAC aerosol models
WCP	World Climate Programme
WINOM	water-insoluble organic matter
WS	water-soluble component of basic aerosol in the 6S code
WSIM	water-soluble inorganic matter
WSOM	water-soluble organic matter

List of Symbols

Latin

a	diameter of a spherical particle
a_c	mode diameter of an aerosol size-distribution curve
a_m	geometrical mean diameter of an aerosol size-distribution curve
$g(\lambda)$	asymmetry factor of an aerosol particle polydispersion at wavelength λ

$K_{ext}(r, \lambda)$, $K_{sca}(r, \lambda)$, $K_{abs}(r, \lambda)$	extinction, scattering, and absorption efficiency factors of an aerosol particle as a function of particle radius r and wavelength λ
$m(\lambda)$	complex refractive index of particulate matter at wavelength λ, equal to $n(\lambda) - i\chi(\lambda)$
$M(a)$	size-distribution function of aerosol mass concentration
M_{DL}	mass fraction of the 6S dust-like (DL) aerosol component
M_i	mass fraction of the ith M-type (RH = 50%) aerosol model
M_{ins}	mass concentration of insoluble substances in the OPAC particulate matter (usually measured in $\mu g\,m^{-3}$)
M_{LW}	mass fraction of liquid water in the M-type (RH = 50%) aerosol models
M_{md}	mass fraction of mineral dust containing quartz and silicates in the FT and BL aerosol models
M_{OC}	mass fraction of the 6S oceanic (OC) aerosol component
M_p	overall mass concentration of atmospheric particulate matter (usually measured in μg per cubic meter of air)
M_{so}	mass fraction of soot substances in the FT and BL aerosol models
M_{SO}	mass fraction of the 6S soot (SO) aerosol component
M_{soo}	mass concentration of soot substances in the OPAC particulate matter (usually measured in $\mu g\,m^{-3}$)
M_{ss}	mass fraction of sea salt particles in the FT and BL aerosol models
M_{WS}	mass fraction of the 6S water-soluble (WS) aerosol component
M_{ws}	mass fraction of water-soluble substances in the FT and BL aerosol models
M_{wss}	mass concentration of water-soluble substances in the OPAC particulate matter (usually measured in $\mu g\,m^{-3}$)
$n(\lambda)$	real part of particulate matter refractive index at wavelength λ
$N(a)$, $N(r)$	aerosol particle number density size-distribution curve given as a function of diameter a or radius r
N_0	ground-level aerosol number concentration
N_i	particle number concentration of the ith mode of a multimodal aerosol lognormal size-distribution curve
N_j	particle number concentration of the jth OPAC basic component
N_T	overall particle number concentration of a multimodal aerosol size-distribution curve (measured in cm^{-3})
N_{tot}	total particle number concentration of aerosol particles per cubic centimeter of air
$p(\cos\theta)$	phase function of an aerosol particle population, normalized to the average energy in all directions and represented as a function of the cosine of scattering angle θ
r	radius of a spherical aerosol particle
r_c	mode radius of a lognormal size-distribution curve (measured in μm)
RH	relative humidity of air

S_T	overall particle surface area of a lognormal size distribution of dry-air aerosol particles
V_g	overall particle volume of a lognormal size distribution of aerosol particles grown by water condensation
V_p	volume percentage of dry particulate matter in wet particles grown by water condensation
V_T	overall particle volume of a lognormal size distribution of dry-air aerosol particles
V_w	volume percentage of liquid water in wet particles grown by water condensation
WS	wind speed (measured in m s^{-1})
x	Mie size parameter equal to $2\pi r/\lambda$

Greek

ϑ_0	apparent solar zenith angle
ρ	dry-air particulate mass density
ω	mean single-scattering albedo of dry-air aerosol particles
θ	scattering angle
λ	wavelength
$\chi(\lambda)$	imaginary part of particulate matter refractive index at wavelength λ
$\omega(\lambda)$	single-scattering albedo of an aerosol particle population at wavelength λ
$\alpha(0.40-0.86\ \mu m)$	mean Ångström's exponent calculated for the values of $\beta_{ext}(\lambda)$ determined over the wavelength range from 0.40 to 0.86 μm
$\omega(0.40-3.75\ \mu m)$	mean single-scattering albedo of the 10 OPAC (RH = 50%) aerosol models and SF aerosol models calculated over the 0.40–3.75 μm wavelength range
$\beta_{ext}(\lambda)$	volume extinction coefficient of an aerosol particle population at wavelength λ
$\beta_{sca}(\lambda)$	volume scattering coefficient of an aerosol particle population at wavelength λ
$\beta_{abs}(\lambda)$	volume absorption coefficient of an aerosol particle population at wavelength λ
Γ_j	mass mixing ratio of the jth OPAC basic component
Γ_p	mass fraction of dry particulate matter in the 6S basic components
φ_r	growth factor of particle radius r by water condensation for the four 6S basic components
ΔV	increase of the overall particle volume of a lognormal size distribution of aerosol particles due to water condensation growth
$\Delta V/V$	increase ratio of the overall particle volume V of a lognormal size distribution of aerosol particles due to water condensation growth
Γ_w	mass fraction of liquid water in the 6S basic components (due to water condensation occurring for RH = 50%)
ρ_w	wet particulate mass density (usually considered at RH = 50%)
σ_g	geometric standard deviation of a lognormal aerosol size-distribution curve

References

Ångström, A. (1964) The parameters of atmospheric turbidity. *Tellus*, **16** (1), 64–75. doi: 10.1111/j.2153-3490.1964.tb00144.x

Burley, J.D. and Johnston, H.S. (1992) Ionic mechanisms for heterogeneous stratospheric reactions and ultraviolet photoabsorption cross sections for NO_2^+, HNO_3, and NO_3^- in sulfuric acid. *Geophys. Res. Lett.*, **19** (13), 1359–1362. doi: 10.1029/92GL01115

Bush, B.C. and Valero, F.P.J. (2002) Spectral aerosol radiative forcing at the surface during the Indian Ocean Experiment (INDOEX). *J. Geophys. Res.*, **107** (D19), 8003. doi: 10.1029/2000JD000020

Carbone, C., Decesari, S., Mircea, M., Giulianelli, L., Finessi, E., Rinaldi, M., Fuzzi, S., Marinoni, A., Duchi, R., Perrino, C., Sargolini, T., Vardè, M., Sprovieri, F., Gobbi, G.P., Angelini, F., and Facchini, M.C. (2010) Size-resolved aerosol chemical composition over the Italian peninsula during typical summer and winter conditions. *Atmos. Environ.*, **44** (39), 5269–5278. doi: 10.1016/j.atmosenv.2010.08.008

Carlson, T.N. and Benjamin, S.G. (1980) Radiative heating rates for Saharan dust. *J. Atmos. Sci.*, **37** (1), 193–213. doi: 10.1175/1520-0469(1980)037<0193:RHRFSD>2.0.CO;2

Carlson, T.N. and Caverly, R.S. (1977) Radiative characteristics of Saharan dust at solar wavelengths. *J. Geophys. Res.*, **82** (21), 3141–3152. doi: 10.1029/JC082i021p03141

Carr, S.B. (2005) The Aerosol Models in MODTRAN: Incorporating Selected Measurements from Northern Australia. Technical Report of the Defence Science and Technology Organisation, No. DSTO-TR-1803, Defence Science and Technology Organisation, Edinburgh, South Australia, 67 pp.

d'Almeida, G.A., Koepke, P., and Shetlle, E.P. (eds) (1991) *Atmospheric Aerosols: Global Climatology and Radiative Characteristics*, A. Deepak Publishing, Hampton, VA, 561 pp.

Deepak, A. and Gerber, H.E. (eds) (1983) *Report of the Experts Meeting on Aerosols and Their Climatic Effects*, WCP-55, 107 pp. (available from World Meteorological Organization, Case Postale No. 5, CH-1211 Geneva, Switzerland).

Deirmendjian, D. (1969) *Electromagnetic Scattering on Spherical Polydispersions*, Elsevier, New York, 290 pp.

Ferrare, R., Feingold, G., Ghan, S., Ogren, J., Schmid, B., Schwartz, S.E., and Sheridan, P. (2006) Preface to special section: Atmospheric Radiation Measurement Program May 2003 Intensive Operations Period examining aerosol properties and radiative influences. *J. Geophys. Res.*, **111**, D05S01. doi: 10.1029/2005JD006908

Hale, G.M. and Querry, M.R. (1973) Optical constants of water in the 200-nm to 200-μm wavelength region. *Appl. Opt.*, **12** (3), 555–563. doi: 10.1364/AO.12.000555

Hänel, G. (1968) The real part of the mean complex refractive index and the mean density of samples of atmospheric aerosol particles. *Tellus*, **20** (3), 371–379. doi: 10.1111/j.2153-3490.1968.tb00378.x

Hänel, G. (1972) Computation of the extinction of visible radiation by atmospheric aerosol particles as a function of the relative humidity, based upon measured properties. *J. Aerosol Sci.*, **3** (5), 377–386. doi: 10.1016/0021-8502(72)90092-4

Hänel, G. (1976) The properties of atmospheric aerosol particles as functions of the relative humidity at thermodynamic equilibrium with the surrounding moist air. *Adv. Geophys.*, **19**, 73–188.

Hänel, G. and Bullrich, K. (1978) Physico-chemical property models of tropospheric aerosol particles. *Beitr. Phys. Atmos.*, **51**, 129–138.

Hänel, G. and Zankl, B. (1979) Aerosol size and relative humidity: water uptake by mixtures of salts. *Tellus B*, **31** (6), 478–486. doi: 10.1111/j.2153-3490.1979.tb00929.x

Hess, M., Koepke, P., and Schult, I. (1998) Optical properties of aerosols and clouds: the software package OPAC. *Bull. Am. Meteorol. Soc.*, **79** (5), 831–844. doi: 10.1175/1520-0477(1998)079<0831:OPOAAC>2.0.CO;2

Hobbs, P.V., Bowdle, D.A., and Radke, L.F. (1985) Particles in the lower troposphere over the high plains of the

United States. Part 1. Size distributions, elemental compositions, and morphologies. *J. Clim. Appl. Meteorol.*, **24** (12), 1344–1356. doi: 10.1175/1520-0450(1985)024<1344:PITLTO>2.0.CO;2

Hofmann, D.J. and Rosen, J.M. (1983a) Stratospheric sulfuric acid fraction and mass estimate for the 1982 volcanic eruption of El Chichón. *Geophys. Res. Lett.*, **10** (4), 313–316. doi: 10.1029/GL010i004p00313

Hofmann, D.J. and Rosen, J.M. (1983b) Sulfuric acid droplet formation and growth in the stratosphere after the 1982 eruption of El Chichón. *Science*, **222** (4621), 325–327. doi: 10.1126/science.222.4621.325

Hummel, J.R., Shettle, E.P., and Longtin, D.R. (1988) A New Background Stratospheric Aerosol Model for Use in Atmospheric Radiation Models. Scientific Report No. 8, AFGL-TR-88-0166, 30 July 1988, Air Force Geophysics Laboratory, Hanscom Air Force Base, Massachusetts.

IPCC (2001) Climate Change *2001*. The Scientific Basis (eds J.T. Houghton, Y. Ding, D.J. Griggs, M. Noguer, P.J. van der Linden, X. Dai, K. Maskell, and C.A. Johnson), Cambridge University Press, Cambridge, New York, 881 pp.

Jaenicke, R. (1993) in *Aerosol-Cloud-Climate Interactions* (ed. P.V. Hobbs), Academic Press, San Diego, CA, pp. 1–31.

Jaenicke, R. and Schutz, L. (1978) Comprehensive study of physical and chemical properties of the surface aerosol in the Cape Verde Islands region. *J. Geophys. Res.*, **83** (C7), 3583–3599. doi: 10.1029/JC083iC07p03585

Kaufman, Y.J., Tanré, D., and Boucher, O. (2002) A satellite view of aerosols in the climate system. *Nature*, **419** (6903), 215–223. doi: 10.1038/nature01091

King, M.D., Harshvardhan, and Arking, A. (1984) A model of the radiative properties of the El Chichón stratospheric aerosol layer. *J. Clim. Appl. Meteorol.*, **23** (7), 1121–1137. doi: 10.1175/1520-0450(1984)023<1121:AMOTRP>2.0.CO;2

Koepke, P., Hess, M., Schult, I., and Shettle, E.P. (1997) Global Aerosol Data Set. Report No. 243, Max Planck Institut für Meteorologie, Hamburg, 44 pp.

McClatchey, R.A., Bolle, H.J., and Kondratyev, K.Ya. (1980) Report of the IAMAP Radiation Commission working group on a Standard Radiation Atmosphere, WMO/IAMAP, 33 pp. (available from AFGL, Hanscom Air Force Base, MA 01731).

Mielonen, T., Aaltonen, V., Lihavainen, H., Hyvärinen, A.-P., Arola, A., Komppula, M., and Kivi, R. (2013) Biomass burning aerosols observed in Northern Finland during the 2010 wildfires in Russia. *Atmosphere*, **4** (1), 17–34. doi: 10.3390/atmos4010017

Palmer, K.F. and Williams, D. (1975) Optical constants of sulfuric acid; application to the clouds of Venus? *Appl. Opt.*, **14** (1), 208–219. doi: 10.1364/AO.14.000208

Patterson, E.M. (1977) Atmospheric extinction between 0.55 μm and 10.6 μm due to soil-derived aerosols. *Appl. Opt.*, **16** (9), 2414–2418. doi: 10.1364/AO.16.002414

Patterson, E.M. (1981) Optical properties of the crustal aerosol: relation to chemical and physical characteristics. *J. Geophys. Res.*, **86** (C4), 3236–3246. doi: 10.1029/JC086iC04p03236

Patterson, E.M., Gillette, D.A., and Stockton, B.H. (1977) Complex index of refraction between 300 and 700 nm for Saharan aerosols. *J. Geophys. Res.*, **82** (21), 3153–3160. doi: 10.1029/JC082i021p03153

Perrone, M.R., Santese, M., Tafuro, A.M., Holben, B., and Smirnov, A. (2005) Aerosol load characterization over south-east Italy by one year of AERONET sun-photometer measurements. *Atmos. Res.*, **75** (1–2), 111–133. doi: 10.1016/j.atmosres.2004.12.003

Pueschel, R.F., Kinne, S.A., Russell, P.B., Snetsinger, K.G., and Livingston, J.M. (1993) in IRS '92: Current Problems in Atmospheric Radiation, Proceedings of the International Radiation Symposium, Tallinn (Estonia), August 2–8, 1992 (eds S. Keevallik and O. Kärner), A. Deepak Publishing, Hampton, VA, pp. 183–186.

Pueschel, R.F., Snetsinger, K.G., Goodman, J.K., Toon, O.B., Ferry, G.V., Oberbeck, V.R., Livingston, J.M., Verma, S., Fong, W., Starr, W.L., and Chan, K.R. (1989) Condensed nitrate, sulfate, and chloride in Antarctic stratospheric aerosols. *J. Geophys. Res.*, **94** (D9), 11271–11284. doi: 10.1029/JD094iD09p11271

Remer, L.A., Kaufman, Y.J., Holben, B.N., Thompson, A.M., and McNamara, D. (1998) Biomass burning aerosol size distribution and modeled optical properties. *J. Geophys. Res.*, **103** (D24), 31879–31891. doi: 10.1029/98JD00271

Seinfeld, J.H. and Pandis, S.N. (2006) Atmospheric Chemistry and Physics, from Air Pollution to Climate Change, 2nd edn, John Wiley & Sons, Inc., Hoboken, NJ, 1225 pp.

Shettle, E.P. (1984) in IRS *'84*: Current Problems in Atmospheric Radiation, Proceedings of the International Radiation Symposium, Perugia, Italy, August 21–28, 1984 (ed. G. Fiocco), A. Deepak Publishing, Hampton, VA, pp. 74–77.

Shettle, E.P. and Fenn, R.W. (1979) Models for the Aerosols of the Lower Atmosphere and the Effects of Humidity Variations on Their Optical Properties. Environmental Research Papers, No. 676, AFGL-Technical Report 79–0214, Air Force Geophysics Laboratory, Hanscom AFB, MA, 94 pp.

Stone, R.S., Anderson, G.P., Andrews, E., Dutton, E.G., Shettle, E.P., and Berk, A. (2007) Incursions and radiative impact of Asian dust in northern Alaska. *Geophys. Res. Lett.*, **34**, L14815. doi: 10.1029/2007GL029878

Takemura, T., Nozawa, T., Emori, S., Nakajima, T.Y., and Nakajima, T. (2005) Simulation of climate response to aerosol direct and indirect effects with aerosol transport-radiation model. *J. Geophys. Res.*, **110**, D02202. doi: 10.1029/2004JD005029

Tomasi, C., Lanconelli, C., Lupi, A., and Mazzola, M. (2013) in *Light Scattering Reviews*, Springer-Praxis Books in Environmental Sciences, vol. **8**, Radiative Transfer and Light Scattering (ed. A.A. Kokhanovsky), Springer-Verlag, Berlin, Heidelberg, pp. 505–626. doi: 10.1007/978-3-642-32106-1

Tomasi, C., Lanconelli, C., Lupi, A., and Mazzola, M. (2015a) in *Light Scattering Reviews*, Springer-Praxis Books in Environmental Sciences, vol. **9**, Light Scattering and Radiative Transfer (ed. A.A. Kokhanovsky), Springer-Verlag, Berlin, Heidelberg, pp. 297–425. doi: 10.1007/978-3-642-37985-7

Tomasi, C., Kokhanovsky, A.A., Lupi, A., Ritter, C., Smirnov, A., O'Neill, N.T., Stone, R.S., Holben, B.N., Nyeki, S., Wehrli, C., Stohl, A., Mazzola, M., Lanconelli, C., Vitale, V., Stebel, K., Aaltonen, V., de Leeuw, G., Rodriguez, E., Herber, A.B., Radionov, V.F., Zielinski, T., Petelski, T., Sakerin, S.M., Kabanov, D.M., Xue, Y., Mei, L., Istomina, L., Wagener, R., McArthur, B., Sobolewski, P.S., Kivi, R., Courcoux, Y., Larouche, P., Broccardo, S., and Piketh, S.J. (2015b) Aerosol remote sensing in polar regions. *Earth Sci. Rev.*, **140**, 108–157. doi: 10.1016/j.earscirev.2014.11.001

Tomasi, C., Lupi, A., Mazzola, M., Stone, R.S., Dutton, E.G., Herber, A., Radionov, V.F., Holben, B.N., Sorokin, M.G., Sakerin, S.M., Terpugova, S.A., Sobolewski, P.S., Lanconelli, C., Petkov, B.H., Busetto, M., and Vitale, V. (2012) An update on polar aerosol optical properties using POLAR-AOD and other measurements performed during the International Polar Year. *Atmos. Environ.*, **52**, 29–47. doi: 10.1016/j.atmosenv.2012.02.055

Tomasi, C., Prodi, F., and Tampieri, F. (1979) Atmospheric turbidity variations caused by layers of Sahara dust particles. *Beitr. Phys. Atmos.*, **52** (3), 215–228.

Tomasi, C. and Tampieri, F. (1977) Size distributions of tropospheric particles in terms of the modified gamma function and relationships between skewness and mode radius. *Tellus*, **29** (1), 66–74. doi: 10.1111/j.2153-3490.1977.tb00710.x

Tomasi, C., Vitale, V., Lupi, A., Cacciari, A., Marani, S., and Bonafè, U. (2003) Marine and continental aerosol effects on the upwelling solar radiation flux in Southern Portugal during the ACE-2 experiment. *Ann. Geophys.*, **46** (2), 467–479. doi: 10.4401/ag-3420

Tomasi, C., Vitale, V., and Tarozzi, L. (1997) Sun-photometric measurements of atmospheric turbidity variations caused by the Pinatubo aerosol cloud in the Himalayan region during the summer periods of 1991 and 1992. *Il Nuovo Cimento C*, **20** (1), 61–88.

Turco, R.P., Whitten, R.C., and Toon, O.B. (1982) Stratospheric aerosols: observation and theory. *Rev. Geophys. Space Phys.*, **20** (2), 233–279. doi: 10.1029/RG020i002p00233

Twitty, J.T. and Weinman, J.A. (1971) Radiative properties of carbonaceous aerosols. *J. Appl. Meteorol.*, **10** (4), 725–731. doi: 10.1175/1520-0450(1971)010<0725:RPOCA>2.0.CO;2

Vermote, E.F., Tanré, D., Deuzé, J.L., Herman, M., and Morcrette, J.-J. (1997a) Second simulation of the satellite signal in the solar spectrum (6S): an overview. *IEEE Trans. Geosci. Remote Sens.*, **35** (3), 675–686. doi: 10.1109/36.581987

Vermote, E., Tanré, D., Deuzé, J.L., Herman, M., and Morcrette, J.J. (1997b) Second Simulation of the Satellite Signal in the Solar Spectrum *(6S)*, 6S User Guide Version 2, July 1997, University of Lille, (France), 218 pp.

Volz, F.E. (1972a) Infrared absorption by atmospheric aerosol substances. *J. Geophys. Res.*, **77** (6), 1017–1031. doi: 10.1029/JC077i006p01017

Volz, F.E. (1972b) Infrared refractive index of atmospheric aerosol substances. *Appl. Opt.*, **11** (4), 755–759. doi: 10.1364/AO.11.000755

Volz, F.E. (1973) Infrared optical constants of ammonium sulfate, Sahara dust, volcanic pumice, and fly ash. *Appl. Opt.*, **12** (3), 564–568. doi: 10.1364/AO.12.000564

WCP-112 (1986) A Preliminary Cloudless Standard Atmosphere for Radiation Computation, World Climate Programme (WCP)/International Association of Meteorology and Atmospheric Physics (IAMAP) Radiation Commission, WCP-112-TD-24, World Meteorological Organization (WMO), Geneva, 53 pp.

Whitby, K.T. (1978) The physical characteristics of sulfur aerosols. *Atmos. Environ.*, **12** (1–3), 135–159. doi: 10.1016/0004-6981(78)90196-8

WMO (1983) World Meteorological Organization (WMO/CAS)/Radiation Commission of IAMAP Meeting of Experts on Aerosols and Their Climatic Effects, WCP 55, Williamsburg, VA, March 28–30, 1983.

7
Remote Sensing of Atmospheric Aerosol

Alexander A. Kokhanovsky, Claudio Tomasi, Boyan H. Petkov, Christian Lanconelli, Maurizio Busetto, Mauro Mazzola, Angelo Lupi, and Kwon H. Lee

7.1
Introduction

Atmospheric aerosol is a global phenomenon. Every cubic centimeter of air contains hundreds or even thousands of small liquid and solid particles of different chemical composition, internal structure, and shape (dust, soot, etc.). The aerosol concentration depends on the location around the globe and the altitude being generally (but not always!) smaller at higher levels in atmosphere. Atmospheric aerosol influences precipitation, weather and climate change, cloud formation and optical properties, air quality, atmospheric radiative transfer, dynamics, and thermodynamics. Therefore, it is of paramount importance to develop techniques for monitoring aerosol properties on a global (say, using satellites) and local (e.g., particulate matter concentrations in cities and at background sites) scales. The main targets for monitoring are:

1. Number concentration of particles
2. Particulate matter mass concentration
3. Aerosol particle size distribution or effective radius
4. Shape and internal structure of aerosol particles
5. Chemical composition
6. Complex refractive index
7. Local optical properties (light absorption, scattering and extinction coefficients, single scattering albedo, optical thickness, asymmetry parameter, polarization characteristics, and phase matrix/function)
8. Vertical profiles of various aerosol parameters (say, spectral backscattering coefficient and depolarization ratio as measured by a *li*ght *d*etection *a*nd *r*anging (LIDAR) or *in situ* observations).

Atmospheric Aerosols: Life Cycles and Effects on Air Quality and Climate, First Edition.
Edited by Claudio Tomasi, Sandro Fuzzi, and Alexander Kokhanovsky.

Optical methods are most suitable for the aerosol remote sensing because the main fraction of aerosol has sizes close to that of a visible (VIS)/shortwave infrared (SWIR) light wavelength. Therefore, aerosols effectively scatter light, and their properties can be derived from various light scattering (and also light absorption/extinction) experiments. This is done by employing various optical instrumentation placed on satellites, ships, and aircrafts. Several ground networks of optical instruments for aerosol monitoring exist (such as AERONET, SKYNET, and EARLINET, to name a few). The main aim of this chapter is to review main methods and recent results obtained in a broad area of aerosol remote sensing using optical methods. Ground-based, airborne, and satellite techniques are presented.

7.2
Ground-Based Aerosol Remote Sensing Measurements

Ground-based remote sensing techniques are commonly used in aerosol studies to determine the main optical parameters of aerosol particles by employing various instruments, such as multispectral Sun photometers and Sun/sky radiometers, nephelometers, aethalometers, absorption photometers, LIDARs, and optical particle counters (OPC):

A. Measurements of aerosol optical thickness (AOT) $\tau_a(\lambda)$ are performed at numerous ultraviolet (UV), VIS, and near-infrared (NIR) wavelengths using a multispectral Sun photometer or a Sun/sky radiometer, such as the Cimel CE318 model manufactured by Cimel Electronique (France) and used in the AERONET network (Holben *et al.*, 1998; Eck *et al.*, 1999) and the Prede POM-01L and POM-02L models manufactured by Prede Incorporated (Japan) and employed in the SKYNET network (Nakajima *et al.*, 1996, 2003). Examining the spectral series of $\tau_a(\lambda)$ measured over the 0.400–0.870 µm spectral range, the best-fit values of Ångström (1964) wavelength exponent α and atmospheric turbidity parameter $\beta(\lambda = 1\,\mu m)$ can be easily calculated. The fine particle fraction $\eta_f(0.50\,\mu m)$, giving the fraction of $\tau_a(\lambda)$ due to fine particle extinction, can be determined by applying the O'Neill, Dubovik, and Eck (2001) procedure to the spectral series of $\tau_a(\lambda)$ derived from the Sun-photometer measurements over the wavelength range from 0.40 to 0.87 µm. This procedure allows one to exploit the spectral curvature information in $\tau_a(\lambda)$ to permit a direct estimation of the fine-mode Ångström exponent α_f and the optical fraction η_f of $\tau_a(\lambda)$ due to the extinction effects produced by the fine particle mode. More details on the O'Neill, Dubovik, and Eck (2001) procedure are provided in Section 7.2.1.4. In addition to the aforementioned solar irradiance measurements, the Cimel Sun-photometer models equipped with polarization filters (in three directions) are often used to take daily measurements of the linear polarization produced by aerosol particles on the principal plane horizon, these measurements being performed for various solar zenith angles throughout the measurement day (Vermeulen *et al.*, 2001). The most recent versions of this instrument are equipped with triplets of polarizing filters peaked at UV (340 and 380 nm), VIS (440, 500, and 675 nm), and infrared (IR) (870, 1020, and 1640 nm) wavelengths.

B. Using the aforementioned AERONET and SKYNET Sun/sky radiometers, routine measurements of sky brightness in the solar almucantar (i.e., of solar aureole) can be carried out for values of apparent solar zenith angle ϑ_0 varying from 50° to 80°, together with direct solar irradiance measurements. The solar almucantar plane contains the Sun and the sky strip having the same altitude of the Sun. Measurements of diffuse solar radiation are carried out along such a strip at both sides of the Sun for several angles usually taken in regular steps. Examining these measurements by using the flexible inversion algorithms developed by Nakajima *et al.* (1996) or those of Dubovik and King (2000), different aerosol optical parameters can be retrieved from the Sun/sky-radiometer field measurements, such as the spectral complex refractive index, the asymmetry factor, the single scattering albedo of atmospheric aerosols suspended in the vertical column of the atmosphere, and their aerosol volume size-distribution curves over the particle radius range from 0.05 to 15 μm.

C. The nephelometer is an instrument designed to measure the scattering effects of aerosol particles suspended in air by employing a light beam and a light detector set to one side (often under an angle of 90°) or backward with respect to the source beam, so that particle density is then measured as a function of the light reflected into the detector from the particles. More technical details on this instrument are available at http://www.esrl.noaa.gov/gmd/aero/instrumentation/RR903_manual.pdf. A photograph of the Radiance Research (RR) nephelometer is shown in Figure 7.1.

D. The aethalometer is an instrument designed to measure the concentration of optically absorbing suspended particles. It uses an optical transmission technique to determine the mass concentration of black carbon (BC) soot particles collected from an airstream passing through a filter using a pair of light sources with wavelengths of 370 nm (or 325 nm) for aromatic organic compounds and 880 nm (or 800 nm) for BC. In the more sophisticated models, a set of sources is used having wavelengths ranging from 370 to 950 nm. The Particle Soot Absorption Photometer (PSAP) manufactured by RR and shown in Figure 7.1 can be alternatively used to measure in near real time the aerosol absorption coefficient. The method is based on the integrating plate technique in which the optical transmission change caused by particle deposition is measured, from which the light absorption coefficient of the deposited particles is evaluated in terms of the well-known Bouguer–Lambert–Beer law (as the natural logarithm of the ratio between source intensity and attenuated intensity) (see at http://www.esrl.noaa.gov/gmd/aero/instrumentation/psap.html). Older versions of PSAP instruments operated at one wavelength only (565 nm), while the new versions are usually equipped with three wavelengths (467, 530, and 660 nm).

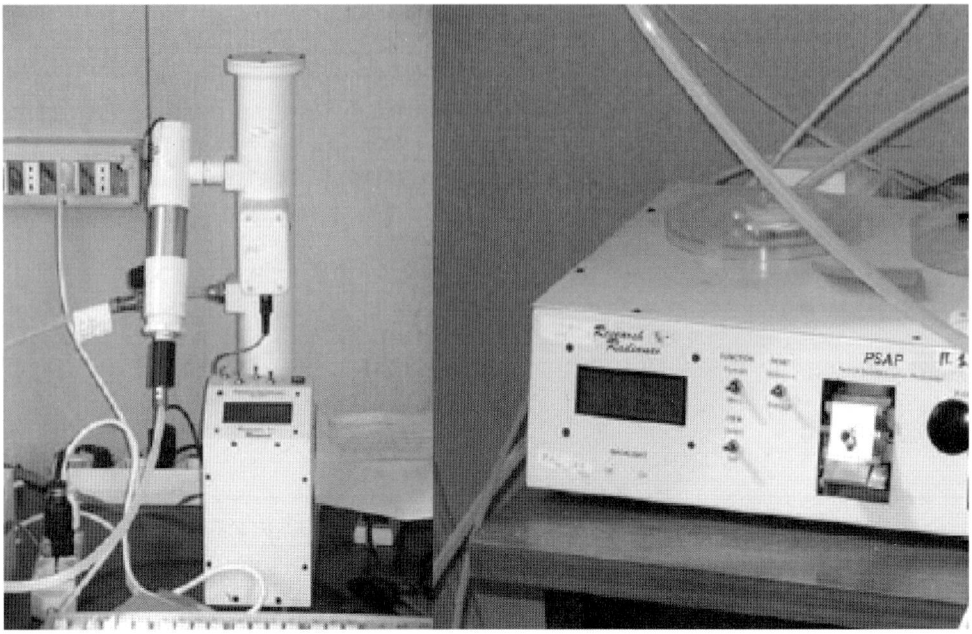

(a) (b)

Figure 7.1 (a) Photograph of the Radiance Research M903 nephelometer employed by the ISAC-CNR group at San Pietro Capofiume (SPC) (Italy) during the AEROCLOUDS experiment campaigns conducted in the 2007–2010 multiyear period to measure the ground-level aerosol scattering coefficient $\beta_{sca}(530\,nm)$ (in m^{-1}). (b) Photograph of the Radiance Research Particle Soot Absorption Photometer (PSAP) measuring at SPC the ground-level aerosol absorption coefficient $\beta_{abs}(573\,nm)$ (in m^{-1}).

E. LIDAR is a remote sensing technology based on the use of a laser and a telescope, the first sending a light pulse through the atmosphere and the second collecting the backscattered LIDAR return signal (see at https://lta.cr.usgs.gov/LIDAR). As the speed of light is known, one can easily calculate the exact atmospheric position from which the LIDAR signal has originated. Therefore, the aerosol vertical profile can be determined from the LIDAR measurements, provided that the scattering properties of aerosol particles are known, at least over the altitude range in which the image of the laser is completely within the field of view (FOV) of the recording telescope. LIDAR pulses can be generated at several UV, VIS, and NIR wavelengths and polarization elements can also be used at the point of light entrance to the telescope to define the shapes of aerosol particles, taking into account that nonspherical particles normally depolarize light while the spherical ones do not depolarize it. Networks of aerosol LIDARs such as the Asian Dust Network (Murayama *et al.*, 2001) and the European Aerosol Research LIDAR Network (Bösenberg *et al.*, 2003) were established to investigate the horizontal and vertical distribution of natural and anthropogenic aerosol plumes in a coherent way on a regional to continental scale.

F. The OPC is an instrument based on the principle of light scattering from particles. It is a real-time instrument that is typically used to measure particles having diameters $a > 0.05\,\mu m$. Therefore, this instrument can provide the size-distribution curve of aerosol particle concentration. A light source (laser) is used to illuminate a stream of aerosol flowing out of a nozzle, and the off-axis from the light beam measures the amount of light scattered from a single particle by refraction and diffraction, both the size and number density of particles being measured at the same time. The size of the particle is figured out by the intensity of the scattered light, provided that the refractive index of the measured aerosol is known. Electronics provides amplification of the low-level signals received from the photodetector and converts each scattered light pulse to a corresponding size category (depending on the pulse height), which is then accumulated in a data logger. Each scattered pulse corresponds to a particle count, and this is incremented in the appropriate size category to obtain particle concentration in a given size interval. For instance, the OPC GRIMM 1.108 "DUSTcheck" model provides the particle number size-distribution curve subdivided into 15 diameter classes from 0.3 μm to 20 μm (Ferrero *et al.*, 2010).

Some examples of field applications of the procedures followed to determine the various optical parameters are presented here, showing the results obtained by applying the previous analysis methods to experimental data sets.

7.2.1
The Multispectral Sun-Photometry Method

Multispectral Sun photometers are instruments specifically designed to carry out ground-level measurements of direct solar irradiance within narrow spectral intervals chosen in the middle of the main windows of the near UV, VIS, and NIR spectrum of the solar radiation passing through the atmosphere and reaching the Earth's surface. In order to obtain these measurements with great accuracy, these instruments need to be equipped with well-defined optical components centered on their optical axis: (i) a quartz window with a circular field stop to prevent dust entering the instrument; (ii) a band-pass filter mounted on the filter wheel, on which a sequence of interference narrowband filters can be positioned step by step, having usually transmission curves with half-bandwidths of 5–10 nm and peak wavelengths carefully chosen within the main windows of the atmospheric transmittance spectrum (those most commonly used are close to around 340, 380, 440, 500, 675, 870, 940 (for precipitable water), 1020, and 1640 nm in the AERONET Cimel CE318 models and to 315, 340, 380, 400, 500, 675, 870, 940 (for precipitable water), 1020, 1600, and 2200 nm in the SKYNET Prede POM-02L models); (iii) an entrance optical condensation system, consisting of a pair of appropriate quartz lenses chosen to avoid appreciable aberration effects; (iv) a frosted quartz diffuser fixed at the focal distance (for green light) from the entrance and combined with a circular diaphragm to achieve an FOV angular diameter ϕ_{FOV} varying from 1° to 1°10′ as λ increases from 315 to 2200 nm, in this way avoiding large errors caused by the diffuse sky radiation entering the instrument; and (v) a stable photodetector with a spectral range from near UV to NIR and a linear responsivity over a broad dynamic range presenting long-term stability characteristics. Both the filter wheel hosting the set of interference filters and the detector system need to be placed inside boxes controlled by thermostat systems capable of maintaining a stable internal temperature within ±1°C during the field measurements to avoid significant variations in the interference filter transmissivity and sensor responsivity performances.

The Sun photometer is usually mounted on a good alt-azimuth system capable of pointing the instrument at the Sun with a precision better than 5′ for FOV diameter $\phi_{\mathrm{FOV}} \approx 1°$. Being the angular diameter of the Sun equal to $\sim 29'$, the sky radiance contribution for $\phi_{\mathrm{FOV}} = 1°$ is totally negligible with respect to the incoming direct solar irradiance at ground level (Tomasi *et al.*, 1999).

7.2.1.1 Calibration of a Sun Photometer Using the Langley Plot Method

Assuming that the output voltage $V(\lambda)$ is provided by a Sun photometer pointed at the Sun for a certain apparent solar zenith angle ϑ_0 and using an interference filter peaked at wavelength λ, it can be stated that the total atmospheric optical thickness $\tau(\lambda)$ is given by the following simple formula based on the Bouguer–Lambert–Beer law:

$$\tau(\lambda) = \frac{1}{m} \ln \left[\frac{F_e V_o(\lambda)}{V(\lambda)} \right], \tag{7.1}$$

where m is the relative optical air mass of the total atmosphere varying as a function of ϑ_0, calibration constant $V_o(\lambda)$ is the output voltage that would be provided by the Sun photometer when pointed at the Sun on a platform located outside the atmosphere (i.e., for null m) for annual mean Earth–Sun distance $D_{\mathrm{E-S}}$, and F_e is the eccentricity correction factor equal to the square power of ratio $D_{\mathrm{E-S}}/D$ between the annual mean Earth–Sun distance $D_{\mathrm{E-S}}$ and the daily Earth–Sun distance D. Using factor F_e in Eq. (7.1), it is possible to take into account that extraterrestrial solar irradiance varies from one day to another throughout the year as a function of distance D. From astronomical calculations, factor F_e is equal to 1.0350 on 1 January and to 1.0351 from 2 to 5 January, subsequently gradually decreases day by day until reaching a value of 0.9666 on 1 July, remains constant until to 8 July, and then starts to increase again until reaching the value of 1.0350 on 30 and 31 December (Iqbal, 1983).

The calibration of a Sun photometer is performed in order to determine the values of constants $V_o(\lambda)$ at all the peak wavelengths λ_p of the interference filters mounted on the instrument. This can be obtained by using standard lamps of known irradiance over the solar radiation spectrum. However, a more precise calibration can be obtained by applying the so-called Langley plot method to Sun-photometer measurements conducted at high-altitude stations outside the planetary boundary layer, where the aerosol load consists of background aerosols only and particulate extinction is stable during the diurnal hours. For instance, AERONET and SKYNET Sun/sky radiometers have been calibrated at the Mauna Loa Observatory (3397 m a.m.s.l., 19°32′N, 155°35′W) in the Hawaii (United States) in the middle of Pacific Ocean (Holben *et al.*, 1998). Intercalibration measurements of various Sun-photometer models have been conducted at the Izaña Observatory (2360 m a.m.s.l., 28°19′N, 16°30′W) in Tenerife (Canary Islands, Spain) in the Atlantic Ocean (Mazzola *et al.*, 2012). Improved applications of the Langley plot method can also be made for low-altitude measurements conducted with a Cimel Sun/sky radiometer (Campanelli *et al.*, 2007). To better illustrate the procedure of the Langley plot method, an application example is shown in Figure 7.2 for the measurements conducted with the UVISIR-1 Sun photometer (Tomasi *et al.*, 1999) at wavelength $\lambda_p = 550\,\text{nm}$, on the top of Sass Pordoi (46°30′N, 11°50′E, 2950 m a.m.s.l..) (Trento, Italy) in the Dolomites (Eastern Alps) on clear-sky days (10 and 11 September, 1985), characterized by stable anticyclonic and cloudless-sky conditions for which values of precipitable water $W < 2.5\,\text{g}\,\text{cm}^{-2}$ were measured during the diurnal hours within the vertical atmospheric column of unit cross section. For such exceptionally low atmospheric turbidity conditions, it is correct to assume that the errors due to the diffuse sky radiance entering the instrument are very low over the whole range of $\vartheta_0 < 82°$ and, hence, for all the values of relative optical air mass $m < 7$.

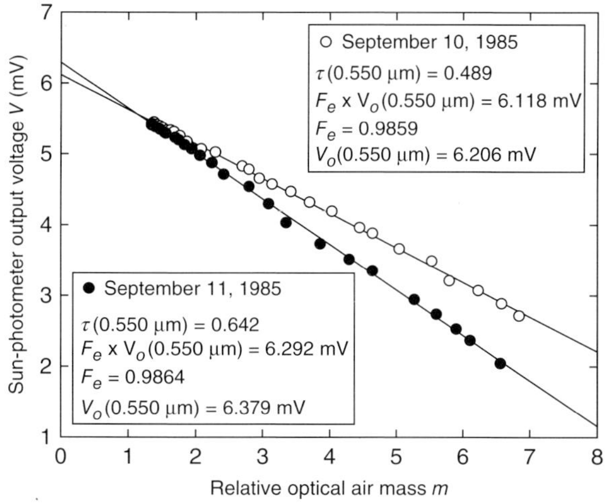

Figure 7.2 Examples of Langley plot best-fit lines determined for the measurements performed during the morning hours of 10 September 1985 (open circles), and September 11 1985 (solid circles), at the top of Sass Pordoi (2950 m a.m.s.l.) (Canazei, Trento, Italy) using the UVISIR-1 sun-photometer (Tomasi, Vitale, and Tagliazucca, 1989; Tomasi *et al.*, 1999) measurements taken at wavelength $\lambda_p = 0.550\,\mu\text{m}$. The values of slope coefficient $\tau(0.550\,\mu\text{m})$ and intercept $F_e V_0(0.550\,\mu\text{m})$ of the two regression lines are given separately for the two calibration days. Both regression lines were obtained for values of regression coefficient r equal to −0.999. An average value of calibration constant $V_0(0.550\,\mu\text{m})$ equal to $6.292(\pm 0.087)\,\text{mV}$ was obtained by applying the Langley plot procedure to the two daily sets of calibration measurements shown in the graph.

Figure 7.2 shows two sequences of output voltages decreasing almost linearly over the range of relative optical air mass m from around 1.5 (at noon) to about 6.8 (after sunrise), yielding regression coefficients very close to 1 in both cases. The two regression lines provide the daily values of intercept $F_e\ V_o(0.550\,\mu m)$ (for $m = 0$) equal to 6.118 on 10 September 1985, and 6.292 on 11 September 1985. Considering that the eccentricity factor F_e is equal to 0.9859 and 0.9864 on such two calibration days, respectively (Iqbal, 1983), the average value of $V_o(0.550\,\mu m)$ was estimated to be equal to $6.292(\pm 0.087)$ mV, therefore for a relative standard deviation equal to 1.4%. When derived from high-altitude calibration measurements conducted for stable atmospheric transparency conditions throughout the observational period, the values of the calibration constant $V_o(\lambda)$ obtained using the Langley plot method can be reliably used in Eq. (7.1) to determine the spectral values of total atmospheric optical thickness $\tau(\lambda)$ at all window wavelengths of the Sun photometer.

7.2.1.2 Determination of Aerosol Optical Thickness

In Eq. (7.1) the product of relative optical air mass m calculated for a molecular atmosphere by total optical thickness $\tau(\lambda)$ yields a measure of the overall extinction affecting the incoming solar radiation along its slant path and is given by the sum of various partial contributions due to Rayleigh scattering, aerosol extinction, and gaseous absorption by water vapor, ozone, nitrogen dioxide and N_2O_4, and oxygen dimer. On this matter, Ångström (1930) pointed out that Fowle (1913) examined a set of atmospheric attenuation measurements conducted with a spectrobolometer at the Astrophysical Observatory at Washington (United States) and was the first to find that the overall atmospheric extinction term $m\ \tau(\lambda)$ given by the Bouguer–Lambert–Beer law can be correctly written as the sum of two terms due to molecular scattering and atmospheric scattering and general absorption, respectively. According to this concept, and considering that the near UV, VIS, and NIR solar radiation is actually attenuated by Rayleigh scattering of air molecules, aerosol extinction (scattering plus absorption), and absorption by the wings of a number of gaseous absorption bands, the product $m\tau(\lambda)$ derived from Eq. (7.1) can be written for wet-air conditions of the atmosphere as the following sum of various partial contributions:

$$m\tau(\lambda) = m\tau_R(\lambda) + m_a\tau_a(\lambda) + m_w\tau_w(\lambda) + m_o\tau_o(\lambda) + m_n\tau_n(\lambda) + m_{od}\tau_{od}(\lambda), \quad (7.2)$$

where (i) suffixes R and a refer to Rayleigh scattering and aerosol particle extinction, respectively, and (ii) suffixes w, o, n, and od refer to water vapor, ozone, nitrogen dioxide (and N_2O_4), and oxygen dimer (O_4), respectively.

In order to calculate the AOT $\tau_a(\lambda)$ from total atmospheric optical thickness $\tau(\lambda)$ derived in terms of Eq. (7.1) from the Sun-photometer measurements, it is of basic importance to determine:

1. The simultaneous partial optical thicknesses due to (i) Rayleigh scattering by air molecules ($\tau_R(\lambda)$) and (ii) absorption by various minor gases of the atmosphere within the near UV, VIS, and NIR windows of the atmospheric

transmission spectrum, produced by water vapor ($\tau_w(\lambda)$), ozone ($\tau_o(\lambda)$), nitrogen dioxide and its dimer N_2O_4 ($\tau_n(\lambda)$), and oxygen dimer O_4 ($\tau_{od}(\lambda)$)
2. The dependence curves of the relative optical air mass parameters m, m_a, m_w, m_o, m_n, and m_{od} used in Eq. (7.2) on angle ϑ_0 by taking into account the main vertical distribution features of the aforementioned atmospheric constituents and the thermodynamic and composition characteristics of the atmosphere in various seasons and for different latitudes.

These calculations are presented in the two following sections.

Calculations of the Partial Optical Thicknesses for Various Atmospheric Constituents
The Rayleigh scattering optical thickness $\tau_R(\lambda)$ can be easily calculated knowing the vertical profiles of air pressure, temperature, and relative humidity (RH) from the surface to the top level of the atmosphere. Optical thickness $\tau_R(\lambda)$ decreases regularly as λ increases being approximately proportional to the inverse of power $\lambda^{4.09}$ at all the near UV, VIS, and NIR wavelengths. Each monochromatic value of $\tau_R(\lambda)$ needs to be accurately calculated in order to correct the total atmospheric optical thickness $\tau(\lambda)$ derived from the Sun-photometer measurements carried out at each window wavelength. In these calculations, one has to take into account that the narrow passband interference filters mounted on these instruments have transmission curves peaked at the selected window wavelengths with (i) half-bandwidths HBW $\approx 0.01\,\mu m$ and ten-bandwidths TBW $\approx 0.02\,\mu m$ for window wavelengths $\lambda < 1.010\,\mu m$, (ii) HBW $\approx 0.04\,\mu m$ and TBW $\approx 0.12\,\mu m$ for the filter peaked at window wavelength $\lambda = 1.600\,\mu m$, and (iii) HBW $\approx 0.05\,\mu m$ and ten-bandwidths TBW $\approx 0.20\,\mu m$ at the window wavelength $\lambda = 2.200\,\mu m$ (Vitale *et al.*, 2000). Therefore, the spectral values of $\tau_R(\lambda)$ given in Table 7.1 were calculated as weighted averages from the spectral values of $\tau_R(\lambda)$ determined by Tomasi *et al.* (2005) at the main window wavelengths of the AERONET and SKYNET Sun/sky radiometers on the basis of the data sets defined by Anderson *et al.* (1986) for six standard atmosphere models over the altitude range from sea level to 120 km by using as weighting functions the transmission curves of the aforementioned narrowband interference filters of the Sun/sky radiometers.

The spectral values of water vapor optical thickness $\tau_w(\lambda)$ within the main windows of the atmospheric transmission spectrum shown in Figure 7.2 can be calculated very simply as the products of precipitable water W by the weak absorption coefficients $k_w(\lambda)$ reported in Table 7.2. These values of $k_w(\lambda)$ were determined by weighting the spectral curves of $k_w(\lambda)$ (determined, for instance, by using the MODTRAN 2/3 code Kneizys *et al.*, 1996) with the previously estimated transmission curves of the various narrowband interference filters mounted on the Sun/sky radiometers. The local measurements of W are routinely conducted by the AERONET and SKYNET sun/sky radiometers and the major part of multispectral sun photometers measuring the so-called hygrometric ratio between the output voltage $V(0.940\,\mu m)$ and one of the nearby window signals $V(0.870\,\mu m)$ or $V(1.020\,\mu m)$, as proposed by Tomasi and Guzzi (1974) and Volz (1974) (see also Tomasi *et al.*, 2000). The spectral values of ozone optical thickness $\tau_o(\lambda)$ can

Table 7.1 Spectral values of the Rayleigh scattering optical thickness $\tau_R(\lambda)$ at the window wavelengths of the AERONET and SKYNET Sun/sky radiometers, as obtained from the calculations made by Tomasi *et al.* (2005) for the six standard atmosphere models defined by Anderson *et al.* (1986).

Wavelength λ (μm)	Model 1 (tropical, 15° N, annual average)	Model 2 (midlatitude winter, 45° N, January)	Model 3 (midlatitude summer, 45° N, July)	Model 4 (subarctic winter, 60° N, January)	Model 5 (subarctic summer, 60° N, July)	Model 6 (US Standard Atmosphere 1976)
0.315	9.912×10^{-1}	9.933×10^{-1}	9.894×10^{-1}	9.878×10^{-1}	9.844×10^{-1}	9.878×10^{-1}
0.340	7.157×10^{-1}	7.173×10^{-1}	7.144×10^{-1}	7.134×10^{-1}	7.108×10^{-1}	7.133×10^{-1}
0.380	4.480×10^{-1}	4.490×10^{-1}	4.472×10^{-1}	4.465×10^{-1}	4.450×10^{-1}	4.465×10^{-1}
0.400	3.616×10^{-1}	3.625×10^{-1}	3.610×10^{-1}	3.604×10^{-1}	3.592×10^{-1}	3.604×10^{-1}
0.500	1.438×10^{-1}	1.442×10^{-1}	1.436×10^{-1}	1.434×10^{-1}	1.429×10^{-1}	1.434×10^{-1}
0.675	4.234×10^{-2}	4.244×10^{-2}	4.225×10^{-2}	4.220×10^{-2}	4.204×10^{-2}	4.220×10^{-2}
0.870	1.518×10^{-2}	1.522×10^{-2}	1.515×10^{-2}	1.514×10^{-2}	1.508×10^{-2}	1.513×10^{-2}
1.020	8.021×10^{-3}	8.040×10^{-3}	8.007×10^{-3}	7.996×10^{-3}	7.967×10^{-3}	7.995×10^{-3}
1.600	1.316×10^{-3}	1.319×10^{-3}	1.313×10^{-3}	1.312×10^{-3}	1.307×10^{-3}	1.311×10^{-3}
2.200	3.678×10^{-4}	3.687×10^{-4}	3.672×10^{-4}	3.667×10^{-4}	3.654×10^{-4}	3.667×10^{-4}

The values of $\tau_R(\lambda)$ have been calculated using the transmission curves of the passband interference filter mounted on the Sun/sky radiometers as weighting functions at various window wavelengths.

Table 7.2 Spectral values of the absorption coefficients of atmospheric water vapor, ozone, and nitrogen dioxide with its dimer, and oxygen dimer optical thickness at the window wavelengths of the AERONET and SKYNET sun/sky radiometers, as obtained from the spectral data shown in Figure 7.2.

Wavelength λ (μm)	Absorption coefficient $k_w(\lambda)$ of atmospheric water vapor (measured in g^{-1} cm^2) (Kneizys *et al.*, 1996)	Absorption coefficient $k_o(\lambda)$ of atmospheric ozone (measured in DU^{-1}) (Inn and Tanaka, 1953; Kneizys *et al.*, 1996)	Absorption coefficient $k_n(\lambda)$ of atmospheric NO_2 and N_2O_4 (measured in (cm STP)$^{-1}$ (Hall and Blacet, 1952; Kneizys *et al.*, 1996)	Oxygen dimer optical thickness $\tau_{od}(\lambda)$ (Michalsky *et al.*, 1999)
0.315	—	1.38×10^{-3}	5.73	—
0.340	—	2.94×10^{-5}	9.82	—
0.380	—	—	1.44	—
0.400	—	1.89×10^{-9}	1.50	—
0.500	8.39×10^{-4}	3.18×10^{-5}	6.34	1.76×10^{-5}
0.675	3.73×10^{-5}	3.98×10^{-5}	—	—
0.870	2.84×10^{-5}	1.46×10^{-6}	—	—
1.020	1.88×10^{-3}	—	—	5.49×10^{-4}
1.600	1.09×10^{-3}	—	—	—
2.200	1.71×10^{-2}	—	—	—

The values of absorption coefficients $k_w(\lambda)$, $k_o(\lambda)$, and $k_n(\lambda)$ and those of $\tau_{od}(\lambda)$ have been calculated using the transmission curves of the passband interference filter mounted on the Sun/sky radiometers as weighting functions at various window wavelengths.

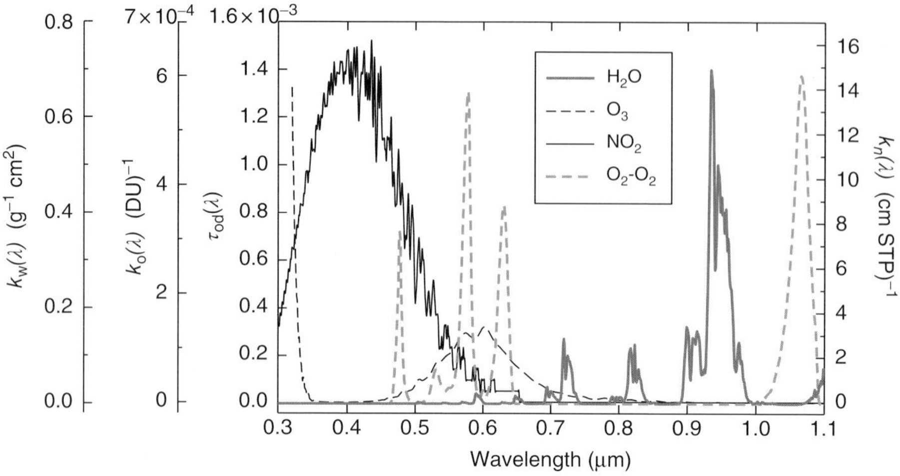

Figure 7.3 Spectral curves of the water vapor absorption coefficient $k_w(\lambda)$, ozone absorption coefficient $k_o(\lambda)$, nitrogen dioxide NO_2 and dimer N_2O_4 absorption coefficient $k_n(\lambda)$, and oxygen dimer (O_4) optical thickness $\tau_{od}(\lambda)$, as derived over the 0.30–1.10 μm wavelength range using the MODTRAN 2/3 computational code and the LOWTRAN 7 model (Kneizys *et al.*, 1996) and additional data available in the literature (Hall and Blacet, 1952; Inn and Tanaka, 1953) for water vapor, ozone, and nitrogen dioxide (+N_2O_4) and the spectral features of $\tau_{od}(\lambda)$ provided by Michalsky *et al.* (1999). Note that NO_2(+N_2O_4) absorption coefficient $k_n(\lambda)$ is measured in (cm STP)$^{-1}$, giving the volume occupied by the NO_2 and N_2O_4 molecules within the vertical atmospheric column of unit cross section, in which standard temperature and pressure (STP) conditions have been assumed at all levels.

be calculated as the products of total-column ozone content by the ozone absorption coefficient $k_o(\lambda)$, which presents rather high values at the wavelengths of 0.315 and 0.340 μm affected by the Huggins band and lower values at the wavelengths ranging from 0.40 to 0.75 μm, which are affected by the weak absorption Chappuis band (Inn and Tanaka, 1953), as can be seen in Figure 7.3.

The spectral values of $k_o(\lambda)$ obtained using the MODTRAN 2/3 code of Kneizys *et al.* (1996) are given in Table 7.2, these values being suitable for the analysis of AERONET and SKYNET data for known values of total ozone. The total atmospheric ozone content is usually measured in Dobson units (DU) when it can be determined from local measurements performed using a ground-based Dobson spectrophotometer (Dobson, 1931) or a Brewer spectrometer (Brewer, 1973) or can be derived from the observations routinely provided above the measurement site by the Ozone Monitoring Instrument (OMI) mounted on the Aura platform to continue the TOMS record for total ozone (Balis *et al.*, 2007).

Nitrogen dioxide NO_2 and its dimer N_2O_4 exhibit a semicontinuum over the $0.32 \le \lambda \le 0.68$ μm wavelength range of the solar spectrum (Hall and Blacet, 1952). The corresponding spectral values of partial optical thickness $\tau_n(\lambda)$ can be calculated by multiplying the NO_2 (and N_2O_4) content (determined by assuming that STP conditions were present along the whole vertical path of the atmosphere) by the overall NO_2 and N_2O_4 absorption coefficient $k_n(\lambda)$ presenting the spectral features shown in Figure 7.3. The spectral values of absorption coefficient $k_n(\lambda)$ produced by these molecular species within the main windows of the atmospheric transmission spectrum are reported in Table 7.2. The total-column content of the two gaseous species can be determined by means of local measurements made with spectrophotometer techniques (Cede *et al.*, 2006) or derived from the OMI/Aura satellite observations conducted over the measurement site (Celarier *et al.*, 2008; Ionov *et al.*, 2008).

The oxygen dimer (O_4) presents a sequence of six absorption bands over the $0.40 \leq \lambda \leq 1.08\,\mu m$ range and two NIR bands centered at the 1.27 and 1.58 μm wavelengths. The spectral values of optical thickness $\tau_{od}(\lambda)$ due to this dimer were directly evaluated by Michalsky *et al.* (1999) within the main windows of the atmospheric transmission spectrum, as shown in Figure 7.3. The values of $\tau_{od}(\lambda)$ obtained within the various windows of the AERONET and SKYNET sun/sky radiometers from the Michalsky *et al.* (1999) study are reported in Table 7.2.

Calculations of the Relative Optical Air Mass Functions for Various Atmospheric Constituents Accurate calculations of $\tau_a(\lambda)$ can be derived by subtracting the correction terms given in Eq. (7.2) for Rayleigh scattering and gaseous absorption from the Sun-photometer measurements of total optical thickness $\tau(\lambda)$ obtained in terms of Eq. (7.1). Taking into account that (i) solar radiation describes a slant path through the atmosphere, which is strongly affected by the refraction effects associated with the specific thermodynamic and composition characteristics of the atmosphere, and (ii) these features vary appreciably with season and latitude, it seems correct to use in Eq. (7.2) a set of relative air mass functions for air molecules (i.e., for Rayleigh scattering, m), aerosols (m_a), atmospheric water vapor (m_w), ozone (m_o), nitrogen dioxide and its dimer (m_n), and oxygen dimer (m_{od}). The relative optical air mass of each atmospheric constituent is calculated as the ratio between the length of the optical path described by the solar light beam along the oblique trajectory and the length of the optical atmospheric path in the zenith direction (Kasten, 1966). According to Thomason, Herman, and Reagan (1983), the relative optical air mass function $m_j(\vartheta_0)$ of the jth atmospheric constituent was calculated for a spherical atmosphere model as a function of ϑ_0. More precisely, $m_j(\vartheta_0)$ is given by the integral of function $k_j(z)$ (describing the vertical profile of the volume extinction coefficient of the jth constituent) calculated from the mean sea level z_o to the atmospheric top level z_{top} (generally assumed to be equal to 120 km) as follows:

$$m_j(\vartheta_0) = (1/K_j) \int_{z_o}^{z_{top}} \frac{k_j(z)dz}{\left[1 - \frac{n_o}{n(z)}(z_o/z)^2 \sin^2\vartheta_0\right]^{-1/2}} \tag{7.3}$$

where (i) K_j is the integral of the volume extinction coefficient $k_j(z)$ of the jth constituent of the atmosphere (measured in per kilometer) over the whole atmosphere, (ii) function $k_j(z)$ represents the vertical profile of the volume extinction coefficient of the jth constituent (chosen among those of wet air, aerosols of various types, water vapor, ozone, nitrogen dioxide, and oxygen dimer) and is used as a weighting function to evaluate the relative Sun path length by taking into account the variations in moist air refractive index $n(z)$ as a function of altitude z, (iii) n_o is the air refractive index at sea level z_o, and (iv) $n(z)$ is the air refractive index varying as a function of z. On the basis of Eq. (7.3), various relative optical air mass functions $m_j(\vartheta_0)$ vary as a function of ϑ_0 presenting features that are mainly influenced by the pressure and temperature conditions of the atmosphere and by the particular vertical distribution features of the jth atmospheric constituent. Therefore, different weighting functions $k_j(z)$ were used in Eq. (7.3) to take into account the variations with height of various extinction or absorption coefficients of aerosol particles and minor atmospheric gases: (i) the volume Rayleigh scattering coefficient to calculate air mass function $m(\vartheta_0)$, (ii) the aerosol volume extinction coefficient $\beta_{ext}(0.55\,\mu m)$ to calculate the aerosol air mass function $m_a(\vartheta_0)$, and (ii) the absorption coefficients of unit concentrations of H_2O, O_3, $NO_2(+N_2O_4)$, and O_4 molecules to calculate the air mass functions $m_w(\vartheta_0)$, $m_o(\vartheta_0)$, $m_n(\vartheta_0)$, and $m_{od}(\vartheta_0)$, respectively.

Reliable calculations of air mass m for a standard atmosphere were made by Kasten and Young (1989) for the International Organization for Standardization (ISO) Atmosphere 1972 model, obtaining the values of m given in Table 7.3. Following the procedure based on Eq. (7.3) and carefully described by Tomasi and Petkov (2014), the values of function $m(\vartheta_0)$ have been here calculated for (i) the six standard atmosphere models of Anderson *et al.* (1986) over the $0 \leq z \leq 120\,km$ range, (ii) the equatorial atmosphere model "Indian Ocean" defined by (Tomasi, 1984), and (iii) the July −75° N and January −75° S atmospheric models defined by Tomasi, Vitale, and De Santis (1998) to represent the thermodynamic characteristics of the Arctic and Antarctic atmospheres during the periods of the year in which sun-photometer measurements are most frequently conducted in the two remote polar regions. The values of $m(\vartheta_0)$ are given in Table 7.3 for thirty selected values of ϑ_0 ranging from 0° to 87°.

The values of relative optical air mass $m_a(\vartheta_0)$ for aerosol extinction were calculated at the same previously selected values of ϑ_0 for average vertical profiles of volume aerosol extinction coefficient $\beta_{ext}(z)$ typical of different latitudinal and seasonal characteristics and different stability conditions of the atmospheric ground layer. Nine particular cases were considered as follows:

1. The first two cases were considered to calculate the relative optical air masses m_{aTs} and m_{aTm} for the thermodynamic conditions of the tropical atmosphere defined by Anderson *et al.* (1986) with vertical profiles of $\beta_{ext}(z)$ similar to that represented by Elterman (1968) and modified within the mixing layer near the surface by assuming that $\beta_{ext}(z)$ decreases exponentially with scale height H_p equal to 1.2 km in the first case and 1.8 km in the second case, in such a way as to represent different vertical mixing cases of the tropical atmosphere, associated with moderate and strong convective motions, respectively.
2. Three cases were considered to calculate the relative optical air masses $m_{aUSs}(\vartheta_0)$, $m_{aUSav}(\vartheta_0)$, and $m_{aUSm}(\vartheta_0)$ defined for thermodynamic conditions

Table 7.3 Values of relative optical air mass *m* of the atmosphere calculated at 30 selected values of apparent solar zenith angle ϑ_0 over the range from 0° to 87° by Kasten and Young (1989) for the International Organization for Standardization (ISO) Standard Atmosphere 1972 model and by Tomasi, Vitale, and De Santis (1998) for the Indian Ocean model of Tomasi (1984), the six atmospheric models of Anderson *et al.* (1986) pertaining to different latitudes and seasons, and the pair of Arctic and Antarctic atmosphere models defined by Tomasi, Vitale, and De Santis (1998).

ϑ_0 (°)	ISO 1972 model	Equatorial Indian Ocean model	Model 1, tropical	Model 2, midlatitude winter	Model 3, midlatitude summer	Model 4, subarctic winter	Model 5, subarctic summer	Model 6, US Standard Atmosphere	July −75°N Arctic model	January −75°S Antarctic model
0	1.0000	1.0000	1.0000	1.0000	1.0000	1.0000	1.0000	1.0000	1.0000	1.0000
10	1.0154	1.0154	1.0154	1.0154	1.0154	1.0154	1.0154	1.0154	1.0154	1.0154
20	1.0640	1.0640	1.0640	1.0640	1.0640	1.0640	1.0640	1.0640	1.0640	1.0640
30	1.1543	1.1542	1.1542	1.1543	1.1543	1.1543	1.1543	1.1543	1.1543	1.1543
40	1.3045	1.3044	1.3044	1.3045	1.3045	1.3045	1.3045	1.3045	1.3045	1.3045
50	1.5535	1.5532	1.5532	1.5535	1.5532	1.5535	1.5534	1.5534	1.5534	1.5534
55	1.7398	1.7393	1.7395	1.7397	1.7395	1.7398	1.7395	1.7395	1.7395	1.7398
60	1.9939	1.9934	1.9934	1.9940	1.9936	1.9942	1.9936	1.9936	1.9938	1.9940
65	2.3552	2.3544	2.3546	2.3555	2.3546	2.3558	2.3551	2.3551	2.3551	2.3555
68	2.6529	2.6518	2.6520	2.6534	2.6520	2.6542	2.6526	2.6526	2.6528	2.6536
70	2.9016	2.9002	2.9005	2.9022	2.9005	2.9031	2.9011	2.9013	2.9013	2.9022
72	3.2054	3.2035	3.2041	3.2067	3.2041	3.2077	3.2051	3.2051	3.2054	3.2067
74	3.5841	3.5815	3.5819	3.5858	3.5823	3.5876	3.5837	3.5840	3.5840	3.5858
75	3.8105	3.8077	3.8081	3.8123	3.8081	3.8146	3.8100	3.8100	3.8108	3.8127
76	4.0682	4.0648	4.0656	4.0705	4.0660	4.0733	4.0676	4.0680	4.0689	4.0713
77	4.3640	4.3599	4.3603	4.3673	4.3612	4.3704	4.3634	4.3638	4.3647	4.3673
78	4.7067	4.7017	4.7022	4.7107	4.7031	4.7149	4.7055	4.7064	4.7078	4.7111
79	5.1081	5.1016	5.1027	5.1134	5.1032	5.1185	5.1067	5.1083	5.1093	5.1139
80	5.5841	5.5758	5.5764	5.5909	5.5781	5.5976	5.5825	5.5836	5.5859	5.5915
81	6.1565	6.1452	6.1464	6.1655	6.1489	6.1754	6.1544	6.1563	6.1594	6.1668
82	6.8568	6.8421	6.8435	6.8689	6.8462	6.8826	6.8545	6.8572	6.8606	6.8709
83	7.7307	7.7101	7.7117	7.7488	7.7163	7.7673	7.7271	7.7318	7.7364	7.7519
84	8.8475	8.8173	8.8200	8.8739	8.8262	8.9023	8.8430	8.8483	8.8562	8.8784
85	10.316	10.269	10.274	10.357	10.284	10.402	10.310	10.318	10.331	10.364
86	12.317	12.231	12.237	12.375	12.256	12.447	12.298	12.310	12.335	12.388
87	15.163	15.030	15.041	15.287	15.077	15.417	15.152	15.169	15.216	15.307

of the US Standard Atmosphere (Anderson *et al.*, 1986), in which the vertical profiles of $\beta_{ext}(z)$ were assumed to (a) decrease with height in an exponential fashion within the mixing layer near the ground – being proportional to the term $\exp(-z/H_p)$ for aerosol scale height H_p equal to (i) 0.8 km to simulate the stability effects due to the presence of a strong thermal inversion near the ground, as it is typical of winter clear-sky days in the midlatitude atmosphere; (ii) 1.1 km to represent average stability conditions of the low troposphere, as it is typical of spring and early autumn cloudless-sky days; and (iii) 1.4 km to represent rather strong convective mixing conditions in the low troposphere, as it is typical of summer days characterized by intense ground heating by solar radiation – and (b) vary at the higher altitudes until to $z_{top} = 50$ km, according to the Elterman (1968) aerosol attenuation model.

3. The last two cases were considered to represent the thermodynamic conditions of the midlatitude summer atmosphere, in which volcanic aerosol particles were formed in the lower stratosphere after a strong volcanic eruption. In these atmospheric aerosol models, the vertical profile of aerosol extinction coefficient $\beta_{ext}(0.55\,\mu m)$ was assumed to be that of Elterman (1968) over the $0 \leq z \leq 50$ km range, with a superimposed peak of $\beta_{ext}(z)$, which was assumed to be similar: (i) in the first case to that of the dense layer of post-Pinatubo volcanic particle layer observed by Russell *et al.* (1993) on 22 January 1992, on board the NASA DC-8 stratospheric airplane at around 19° S latitude, presenting a marked peak of $\beta_{ext}(z)$ at $z \approx 24$ km (formed about seven months after the Mt. Pinatubo eruption of mid-June 1991), for which the air mass $m_{aMSfv}(\vartheta_0)$ was calculated at various apparent solar zenith angles, and (ii) in the second case to that observed by Russell *et al.* (1993) on 20 March 1992, which presents a considerably more attenuated layer of 9-month aged volcanic aerosol particles characterizing the vertical profile of $\beta_{ext}(z)$ with a maximum located at $z \approx 16$ km, for which the relative optical air mass function $m_{aMSav}(\vartheta_0)$ was calculated.
4. Two additional cases were associated with the thermodynamic conditions of the polar atmospheres and background aerosol scenarios, in which the vertical profiles of $\beta_{ext}(z)$ were assumed to be those of Arctic background aerosol and Antarctic background aerosol defined by Tomasi and Petkov (2014) to calculate the relative optical air mass functions $m_{baa}(\vartheta_0)$ and $m_{bAa}(\vartheta_0)$, respectively.

The angular values of the aforementioned eleven relative optical air mass functions for aerosol extinction are given in Table 7.4. On this matter, it is useful mention that Tomasi and Petkov (2014) calculated also the angular values of the following relative optical air mass functions: (i) $m_{avh}(\vartheta_0)$, for an average Arctic haze case; (ii) $m_{AH}(\vartheta_0)$, for a dense Arctic haze case at Ny-Ålesund (Svalbard); (iii) $m_{KV}(\vartheta_0)$, for the Kasatochi volcanic particle layer observed over the Arctic; and (iv) $m_P(\vartheta_0)$, for a 2-year aged Mt. Pinatubo volcanic particle layer over Antarctica.

The angular values of relative optical air mass functions $m_w(\vartheta_0)$ for atmospheric water vapor, $m_o(\vartheta_0)$ for atmospheric ozone, $m_n(\vartheta_0)$ for atmospheric nitrogen dioxide (with its dimer N_2O_4), and $m_{od}(\vartheta_0)$ for atmospheric oxygen dimer were calculated for the US Standard Atmosphere (Model 6) of Anderson *et al.* (1986)) at 30

Table 7.4 Values of relative optical air mass m_a for aerosols calculated at 30 selected values of apparent solar zenith angle ϑ_0 over the range from 0° to 87° for different vertical profiles of volume aerosol extinction coefficient in (a) the tropical atmosphere model of Anderson *et al.* (1986) with different aerosol scale heights H_p equal to 1.2 and 1.8 km; (b) the US Standard Atmosphere model of Anderson *et al.* (1986) with different H_p equal to 0.8, 1.1, and 1.4 km; (c) the two midlatitude summer atmosphere models of Anderson *et al.* (1986) with volcanic aerosol layers peaked at 16 and 24 km, respectively, according to the Russell *et al.* (1993) airborne measurements; (d) the July – 75° N Arctic atmosphere model of Tomasi, Vitale, and De Santis (1998) with the Arctic background aerosol model of Tomasi and Petkov (2014); and (e) the January – 75° S Antarctic atmosphere model of Tomasi, Vitale, and De Santis (1998) with the Antarctic background aerosol model of Tomasi and Petkov (2014).

ϑ_0 (°)	Tropical atmosphere (H_p = 1.2 km, and ground-level visual range VR = 15.44 km), m_{aTs}	Tropical atmosphere (H_p = 1.8 km and ground-level visual range VR = 20.57 km), m_{aTm}	US Standard Atmosphere (H_p = 0.8 km and ground-level visual range VR = 12.37 km), m_{aUSs}	US Standard Atmosphere (H_p = 1.1 km and ground-level visual range VR = 14.63 km), m_{aUSav}	US Standard Atmosphere (H_p = 1.4 km and ground-level visual range VR = 17.11 km), m_{aUSm}	Midlatitude summer atmosphere with volcanic aerosol layer peaked at 16 km, m_{aMSfv}	Midlatitude summer atmosphere with volcanic aerosol layer peaked at 24 km, m_{aMSav}	July –75° N Arctic atmosphere with Arctic background aerosol, m_{baa}	January –75° S Antarctic atmosphere with Antarctic background aerosol, m_{bAa}
0	1.0000	1.0000	1.0000	1.0000	1.0000	1.0000	1.0000	1.0000	1.0000
10	1.0154	1.0154	1.0154	1.0154	1.0154	1.0154	1.0154	1.0154	1.0154
20	1.0641	1.0641	1.0641	1.0641	1.0641	1.0639	1.0639	1.0641	1.0641
30	1.1545	1.1545	1.1545	1.1545	1.1545	1.1541	1.1540	1.1546	1.1546
40	1.3049	1.3048	1.3049	1.3049	1.3049	1.3039	1.3036	1.3052	1.3051
50	1.5545	1.5543	1.5546	1.5546	1.5545	1.5520	1.5515	1.5552	1.5550
55	1.7415	1.7412	1.7417	1.7416	1.7415	1.7375	1.7366	1.7426	1.7423
60	1.9968	1.9963	1.9970	1.9969	1.9967	1.9900	1.9885	1.9986	1.9981
65	2.3604	2.3594	2.3608	2.3606	2.3602	2.3481	2.3456	2.3636	2.3627
68	2.6608	2.6593	2.6614	2.6610	2.6604	2.6424	2.6387	2.6656	2.6642
70	2.9122	2.9102	2.9130	2.9125	2.9116	2.8874	2.8825	2.9186	2.9168
72	3.2200	3.2173	3.2212	3.2204	3.2193	3.1859	3.1792	3.2288	3.2263
74	3.6050	3.6011	3.6067	3.6055	3.6039	3.5564	3.5472	3.6176	3.6140

(*continued overleaf*)

Table 7.4 (Continued).

ϑ_0 (°)	Tropical atmosphere (H_p = 1.2 km, and ground-level visual range VR = 15.44 km), m_{aTs}	Tropical atmosphere (H_p = 1.8 km and ground-level visual range VR = 20.57 km), m_{aTm}	US Standard Atmosphere (H_p = 0.8 km and ground-level visual range VR = 12.37 km), m_{aUSs}	US Standard Atmosphere (H_p = 1.1 km and ground-level visual range VR = 14.63 km), m_{aUSav}	US Standard Atmosphere (H_p = 1.4 km and ground-level visual range VR = 17.11 km), m_{aUSm}	Midlatitude summer atmosphere with volcanic aerosol layer peaked at 16 km, m_{aMSfv}	Midlatitude summer atmosphere with volcanic aerosol layer peaked at 24 km, m_{aMSav}	July −75° N Arctic atmosphere with Arctic background aerosol, m_{baa}	January −75° S Antarctic atmosphere with Antarctic background aerosol, m_{bAa}
75	3.8358	3.8311	3.8378	3.8365	3.8345	3.7769	3.7660	3.8511	3.8467
76	4.0993	4.0935	4.1018	4.1001	4.0977	4.0271	4.0140	4.1180	4.1126
77	4.4027	4.3955	4.4058	4.4038	4.4008	4.3130	4.2972	4.4259	4.4193
78	4.7556	4.7466	4.7596	4.7570	4.7532	4.6426	4.6233	4.7850	4.7766
79	5.1711	5.1594	5.1762	5.1729	5.1680	5.0260	5.0024	5.2088	5.1979
80	5.6669	5.6515	5.6736	5.6692	5.6628	5.4769	5.4478	5.7163	5.7020
81	6.2682	6.2476	6.2773	6.2714	6.2627	6.0135	5.9776	6.3345	6.3153
82	7.0121	6.9834	7.0248	7.0164	7.0044	6.6610	6.6169	7.1037	7.0769
83	7.9547	7.9133	7.9730	7.9609	7.9436	7.4543	7.4016	8.0856	8.0470
84	9.1858	9.1234	9.2134	9.1952	9.1690	8.4438	8.3849	9.3808	9.3225
85	10.858	10.758	10.902	10.873	10.831	9.7034	9.6509	11.164	11.071
86	13.253	13.082	13.330	13.279	13.207	11.348	11.345	13.767	13.605
87	16.956	16.627	17.106	17.006	16.865	13.574	13.754	17.897	17.583

selected values of apparent solar zenith angle ϑ_0 from 0° to 87°, obtaining the values of the four optical air mass functions In the calculations of $m_{od}(\vartheta_0)$, the oxygen dimer concentration was assumed to be proportional to the square power of O_2 concentration, according to Michalsky *et al.* (1999), therefore using as weighting function in Eq. (7.3) the vertical profile of the square of molecular oxygen concentration defined by Tomasi and Petkov (2014). The values of parameters $m_w(\vartheta_0)$, $m_o(\vartheta_0)$, $m_n(\vartheta_0)$, and $m_{od}(\vartheta_0)$ calculated for the US Standard Atmosphere (Model 6) of Anderson *et al.* (1986) and given in Table 7.5 can be confidently used for all the atmospheric conditions, differing only slightly from the values calculated for different latitudes and seasons, that is, for the following eight atmospheric models: (i)

Table 7.5 Values of relative optical air mass functions $m_w(\vartheta_0)$ for atmospheric water vapor, $m_o(\vartheta_0)$ for atmospheric ozone, $m_n(\vartheta_0)$ for atmospheric nitrogen dioxide (and N_2O_4), and $m_{od}(\vartheta_0)$ for atmospheric oxygen dimer, calculated for the US Standard Atmosphere (Model 6) (Anderson *et al.*, 1986) at 30 selected values of apparent solar zenith angle ϑ_0 ranging from 0° to 87°.

ϑ_0 (°)	Atmospheric water vapor air mass $m_w(\vartheta_0)$	Atmospheric ozone air mass $m_o(\vartheta_0)$	Atmospheric nitrogen dioxide (and N_2O_4) air mass $m_n(\vartheta_0)$	Atmospheric oxygen dimer air mass $m_{od}(\vartheta_0)$
0	1.0000	1.0000	1.0000	1.0000
10	1.0154	1.0153	1.0154	1.0154
20	1.0643	1.0637	1.0639	1.0641
30	1.1545	1.1534	1.1537	1.1545
40	1.3051	1.3024	1.3026	1.3048
50	1.5549	1.5485	1.5484	1.5544
55	1.7421	1.7319	1.7318	1.7413
60	1.9980	1.9807	1.9800	1.9963
65	2.3637	2.3316	2.3289	2.3596
68	2.6653	2.6179	2.6139	2.6596
70	2.9179	2.8548	2.8495	2.9105
72	3.2290	3.1412	3.1330	3.2176
74	3.6173	3.4933	3.4818	3.6015
75	3.8515	3.7011	3.6871	3.8315
76	4.1191	3.9347	3.9181	4.0939
77	4.4255	4.1994	4.1789	4.3959
78	4.7856	4.5008	4.4744	4.7469
79	5.2093	4.8464	4.8133	5.1595
80	5.7175	5.2461	5.2035	5.6511
81	6.3360	5.7107	5.6494	6.2460
82	7.1037	6.2553	6.1749	6.9795
83	8.0874	6.8975	6.7909	7.9044
84	9.3783	7.6561	7.5121	9.1030
85	11.156	8.5533	8.3561	10.710
86	13.737	9.6034	9.3298	12.960
87	17.793	10.798	10.423	16.287

the Indian Ocean model of Tomasi (1984), (ii) the five atmospheric models defined by Anderson *et al.* (1986) for different latitudes and seasons (and named tropical, midlatitude winter, midlatitude summer, subarctic winter, and subarctic summer, respectively), and (iii) the July −75° N Arctic and the January −75° S Antarctic models of Tomasi, Vitale, and De Santis (1998). More precisely, (a) the values of $m_w(\vartheta_0)$ determined for the US Standard Atmosphere model were found to differ from those calculated for the other eight aforementioned atmospheric models by less than ±0.03% for $\vartheta_0 < 60°$ and by ±0.03% for $\vartheta_0 = 70°$, ±0.12% for $\vartheta_0 = 80°$, and ±1.2% for $\vartheta_0 = 87°$; (b) the values of $m_o(\vartheta_0)$ calculated for the US Standard Atmosphere model were estimated to differ from those of the other eight atmospheric models by less than ±0.13% for $\vartheta_0 < 60°$ and by no more than ±0.32% for $\vartheta_0 = 70°$, ±1.2% for $\vartheta_0 = 80°$, and ±4.5% for $\vartheta_0 = 87°$; (c) the values of $m_n(\vartheta_0)$ calculated for the US Standard Atmosphere model were estimated to differ from those of the other atmospheric models by less than ±0.03% for $\vartheta_0 < 60°$ and by no more than ±0.07% for $\vartheta_0 = 70°$, ±0.31% for $\vartheta_0 = 80°$, and ±1.2% for $\vartheta_0 = 87°$; and (d) the values of $m_{\text{od}}(\vartheta_0)$ calculated for the US Standard Atmosphere model were estimated to differ from those of the other models by less than ±0.02% for $\vartheta_0 < 60°$ and by no more than ±0.05% for $\vartheta_0 = 70°$, ±0.19% for $\vartheta_0 = 80°$, and ±1.4% for $\vartheta_0 = 87°$.

7.2.1.3 Determination of Aerosol Optical Parameters from Sun-Photometer Measurements

The spectral values of AOT $\tau_a(\lambda)$ can be reliably evaluated at each window wavelength λ in terms of the following expression derived from Eqs. (7.1) and (7.2):

$$\tau_a(\lambda) = (1/m_a)\ln\left[\frac{F_e V_o(\lambda)}{V(\lambda)}\right] - (m/m_a)\tau_R(\lambda) - (m_w/m_a)\tau_w(\lambda) - (m_o/m_a)$$
$$\tau_o(\lambda) - (m_n/m_a)\tau_n(\lambda) - (m_{od}/m_a)\tau_{\text{od}}(\lambda) \quad (7.4)$$

for each output voltage $V(\lambda)$ measured by a Sun photometer for relative optical air mass m and the corresponding relative optical air masses m_a, m_w, m_o, m_n, and m_{od} given in Table 7.5 as a function of ϑ_0. Following this procedure, the AERONET and SKYNET Sun/sky radiometers are used to provide regular measurements of $\tau_a(\lambda)$ at various hours of the cloudless-sky days and various VIS and NIR (VNIR) window wavelengths. The values of $\tau_a(\lambda)$ obtained over the 0.40–0.87 μm spectral range can be examined in terms of the Ångström (1964) formula

$$\tau_a(\lambda) = \beta(1.00\,\mu\text{m})\lambda^{-\alpha}, \quad (7.5)$$

where wavelength λ is measured in micrometer to determine the best-fit values of the atmospheric turbidity parameters α (commonly named Ångström wavelength exponent) and $\beta(1.00\,\mu\text{m})$ (giving in practice the best-fit value of $\tau_a(\lambda)$ at unit wavelength). The Prede POM-02L measurement of $\tau_a(1.020\,\mu\text{m})$ was omitted in the calculations of α and $\beta(1.00\,\mu\text{m})$ because $\tau_a(1.020\,\mu\text{m})$ is lower than 0.05 for high atmospheric transparency conditions and, hence, could cause not negligible uncertainties in calculating the pair of atmospheric turbidity parameters being affected by an instrumental error usually greater than 0.01. Figure 7.4 shows four

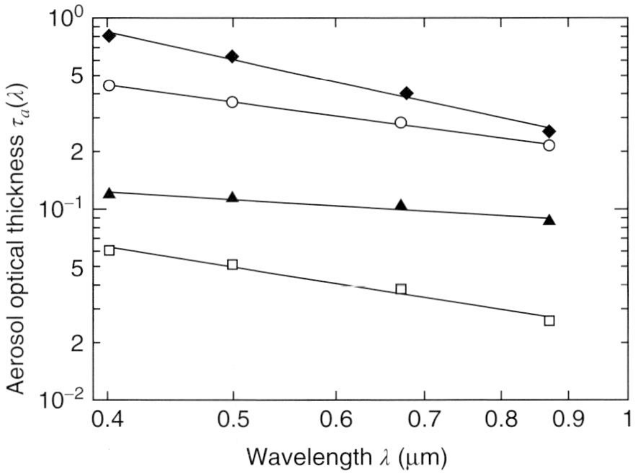

Figure 7.4 Scatter plots of the natural logarithms of aerosol optical thickness $\tau_a(\lambda)$ versus the logarithm of wavelength λ according to the Ångström (1964) formula $\tau_a(\lambda) = \beta\lambda^{-\alpha}$, as obtained over the spectral range from 0.40 to 0.87 μm for four sets of measurements of $\tau_a(\lambda)$ conducted at San Pietro Capofiume using the Prede POM-02L sun/sky radiometer of the SKYNET network at 12:00 local time of the following measurement days: (a) 22 May 2007 (open circles), giving the best-fit values of $\alpha = 0.922$ and $\beta = 0.191$ (with regression coefficient $r_\alpha = 0.998$) for continental aerosol with thin layers of Saharan dust at 2–3 km levels; (b) 23 June 2007 (solid triangles), giving the best-fit values of $\alpha = 0.407$ and $\beta = 0.085$ (with $r_\alpha = 0.965$) ($\alpha = 0.407$, $\beta = 0.085$, $r_\alpha = 0.965$) for anthropogenic–continental (polluted) aerosol with a Saharan dust layer; (c) 8 February 2008 (open squares), giving the best-fit values of $\alpha = 1.085$ and $\beta = 0.023$ (with $r_\alpha = 0.990$) for clear-sky conditions and background continental aerosols from northwestern Europe; and (d) 15 February 2008 (solid diamonds), giving the best-fit values of $\alpha = 1.490$ and $\beta = 0.216$ (with $r_\alpha = 0.995$) for anthropogenic aerosol transported from eastern Europe and northern Russia at all levels.

application examples of the Ångström (1964) formula to field measurements of $\tau_a(\lambda)$ conducted at San Pietro Capofiume (SPC) (44°39′N, 11°39′E, 9 m a.m.s.l.) (hereinafter referred to as SPC), located in the middle of Po Valley (north Italy), using the Prede POM-02L Sun/sky radiometer of the SKYNET network at 12:00 local time (LT) of each day:

- 22 May 2007 ($\alpha = 0.922$, $\beta = 0.191$, regression coefficient $r_\alpha = 0.998$), with optically predominant extinction by continental aerosol particles transported from the Balkans at levels lower than 1.5 km and from southern Italy and Greece at upper levels, containing some thin layers of Saharan dust
- 23 June 2007 ($\alpha = 0.407$, $\beta = 0.085$, $r_\alpha = 0.965$), with particulate extinction mainly produced by anthropogenic–continental (polluted) aerosol that circulated over central and southern Italy during the last days at levels lower than 1 km and a Saharan dust layer at upper altitudes
- 8 February 2008 ($\alpha = 1.085$, $\beta = 0.023$, $r_\alpha = 0.990$), with attenuation mainly produced by background continental aerosols transported from northwestern Europe at levels lower than 3 km and from the Scandinavian regions at upper levels

- 15 February 2008 ($\alpha = 1.490$, $\beta = 0.216$, $r_\alpha = 0.995$), for anthropogenic–continental aerosol transported from eastern Europe and northern Russia at all levels

Spectral sets of Prede POM-02L measurements were regularly performed at the SPC station during the 16-month period from 16 May 2007 to 31 August 2008, using the instrument shown in Figure 7.5. In general, each complete spectral series of direct solar irradiance $V(\lambda)$ was collected within a time interval of $\sim$ 15 minutes and was measured every 30 min throughout each day from sunrise to sunset. On each measurement day, the daily mean values of $\tau_a(\lambda)$ were calculated at all window wavelengths examining the direct solar irradiance measurements for which the daily mean value of exponent $\alpha(0.40-0.87\,\mu m)$ was obtained. The time patterns of the daily average values of $\tau_a(0.50\,\mu m)$ and $\alpha(0.40-0.87\,\mu m)$ determined during the 16-month measurement period are shown in Figure 7.6, together with that of the fine particle fraction $\eta_f(0.50\,\mu m)$ calculated over the VNIR wavelength range by applying the O'Neill, Dubovik, and Eck (2001) procedure. As mentioned earlier, parameter $\eta_f(0.50\,\mu m)$ yields the percentage of $\tau_a(0.50\,\mu m)$ due to the submicrometer aerosol particle component, while the complementary fraction $[1 - \eta_f(0.50\,\mu m)]$ of $\tau_a(0.50\,\mu m)$ is attributed to the residual overall supermicrometer component of the vertical load of aerosol particles.

The spectral series of $\tau_a(\lambda)$ obtained earlier were carefully analyzed to (i) determine the most significant spectral characteristics of aerosol extinction over the Po Valley (Italy) and (ii) evaluate the variations produced by the coarse

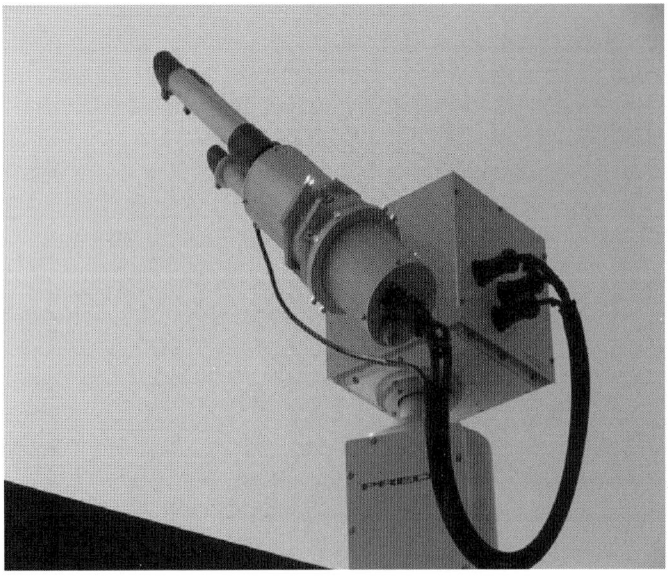

Figure 7.5 Picture of the Prede POM-02L Sun/sky radiometer employed at San Pietro Capofiume (Italy) during the AEROCLOUDS field campaigns from 16 May 2007 to 31 August 2008.

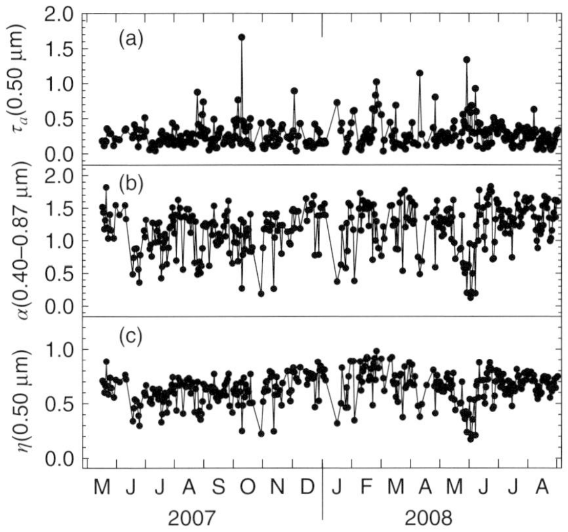

Figure 7.6 Time patterns of the daily mean values of (a) aerosol optical thickness $\tau_a(0.50\,\mu m)$, (b) Ångström (1964) wavelength exponent $\alpha(0.40-0.87\,\mu m)$, and (c) fine particle fraction $\eta_f(0.50\,\mu m)$, as derived from the aerosol optical thickness measurements conducted with the Prede POM-02L Sun/sky radiometer at San Pietro Capofiume (Italy) from 16 May 2007 to 31 August 2008. The values of parameter $\eta_f(0.50\,\mu m)$ have been calculated using the O'Neill, Dubovik, and Eck (2001) procedure.

particles during the frequent Saharan dust transport episodes from north Africa over northern Italy. Figure 7.6a shows the time patterns of the daily mean values of $\tau_a(0.50\,\mu m)$, indicating that the most pronounced peaks of $\tau_a(0.50\,\mu m)$ were recorded in autumn 2007 and in late winter and spring 2008, presumably caused by the presence of Saharan dust layers. Important columnar atmospheric loads of coarse particles are in general present on the Saharan dust transport days, yielding in general rather low values of α, as can be clearly seen in Figure 7.6b. In the other cases, rather high values of $\tau_a(0.50\,\mu m)$ were found to be associated with values of exponent α largely greater than 1, presumably due to particulate extinction effects mainly caused by continental and/or anthropogenic aerosol particles, which consist for the major part of fine particles mainly composed of nitrate and soot substances. The daily mean values of $\tau_a(0.50\,\mu m)$ were found to vary mainly (i) from 0.05 to 0.40 in winter 2008, with an average value of 0.31; (ii) from 0.12 to 0.40 in spring, with average values of 0.22 and 0.29 in spring 2007 and spring 2008, respectively; (iii) from 0.10 to 0.44 in summer, with average values of 0.25 and 0.27 in summer 2007 and summer 2008, respectively; and (iv) from 0.10 to 0.46 in autumn 2007, with an average value of 0.27, as reported in Table 7.6.

In order to give a measure of the extinction effects produced by the fine and coarse particle modes, it is useful to examine the time patterns of the daily mean values of exponent $\alpha(0.40-0.87\,\mu m)$ shown in Figure 7.6b. It can be seen that

Table 7.6 Seasonal mean values of the main optical parameters of aerosol particles in the atmospheric column, as derived from the Prede POM-02L measurements performed at San Pietro Capofiume during the AEROCLOUDS field campaigns conducted in 2007 and 2008.

Season	Aerosol optical thickness τ_a(0.50 μm)	Ångström exponent α(0.40–0.87 μm)	Fine particle fraction η_f(0.50 μm)	Refractive index real part n(0.50 μm)	Refractive index imaginary part χ(0.50 μm)	Asymmetry factor g(0.50 μm)	Single scattering albedo ω(0.50 μm)
Spring 2007	0.22 ± 0.17	1.33 ± 0.36	0.67 ± 0.08	1.46 ± 0.06	0.002 ± 0.001	0.64 ± 0.04	0.98 ± 0.05
Summer 2007	0.25 ± 0.13	1.04 ± 0.37	0.56 ± 0.14	1.47 ± 0.07	0.003 ± 0.002	0.67 ± 0.04	0.95 ± 0.06
Autumn 2007	0.27 ± 0.17	1.09 ± 0.24	0.61 ± 0.12	1.46 ± 0.06	0.005 ± 0.004	0.69 ± 0.05	0.94 ± 0.05
Winter 2008	0.31 ± 0.16	1.29 ± 0.30	0.76 ± 0.23	1.45 ± 0.05	0.006 ± 0.004	0.70 ± 0.05	0.95 ± 0.04
Spring 2008	0.29 ± 0.16	1.15 ± 0.34	0.62 ± 0.16	1.45 ± 0.06	0.004 ± 0.002	0.68 ± 0.04	0.95 ± 0.05
Summer 2008	0.27 ± 0.14	1.32 ± 0.38	0.64 ± 0.13	1.43 ± 0.07	0.003 ± 0.002	0.66 ± 0.04	0.95 ± 0.06

exponent $\alpha(0.40-0.87\,\mu m)$ exhibits large variations from less than 0.2 to about 1.8 throughout the whole measurement period. The daily minima correspond in general to peaks of $\tau_a(0.50\,\mu m)$ produced for the major part by Saharan dust transported over the Mediterranean Sea, these cases occurring most frequently in late winter and early spring and presenting predominant extinction features typical of coarse particles. Conversely, daily mean values of $\alpha(0.40-0.87\,\mu m)$ ranging mainly from 1.3 to 1.7 were measured most frequently in late autumn and early winter, associated with loads of relatively small haze particles grown within the atmospheric ground layer, which may sometimes provide rather high values of $\tau_a(0.50\,\mu m)$. The daily mean values of $\alpha(0.40-0.87\,\mu m)$ were found to appreciably vary from one season to another, varying mainly (i) from 0.70 to 1.60 in winter 2008, with an average value of 1.29 ± 0.30; (ii) from 0.55 to 1.65 in spring, with average values of 1.33 ± 0.36 and 1.15 ± 0.34 in 2007 and 2008, respectively; (iii) from 0.90 to 1.65 in summer, with average values of 1.04 ± 0.37 and 1.32 ± 0.38 in 2007 and 2008, respectively; and (iv) from 0.75 to 1.40 in autumn 2007, with an average seasonal value of 1.09 ± 0.24, as given in Table 7.6.

The time patterns of the daily mean values of fine particle fraction $\eta_f(0.50\,\mu m)$ are shown in Figure 7.6c. They present values mainly ranging from 0.5 to 0.8, with (i) numerous pronounced minima lower than 0.40, clearly indicating that these particulate extinction cases are mainly caused by coarse particles, associated with the transport of thick Saharan dust layers, and yielding high values of $\tau_a(0.50\,\mu m)$, and (ii) rare cases higher than 0.8, due to predominant extinction by fine particles, which often consist of anthropogenic particles containing mainly nitrates and organic carbon (OC). The daily mean values of $\eta_f(0.50\,\mu m)$ were found to vary appreciably with season: (i) from 0.45 to 0.96 in winter 2008, with a seasonal mean value of 0.76 ± 0.23; (ii) from 0.35 to 0.85 in spring 2007 and spring 2008, with seasonal mean values of 0.67 ± 0.08 and 0.62 ± 0.16, respectively; (iii) from 0.38 to 0.75 in summer 2007, yielding a seasonal mean value of 0.56 ± 0.14, and from 0.40 to 0.88 in summer 2008, providing a seasonal mean value of 0.64 ± 0.13; and (iv) from 0.42 to 0.83 in autumn, with a seasonal mean value of 0.61 ± 0.12.

Regular measurements of sky brightness in the almucantar were performed during the 16-month field campaign conducted at SPC. A time interval of about 10 min was generally employed to perform a complete angular scanning of diffuse solar radiation, consisting of measurements taken at six window wavelengths equal to 0.380, 0.400, 0.500, 0.675, 0.870, and 1.020 μm, between one spectral series of direct solar irradiance measurements and the other. These measurements were subsequently examined using the retrieval procedure defined by Nakajima *et al.* (1996, 2003) for the SKYNET data (or the procedure of Dubovik and King, 2000 for the AERONET data). Figure 7.7 shows the time patterns of the daily mean values of the following four optical parameters of aerosols: (a) the real part $n(0.50\,\mu m)$; (b) the imaginary part $\chi(0.50\,\mu m)$; (c) the asymmetry factor $g(0.50\,\mu m)$, giving the measure of the degree of anisotropy of particulate scattering diagram; and (d) the single scattering albedo $\omega(0.50\,\mu m)$, given by the ratio between scattering efficiency and total extinction (scattering + absorption)

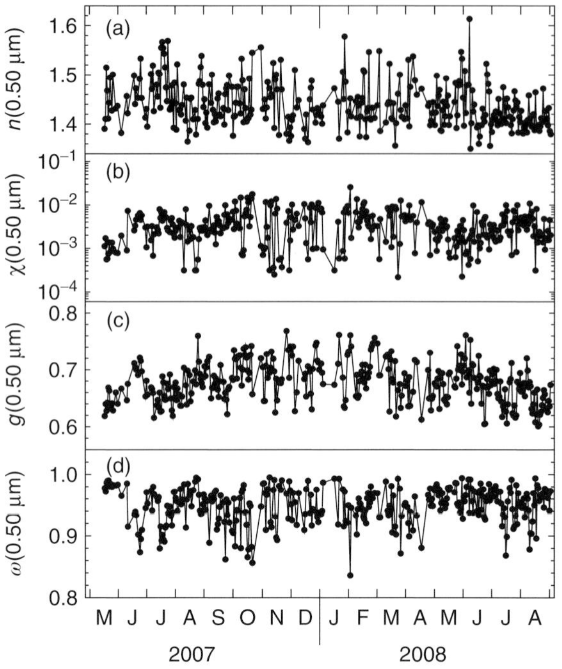

Figure 7.7 Time patterns of the daily mean values of (a) real part $n(0.50\,\mu m)$ of columnar aerosol refractive index, (b) imaginary part $\chi(0.50\,\mu m)$ of columnar aerosol refractive index, (c) asymmetry factor $g(0.50\,\mu m)$ of columnar aerosols, and (d) single scattering albedo $\omega(0.50\,\mu m)$ of columnar aerosols, as derived from the sky-brightness measurements in the almucantar conducted at the 0.50 μm wavelength with the Prede POM-02L Sun/sky radiometer at San Pietro Capofiume (northern Italy) from 16 May 2007 to 1 September 2008, using the SkyRad 4.2 inversion code (Nakajima, Tanaka, and Yamauchi, 1983; Nakajima *et al.*, 1996).

efficiency of the aerosol particle polydispersion. Such four optical parameters fully characterize the optical properties of aerosol particles, defining also their absorption characteristics.

Figure 7.7a shows the time patterns of the daily mean values of real part $n(0.50\,\mu m)$ of particulate matter refractive index derived by using the version 4.2 (Campanelli *et al.*, 2007) of the SkyRad inversion code defined by Nakajima, Tanaka, and Yamauchi (1983); Nakajima *et al.* (1996) for the Prede POM-02L sky-brightness measurements conducted at SPC from mid-May 2007 to the end of August 2008. The daily mean values of $n(0.50\,\mu m)$ were found to vary mainly between 1.35 and 1.55 throughout the year, with higher values in summer and autumn 2007 and in spring 2008 caused for the most part by Saharan dust particle loads (containing in general high concentrations of aluminite and other quartz compounds). Conversely, rather low values of $n(0.50\,\mu m)$ were measured in autumn 2007 and winter 2008, when particle growth processes by condensation have caused a strong uptake of liquid water by aerosol particles leading the real part $n(\lambda)$ to approximate the liquid water value $n = 1.33$. Thus, the daily mean values of $n(0.50\,\mu m)$ were found to range (i) from 1.38 to 1.55 in winter, (ii) from 1.40 to 1.52 in spring, (iii) from 1.39 to 1.52 in summer, and (iv) from 1.41 to 1.52 in autumn, with the seasonal mean values given in Table 7.6.

The time patterns of the daily mean values of $\chi(0.50\,\mu m)$ are shown in Figure 7.7b, estimated to range mainly from 0.0001 to 0.05 throughout the year. The seasonal data sets provided seasonal mean values ranging between 0.002 (in spring 2007) and 0.006 (in winter 2008), as reported in Table 7.6.

The asymmetry factor $g(\lambda)$ was determined at different window wavelengths to estimate how much the forward lobe of the aerosol scattering diagram is more pronounced than the backward lobe. This parameter varies as a function of complex particulate matter refractive index and depends on the particle size-distribution curve shape parameters according to the Mie (1908) theory. The time patterns of the daily mean values of $g(0.50\,\mu m)$ measured at SPC during the field campaign are shown in Figure 7.7c. It can be noted that $g(0.50\,\mu m)$ varied from 0.6 to 0.8 during the SPC observational period, presenting rather high values in autumn 2007, greater than 0.70, and lower values in May and July 2007 and in summer 2008, during the periods of most frequent Saharan dust transport occurrence, varying with season: (i) from 0.65 to 0.75 in winter, giving a mean seasonal value $g_{ms} = 0.70 \pm 0.04$; (ii) from 0.63 to 0.72 in spring ($g_{ms} = 0.66 \pm 0.06$); (iii) from 0.62 to 0.71 in summer ($g_{ms} = 0.66 \pm 0.06$); and (iv) from 0.64 to 0.73 in autumn ($g_{ms} = 0.69 \pm 0.05$). The seasonal mean values of $g(0.50\,\mu m)$ measured from spring 2007 to summer 2008 are given in Table 7.6.

The single scattering albedo $\omega(\lambda)$ of an aerosol particle polydispersion provides the percentage of monochromatic incoming solar radiation attenuated by aerosol scattering during its passage through the atmosphere. Therefore, the difference between unit and $\omega(\lambda)$ provides the fraction of incoming solar radiation absorbed by airborne aerosol particles. The time patterns of the daily mean values of $\omega(0.50\,\mu m)$ measured at SPC in 2007 and 2008 are shown in Figure 7.7d. The results indicate that the major part of daily mean values of $\omega(0.50\,\mu m)$ ranged between 0.8 and nearly 1.0 over the whole measurement period. Values greater than 0.95 indicated the presence of aerosol loads characterized by intense scattering and low absorption (as in the cases of sea-salt particles transported from the Atlantic Ocean and Mediterranean Sea), while those ranging from 0.8 to 0.9 were in general associated with aerosol particles having industrial–anthropogenic origin. The rare cases with $\omega(0.50\,\mu m) < 0.8$ were caused by aerosol particle loads containing relatively high concentrations of soot substances from biomass burning (BB) activities and combustion processes in urban areas. This important optical parameter was estimated to vary with season: (i) from 0.92 to 0.98 in winter, yielding a mean seasonal value $\omega_{ms} = 0.95 \pm 0.04$; (ii) from 0.91 to 0.99 in spring ($\omega_{ms} = 0.95 \pm 0.05$); (iii) from 0.90 to 0.98 in summer ($\omega_{ms} = 0.95 \pm 0.05$); and (iv) from 0.89 to 0.98 in autumn ($\omega_{ms} = 0.94 \pm 0.05$), as reported in Table 7.6.

Knowing the spectral characteristics of the previously examined aerosol optical parameters $\tau_a(\lambda)$, $n(\lambda)$, $\chi(\lambda)$, $g(\lambda)$, and $\omega(\lambda)$ as well as those of phase function $p(\theta)$ defined for the scattering angle θ, each spectral series of $\tau_a(\lambda)$ was analyzed with the SkyRad 4.2 inversion algorithm to retrieve the volume size-distribution curve $\Gamma(r) = d\Gamma/d(\ln r)$ of the aerosol particles suspended in the vertical column of unit cross section. Five examples of size-distribution curves pertaining to Prede POM-02L measurements conducted on the five measurement days listed in Table 7.7 are shown in Figure 7.8 as a function of particle radius r. Two days regarded continental aerosol loads transported from the Balkans on 22 May 2007, and from northwestern Europe and Scandinavia on 8 February 2008. The first size-distribution curve determined on 22 May 2007, was found to be clearly characterized by three particle modes, with a first large mode of fine particles centered at $r_c \approx 0.10\,\mu m$ (mainly consisting of background continental aerosols), a second mode of accumulation/coarse particles with mode radius $r_c \approx 2\,\mu m$ (mainly due to natural/agricultural sources in the Balkans, Greece, and southern Italy), and a third large mode of coarse particles with $r_c \approx 12\,\mu m$ (probably due to the presence of thin layers of Saharan dust at 2–3 km levels).

On 8 February 2008, when an atmospheric load of pure continental aerosols moved from remote unpolluted areas and the overall aerosol size-distribution curve was found to exhibit three marked modes centered at $r_c \approx 0.12\,\mu m$ (fine particles), $r_c \approx 1.8\,\mu m$ (accumulation/coarse particles), and $r_c \approx 10\,\mu m$ (coarse particles), respectively, the marked fine particle mode and the less pronounced modes of accumulation and coarse particles are associated with the particle transport over long distances from the remote Arctic regions and northwestern Europe.

Two other cases among those shown in Figure 7.8 regarded transport of anthropogenic–continental aerosols over northern Italy, the first observed on 23 June 2007, and related to local circulation motions and the latter measured on 15 February 2008, associated with particulate transport from eastern Europe and northern Russia. The retrieved size-distribution curve of 23 June 2007, was found to exhibit three modes centered at radii $r_c \approx 0.2\,\mu m$ (fine and accumulation particles, associated with the low pollution characteristics of the aerosol sources located in central and southern Italy), $r_c \approx 1.5\,\mu m$ (accumulation/coarse particles), and $r_c \approx 5\,\mu m$ (coarse particles), which were also due to residual Saharan dust layers suspended at upper tropospheric levels.

Similar results were retrieved on 15 February 2008, when the size-distribution curve was found to show a first mode of fine and accumulation particles (with $r_c \approx 0.2\,\mu m$), a second mode of accumulation/coarse particles (with $r_c \approx 1.6\,\mu m$), and a third mode of coarse particles (with $r_c \approx 5.5\,\mu m$). In this case of predominant transport of anthropogenic aerosol from eastern Europe and northern Russia at all levels, the fine particle mode exhibited a relatively high content of fine particles of continental origins and was markedly separated from the second mode, which was found to present concentration values comparable with those of the coarse particles, due for the most part to natural mobilization processes by winds over the agricultural and mountain areas.

The fifth example shown in Figure 7.8 was recorded on 17 July 2007, regarding (a) anthropogenic–continental aerosols originated from local sources at levels lower than 1.5 km, (b) continental aerosols from central Europe at altitudes varying from 1.5 to 2.5 km, and (c) Saharan dust at levels ranging from 3 to 5 km. This

Table 7.7 Optically predominant aerosol type during the five measurement days considered in Figure 7.7 and characterized by multimodal features of the aerosol particle size-distribution curves in the atmospheric vertical column.

Measurement day	Optically predominant aerosol type	Daily mean aerosol optical thickness $\tau_a(0.50\,\mu m)$	Daily mean Ångström wavelength exponent $\alpha(0.40-0.87\,\mu m)$	Daily mean fine particle fraction $\eta_f(0.55\,\mu m)$
22 May 2007	Continental aerosol from the Balkans at levels < 1.5 km and from southern Italy and Greece at upper levels, with thin layers of Saharan dust at 2–3 km levels	0.37	0.91	0.52
23 June 2007	Anthropogenic–continental polluted aerosol circulating over central and southern Italy	0.11	0.41	0.62
17 July 2007	Anthropogenic–continental aerosol at levels <1.5 km, continental aerosol from central Europe at middle levels, and Saharan dust at the 4–5 km levels	0.31	0.73	0.42
8 February 2008	Continental aerosol from northwestern Europe at levels < 3 km and from Scandinavia at upper levels	0.05	1.08	0.47
15 February 2008	Anthropogenic–continental aerosol from eastern Europe and northern Russia at all levels	0.22	1.55	0.82

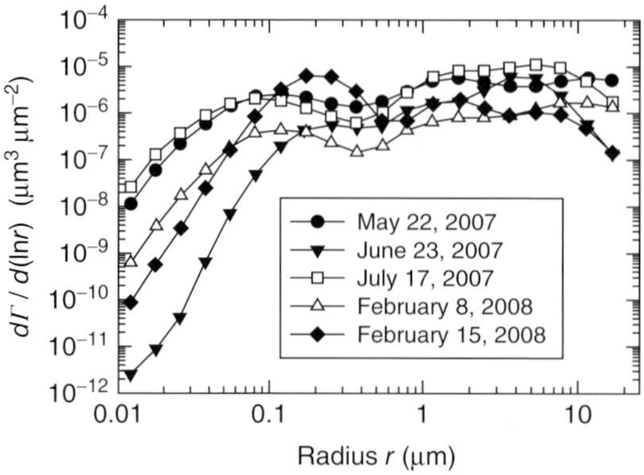

Figure 7.8 Size-distribution curves of the overall volume Γ occupied by the aerosol particles in the vertical atmospheric column $d\Gamma/d(\ln r)$ (measured in cubic micrometer per unit cross section (equal to 1 μm^2) of the atmospheric column), as determined at various local times (LT) of the following days: (1) 22 May 2007 (08:27 LT), for continental aerosols; (2) 23 June 2007 (11:45 LT), for anthropogenic–continental aerosols; (3) 17 July 2007 (17:27 LT), for anthropogenic–continental aerosols and Saharan dust; (4) 8 February 2008 (10:45 LT), for continental aerosols; and (5) 15 February 2008 (11:00 LT), for anthropogenic–continental aerosols.

mixed load of aerosol particles contributed efficiently to give daily mean values of $\tau_a(0.50\,\mu m) > 0.3$, $\alpha(0.40-0.87\,\mu m) < 0.75$, and $\eta_f(0.50\,\mu m) < 0.45$, because of the marked extinction effects produced by the coarse particles mobilized over the north African desert regions. The size-distribution curve was estimated to consist of three modes during such a complex transport case, with a first rather weak mode of fine (presumably man-made) particles centered at $r_c \approx 0.08\,\mu m$, a second more marked mode of accumulation/coarse particles (mainly of continental origin) with $r_c \approx 1.8\,\mu m$, and a third well-pronounced mode of coarse particles centered at $r_c \approx 5.5\,\mu m$ and mainly due to desert dust particles.

7.2.1.4 Relationship between the Fine Particle Fraction and Ångström Wavelength Exponent

The Ångström wavelengthexponent α is an operationally robust optical parameter containing size information on all the optically active aerosols in the FOV of a Sun photometer. Assuming that the optical effects of a typical particle size distribution can be separated into submicrometer (fine) and supermicrometer (coarse) components, O'Neill, Dubovik, and Eck (2001) showed that one can exploit the spectral curvature information in the measured spectral series of $\tau_a(\lambda)$ to achieve the estimates of the fine-mode Ångström exponent $\alpha_f(\lambda)$ and optical fraction $\eta_f \tau_a(\lambda)$ attributed to the fine particle mode. In the O'Neill, Dubovik, and Eck (2001) analysis, exponent $\alpha(\lambda)$ was determined as the derivative $d[\ln \tau_a(\lambda)]/d(\ln \lambda)$, whose dependence on wavelength can be explicitly expressed in terms of the analytical form

$$\alpha(\lambda) = [\alpha_f(\lambda)\tau_a(\lambda) + \alpha_c(\lambda)\tau_a(\lambda)]/\tau_a(\lambda)]. \tag{7.6}$$

Assuming that

$$\eta_f(\lambda) = [\alpha(\lambda) - \alpha_f(\lambda)]/[\alpha_f(\lambda) - \alpha_c(\lambda)], \tag{7.7}$$

exponent $\alpha(\lambda)$ defined in Eq. (7.6) can be written in the form

$$\alpha(\lambda) = \alpha_f(\lambda)\eta_f(\lambda) + \alpha_c(\lambda)[1 - \eta_f(\lambda)], \tag{7.8}$$

which explicitly contains the fine particle fraction $\eta_f(\lambda)$. Differentiating Eq. (7.8), the following spectral derivative is obtained:

$$\alpha'(\lambda) = \alpha_f'(\lambda)\eta_f(\lambda) + \alpha'(\lambda)[1 - \eta_f(\lambda)] - \eta_f(\lambda)[1 - \eta_f(\lambda)][\alpha_f'(\lambda) - \alpha_c(\lambda)]^2. \tag{7.9}$$

Taking into account the definition of $\eta_f(\lambda)$ given in Eq. (7.7), the difference $[\alpha_f(\lambda) - \alpha_c(\lambda)]$ can be represented in the following form:

$$\alpha_f(\lambda) - \alpha_c(\lambda) = (1/2)\left\{ y + \left[y^2 + 4\left(\alpha_f' - \alpha_c'\right)\right]^{1/2} \right\}, \tag{7.10}$$

in which parameter y is given by

$$y = (\alpha - \alpha_c) - [(\alpha' - \alpha_c')/(\alpha - \alpha_c)] = (\alpha_f - \alpha_c) - [(\alpha_f' - \alpha_c')/(\alpha_f - \alpha_c)]. \tag{7.11}$$

Applying the previous procedure to the field measurements conducted at SPC with the Prede POM-02L Sun/sky radiometer, parameters α_f and α_c were determined from the measurements of $\tau_a(\lambda)$ performed at the 0.400, 0.500, 0.675, and 0.870 μm wavelengths and plotted versus wavelength on a bilogarithmic scale, according to the best-fit examples obtained using the Ångström (1964) formula and shown in Figure 7.4.

A first example of scatter plot of parameter $\eta_f(0.50\,\mu\text{m})$ calculated using the O'Neill, Dubovik, and Eck (2001) algorithm is shown in Figure 7.9 versus exponent $\alpha(0.44-0.87\,\mu\text{m})$ for the whole set of the Prede POM-02L measurements conducted during the last 2 weeks of May 2007 at different hours of cloudless-sky days (and, hence, for various solar zenith angles). The data substantially refer to two different predominant aerosol types, the first characterized by predominant rural-continental aerosol extinction and the second by prevailing Saharan dust extinction effects. The scatter plot of Figure 7.9a shows that low values of η_f are associated with relatively low values of α in cases where Saharan dust extinction predominates on fine particle attenuation, mainly because of its relatively high coarse particle content.

In fact, parameter η_f found for Saharan dust increases gradually from about 0.2 to around 1.0 as α increases from 0.2 to 1.5, showing a gradually larger dispersion with increasing α. These features provide evidence that a biunivocal relationship does not exist between η_f and α, since both parameters exhibit a large variability arising from the different multimodal size-distribution characteristics of the particle size distribution, in which desert dust is mixed with a less important content of rural-continental aerosols. Predominant contents of rural-continental aerosol particles were found to yield values of η_f mainly ranging from 0.8 to 1, while exponent α increases gradually from about 1.5 to beyond 2.0. As a whole, Figure 7.9a shows a measure of the large dispersion of fine fraction η_f when plotted versus

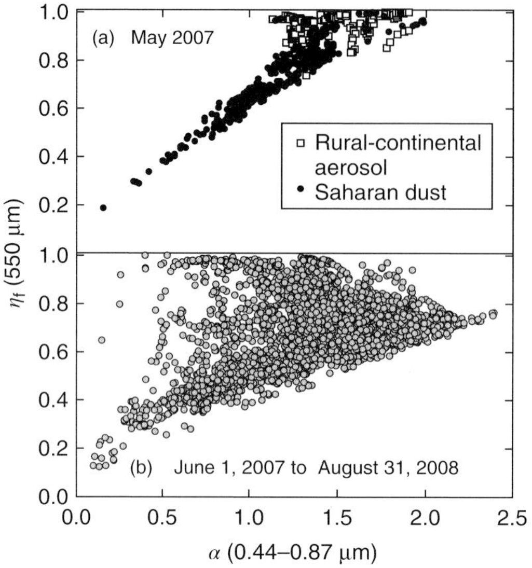

Figure 7.9 (a) Scatter plot of fine particle fraction $\eta_f(0.50\,\mu m)$ calculated using the algorithm of O'Neill, Dubovik, and Eck (2001) represented as a function of simultaneous Ångström's wavelength exponent $\alpha(0.44 - 0.87\,\mu m)$, as obtained by examining the Prede POM-02L Sun/sky-radiometer measurements conducted at SPC during the last 2 weeks of May 2007 at different hours from sunrise to sunset. Open squares refer to the measurements performed for atmospheric loads with predominant rural-continental aerosol particles, and small solid circles to those performed for loads with prevailing content of Saharan dust. (b) As in part (a), for the whole set of daily mean values of parameters $\eta_f(0.50\,\mu m)$ and $\alpha(0.44\text{–}0.87\,\mu m)$ derived from the Prede POM-02L Sun/sky-radiometer measurements conducted for all aerosol types (gray circles) and for cloudless-sky conditions during the period of the 15-month AEROCLOUDS field campaign conducted from 1 June 2007 to 31 August 2008.

$\alpha(0.44-0.87\,\mu m)$, such a variability depending substantially on the large variations of the efficiencies with which fine and coarse particles contribute to the overall particulate extinction. Finally, parameter η_f approximates the unit value as the fine particle extinction effects prevail more and more markedly on those produced by coarse particles, while exponent α increases until reaching an upper limit exceeding 2.4.

These features can be more clearly seen in Figure 7.9b, which presents a 15-month scatter plot of the daily mean values of η_f versus those of α, such a very large data set being collected for the measurements carried out from 1 June 2007 to 31 August 2008. The dispersion characteristics of this scatter plot are similar to those of the first example of May 2007 but seem to be largely more dispersed because of the wider variety of atmospheric turbidity conditions examined during such a long measurement period. The results confirm that the dispersion features of η_f increase gradually as α increases passing from cases with prevailing extinction effects due to coarse particles to extinction cases dominated by the fine particle mode. Figure 7.8b defines in practice a large domain for such a high number of cases, limited in its lower part by a well-defined curve and presenting more dispersed values in its upper part for values of α mainly greater than 1.0. The daily mean values of α determined during the AEROCLOUDS project 2007–2008 campaign for predominant extinction conditions given by continental particles mixed with local anthropogenic aerosol were estimated to mainly range from 0.90 to 1.60, presenting a large variety of linear combinations of fine and coarse particle modes leading parameter η_f to assume corresponding values ranging from less than 0.4 to nearly unit. Cases of mixed marine/continental aerosol loads were often observed during the AEROCLOUDS campaign, with daily mean values of α ranging mainly between 0.70 and 1.30 and gradually increasing as the relative contribution by continental particles increased on that of marine aerosols and parameter η_f correspondingly increases to approach values indicating the optical predominance of fine particles.

7.2.2 Measurements of Volume Extinction, Scattering, and Absorption Coefficients at Ground Level Using Nephelometer and PSAP Techniques

Ground-level measurements of the volume aerosol scattering coefficient $\beta_{sca}(0.530\,\mu m)$ and volume aerosol absorption coefficient $\beta_{abs}(0.573\,\mu m)$ were conducted at the SPC station during the AEROCLOUDS field campaign from mid-May 2007 to late September 2010 using a M903 nephelometer and PSAP, both manufactured by RR. The M903 nephelometer is a portable low-power instrument designed for environmental monitoring, equipped with a light source having wavelength $\lambda = 0.530\,\mu m$. It provides measurements of the backscattering coefficient for particulate mass concentrations varying from ~ 1 to $10^3\,\mu g\,m^{-3}$. The instrument is equipped with an internal data logger to store the scattering coefficient averages and the parameters used to estimate the backscattering coefficient and provide a measurement of the average light absorption coefficient. The method is based on the integrating plate technique, in which the change in the optical transmission of a filter caused by particle deposition is related to the light absorption coefficient of the deposited particles determined in terms of the Bouguer–Lambert–Beer law for light attenuation. The PSAP model employed at SPC operated at the 0.573 μm wavelength, while the more recent versions operate with a triplet of wavelengths (0.467, 0.530, and 0.660 μm).

In place of PSAPs, various aethalometer models can be employed in field studies to measure the aerosol absorption coefficient $\beta_{abs}(\lambda)$ at a visible wavelength. The aethalometer is an instrument where a particulate sample is optically analyzed to determine the mass concentration of BC particles collected from an airstream passing through a filter and operates at two wavelengths (0.370 μm for aromatic organic compounds and 0.880 μm for BC) or seven wavelengths ranging from 0.370 to 0.950 μm.

The aforementioned M903 nephelometer and PSAP measurements were conducted at SPC on each measurement day using the two previously described

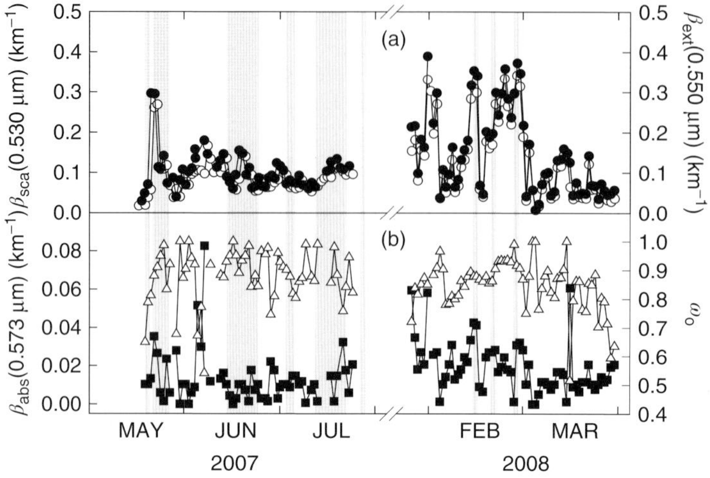

Figure 7.10 (a) Time patterns of the daily mean values of aerosol volume scattering coefficient $\beta_{sca}(0.530\,\mu m)$ (open circles) derived from the nephelometer measurements and approximate volume extinction coefficient $\beta_{sca}(0.550\,\mu m)$ determined as equal to the sum of $\beta_{sca}(0.530\,\mu m)$ and $\beta_{abs}(0.573\,\mu m)$ (solid squares), recorded at SPC (Italy) over the two periods from 16 May to 27 July 2007, and from 24 January to 31 March 2008. (b) As in part (a), for the daily mean values of aerosol volume absorption coefficient $\beta_{abs}(0.573\,\mu m)$ (solid squares) derived from the PSAP measurements and surface-level aerosol single scattering albedo ω_o in the visible (open upward triangles), estimated in terms of ratio $\beta_{sca}(0.530\,\mu m)/\beta_{ext}(0.550\,\mu m)$ determined examining the nephelometer and PSAP measurements. The measurement days characterized by the presence of Saharan dust in the low troposphere during the two seasonal field campaigns are marked with gray thin vertical strips.

instruments over a 20 min period every 2 h to determine the daily average value of $\beta_{sca}(0.530\,\mu m)$ and $\beta_{abs}(0.573\,\mu m)$, respectively. The time patterns of the daily average values of these two coefficients are shown in Figure 7.10 over the two measurement periods from 16 May to 27 July 2007, and from 24 January to 31 March 2008, together with the corresponding time patterns of volume extinction coefficient $\beta_{ext}(0.550\,\mu m)$ calculated as the arithmetic sum of $\beta_{sca}(0.530\,\mu m)$ and $\beta_{abs}(0.573\,\mu m)$. It can be noted in Figure 7.10 that numerous measurement days of spring/summer 2007 were affected by Saharan dust transport episodes from northern Africa (marked in gray color). It is worth noting that they coincide with the major part of the peaks characterizing the time patterns of $\beta_{ext}(0.550\,\mu m)$ and $\beta_{sca}(0.530\,\mu m)$, providing evidence of the importance of Saharan dust deposition to cause the ground-level particulate extinction increase on such particular days. Only a few days with Saharan dust transport were observed during the winter 2008 campaign, which were associated with rather high values of $\beta_{ext}(0.550\,\mu m)$ and $\beta_{sca}(0.530\,\mu m)$ with respect to the average aerosol extinction features observed in the winter season over the Po Valley (Italy) and mainly due to anthropogenic aerosol.

The time patterns of the daily mean values of volume absorption coefficient $\beta_{abs}(0.573\,\mu m)$ directly derived from the PSAP measurements performed during the two campaigns are shown in Figure 7.10b. These values were in general lower than $0.04\,km^{-1}$ during both the field campaigns, a large part of the higher values being recorded in concomitance with the Saharan dust transport episodes. Combining the field measurements of $\beta_{sca}(0.530\,\mu m)$ with the previously assumed values of $\beta_{ext}(0.550\,\mu m) = \beta_{sca}(0.530\,\mu m) + \beta_{abs}(0.573\,\mu m)$, it is possible to estimate with good approximation the daily mean values of ground-level aerosol single scattering albedo ω_o in the VIS which assumed to be equal to the ratio $\beta_{sca}(0.530\,\mu m)/\beta_{ext}(0.550\,\mu m)$. The time patterns of the daily mean values of ω_o shown in Figure 7.10b indicate that ω_o varied mainly between 0.80 and 0.98 in spring/summer of 2007 and between 0.70 and 0.95 during the winter 2008, presenting a sharp decrease during the last week of March 2008, which was frequently affected by the presence of desert dust particles characterized by particularly strong absorption features.

7.2.3
Vertical Profiles of Backscatter and Extinction Coefficients from LIDAR Measurements

The LIDAR is an instrument conceptually similar to the radar, which allows one to carry out remote sensing measurements of atmospheric aerosols. In the LIDAR technique, a UV, VIS, or IR laser light beam is emitted by a laser source to sound the atmosphere with a high vertical resolution. Optical telescopes and photomultipliers (or photodiodes) are used to collect and detect the light backscattered by the atmosphere toward the receiver at the ground (for details, see Measures, 1992). Combined with appropriate correction procedures for the Rayleigh scattering effects by air molecules, the LIDAR technique constitutes the most efficient tool to observe the vertical distribution of atmospheric aerosols, with height resolution of a few meters. A quantitative inversion of the LIDAR signal is performed for this purpose, provided that the relationship existing between the volume aerosol extinction coefficient $\beta_{ext}(\lambda_o)$ at LIDAR wavelength λ_o and the corresponding backscatter coefficient $\beta_{bs}(\lambda_o)$ is known within the various tropospheric altitude ranges. Coefficient $\beta_{ext}(\lambda_o)$ is usually measured in per kilometer and given by the integral of the Mie (1908) extinction efficiency multiplied by (i) the cross section πr^2 of the aerosol particle having radius r and (ii) the multimodal size-distribution $n(r) = dN/dr$ of aerosols over the whole particle radius range observed in the atmosphere. Backscatter coefficient $\beta_{bs}(\lambda_o)$ (measured in steradian per kilometer) is given by the integral of the backscatter efficiency defined by the Mie (1908) theory and multiplied by (i) the cross section πr^2 of the spherical aerosol particle having radius r and (ii) the multimodal aerosol size-distribution curve $n(r)$ (Collis and Russell, 1976; Klett, 1985). The aerosol extinction cross section is related to the backscatter aerosol cross section through a phase function, which closely depends on the aerosol microphysical and optical properties (shape, size distribution, and refractive index). Therefore, in order to calculate realistically the vertical profiles of coefficients $\beta_{ext}(\lambda_o)$ and $\beta_{bs}(\lambda_o)$, it is crucial to know (i) the aerosol composition at various altitudes and, hence, determine with good accuracy the real and imaginary parts of particulate matter refractive index and (ii) the shape parameters of the multimodal size-distribution curve of the aerosol particles that most efficiently scatter the LIDAR light beam, at least over the radius range from 10^{-2} to 20 μm. Therefore, the procedure followed to analyze the LIDAR data substantially includes the definition of the aerosol size distribution and complex refractive index of particulate matter, calculations of aerosol properties, and the study of the average behavior of the aerosol optical parameters, evaluated on the basis of known particle size distributions and refractive index. For this reason, the LIDAR measurements are commonly conducted for a large number of runs to extract significant average experimental data and achieve a statistically correct description of the aerosol optical properties.

At each altitude, the LIDAR ratio $R_\ell(\lambda_o)$ at wavelength λ_o is equal to the ratio between the aerosol extinction coefficient $\beta_{ext}(\lambda_o)$ and the aerosol backscatter coefficient $\beta_{bs}(\lambda_o)$, which depend both on the chemical composition of aerosol particles and on the shape of aerosol particles and the multimodal features of the aerosol size-distribution curve. On the basis of calculations made according to the Mie (1908) theory for various aerosol types, Ackermann (1998) estimated that LIDAR ratio $R_\ell(0.532\,\mu m)$ pertaining to aerosol particles grown for air RH equal to ∼ 50% was equal to 27 sr for maritime aerosol, about 55 sr for continental aerosol, and about 19 sr for desert dust. The LIDAR ratio can vary strongly with height, especially when marine, anthropogenic (urban, BB), and/or desert dust particles or mixtures of these basic aerosol types are present in layers suspended one above the other (Ansmann *et al.*, 2001, 2002). Typical values of $R_\ell(0.532\,\mu m)$ determined with a Raman LIDAR for different aerosol types were estimated to vary for RH conditions ranging between 30% and more than 80%: (i) from 20 to 35 sr for marine aerosols (Ansmann *et al.*, 2001; Franke *et al.*, 2001), (ii) from 50 to 80 sr for Saharan dust (Mattis *et al.*, 2002), (iii) from 35 to 70 sr for less absorbing urban particles (Ansmann *et al.*, 2001; Franke *et al.*, 2001), and (iv) from 70 to 100 sr for absorbing particles originated from BB (Franke *et al.*, 2001; Wandinger *et al.*, 2002). Therefore, considering that $R_\ell(0.532\,\mu m)$ is expected to exhibit large variations (for instance, from 20 to 100 sr), it seems in practice impossible to estimate trustworthy the aerosol extinction profiles in all cases where aerosols have mixed origins.

LIDAR measurements were conducted by Angelini *et al.* (2009) at the Torre Sarca experimental station (45°31′N, 9°13′E, 125 m a.m.s.l.) of the Milano-Bicocca University, located in the northern side of the Milan urban area, on 200 measurement days from January 2007 to February 2008, as a part of the field activities conducted during the QUITSAT project (Di Nicolantonio, Cacciari, and Tomasi, 2009). The automated Vaisala LD-40 LIDAR ceilometer operating at the 0.855 μm wavelength (and shown in Figure 7.11) was used to study the evolutionary patterns of the mixing layer height (MLH), having the technical characteristics described by Emeis, Schäfer, and Münkel (2009). In order to achieve a better signal-to-noise

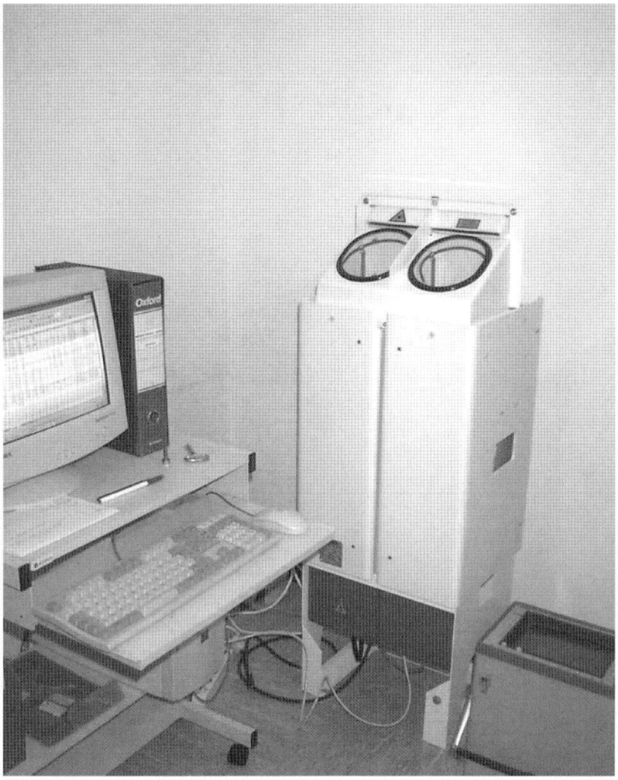

Figure 7.11 Automated Vaisala LD-40 ceilometer used by Angelini *et al.* (2009) at the Torre Sarca (Milano-Bicocca University) station in the center of Milan (Italy) during the QUITSAT field campaigns of summer 2007 and winter 2008. (Reproduced with kind permission of G. P. Gobbi and F. Angelini, ISAC-CNR, Rome Tor Vergata, Via del Fosso del Cavaliere 100, 00133 Roma, RM Lazio, Italy).

ratio, the aerosol backscatter profiles were regularly acquired by Angelini *et al.* (2009) at Torre Sarca (Milan) every 15 s and averaged over 15 min periods.

Adopting these selection criteria, a set of 96 average vertical profiles of LIDAR signals was collected on each measurement day on average. Due to the instrumental limitations, the lowest height monitored by the automated Vaisala LD-40 ceilometer was about 60 m above the ground, and the top level was around 4 km. The analysis of the LIDAR measurements was carried out to retrieve the regular vertical profiles of coefficient $\beta_{ext}(0.855\,\mu m)$ and estimate the evolutionary patterns of the MLH during each measurement day (Eresmaa *et al.*, 2006). Six vertical profiles of $\beta_{ext}(0.855\,\mu m)$ are shown in Figure 7.12, obtained on 19 July 2007, at six distinct hours: three in the early morning, one in late morning, one at noon, and the last in early afternoon. In the early morning, the MLH slowly increased from about 0.2 km to more than 0.5 km as a result of the gradual more intense heating of the surface by solar radiation, while a dense layer of aerosol particles was

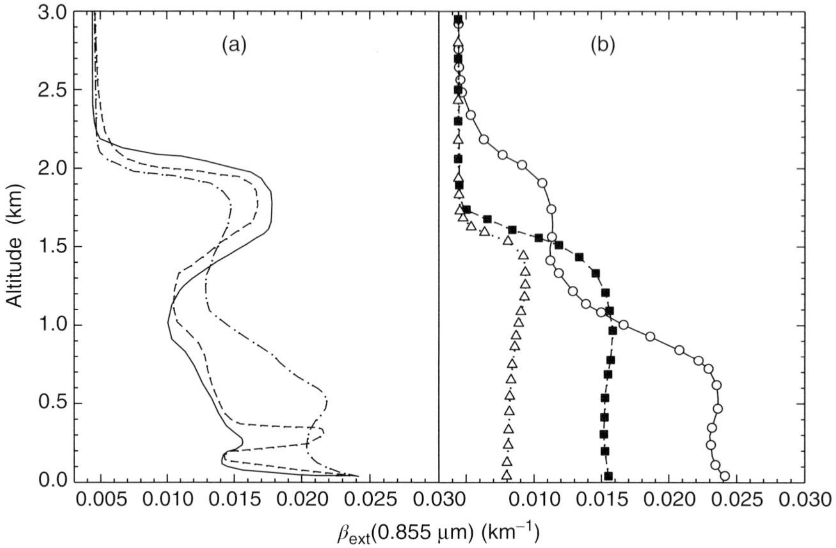

Figure 7.12 (a) Vertical profiles of the volume extinction coefficient $\beta_{ext}(0.855\,\mu m)$ derived over the first 4 km of the atmosphere from the automated Vaisala LD-40 ceilometer measurements performed by Angelini *et al.* (2009) as a part of the national QUITSAT project (Di Nicolantonio, Cacciari, and Tomasi, 2009) on 19 July 2007, above the urban site of Torre Sarca in Milan (45°31′N, 9°13′E, Milano-Bicocca University) at three hours of early morning: (i) 04:03 UTC (solid curve), (i) 05:15 UTC (dashed curve), and 06:40 UTC (dashed and dotted curve). (b) As in the left-hand part, for the Vaisala LD-40 ceilometer measurements conducted during the warmest part of 19 July 2007, at (i) 10:00 UTC (solid curve with open circles), (ii) 12:00 UTC (dashed curve with solid squares), and (iii) 14:00 UTC (dotted curve with open triangles).

observed at altitudes of 1.6–2.0 km, formed by the intense convective motions generated during the warmest hours of the previous days. In the late morning, the surface warming by solar radiation gradually increased, and the MLH lifted to 0.6 km at 10:00 UTC, while the extinction intensity of the upper layer gradually decreased to less than 50% its nighttime intensity. The MHL increased to about 1.2 km at 12:00 UTC and to ∼ 1.4 km at 14:00 UTC, while the higher thick layer of aerosol particles totally vanished as a result of an extended vertical mixing involving the whole boundary layer and due to the strong surface heating by solar radiation.

7.2.4
Measurements of the Aerosol Size Distribution Using an Optical Particle Counter

Vertical profiles of particulate matter size-distribution curves were achieved at the Torre Sarca (Milan) station during the QUITSAT experiment by using an OPC (GRIMM 1.108 "DUSTcheck") model (shown in Figure 7.13) on board the Milano-Bicocca University-tethered balloon. This instrument is typically designed

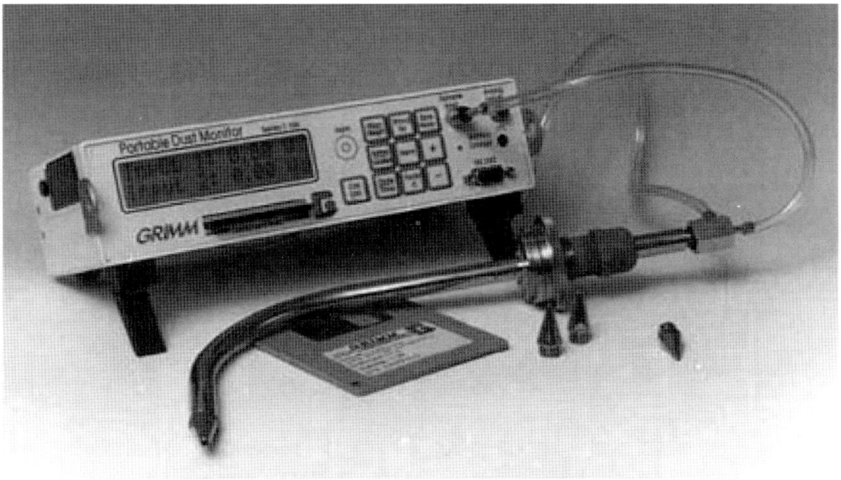

Figure 7.13 Picture of the optical particle counter (GRIMM 1.108 "DUSTcheck") model mounted on board the Milano-Bicocca University-tethered balloon and used at the Torre Sarca (Milano-Bicocca University) station in the center of Milan (Italy) during the 3 years from 2005 to 2008 to measure the aerosol size spectra at various levels of the atmospheric boundary layer. (Reproduced with kind permission of Ezio Bolzacchini and Luca Ferrero, Department of Earth and Environmental Sciences, University of Milano-Bicocca, Piazza della Scienza 1, 20126, Milano, Italy).

to measure in real time both fine and coarse particle light scattering. A light source is used, and generally laser or laser diode is adjusted to illuminate a stream of aerosol flowing out of a nozzle. The off-axis from the light beam measures the amount of light scattered from a single particle, the number density and particle size being estimated on the basis of the scattered light intensity. Using this technique, it provides the particle number concentrations within 15 size classes over the $0.3 \leq a \leq 20\,\mu m$ size range.

Ground-based OPC measurements were conducted and analyzed by Ferrero *et al.* (2010) to define the aerosol size-distribution curves at ground level. Figure 7.14 shows the seasonal mean size-distribution curves of aerosol number concentration obtained by averaging the two following sets of OPC ground-level measurements consisting of (i) 142 size-distribution curves measured during the winter months of the three years from 2005 to 2008 and (ii) 72 size-distribution curves measured during the summer months from 2005 to 2008. The comparison provides evidence of the multimodal characteristics of the two seasonal particle size-distribution curves determined over the $0.3-1.5\,\mu m$ and $2.0-28\,\mu m$ size ranges. It can be noted in Figure 7.14 that the number concentration of the fine particle mode measured in winter is considerably higher by nearly one order of magnitude than that measured in summer. More similar values of coarse particle concentration were measured during the two seasonal periods, obtaining only slightly higher values in summer over the size range $a > 10\,\mu m$.

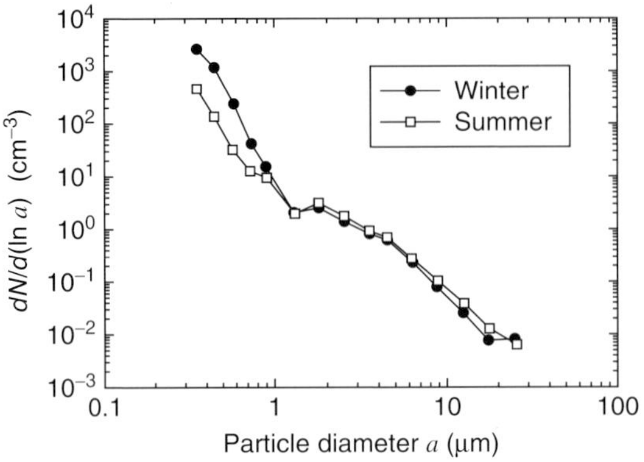

Figure 7.14 Comparison between the seasonal mean size-distribution curves of aerosol number concentration $dN/d(\ln a)$ obtained as a function of the natural logarithm of particle diameter a from (i) an overall number of 142 size-distribution curves measured at the Torre Sarca (Milano-Bicocca University) station in the center of Milan (Italy) during the winter months of 3 years from 2005 to 2008 (solid circles) and (ii) a set of 72 ground-level size-distribution curves measured during the summer months from 2005 to 2008 (open squares). (This graph was prepared using the data sets of Ferrero *et al.* (2010).)

Vertical profiles of the particle size-distribution curves were measured at Torre Sarca (Milan) station over the first hundred meters above the ground by using the GRIMM 1.108 "DUSTcheck" OPC mounted on a spherical helium-filled tethered balloon (PU balloon, 4 m in diameter, 33.5 m^3 volume) (for details, see Ferrero *et al.*, 2010). The OPC system acquired data with a time resolution of 6 s to reach the highest spatial resolution during the balloon's flight. In order to avoid any wrong information in determining the original particle size-distribution characteristics that could be caused by too high RH conditions, Ferrero *et al.* (2010) carried out the vertical profiles of particle mass concentration only for RH < 65%. The maximum height reached during each balloon launch varied with the atmospheric conditions occurring on each measurement day, the most usual height range being of 300–600 m above the ground. Figure 7.15 shows that the fine particle number concentration varied mainly between 50 and about 190 cm^{-3} during the three morning sampling periods of 19 July 2007, presenting a marked peak at around 0.2 km height at about 06:15 UTC. Conversely, the vertical profiles of coarse particle number density showed almost stable values over the whole ground layer below the 0.6 km height at all three sampling hours.

7.3
Airborne Remote Sensing Measurements of Aerosol Optical Properties

Airborne measurements of aerosol microphysical and optical parameters are conducted using a variety of *in situ* and remote sensing techniques suitable for

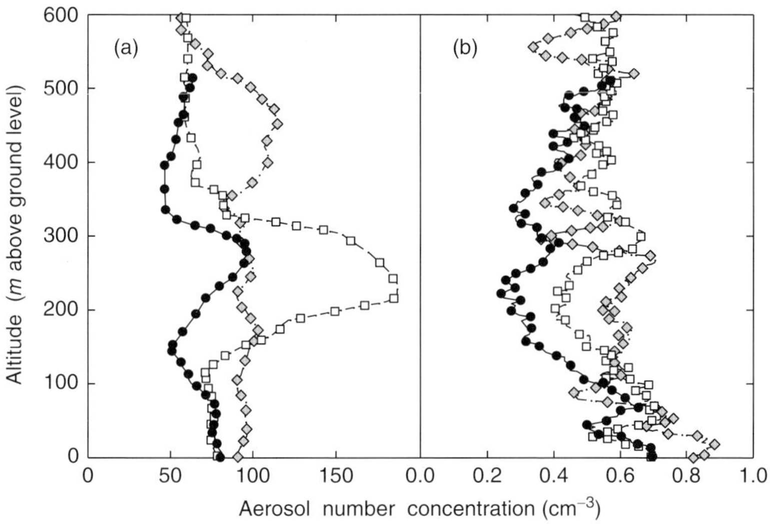

Figure 7.15 Vertical profiles of the aerosol particle number concentration within the mixing layer of the urban atmosphere with (a) sizes varying between 0.3 and 1.6 μm (fine particle mode) and (b) with sizes > 1.6 μm (coarse particle mode), as measured over three different periods of 19 July 2007, using an optical particle counter (OPC GRIMM 1.108 "DUSTcheck" model) mounted on board the tethered balloon of the Milano-Bicocca University (Ferrero *et al.*, 2010) above the urban site of Torre Sarca (Milan) (45°31′N, 9°13′E). The measurement periods were (i) from 03:55 UTC to 04:12 UTC (solid circles), (ii) from 05:05 UTC to 05:25 UTC (open squares), and (iii) from 06:25 UTC to 06:55 UTC (gray diamonds).

sampling, analyzing, and monitoring aerosol particles. The airborne remote sensing instruments are based on both passive techniques, analyzing the interference effects of aerosol particles with natural radiation, and active techniques, in which an artificial radiation source is employed (as made, for instance, in the LIDAR systems). Numerous experiments have been conducted over the past decades aboard aircraft equipped with remote sensing instruments to study the microphysical and optical characteristics of aerosol particles, the spectral dependence features of AOT, and the vertical profiles of particulate extinction, scattering, and absorption coefficients. Some experimental campaigns have been examined here to provide an exhaustive information on the role of aerosols in the radiation exchange processes that regulate the radiation budget of the surface–atmosphere system and better describe the optical properties of natural and man-made aerosols. Table 7.8 presents the list of the eight projects considered in this study to illustrate some significant results obtained from airborne remote sensing measurements and describing the characteristics of various aerosol types and provides the measurement period and the remote sensing instruments mounted aboard the aircraft in such selected field experiments.

Table 7.8 List of the experimental projects and measurement periods, during which airborne remote sensing measurements were conducted on board aircraft.

Project and measurement period	Aircraft	Airborne remote sensing instruments
AASE-II (Russell *et al.*, 1993), January–March 1992	NASA DC-8	Ames Airborne Tracking Sunphotometer AATS-6 (Matsumoto *et al.*, 1987) and multiwavelength LIDAR system (Browell *et al.*, 1993).
ALIVE (Gunter *et al.*, 1993), February 1989–June 1990	NOAA King Air	3-wavelength nephelometer, 1-wavelength aethalometer, Forward-Scattering Spectrometer Probe (FSSP), and Active-Scattering Aerosol Spectrometer Probe (ASASP) (Kim *et al.*, 1993).
SCAR-A (Remer *et al.*, 1997), July 1993	University of Washington Convair C-131A	2-wavelength Nd:YAG laser (Hobbs *et al.*, 1991), 3-wavelength integrating nephelometer, 1-wavelength Radiance Research PSAP photometer, and Passive Cavity Aerosol Spectrometer Probe (PCASP) over the $0.1 \le a \le 3.0\,\mu m$ size range.
TARFOX (Russell, Hobbs, and Stowe, 1999a), July 1996	University of Washington Convair C-131A	Sun photometer, scanning radiometers (Russell *et al.*, 1999b)
ACE-2 (Russell and Heintzenberg, 2000), June–July 1997	MRF C-130 Hercules	PCASP over the size range $a > 100\,nm$, FSSP over the size range $a > 500\,nm$, and broadband radiometers.
	Modified Cessna 337 Skymaster (Pelican)	14-channel NASA Ames Airborne Tracking Sunphotometer AATS-14, 3-wavelength integrating nephelometer (TSI 3563 model), Radiance Research PSAP photometer, a pair of Radiance Research M903 ($\lambda = 0.530\,\mu m$) nephelometers for both wet and dry RH conditions, and a pair of optical particle counters (OPCs) over the size range $5 \times 10^{-3} \le a \le 8\,\mu m$ (Livingston *et al.*, 2000; Schmid *et al.*, 2000).

PRIDE (Reid *et al.*, 2003), 28 June–24 July 2000	Piper Navajo	6-channel NASA Ames Airborne Tracking Sunphotometer AATS-6 (Matsumoto *et al.*, 1987), NASA Ames Solar Spectral Flux Radiometer (SSFR), a pair of upwelling and downwelling hyperspectral radiometers, PCASP (PCASP-100X), and FSSP (FSSP-100X) (Collins *et al.*, 2000)
ARCTAS/ARCPAC (Jacob *et al.*, 2010; Brock *et al.*, 2011), spring 2008	NASA DC-8	A pair of Single Particle Soot Photometers (SP2) to detect BC (soot) mass over the 100–600 nm size range, high-resolution time-of-flight aerosol mass spectrometers (HR-ToF-AMS) for the aerosol chemical analysis, TSI 3-λ nephelometer, 3-λ Radiance Research PSAP photometer, UHSAS2 sizer over the 0.08–3.0 μm size range, and TSI APS3 spectrometer model measuring the aerosol size distribution over the 0.78–20.0 μm size range.
	NASA P-3B	HR-ToF-AMS to measure submicrometer nonrefractory aerosol chemistry, TSI 3-λ nephelometer, 3-λ Radiance Research PSAP, a pair of custom TSI Differential Mobility Analyzers providing the aerosol size distribution over the 0.01–0.20 μm and 0.01–0.50 μm size ranges, OPC over the 0.15–8.0 μm size range, and TSI Aerodynamic Particle Sizer (APS) spectrometer over the 0.78–20.0 μm size range.

(*continued overleaf*)

Table 7.8 (Continued)

Project and measurement period	Aircraft	Airborne remote sensing instruments
	NOAA WP-3D	SP2 photometer, HR-ToF-AMS, photoacoustic apparatus to measure submicrometer aerosol extinction, Radiance Research PSAP photometer, and some supplemental OPCs.
PAMARCMIP (Stone *et al.*, 2010), 1–25 April, 2009	Polar-5 (AWI Institute, Germany)	8-channel sun photometer manufactured by NOAA ESRL GMD (Boulder, United States) in cooperation with ISAC-CNR (Bologna, Italy), Airborne Mobile Aeroscl LIDAR (AMALi) system (AWI Institute, Potsdam, Germany), SP2, and Ultra-High Sensitivity Aerosol Spectrometer (UHSAS).

7.3.1 Main Results Derived from the Second Airborne Arctic Stratospheric Expedition (AASE-II) Measurements

A number of NASA DC-8 flights were performed at stratospheric altitudes in January–March 1992 as a part of the Second Airborne Arctic Stratospheric Expedition (AASE-II) by carrying out measurements of AOT at VNIR wavelengths produced by the Mt. Pinatubo volcanic aerosol particles using the Ames Airborne Tracking Sunphotometer (Matsumoto *et al.,* 1987). Vertical profiles of the backscattering coefficient $\beta_{bs}(\lambda)$ of volcanic aerosol were also measured at altitudes z ranging from 10 to nearly 30 km and at latitudes varying from 18° S to 20° S using an airborne LIDAR during some DC-8 flights (Browell *et al.,* 1993), as shown in Figure 7.16. Three vertical profiles of volcanic aerosol backscattering coefficient $\beta_{bs}(\lambda)$ were derived by Russell *et al.* (1993) analyzing the LIDAR measurements recorded aboard the NASA DC-8 aircraft using a multiwavelength LIDAR at latitudes of 18–20° S and altitudes ranging from $\sim$ 10 to 30 km: (i) a first profile of $\beta_{bs}(1.064\,\mu m)$ was measured on 19 January 1992, presenting a pronounced peak at $\sim$ 15 km height, and some secondary layers at the $20 \leq z \leq 24$ km altitudes; (ii) a profile of $\beta_{bs}(0.604\,\mu m)$ measured on 28 January 1992, with a large maximum centered at $z \approx 24$ km; and (iii) a profile

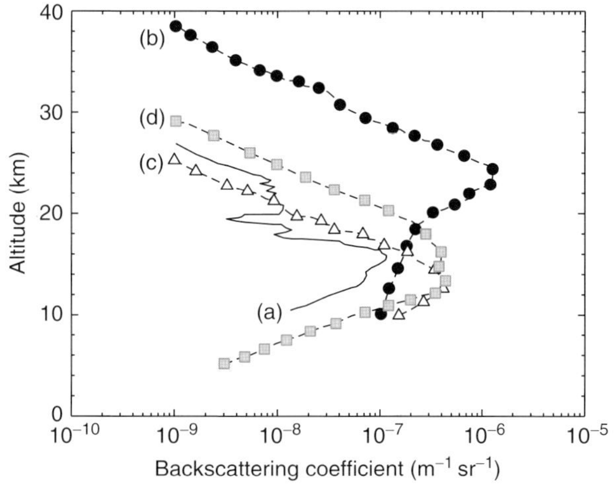

Figure 7.16 Vertical profiles of aerosol backscattering coefficient derived from the multiwavelength LIDAR measurements conducted on board the NASA DC-8 aircraft during the Second Airborne Arctic Stratospheric Expedition (AASE-II) at altitudes ranging from about 10 to 30 km on different measurement days: (a) 19 January 1992, at wavelength $\lambda = 1.064\,\mu m$ (solid curve); (b) 28 January 1992, at wavelength $\lambda = 0.604\,\mu m$ (dashed curve with solid circles); (c) 14 March 1992, at wavelength $\lambda = 0.604\,\mu m$ (dashed curve with open triangles); and (d) 20 March 1992, at wavelength $\lambda = 0.604\,\mu m$ (dashed curve with gray squares). (Graph prepared using the AASE-II data of Russell *et al.* (1993).)

of β_{bs}(0.604 μm) measured on 20 March 1992, extending from 9 to 25 km levels, with a large maximum occupying the $10 \leq z \leq 20$ km altitude range.

7.3.2 Airborne Remote Sensing Measurements during the Army LIDAR Verification Experiment (ALIVE)

The Army LIDAR Validation Experiment (ALIVE) was conducted over the south-central part of New Mexico (United States) during four intensive periods in February 1989, July 1989, December 1989, and May–June 1990. Optical measurements were carried out by Gunter *et al.* (1993) using various instruments mounted on the National Oceanic and Atmospheric Administration (NOAA) King Air research aircraft. In particular, a three-wavelength (0.449, 0.536, and 0.690 μm) nephelometer and an aethalometer were used during the four aforementioned field periods to assess the effects of aerosol scattering and absorption on shortwave radiation and define the vertical features of particulate extinction. The aerosol vertical profiles and size-distribution features were investigated by Kim *et al.* (1993) using the airborne nephelometer, aethalometer, Forward-Scattering Spectrometer Probe (FSSP), and Active-Scattering Aerosol Spectrometer Probe (ASASP), the last two instruments being mounted on the wings of the NOAA King Air research aircraft. The vertical distribution profiles of background atmospheric aerosols were determined over the $0.2 \leq z \leq 4.5$ km altitude range, showing bimodal features typical of the continental atmospheric aerosols over the whole $0.1 \leq a \leq 32$ μm size range, therefore including both fine and coarse particle modes, whose average aerosol size-distribution curves were well approximated by lognormal curves. The free troposphere aerosol size distributions were measured in December 1989 (ALIVE III) and June 1990 (ALIVE IV), resembling often the Junge power-law distribution characteristics.

7.3.3 Airborne Measurements Performed during the Sulfate Clouds and Radiation–Atlantic (SCAR-A) Experiment

The Sulfate Clouds and Radiation – Atlantic (SCAR-A) experiment was conducted in the mid-Atlantic region of the eastern United States in July 1993, carrying out both airborne and ground-based sun-radiometer measurements to better define the aerosol composition and particulate matter radiative parameters. Airborne *in situ* measurements were carried out on board the University of Washington Convair C-131A research aircraft during 13 flights using *in situ* samplers with which the concentrations and size distributions of aerosol particles were measured continuously over the size range $> 5 \times 10^{-2}$ μm (Hegg, Ferek, and Hobbs, 1993) and the remote sensing instruments reported in Table 7.8. In particular, a dual-wavelength Nd:YAG laser LIDAR mounted within the aircraft fuselage and pointed directly upward through a window was used during the flights, having as receiver a 33.6 cm Cassegrain telescope coupled with a photomultiplier tube detector at $\lambda = 0.532\,\mu m$ and an avalanche photodiode detector at $\lambda = 1.064\,\mu m$ (Hobbs *et al.*, 1991). Examining the particulate sampling data and comparing them with the LIDAR measurements and the spectral estimates of AOT $\tau_a(\lambda)$ derived from the Sun/sky-radiometer measurements conducted simultaneously at four ground stations in the experimental area, Remer *et al.* (1997) determined the microphysical and radiative characteristics of the aerosol particles at various altitudes and defined the size-distribution shape parameters and the scattering features of atmospheric aerosol particles, which were found to be predominantly associated with urban/industrial pollution particles.

The daily mean values of the scattering component $\tau_{sca}(0.450\,\mu m)$ of AOT obtained from the nephelometer measurements were estimated to vary from 0.16 to 0.45 during the experiment and the absorption component $\tau_{abs}(0.450\,\mu m)$ to range from 0.0005 to 0.005, while the daily mean values of $\tau_a(0.440\,\mu m)$ derived from the ground-based Sun radiometers varied from 0.24 to 0.76. The comparison between the estimates of AOT $\tau_a(0.670\,\mu m)$ derived from the ground-based Sun-radiometer measurements and those of $\tau_a(0.670\,\mu m)$ derived within 6 h from the airborne data collected at altitudes $z < 30\,km$ showed a very good agreement with relative discrepancies ranging from 5% to 17%. The single scattering albedo $\omega(0.440\,\mu m)$ was estimated to range mainly between 0.980 and 0.999 during the summer, showing that aerosol particles were mainly composed of sulfate substances (by more than 90%) and contained only minor fractions of sea salt, nitrate substances, and organics (Hegg *et al.*, 1995). The real part of the refractive index was evaluated to vary from 1.52 (ammonium sulfate) to 1.47 (ammonium bisulfate), resulting in residual differences between observed and calculated scattering coefficient not exceeding 10%.

The analysis of the particle size spectra obtained using the two aforementioned different methodologies provided mainly bimodal features of the volume aerosol size distributions at various altitudes from surface level to 2 km, presenting often a fine particle mode with mode diameter $a_c \approx 0.4\,\mu m$ and a coarse particle mode with a_c ranging mainly from 4 to 6 μm. The vertical profiles of light scattering coefficient $\beta_{sca}(0.450\,\mu m)$ derived from both airborne nephelometer and LIDAR measurements provided evidence of the multilayer characteristics of the aerosol vertical distribution within cloud-free air masses in the low troposphere, where these profiles spanned vertical layers of variable optical thickness located at altitudes ranging from 0.6 to 2.5 km.

7.3.4
Airborne Measurements Conducted during the Tropospheric Aerosol Radiative Forcing Observational Experiment (TARFOX)

The Tropospheric Aerosol Radiative Forcing Observational Experiment (TARFOX) was conducted from 10 July to 31 July 1996, to study the aerosol effects on atmospheric radiation budget over the eastern US coastal regions, where urban/industrial haze plumes move in general from the continent to the Atlantic Ocean. The intensive field campaign included remote sensing measurements carried out at various land sites, onboard ships, and four satellites (GOES-8, NOAA-14, ERS-2, and Landsat) and aboard four airplanes (ER-2, C-130, C-131A, and a modified Cessna (Pelican)) (Russell, Hobbs, and Stowe, 1999a; Hobbs, 1999). Upward and downward solar radiation fluxes were measured on board the UK Meteorological Research Flight (MRF) C-130 aircraft over the 0.3–3.0 μm and 0.7–3.0 μm wavelength ranges (Hignett *et al.*, 1999), together with additional radiometric measurements of the angular distribution of sky radiance from mid-VIS to the IR and supplemental data providing the aerosol particle size-distribution curves, particulate matter chemical composition, and absorption coefficients. Spectral series of $\tau_a(\lambda)$, backscatter profiles, aerosol and cloud absorption and scattering coefficients, aerosol hygroscopic growth factors, and cloud/fog optical properties were measured on board the C-131 A aircraft of the University of Washington (Russell *et al.*, 1999b).

Examining the previous measurements, the AOT τ_a(0.55 μm) was estimated to vary mainly from 0.07 to 0.55 over the observed region, giving an overall mean value of 0.25, and exponent α to range from 1.0 to 2.3, yielding an overall mean value of 1.66. In particular, Redemann *et al.* (2000a) estimated that the real part n(0.815 μm) of the particulate matter refraction index varied from 1.33 (liquid water) to 1.62 (sand) and the imaginary part χ(0.815 μm) from the very low values close to 0.4, as it is typical of sea salt contaminated by high concentration of carbonaceous soot substances. Aerosol single scattering albedo ω(0.45 μm) was estimated by Redemann *et al.* (2000b) to range from 0.900 to 0.985, with an average value of 0.96, while asymmetry factor g(0.45 μm) varied from 0.60 to 0.80. Figure 7.17 shows two examples of the vertical profiles of aerosol extinction coefficient and asymmetry factor derived by Redemann *et al.* (2000b) using the Fu and Liou (1992) radiative transfer model over the 0.2–0.7 μm wavelength range for the data sets recorded on 17 July and 24 July 1996, respectively. For these aerosol optical properties, the AOT $\tau_a(\lambda)$ was measured at various VNIR wavelengths from 0.30 to 0.70 μm spectral range and found to vary mainly between 0.05 and 0.40 during the TARFOX intensive period.

7.3.5 The Aerosol Characterization Experiment 2 (ACE-2) Airborne Remote Sensing Measurements

The Aerosol Characterization Experiment 2 (ACE-2) experiment was an important part of the International Global Atmospheric Chemistry (IGAC) program conducted from 16 June to 25 July 1997, over the subtropical northeast Atlantic Ocean sector limited by Portugal, Azores Islands, and Canary Islands (Raes *et al.*, 2000; Russell and Heintzenberg, 2000). The experiment included 60 coordinated aircraft missions (with 6 instrumented airplanes for a total of 450 flight hours) and numerous additional *in situ* particle sampling and Sun-photometer/radiometer measurements performed on board the R/V Vodyanitsky ship and at ten ground stations in southern Portugal, Tenerife island (Canary Islands, Spain), and Madeira (Portugal), with the main goal of studying the aerosol radiative properties and their

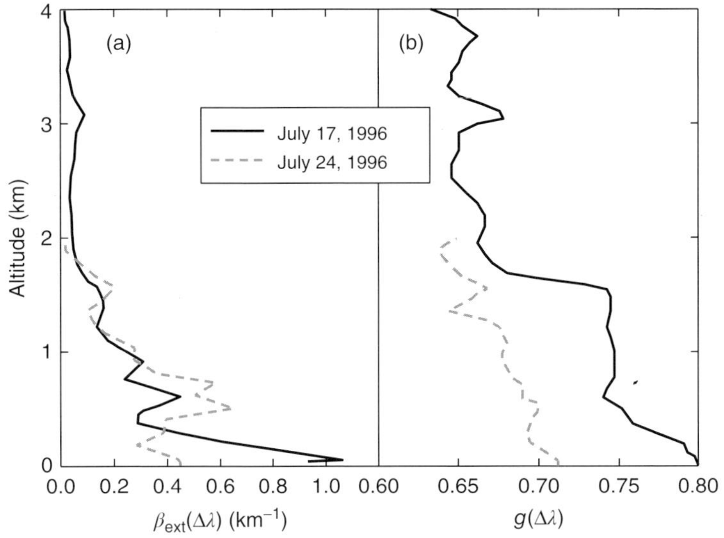

Figure 7.17 (a) Vertical profiles of aerosol volume extinction coefficient $\beta_{ext}(\Delta\lambda)$ obtained by Redemann *et al.* (2000b) for the first band of the Fu and Liou (1992) radiative transfer model applied over the wavelength range $\Delta\lambda$ from 0.2 to 0.7 μm to the field data collected on the two TARFOX measurement days of 17 July 1996 (solid curve), and 24 July 1996 (gray dashed curve). (b) As in the left-hand side, for the vertical profiles of asymmetry factor $g(\Delta\lambda)$ calculated by averaging the Fu and Liou (1992) data over the wavelength range $\Delta\lambda$ from 0.2 to 0.7 μm.

effects on the radiation exchange processes occurring in this maritime region and mainly due to background marine, anthropogenic polluted aerosol from Europe, and mineral dust (MD) transported from Saharan regions.

Four instrumented airplanes were used over the ACE-2 observation area. In particular, aerosol remote sensing measurements were conducted on board the two following aircraft carrying the instruments listed in Table 7.8:

1. The C-130 Hercules aircraft of the MRF, Farnborough (United Kingdom), measuring (i) the aerosol size spectra with $a > 3$ nm, with $a > 100$ nm, and with $a > 500$ nm, (ii) the heated/ambient aerosol size-distribution curves with a Scanning Mobility Particle Sizer (SMPS), (iii) some broadband radiometers, (iv) the aerosol scattering and absorption coefficients, and (v) CCN spectra
2. The modified Cessna 337 Skymaster (Pelican) aircraft operated by the Center for Interdisciplinary Remotely Piloted Aircraft Studies (CIRPAS), US Navy (United States), which was equipped with the instruments listed in Table 7.8 and used to measure the aerosol chemical, physical, and optical properties from sea level to ~ 4 km altitude (Livingston *et al.*, 2000)

The AATS-14 sun photometer mounted on the Pelican aircraft acquired data of good quality during 14 of the 21 ACE-2 flights (Schmid *et al.*, 2000). Vertical

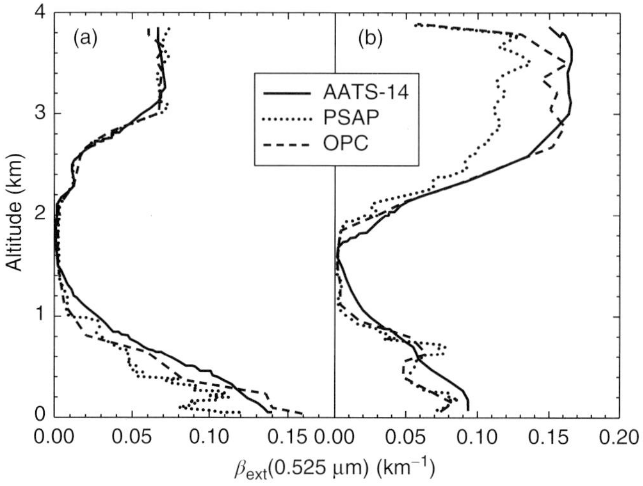

Figure 7.18 (a) Vertical profiles of volume extinction coefficient $\beta_{ext}(0.525\,\mu m)$ (solid curve) retrieved from the AATS-14 sun-photometer measurements performed aboard the Pelican aircraft during the tf15 flight (8 July 1997) and the vertical profiles derived from the nephelometer and PSAP airborne measurements (dotted curve) and the Caltech OPC airborne measurements (black dashed curve). (b) As in the left-hand side for the AATS-14, nephelometer and PSAP and Caltech OPC airborne measurements performed aboard the Pelican aircraft during the tf20 flight (17 July 1997). (Graph prepared using the ACE-2 data sets of Schmid *et al.* (2000).)

profiles of $\tau_a(\lambda)$ and $\beta_{ext}(\lambda)$ were retrieved from the AATS-14 measurements conducted at the four wavelengths $\lambda = 0.380$, 0.525, 0.864, and 1.556 μm, among which those of $\beta_{ext}(\lambda)$ obtained during two flights conducted on 8 and 17 July 1997, are shown in Figure 7.18a. These results clearly indicate that aerosol extinction decreased rapidly with height from ground level to 1.5–2 km altitudes and then increased again to reach a maximum at ~ 3.2 km altitude, where a Saharan dust layer subsided for several days. These vertical profiles were estimated to agree very closely with those of coefficient $\beta_{ext}(\lambda)$ calculated using the size-distribution measurements made with the Caltech OPC at $\lambda = 0.525\,\mu m$ and derived by combining scattering and absorption measurements performed with the aforementioned nephelometers and PSAP instrument, as can be seen in Figure 7.18a.

During the flight of 17 July 1997, the AATS-14 and LIDAR measurements provided evidence of the presence of a 3-layer structure above the observation area, consisting of (i) a marine boundary layer (MBL) near the ground, where $\tau_a(0.525\,\mu m)$ decreased from 0.37 to 0.32; (ii) a clean layer at altitudes between 1 and 2 km, in which $\tau_a(0.525\,\mu m)$ remained virtually constant with height; and (iii) an elevated layer occupying the altitude range from about 3 to 5 km, where $\tau_a(0.525\,\mu m)$ decreased from 0.3 to less than 0.02 (Schmid *et al.*, 2000). The AATS-14 data were found to agree very closely with those taken with airborne OPC, nephelometer, and PSAP instruments also in this case, as shown in Figure 7.18b:

(a) the aerosol size distribution sampled within the MBL was characterized by bimodal features with a first mode centered at mode diameter $a_c \approx 0.3\,\mu m$ and a second mode of coarse particles with $a_c \approx 5\,\mu m$, and (b) the dust layer suspended at the 2–4 km altitudes presented a multimodal surface size-distribution curve with maxima centered at around 1.5 μm and 4 μm diameters.

7.3.6 Airborne Remote Sensing Measurements during the Puerto Rico Dust Experiment (PRIDE)

The Puerto Rico Dust Experiment (PRIDE) was conducted during four weeks from 28 June to 24 July 2000, by a group of scientists from US Navy, NASA, and various university institutions. The field activities included ground-based remote sensing measurements, multispectral measurements of $\tau_a(\lambda)$ conducted with a pair of AERONET sun photometers at two sites located in the experimental area, satellite-based observations, and *in situ* and remote sensing measurements performed on board the research vessel Chapman and the Piper Navajo and Cessna 172 research aircraft (Reid *et al.*, 2003). The Piper Navajo was used to collect 61 h of data during 21 flights made in proximity of the islands of Puerto Rico, St. Thomas, and St. Croix, using the instruments listed in Table 7.8, from which airborne aerosol observations were collected for both clean marine aerosol particles (8–19 July) and predominant extinction by Saharan dust (20–24 July).

For the clean marine conditions observed during the flights of 8, 11, 17, and 19 July, the AOT $\tau_a(\lambda)$ measured at visible wavelengths assumed values lower than 0.08 in the vicinity of Puerto Rico, with rather low daily mean values of exponent α varying between 0.25 and 0.47 on the four days, indicating that the coarse sea-salt particle mode produced the dominant part of aerosol extinction. Dry sea-salt concentrations of 15–20 $\mu g\,m^{-3}$ were estimated on those days, compared with dust concentration values smaller than 5 $\mu g\,m^{-3}$ measured aboard the aircraft. The vertical profiles of fine-mode mass concentration were found to decrease gradually with the height across the MBL and the upper layers at 2–3 km altitudes. The vertical profiles of $\tau_a(\lambda)$ defined by Livingston *et al.* (2003) examining the AATS-6 airborne sun-photometer measurements were estimated to present a very rapid decrease up to 2–3 km altitudes and slower variations at the upper levels at all window wavelengths. The vertical profiles of $\beta_{ext}(\lambda)$ were then calculated on the basis of the variations of $\tau_a(\lambda)$ with height. An example of the vertical profile of $\beta_{ext}(0.526\,\mu m)$ obtained from the AATS-6 measurements made on 6 July 2000, is shown in Figure 7.19, providing evidence of the gradual decrease of marine aerosol extinction with height over the first 4 km of the atmosphere.

Only over the last 5 days of the measurement campaign, from 20 to 24 July 2000, it was possible to monitor the vertical profiles of Saharan dust particle concentration aboard the Piper Navajo, during the transport of these particles into the Caribbean region. The first signs of dust event appeared at Puerto Rico on the late evening of 19 July and became markedly evident during the morning flight of 20 July, when $\tau_a(\lambda)$ increased from 0.06 to 0.27 at the mid-VIS wavelengths. The

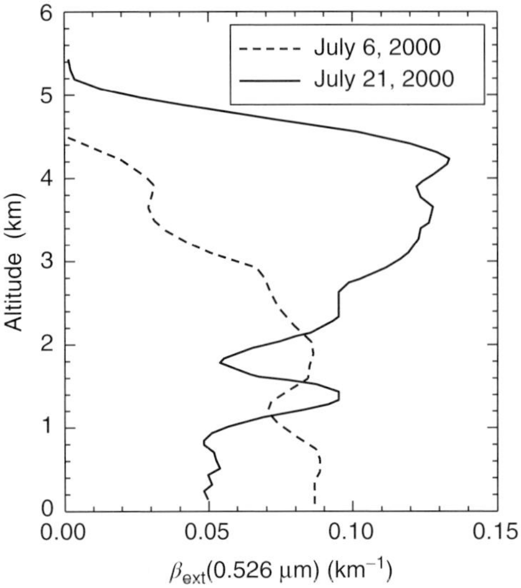

Figure 7.19 Examples of vertical profiles of the aerosol extinction coefficient $\beta_{ext}(0.526\,\mu m)$ derived from the measurements of aerosol optical thickness $\tau_a(0.526\,\mu m)$ performed with the AATS-6 sun-photometer on board the SSC San Diego Navajo aircraft during the flights of (i) 6 July 2000 (ascent flight, 14:42–15:30 UTC), for a predominant load of clean marine aerosol in the low troposphere from sea level to 6 km altitude (dashed curve) and (ii) 21 July 2000 (descent flight, 13:51–14:51 UTC), for a dense layer of Saharan dust (SAL) extended from about 1 to 4.6 km altitudes (solid curve). (Graph prepared using the PRIDE data sets of Livingston *et al.* (2003).)

Saharan Air Layer (SAL) exhibited a subtropical subsidence inversion top level located at 5.6 km height and presented a clear optical dominance of dust particles between 2.3 and 4.5 km levels. Atmospheric dust was clearly monitored over the FSSP 1.5–3 μm and PCASP 0.3–1.1 μm size ranges, showing only slight shifts in particle sizes with altitude. Values of dust optical thickness $\tau_a(\lambda) \approx 0.35$ were measured in the mid-VIS during the flight of 21 July, while dust concentrations were found to be slightly lower than those recorded on the previous day, and accumulation particle mode showed an increase in the SAL, supporting the presence of some minor anthropogenic aerosol components. The vertical profile of $\beta_{ext}(\lambda)$ derived by Livingston *et al.* (2003) analyzing the AATS-6 vertical profiles of $\tau_a(\lambda)$ of 21 July (in the afternoon) is shown in Figure 7.19, clearly indicating that a dense layer of Saharan dust was present at altitudes ranging from 1 to more than 4.4 km, with a maximum centered at ~ 4 km.

7.3.7
The ARCTAS/ARCPAC Airborne Remote Sensing Measurements in the Western Arctic

The Arctic Research of the Composition of the Troposphere from Aircraft and Satellites (ARCTAS) field campaign was funded by NASA in the spring of 2008

(Jacob *et al.*, 2010), in conjunction with the NOAA-funded Aerosol, Radiation, and Cloud Processes affecting Arctic Climate (ARCPAC) project (Brock *et al.*, 2011). The NASA DC-8 and P-3B and NOAA WP-3D aircraft conducted over 160 h of *in situ* sampling and remote sensing measurements over the $0.1 \leq z \leq 12$ km altitude range throughout the Western Arctic north (i.e., Alaska to Greenland) at latitudes higher than 55° N, the three aircraft being equipped with (i) various instruments providing trace gas and aerosol chemistry measurements and *in situ* direct measurements of the aerosol size distribution and BC mass concentration and (ii) the remote sensing instruments indicted in Table 7.8.

The NASA DC-8 and P-3B sampled air masses at different altitudes over the Western Arctic between 31 March and 19 April 2008, including also three sorties from Alaska to Greenland with one DC-8 flight via the North Pole. The NOAA WP-3D flights were conducted between 11 and 24 April 2008, over Alaska and the surrounding Arctic Ocean areas. All the three platforms encountered air masses containing anthropogenic urban/industrial emissions, BB, and MD particulate loads. Examining these measurements, McNaughton *et al.* (2011) estimated that:

1. The vertical profiles of total and submicrometer dry aerosol extinction presented episodically some pronounced peaks in the middle troposphere, at altitudes ranging from ∼ 2 to 7 km, during MD transport episodes over the Western Arctic, with values of submicrometer aerosol extinction coefficient increasing from less than $0.02\,km^{-1}$ near the surface to $0.03-0.06\,km^{-1}$ at the 3–6 km altitudes.
2. The BC mass concentration was found to exhibit various peaks in the middle troposphere, with values ranging from 200 to $300\,ng\,s\,m^{-3}$ (where second per cubic meter indicates standard cubic meter, i.e., the unit air volume calculated for standard air pressure and temperature conditions) and values four times lower at the surface (∼ $50\,ng\,s\,m^{-3}$). A high variability of this parameter was also found from one site to another at midtropospheric levels, the BC mass concentration varying from 1 to $5\,ng\,s\,m^{-3}$ in the less polluted cases and from 1500 to $1800\,ng\,s\,m^{-3}$ within the forest fire plumes.
3. The ratio of organic aerosol to sulfate aerosol mass concentrations measured with the aerosol mass spectrometers for the accumulation particle mode was found to vary from 0.1 to as much as 20 in air masses strongly influenced by BB emissions. It was found that only 11–14% of the samples collected on board the three aircraft were dominated by BB emissions, the Western Arctic air masses being influenced on average by BB for about 42–47% of the total BC burden.
4. A key parameter linking BC emissions to their light absorption properties is the mass absorption efficiency (MAE), usually measured in square meter per gram. It was estimated by McNaughton *et al.* (2011) by plotting total aerosol absorption measured by the airborne PSAPs versus the BC content of Arctic aerosol to determine the regression lines and the slope coefficient (i.e., MAE). Despite some complications due to spurious absorption effects produced by

OC (present in humic-like substances and MD) and other sampling errors, McNaughton *et al.* (2011) estimated that (i) MAE parameter at $\lambda = 0.470\,\mu m$ was equal to $15.2 \pm 0.3\,m^2\,g^{-1}$ for the PSAP measurements recorded aboard the DC-8 aircraft and $14.8\,m^2\,g^{-1}$ for the PSAP measurements taken aboard the P-3B aircraft; (ii) the MAE's values at $\lambda = 0.530\,\mu m$ were equal to 9.5 ± 0.2 and $10.3 \pm 0.4\,m^2\,g^{-1}$ on board the DC-8 and P-3B aircraft, respectively; and (iii) the MAE's values at $\lambda = 0.660\,\mu m$ were also found to agree closely, being equal to 6.7 ± 0.1 and $6.9 \pm 0.3\,m^2\,g^{-1}$, respectively. The previous estimates of MAE represent the total absorption efficiency, which includes absorption by non-BC carbonaceous species, MD, and possible amplification effects due to coatings. A careful analysis of these measurements demonstrated that the MAE's variations for total aerosol are very small compared to those calculated through the bulk analysis, indicating that total absorption is dominated by light-absorbing carbon (LAC).

5. Taking into account that (i) the wavelength dependence of BC is relatively weak and expected to vary as the inverse of wavelength and (ii) absorption by brown carbon (BrC) is typically very small at $0.660\,\mu m$, where it is primarily due to BC, the excess light absorption at the 0.470 and $0.530\,\mu m$ wavelengths with respect to that measured at $\lambda = 0.660\,\mu m$ was attributed by McNaughton *et al.* (2011) to the BrC amount producing a wavelength dependence in excess of λ^{-1}. They also determined the submicrometer refractory volume and converted it to an estimate of non-BC refractory mass (M_{ref}) by assuming a particulate density of $1.3\,g\,cm^{-1}$ and subtracting the mass of BC measured concurrently with the SP2 photometers mounted on board the DC-8 and WP-3D aircraft. Following this procedure, the MAE for the refractory mass was estimated to be equal to 0.83 ± 0.15 and $0.27 \pm 0.08\,m^2\,g^{-1}$ at the $0.470\,\mu m$ and $0.530\,\mu m$ wavelengths, respectively. In these calculations, the actual constituents of M_{ref} include presumably both refractory organic matter and nonorganic submicrometer refractory aerosol from urban/industrial and natural sources. Assuming that M_{ref} can be totally attributed to BrC, ARCTAS estimates of MAE_{BC} were obtained by McNaughton *et al.* (2011) equal to 11.2 ± 0.8, 9.5 ± 0.6, and $7.4 \pm 0.7\,m^2\,g^{-1}$ at the $0.470\,\mu m$, $0.530\,\mu m$, and $0.660\,\mu m$ wavelengths, respectively.
6. Intense plumes of coarse particles were sometimes monitored at 2–6 km altitudes during Asian dust transport episodes. The ARCTAS/ARCPAC measurements supported the hypothesis that Western Arctic air masses are heavily influenced by emissions of Asian dust transported across the Pacific Ocean after lofting by midlatitude cyclones over Central Asia, East Asia, and eastern North Pacific (Stohl, 2006). The DC-8 spent the most time at high altitude and in the High Arctic, north of 70° N, whereas the WP-3D measurements were mainly conducted in the lower part of the troposphere at latitudes close to 55° N over Alaska. These airborne measurements showed that the supermicrometer MD concentration presented in general rather low values near the surface, values $< 0.5\,\mu g\,s\,m^{-3}$ at the 2 km level,

and slowly increasing values with height at upper levels until reaching values of 2.8 μg s m^{-3} above 4 km. At the 6 km level, dust concentrations reached in general average values as high as 2.1−3.4 μg s m^{-3}. Total light absorption was found to be dominated by LAC, although its concentration usually decreases above 6 km. Therefore, absorption by MD accounted for only 4−12% and 3−9% of total absorption at the 0.470 μm and 0.530 μm wavelengths, respectively.

An attempt was made by McNaughton *et al.* (2011) to estimate the MAE for supermicrometer dust by following a procedure similar to that undertaken for LAC. The dust data were selected by choosing only the data presenting a supermicrometer mass fraction > 0.75 or values of scattering Ångström exponent <1.0. Further, sea salt was eliminated by restricting the analysis to data with values of absorption Ångström exponent > 1.5. The regression lines drawn for such a limited subset of data provided MAE estimates of 0.034 $m^2 g^{-1}$ at $\lambda = 0.470$ μm and 0.017 $m^2 g^{-1}$ at $\lambda = 0.530$ μm, determined with very low regression coefficients.

7. A characterization of the vertical profiles of mass concentration was made by McNaughton *et al.* (2011) for different aerosol types, considering that radiative forcing efficiency of BC, BrC, and MD over the polar regions is appreciably influenced by the aerosol vertical distribution features. BC is the dominant absorber and exhibits vertical distributions differing from those of MD and nonabsorbing sea salt. The ARCTAS/ARCPAC measurements were analyzed by taking into account the wavelength dependence features of such aerosol components, finding that (i) BC presents the most pronounced peaks in the middle troposphere, with values greater than 0.1 μg s m^{-3} at altitudes ranging from 2 to 6 km; (ii) the mean mass concentration of supermicrometer MD increases from levels near the surface to 4−6 km heights, showing in general a multilayered profile; and (iii) the mean sea-salt mass concentration decreases slowly with height until reaching background low values at ~ 3 km.

   The mean vertical profiles of aerosol mass concentrations (measured in microgram per cubic meter) during the ARCTAS/ARCPAC for BC, submicrometer mass (assumed to have mass density $\rho = 1.6\,g\,cm^{-3}$), submicrometer refractory mass (for mass density $\rho = 1.3\,g\,cm^{-3}$), supermicrometer MD (for dry bulk density $\rho = 2.06\,g\,cm^{-3}$), and sea salt (for dry bulk density $\rho = 2.20\,g\,cm^{-3}$) are shown in Figure 7.20, as obtained by McNaughton *et al.* (2011).

   The vertical profile of BrC mass can be obtained from these profiles by subtracting BC mass from submicrometer refractory mass. Figure 7.20 shows that the peak altitudes are equal to 3.6 km for BC, 4.4 km for submicrometer mass, 3.6 km for submicrometer refractory mass, and 4.4 km for MD, while sea-salt mass concentration decreases gradually as height increases to levels located above the MBL. The plumes of BB particles were sampled repeatedly by the NOAA WP-3D after 19 April 2008, and only for a short time period by the NASA DC-8 and P-3B aircraft on 19 April. Examining the results derived

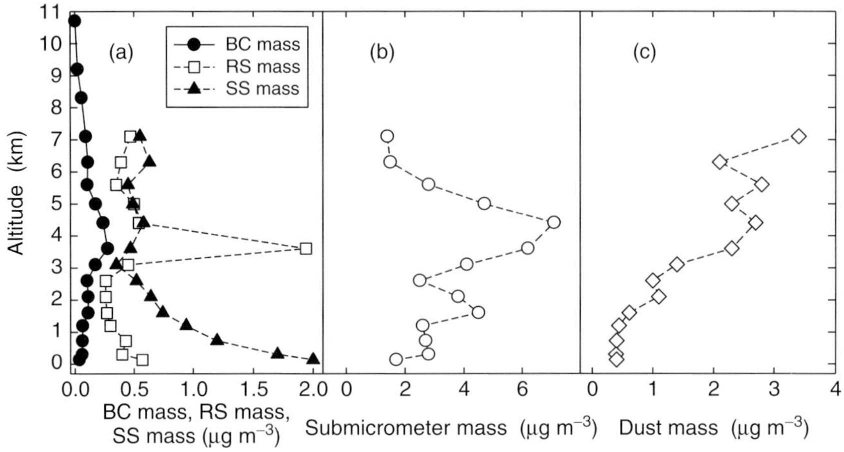

Figure 7.20 (a) Vertical profiles of (i) mean black carbon (BC) mass concentration (solid curve with solid circles), (ii) mean refractory submicrometer (RS) aerosol mass concentration (for mass density $\rho = 1.3\,\mathrm{g\,cm^{-3}}$) (dashed curve with open squares), and (iii) mean sea-salt (SS) mass concentration (dashed curve with solid triangles). (b) Vertical profiles of mean submicrometer aerosol mass concentration (for mass density $\rho = 1.6\,\mathrm{g\,cm^{-3}}$) (dashed curve with open circles). (c) Vertical profiles of mean dust mass concentration (for dry bulk density $\rho = 2.06\,\mathrm{g\,cm^{-3}}$) (dashed curve with open diamonds). All the mass concentrations are measured in microgram per cubic meter. The graphs were prepared using the ARCTAS/ARCPAC data published by McNaughton *et al.* (2011) in Table 5.

from these aircraft data, the light absorption over the Arctic was estimated to be predominantly due to light-absorbing carbonaceous species derived from urban/industrial activities and BB fires (by 88–99% of the whole), and BB aerosols were evaluated to account for just 11–14% of the air volume sampled onboard aircraft.

8. The average vertical profiles of total light extinction coefficients $\beta_{ext}(\lambda)$ at wavelengths $\lambda = 0.470\,\mu m$, $\lambda = 0.530\,\mu m$, and $\lambda = 0.660\,\mu m$ were determined by examining the airborne aerosol light scattering and absorption coefficients derived from the nephelometer and PSAP measurements performed on board the DC-8, P-3B, and WP-3D aircraft. The vertical profiles of $\beta_{ext}(\lambda)$ are shown for all the three wavelengths in Figure 7.21a, presenting some marked peaks at the 3.6 km level. The vertical profiles of total single scattering albedo $\omega(\lambda)$ derived at the three nephelometric wavelengths from the airborne nephelometer and PSAP measurements are shown in Figure 7.21b, indicating that they are rather stable with height ranging mainly between 0.93 and 0.98 at tropospheric levels.

   The results presented in Figure 7.21 were also integrated along the vertical path of the atmosphere to determine the submicrometer AOT $\tau_a(0.530\,\mu m)$, found to be on average equal to 0.12, and the average value of weighted column single scattering albedo $\omega(0.530\,\mu m) = 0.94 \pm 0.5$, while the values

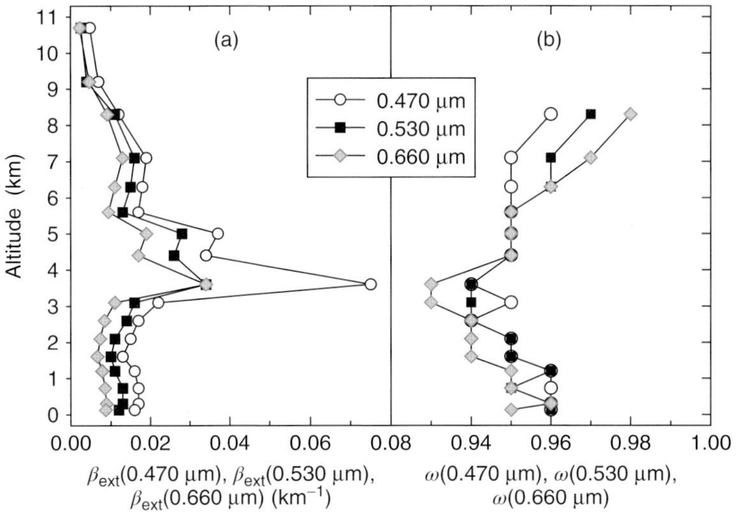

Figure 7.21 (a) Vertical profiles of the mean values of total light extinction coefficients $\beta_{ext}(0.470\,\mu m)$ (open circles), $\beta_{ext}(0.530\,\mu m)$ (solid squares), and $\beta_{ext}(0.660\,\mu m)$ (gray diamonds), as measured in per kilometer by McNaughton *et al.* (2011) examining the 1 min averaged data measured on board the NASA DC-8 and the P-3B aircraft during the ARCTAS airborne campaign. (b) As in (a) for the mean values of aerosol single scattering albedo $\omega(0.470\,\mu m)$ (open circles), $\omega(0.530\,\mu m)$ (solid squares), and $\omega(0.660\,\mu m)$ (gray diamonds) obtained as ratios between the respective aerosol scattering and aerosol extinction coefficients. Both graphs were prepared using the ARCTAS/ARCPAC data published by McNaughton *et al.* (2011) in Table 6.

of submicrometer $\tau_a(0.530\,\mu m)$ and $\omega(0.530\,\mu m)$ calculated from the data collected aboard the DC-8 aircraft were estimated to be equal to 0.15 and 0.95 ± 0.3, respectively. The presence of the intense but episodic BB plumes observed during the ARCTAS/ARCPAC flights has contributed to increase the values of $\tau_a(\lambda)$ by $\sim 1/3$ and those of $\omega(\lambda)$ by 0.01. Low values of scattering Ångström exponent but high values of absorbing Ångström exponent were found for Asian dust at high altitude levels between 6 and 8 km, the contribution of MD to total light absorption being significant only above the 6 km level, since these particles contribute to provide absorption percentages of 4–12% and 3–8% at the 0.470 μm and 0.530 μm wavelengths, respectively. At these altitudes, light absorption by BrC from urban/industrial emissions is on average responsible for 9% (at $\lambda = 0.470\,\mu m$) and 4% (at $\lambda = 0.530\,\mu m$) of total light absorption, whereas absorption by BrC from BB emissions was estimated to account for 12% and 6% of total light absorption, respectively. These estimates indicate that light absorption by anthropogenic BrC is responsible for a higher percentage of total light absorption than that attributed to the natural sources of MD. In addition, LAC from urban/industrial sources was found to yield 51–54% of total light absorption, while LAC from BB emissions accounts for 42–45% of aerosol light absorption. Therefore, in spite of the limitations imposed by PSAP instrument noise, the 2008 ARCTAS/ARCPAC results indicated that anthropogenic LAC accounts for 94–99% of total light absorption at $\lambda = 0.470\,\mu m$ and 96–99% of total light absorption at $\lambda = 0.530\,\mu m$ in the Western Arctic atmosphere below the 6 km altitude.

9. Additional airborne multispectral sun-photometer measurements were conducted in the Arctic region during the ARCTAS project in March and April 2008 over Alaska (United States) using the NASA P-3 aircraft, on which the NASA Ames 14-channel Airborne Tracking Sunphotometer (AATS-14) and other instruments were deployed to measure the microphysical and optical properties of sampled boreal forest fire smoke, Asian dust, and emissions from mining sites (Shinozuka *et al.*, 2011). The NASA P-3 aircraft carried the following instruments: (i) the AATS-14 sun-photometer, providing measurements of AOT $\tau_a(\lambda)$ at 14 selected wavelengths ranging from $0.353\,\mu m$ (with half-bandwidth HBW = 2 nm) to $2.14\,\mu m$ (with HBW = $17.3\,\mu m$); (ii) two TSI (model 3563) integrating nephelometers providing continuous measurements of total and submicrometer aerosol scattering coefficients at the 0.450, 0.550, and $0.700\,\mu m$ wavelengths in time steps of ~ 5 s; (iii) a pair of 3-λ RR PSAP photometers, with which the aerosol light absorption was continuously measured; (iv) a pair of single-wavelength RR (M903 model) nephelometers, which were operated in parallel to measure the RH effects on aerosol scattering; (v) an Aerodyne High-Resolution Time-of-Flight Aerosol Mass Spectrometer (AMS) used to characterize volatile ionic and organic aerosol components over the 50–700 nm size range; and (vi) a Single Particle Soot Photometer (SP2), providing measurements of BC (soot) mass over the 100–600 nm size range. One of the TSI nephelometers and one of the PSAPs were operated behind a $1\,\mu m$ (aerodynamic diameter) impactor to measure the submicron fraction of scattering and absorption coefficients. The previously listed *in situ* measurements were also integrated vertically to yield the partial values of $\tau_a(\lambda)$ over the most significant particulate layers. Shinozuka *et al.* (2011) examined the AATS-14 data set collected during 25 ARCTAS flights over Alaska, the Arctic Ocean, and Greenland obtaining values of $\tau_a(0.499\,\mu m)$ constantly decreasing from 0.07 near the surface to 0.01 at the 7.5 km level. A modified Ångström wavelength exponent (α_λ) was determined examining the wavelength dependence features of $\tau_a(\lambda)$, finding (i) values as low as 0.7 above the Arctic Sea covered by ice on those observation days, presumably due to predominant extinction by either dust or ice particles rather than sea-salt particles; (ii) high values equal to 1.7 ± 0.1 on average, at altitudes varying between 2 and 4 km on 13 and 15 April, presumably associated with predominant extinction by forest fire particles; and (iii) values equal on average to 1.4 ± 0.3 in other cases, in which the aerosol extinction was mainly due to anthropogenic pollution (Arctic haze).

7.3.8 The Airborne Measurements Conducted during the Pan-Arctic Measurements and Arctic Regional Climate Model Intercomparison Project (PAM-ARCMIP)

The airborne campaign Pan-Arctic Measurements and Arctic Regional Climate Model Intercomparison Project (PAM-ARCMIP) was conducted in the Arctic region from 1 to 25 April 2009, organized by the Alfred Wegener Institute for Polar and Marine Research (AWI) (Germany). The Polar-5 research aircraft was used during the campaign, on board with a series of Arctic circumnavigations being conducted from the inner regions of Alaska (United States) to the Svalbard Archipelago (Norway) across the coastal regions of Northwest Territories and Nunavut (Canada) and northern Greenland, at latitudes ranging from 65° N to 78° N, equipped with the instruments described in Table 7.8. The general goal of the project was to collect comprehensive data sets in order to better understand the Arctic climate system and in particular the radiative effects produced by Arctic aerosol. On command, the 8-channel NOAA-CNR Sun-photometer system was pointed at the Sun, and the data acquisition system connected with it recorded spectral irradiance at the eight peak wavelengths λ_p = 368, 412, 500, 610, 675, 778, 862, and 1050 nm, with 1 s time resolution, along with GPS position and altitude along the flight track.

Nearly eighty-eight thousand individual spectra of direct solar irradiance were recorded during the campaign, from which as many spectral series of $\tau_a(\lambda)$ were calculated. The sun-photometer observations were made with a horizontal resolution of about 50 m and vertical resolution of $\sim$ 5 m, from near the surface (at a height of $\sim$ 60 m above the ground) to $z \approx 4$ km. Clear-sky observations were conducted above various areas where some ground-based sun photometers were simultaneously operated: Ny-Ålesund (Svalbard) from 3 to 6 April, Station Nord (Greenland) on 8 April, Alert in the Qikiqtaaluk region (Nunavut, Canada) on 9 and 11 April, North Pole 36 on 10 April, Eureka (on the Fosheim Peninsula (Qikiqtaaluk region, Nunavut, Canada) on 13 April, Resolute in the Cornwallis Island (Nunavut, Canada) on 14 and 15 April, Sachs Harbor in the Inuvik region (Northwest Territories, Canada) on 16 and 17 April, and Barrow (Alaska) on 24 and 25 April. Before being analyzed to determine the spectral values of $\tau_a(\lambda)$, all the above Sun-photometer measurements of direct solar irradiance were first screened to minimize the attenuation effects of thin clouds by following the two different procedures defined independently by the NOAA ESRL GMD and ISAC-CNR groups (Stone *et al.*, 2010). The values of total atmospheric optical thickness $\tau(\lambda)$ derived at various wavelengths in terms of the Bouguer–Lambert–Beer law applied to the direct solar irradiance measurements were carefully corrected for the Rayleigh scattering and gaseous absorption effects produced by ozone, nitrogen dioxide, water vapor, and oxygen dimer. The vertical profiles of $\tau_a(\lambda)$ were then determined for each flight at various window wavelengths, and the corresponding values of Ångström exponent $\alpha(412/675)$ were calculated directly by using the monochromatic values of $\tau_a(0.412\,\mu m)$ and $\tau_a(0.675\,\mu m)$ derived from the airborne Sun-photometer measurements. Some examples of the vertical

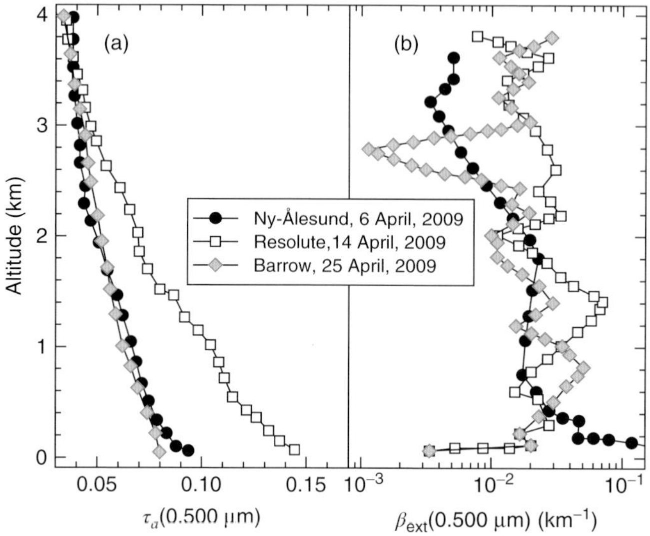

Figure 7.22 Vertical profiles over the altitude range from surface to 4 km of (a) aerosol optical thickness $\tau_a(0.500\,\mu m)$ measured on board the Polar-5 aircraft of the Alfred Wegener Institute for Polar and Marine Research (Germany) using the 8-channel Sun-photometer system developed by the NOAA ESRL GMD (Boulder, Colorado, USA) and ISAC-CNR (Bologna, Italy) at (i) Ny-Ålesund on 6 April 2009 (solid circles), (ii) Resolute on 14 April 2009 (open squares), and (iii) Barrow on 25 April 2009 (gray diamonds); and (b) volume extinction coefficient $\beta_{ext}(0.500\,\mu m)$ derived from the Sun-photometer measurements of $\tau_a(0.500\,\mu m)$ shown in (a) and conducted in the atmospheric columns above the same sites. (Graph prepared using the PAM-ARCMIP data of Stone *et al.* (2010).)

profiles of $\tau_a(0.500\,\mu m)$ derived from the Sun-photometer measurements of $\tau_a(0.500\,\mu m)$ are shown in Figure 7.22a, as obtained by analyzing the airborne Sun-photometer measurements recorded at Ny-Ålesund on 6 April 2009, Resolute on 14 April 2009, and Barrow on 25 April 2009. Figure 7.22a shows that the values of $\tau_a(0.500\,\mu m)$ exceeded the background surface-level values recorded in the previous years and decreased gradually with height until reaching values lower than 0.1 at $z \approx 4\,km$ above the three stations, such relatively high values being probably due to extinction by volcanic aerosols generated by the March eruptions of Mount Redoubt in Alaska.

The vertical profiles of $\beta_{ext}(0.500\,\mu m)$ were derived from the vertical profiles of $\tau_a(0.500\,\mu m)$. They show that $\beta_{ext}(0.500\,\mu m)$ is subject to large variations with height due to the presence of multilayered structures of aerosol particles but decreases on average with height in spite of such sharp changes. In particular, strong variations of $\beta_{ext}(0.500\,\mu m)$ with height were detected within the thermal inversion layer (having a depth of $\sim 0.65\,km$ on average), presumably because of the presence of haze particle layers above the ground level. As can be seen in Figure 7.22b, $\beta_{ext}(0.500\,\mu m)$ was estimated to slowly decrease with height until reaching values ranging from 0.01 to $0.02\,km^{-1}$ at heights $z > 3\,km$.

The vertical profiles of Ångström wavelength exponent α (calculated for the pair of AOT values at the 412 and 675 nm wavelengths) were found to increase in general with height from surface level up to $z \approx 3$ km, assuming average values equal to ~ 1.3 at the surface and to ~ 1.6 at $z = 2$ km and then remaining almost constant through the whole middle troposphere. These variations are certainly due to pronounced changes in the aerosol size-distribution curve occurring from one level to another, such an increase in α being clearly associated with a marked decrease of particle sizes, which is fully confirmed by the *in situ* values of geometric mean diameter calculated examining the airborne EC Ultra-High Sensitivity Aerosol Spectrometer (UHSAS) measurements conducted within the 60–1000 nm size range.

An important contribution to the knowledge of the vertical profiles of aerosol extinction was given by the Airborne Mobile Aerosol LIDAR (AMALi) system measurements conducted during the PAM-ARCMIP aircraft mission. This LIDAR system was developed by AWI–Potsdam and operates at the two wavelengths $\lambda = 0.355\,\mu m$ and $\lambda = 0.532\,\mu m$, utilizing a Nd:YAG pulsed laser and providing 7.5 m vertically resolved profiles of (i) volume extinction coefficient $\beta_{ext}(\lambda)$ measured in per kilometer at LIDAR wavelength $\lambda = 0.532\,\mu m$ and (ii) linear depolarization ratio $\delta(0.532\,\mu m)$. This airborne LIDAR was operated in either nadir or zenith configuration. On 4 April 2009, the LIDAR data were recorded above Ny-Ålesund (Svalbard) in the nadir position, with a vertical resolution of 7.5 m and horizontal resolution varying with the aircraft speed, so that data averaged over 15 s were in practice related to a horizontal resolution of ~ 750 m on average. In general, profiles of backscatter coefficient $\beta_{bs}(0.532\,\mu m)$ and linear volume depolarization ratio $\delta(0.532\,\mu m)$ were selected for cloud-free conditions of the atmosphere, between the ground level and 250 m below the cruising altitude, that is, generally below $z = 2.8$ km. The AMALi backscatter coefficient $\beta_{bs}(0.532\,\mu m)$ was calculated by using iteratively the Klett (1981) algorithm for a value of LIDAR ratio $R_\ell(0.532\,\mu m) = 20$ sr, typical of marine aerosols (Ansmann *et al.*, 2001). The volume depolarization ratio $\delta(0.532\,\mu m)$ was calculated as the ratio of the perpendicular and parallel LIDAR signals to measure the asphericity of the scattering particles.

Figure 7.23 shows some examples of the vertical profiles of (i) $\beta_{bs}(0.532\,\mu m)$, (ii) LIDAR ratio $R_\ell(0.532\,\mu m)$, obtained from altitude step-by-step comparisons between LIDAR and sun-photometer data, (iii) linear volume depolarization ratio $\delta(0.532\,\mu m)$ measured with the AMALi LIDAR, (iv) extinction coefficient $\beta_{ext}(0.532\,\mu m)$ calculated as the product $\beta_{bs} \times R_\ell(0.532\,\mu m)$, and (v) AOT $\tau_a(0.532\,\mu m)$ derived by smoothing and integrating the values of $\beta_{ext}(0.532\,\mu m)$ derived from the airborne sun-photometer measurements conducted above Ny-Ålesund in early April 2009. The vertical profiles of $\beta_{bs}(0.532\,\mu m)$ clearly indicate that the most pronounced aerosol scattering effects observed above Ny-Ålesund were produced on those days within the ground layer of 1.5 km depth, while low backscattering features were caused at upper levels by background aerosol loads presenting LIDAR ratios typical of mixed marine/continental particle loads. The vertical profiles of LIDAR ratio $R_\ell(0.532\,\mu m)$ showed values lower than 20 sr

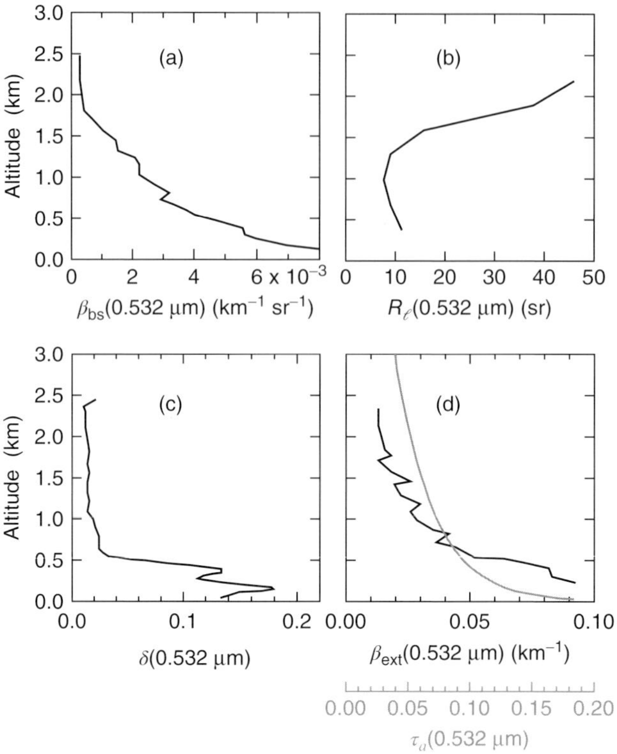

Figure 7.23 Average vertical profiles derived from the AMALi measurements conducted above Ny-Ålesund on 3–6 April 2009: (a) backscatter coefficient $\beta_{bs}(0.532\,\mu m)$ calculated by using iteratively the Klett (1981) algorithm, (b) LIDAR ratio $R_\ell(0.532\,\mu m)$ derived from altitude step-by-step comparisons with Sun-photometer data, (c) linear volume depolarization ratio $\delta(0.532\,\mu m)$ measured with the AMALi LIDAR, and (d) volume extinction coefficient $\beta_{ext}(0.532\,\mu m)$ calculated as the product $\beta_{bs}(0.532\,\mu m) \times R_\ell(0.532\,\mu m)$ (solid curve), which is compared in part (d) with the vertical profile of aerosol optical thickness $\tau_a(0.532\,\mu m)$ derived by smoothing and integrating the values of $\beta_{ext}(0.532\,\mu m)$ obtained from the airborne Sun-photometer measurements (gray solid curve). The graphs were prepared using the PAM-ARCMIP data of Hoffmann *et al.* (2012).

within the ground layer of 1.5 km depth (marine aerosol) and values increasing with height until to exceed 40 sr at altitudes $z > 2.2$ km, which more typically represent loads of mixed marine/continental aerosol particles. The vertical profile of volume depolarization ratio $\delta(0.532\,\mu m)$ was found to range between 0.1 and 0.2 within the ground layer (with $z < 0.5$) and to decrease with height until reaching values lower than 2×10^{-2} for $z > 0.7$ km. Since this parameter depends on size, chemical composition, and shape of the aerosol particles, the above variations with altitude combined with the low LIDAR ratios indicate reasonably that (i) the major part of the observed aerosol particles had nearly spherical shapes and (ii) only a few ice crystals were mixed with the Arctic haze particles within the ground layer, generally presenting only a negligible absorption component. The vertical profiles of extinction coefficient $\beta_{ext}(0.532\,\mu m) = \beta_{bs}(0.532\,\mu m) \times R_{\ell}(0.532\,\mu m)$ calculated from the AMALi measurements are shown in Figure 7.23d, showing that this parameter assumed values close to $10^{-1}\,km^{-1}$ near the surface and decreased rapidly with height until becoming smaller than $2 \times 10^{-2}\,km^{-1}$ at altitude $z = 1\,km$. This profile obtained by Hoffmann *et al.* (2012) is compared in Figure 7.23d with that of $\tau_a(0.532\,\mu m)$ derived by smoothing and integrating the values of aerosol extinction coefficient calculated on the basis of the airborne Sun-photometer measurements described earlier, showing that the columnar $\tau_a(0.532\,\mu m)$ decreased from about 0.18 near the surface to ~ 0.10 at $z = 0.5\,km$ and became smaller than 0.05 at $z = 3\,km$, presenting particularly marked multilayered features within the altitude range $z < 1\,km$.

7.4 Satellite-Borne Aerosol Remote Sensing Measurements

Being a global phenomenon aerosol requires global means of observation. Probably, the best possibility to observe aerosols is from geostationary platforms with a high temporal resolution (say, 15 min). The correspondent geostationary systems exist over the United States, Europe, Japan, India, and some other locations. However, more geostationary satellites must be launched to cover the globe without gaps. Another question is the equipment of the corresponding platforms. They require multiangular spectral measurements combined with LIDAR and polarimetric observations. The LIDARs and multiangular polarimeters are usually used on the polar-orbiting systems and not on the geostationary platforms. However, this may change in the future due to the need of humankind for a better understanding of atmospheric pollution and climate change, where atmospheric aerosol is a key player. In this section we discuss main instrumentation and methods to retrieve atmospheric aerosol using satellite observations.

7.4.1 Satellite Instrumentation

In this section we describe the selected instruments (see also Table 7.9), which are used for the aerosol remote sensing from space.

Moderate-Resolution Imaging Spectroradiometer (MODIS) is a payload scientific instrument launched into the Earth's orbit by NASA in 1999 on board the Terra (EOS AM) satellite and in 2002 on board the Aqua (EOS PM) satellite. The MODIS captures data in 36 spectral bands ranging in wavelength from 0.40 to 14.4 μm and at varying spatial resolutions (2 bands at 250 m, 5 bands at 500 m, and 29 bands at 1 km). Together the instruments on Terra and Aqua image the entire Earth every 1 to 2 days. They are designed to provide measurements in large-scale global dynamics including changes in Earth's cloud cover, radiation budget, and processes occurring in the oceans, on land, and in the lower atmosphere.

Table 7.9 The history of platforms and sensors used to derive aerosol properties from space.

Launch	End	Platform	Instrument	Number of bands (wavelengths (μm))	Accuracy	References[a)]
1972	1978	Landsat (ERTS-1)	MSS	4(0.5–1.1)	$\tau(10\%)$	Griggs (1975)
1974	1981	SMS-1, SMS-2	VISSR	5(0.65–12.5)		
1975	Present	GOES-1–12	VISSR	5(0.65–12.5)	$\tau(18-34\%)$ [b)]	Knapp, Vonder Harr, and Kaufman (2002)
1975	1975	Apollo-Soyuz	SAM	0.83	—	McCormick *et al.* (1979)
1977	2005	GMS-1–5	VISSR	4(0.45–12.5)	—	—
1978	1980	TIROS-N	AVHRR	4(0.58–11.5)	—	—
1978	1993	Nimbus-7	SAM-2, CZCS, TOMS	1 6(0.443–11.5) 6(0.312–0.380)	$\sigma_{ext}(10\%)$ — —	McCormick *et al.* (1979) — —
1979	1981	AEM-B	SAGE	4(0.385,0.45,0.6,1.0)	$\sigma_{ext}(10\%)$	Chu and McCormick (1979)
1979	Present	NOAA-6–16	AVHRR	5(0.58–12)	$\tau(10\%)$[c)], $\tau(3.6\%)$[d)]	Stowe, Ignatov, and Singh (1997); Mishchenko *et al.* (1999)
1984	2005	ERBS	SAGE-2	4(0.386–1.02)	$\sigma_{ext}(10\%)$	Chu *et al.* (1989)
1997	Present	TRMM	VIRS	5(0.63–12)	$\tau(35\%)$, $\alpha(\pm 0.5)$	Ignatov and Stowe (2000)
1991	1996	SPOT-3	POAM-2	9(0.353–1.060)	$\sigma_{ext}(\sim 20\%)$	Randall *et al.* (1996)
1991	1999	ERS-1	ATSR, GOME	4(1.6, 3.7, 11, 12) 4(0.24–0.79)	— —	— Torricella *et al.* (1999)
1992	2005	UARS	HALOE	8(2.45–10.01)	$r_{eff}(\pm 15\%)$, $\sigma_{ext}(\pm 5\%)$	Hervig, Deshler, and Russell (1998)
1994	1994	SSD	LITE	3(0.355, 0.532, 1.064)	$\beta(\lambda_1)/\beta(\lambda_2)(< 5\%)$	Gu *et al.* (1997)
1995	Present	ERS-2	ATSR-2, GOME	7(0.55–12) 0.24–0.79	$\tau(< 0.03)$, $\tau(30\%)$	Veefkind *et al.* (1999)
1996	Present	Earth Probe	TOMS	6(0.309–0.360)	$\tau(20-30\%)$ [e)]	Torres *et al.* (2002)
1996	1997	ADEOS	POLDER, ILAS, OCTS	9(0.443–0.910) 2(0.75–0.78, 6.21–11.77) 7(0.412–0.865)	$\tau(20-30\%)$[f)], — —	Herman *et al.* (1997) — —

1997	Present	OrbView-2	SeaWiFS	8(0.412–0.865)	τ(5–10%)	Gordon and Wang (1994)
1998	Present	SPOT-4	POAM-3	9(0.354–1.018)	σ_{ext}(±30%)	Randall *et al.* (2001)
1999	Present	Terra	MODIS,	36 (0.4–14.4)	τ(5–15%)[g], τ(10–20%)	Remer *et al.* (2005)
			MISR	4 (0.45–0.87)		Kahn *et al.* (2005)
2001	2005	Meteor-3M	SAGE-3	9(0.385–1.545)	σ_{ext}(5%), τ(5%)	Thomason, Poole, and Randall (2007)
2001	Present	PROBA	CHRIS	62(0.4–1.05)	—	Barnsley *et al.* (2004)
2001	Present	Odin	OSIRIS	0.274–0.810	σ_{ext}(15%)	Bourassa *et al.* (2007)
2002	Present	Aqua	MODIS	—	—	—
2002	Present	ENVISAT	AATSR, MERIS,	7(0.55–12.0)	τ(0.16), τ(∼ 0.2), AI(∼ 0.4)	Grey, North, and Los (2006);
			SCIAMACHY	15 (0.4–1.05)		Vidot, Santer, and Aznay (2008);
				0.24–2.4		de Graaf and Stammes (2005)
2002	2003	ADEOS-II	POLDER-2,	9(0.443–0.910)	— —, τ(∼ 0.1)	—
			ILAS-2,	4(0.75–12.85)		Zasetsky and Sloan (2005)
			GLI	36(0.38–12)		Murakami *et al.* (2006)
2002	Present	MSG-1	SEVIRI	12(0.6–13.4)	T(0.08)	Popp *et al.* (2007)
2003	2003-	ICESat	GLAS	2(0.532, 1.064)	σ_{ext}(10%), τ(20%)	Palm *et al.* (2002)
2004	Present	Aura	OMI,	3(0.27–0.5)	τ(30%), σ_{ext}(5–25%)	Torres *et al.* (2007);
			HIRDLS	21(6–18)		Froidevaux and Douglass (2001)
2004	Present	PARASOL	POLER-3	8(0.44–0.91)	—	—
2006	Present	CALIPSO	CALIOP	2(0.532, 1.064)	τ(∼ 25%)	Omar *et al.* (2013)
2010	Present	COMS	MI	5(0.45–12.5)	— —	Lee, Lee, and Chung (2016)
			GOCI	8(0.412–0.865)		Lee *et al.* (2010); Lee *et al.* (2012)
2014	Present	Himawari-8	AHI	16(0.47–13.3)	—	—

a) References of the validation study for accuracy are listed here.
b) Accuracy for operational GOES aerosol retrieval may apply for other GOES series.
c) Accuracy for single-channel AVHRR aerosol retrieval algorithm may apply for other AVHRR series.
d) Accuracy for two-channel AVHRR aerosol retrieval algorithm may apply for other AVHRR series.
e) Accuracy for TOMS AOT retrieval from Nimbus-7 to Earth Probe.
f) May apply for the POLDER-2 and POLDER-3.
g) Same to the MODIS/Aqua.

Three onboard calibrators (a solar diffuser combined with a solar diffuser stability monitor, a spectral radiometric calibration assembly, and a black body) provide in-flight calibration.

Visible Infrared Imaging Radiometer Suite (VIIRS) is a sensor on board the Suomi National Polar-orbiting Partnership weather satellite. VIIRS is a scanning radiometer which collects imagery and radiometric measurements of the land, atmosphere, cryosphere, and oceans in the VIS and IR bands of the electromagnetic spectrum. VIIRS is an advanced, modular multichannel imager and radiometer (of OLS, AVHRR/3, MODIS, and SeaWiFS heritage) with the objective to provide global observations (moderate spatial resolution) of land, ocean, and atmosphere parameters at high temporal resolution (daily).

VIIRS is a multispectral (22-band) optomechanical radiometer, employing a cross-track rotating telescope fore-optics design (operating on the whisk broom scanner principle), to cover a wide swath. The rotating telescope assembly (RTA) (of 20 cm diameter) concept of SeaWiFS heritage allows a low straylight performance. An observation scene is imaged onto three focal planes, separating the VNIR, SWIR/MWIR, and TIR energy – covering a spectral range of 0.4–12.5 μm (see Table 7.10). The VNIR focal plane array (FPA) has nine spectral bands, the SWIR/MWIR FPA has eight spectral bands, and the TIR FPA four spectral bands. The integral day/night band (DNB) capability provides a very large dynamic range low-light capability in all VIIRS orbits. The detector line arrays (16 detectors in each array for the SWIR/MWIR and TIR bands, 32 detectors in the array for the VNIR and DNB (Pan) bands) of the whisk broom scanner are oriented in the along-track direction.

Typical data products (types) of VIIRS include atmospheric, clouds, Earth's radiation budget, clear-air land/water surfaces, sea surface temperature (SST), ocean color, and low-light VIS imagery. A swath width of 3000 km is provided (corresponding to FOV = $\pm 55.84°$) with a spatial resolution for imagery-related products of no worse than 0.4to 0.8 km (nadir to edge of scan). The radiometric bands provide a resolution about twice in size to the imagery bands. The swath widths and locations are individually programmable by band (improved resolution views of selected target near nadir).

Advanced Along-Track Scanning Radiometer (AATSR) is a multichannel imaging radiometer with the principal objective of providing data concerning global SST to the high levels of accuracy and stability required for monitoring and carrying out research into the behavior of the Earth's climate. AATSR can measure Earth's surface temperature to a precision of 0.3 °K (0.5 °F) for climate research. Among the secondary objectives of AATSR is the observation of environmental parameters such as aerosols, clouds, fires, gas flares, water content, biomass, and vegetal health and growth. AATSR is the successor of ATSR-1 and ATSR-2, payloads of ERS-1 and ERS-2. AATSR has operated from ENVISAT platform (2002–2012). Its successor (Sea and Land Surface Temperature Radiometer (SLSTR)) performs measurements from Sentinel-3A, Sentinel-3B, and Sentinel-3C platforms (launch years: 2016, 2017, 2020) at the following wavelengths: 0.555, 0.659, 0.865, 1.375, 1.61, 2.25, 3.74, 10.85, and 12 μm. The Sentinel-3 mission is

Table 7.10 Bands of VIIRS.

Band	Center wavelength (μm)	Bandwidth (μm)	Comment (driving EDR observation requirements)
Visible and near-infrared (VNIR) spectral region, use of Si detectors in FPA			
DNB	0.70	0.40	Day/night band, broad bandwidth maximizes signal (essential nighttime reflected band)
M1	0.412	0.02	Ocean color, suspended matter, net heat flux, mass loading
M2	0.445	0.018	Ocean color, suspended matter, net heat flux, mass loading
M3	0.488	0.02	Ocean color EVI, surface type, aerosols, suspended matter, net heat flux, mass loading
M4	0.555	0.02	Ocean color, surface type, suspended matter, net heat flux, mass loading
I1	0.640	0.05	Imagery, NDVI, cloud mask/cover, cloud optical properties, surface type, albedo, snow/ice, soil moisture
M5	0.672	0.02	Ocean color, aerosols, suspended matter, net heat flux, littoral transport, mass loading
M6	0.746	0.015	Ocean color, mass loading
I2	0.865	0.039	Imagery NDVI (NDVI heritage band), snow/ice, surface type, albedo
M7	0.865	0.039	Ocean color, cloud mask/cover, aerosols, soil moisture, net heat flux, mass loading
Shortwave infrared (SWIR) spectral region, use of photovoltaic (PV) HgCdTe detectors			
M8	1.24	0.02	Cloud optical properties (essential over snow/ice), active fires
M9	1.378	0.015	Cloud mask/cover (thin cirrus detection), aerosols, net heat flux
M10	1.61	0.06	Aerosols, cloud optical properties, cloud mask/cover (cloud/snow detection), active fires, soil moisture, net heat flux
I3	1.61	0.06	Imagery snow/ice (cloud/snow differentiation), surface type, albedo
M11	2.25	0.05	Aerosols (optimal aerosol optical thickness over land), cloud optical properties, surface type, active fires, net heat flux
Midwave infrared (MWIR) spectral region, use of PV HgCdTe detectors			
I4	3.74	0.38	Imagery (identification of low and dark stratus), active fires
M12	3.70	0.18	Sea surface temperature (SST), cloud mask/cover, cloud EDRs, surface type, land/ice surface temperature, aerosols
M13	4.05	0.155	SST (essential for skin SST in tropics and during daytime), land surface temperature, active fires, precipitable water
Thermal infrared (TIR) spectral region, use of PV HgCdTe detectors			
M14	8.55	0.3	Cloud mask/cover (pivotal for cloud phase detection at night, cloud optical properties
M15	10.763	1.00	SST, cloud EDRs and SDRs (Science Data Records), land/ice surface temperature, surface type
I5	11.450	1.9	Imagery (nighttime imagery band)
M16	12.013	0.95	SST, cloud mask/cover, land/ice surface temperature, surface type

one of the elements of the new European Commission's Earth Observation Programme Copernicus, previously known as Global Monitoring for Environment and Security (GMES). The swath of this instrument will be equal to 1400 km for the nadir view and 740 km for the dual view.

Multiangle Imaging SpectroRadiometer (MISR) consists of nine cameras arranged to view along track that acquire image data with nominal view zenith angles relative to the surface reference ellipsoid of 0.0°, ±26.1°, ±45.6°, ±60.0°, and ±70.5° (forward and afterward of the Terra satellite) in four spectral bands (446, 558, 672, and 866 nm). In the global mode, the 672 nm (red) band images are acquired with a nominal maximum cross-track ground spatial resolution of 275 m in all nine cameras, and information from all bands is provided at this resolution in the nadir camera as well (Diner *et al.*, 1998, 2001). The width of the overlapped swath of MISR is 360 km, providing global multiangle coverage of the entire Earth in 9 days at the equator and in 2 days near the poles.

POLarization and Directionality of the Earth's Reflectances (POLDER) instrument is a passive optical imaging radiometer and polarimeter developed by the French space agency CNES. POLDER utilizes a push broom scanner. The device's optical system uses a telecentric lens and a charge-coupled device matrix with a resolution of 242 × 548 pixels. The focal length is 3.57 mm (0.141 in.) with a focal ratio of 4.6. The FOV ranges from ±43° to ±57°, depending on the tracking method. The device scans between 443 and 910 nm FWHM, depending on the objective of the measurement. The shorter wavelengths (443–565 nm) typically measure ocean color, whereas the longer wavelengths (670–910 nm) are used to study vegetation and water vapor content. The rotating wheels with respective polarizers allow for the measurements of the three first components of the Stokes vector of the reflected solar light. This enables the determination of the intensity, degree of linear polarization, and the angle of polarization plane (or the direction of the oscillations of the electric vector in the reflected solar light beam at the top of atmosphere (TOA)). Since 1996, three POLDER instruments have flown: POLDER-1 on board JAXA ADEOS-I (November 1996–June 1997), POLDER-2 on board JAXA ADEOS-II (April 2003–October 2003), and PARASOL on board CNES/Myriade microsatellite within the A-train constellation (December 2004–2013). The ICARE Data and Services Center is in charge of the POLDER/PARASOL-derived atmospheric products, including level 2 and 3 algorithm development, data production and distribution, and imagery processing. All the PARASOL or POLDER data (level 1, 2, or 3) are available online at ICARE using ftp (ftp.icare.univ-lille1.fr) or web interface without any delay.

Multiviewing, Multichannel, Multipolarization Imager (3MI) is a two-dimensional push broom radiometer dedicated to aerosol and cloud characterization for climate monitoring, air quality forecasting, and numerical weather prediction (Marbach *et al.*, 2013, 2015). It will fly together with eight other instruments (METimage, ICI, Sentinel-5 (S-5), IASI-NG, MWS, MWI, RO, and SCA) on the MetOp-SG (2021). The synergetic use with other instruments (especially METimage, IASI-NG, and S-5) could be of advantage for the 3MI retrievals of the atmospheric state (aerosol, clouds, trace gases). The planned mission time framework is 20 years (till 2040). The satellite will fly in a sun-synchronous low Earth orbit at 832 km altitude (09:30 LT of the descending mode). As a matter of fact, the 3MI will operate from the MetOp-SG-A satellite. The MetOp-SG-B satellite will fly in the same orbit separated by 180 degrees having on board ICI, MWI, and RO. The RO instrument will be placed on both satellites. The 3MI swath is 2200 km, and spatial sampling is 4 km at nadir. The first three components of the Stokes vector of the reflected light (I, Q, and U) will be measured at nine channels (see Table 7.11) in the spectral range from 410 to 2130 nm for up to 14 observation directions. Just intensity measurements will be performed at 763, 765, and 910 nm. The 3MI design consists basically of a filter and polarizer wheel rotating in front of the detectors. For design purpose the spectral channels have been split into VNIR and SWIR filters and polarizers with dedicated detectors and optical heads. The multipolarization (3 acquisitions within 1 s for the polarized channel) and multispectral acquisitions are done within a wheel rotation of less than 7 s. 3MI binary detector readouts will be calibrated to radiances using both on-ground calibration data (level 1b1) and using vicarious calibration methods from in-flight data (level 1b). The multiviewing capability will be achieved by successive images of the same spectral band observing the scene under different angles, allowing up to 14 views per target (see Figure 7.24).

The measurements are performed in a similar way as it was designed for the POLDER. This enables the determination of the degree of linear polarization of reflected light and also the direction of the oscillations of the electric vector in the light beam. The spectral channels have been split into VNIR (410, 443, 490, 555,

Table 7.11 3MI channels (channels 1–9 for the VNIR optical head and the channels 9–12 for the SWIR optical head).

Channel number	Channel center (nm)	Channel width (nm)	Measured components of the Stokes vector
1	410	20	I, Q, U
2	443	20	I, Q, U
3	490	20	I, Q, U
4	555	20	I, Q, U
5	670	20	I, Q, U
6	763	10	I
7	765	40	I
8	865	40	I, Q, U
9	910	20	I
10	1370	40	I, Q, U
11	1650	40	I, Q, U
12	2130	40	I, Q, U

The range of viewing zenith angles is different for different optical heads being larger for the VNIR optical head.

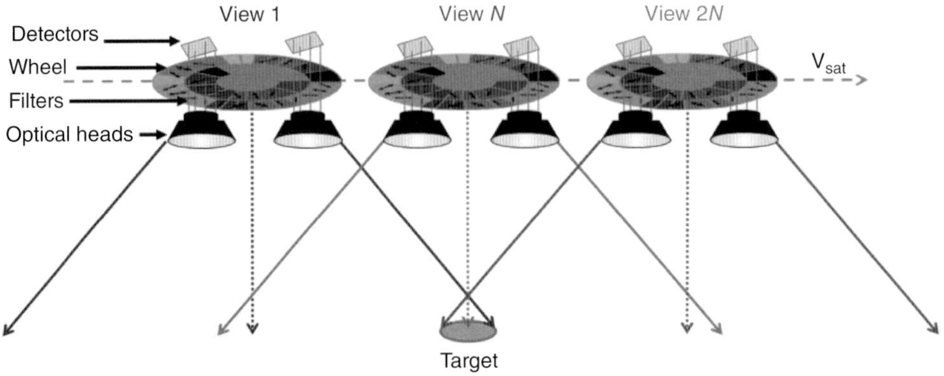

Figure 7.24 The 3MI concept, multiview, multispectral, and multipolarization sampling (Marbach *et al.*, 2015).

670, 763, 765, 865, and 910 nm) and SWIR (910, 1370, 1650, and 2130 nm) filters and polarizers with dedicated detectors and optical heads.

There is no onboard calibration. Therefore, the 3MI calibration will be performed by using other instruments from the MetOp-SG payload and also other instruments with similar channels flying at the same time in space. The instrument spectral response functions for 3MI are not known at the moment. It will be assumed that they can be modeled by the Gaussian distribution with the width at the half maximum equal to the channel width given in Table 7.10.

Cloud-Aerosol LIDAR with Orthogonal Polarization (CALIOP) is a two-wavelength polarization-sensitive LIDAR that provides high-resolution vertical profiles of aerosols and clouds. CALIOP utilizes three receiver channels: one measuring the 1064 nm backscatter intensity and two channels measuring orthogonally polarized components of the 532 nm backscattered signal. Dual 14-bit digitizers on each channel provide an effective 22-bit dynamic range. The receiver telescope is 1 m in diameter. A redundant laser transmitter is included in the payload. The continuity of atmospheric LIDAR measurements from space will be provided by the ATmospheric LIDar (ATLID) to be launched on board Earth Cloud, Aerosol and Radiation Explorer (EarthCARE). It gives vertical profiles of aerosols and thin clouds. The ATLID operates at a wavelength of 355 nm and has a high spectral resolution receiver and depolarization channel. The EarthCARE, due for launch in 2012 (ESA, 2004), is a joint European–Japanese mission addressing the need for a better understanding of the interactions between cloud, radiative, and aerosol processes that play a role in climate regulation. Four distinct instruments are planned for deployment on EarthCARE: the Backscatter LIDAR (ATLID), the Cloud Profiling Radar (CPR), the Multispectral Imager (MSI), and the Broadband Radiometer (BBR). The goal of this mission is to improve the representation and understanding of the Earth's radiative balance in climate and numerical weather forecast models by acquiring vertical profiles of clouds and aerosols, as well as the radiances at the TOA.

7.4.2
Methods

The AOT can be determined using ground-based radiometers as discussed in the beginning of this section. The corresponding measurement procedures are known at least for a century (Fowle, 1913). The history of satellite aerosol remote sensing is considerably shorter. The first paper on the AOT determination from space was published 40 years ago (Griggs, 1975). The retrieval has been performed over ocean, where the increase of aerosol load leads to the increase of the TOA reflectance. Since then a lot of methods have been developed for the retrieval of aerosol properties over both ocean and land surfaces using various instrumentation (Kokhanovsky and de Leeuw, 2009). The retrieval techniques depend on the instrument used. They can be characterized in the following ways:

Single-view spectral observations (see Section 7.4.3.1)
Double-view spectral observations (see Section 7.4.3.2)
Multiview spectral observations (see Section 7.4.3.3)
Multiview spectral and polarimetric observations (see Section 7.4.3.4).

We will consider the selected algorithms now.

7.4.2.1 The Algorithms Based on the Single-View Spectral Observations

The typical example of such an algorithm is that developed for MODIS. The algorithm is robust over ocean. However, it has reduced accuracy for retrievals over land. In the framework of the MODIS algorithm, the lookup tables (LUTs) are used. They provide the reflectance derived with the use of the vector radiative transfer model for various solar zenith angles, viewing zenith angles, relative azimuths, and AOTs for several fine and coarse aerosol modes. The inversion first interpolates the LUT to match the light incidence/observation geometry. The inversion determines the total spectral AOT and the fine-mode weighting (FMW) f that minimizes the least squares difference error between the modeled and observed spectral reflectance. The value of FMW is defined via the following equation:

$$f = \frac{\tau_f}{\tau_a} \tag{7.12}$$

Here $\tau_a = \tau_f + \tau_c$ is the AOT, τ_f is the contribution of the fine-mode aerosol to the AOT, and τ_c is the contribution of the coarse-mode aerosol to the AOT. The possibility for the determination of the FMW is due to the fact that one can present the TOA reflectance R as

$$R(\tau_a) = fR_f(\tau_a) + (1-f)R_c(\tau_a). \tag{7.13}$$

This relationship exactly holds in the single scattering approximation for non-absorbing aerosols. Indeed, it follows in the single scattering approximation for the atmospheric aerosol composed of two modes (fine and coarse):

$$R(\tau_a) = (A\tau_f p_f(\theta) + A\tau_c p_c(\theta))/\tau_a \equiv fR_f + (1-f)R_c \tag{7.14}$$

where the constant A does not depend on the properties of aerosol, $p(\theta)$ is the aerosol phase function (for fine and coarse modes), $R_{f(c)}(\tau_a) = A\tau_a p_{f(c)}(\theta)$, and θ is the scattering angle. The accuracy of this approximation decreases for larger values of the AOT and also for absorbing aerosols. Then one must use LUTs calculated for the external aerosol mixture with the following parameters:

$$p(\theta) = \frac{\tau_f p_f(\theta) + \tau_c p_c(\theta)}{\tau_a}, \quad \omega_0 = \frac{\tau_f \omega_{0f} + \tau_c \omega_{0c}}{\tau_a}, \quad \tau_a = \tau_f + \tau_c. \tag{7.15}$$

Clearly, in this case, the size of LUTs increases considerably. Alternatively, the differences between two approaches can be parameterized, and corrections to Eq. (7.13) can be used (Abdou *et al.*, 2005). The approach based on Eq. (7.13) works quite well for oceanic aerosol because it is nonabsorbing in most of cases. Also AOT is generally quite low (0.1 or so) over oceans. The errors increase for the retrievals over land both because aerosol load can be substantial and because absorbing aerosol particles (say, smoke, dust) can present, which reduces the single scattering albedo (down to 0.8 or so). Another problem with aerosol retrievals over land that the surface reflectance can be quite large. This problem is usually avoided using the measurements in the NIR (say, at 1.6 or 2.1 μm), where the contribution of aerosol is low and the TOA reflectance is almost coincident with the surface reflectance r. Various assumptions on the correlation of the surface reflectance in the NIR and VIS can be used. This enables the estimation of the surface reflectance in the VIS needed for the retrieval routine. Alternatively, the measurements in the UV can be used because most of surfaces (but not ice and snow) are only weekly reflective there.

7.1.2.2 Double-View Spectral Observations

The double view enables the elimination of surface in the retrieval process. It can be done in the following way. The TOA reflectance over a reflective Lambertian ground can be presented in the following form:

$$R = R_0 + \frac{AT}{1 - \mathrm{Ar}}, \tag{7.16}$$

where A is the surface albedo, r is the spherical albedo of atmospheric layer, and T is its transmittance. Let us imagine that the observations are performed at two angles for the same pixel characterized by the surface albedo A. Then one can derive from Eq. (7.16)

$$\frac{R_1 - R_{01}}{R_2 - R_{02}} = \frac{T_1}{T_2}, \tag{7.17}$$

where indices 1 and 2 signify the corresponding observation angle. Equation (7.17) does not contain the surface albedo and can be used to determine the AOT, aerosol type, and FMW using the minimization procedure. For this the LUTs over nonreflecting surfaces (R_0) must be used. In addition, the LUTs for the transmittance T must be constructed.

The approximate account for the non-Lambertian nature of real-world surfaces is usually performed in the following way. First of all it is assumed that A in

Eq. (7.16) depends on the angle. Then it follows from Eq. (7.16) at a given wavelength λ that

$$\frac{R_1 - R_{01}}{R_2 - R_{02}} = k\frac{T_1}{T_2}. \tag{7.18}$$

The value of k is equal to 1 for Lambertian surfaces (see Eq. (7.17)). One may assume that the factor k is almost spectrally neutral. Then it can be obtained from the measurements in the NIR (say, at 1.6 or 2.1 micrometers) taking into account that

$$k = \frac{R_1}{R_2}. \tag{7.19}$$

This relationship holds approximately in the NIR taking into account that the transmittances are close to one and atmospheric reflectance is close to zero in the NIR. Having derived k using satellite measurements via Eq. (7.19), one can use Eq. (7.18) (with known value of k) to derive AOT for several wavelengths. The aerosol model, which provides the best fit of spectral and double-view measurements, is selected.

7.4.2.3 Multiview Spectral Observations

The multiview enables both the elimination of the surface in the retrieval process and better selection of aerosol models as compared to just double-view measurements. It follows from Eq. (7.18) that

$$k = \frac{R_1 - R_{01}}{R_2 - R_{02}} \times \frac{T_2}{T_1}. \tag{7.20}$$

Let assume that there are j angular measurements at a given wavelength λ. Then it follows from Eq. (7.20) that

$$k_i = \frac{R_1 - R_{01}}{R_i - R_i} \times \frac{T_i}{T_1}, i = 2, \ldots j. \tag{7.21}$$

This factor must be spectrally neutral. The aerosol model and spectral optical thickness, which provides the spectral neutrality of the factor k, is used as an output of the algorithm. Sometimes not factor k_j itself but the ratio of the factor to the average value of this factor for all views for a given spectral band is used.

7.4.2.4 Multiview Spectral and Polarimetric Observations

The reflected solar light is characterized by the Stokes vector, which depends on the wavelength, the geometry of illumination/observation, and also the properties of underlying surface and atmosphere including atmospheric aerosol. Therefore, the maximum information on the aerosol underlying system can be achieved, if the spectral multiangular measurements of the Stokes vector (and not just intensity) of the reflected (or transmitted) light are performed. Such measurements have been performed by POLDER 1, POLDER 2, and POLDER 3 instruments. Currently there is no polarimeter in space. However, it is planned that they appear in 2021 (e.g., 3MI on board EUMETSAT PS-SG mission).

7.4.2.5 Retrievals over Ocean Using Multiangle Polarimetric Observations

The retrievals are performed by selecting the aerosol model, which provides best fits of the angular measurements of the reflectance and degree of polarization of reflected solar light at selected channels with the precalculated LUTs.

The best fit is found by minimizing the following function:

$$\Delta = \sum_{k=1}^{K} \sum_{l=1}^{L} \{\delta_R^2 + \delta_q^2\}, \tag{7.22}$$

where

$$\delta_R = R(\lambda_k, \vartheta_0, \vartheta_l, \varphi_l) - R_{\text{meas}}(\lambda_k, \vartheta_0, \vartheta_l, \varphi_l),$$
$$\delta_q = q(\lambda_k, \vartheta_0, \vartheta_l, \varphi_l) - q_{\text{meas}}(\lambda_k, \vartheta_0, \vartheta_l, \varphi_l).$$

Here ϑ_l is the viewing zenith angle, and φ_l is the relative azimuthal angle. Only the wavelengths $\lambda_k = 670, 865, 1650$, and 2130 nm are used in the retrievals over the ocean. Therefore, $K = 4$. The shorter wavelengths are not used in the retrieval process over ocean because the characterization/retrieval of the surface reflectance is needed then. This leads to the biases in the retrieved aerosol properties. For the wavelengths above 670 nm, ocean is a blackbody in a good approximation. Therefore, the retrievals over ocean are often performed for the LUTs calculated at the surface albedo equal to zero.

7.4.2.6 Retrievals over Land

The channels at 410, 443, 490, 555, 670, and 2130 nm are used for AOT retrievals over land. The information on the surface reflectance is taken from the measurements at 2130 nm neglecting aerosol contribution at this channel.

The aerosol parameters are found by minimizing the functional

$$\Lambda = \sum_{l=1}^{L} \sum_{j=1}^{14} \left\| k_j - \frac{R_1 - R_{01}}{R_{j-1} - R_{0j-1}} \times \frac{T_2}{T_1} - s_j \right\|^2 \to 0,$$

where s_j is the part of the functional related to the polarization measurements.

The accuracy of retrievals using different algorithms is shown in Figure 7.25.

7.4.2.7 Aerosol Retrieval Using an Artificial Neural Network Technique

The neural network (hereafter NN), developed with an artificial intelligence (AI) or a machine learning technology, has been applied in aerosol retrieval applications based on different satellite platforms and achieved promising results. Because NN can learn complex linear or nonlinear relationships in the radiometric data, it has been proven to be a useful tool to solve ill-posed problems as the retrieval of aerosol from satellite radiance measurements. For examples, a commercial NN package with multithreshold techniques was used to identify BB aerosols using the AVHRR imagery data (Li et al., 2001), and the results showed reasonable correspondence with TOMS AI (Li *et al.*, 2009). As a NN technique with data mining scheme, LUTs storing TOA reflectance values with various atmosphere–ocean conditions are used for tuning the NN. This LUT-based NN technique was used for aerosol retrieval from reflectances at two ADEOS/OCTS

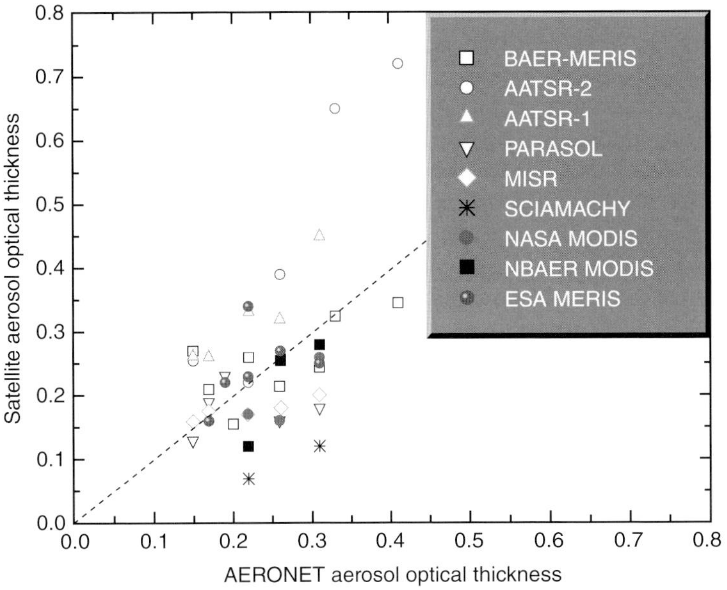

Figure 7.25 Comparison of satellite and ground measurements of AOT at 0.55 μm (Kokhanovsky *et al.*, 2007).

bands (0.67 and 0.865 μm) (Okada, Mukai, and Sano, 2001). Automatic dust storm detection was available by using NN for the MODIS images (Rivas-Perea *et al.*, 2010; El-ossta, Qahwaji, and Ipson, 2013) and for AOT retrieval from the Atmospheric Infrared Sounder (AIRS) (Han and Sohn, 2013).

More recently, simultaneous dust storm detection and retrieval of AOT based on the multilayer perceptron (MLP) NN model with a back-propagation (BP) learning algorithm (hereafter MLP-BP) for the geostationary satellite images were introduced (Xiao *et al.*, 2015). As a feed-forward NN model, the MLP-BP has at least one layer of "hidden" neurons whose nonlinear functions are able to operate as universal function approximations. When enough hidden neurons and training data are used as input, NNs are capable of learning the mathematical relation between inputs and outputs (see Figure 7.26). The NN technique has proven to reduce processing times and is promising for effective aerosol retrievals from satellite remote sensing. This automated method can be used to near real-time warning of dust storms for both environmental authorities and public.

7.4.3 Examples of Aerosol Retrievals

7.4.3.1 Global View of Aerosol Distribution from Passive Sensor

Aerosol products have been retrieved from Earth observation satellites. Especially, aerosol records from the single- or dual-band observation sensors, such as

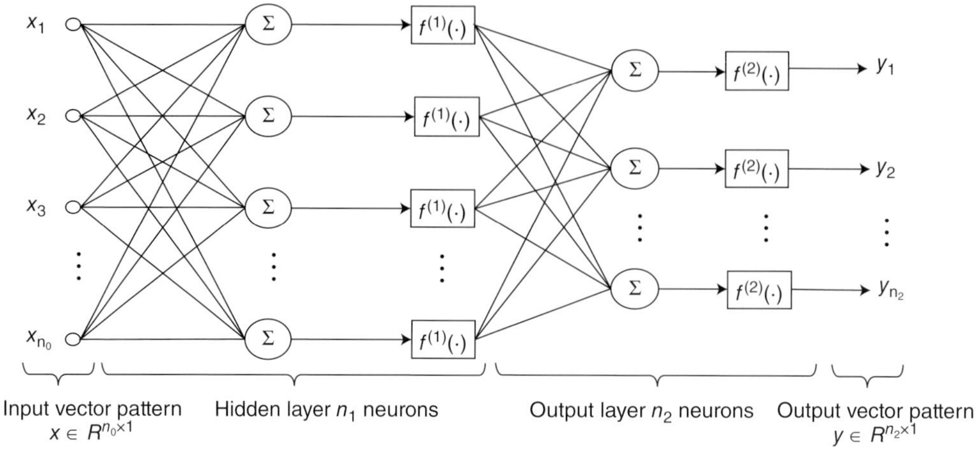

Figure 7.26 A multilayer perceptron with one hidden layer (after Xiao *et al.*, 2015).

AVHRR and TOMS measurements, are the longest records of over 37 years. Multiwavelength sensors – for example, the SeaWiFS from 1997, MODIS, and MISR from 2000 – were providing improved aerosol products. Those aerosol products can be used for understanding climate change and studies of atmospheric pollution.

In Figure 7.27, we show the distribution of AOT as derived from space. In this figure, 16 years MODIS retrieved AOT data (http://ladsweb.nascom.nasa.gov) are shown. The MODIS retrieved AOT was previously validated by ground-based sun photometer using a spatiotemporal approach (Ichoku et al. 2002) and retrieval errors were reported as $\Delta\tau = \pm 0.05 \pm 0.2\tau$ over ocean (Remer et al. 2005). Aerosols near source regions are particularly near sources of smoke and dust. It follows that the AOT is large near source regions of smoke and dust, such as deserts and populated areas. For example, so-called dust belt from the Sahara eastward across Middle East and into central China is clearly evident as shown in Figure 7.28. Transport of desert dust from the Sahara westward across the Atlantic Ocean is recorded. However, large aerosol pollution is seen over parts over Asia and transport of finer polluted aerosols from Asia across the Northern Pacific is evident.

7.4.3.2 Aerosol Retrieval from Different Sensors and Retrieval Algorithms

General view of aerosol distribution from satellite observations shows similar pattern as explained in the former section. However, aerosol intensity derived by the algorithm for the satellite sensor may have discrepancy. Inherent uncertainty contained in the retrieval algorithm can cause error in the retrieved aerosol products. Comparison of global mean AOT over ocean derived from different satellites with their own retrieval algorithms is shown in Figure 7.29. All data sets show strong peaks of AOT by volcano eruptions of El Chichón in March 1982 and Mt. Pinatubo

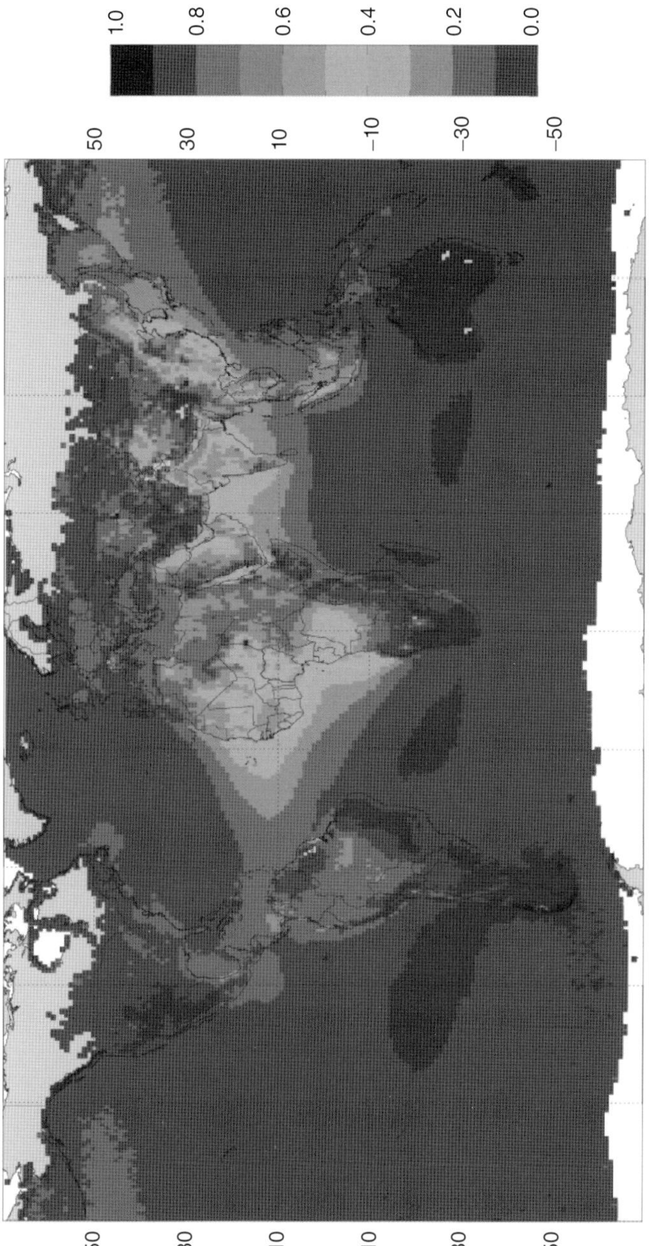

Figure 7.27 Mean aerosol optical thickness from the Terra MODIS collection 6 data, March 2001–October 2015.

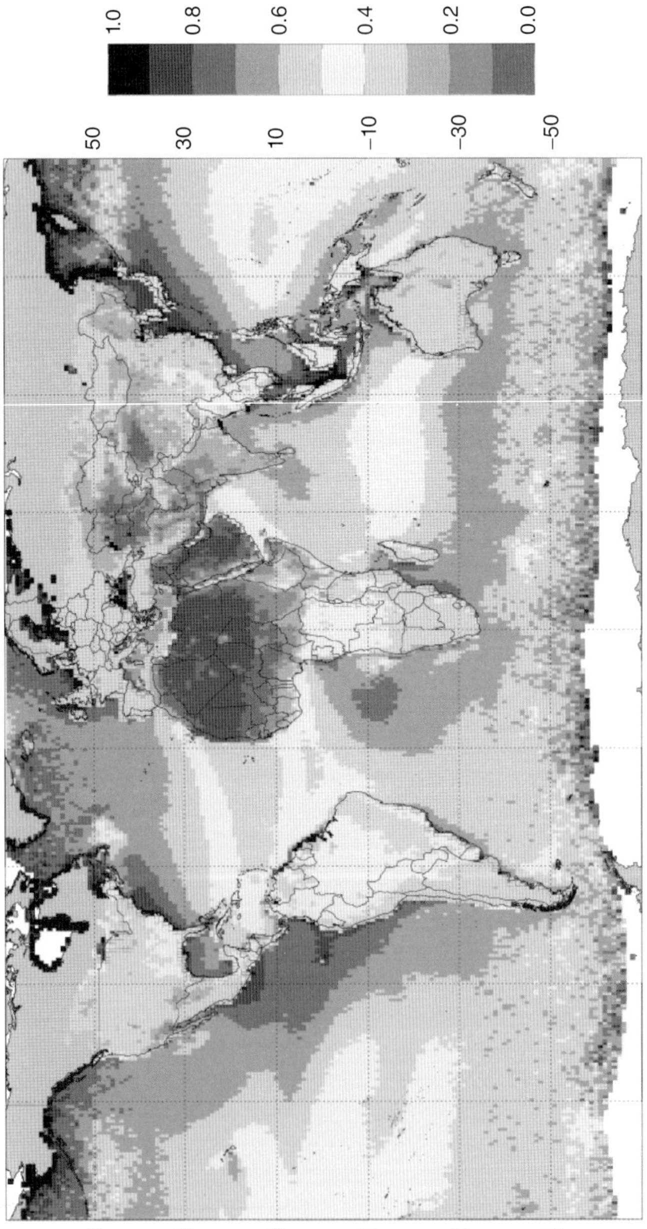

Figure 7.28 Mean fine fraction of aerosol optical thickness over ocean and normalized Ångström exponent over land from the Terra MODIS collection 6 data, March 2001 – October 2015.

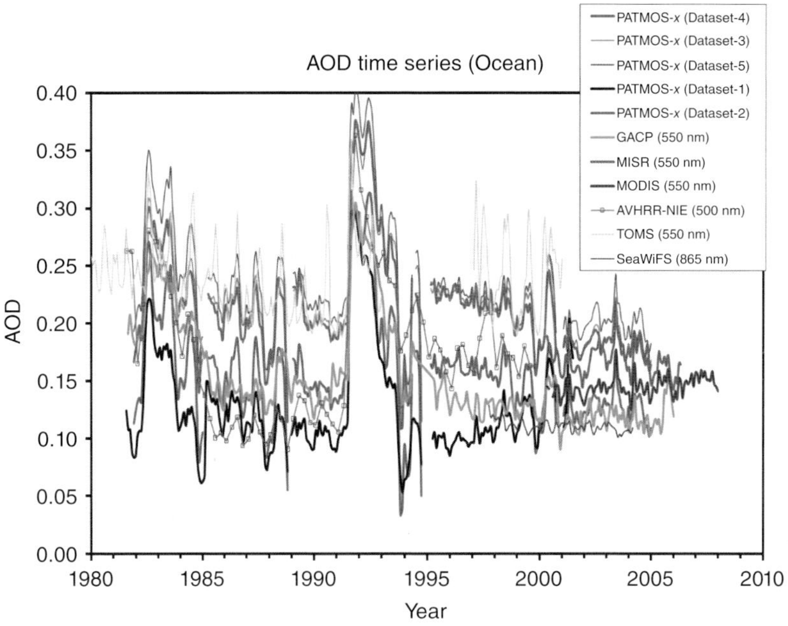

Figure 7.29 Comparison of global mean AOD computed from five versions of PATMOS experimental products, with reference to the TOMS/OMI, SeaWiFS, MODIS, and MISR (after Li *et al.*, 2009).

in June 1991. Apart from these two strong global aerosol events, long-term variations could be seen. However, largest systematic discrepancy is found among the different data sets. Sensor's radiometric calibration, cloud contamination, correction of the surface effect, atmospheric absorption, and assumption of aerosol properties can be arise this discrepancy (Li *et al.*, 2009).

7.4.3.3 Time-Resolved Observation from Geostationary Platform

Aerosol retrieval from the geostationary satellite observations provides timely variation of aerosol spatial distribution and optical property changes. Geostationary satellites have been provided with AOTs as well as aerosol type which are used to investigate spatial and temporal patterns of severe aerosol events such as desert dust storm, BB, or smog pollution. For example, Figure 7.30 shows AOT maps by the Geostationary Ocean Color Imager (GOCI) on board the Communication, Ocean and Meteorological Satellite (COMS) on dusty days of 1–2 May 2011. The Asian dust is clearly seen in the hourly GOCI RGB color composite images, and their moving pattern through the Yellow Sea can be recognized by the very well-defined and high AOT values. Figure 7.30 also shows the same area covered by MODIS AOT data (level 2, 10 km spatial resolution), which have similar spatial patterns for aerosol. However, GOCI-retrieved AOT with the 500 m spatial resolution shows a higher spatial resolution, sufficient to identify the extent of the dust plume.

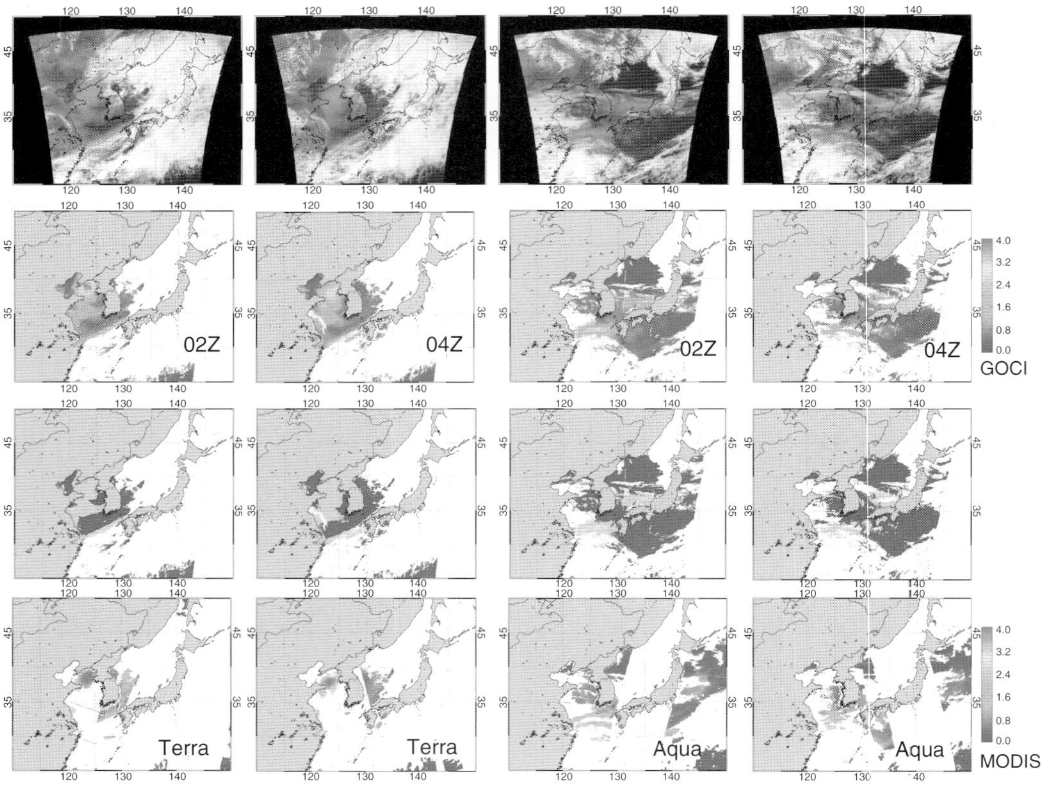

Figure 7.30 GOCI image (top line), retrieved GOCI AOT at 550 nm (second line) with selected aerosol types (third line), and MODIS L2 AOT (bottom line) on 1–2 May 2011. Aerosol types are given by dark grey (maritime clean) and light grey (desert dust) (after Lee *et al.*, 2012).

7.4.3.4 Atmospheric Anatomy from the Active Sensing Platform

Backscatter signal observed from the spaceborne LIDAR system is providing unique three-dimensional spatial distribution as well as temporal variations for atmospheric aerosols. The CALIOP, carried on the Cloud-Aerosol LIDAR and Infrared Pathfinder Satellite Observation (CALIPSO) satellite, has carried out continuous observations of aerosol profiles. Typically, CALIOP can derive the particulate extinction coefficient and depolarization ratio for each altitude according to time and space. For example, Figure 7.31 shows the variation of optical properties of aerosols over Northeast Asia (E 110–140°, N 20–50°). In these geostatistical results, the most frequent altitudes of aerosols are clearly

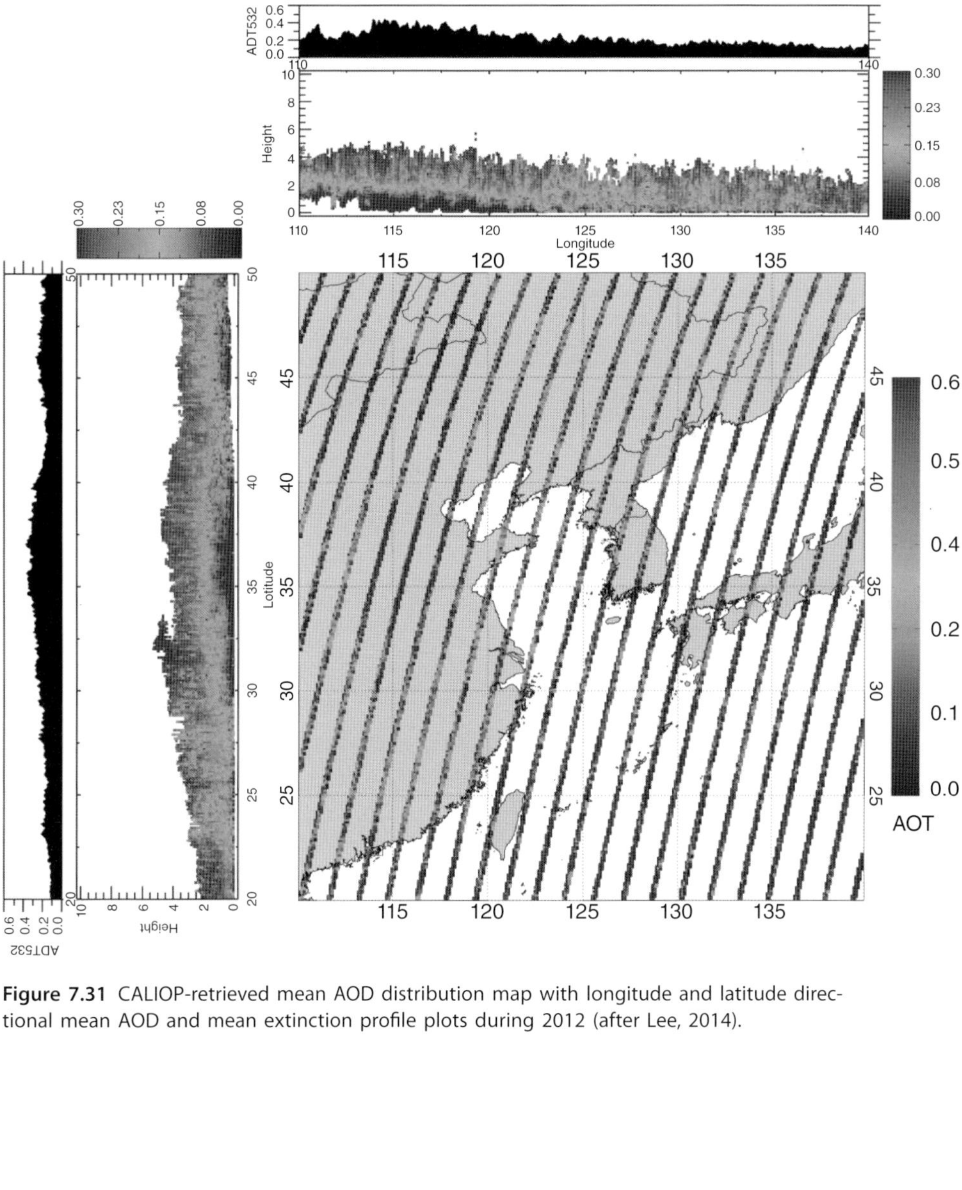

Figure 7.31 CALIOP-retrieved mean AOD distribution map with longitude and latitude directional mean AOD and mean extinction profile plots during 2012 (after Lee, 2014).

identified, and mean aerosol profiles vary with location. Since relatively high aerosol loads are found over urbanized area, it is considered that aerosols mixed with pollution mainly exist over area. These regional three-dimensional aerosol information will be helcpful to understand aerosol climatology and effects on climate.

Abbreviations

AMS	aerosol mass spectrometer
BC	black carbon
BB	biomass burning
BrC	brown carbon
CBL	convective boundary layer
CNs	condensation nuclei
CCNs	cloud condensation nuclei
CNC	condensation nucleus counter
DMA	differential mobility analyzer
EC	elemental carbon
FMW	fine mode weighting
FOV	field of view
HBW	half-bandwidth of an interference filter mounted on a Sun photometer
HR-ToF-AMS	high-resolution time-of-flight aerosol mass spectrometer
IR	infrared
LAC	light-absorbing carbon
LUTs	lookup tables
MLH	mixing layer height
MAE	mass absorption efficiency
MAE_{BC}	black carbon mass absorption efficiency
MBL	marine boundary layer
MD	mineral dust
NIR	near-infrared
OC	organic carbon
OPC	optical particle counter
PSAP	particle soot absorption photometer
RS	refractory submicrometer
SAL	Saharan air layer
SS	sea salt
SST	sea surface temperature
TBW	ten-bandwidth of an interference filter used in Sun photometry
TOA	top of the atmosphere
UHSAS	ultrahigh sensitivity aerosol spectrometer
UV	ultraviolet
VIS	visible

List of Symbols

Latin

a	aerosol particle diameter
a_c	mode diameter
A	surface albedo
Ar	spherical albedo of an atmospheric layer
τ_a	aerosol optical thickness (AOT)
a_{oe}	aerosol particle diameter measured by an optical particle counter or other similar remote sensing instruments
$D_{\mathrm{E-S}}$	annual mean Earth–Sun distance
$dN/d(\ln a)$	size distribution of aerosol particles as a function of particle diameter a plotted on a logarithmic scale
D	daily Earth–Sun distance
F_e	eccentricity correction factor equal to $(D_{\mathrm{E-S}}/D)^2$
$g(\lambda)$	asymmetry factor of aerosol particles at wavelength λ
g_{ms}	mean seasonal value of aerosol asymmetry factor
H_p	aerosol scale height defining the vertical profile of aerosol volume extinction coefficient decreasing in an exponential fashion with height
K_j	integral of the volume extinction coefficient $k_j(z)$ over the vertical atmospheric path from z_o to z_{top}
$k_j(z)$	volume extinction (or absorption) coefficient of the jth constituent varying as a function of altitude z
$k_n(\lambda)$	absorption coefficient of atmospheric NO_2 (+N_2O_4) content at wavelength λ
$k_o(\lambda)$	absorption coefficient of atmospheric ozone content at wavelength λ
$k_w(\lambda)$	absorption coefficient of atmospheric water vapor content at wavelength λ
$m(\vartheta_0)$	relative optical air mass (varying as a function of ϕ) for a standard atmosphere
$m_a(\vartheta_0)$	relative optical air mass for aerosol extinction (varying as a function of ϕ)
$m_{\mathrm{AH}}(\vartheta_0)$	relative optical air mass for a dense Arctic haze case
$m_{\mathrm{aMSav}}(\vartheta_0)$	relative optical air mass of the Mid-latitude Summer Atmosphere model with a stratospheric volcanic aerosol layer peaked at altitude $z = 24$ km
$m_{\mathrm{aMSfv}}(\vartheta_0)$	relative optical air mass of the Mid-latitude Summer Atmosphere model with a stratospheric volcanic aerosol layer peaked at altitude $z = 16$ km
$m_{\mathrm{aTm}}(\vartheta_0)$	relative optical air mass of the Tropical Atmosphere model with $H_p = 1.8$ km

$m_{\mathrm{aTs}}(\vartheta_0)$	relative optical air mass of the Tropical Atmosphere model with $H_p = 1.2$ km
$m_{\mathrm{aUSav}}(\vartheta_0)$	relative optical air mass of the U. S. Standard Atmosphere model with $H_p = 1.1$ km
$m_{\mathrm{aUSm}}(\vartheta_0)$	relative optical air mass of the U. S. Standard Atmosphere model with $H_p = 1.4$ km
$m_{\mathrm{aUSs}}(\vartheta_0)$	relative optical air mass of the U. S. Standard Atmosphere model with $H_p = 0.8$ km
$m_{\mathrm{avh}}(\vartheta_0)$	relative optical air mass for an average Arctic haze case
$m_{\mathrm{bAa}}(\vartheta_0)$	relative optical air mass of the January −75°S Antarctic atmosphere with Antarctic background aerosol
$m_{\mathrm{baa}}(\vartheta_0)$	relative optical air mass of the July −75°N Arctic atmosphere with Arctic background aerosol
$m_j(\vartheta_0)$	relative optical air mass function of the *j*th atmospheric constituent as a function of ϑ_0
$m_{\mathrm{KV}}(\vartheta_0)$	relative optical air mass for the Kasatochi volcanic particle layer over the Arctic
M_{ref}	non-BC refractory mass
$m_n(\vartheta_0)$	relative optical air mass for nitrogen dioxide absorption as a function of ϑ_0
$m_o(\vartheta_0)$	relative optical air mass for ozone absorption as a function of ϑ_0
$m_{\mathrm{od}}(\vartheta_0)$	relative optical air mass for oxygen dimer absorption as a function of ϑ_0
$m_P(\vartheta_0)$	relative optical air mass for the Mt. Pinatubo volcanic particle layer over Antarctica
$m_w(\vartheta_0)$	relative optical air mass for water vapor absorption as a function of ϑ_0
N	total number of aerosol particles in the vertical atmospheric column of unit cross section
$n(\lambda)$	real part of particulate refractive index at wavelength λ
$n(r) = dN/dr$	size distribution of aerosol particles as a function of particle radius r
$n(z)$	air refractive index as a function of altitude z
$n(z)$	moist air refractive index as a function of altitude z
n_o	air refractive index at sea level z_o
$p(\theta)$	aerosol phase function
$p_f(\theta)$	part of the aerosol phase function due to the fine particle mode
$p_c(\theta)$	part of the aerosol phase function due to the coarse particle mode
r	particle radius
r_α	regression coefficient of the best-fit solution found applying the Ångström formula to spectral series of $\tau_a(\lambda)$
r_c	mode radius of a fine particle mode, an accumulation particle mode, or a coarse particle mode
r_{eff}	mean effective radius of a particle mode

$R_\ell(\lambda_o)$	LIDAR ratio at wavelength λ_o
$R_\ell(0.532\,\mu m)$	LIDAR ratio measured by the AMALi LIDAR system
R	reflectance at the TOA level
R_s	surface reflectance
R_f	part of top-of-atmosphere reflectance due to the fine particle mode
R_c	part of top-of-atmosphere reflectance due to the coarse particle mode
RH	relative humidity of air
$s\ m^{-3}$	standard cubic meter of air
T	transmittance of an atmospheric layer
$V(\lambda)$	output voltage of a sun photometer at wavelength λ (usually measured in mV)
$V(0.870\,\mu m)$, $V(1.020\,\mu m)$	output voltages of a Sun photometer within the atmospheric transmittance windows limiting the $\rho\sigma\tau$ absorption band by water vapor (usually measured in mV)
$V(0.940\,\mu m)$	output voltage of a Sun photometer at a mid-wavelength of the $\rho\sigma\tau$ absorption band by water vapor (usually measured in millivolt
$V_o(\lambda)$	calibration constant of a Sun photometer at wavelength λ, giving the extraterrestrial value of $V(\lambda)$ for annual mean Earth–Sun distance (usually measured in millivolt)
VR	ground-level visual range
W	precipitable water for standard pressure and temperature (STP) conditions, also labelled "atmospheric water vapor content"
y	parameter defined in Eq. (2.11) and used to develop the geometrical series of the difference $\alpha_f - \alpha_c$
z	altitude, height
z_o	mean sea level to the atmosphere
z_{top}	top level of the atmosphere (here assumed to be equal to 120 km

Greek

$\alpha(\lambda)$	wavelength dependence function of Ångström exponent
α	Ångström wavelength exponent
α_λ	modified Ångström exponent at wavelength λ
$\alpha(0.40-0.87\,\mu m)$	Ångström wavelength exponent determined over the 0.40–0.87 μm spectral range
α'	spectral first derivative of $\alpha(\lambda)$ with respect to λ
$\alpha_c(\lambda)$	wavelength dependence function of Ångström exponent relative to coarse particles
α_c'	spectral first derivative of $\alpha_c(\lambda)$ with respect to λ
$\alpha_f(\lambda)$	wavelength dependence function of Ångström exponent relative to fine particles

α_f'	spectral first derivative of $\alpha_f(\lambda)$ with respect to λ
β	second atmospheric turbidity parameter giving the best-fit value of $\tau_a(\lambda)$ at the 1 μm wavelength
$\beta_{abs}(0.573\,\mu m)$	volume absorption coefficient measured with the RR PSAP photometer
$\beta_{bs}(\lambda)$	volume aerosol backscattering coefficient at wavelength λ
$\beta_{bs}(0.532\,\mu m)$	backscatter coefficient measured by the AMALi LIDAR system
$\beta_{sca}(\lambda)$	volume aerosol scattering coefficient at wavelength λ
$\beta_{sca}(0.530\,\mu m)$	volume scattering coefficient measured with the RR M903 nephelometer
$\beta_{ext}(\lambda)$	volume aerosol extinction coefficient at wavelength λ
$\beta_{ext}(0.550\,\mu m)$	average extinction coefficient calculated at wavelength $\lambda = 0.55\,\mu m$ as the sum of $\beta_{sca}(0.530\,\mu m)$ and $\beta_{abs}(0.573\,\mu m)$
$\beta_{ext}(z)$	volume aerosol extinction coefficient at altitude z
$\beta_{ext}(\lambda_o)$	volume aerosol extinction coefficient at LIDAR wavelength λ_o
$\beta_{ext}(0.532\,\mu m)$	volume extinction coefficient calculated from the AMALi LIDAR system measurements
$\delta(0.532\,\mu m)$	volume depolarization ratio measured by the AMALi LIDAR system
Φ_{FOV}	diameter of the circular field of view (FOV) of a Sun photometer
λ	wavelength (usually measured in μm or in nm)
λ_p	peak wavelength of a narrow-band interference filter
ρ	mass density or dry bulk density of particulate matter
ϑ_0	apparent solar zenith angle
$\eta_f(\lambda)$	fine particle fraction, giving the fraction of $\tau_a(\lambda)$ due to fine particle extinction
$\eta_f(0.50\,\mu m)$	fine particle fraction determined for a chosen separation wavelength $\lambda = 0.50\,\mu m$
θ	scattering angle
ϑ_l	viewing zenith angle
φ_l	relative azimuthal angle
$\chi(\lambda)$	imaginary part of particulate refractive index at wavelength λ
k	proportionality factor between reflectances pertaining to two incoming beams having different incidence angles
$\Gamma(r)$	volume size-distribution curve of aerosol particles (suspended in the vertical column of unit cross section) as a function of particle radius
$\tau(\lambda)$	total atmospheric optical thickness at wavelength λ
$\tau_a(\lambda)$	aerosol optical thickness at wavelength λ
$\tau_{abs}(\lambda)$	absorption component of aerosol optical thickness $\tau_a(\lambda)$
$\tau_{sca}(\lambda)$	scattering component of aerosol optical thickness $\tau_a(\lambda)$
$\tau_n(\lambda)$	nitrogen dioxide (and N_2O_4) optical thickness at wavelength λ
$\tau_o(\lambda)$	ozone optical thickness at wavelength λ
$\tau_{od}(\lambda)$	oxygen dimer optical thickness at wavelength λ

$\tau_R(\lambda)$	Rayleigh scattering optical thickness at wavelength λ
$\tau_w(\lambda)$	water vapor optical thickness at wavelength λ
τ_f	fraction of aerosol optical thickness due to the fine particle mode
τ_c	fraction of aerosol optical thickness due to the coarse particle mode
$\omega(\lambda)$	single-scattering albedo of aerosol particles at wavelength λ
ω_{ms}	mean seasonal value of aerosol single-scattering albedo
ω_o	ground-level aerosol single-scattering albedo
ω_0	single-scattering albedo of columnar aerosol particles

References

Abdou, W.A., Diner, D.J., Martonchik, J.V., Bruegge, C.J., Kahn, R.A., Gaitley, B.J., Crean, K.A., Remer, L.A., and Holben, B. (2005) Comparison of coincident Multiangle Imaging Spectroradiometer and Moderate Resolution Imaging Spectroradiometer aerosol optical depths over land and ocean scenes containing Aerosol Robotic Network sites. *J. Geophys. Res.*, **110**, D10S07. doi: 10.1029/2004JD004693

Ackermann, J. (1998) The extinction-to-backscatter ratio of tropospheric aerosol: a numerical study. *J. Atmos. Oceanic Technol.*, **15** (4), 1043–1050. doi: 10.1175/1520-0426(1998)015b1043:TETBRON2.0.CO;2

Anderson, G.P., Chetwind, J.H., Clough, S.A., Shettle, E.P., and Kneizys, F.X. (1986) AFGL Atmospheric Constituent Profiles (0–120 km). Environmental Research Papers, No. 954, AFGL-TR-86-0110. Optical Physics Division (Air Force Geophysics Laboratory), Hanscom Air Force Base, MA.

Angelini, F., Barnaba, F., Landi, T.C., Caporaso, L., and Gobbi, G.P. (2009) Study of atmospheric aerosols and mixing layer by LIDAR. *Radiat. Prot. Dosim.*, **137** (3-4), 275–279. doi: 10.1093/rpd/ncp219

Ångström, A. (1930) On the atmospheric transmission of Sun radiation. II. *Geogr. Ann.*, **12**, 130–159.

Ångström, A. (1964) The parameters of atmospheric turbidity. *Tellus*, **16** (1), 64–75. doi: 10.1111/j.2153-3490.1964.tb00144.x

Ansmann, A., Wagner, F., Althausen, D., Müller, D., Herber, A., and Wandinger, U. (2001) European pollution outbreaks during ACE 2: lofted aerosol plumes observed with Raman lidar at the Portuguese coast. *J. Geophys. Res.*, **106** (D18), 20725–20733. doi: 10.1029/2000JD000091

Ansmann, A., Wagner, F., Müller, D., Althausen, D., Herber, A., von Hoyningen-Huene, W., and Wandinger, U. (2002) European pollution outbreaks during ACE 2: optical particle properties inferred from multiwavelength lidar and star-Sun photometry. *J. Geophys. Res.*, **107** (D15, 4259), AAC 8-1–AAC 8-14. doi: 10.1029/2001JD001109

Balis, D., Kroon, M., Koukouli, M.E., Brinksma, E.J., Labow, G., Veefkind, J.P., and McPeters, R.D. (2007) Validation of Ozone Monitoring Instrument total ozone column measurements using Brewer and Dobson spectrophotometer ground-based observations. *J. Geophys. Res.*, **112**, D24S46. doi: 10.1029/2007JD008796

Barnsley, M.J., Settle, J.J., Cutter, M.A., Lobb, D.R., and Teston, F. (2004) The PROBA/CHRIS mission: a low-cost smallsat for hyperspectral, multi-angle, observations of the Earth surface and atmosphere. *IEEE Trans. Geosci. Remote Sens.*, **42** (7), 1512–1520. doi: 10.1109/TGRS.2004.827260

Bösenberg, J., Matthias, V., Amodeo, A. *et al.* (2003) EARLINET: A European Aerosol Research Lidar Network to Establish an Aerosol Climatology. MPI-Report 348, Max-Planck-Institut für Meteorologie, Hamburg, 192 pp.

Bourassa, A.E., Degenstein, D.A., Gattinger, R.L., and Llewellyn, E.J. (2007) Stratospheric aerosol retrieval with optical spectrograph and infrared imaging system limb scatter measurements. *J. Geophys. Res.*, **112**, D10217. doi: 10.1029/2006JD008079

Brewer, A.W. (1973) A replacement for the Dobson spectrophotometer? *Pure Appl. Geophys.*, **106–108** (1), 919–927. doi: 10.1007/BF00881042

Brock, C.A., Cozic, J., Bahreini, R., Froyd, K.D., Middlebrook, A.M., McComiskey, A., Brioude, J., Cooper, O.R., Stohl, A., Aikin, K.C., de Gouw, J.A., Fahey, D.W., Ferrare, R.A., Gao, R.-S., Gore, W., Holloway, J.S., Hübler, G., Jefferson, A., Lack, D.A., Lance, S., Moore, R.H., Murphy, D.M., Nenes, A., Novelli, P.C., Nowak, J.B., Ogren, J.A., Peischl, J., Pierce, R.B., Pilewskie, P., Quinn, P.K., Ryerson, T.B., Schmidt, K.S., Schwarz, J.P., Sodemann, H., Spackman, J.R., Stark, H., Thomson, D.S., Thornberry, T., Veres, P., Watts, L.A., Warneke, C., and Wollny, A.G. (2011) Characteristics, sources, and transport of aerosols measured in spring 2008 during the aerosol, radiation, and cloud processes affecting Arctic Climate (ARCPAC) Project. *Atmos. Chem. Phys.*, **11**, 2423–2453. doi: 10.5194/acp-11-2423-2011

Browell, E.V., Butler, C.F., Fenn, M.A., Grant, W.B., Ismail, S., Schoeberl, M., Toon, O.B., Loewenstein, M., and Podolske, J. (1993) Ozone and aerosol changes during the 1991-1992 Airborne Arctic Stratospheric Expedition. *Science*, **261** (5125), 1155–1158. doi: 10.1126/science.261.5125.1155

Campanelli, M., Estellés, V., Tomasi, C., Nakajima, T., Malvestuto, V., and Martìnez-Lozano, J.A. (2007) Application of the SKYRAD Improved Langley plot method for the in situ calibration of CIMEL Sun-sky photometers. *Appl. Opt.*, **46** (14), 2688–2702. doi: 10.1364/AO.46.002688

Cede, A., Herman, J.R., Richter, A., Krotkov, N., and Burrows, J.P. (2006) Measurements of nitrogen dioxide total column amounts using a Brewer double spectrophotometer in direct Sun mode. *J. Geophys. Res.*, **111** (D5). doi: 10.1029/2005JD006585

Celarier, E.A., Brinksma, E.J., Gleason, J.F., Veefkind, J.P., Cede, A., Herman, J.R., Ionov, D., Goutail, F., Pommereau, J.-P., Lambert, J.-C., van Roozendael, M., Pinardi, G., Wittrock, F., Schönhardt, A., Richter, A., Ibrahim, O.W., Wagner, T., Bojkov, B., Mount, G., Spinei, E., Chen, C.M., Pongetti, T.J., Sander, S.P., Bucsela, E.J., Wenig, M.O., Swart, D.P.J., Volten, H., Kroon, M., and Levelt, P.F. (2008) Validation of Ozone Monitoring Instrument nitrogen dioxide columns. *J. Geophys. Res.*, **113**, D15S15. doi: 10.1029/2007JD008908

Chu, W.P. and McCormick, M.P. (1979) Inversion of stratospheric aerosol and gaseous constituents from spacecraft solar extinction data in the 0.38-1.0-μm wavelength region. *Appl. Opt.*, **18** (9), 1404–1413. doi: 10.1364/AO.18.001404

Chu, W.P., McCormick, M.P., Lenoble, J., Brognoiz, C., and Pruvost, P. (1989) SAGE II inversion algorithm. *J. Geophys. Res.*, **94** (D6), 8339–8351. doi: 10.1029/JD094iD06p08339

Collins, D.R., Jonsson, H.H., Seinfeld, J.H., Flagan, R.C., Gassó, S., Hegg, D.A., Russell, P.B., Schmid, B., Livingston, J.M., Öström, E., Noone, K.J., Russell, L.M., and Putaud, J.P. (2000) In situ aerosol-size distributions and clear-column radiative closure during ACE-2. *Tellus B*, **52** (2), 498–525. doi: 10.1034/j.1600-0889.2000.00008.x

Collis, R.T.H. and Russell, P.B. (1976) Lidar measurement of particles and gases by elastic backscattering and differential absorption, in *Laser Monitoring of the Atmosphere* (ed. E.D. Hinkley), Springer-Verlag, New York, pp. 71–151.

de Graaf, M. and Stammes, P. (2005) SCIAMACHY Absorbing Aerosol Index – calibration issues and global results from 2002–2004. *Atmos. Chem. Phys.*, **5**, 2385–2394, SRef-ID: 1680-7324/acp/2005-5-2385.

Diner, D.J., Abdou, W.A., Bruegge, C.J., Conel, J.E., Crean, K.A., Gaitley, B.J., Helmlinger, M.C., Kahn, R.A., Martonchik, J.V., Pilorz, S.H., and Holben, B.N. (2001) MISR aerosol optical depth retrievals over southern Africa during the SAFARI-2000 dry season campaign. *Geophys. Res. Lett.*, **28** (16), 3127–3130. doi: 10.1029/2001GL013188

Diner, D.J., Beckert, J.C., Reilly, T.H., Bruegge, C.J., Conel, J.E., Kahn, R.A., Martonchik, J.V., Ackerman, T.P., Davies,

R., Gerstl, S.A.W., Gordon, H.R., Muller, J.-P., Myneni, R.B., Sellers, P.J., Pinty, B., and Vertraete, M.M. (1998) Multi-angle Imaging SpectroRadiometer (MISR) instrument description and experiment overview. *IEEE Trans. Geosci. Remote Sens.*, **36** (4), 1072–1087.

Di Nicolantonio, W., Cacciari, A., and Tomasi, C. (2009) Particulate matter at surface: Northern Italy monitoring based on satellite remote sensing, meteorological fields, and in-situ samplings. *IEEE J. Sel. Top. Appl. Earth Observ. Remote Sens.*, **2** (4), 284–292. doi: 10.1109/JSTARS.2009.2033948

Dobson, G.M.B. (1931) A photoelectric spectrophotometer for measuring atmospheric ozone. *Proc. Phys. Soc.*, **43**, 324–328.

Dubovik, O. and King, M.D. (2000) A flexible inversion algorithm for the retrieval of aerosol optical properties from Sun and sky radiance measurements. *J. Geophys. Res.*, **105** (D16), 20673–20696. doi: 10.1029/2000JD900282

Eck, T.F., Holben, B.N., Reid, J.S., Dubovik, O., Smirnov, A., O'Neill, T., Slutsker, I., and Kinne, S. (1999) Wavelength dependence of the optical depth of biomass burning, urban, and desert dust aerosols. *J. Geophys. Res.*, **104** (D24), 31333–31349. doi: 10.1029/1999JD200923

El-ossta, E., Qahwaji, R., and Ipson, S.S. (2013) Detection of dust storms using MODIS reflective and emissive bands. *IEEE J. Sel. Top. Appl. Earth Observ. Remote Sens.*, **6** (6), 2480–2485. doi: 10.1109/JSTARS.2013.2248131

Elterman, L. (1968) UV, Visible, and IR Attenuation for Altitudes to 50 km, 1968. Environmental Research Papers, No. 285, AFCRL-68-0153, April 1968, Air Force Cambridge Research Laboratories, L. G. Hanscom Field, Bedford, MA.

Emeis, S., Schäfer, K., and Münkel, C. (2009) Observation of the structure of the urban boundary layer with different ceilometers and validation by RASS data. *Meteorol. Z.*, **18** (2), 149–154. doi: 10.1127/0941-2948/2009/0365

Eresmaa, N., Karppinen, A., Joffre, S.M., Räsänen, J., and Talvitie, H. (2006) Mixing height determination by ceilometer. *Atmos. Chem. Phys.*, **6** (6), 1485–1493. doi: 10.5194/acp-6-1485-2006

European Space Agency (ESA) (2004) Reports for Mission Selection, The Six Candidate Earth Explorer Missions: EarthCARE, ESA SP-1279(1).

Ferrero, L., Perrone, M.G., Petraccone, S., Sangiorgi, G., Ferrini, B.S., Lo Porto, C., Lazzati, Z., Cocchi, D., Bruno, F., Greco, F., Riccio, A., and Bolzacchini, E. (2010) Vertically-resolved particle size distribution within and above the mixing layer over the Milan metropolitan area. *Atmos. Chem. Phys.*, **10** (8), 3915–3932. doi: 10.5194/acp-10-3915-2010

Fowle, F.E. (1913) The non-selective transmissibility of radiation through dry and moist air. *Astrophys. J.*, **37**, 392–406.

Franke, K., Ansmann, A., Müller, D., Althausen, D., Wagner, F., and Scheele, R. (2001) One-year observations of particle lidar ratio over the tropical Indian Ocean with Raman lidar. *Geophys. Res. Lett.*, **28** (24), 4559–4562. doi: 10.1029/2001GL013671

Froidevaux, L. and Douglass, A (2001) *Earth Observing System (EOS) Aura Science Data Validation Plan*, (see https://aura.gsfc.nasa.gov/images/aura_validation_v1.0.pdf)

Fu, Q. and Liou, K.N. (1992) On the correlated k-distribution method for radiative transfer in nonhomogeneous atmospheres. *J. Atmos. Sci.*, **49** (22), 2139–2156. doi: 10.1175/1520-0469(1992)049<2139:OTCDMF>2.0.CO;2

Gordon, H.R. and Wang, M. (1994) Retrieval of water-leaving radiance and aerosol optical thickness over the oceans with SeaWiFS: a preliminary algorithm. *Appl. Opt.*, **33** (3), 443–452. doi: 10.1364/AO.33.000443

Grey, W.M.F., North, P.R.J., and Los, S.O. (2006) Computationally efficient method for retrieving aerosol optical depth from ATSR-2 and AATSR data. *Appl. Opt.*, **45** (12), 2786–2795. doi: 10.1364/AO.45.002786

Griggs, M. (1975) Measurements of atmospheric aerosol optical thickness over water using ERTS-1 data. *J. Air Pollut. Control Assoc.*, **25**, 622–626.

Gu, Y.Y., Gardner, C.S., Castleberg, P.A., Papen, G.C., and Kelley, M.C. (1997) Validation of the lidar in-space technology experiment: stratospheric temperature and aerosol measurements. *Appl. Opt.*, **36** (21), 5148–5157. doi: 10.1364/AO.36.005148

Gunter, R.L., Hansen, A.D.A., Boatman, J.F., Bodhaine, B.A., Schnell, R.C., and Garvey, O.M. (1993) Airborne measurements of aerosol optical properties over South-central New Mexico. *Atmos. Environ. Part A*, **27** (8), 1363–1368. doi: 10.1016/0960-1686(93)90262-W.

Hall, T.C. Jr., and Blacet, F.E. (1952) Separation of the absorption spectra of ~ 5 s and λ in the range of 2400 – 5000 Å. *J. Chem. Phys.*, **20** (11), 1745–1749.

Han, H.-J. and Sohn, B.J. (2013) Retrieving Asian dust AOT and height from hyperspectral sounder measurements: an artificial neural network approach. *J. Geophys. Res. Atmos.*, **118**, 837–845. doi: 10.1002/jgrd.50170

Hegg, D.A., Ferek, R.J., and Hobbs, P.V. (1993) Light scattering and cloud condensation nucleus activity of sulfate aerosol measured over the northeast Atlantic Ocean. *J. Geophys. Res.*, **98** (D8), 14887–14894. doi: 10.1029/93JD01615

Hegg, D.A., Hobbs, P.V., Ferek, R.J., and Waggoner, A.P. (1995) Measurements of some aerosol properties relevant to radiative forcing on the east coast of the United States. *J. Appl. Meteorol.*, **34** (10), 2306–2315. doi: 10.1175/1520-0450(1995)034<2306:MOSAPR>2.0.CO;2

Herman, J.R., Bhartia, P.K., Torres, O., Hsu, C., Seftor, C., and Celarier, E. (1997) Global distribution of UV-absorbing aerosols from Nimbus 7/TOMS data. *J. Geophys. Res.*, **102** (D14), 16911–16922. doi: 10.1029/96JD03680

Hervig, M.-E., Deshler, T., and Russell, J.M. III, (1998) Aerosol size distributions obtained from HALOE spectral extinction measurements. *J. Geophys. Res.*, **103** (D1), 1573–1583. doi: 10.1029/97JD03081

Hignett, P., Taylor, J.P., Francis, P.N., and Glew, M.D. (1999) Comparison of observed and modeled direct aerosol forcing during TARFOX. *J. Geophys. Res.*, **104** (D2), 2279–2287. doi: 10.1029/98JD02021

Hobbs, P.V. (1999) An overview of the University of Washington airborne measurements and results from the Tropospheric Aerosol Radiative Forcing Observational Experiment (TARFOX). *J. Geophys. Res.*, **104** (D2), 2233–2238. doi: 10.1029/98JD02283

Hobbs, P.V., Radke, L.F., Lyons, J.H., Ferek, R.J., and Coffman, D.J. (1991) Airborne measurements of particle and gas emissions from the 1990 volcanic eruptions of Mount Redoubt. *J. Geophys. Res.*, **96** (D10), 18735–18752. doi: 10.1029/91JD01635

Hoffmann, A., Osterloh, L., Stone, R., Lampert, A., Ritter, C., Stock, M., Tunved, P., Hennig, T., Böckmann, C., Li, S.-M., Eleftheriadis, K., Maturilli, M., Orgis, T., Herber, A., Neuber, R., and Dethloff, K. (2012) Remote sensing and in-situ measurements of tropospheric aerosol, a PAMARCMiP case study. *Atmos. Environ.*, **52 C**, 56–66. doi: 10.1016/j.atmosenv.2011.11.027

Holben, B.N., Eck, T.F., Slutsker, I., Tanré, D., Buis, J.P., Setzer, A., Vermote, E., Reagan, J.A., Kaufman, Y.J., Nakajima, T., Lavenu, F., Jankowiak, I., and Smirnov, A. (1998) AERONET – A federated instrument network and data archive for aerosol characterization. *Remote Sens. Environ.*, **66** (1), 1–16. doi: S0034-4257(98)00031-5

Ichoku, C., Chu, D.A., Mattoo, S., Kaufman, Y.J., Remer, L.A., Tanré, D., Slutsker, I. and Holben, B.N. (2002). A spatio-temporal approach for global validation and analysis of MODIS aerosol products. *Geophys. Res. Lett.*, **29** (12), MOD1-1 – MOD1-4, doi: 10.1029/2001GL013206

Ignatov, A. and Stowe, L. (2000) Physical basis, premises, and self-consistency checks of aerosol retrievals from TRMM VIRS. *J. Appl. Meteorol.*, **39** (12), 2259–2277. doi: 10.1175/1520-0450(2001)040<2259:PBPASC>2.0.CO;2

Inn, E.C.Y. and Tanaka, Y. (1953) Absorption coefficient of ozone in the ultraviolet and visible regions. *J. Opt. Soc. Am.*, **43** (10), 870–873.

Ionov, D.V., Timofeyev, Y.M., Sinyakov, V.P., Semenov, V.K., Goutail, F., Pommereau, J.-P., Bucsela, E.J., Celarier, E.A., and Kroon, M. (2008) Ground-based validation of EOS-Aura OMI 50–700 nm vertical column data in the midlatitude mountain ranges of Tien Shan (Kyrgyzstan) and Alps (France). *J. Geophys. Res.*, **113**, D15S08. doi: 10.1029/2007JD008659

Iqbal, M. (1983) *An Introduction to Solar Radiation*, Academic Press, Toronto, p. 390.

Jacob, D.J., Crawford, J.H., Maring, H., Clarke, A.D., Dibb, J.E., Emmons, L.K., Ferrare, R.A., Hostetler, C.A., Russell, P.B., Singh, H.B., Thompson, A.M., Shaw, G.E., McCauley, E., Pederson, J.R., and Fisher, J.A. (2010) The Arctic Research of the Composition of the Troposphere from Aircraft and Satellites (ARCTAS) mission: design, execution, and first results. *Atmos. Chem. Phys.*, **10**, 5191–5212. doi: 10.5194/acp-10-5191-2010

Kahn, R.A., Gaitley, B.J., Martonchik, J.V., Diner, D.J., Crean, K.A., and Holben, B. (2005) Multiangle Imaging Spectroradiometer (MISR) global aerosol optical depth validation based on 2 years of coincident Aerosol Robotic Network (AERONET) observations. *J. Geophys. Res.*, **110**, D10S04. doi: 10.1029/2004JD004706

Kasten, F. (1966) A new table and approximation formula for the relative optical air mass. *Arch. Meteorol. Geophys. Bioklimatol., Ser. B*, **14** (2), 206–223.

Kasten, F. and Young, A.T. (1989) Revised optical air mass tables and approximation formula. *Appl. Opt.*, **28** (22), 4735–4738. doi: 10.1364/AO.28.004735

Kim, Y.J., Boatman, J.F., Gunter, R.L., Wellman, D.L., and Wilkison, S.W. (1993) Vertical distribution of atmospheric aerosol size distribution over South-central New Mexico. *Atmos. Environ.*, **27A** (8), 1351–1362. doi: 10.1016/0960-1686(93)90261-V

Klett, J. (1981) Stable analytical inversion solution for processing lidar returns. *Appl. Opt.*, **20** (2), 211–220. doi: 10.1364/AO.20.000211

Klett, J.D. (1985) Lidar inversion with variable backscatter/extinction ratios. *Appl. Opt.*, **24** (11), 1638–1643. doi: 10.1364/AO.24.001638

Knapp, K.R., Vonder Harr, T.H., and Kaufman, Y.J. (2002) Aerosol optical depth retrieval from GOES-8: uncertainty study and retrieval validation over South America. *J. Geophys. Res.*, **107** (D7), 4055. doi: 10.1029/2001JD000505

Kneizys, F.X., Abreu, L.W., Anderson, G.P., Chetwind, J.H., Shettle, E.P., Berk, A., Bernstein, L.S., Robertson, D.C., Acharya, P., Rothman, L.S., Selby, J.E.A., Gallery, W.O., and Clough, S.A. (1996) in *The MODTRAN 2/3 Report and LOW-TRAN 7 Model* (eds L.W. Abreu and G.P. Anderson), Contract F19628-91-C-0132, Phillips Laboratory, Geophysics Directorate, PL/GPOS, Hanscom AFB, MA, p. 261.

Kokhanovsky, A.A. and de Leeuw, G. (eds) (2009) *Satellite Aerosol Remote Sensing Over Land*, Springer-Praxis, Berlin, p. 388.

Kokhanovsky, A.A., Breon, F.M., Cacciari, A., Carboni, E., Diner, D., Di Nicolantonio, W., Grainger, R.G., Grey, W.M.F., Höller, R., Lee, K.-H., Li, Z., North, P.R.J., Sayer, A.M., Thomas, G.E., and von Hoyningen-Huene, W. (2007) Aerosol remote sensing over land: a comparison of satellite retrievals using different algorithms and instruments. *Atmos. Res.*, **85** (3-4), 372–394. doi: 10.1016/j.atmosres.2007.02.008

Lee, K.H. (2014) 3-D perspectives of atmospheric aerosol optical properties over northeast Asia using LIDAR onboard the CALIPSO satellite. *Korean J. Remote Sens.*, **30** (5), 559–570. doi: 10.7780/kjrs.2014.30.5.2

Lee, J., Kim, J., Song, C.H., Ryu, J.-H., Ahn, Y.-H., and Song, C.K. (2010) Algorithm for retrieval of aerosol optical properties over the ocean from the Geostationary Ocean Color Imager. *Remote Sens. Environ.*, **114** (5), 1077–1088. doi: 10.1016/j.rse.2009.12.021

Lee, K.H., Lee, K.-T., and Chung, S.R. (2016) Time-resolved observation of volcanic ash using COMS/MI: a case study from the 2011 Shinmoedake eruption. *Remote Sens. Environ.*, **173**, 122–132. doi: 10.1016/j.rse.2015.11.014

Lee, K.H., Ryu, J.H., Ahn, J.H., and Kim, Y.J. (2012) First retrieval of data regarding spatial distribution of Asian dust aerosol from the Geostationary Ocean Color Imager. *Ocean Sci. J.*, **47** (4), 465–472. doi: 10.1007/s12601-012-0042-2

Li, Z., Khananian, A., Fraser, R., and Cihlar, J. (2001) Automatic detection of fire smoke using artificial neural networks and threshold approaches applied to AVHRR imagery. *IEEE Trans. Geosci. Remote Sens.*, **39** (9), 1859–1870.

Li, Z., Zhao, X., Kahn, R., Mishchenko, M., Remer, L., Lee, K.-H., Wang, M., Laszlo, I., Nakajima, T., and Maring, H. (2009) Uncertainties in satellite remote sensing of aerosols and impact on monitoring its long-term trend: a review and perspective. *Ann. Geophys.*, **27** (7), 2755–2770. doi: 10.5194/angeo-27-2755-2009

Livingston, J.M., Kapustin, V.N., Schmid, B., Russell, P.B., Quinn, P.K., Bates, T.S., Durkee, P.A., Smith, P.J., Freudenthaler, V., Wiegner, M., Covert, D.S., Gassó, S., Hegg, D., Collins, D.R., Flagan, R.C., Seinfeld, J.H., Vitale, V., and Tomasi, C. (2000) Shipboard sunphotometer measurements of aerosol optical depth spectra and columnar water vapor during ACE-2, and comparison with selected land, ship, aircraft, and satellite measurements. *Tellus B*, **52** (2), 594–619. doi: 10.1034/j.1600-0889.2000.00045.x

Livingston, J.M., Russell, P.B., Reid, J.S., Redemann, J., Schmid, B., Allen, D.A., Torres, O., Levy, R.C., Remer, L.A., Holben, B.N., Smirnov, A., Dubovik, O., Welton, E.J., Campbell, J.R., Wang, J., and Christopher, S.A. (2003) Airborne Sun photometer measurements of aerosol optical depth and columnar water vapor during the Puerto Rico Dust Experiment and comparison with land, aircraft, and satellite measurements. *J. Geophys. Res.*, **108** (D19), 8588. doi: 10.1029/2002JD02520

McCormick, M.P., Hamill, P., Chu, W.P., Swissler, T.J., McMaster, L.R., and Pepin, T.J. (1979) Satellite studies of the stratospheric aerosol. *Bull. Am. Meteorol. Soc.*, **60** (9), 1038–1046. doi: 10.1175/1520-0477(1979)060<1038:SSOTSA>2.0.CO;2

McNaughton, C.S., Clarke, A.D., Freitag, S., Kapustin, V.N., Kondo, Y., Moteki, N., Sahu, L., Takegawa, N., Schwarz, J.P., Spackman, J.R., Watts, L., Diskin, G., Podolske, J., Holloway, J.S., Wisthaler, A., Mikoviny, T., de Gouw, J., Warneke, C., Jimenez, J., Cubison, M., Howell, A., Middlebrook, G., Bahreini, R., Anderson, B.E., Winstead, E., Thornhill, K.L., Lack, D., Cozic, J., and Brock, C.A. (2011). Absorbing aerosol in the troposphere of the Western Arctic during the 2008 ARCTAS/ARCPAC airborne field campaigns. *Atmos. Chem. Phys.*, **11**, 7561–7582. doi: 10.5194/acp-11-7561-2011

Marbach, T., Phillips, P., Lacan, A., and Schlüssel, P. (2013) The Multi-Viewing, -Channel, -Polarisation Imager (3MI) of the EUMETSAT Polar System - Second Generation (EPS-SG) dedicated to aerosol characterisation. Proceedings SPIE 8889, Sensors, Systems, and Next-Generation Satellites XVII, 88890I, October 16, 2013, doi: 10.1117/12.2028221.

Marbach, T., Riedi, J., Lacan, A., and Schlüssel, P. (2015) The 3MI mission: multi-viewing-channel-polarisation imager of the EUMETSAT polar system: second generation (EPS-SG) dedicated to aerosol and cloud monitoring. Proceedings SPIE 9613, Polarization Science and Remote Sensing VII, 961310, September 1, 2015, doi: 10.1117/12.2186978.

Matsumoto, T., Russell, P.B., Mina, C., Van Ark, W., and Banta, V. (1987) Airborne tracking sunphotometer. *J. Amos. Oceanic Technol.*, **4** (2), 336–339. doi: 10.1175/1520-0426(1987)004<0336:ATS>2.0.CO;2

Mattis, I., Ansmann, A., Müller, D., Wandinger, U., and Althausen, D. (2002) Dual-wavelength Raman lidar observations of the extinction-to-backscatter ratio of Saharan dust. *Geophys. Res. Lett.*, **29** (9, 1306,), 20-1–20-4. doi: 10.1029/2002GL014721

Mazzola, M., Stone, R.S., Herber, A., Tomasi, C., Lupi, A., Vitale, V., Toledano, C., Cachorro, V.E., Torres, B., Berjon, A., Ortiz, J.P., O'Neill, N.T., Masataka, S., Stebel, K., Aaltonen, V., Zielinski, T., Petelski, T., Goloub, P., Blarel, L., Li, Z., Abboud, I., Cuevas, E., Stock, M., Schulz, K.-H., and Virkkula, A. (2012) Evaluation of Sun photometer capabilities for the retrievals of aerosol optical depth at high latitudes: the POLAR-AOD intercomparison campaigns. *Atmos. Environ.*, **52 C**, 4–17. doi: 10.1016/j.atmosenv.2011.07.042

Measures, R.M. (ed.) (1992) *Laser Remote Sensing Fundamentals and Applications*, Krieger Publisher Company, Malabar, FL, p. 524.

Michalsky, J., Beauharnois, M., Berndt, J., Harrison, L., Kiedron, P., and Min, Q. (1999) 100–600 nm absorption band identification based on optical depth spectra of the visible and near-infrared. *Geophys. Res. Lett.*, **26** (11), 1581–1584. doi: 10.1029/1999GL900267

Mie, G. (1908) Beiträge zur Optik trüber Medien, speziell kolloidaler Metallösungen. *Ann. Phys., Vierte Folge*, **25** (3), 377–445.

Mishchenko, M.I., Geogdzhayev, I.V., Cairns, B., Rossow, W.B., and Lacis, A.A. (1999) Aerosol retrievals over the ocean by use of channels 1 and 2 AVHRR data: sensitivity analysis and preliminary results. *Appl. Opt.*, **38** (36), 7325–7341. doi: 10.1364/AO.38.007325

Murakami, H., Sasaoka, K., Hosoda, K., Fukushima, H., Toarani, M., Frouin, R., Mitchell, B.G., Kahru, M., Deschamps, P.Y., Clark, D., Flora, S., Kishino, M., Saitoh, S., Asanuma, I., Tanaka, A., Sasaki, H., Yokouchi, K., Kiyomoto, Y., Saito, H., Dupouy, C., Siripong, A., Matsumura, S., and Ishizaka, J. (2006) Validation of ADEOS-II GLI ocean color products using in-situ observations. *J. Oceanogr.*, **62** (3), 373–393. doi: 10.1007/s10872-006-0062-6

Murayama, T., Sugimoto, N., Uno, I., Kinoshita, K., Aoki, K., Hagiwara, N., Liu, Z., Matsui, I., Sakai, T., Shibata, T., Arao, K., Sohn, B.-J., Won, J.-G., Yoon, S.-C., Li, T., Zhou, J., Hu, H., Abo, M., Iokibe, K., Koga, R., and Iwasaka, Y. (2001) Ground-based network observation of Asian dust events of April 1998 in east Asia. *J. Geophys. Res.*, **106** (D16), 18345–18359. doi: 10.1029/2000JD900554

Nakajima, T., Sekiguchi, M., Takemura, T., Uno, I., Higurashi, A., Kim, D., Sohn, B.J., Oh, S.-N., Nakajima, T.Y., Ohta, S., Okada, I., Takemura, T., and Kawamoto, K. (2003) Significance of direct and indirect radiative forcings of aerosols in the East China Sea region. *J. Geophys. Res.*, **108** (D23), 8658. doi: 10.1029/2002JD003261

Nakajima, T., Tanaka, M., and Yamauchi, T. (1983) Retrieval of the optical properties of aerosols from aureole and extinction data. *Appl. Opt.*, **22** (19), 2951–2959. doi: 10.1364/AO.22.002951

Nakajima, T., Tonna, G., Rao, R., Boi, P., Kaufman, Y., and Holben, B.N. (1996) Use of sky brightness measurements from ground for remote sensing of particulate polydispersions. *Appl. Opt.*, **35** (15), 2672–2686.

Okada, Y., Mukai, S., and Sano, I. (2001) Neural network approach for aerosol retrieval. International Geoscience and Remote Sensing Symposium, IEEE/IGARSS, Vol. **4**, pp. 1716–1718.

Omar, A.H., Winker, D.M., Tackett, J.L., Giles, D.M., Kar, J., Liu, Z., Vaughan, M.A., Powell, K.A., and Trepte, C.R. (2013) CALIOP and AERONET aerosol optical depth comparisons: one size fits none. *J. Geophys. Res. Atmos.*, **118**, 4748–4766. doi: 10.1002/jgrd.50330

O'Neill, N.T., Dubovik, O., and Eck, T.F. (2001) Modified Ångström exponent for the characterization of submicrometer aerosol. *Appl. Opt.*, **40** (15), 2368–2375. doi: 10.1364/AO.40.002368

Palm, S., Hart, W., Hlavka, D., Welton, E.J., Mahesh, A., and Spinhirne, J. (2002) *GLAS Atmospheric Data Products. GLAS Algorithm Theoretical Basis Document (ATBD). Version 4.2*, Science Systems and Applications, Inc, Lanham, MD.

Popp, C., Hauser, A., Foppa, N., and Wunderle, S. (2007) Remote sensing of aerosol optical depth over central Europe from MSG-SEVIRI data and accuracy assessment with ground-based AERONET measurements. *J. Geophys. Res.*, **112**, D24S11. doi: 10.1029/2007JD008423

Raes, F., Bates, T., McGovern, F., and Van Liedekerke, M. (2000) The 2nd Aerosol Characterization Experiment (ACE-2): general overview and main results. *Tellus B*, **52** (2), 111–125. doi: 10.1034/j.1600-0889.2000.00124.x

Randall, C.E., Bevilacqua, R.M., Lumpe, J.D., and Hoppel, K.W. (2001) Validation of POAM III aerosols: comparison to SAGE II and HALOE. *J. Geophys. Res.*, **106** (D21), 27525–27536. doi: 10.1029/2001JD000528

Randall, C.E., Rusch, D.W., Olivero, J.J., Bevilacqua, R.M., Poole, L.R., Lumpe, J.D., Fromm, M.D., Hoppel, K.W., Hornstein, J.S., and Shettle, E.P. (1996) An overview of POAM II aerosol measurements at 1.06 μm. *Geophys. Res. Lett.*, **23** (22), 3195–3198. doi: 10.1029/96GL02921

Redemann, J., Turco, R.P., Liou, K.N., Russell, P.B., Bergstrom, R.W., Schmid, B., Livingston, J.M., Hobbs, P.V., Hartley, W.S., Ismail, S., Ferrare, R.A., and Browell,

E.V. (2000a) Retrieving the vertical structure of the effective aerosol complex index of refraction from a combination of aerosol in situ and remote sensing measurements during TARFOX. *J. Geophys. Res.*, **105** (D8), 9949–9970. doi: 10.1029/1999JD901044

Redemann, J., Turco, R.P., Liou, K.N., Hobbs, P.V., Hartley, W.S., Bergstrom, R.W., Browell, E.V., and Russell, P.B. (2000b) Case studies of the vertical structure of the direct shortwave aerosol radiative forcing during TARFOX. *J. Geophys. Res.*, **105** (D8), 9971–9979. doi: 10.1029/1999JD901042

Reid, J.S., Kinney, J.E., Wesphal, D.L., Holben, B.N., Welton, E.J., Tsay, S.-C., Eleuterio, D.P., Campbell, J.R., Christopher, S.A., Colarco, P.R., Jonsson, H.H., Livingston, J.M., Maring, H.B., Meier, M.L., Pilewskie, P., Prospero, J.M., Reid, E.A., Remer, L.A., Russell, P.B., Savoie, D.L., Smirnov, A., and Tanré, D. (2003) Analysis of measurements of Saharan dust by airborne and ground based remote sensing methods during the Puerto Rico Dust Experiment (PRIDE). *J. Geophys. Res.*, **108** (D19), 8586. doi: 10.1029/2002JD002493

Remer, L.A., Gassó, S., Hegg, D.A., Kaufman, Y.J., and Holben, B.N. (1997) Urban/industrial aerosol: ground-based Sun/sky radiometer and airborne in-situ measurements. *J. Geophys. Res.*, **102** (D14), 16849–16859. doi: 10.1029/96JD01932

Remer, L.A., Kaufman, Y.J., Tanré, D., Mattoo, S., Chu, D.A., Martins, J.V., Li, R.-R., Ichoku, C., Levy, R.C., Kleidman, R.G., Eck, T.F., Vermote, E., and Holben, B.N. (2005) The MODIS aerosol algorithm, products and validation. *J. Atmos. Sci.*, **62** (4), 947–973. doi: 10.1175/JAS3385.1

Rivas-Perea, P., Rosiles, J., Murguia, M., and Tilton, J. (2010) Automatic dust storm detection based on supervised classification of multispectral data, in *Soft Computing Recognition Based on Biometrics*, vol. **312** (eds P. Melin, J. Kacprzyk, and W. Pedrycz), Springer-Verlag, Berlin/Heidelberg, pp. 443–454.

Russell, P.B. and Heintzenberg, J. (2000) An overview of the ACE 2 clear sky column closure experiment (CLEARCOLUMN). *Tellus B*, **52** (2), 463–483. doi: 10.1034/j.1600-0889.2000.00013.x

Russell, P.B., Livingston, J.M., Pueschel, R.F., Reagan, J.A., Browell, E.V., Toon, G.C., Newman, P.A., Schoeberl, M.R., Lait, L.R., Pfister, L., Gao, Q., and Herman, B.M. (1993) Post-Pinatubo optical depth spectra vs. latitude and vortex structure: airborne tracking sunphotometer measurements in AASE II. *Geophys. Res. Lett.*, **20** (22), 2571–2574. doi: 10.1029/93GL03006

Russell, P.B., Hobbs, P.V., and Stowe, L.L. (1999a) Aerosol properties and radiative effects in the United States East Coast haze plume: an overview of the Tropospheric Aerosol Radiative Forcing Observational Experiment (TARFOX). *J. Geophys. Res.*, **104** (D2), 2213–2222. doi: 10.1029/1998JD200028

Russell, P.B., Livingston, J.M., Hignett, P., Kinne, S., Wong, J., Chien, A., Bergstrom, R., Durkee, P., and Hobb, P.V. (1999b) Aerosol-induced radiative flux changes off the United States mid-Atlantic coast: comparison of values calculated from sunphotometer and in situ data with those measured by airborne pyranometer. *J. Geophys. Res.*, **104** (D2), 2289–2307. doi: 10.1029/1998JD200025

Schmid, B., Livingston, J.M., Russell, P.B., Durkee, P.A., Jonsson, H.H., Collins, D.R., Flagan, R.C., Seinfeld, J.H., Gassó, S., Hegg, D.A., Öström, E., Noone, K.J., Welton, E.J., Vos, K.J., Gordon, H.R., Formenti, P., and Andreae, M.O. (2000) Clear-sky closure studies of lower tropospheric aerosol and water vapor during ACE-2 using airborne sunphotometer, airborne in-situ, space-borne, and ground-based measurements. *Tellus B*, **52** (2), 568–593. doi: 10.1034/j.1600-0889.2000.00009.x

Shinozuka, Y., Redemann, J., Livingston, J.M., Russell, P.B., Clarke, A.D., Howell, S.G., Freitag, S., O'Neill, N.T., Reid, E.A., Johnson, R., Ramachandran, S., McNaughton, C.S., Kapustin, V.N., Brekhovskikh, V., Holben, B.N., and McArthur, L.J.B. (2011) Airborne observation of aerosol optical depth during ARCTAS: vertical profiles, inter-comparison and fine-mode fraction. *Atmos. Chem.*

Phys., **11** (8), 3673–3688. doi: 10.5194/acp-11-3673-2011

Stohl, A. (2006) Characteristics of atmospheric transport into the Arctic troposphere. *J. Geophys. Res.*, **111**, D11306. doi: 10.1029/2005JD006888

Stone, R.S., Herber, A., Vitale, V., Mazzola, M., Lupi, A., Schnell, R.C., Dutton, E.G., Liu, P.S.K., Li, S.-M., Dethloff, K., Lampert, A., Ritter, C., Stock, M., Neuber, R., and Maturilli, M. (2010) A three-dimensional characterization of Arctic aerosols from airborne Sun photometer observations; PAM-ARCMIP, April 2009. *J. Geophys. Res.*, **115**, D13203. doi: 10.1029/2009JD013605

Stowe, L.L., Ignatov, A.M., and Singh, R.R. (1997) Development, validation, and potential enhancements to the second-generation operational aerosol product at the National Environmental Satellite, Data, and Information Service of the National Oceanic and Atmospheric Administration. *J. Geophys. Res.*, **102** (D14), 16923–16934. doi: 10.1029/96JD02132

Thomason, L.W., Herman, B.M., and Reagan, J.A. (1983) The effect of atmospheric attenuators with structured vertical distributions on air mass determinations and Langley plot analyses. *J. Atmos. Sci.*, **40** (7), 1851–1854. doi: 10.1175/1520-0469(1983)040<1851:TEOAAW>2.0.CO;2

Thomason, L.W., Poole, L.R., and Randall, C.E. (2007) SAGE III aerosol extinction validation in the Arctic winter: comparisons with SAGE II and POAM III. *Atmos. Chem. Phys.*, **7**, 1423–1433. doi: 10.5194/acp-7-1423-2007

Tomasi, C. (1984) Vertical distribution features of atmospheric water vapor in the Mediterranean, Red Sea and Indian Ocean. *J. Geophys. Res.*, **89** (D2), 2563–2566. doi: 10.1029/JDo89iD02p02563

Tomasi, C. and Guzzi, R. (1974) High precision atmospheric hygrometry using the solar infrared spectrum. *J. Phys. E: Sci. Instrum.*, **7** (8), 647–649. doi: 10.1088/0022-3735/7/8/018

Tomasi, C. and Petkov, B.H. (2014) Calculations of relative optical air masses for various aerosol types and minor gases in Arctic and Antarctic atmospheres. *J. Geophys. Res. Atmos.*, **119** (3), 1363–1385. doi: 10.1002/2013JD020600

Tomasi, C., Marani, S., Vitale, V., Wagner, F., Cacciari, A., and Lupi, A. (2000) Precipitable water evaluations from infrared sun-photometric measurements analyzed using the atmospheric hygrometry technique. *Tellus B*, **52** (2), 734–749. doi: 10.1034/j.1600-0889.2000.00032.x

Tomasi, C., Vitale, V., and De Santis, L.V. (1998) Relative optical mass functions for air, water vapour, ozone and nitrogen dioxide in atmospheric models presenting different latitudinal and seasonal conditions. *Meteorol. Atmos. Phys.*, **65** (1-2), 11–30. doi: 10.1007/BF01030266

Tomasi, C., Vitale, V., Lupi, A., Cacciari, A., and Marani, S. (1999) Use of multi-wavelength sun-radiometers for precise ground-based measurements of the aerosol optical thickness. IGARSS 99 Proceedings (IEEE 1999 International Geoscience and Remote Sensing Symposium), Remote Sensing of the System Earth, A Challenge for the 21st Century, Vol. I, Congress Centrum Hamburg (Germany), June 28 - July 2, 1999, pp. 354–358, doi: 10.1109/IGARSS.1999.773496.

Tomasi, C., Vitale, V., Petkov, B., Lupi, A., and Cacciari, A. (2005) Improved algorithm for calculations of Rayleigh-scattering optical depth in standard atmospheres. *Appl. Opt.*, **44** (16), 3320–3341. doi: 10.1364/AO.44.003320

Tomasi, C., Vitale, V., and Tagliazucca, M. (1989) Atmospheric turbidity measurements at Terra Nova Bay during January and February 1988. SIF - Conference Proceedings, Vol. 20, pp. 67–77.

Torres, O., Bhartia, P.K., Herman, J.R., Sinyuk, A., Ginoux, P., and Holben, B. (2002) A long-term record of aerosol optical depth from TOMS observations and comparison to AERONET measurements. *J. Atmos. Sci.*, **59** (3), 398–413. doi: 10.1175/1520-0469(2002)059<0398:ALTROA>2.0.CO;2

Torres, O., Bhartia, P.K., Sinyuk, A., Welton, E.J., and Holben, B. (2007) Total Ozone Mapping Spectrometer measurements of aerosol absorption from space: comparison to SAFARI 2000 ground-based observations. *J. Geophys. Res.*, **110**, D10. doi: 10.1029/2004JD004611

Torricella, F., Cattani, E., Cervino, M., Guzzi, R., and Levoni, C. (1999) Retrieval of aerosol properties over the ocean using Global Ozone Monitoring Experiment measurements: method and applications to test cases. *J. Geophys. Res.*, **104** (D10), 12085–12098. doi: 10.1029/1999JD900040

Vermeulen, A., Devaux, C., Tanré, D., Herman, M., Holben, B., Blarel, L., Chatenet, B., and Pietras, C. (2001) Aeronet polarization measurements. American Geophysical Union, Fall Meeting 2001, abstract #A41B-0024, bibliographic code: 2001AGUFM.A41B0024V.

Vitale, V., Tomasi, C., von Hoyningen-Huene, W., Bonafè, U., Marani, S., Lupi, A., Cacciari, A., and Ruggeri, P. (2000) Spectral measurements of aerosol particle extinction in the 0.4–3.7 μm wavelength range, performed at Sagres with the IR-RAD sun-radiometer. *Tellus B*, **52** (2), 716–733. doi: 10.1034/j.1600-0889.2000.00028.x

Veefkind, J.P., de Leeuw, G., Durkee, P.A., Russell, P.B., Hobbs, P.V., and Livingston, J.M. (1999) Aerosol optical depth retrieval using ATSR-2 and AVHRR data during TARFOX. *J. Geophys. Res.*, **104** (D2), 2253–2260. doi: 10.1029/98JD02816

Vidot, J., Santer, R., and Aznay, O. (2008) Evaluation of the MERIS aerosol product over land with AERONET. *Atmos. Chem. Phys.*, **8**, 7603–7617. doi: 10.5194/acp-8-7603-2008

Volz, F.E. (1974) Economical multispectral sun-photometer for measurements of aerosol extinction from 0.44 μm to 1.6 μm and precipitable water. *Appl. Opt.*, **13** (8), 1732–1733. doi: 10.1364/AO.13.001732

Wandinger, U., Müller, D., Böckmann, C., Althausen, D., Matthias, V., Bösenberg, J., Weiß, V., Fiebig, M., Wendisch, M., Stohl, A., and Ansmann, A. (2002) Optical and microphysical characterization of biomass-burning and industrial-pollution aerosols from multiwavelength lidar and aircraft measurements. *J. Geophys. Res.*, **107** (D21, 8125,), LAC 7-1–LAC 7-4. doi: 10.1029/2000JD000202

Xiao, F., Wong, M.S., Lee, K.H., Campbell, J.R., and Shea, Y. (2015) Retrieval of dust storm aerosols using an integrated Neural Network model. *Comput. Geosci.*, **85**, 104–114. Online publication date: 1 December 2015.

Zasetsky, A.Y. and Sloan, J.J. (2005) Monte Carlo approach to identification of the composition of stratospheric aerosols from infrared solar occultation measurements. *Appl. Opt.*, **44** (22), 4785–4790. doi: 10.1364/AO.44.004785

8
Aerosol and Climate Change: Direct and Indirect Aerosol Effects on Climate

Claudio Tomasi, Christian Lanconelli, Mauro Mazzola, and Angelo Lupi

8.1
Introduction

Airborne aerosols exhibit overall number concentrations usually varying from a few hundreds per cubic centimeter of air in the most remote areas of the Earth to more than $10^4\,cm^{-3}$ in the most polluted urban areas over the size range from 10^{-2} to more than 50 μm. They originate from both primary sources or through secondary processes. Primary aerosols are emitted from both natural and anthropogenic sources. The natural particles mainly consist of mineral dust mobilized in desert and semiarid regions, sea-salt particles generated over the oceans, volcanic dust injected into the low stratosphere by violent eruptions of debris and gases, biogenic aerosols (like viruses, bacterial cells, fungi, and spores from plants and animals), and smokes from spontaneous combustion forest fires. Anthropogenic aerosols mainly consist of industrial dust, soil dust mobilized by strong winds in agricultural areas, and smokes generated through fossil fuel combustion and waste and biomass burning. Secondary aerosols are formed in the atmosphere through chemical (mainly heterogeneous) reactions involving sulfur dioxide, nitrogen oxides, biogenic volatile organic compounds (VOCs), and other chemical species emitted from both natural and anthropogenic activities (Seinfeld and Pandis, 2006).

The aerosol particle sizes are comparable with the incoming solar radiation wavelengths, mainly ranging from 0.3 to 4.0 μm, and in general smaller than those of terrestrial radiation, mainly varying between 4.0 and 25 μm. Therefore, as clearly stated by the Mie (1908) scattering theory, aerosols interact very strongly with the solar (shortwave) radiation and rather weakly with the terrestrial (longwave) radiation. Because of such strong interactions, aerosols induce important scattering and absorption effects on the radiation budget of the land–atmosphere–ocean system, extinguishing and scattering backward the shortwave radiation very efficiently and attenuating more weakly the

Atmospheric Aerosols: Life Cycles and Effects on Air Quality and Climate, First Edition.
Edited by Claudio Tomasi, Sandro Fuzzi, and Alexander Kokhanovsky.

longwave radiation emitted upward by the terrestrial surface and the various atmospheric layers. Consequently, intense direct climatic effects are induced by atmospheric aerosols through scattering and absorption of incident solar radiation, generally causing a marked decrease in the flux density of direct shortwave radiation at the surface and an increase in the solar radiation fraction reflected back to space. At the same time, scattering by aerosols causes an increase in the diffuse solar radiation flux reaching the surface and in the fraction of shortwave radiation scattered upward, therefore contributing to enhance the surface–atmosphere system albedo (Haywood and Boucher, 2000). These radiative effects were evaluated over the last decades by means of both ground-based and *in situ* measurements of the aerosol optical parameters and satellite-borne data analyses (King *et al.,* 1999; Remer and Kaufman, 2006). The absorption of incoming solar radiation by aerosols is particularly marked in cases where particulate matter contains significant fractions of soot substances containing black carbon (BC) and/or elemental carbon (EC) and therefore induces appreciable warming effects in the lower part of the troposphere (Bellouin *et al.,* 2003).

The scattering and absorption of longwave terrestrial radiation by atmospheric aerosols can modify the cooling rate of the atmospheric boundary layer (BL), especially within dense layers of haze particles or desert dust (Zhang and Christopher, 2003). However, aerosols are in general estimated to scatter and absorb rather weakly the infrared (IR) radiation on a global scale, contributing only slightly to enhance the greenhouse effect of the atmosphere, which is mainly generated for cloud-free sky conditions by the thermal radiation absorption due to the numerous bands of H_2O, CO_2, CH_4, N_2O, and CFCs (Hansen, Sato, and Ruedy, 1997).

Together with the aforementioned studies, numerous other field and satellite-based investigations were conducted by the scientific community over the past 40 years (McCormick and Ludwig, 1967; Kaufman, Tanré, and Boucher, 2002), clearly showing that the direct aerosol-induced radiative forcing (hereinafter referred to as DARF) effects are particularly intense in the midlatitude industrial and densely populated regions, where anthropogenic aerosol emissions are particularly strong. Conversely, the DARF effects were estimated to be relatively weak in the remote oceanic areas (Takemura *et al.,* 2005) and in the polar regions (Tomasi *et al.,* 2015a,b), due to the fact that the background number and mass concentrations of particles are usually much lower in these regions than at midlatitudes. The DARF effects induced by airborne aerosols on the radiation budget of the surface–atmosphere system are usually determined by considering only the changes caused in the solar radiation budget over the spectral range from the ultraviolet (UV) to the near infrared (NIR) and neglecting the effects of the atmospheric particulate matter on the incoming and outgoing fluxes of terrestrial radiation. This choice is due to the fact that the aerosol scattering effects on the longwave radiation are less intense than those affecting solar radiation (Mie, 1908). On this matter, Tafuro *et al.* (2007) estimated that the monthly mean values of longwave DARF at the top-of-atmosphere (ToA) level vary between 0.1% and 2% of the shortwave DARF evaluated throughout the year and the longwave DARF at the surface level varies between 10% and 25% of the corresponding shortwave DARF at the BoA level. The net flux of shortwave radiation at the ToA level is given by the difference between the downwelling and upwelling fluxes. Because of the intense scattering and absorption effects produced by aerosol particles, a significant change in the net flux of shortwave radiation is usually measured, which provides the amount of energy available for "governing" the Earth's climate per unit surface and unit time. Since such a net flux variation yields the DARF effects at the ToA level, it causes a cooling (or warming) effect, depending on the negative (or positive) sign of this net flux change (Chylek and Coakley, 1974). Calculations of the DARF effects are made in the present study for various aerosol types.

8.2 The Instantaneous DARF Effects at the ToA and BoA Levels and in the Atmosphere

The perturbation induced by aerosols on the radiation budget of the surface–atmosphere system is evaluated by assuming that it is due to a change in the overall vertical atmospheric content and/or in the optical characteristics of aerosol particles, while all the other atmospheric constituents are kept unperturbed. Therefore, the DARF at the ToA level can be calculated as the change in the net flux of solar radiation of the climate system outside the atmosphere, which causes an energy deficit or a surplus (Hansen, Sato, and Ruedy, 1997). At any given time, the change depends on (i) the optical parameters of airborne aerosols; (ii) the spectral intensity of incoming shortwave radiation $I{\downarrow}(\lambda)$ crossing the atmosphere, which varies as a function of solar zenith angle ϑ_0 throughout the day; and (iii) the angular and spectral characteristics of surface reflectance properties. Similar concepts are also valid in the calculations of the DARF term at the BoA level, due to the perturbation in the shortwave radiation field at the surface caused by aerosol scattering and absorption of incoming solar radiation. Correspondingly, the net flux change in the atmosphere can be calculated as the difference between the DARF at the ToA level and that simultaneously induced at the BoA level (Ramanathan *et al.*, 2001a). Therefore, before evaluating the instantaneous DARF effects at the ToA and BoA levels and within the atmosphere, it is useful to describe the main spectral features of the incoming solar radiation at the ToA and BoA levels in a standard atmosphere.

8.2.1 The Spectral Characteristics of Solar Radiation

Let us consider a beam of the incoming radiation passing through an arbitrarily thin layer of the atmosphere along a specific path crossing the atmosphere. For each kind of gas molecule and particle encountered by the beam, its monochromatic intensity is decreased by the increment

$$dI \downarrow (\lambda) = -I \downarrow (\lambda) K_{ext}(\lambda) N_{tot} \sigma_{ext}\, ds, \tag{8.1}$$

where N_{tot} is the total number of particles per unit volume of air, σ_{ext} is the extinction cross section of each particle, $K_{ext}(\lambda)$ is the extinction efficiency, and ds is the differential path length along the ray path of the incident radiation, where parameter $K_{ext}(\lambda)$ represents the combined effects of scattering and absorption in extinguishing the radiation passing through the atmospheric layer. The product $K_{ext}(\lambda)\, N_{tot}\, \sigma_{ext}$ in Eq. (8.1) is called the extinction cross section. In the case of a gaseous atmospheric constituent, it is sometimes convenient to express the rate of absorption in the form

$$dI \downarrow (\lambda) = -I(\lambda) \rho_a M_j k_j(\lambda) ds \tag{8.2}$$

where ρ_a is the density of the air, M_j is the mass of the jth absorbing gas per unit mass of air, and $k_j(\lambda)$ is the mass absorption coefficient of the jth absorbing gas as a function of wavelength, which has dimensions of surface per unit mass. In both Eqs. (8.1) and (8.2), the products $K_{ext}(\lambda) N_{tot}$ and $\rho_a \mathrm{M_j k_j}(\lambda)$ are the aerosol extinction coefficient and the gaseous absorption coefficient, respectively, both defined per unit length path. The contributions of the various gases and particles are additive in the radiative transfer evaluations, since they are the contributions of scattering and absorption due to aerosols and/or gases in attenuating the incident beam of radiation, that is,

$$K_{ext}(\lambda) = K_{sca}(\lambda) + K_{abs}(\lambda) \tag{8.3}$$

where parameter $K_{ext}(\lambda)$ provided by aerosols includes at any given place and time the various contributions due to the wide variety of shapes and sizes, as well as cloud droplets and ice crystals that may be present in cloudy atmosphere cases. Nonetheless, considering only the cloud-free sky cases, it is useful to consider the case of scattering by a spherical particle of diameter a, for which the extinction, scattering, or absorption efficiency parameters $K_{ext}(\lambda)$, $K_{sca}(\lambda)$, and $K_{abs}(\lambda)$ used in Eq. (8.3) can be prescribed on the basis of theory, as a function of the dimensionless Mie size parameter $x = \pi \mathrm{a}/\lambda$ and the particulate complex index of refraction $m(\lambda)$ at wavelength λ, whose real part $n(\lambda)$ is the ratio of the speed of light in a vacuum to the speed at which light travels when it is passing through the particle and imaginary part $\chi(\lambda)$ gives a measure of the light absorption by particles.

For the scattering of radiation in the visible part of the spectrum, x ranges from much less than 1 for air molecules to ~1 for haze and smoke particles and to ≫1 for raindrops. Aerosol particles with $x \ll 1$ are relatively ineffective at scattering radiation. Within this so-called "Rayleigh scattering" regime, scattering efficiency $K_{sca}(\lambda)$ is roughly proportional to $\lambda^{-R(\lambda)}$ (being λ measured in micrometer), and the scattering is divided evenly between the forward and backward hemispheres. Exponent $R(\lambda)$ assumes best-fit values slowly decreasing with wavelength equal to 4.759 ± 0.039 within the 0.20–0.25 μm spectral range and to 4.091 ± 0.001 within the 0.50–0.55 μm range and varying more slowly at the upper wavelengths until reaching a value of 4.002 within the 3.5–4.0 μm range (Tomasi *et al.*, 2005). This implies that the violet component of the visible sunlight is more intensely scattered than the red component, causing the diffuse light of the sky to appear as blue to our eyes. For values of Mie parameter x comparable to or greater than 1, the scattered radiation is directed mainly into the forward hemisphere. This is the case of aerosol particles, which cause an overall extinction coefficient approximately proportional to $\lambda^{-\alpha}$ over the visible and NIR spectral range, with the Ångström (1964) exponent α mainly ranging from nearly zero to more than 2 for the different size-distribution characteristics of aerosols.

The shortwave radiative energy emitted by the Sun and the longwave energy emitted by the Earth have spectral distribution curves similar to those of two blackbodies having temperatures close to 6000 K and 250 K. Figure 8.1 shows the spectral distribution curve of extraterrestrial solar irradiance $I_{\text{ToA}}\downarrow(\lambda)$ over the overall 0.20–4.00 μm spectral range. The integral of $I_{\text{ToA}}\downarrow(\lambda)$ over the whole wavelength range is equal to the so-called solar constant recently evaluated to be equal 1365 W m^{-2}, which was estimated to be equivalent to the energy emitted

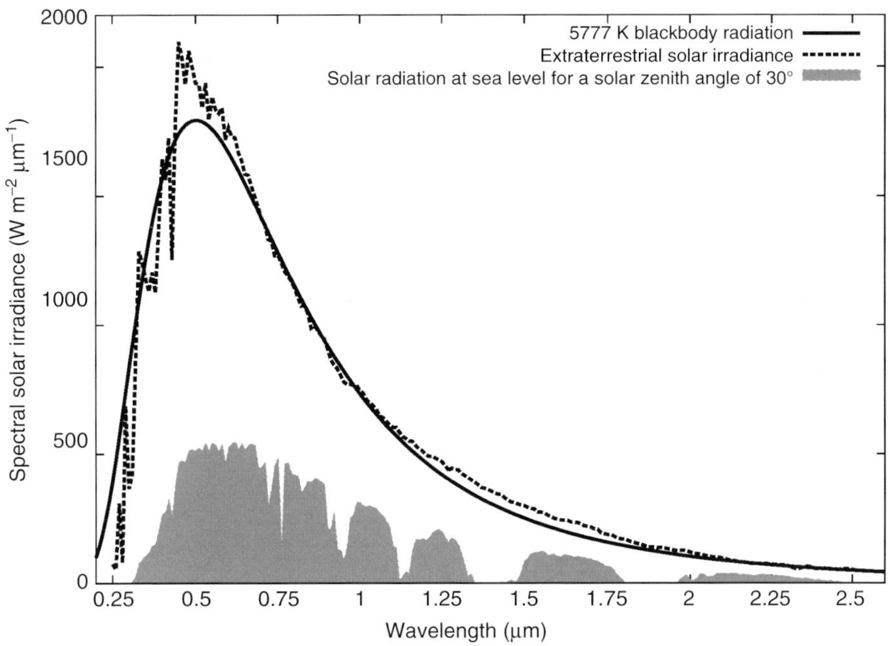

Figure 8.1 Comparison between the spectral curve of a blackbody having an emission temperature of 5777 K for a mean Earth–Sun distance (solid curve) (Fröhlich, 2013) and the spectral curve of solar irradiance $I_{\text{ToA}}(\lambda)$ outside the atmosphere (dashed curve). The gray area represents the sea-level spectrum of solar radiation which has crossed the standard atmosphere for relative optical air mass $\mu = 2$ (i.e., for solar zenith angle $\vartheta_0 = 30°$), which shows a number of absorption bands by water vapor, carbon dioxide, oxygen, and ozone.

by a blackbody having an effective emission temperature $T_e = 5777$ K (Fröhlich, 2013). For such a Sun emission temperature, Wien's displacement law provides a maximum value of the spectral distribution curve of solar radiation flux density located at wavelength $\lambda_{max} \approx 0.48$ μm. Figure 8.1 also shows that about 7% of the extraterrestrial solar radiation flux density $I_{ToA}\downarrow(\lambda)$ belongs to the UV spectral range ($\lambda \leq 0.38$ μm), 45% to the visible spectral range ($0.38 < \lambda \leq 0.76$ μm), 43% to the NIR wavelength range ($0.76 < \lambda \leq 2.2$ μm), and 5% only to the IR range ($2.2 < \lambda \leq 25$ μm) (Iqbal, 1983). The amount of total solar energy received over the entire wavelength range by the unit area and per unit time at the ToA for mean Sun–Earth distance is called *solar constant*, although it is somewhat misleading since the incoming solar irradiance varies with the 11-year solar activity cycle and more generally over centennial time scales with pronounced cycles of 35–40 and 80–90 years. Throughout the year, the extraterrestrial solar irradiance is subject to vary regularly from one day to another because of the Earth–Sun distance changes, leading to a cyclic variation of ±3.34% around the annual mean value.

During its passage through the cloudless atmosphere, the incoming solar radiation is attenuated by Rayleigh scattering, aerosol extinction, and absorption by atmospheric gases:

A. Rayleigh scattering is produced by the interactions between the electromagnetic radiation and air molecules. The intensity of radiant energy scattered by a unit air volume in a certain direction within the unit solid angle is given by the angular function $(1 + \cos^2 \theta)$, where parameter θ is the scattering angle between the direction of incidence and the scattering direction. The shape of the Rayleigh scattering angular diagram consists of two similar forward and backward lobes, indicating that about the half part of scattered solar radiation moves forward and the remaining half part moves backward.
B. Aerosol particles have sizes ranging from a few tens of Ångström (molecular clusters) to several tens of micrometers (cloud and fog droplets). They are generally divided into three principal size classes: (i) small particles with diameters $a < 0.2$ μm, called ultrafine particles or nuclei; (ii) large particles, with a ranging from 0.2 to ~2.5 μm, usually called accumulation particles; and (iii) large particles having diameters $a > 2.5$ μm, commonly called coarse particles. The Mie (1908) electromagnetic theory describes the scattering and absorption of an electromagnetic wave by a spherical particle assuming that the radiation field of the particle is the sum of various fields created by quadrupoles, octupoles, and multipoles of higher orders. It shows that aerosol scattering and absorption of incoming solar radiation are both closely related to the particle size, ratio a/λ, and complex refractive index $m(\lambda)$ of particulate matter, which varies considerably as a function of λ and closely depends on the aerosol physicochemical properties. The angular-distribution function of the scattered intensity usually presents a pronounced forward lobe, a less marked backward lobe, and numerous peaks and troughs on the sides and varies widely as a function of a/λ. An atmospheric particle layer produces the extinction, scattering, and

absorption coefficients $K_{ext}(\lambda)$, $K_{sca}(\lambda)$, and $K_{abs}(\lambda)$ (commonly measured per unit path length), which depend on the shape of the particle size-distribution curve, particle number concentration, and particle refractive index. Therefore, the aforementioned three coefficients vary with altitude, closely depending on the vertical profiles of aerosol concentration. Dense aerosol layers can cause a strong extinction of incoming solar radiation at visible and NIR wavelengths, producing an appreciable increase in the albedo of the Earth–atmosphere system and a strong reduction of the solar radiation flux reaching the Earth's surface, while aerosol absorption of solar radiation contributes in general to reduce the overall albedo and release additional heat in the atmosphere.

C. Atmospheric gases absorb very strongly the solar radiation passing through the atmosphere. In the UV spectral region ($0.20 \leq \lambda \leq 0.38$ μm), the solar radiation is strongly absorbed by molecular oxygen (Schumann–Range and Herzberg continuum, Schumann–Runge bands, Herzberg bands), molecular nitrogen (Lyman–Birge–Hopfield bands), and molecular ozone (Hartley bands and Huggins bands). In the visible, weak absorption bands are due to the molecular oxygen bands centered at $\lambda = 0.6884$ μm and $\lambda = 0.7621$ μm (as shown in Figure 8.1), the ozone Chappuis band over the $0.45 \leq \lambda \leq 0.78$ μm range, some relatively weak absorption bands of water vapor over the $0.54 \leq \lambda \leq 0.74$ μm range, and a semicontinuum together with a nitrogen dioxide absorption band over the $0.25 \leq \lambda \leq 0.58$ μm range. In the IR spectrum, a number of water vapor absorption bands are characterized by intensities gradually increasing with λ (see Figure 8.1). They are in general identified by using by Greek letters: band α occupies the 0.70–0.74 μm wavelength range; band called "0.8 μm" extends from 0.796 to 0.847 μm; bands ρ, σ, and τ cover the wavelength range from 0.870 to 0.990 μm; band Φ occupies the 1.08–1.20 μm spectral range; band Ψ covers the 1.25–1.54 μm range; band Ω covers the 1.69–2.08 μm range; band X extends from 2.27 to 2.99 μm; and band X′ occupies the 2.99–3.57 μm wavelength range.

8.2.2
Vertical Features of Aerosol Volume Extinction Coefficient

The vertical profile of aerosol mass concentration typically shows an exponential decrease with altitude z up to a certain height and a rather constant profile above that altitude (Gras, 1991). The aerosol mass concentration as a function of height $M(z)$ can then be expressed as

$$M(z) = M(0)\exp(-z/H_p), \tag{8.4}$$

where $M(0)$ is the surface concentration of particulate mass and H_p *is* the so-called scale height. Evaluations of H_p were obtained by (Tomasi, 1982) examining a large set of sun-photometer measurements of aerosol optical thickness (AOT) at visible and NIR wavelengths conducted at some sites of the Po Valley (Italy), finding that this parameter ranges mainly from 0.5 to 1.3 km in winter for continental-polluted aerosol loads trapped within the thermal inversion layer near the ground and exceeds 2 km on summer days, due to the intense convective motions favoring the vertical transport of these continental and rural aerosols. Various exponential vertical profiles have been determined by Jaenicke (1993) for several aerosol types, obtaining long-term average values of $H_p = 0.9$ km for the maritime aerosol, $H_p = 0.73$ km for the remote continental aerosol, H_p =2.0 km for desert dust, and $H_p = 30$ km for background polar aerosol. Significant variations of H_p were observed by Jaenicke (1993) in anthropogenic plumes, strong pollution areas, or during nucleation events in the free troposphere (FT). Therefore, aerosol particles can cause significant attenuation effects on the solar radiation flux during its passage though the atmosphere, especially within the ground layer. Because of these extinction effects, the spectral distribution of solar irradiance $I_{\mathrm{BoA}\downarrow}(\lambda)$ reaching the Earth's surface is more or less intensely modified by the continuous spectral extinction by airborne aerosols. When reflected upward by the surface, solar irradiance $I_{\mathrm{BoA}\downarrow}(\lambda)$ assumes a new spectral distributions strongly influenced by the spectral and angular characteristics of surface reflectance, which is subsequently attenuated by the atmosphere before reaching the ToA level and the outer space. The spectral and directional changes produced by the Earth's surface can vary largely as a function of the surface reflectance characteristics. The theoretical aspects associated with the aerosol optical characteristics and the surface reflectance properties are examined in the following two paragraphs.

8.2.3 Aerosol Extinction Models and Optical Characteristics

The aerosol optical parameters can vary considerably from one aerosol type to another since the size-distribution shape parameters and the chemical composition features of particulate matter closely depend on the particle origins and the growth and long-range transport processes involving the particles during their atmospheric life. The characterization of the aerosol size-distribution curves and optical parameters was made here by using the size-distribution curves and the optical characteristics described in Chapter 6 of this book for the following aerosol extinction models:

1. The aerosol models 6S-C (continental), 6S-M (maritime), and 6S-U (urban) defined by Vermote *et al.* (1997a,1997b) and the background desert dust model BDD (Shettle, 1984), the El Chichón stratospheric volcanic aerosol model ESV (King, Harshvardhan, and Arking, 1984), the biomass burning smoke (BBS) model (Remer *et al.*, 1998) (all defined by Tomasi *et al.* (2013)), and the winter continental-polluted aerosol model WPV defined by Tomasi *et al.* (2015a)
2. The four aerosol extinction models defined for background desert dust (BDD) by Shettle (1984), El Chichón stratospheric volcanic (ESV) aerosol by King, Harshvardhan, and Arking (1984), BBS by us on the basis of the Remer *et al.* (1998) data sets, and winter continental-polluted aerosol (PVW) by us for the

size-distribution and composition data sets collected by Carbone *et al.* (2010) over the Po Valley (Italy) in the winter season

3. The eight modified (M-type) aerosol models defined by Tomasi *et al.* (2013) using the 6S basic components to simulate the wet-air (relative humidity RH = 50%) aerosol radiative properties by combining the 6S-M and 6S-C maritime and continental aerosol models
4. The 10 OPAC aerosol models determined by Hess, Koepke, and Schult (1998) for RH = 50% to represent aerosols of different origins
5. The four wet-air (RH = 50%) aerosol models SF (Shettle and Fenn, 1979) for rural, maritime, tropospheric, and urban aerosols
6. The seven additional aerosol models defined by Tomasi *et al.* (2013) to represent two Saharan dust (SD-1 and SD-2) models (Tomasi, Prodi, and Tampieri, 1979), two BBS models based on the Carr (2005) field measurements conducted in the FT and within the atmospheric BL, and three pre- and post-Pinatubo volcanic particle models (PV-1, PV-2, PV-3) observed in the low stratosphere (Pueschel *et al.*, 1993; Tomasi, Vitale, and Tarozzi, 1997)
7. The seven polar aerosol models defined by Tomasi *et al.* (2015a) for the Arctic haze (AH), Arctic summer background (ASB), Asian dust (AD), boreal forest fire smoke (BFFS) particle models, and the Antarctic austral summer coastal (AASC), austral summer Antarctic Plateau (ASAP), and Antarctic austral winter coastal (AAWC) aerosol models.

For all the aforementioned 43 aerosol extinction models, the following optical parameters have been defined in Chapter 6: (i) the complex refractive index $m(\lambda) = n(\lambda) - i\chi(\lambda)$; (ii) the volume extinction, scattering, and absorption coefficients $\beta_{ext}(\lambda)$, $\beta_{sca}(\lambda)$, and $\beta_{abs}(\lambda)$ at 11 selected wavelengths from 0.30 to 3.75 μm; (iii) the spectral single scattering albedo $\omega(\lambda)$ (hereinafter also referred to as SSA), calculated for the various aerosol particle size-distribution curves at the same 11 wavelengths; (iv) the average SSA ω calculated over the 0.30–3.75 μm wavelength range; and the asymmetry factor $g(\lambda)$ determined at the 11 wavelengths selected earlier. Some examples of the chemical composition of fine and accumulation particles are shown in Figure 8.2 for different aerosol types sampled in urban, continental, and remote continental air masses over central Europe and during the AEROCLOUDS field campaigns conducted at four sites of the Po Valley (Italy) during the May–July 2007 and January–March 2008 periods, giving evidence of the marked variations occurred in the mass fractions of the various components, which have led to pronounced changes in the particulate optical characteristics. The examples shown in Figure 8.2 clearly indicate that atmospheric aerosols contain sulfates, nitrates, ammonium ions, organic matter (OM), EC, various crustal species, sea salt (prevailing in maritime particles), metal oxides, hydrogen ions, and liquid water. Substances such as sulfate, ammonium, OM, EC, and certain transition metals are predominantly present in the fine particle fraction. Nitrate can be often present in both fine and coarse particles, the nitrate found in the fine particles being usually produced by the nitric acid/ammonia reaction, leading to the formation of ammonium nitrate, and that found in the coarse particles by

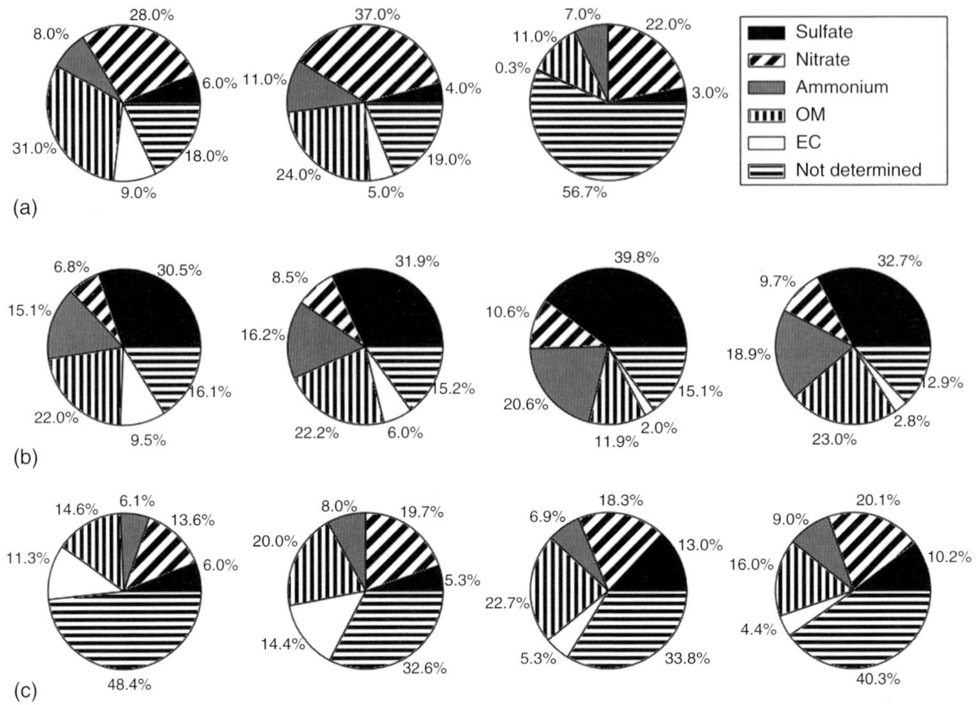

Figure 8.2 Average composition diagrams of particulate matter sampled at various sites of central Europe and northern Italy in different seasonal periods, giving the particle mass concentration of each aerosol component (OM = organic matter, EC = elemental carbon): (a) from left to right, fine aerosol samples (with diameter $a < 2.5\ \mu m$) collected at urban, nonurban continental, and remote sites (Heintzenberg, 1994); (b) fine continental-polluted aerosol ($a < 2.5\ \mu m$) sampled in winter at the Milan (urban), Bologna (urban), San Pietro Capofiume (rural), and Oasi Bine (rural) sites in the Po Valley (northern Italy); and (c) fine continental aerosol ($a < 2.5\ \mu m$) sampled in summer at the Milan (urban), Bologna (urban), San Pietro Capofiume (rural), and Oasi Bine (rural) sites in the Po Valley (northern Italy) (E. Bolzacchini, private communication).

coarse particle/nitric acid reactions. Crustal materials containing silicon, calcium, magnesium, aluminum, and iron and biogenic organic particles (pollen, spores, plant fragments) are usually most sampled in coarse particles. The analyses of samples of typical urban aerosols indicate that sulfate, nitrate, and ammonium ions exhibit in general two modes over the $0.1 - 1.0\ \mu m$ size range, together with a third mode coarse particles: (i) the first mode is generally peaked at mode diameter $a_c \approx 0.2\ \mu m$, being originated by condensation of secondary aerosol components from the gas phase; (ii) the second mode is centered at $a_c \approx 0.7\ \mu m$, its formation being mainly attributed to heterogeneous aqueous-phase reactions; and (iii) the third mode of supermicron particles usually consists of nitrates by more than 50%, and NaCl for a lower percentage, the nitrate being due to reactions of nitric acid with sea salt or crustal matter, constituting an interesting case where secondary aerosol matter (nitrate) is formed through the reaction of a naturally produced material (sea salt or dust) and an anthropogenic pollutant (nitric acid). Relatively low mass fractions of primary carbonaceous particles are in general found in urban and industrial aerosols, mainly produced by combustion (pyrogenic), chemical (commercial products), geologic (fossil fuels), and natural (biogenic) sources.

The mass concentration size-distribution curve of primary particles emitted from boilers, fireplaces, automobiles, diesel trucks, and meat cooking operations is dominated by a mode centered at a_c equal to around 0.1–0.2 μm. Most of the observed mass size distributions are unimodal, and 90% or more of the emitted mass is in the form of submicron particles. In particular, the mass size-distribution curve of EC emitted by automobiles in urban areas is unimodal with a peak around 0.1 μm diameter. The ambient distribution of EC in polluted areas is often bimodal with peaks over the 0.05–0.12 μm (Aitken nuclei) size range and the 0.5–1.0 μm (accumulation) size range. In polluted urban areas the nuclei mode usually dominates, containing almost 75% of the total EC, being the result of the emissions from primary EC sources, and the second mode is mainly due to the accumulation of secondary aerosol products and their subsequent growth by condensation. The aforementioned variations in the aerosol chemical composition characteristics generally cause significant changes in SSA. For instance, a predominant sea-salt particles yields a value of complex refractive index $m(\lambda)$ characterized by very weak absorption and, hence, to spectral values of SSA very close to unit at all the visible and NIR wavelengths and spectral values of SSA very close to unit at all visible and NIR wavelengths, whereas a value of SSA = 0.94 was found for the mixed maritime–continental model M-4 of Tomasi *et al.* (2013) containing only a few percents of soot substances. More generally, (i) rural and continental aerosols were estimated in Chapter 6 to yield values of $\omega(\lambda)$ ranging mainly from 0.80 (model M-9) to 0.95 (model SF-T), depending closely on their relative content of polluted absorbing substances; (ii) desert dust was estimated to give values of $\omega(\lambda)$ varying from ~0.70 (model SD-2) to 0.93 (model BDD); (iii) urban aerosols provide values $\omega(\lambda)$ varying between 0.60 (models 6S-U, SF-U, and M-14) and around 0.75 (models M-11 and OPAC Urban); and (iv) stratospheric volcanic aerosols, consisting of sulfuric acid and liquid water were estimated to give stable values of $\omega(\lambda)$ very close to unit at all the visible and NIR wavelengths (as found for models ESV, PV-1, PV-2, and PV-3).

8.2.4
Modeling the Underlying Surface Reflectance Characteristics

Using the two-stream approximation procedure (Coakley and Chylek, 1975) to simulate the radiative transfer processes occurring in a plane atmosphere containing an aerosol particle layer with well-defined absorptance and reflectance characteristics, Chylek and Coakley (1974) demonstrated that (i) aerosol particles can induce cooling or warming effects in the atmosphere, depending on the surface albedo conditions for any aerosol particle ratio between absorptance and reflectance, and (ii) DARF at the ToA level tends to change sign passing from cooling to warming for all the initial surface albedo characteristics as the absorptance properties of particulate matter increase with respect to reflectance, until exceeding a critical value of the absorptance/reflectance ratio. This change in sign of the DARF effects induced at the ToA level is expected to occur, for instance, when aerosols suspended over an oceanic area with low surface albedo are transported over surfaces presenting higher surface albedo conditions. In order to investigate the dependence features of the DARF effects on the surface reflectance properties, as they take place at both ToA and BoA levels of the atmosphere, a set of 16 surface reflectance models were prepared by Tomasi *et al.* (2013) to improve the representation of the geometrical and spectral characteristics of surface reflectance defined in the 6S code (Vermote *et al.*, 1997a), subdivided into four classes of bidirectional reflectance distribution functions (BRDF) according to Kriebel (1978). Each BDRF function is determined in terms of ratio $f_e = dL\uparrow(\eta_{pz}, \phi)/dF\downarrow(\vartheta_0, \phi_o)$ between the radiance $dL\uparrow$ reflected upward by the surface in the direction individuated by the pair of polar zenith and azimuth angles η_{pz} and ϕ and the incident irradiance $dF\downarrow$ coming from the direction (ϑ_0, ϕ_o). It is measured in per steradian and assumes a constant value equal to $1/\pi$ for an ideal Lambertian reflector. In order to improve these BRDF representations, the concept of bidirectional reflectance factor $R(\lambda, \vartheta_0, \phi_o, \eta_{pz}, \phi)$ was adopted. It represents the ratio between the real BRDF surface reflectance and the BRDF reflectance of an ideal (100%) Lambertian reflector, as a function of wavelength λ and of the four angular coordinates of the Sun-surface-external viewer system (i.e., the nadir angle η_{pz} and azimuth angle ϕ of the observer and solar zenith angle ϑ_0 and azimuth angle ϕ_o of the Sun). Actually, factor $R(\lambda, \vartheta_0, \phi_o, \eta_{pz}, \phi)$ provides the ratio between the upwelling irradiance $F\uparrow(\eta_{pz}, \phi)$ reflected by the surface in a certain upwelling direction and the incident flux of a collimated beam of incoming solar radiation having direction (ϑ_0, ϕ_o). The plane perpendicular to the surface, containing both the Sun and the ground-level reference spot, defines the principal plane of reflection. Factor $R(\lambda, \vartheta_0, \phi_o, \eta_{pz}, \phi)$ is commonly assumed to exhibit a cylindrical symmetry with respect to the principal plane of reflection. Thus, its mathematical representation is in general made considering only the difference between the two angles ϕ_o and ϕ, which will be hereinafter referred to as the angular difference ϕ'.

To obtain precise calculations of DARF at the ToA level for each geometrical configuration of the surface–atmosphere system defined by the angular parameters ϑ_0, η_{pz}, and ϕ', we decided to define a set of BRDF models based on rigorous physical concepts, allowing us to perform the calculations of the DARF terms planned in the present study. For this purpose, we carried out realistic simulations of the deep surface–atmosphere coupling effects induced by the complex radiative transfer processes occurring inside the surface–air interface layer. Various parameterization criteria were adopted, based on both hyperspectral (H-type) and nonhyperspectral (N-H-type) models, where the reflected radiance fields generated by surfaces were represented at each wavelength over the 0.30–4.00 μm spectral range as a function of the aforementioned angular coordinates, to totally cover the 2π upward solid angle. To represent the surface reflectance properties characterized by various spectral and angular dependence features, the sixteen BRDF models were determined in terms of the following functions typical of each surface:

The spectral directional hemispherical reflectance $R_{bs}(\lambda, \vartheta_0)$ is commonly called "black-sky albedo" and is calculated by integrating the function $R(\lambda, \vartheta_0, \phi_o = 0°, \eta, \phi)$ over the 2π upward solid angle to represent the spectral curve of surface reflectance as a function of solar zenith angle ϑ_0 only (Román *et al.*, 2010). This spectral function is considered to be valid in the ideal case in which the diffuse component of the global (e.g., direct + diffuse) solar radiation field is assumed to be null and provides the spectral curve of the black-sky albedo for $\vartheta_0 = 0°$, expressed in terms of the following analytical form:

$$R_{bs}(\lambda, \vartheta_o = 0°) = \frac{1}{\pi}\int_0^{2\pi}\int_0^{\pi/2} R(\lambda, \vartheta_o, \phi_o, \eta, \phi)\cos\eta\sin\eta\, d\eta\, d\phi. \quad (8.5)$$

The spectral bihemispherical reflectance $R_{ws}(\lambda)$ is called "white-sky albedo" and is determined by integrating the function $R_{bs}(\lambda, \theta_o)$ over the 2π upward solid angle and assuming that the incoming solar radiation field consists only of the diffuse isotropic component $D\downarrow(\lambda)$. Therefore, function $R_{ws}(\lambda)$ represents in practice the spectral curve of surface albedo relative to the diffuse component of incoming solar radiation:

$$R_{ws}(\lambda) = \frac{1}{\pi}\int_0^{2\pi}\int_0^{\pi/2} R_{bs}(\lambda, \vartheta_o)\cos\vartheta_o\sin\vartheta_o d\vartheta_o d\phi_o. \quad (8.6)$$

Bearing in mind that the white-sky albedo is given in Eq. (8.6) by the double integral of the black-sky albedo $R_{bs}(\lambda, \theta_o)$ over the whole intervals of the two downwelling polar angles and that $R_{bs}(\lambda, \theta_o)$ is obtained in Eq. (8.5) through the double hemispherical integration of the bidirectional reflectance factor $R(\lambda, \vartheta_0, \phi_o, \eta, \phi)$ over the entire ranges of the two upwelling polar angles, it is evident that $R_{ws}(\lambda)$ does not depend on the geometrical configuration of the Sun-surface-external viewer system. In cases of isotropic surface reflectance conditions, the white-sky albedo can be assumed to be equal to that of an equivalent Lambertian reflector.

The spectral curve of surface albedo $R_L(\lambda, \theta_o)$ is obtained as the weighted average of the spectral surface reflectance contributions associated with the black-sky and white-sky albedo conditions of the solar radiation field, respectively. This approximate function can be calculated in terms of the analytical form of the Lewis (1995) function defined by the following equation:

$$R_L(\lambda, \vartheta_o) = R_{bs}(\vartheta_o)[1 - D\downarrow(\lambda)] + R_{ws}(\vartheta_o)D\downarrow(\lambda), \quad (8.7)$$

where $D\downarrow(\lambda)$ is the spectral curve of the diffuse fraction of downwelling (global) solar radiation $I\downarrow(\lambda)$ reaching the surface, which can be calculated as a function of solar zenith angle ϑ_0 using the 6S code (Vermote *et al.*, 1997b) for any atmospheric aerosol content.

In order to obtain realistic evaluations of the DARF effects occurring inside the surface–atmosphere system, we determined a set of BRDF models over the 0.30–4.0 μm spectral range of incoming solar radiation and for a range of surface albedo increasing from less than 0.1 (over the oceans) to more than 0.8 (over ice-covered areas). More precisely, four surface reflectance classes were defined for seawater areas (ocean surfaces (OS)), vegetation-covered (VS) and agricultural land areas, bare soil (BS) and arid areas, and snow- and ice-covered polar surfaces (PS), having the following spectral and angular characteristics:

The OS class consists of four BRDF OS reflectance models representing the typical oceanic surface reflectance conditions described by the OCEAN hyperspectral model (Morel, 1988) and developed using the OCEAN subroutine given in the 6S code (Vermote *et al.*, 1997b). In defining the four OS models, we included the sun glint reflectance effects (Cox and Munk, 1954) and the effects due to Fresnel's reflection (Born and Wolf, 1975) as well as the whitecaps model characteristics (Koepke, 1984) and the recently improved spectral reflectance modeling features of whitecaps (Kokhanovsky, 2004). The OCEAN subroutine was used to calculate the BRDF function curves (a) for numerous triplets of angles ϑ_0, η_{pz}, and $\phi' = \phi_0 - \phi$, (b) as a function of the wind speed WS, and (c) for various sets of selected values of the following environment parameters: (i) wind direction *WD*, assumed to lie on the vertical plane $\phi_o = 0°$ for all the OS BRDF models; (ii) seawater pigment concentration C_p, equal to 10^{-4} g m^{-3}, this assumption being made considering that variations in C_p of more than four orders of magnitude can cause only relatively small changes in the sea surface reflectance, which is <10% even in extremely polluted cases; and (iii) seawater salt concentration C_{ss} equal to 34.3 ppt, an increase in C_{ss} from 0 to 48 ppt being estimated to yield sea surface reflectance changes ≪1%. For the previous characteristics, the following four OS models were obtained: (i) OS1 for WS = 2 m s^{-1}, (ii) OS2 for WS = 5 m s^{-1}, (3) OS3 for WS = 10 m s^{-1}, and (4) OS4 for WS = 20 m s^{-1}. The corresponding spectral curves of surface albedo $R_L(\lambda, \vartheta_0 = 60°)$ are shown in Figure 8.3, presenting nearly continuous features over the whole wavelength range from 0.4 to 2.5 μm for the first three OS models and some discontinuities at wavelengths around 1.5 and 2.0 μm for the OS4 model. The corresponding monochromatic values of white-sky albedo $R_{ws}(\lambda)$ are given in Table 8.1 for the four OS models, as calculated at 21 selected wavelengths over the spectral range from 0.40 to 2.50 μm, and found to vary slowly as a function of wavelength. All the results are reported in Table 8.1, as determined for the M-8 continental aerosol model and AOT = 0.10 at wavelength $\lambda = 0.55$ μm, showing that some minor variations in broadband albedo $A(\vartheta_0)$ can arise from small changes in the solar diffuse radiation flux $I_d\downarrow(\lambda, \vartheta_0)$ defined in Eq. (8.7), which are caused by variations in the vertical atmospheric contents of particulate constituents

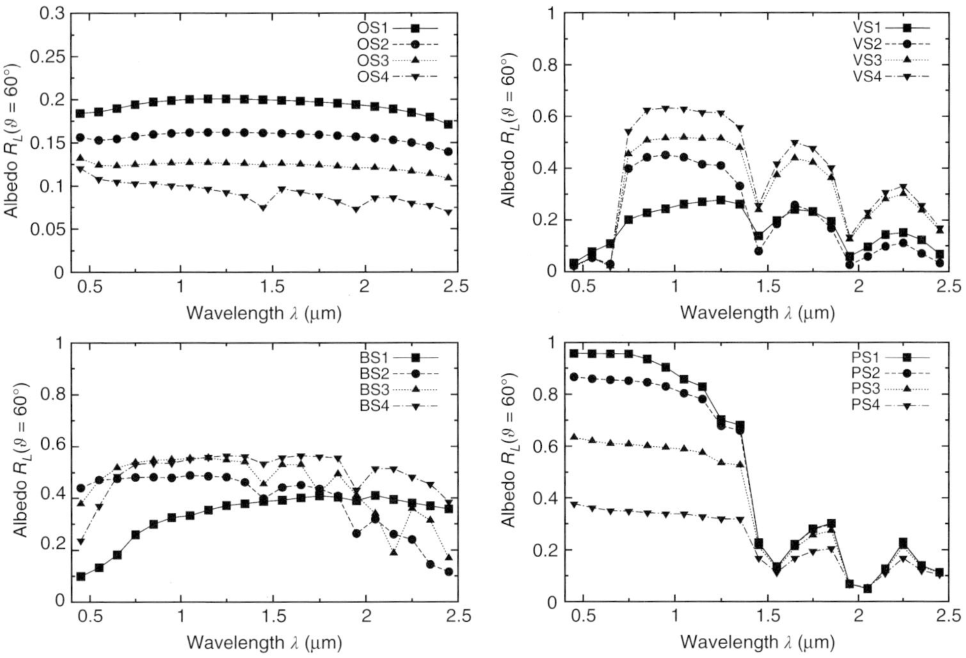

Figure 8.3 Spectral curves of surface albedo $R_L(\lambda)$ (Lewis and Barnsley, 1994), as defined in Eq. (8.7) over the 0.40–2.50 μm wavelength range for the four OS, VS, BS, and PS classes of BRDF surface reflectance models considered in the present study. All the reflectance models were determined for (i) the optical characteristics of the US62 atmosphere model (Dubin, Sissenwine, and Teweles, 1966), (ii) the scattering and absorption properties of the M-8 aerosol model consisting of pure continental particles (Tomasi *et al.* 2013), (iii) aerosol optical thickness $\tau_a(0.55\ \mu m) = 0.10$, and (iv) solar zenith angle $\vartheta_o = 60°$. Note that the range of $R_L(\lambda)$ is from 0 to 1 for the VS, BS, and PS BRDF models and from 0 to 0.3 for the OS BRDF models.

capable of scattering the shortwave radiation. Table 8.2 provides the values of black-sky albedo $R_{bs}(\vartheta_0 = 0°)$ at null zenith illumination angle and white-sky albedo R_{ws}, together with those of broadband albedo $A(\vartheta_0)$ calculated for nine values of ϑ_0 increasing in steps of 10° over the 0–80° range, using the following analytical form:

$$A(\vartheta_0) = \int_{0.40\ \mu m}^{2.50\ \mu m} R_L(\lambda, \vartheta_0) I \downarrow (\lambda, \vartheta_0)\, d\lambda \Big/ \int_{0.40\ \mu m}^{2.50\ \mu m} I \downarrow (\lambda, \vartheta_0)\, d\lambda, \qquad (8.8)$$

where the spectral quantities $R_L(\lambda, \vartheta_0) \times I\downarrow(\lambda, \vartheta_0)$ and $I\downarrow(\lambda, \vartheta_0)$ were integrated over the 0.40–2.50 μm wavelength range. In these calculations, some minor variations in the broadband albedo $A(\vartheta_0)$ can arise from the changes affecting the diffuse fraction $D\downarrow$ of incoming solar irradiance considered in Eq. (8.7), which are associated with variations in the atmospheric aerosol contents. The values of $A(\vartheta_0)$ given in Table 8.2 for the four OS models

Table 8.1 Monochromatic values of white-sky albedo $R_{ws}(\lambda)$ shown in Figure 8.5, as obtained at 21 wavelengths selected over the 0.45–2.45 μm spectral range for the 16 BRDF surface reflectance models subdivided into the OS, VS, BS, and PS classes.

Wave length λ (μm)	OS1	OS2	OS3	OS4	VS1	VS2	VS3	VS4	BS1	BS2	BS3	BS4	PS1	PS2	PS3	PS4
0.45	0.184	0.156	0.132	0.120	0.034	0.023	0.019	0.021	0.099	0.439	0.378	0.237	0.957	0.866	0.633	0.377
0.55	0.186	0.153	0.124	0.108	0.077	0.054	0.053	0.056	0.133	0.471	0.470	0.370	0.956	0.859	0.620	0.362
0.65	0.190	0.155	0.123	0.105	0.108	0.029	0.022	0.021	0.182	0.476	0.519	0.486	0.955	0.855	0.609	0.350
0.75	0.1945	0.158	0.125	0.103	0.200	0.399	0.455	0.543	0.260	0.481	0.538	0.531	0.956	0.852	0.607	0.348
0.85	0.197	0.160	0.126	0.103	0.228	0.442	0.507	0.624	0.300	0.482	0.548	0.536	0.935	0.845	0.601	0.343
0.95	0.199	0.161	0.127	0.101	0.244	0.451	0.516	0.632	0.325	0.480	0.550	0.536	0.903	0.830	0.594	0.338
1.05	0.200	0.162	0.127	0.100	0.262	0.442	0.519	0.629	0.333	0.489	0.554	0.551	0.857	0.803	0.588	0.338
1.15	0.201	0.162	0.127	0.097	0.271	0.415	0.514	0.615	0.354	0.486	0.556	0.559	0.828	0.780	0.574	0.327
1.25	0.201	0.162	0.127	0.093	0.278	0.411	0.515	0.614	0.372	0.483	0.547	0.565	0.702	0.678	0.535	0.319
1.35	0.200	0.162	0.126	0.088	0.262	0.332	0.479	0.558	0.379	0.462	0.540	0.562	0.680	0.660	0.527	0.318
1.45	0.200	0.161	0.124	0.075	0.138	0.079	0.239	0.258	0.387	0.398	0.454	0.534	0.227	0.226	0.211	0.167
1.55	0.199	0.161	0.126	0.097	0.199	0.184	0.374	0.418	0.392	0.442	0.530	0.558	0.134	0.134	0.129	0.112
1.65	0.198	0.160	0.125	0.093	0.241	0.258	0.439	0.500	0.401	0.451	0.530	0.564	0.220	0.219	0.206	0.167
1.75	0.197	0.159	0.124	0.089	0.233	0.233	0.422	0.477	0.409	0.436	0.422	0.561	0.280	0.278	0.255	0.193
1.85	0.196	0.158	0.123	0.082	0.195	0.167	0.363	0.403	0.403	0.408	0.494	0.556	0.302	0.299	0.273	0.204
1.95	0.194	0.157	0.121	0.073	0.059	0.027	0.125	0.135	0.390	0.264	0.410	0.431	0.068	0.068	0.068	0.068
2.05	0.192	0.155	0.121	0.086	0.096	0.059	0.213	0.232	0.410	0.319	0.340	0.515	0.050	0.050	0.051	0.054
2.15	0.189	0.153	0.119	0.086	0.143	0.098	0.281	0.307	0.395	0.261	0.188	0.514	0.126	0.125	0.121	0.109
2.25	0.185	0.150	0.117	0.080	0.150	0.111	0.301	0.331	0.381	0.240	0.359	0.482	0.229	0.227	0.211	0.167
2.35	0.179	0.146	0.114	0.078	0.122	0.070	0.237	0.257	0.370	0.144	0.313	0.454	0.139	0.138	0.134	0.119
2.45	0.171	0.140	0.109	0.070	0.067	0.033	0.156	0.169	0.358	0.116	0.168	0.385	0.113	0.113	0.111	0.105

Table 8.2 Values of directional hemispherical reflectance $R_{bs}(0°)$ (black-sky albedo) at null zenith illumination angle, bihemispherical reflectance R_{ws} (white-sky albedo, which does not depend on solar zenith angle ϑ_0), and broadband albedo $A(\vartheta_0)$ calculated for nine values of solar zenith angle ϑ_0 (increasing from 0° to 80° in steps of 10° in steps of 10°), as obtained for the 16 surface reflectance models listed in the first column and the spectral characteristics of the US62 standard atmosphere model (Dubin, Sissenwine, and Teweles, 1966) and the M-8 (pure continental) aerosol model of Tomasi *et al.* (2013).

Model	$R_{bs}(0°)$	R_{ws}	$A(0°)$ (1.0000)	$A(10°)$ (1.0148)	$A(20°)$ (1.0634)	$A(30°)$ (1.1536)	$A(40°)$ (1.3037)	$A(50°)$ (1.5526)	$A(60°)$ (1.9928)	$A(70°)$ (2.8999)	$A(80°)$ (5.5803)
OS1	0.026	0.070	0.030	0.030	0.030	0.035	0.048	0.083	0.193	0.601	2.098
OS2	0.026	0.069	0.030	0.030	0.030	0.033	0.043	0.071	0.158	0.425	1.202
OS3	0.028	0.069	0.031	0.031	0.032	0.034	0.042	0.065	0.127	0.284	0.714
OS4	0.044	0.081	0.048	0.048	0.049	0.051	0.057	0.072	0.105	0.180	0.385
VS1	0.134	0.153	0.135	0.134	0.133	0.139	0.141	0.149	0.155	0.175	0.202
VS2	0.170	0.203	0.172	0.450	0.450	0.178	0.450	0.450	0.210	0.450	0.450
VS3	0.201	0.243	0.203	0.202	0.204	0.213	0.223	0.239	0.258	0.287	0.340
VS4	0.250	0.292	0.253	0.253	0.256	0.264	0.274	0.289	0.306	0.333	0.386
BS1	0.225	0.237	0.226	0.225	0.224	0.228	0.229	0.235	0.240	0.256	0.278
BS2	0.431	0.457	0.434	0.431	0.429	0.438	0.441	0.453	0.461	0.487	0.510
BS3	0.456	0.484	0.459	0.456	0.455	0.463	0.467	0.480	0.489	0.519	0.549
BS4	0.424	0.450	0.426	0.424	0.422	0.430	0.433	0.446	0.455	0.485	0.521
PS1	0.824	0.847	0.827	0.827	0.830	0.834	0.84	0.847	0.854	0.862	0.865
PS2	0.720	0.761	0.726	0.727	0.732	0.739	0.749	0.761	0.775	0.789	0.796
PS3	0.461	0.536	0.472	0.475	0.483	0.496	0.514	0.537	0.564	0.591	0.608
PS4	0.214	0.296	0.223	0.226	0.234	0.249	0.269	0.296	0.329	0.365	0.390

The values of relative optical air mass m given in brackets for the various values of $A(\vartheta_0)$ were evaluated by Tomasi *et al.* (1997). The values of $A(80°)$ given in the last column are in general affected by high uncertainties, due to the fact that the four OC models are quite failing for $\vartheta_0 > 75°$.

were found to slowly increase passing from OS1 to OS4 for low values of ϑ_0 and decrease appreciably for $\vartheta_0 \geq 30°$, presenting even more rapid variations as one passes from the OS1 model to the OS4 one, as ϑ_0 assumes gradually higher values. This behavior is presumably due to an increasing specular reflectance effect occurring for calm wind conditions (in our case for WS = 2 m s^{-1}).

The vegetated surfaces (VS) class, which consists of four BRDF reflectance models representing the albedo characteristics of vegetation-covered surfaces, as derived from the AK subroutine reported by the 6S code (Vermote *et al.*, 1997b), used to evaluate the reflectance properties of vegetated surfaces in terms of the hyperspectral (Kuusk, 1994) model simulations. This subroutine utilizes the PROSPECT code (Jacquemoud and Baret, 1990) to simulate the chlorophyll absorption features and the Nilson and Kuusk (1989) algorithm to represent the reflectance anisotropy characteristics of the one-layer canopy coverage. The angular BRDF values were calculated as a function of numerous parameters characterizing the physical, optical, biological, and leaf parameters of the vegetation coverage, such as the chlorophyll content C_{AB}, leaf water equivalent thickness C_{lw}, effective number N_{eff} of elementary layers inside a leaf, ratio C_n of refractive indices of the leaf surface wax and internal material, and weight S_1 of the first Price (1990) function for the soil reflectance, leaf area index LAI, elliptical eccentricity E_{ell} of the leaf angle distribution, modal inclination Q_m of the leaf distribution, and relative leaf size S_L with respect to the canopy depth. The four VS models were obtained by varying the aforementioned parameters, the first two models being based on the Kuusk (1994) corn model and the third and fourth models based on the Kuusk (1994) soybean model, for the values of vegetation coverage and leaf structural parameters given in Table 8.3. The spectral curves of surface albedo $R_L(\lambda, \vartheta_0)$ obtained for the four VS models are shown in Figure 8.3 over the 0.40–2.50 μm wavelength range, while the values of broadband albedo $A(\vartheta_0)$ are given in Table 8.2 for each model, as obtained for values of ϑ_0 increasing in steps of 10° from 0° to 80°. The typical spectral reflectance features of a vegetation-covered surface, presenting a sharp reflectance increase at ~0.70 μm wavelength, and commonly called "red edge," is well reproduced by the Kuusk (1994) model. Model VS1 appears to be suitable for representing a canopy clearly affected by drought conditions, for which the soil spectral signature emerges from the background, while the other three VS models can be more confidently used to represent vegetation coverages presenting increasing values of LAI and gradually more enhanced features of the so-called red edge. The monochromatic values of surface albedo $R_L(\lambda, \vartheta_0 = 60°)$ obtained for the four VS models are given in Table 8.1 at the 21 selected wavelengths from 0.40 to 2.50 μm. As shown in Table 8.2, the white-sky albedo $R_{ws}(\lambda)$ increases from 0.15 to nearly 0.30 passing from the VS1 model to the last one.

The BS class consists of four BRDF surface reflectance models derived from the nonhyperspectral model proposed by Rahman, Pinty, and Verstraete (1993)

Table 8.3 Values of leaf parameters assumed to give form to the four vegetated surface (VS) BRDF reflectance models used in the present study.

BRDF model	Chlorophyll content C_{AB} (μg cm^{-2})	Thickness C_{lw} (cm)	Effective number N_{eff}	Ratio C_n	Weight S_1	Leaf area index LAI	Elliptical eccentricity E_{ell}	Modal inclination Q_m (degrees)	Relative leaf size S_L
VS1	100	4.0×10^{-2}	1.09	0.9	0.213	0.1	0.972	10.7	0.1
VS2	100	2.6×10^{-2}	1.09	0.9	0.213	1.5	0.972	10.7	0.1
VS3	82.2	5.0×10^{-3}	1.24	0.9	0.225	2.5	0.965	45.8	0.1
VS4	82.2	5.0×10^{-3}	1.24	0.9	0.225	5.0	0.965	45.8	0.1

and used as subroutine to implement the 6S code (Vermote *et al.*, 1997a). These models are (i) the dry sand BS1 model, directly taken from the 6S code; (ii) the illite BS2 model, represented by a mixing of clay minerals having crystal structures very similar to that of muscovite; (iii) the alunite BS3 model, obtained for the spectral reflectance characteristics of a mineral consisting of a hydrous potassium aluminum sulfate and presenting massive forms mixed with rhombohedral crystals; and (iv) the montmorillonite BS4 model, represented by a soft clay mineral consisting mainly of a hydrous aluminum silicate in which aluminum was exchanged abundantly with magnesium and other bases. The last three spectral surface reflectance curves were taken from the USGS Library (http://speclab.cr.usgs.gov), giving the wavelength dependence features of surface reflectance $R_L(\lambda, \vartheta_0 = 60°)$ shown in Figure 8.3 over the 0.40–2.50 μm spectral range. The angular dependence characteristics of the aforementioned BRDF surface reflectance functions were defined by assuming that they are similar to those determined by Rahman, Pinty, and Verstraete (1993) and using the parameter An to define the anisotropy degree of surface reflectance and parameter As to evaluate its asymmetry degree and regulate the relative intensities of forward and backward scattering. More precisely, An was assumed to be equal to 0.648 within the wavelength interval $\lambda < 0.630$ μm and to 0.668 for $\lambda > 0.915$ μm, respectively, while As was kept as equal to −0.290 and −0.268 within the two previously defined wavelength intervals, respectively, according to the Rahman, Pinty, and Verstraete (1993) calculations made for a plowed field surface. The spectral values of An and As over the 0.630–0.915 μm wavelength range were calculated using a linear interpolation procedure in wavelength between the values assumed earlier. The spectral features of surface albedo $R_L(\lambda)$ defined in Eq. (8.7) for the four BS models are shown in Figure 8.3. The monochromatic values of $R_L(\lambda, \vartheta_0 = 60°)$ determined at the previously chosen wavelengths for the four BS models are given in Table 8.1, and the corresponding values of broadband albedo $A(\vartheta_0)$ obtained for the nine values of ϑ_0 using the values of parameters *An* and *As* determined earlier are given in Table 8.2. For the previous assumptions, the white-sky albedo $R_{ws}(\lambda)$ was estimated to be approximately equal to 0.23 for the BS1 model, 0.43 for the BS2 model, 0.46 for the BS3 model, and 0.42 for the BS4 model.

The PS class consists of four surface reflectance models representing various surfaces covered by snow fields and/or glaciers, which were derived using the hyperspectral surface reflectance model of the direct component of black-sky albedo $R_{bs}(\lambda, \vartheta_0)$ defined in Eq. (8.5) and the diffuse component of white-sky albedo $R_{ws}(\lambda, \vartheta_0)$ described in Eq. (8.6). The main parameters of the PS models are the size distributions of the snow and BC grains, represented by means of lognormal size-distribution functions, and the concentration of dust and/or BC in the snow surface layer (Wiscombe and Warren, 1980; Warren and Wiscombe, 1980). In modeling such surface reflectance data, we assumed that the reflectance effects produced by a volume concentration of BC particles equal to 1 ppb are comparable with those given by a volume concentration of dust particles equal to 100 ppm. Both the particle types were observed to cause a relevant reduction of albedo over the spectral range $\lambda < 1\,\mu m$, where soot particles cause a decrease in reflectance and dust particles correspondingly cause an appreciable increase at the 0.60–0.70 μm wavelengths. Snow grains with sizes varying between 50 and 500 μm were found to produce only weak variations at visible wavelengths. These features agree in general with the recent results obtained by Kokhanovsky and Breon (2012), who investigated the anisotropic characteristics of the snow BRDF reflectance and defined a semiempirical spectral model based on detailed representations of the forward scattering maximum dependence on view zenith angle and azimuthal reflectance variations. Various surface albedo simulations were made by assuming that the main component of the surface layer is constituted by snow grains having radii equal to $10^2\,\mu m$ and additional concentrations are given by soot particles, presenting lognormal size-distribution curves centered at the $10^{-1}\,\mu m$ radius. A Mie algorithm was defined to calculate the spectral values of $\omega(\lambda)$ and asymmetry factor $g(\lambda)$ of such a snow grain size distribution at 217 selected wavelengths over the 0.40–2.50 μm range, according to Warren and Wiscombe (1980), using the spectral values of complex refractive index defined by Warren (1984) for snow grains together with those of BC provided by the 6S aerosol soot component (Vermote *et al.*, 1997b). Following this procedure, the values of the real part were estimated to range between 1.75 and 1.90, and those of the imaginary part between 0.43 and 0.57 over the entire solar spectrum. The direct and diffuse solar radiation components were calculated following the semiempirical parameterization method of Wiscombe and Warren (1980), based on the use of the values of $\omega(\lambda)$ and $g(\lambda)$ calculated previously for bimodal size distributions consisting of soot particles and snow grains. Four models have been obtained following the previous procedure, assuming gradually increasing values of the soot volume concentration C_{soot}: (i) model PS1 for almost pure snow ($C_{soot} = 2 \times 10^{-3}$ ppm), (ii) model PS2 for slightly contaminated snow ($C_{soot} = 4 \times 10^{-2}$ ppm), (iii) model PS3 for a snow coverage with $C_{soot} = 4 \times 10^{-1}$ ppm, and (iv) model PS4 for heavy carbon-contaminated snow ($C_{soot} = 2$ ppm). The spectral curves of surface albedo $R_L(\lambda, \vartheta_0 = 60°)$ are shown in Figure 8.3, while the monochromatic values of this parameter are given in Table 8.1 for the four PS models and at numerous selected wavelengths from 0.40 to 2.50 μm. The corresponding nine values of $A(\vartheta_0)$ are given in Table 8.2. It can be seen in Figure 8.3 that the spectral signature of the snow surface is characterized by an appreciable decrease in reflectance for $\lambda > 1\,\mu m$, presenting well-defined minima corresponding to the water absorption bands Ψ (centered at $\lambda \approx 1.40\,\mu m$) and Ω (centered at $\lambda \approx 1.87\,\mu m$). Broadband white-sky albedo was estimated to vary between 0.30 (for polluted snow cover model PS4) to nearly 0.85 (for clean snow cover model PS1).

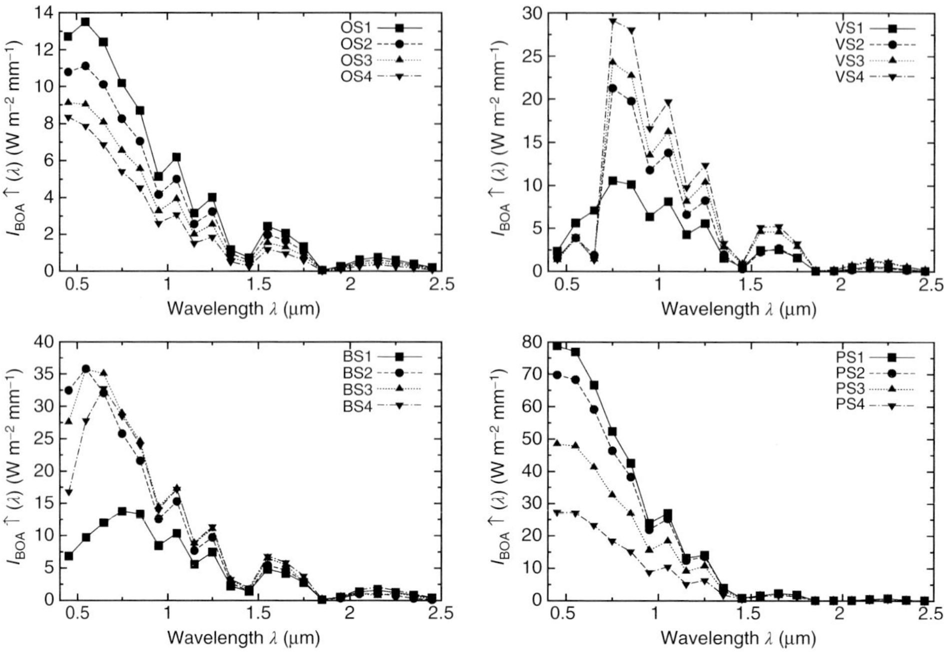

Figure 8.4 Spectral curves of the upwelling irradiance $I_{BoA}\uparrow(\lambda)$ at the BoA level, given by the product of spectral surface albedo $R_L(\lambda, \vartheta_0 = 60°)$ by the incoming global (direct + diffuse) irradiance $I_{BoA}\downarrow(\lambda, \vartheta_0 = 60°)$ at the surface, as obtained for the four OS, VS, BS, and PS classes of BRDF surface reflectance models considered in the present work. All the BRDF reflectance models were determined for (i) the optical characteristics of the US62 atmosphere model (Dubin, Sissenwine, and Teweles, 1966), (ii) the scattering and absorption properties of the M-8 aerosol model consisting of pure continental particles (Tomasi *et al.* 2013), and (iii) aerosol optical thickness $\tau_a(0.55\,\mu m) = 0.10$. Note that reflected irradiance $I_{BoA\uparrow}(\lambda, \vartheta_0 = 60°)$ is presented over different ranges for the various OS, VS, BS, and PS BRDF models.

In order to calculate the DARF effects, it is necessary to calculate the differences between the irradiance relative to pristine atmospheric transparency conditions and the irradiance relative to a turbid atmosphere. Therefore, it is useful to have a clear picture of the spectral dependence features of the upwelling irradiance $I_{BoA}\uparrow(\lambda)$ reflected at the surface, arising from the combined effects of atmospheric attenuation on the incoming solar irradiance reaching the surface and spectral surface reflectance. The spectral curves of the upwelling irradiance $I_{BoA}\uparrow(\lambda)$ are shown in Figure 8.4 for the four sets of surface reflectance models OS, VS, BS, and PS, each obtained in terms of the product of the global downwelling irradiance by surface albedo $R_o(\lambda, \vartheta_0)$. Figure 8.5 shows the spectral curves of white-sky albedo $R_{ws}(\lambda)$, which can be used to determine the reflectance conditions of the equivalent Lambertian reflectance models belonging to the OS, VS, BS, and PS classes. The monochromatic values of $R_{ws}(\lambda)$ obtained over the 0.40–2.50 μm spectral

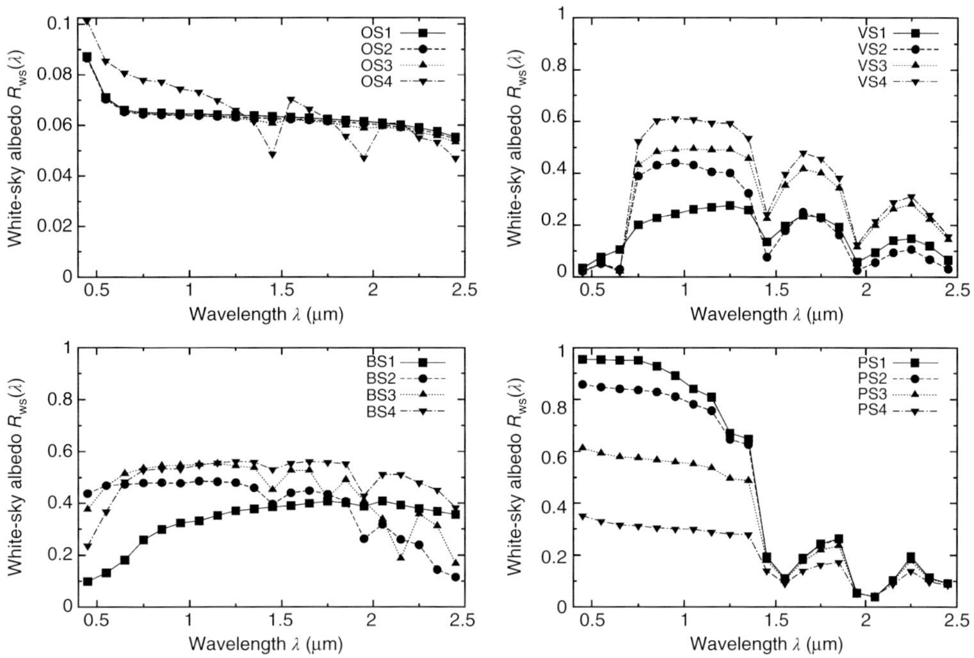

Figure 8.5 Spectral curves of white-sky albedo $R_{ws}(\lambda)$ defined in Eq. (8.6) over the 0.40–2.50 μm spectral range for the four OS, VS, BS, and PS classes of BRDF surface reflectance models. Note that the range of $R_{ws}(\lambda)$ is from 0 to 1 for the VS, BS, and PS BRDF models and from 0 to 0.1 for the OS BRDF models.

range are presented in Table 8.2 for the sixteen BRDF surface reflectance models, indicating that white-sky albedo $R_{ws}(\lambda)$ assumes values very close to those of $R_{bs}(\lambda)$ calculated for $\vartheta_0 = 60°$, as it is typical of natural surfaces.

The non-Lambertian surface reflectance curves illustrated in Figure 8.6 for the OS3, VS3, BS3, and PS3 surface models indicate that the values of $R_{ws}(\lambda)$ were intermediate between the two black-sky albedo curves obtained for $\vartheta_0 = 50°$ and $\vartheta_0 = 60°$. In order to weight the SSA effects on solar radiation, the scattering and absorption features occurring at visible and NIR wavelengths were taken into account, bearing in mind that aerosols are mostly present within the lower part of the troposphere. Accordingly, we used the average ω as SSA key parameter to evaluate the intensity of the DARF effects induced by airborne aerosols.

8.2.5
Calculations of Instantaneous DARF Terms at the ToA and BoA Levels and within the Atmosphere

The net flux of shortwave radiation at the ToA level (or at another level close to the tropopause) is given by the difference between the incoming flux $F\downarrow$ and the outgoing flux $F\uparrow$ of solar radiation. Since these fluxes vary as a function of the

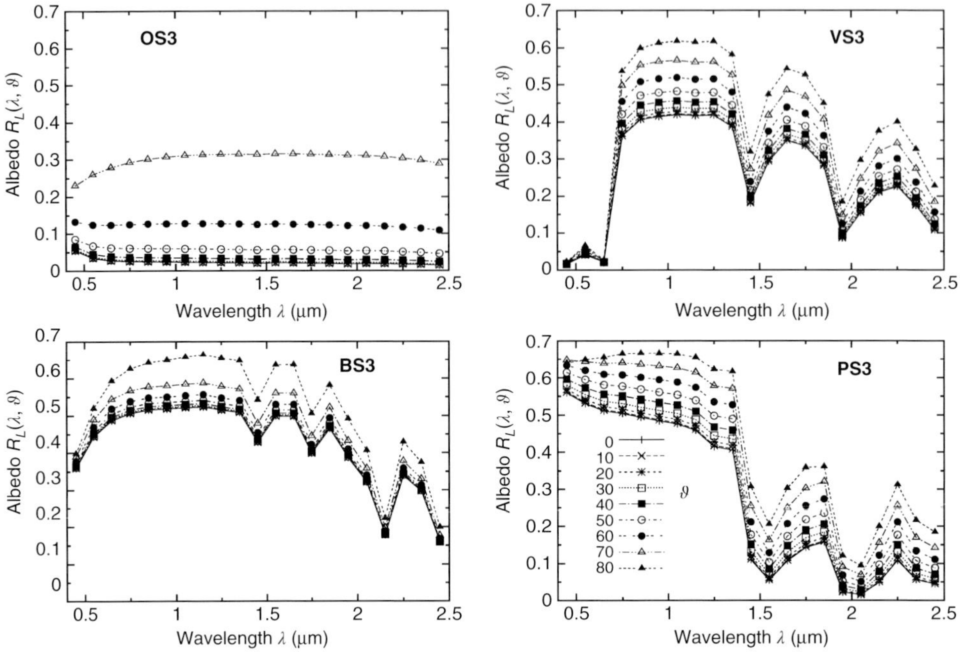

Figure 8.6 Spectral curves of surface albedo $R_L(\lambda, \vartheta_0)$ defined in Eq. (8.7) over the 0.40–2.50 μm spectral range for incoming global (direct + diffuse) solar irradiance at (i) different solar zenith angles ϑ_0 taken in steps of 10° from 0° to 80° and for (ii) the OS3, VS3, BS3, and PS3 BRDF surface reflectance models. In this surface–atmosphere system, the atmosphere was assumed to have (i) the optical characteristics of the US62 atmosphere model (Dubin, Sissenwine, and Teweles, 1966), (ii) the scattering and absorption properties of the M-8 aerosol model consisting of pure continental particles (Tomasi *et al.* (2013)), and (iii) aerosol optical thickness $\tau_a(0.55\,\mu m) = 0.10$. Note that the spectral curve of $R_L(\lambda, \vartheta_0)$ calculated for the OS3 BRDF reflectance model at $\vartheta_0 = 80°$ assumes values higher than 0.9 and therefore is not reported in the OS3 graph.

radiation scattering and absorption effects due to the atmospheric constituents, the net flux change gives a measure of the radiative forcing (RF) acting on the surface–atmosphere system. This concept implies that the net flux change can assume a negative (cooling) or positive (warming) sign at the ToA level of the atmosphere. If the concentrations of all the atmospheric constituents other than aerosols are kept constant, the above change can be attributed to airborne particles only (IPCC, 1996) and to depend on various parameters related to the optical characteristics and vertical profiles of aerosol mass concentration (Hansen, Sato, and Ruedy, 1997). The DARF effects on the climate system are commonly evaluated by considering only the shortwave (solar) radiation flux change and neglecting the radiative effects produced by aerosols on the longwave radiation. They are generally evaluated instantaneously for various prefixed hours of the day and then calculated as diurnal averages over the 24 hour period (Bush and Valero, 2003). On the basis of these remarks, the instantaneous DARF term $\Delta F_{\mathrm{ToA}}(t)$ occurring at a certain time t can be simply determined at the ToA level as the difference between (i) the net radiative flux for the turbid atmosphere containing a certain columnar content of aerosol particles and (ii) the same quantity in a pristine atmosphere without aerosols. Thus, the instantaneous forcing $\Delta F_{ToA}(t)$ induced by aerosol particles can be represented in terms of the following formula:

$$\Delta F_{ToA}(t) = F_{net} \downarrow (t) - F^{*}{}_{net} \downarrow (t), \tag{8.9}$$

where (i) the ToA level net flux $F_{net}{\downarrow}(t)$ is given by the difference

$$F_{net} \downarrow (t) = F \downarrow (t) - F \uparrow (t), \tag{8.10}$$

between the instantaneous shortwave downwelling flux $F{\downarrow}(t)$ (affected by all the attenuation processes occurring in the atmosphere) and the instantaneous shortwave upwelling flux $F{\uparrow}(t)$, both determined at the ToA level for an atmosphere including all its constituents, and (ii) the net flux $F^{*}{}_{net}{\downarrow}(t)$ at the ToA level is given by the difference between the corresponding shortwave downwelling flux $F^{*}{\downarrow}(t)$ and the shortwave upwelling flux $F^{*}{\uparrow}(t)$, both calculated in the pristine atmosphere without aerosols. The previous procedure provides the following final equation:

$$F^{*}{}_{net}(t) = F^{*} \downarrow (t) - F^{*} \uparrow (t), \tag{8.11}$$

where the instantaneous downwelling flux $F^{*}{\downarrow}(t)$ is not altered by atmospheric aerosols. Therefore, instantaneous flux $F^{*}{\downarrow}(t)$ in Eq. (8.11) is equal to flux $F{\downarrow}(t)$ given in Eq. (8.10). Consequently, the instantaneous term $\Delta F_{ToA}(t)$ calculated in terms of Eq. (8.9) is directly given by the difference

$$\Delta F_{ToA}(t) = F^{*} \uparrow (t) - F \uparrow (t), \tag{8.12}$$

which shows that this DARF term can be correctly evaluated by simply subtracting the upward solar radiation flux leaving the real atmosphere with aerosols from the upwelling solar radiation flux emerging from the pristine atmosphere without aerosols. According to the conventional definition of RF in the atmosphere (IPCC, 1996), negative values of $\Delta F_{ToA}(t)$ indicate that aerosols cause an increase in the upwelling flux of solar radiation and hence an increase in the surface–atmosphere system albedo, therefore causing direct cooling effects on the climate system. Conversely, positive values of $\Delta F_{ToA}(t)$ imply that lower upwelling solar radiation fluxes are induced by aerosols, causing a decrease in the overall albedo and, consequently, producing significant atmospheric warming effects (Zhao *et al.*, 2008).

The instantaneous aerosol RF $\Delta F_{BoA}(t)$ at the surface (i.e., at the BoA level) gives a measure of the perturbation induced by airborne aerosols in the net flux reaching the surface. It can be calculated as the difference between the net shortwave flux at surface level in the atmosphere with aerosols and the net shortwave flux at surface level in the same atmosphere assumed to be without aerosols (Bush and Valero, 2002, 2003). Therefore, it can be expressed at a given time t as the difference

$$\Delta F_{BoA}(t) = \Phi_{net}(t) - \Phi^{*}{}_{net}(t), \tag{8.13}$$

where the net flux $\Phi_{net}(t)$ at the surface is given by the difference between the downwelling flux $\Phi\downarrow(t)$ and the upwelling flux $\Phi\uparrow(t)$ according to the following the formula:

$$\Phi_{net}(t) = \Phi \downarrow (t) - \Phi \uparrow (t). \tag{8.14}$$

Assuming that A is the average surface albedo determined over the shortwave radiation spectral range, the instantaneous upwelling flux $\Phi\uparrow(t)$ at the surface can be expressed with good approximation as the product

$$\Phi_{net}(t) = A \times \Phi \downarrow (t). \tag{8.15}$$

Thus, combining Eqs. (8.14) and (8.15), the net flux $\Phi_{net}(t)$ at the surface can be written in the following form:

$$\Phi_{net}(t) = (1 - A) \quad \Phi \downarrow (t). \tag{8.16}$$

To calculate the net flux $\Phi_{net}(t)$ as well as the downwelling flux $\Phi\downarrow(t)$ of solar radiation reaching the surface level after its passage through the pristine atmosphere without aerosols and the upwelling flux $\Phi\uparrow(t)$ passing through the turbid atmosphere, the most commonly used radiative transfer codes such as the 6S code of Vermote *et al.* (1997a) can be used, applied to the most common atmospheric models available in the literature or to sets of vertical profiles of temperature, pressure, and water vapor partial pressure derived from local meteorological radiosounding measurements. In these calculations, it is important to take into account that the field measurements of aerosol composition, optical, and size-distribution parameters are affected by not negligible errors, which can cause important errors in the calculations of the instantaneous DARF effects made using clear-sky radiative transfer codes (Valero and Bush, 1999).

The DARF effects occurring at the ToA and BoA levels imply that an instantaneous aerosol thermodynamic forcing $\Delta F_{Atm}(t)$ is produced by aerosols within the atmosphere. It can be evaluated as the difference between the term $\Delta F_{ToA}(t)$ defined in Eq. (8.12) and the term $\Delta F_{BoA}(t)$ given in Eq. (8.13):

$$\Delta F_{Atm}(t) = \Delta F_{ToA}(t) - \Delta F_{BoA}(t). \tag{8.17}$$

The RF term $\Delta F_{Atm}(t)$ is the instantaneous change in the atmospheric energy budget, which is not explicitly caused by aerosol-induced RF but is more likely due to an internal redistribution of the energy surplus, which modifies the amount of latent heat released by aerosols both directly and during the cloud formation processes (Ramanathan *et al.*, 2001b). Because of these exchanges, the atmospheric stability conditions may be considerably modified, exerting their influence on the heating rates, surface temperatures, and cloud formation and persistence, all contributing to induce further changes in the local atmospheric cooling and warming processes.

As mentioned earlier, large uncertainties still exist regarding the role of atmospheric aerosols in causing climate change effects. In particular, large knowledge gaps still remain on (i) the dependence of DARF on the microphysical and composition characteristics of aerosols, (ii) the multimodal characteristics of the particle size-distribution curves and their variations with altitude, (iii) the most common vertical distribution features of aerosol number and mass concentrations and aerosol optical parameters, and (iv) the spectral and directional characteristics of surface reflectance in the various areas of the Earth. The last point is of crucial importance, since the DARF effects strongly depend not only on the optical parameters of atmospheric aerosols but also on the surface reflectance properties (Chylek and Coakley, 1974). With regard to this, it is worth mentioning that the Lambertian surface reflectance models most commonly used in the literature to calculate the energy budget of the surface – atmosphere system do not realistically describe the reflectance characteristics of the oceanic and land regions, especially for high values of solar zenith angle ϑ_0. For this reason, it seems more appropriate to use the BRDF models described in the previous section to represent more realistically the surface reflectance characteristics, which are in general non-Lambertian in most cases. The use of the BRDF models is also crucial when the spectral albedo curves of the various surfaces are evaluated as the sum of the so-called black-sky and white-sky albedo terms defined earlier in Eqs. (8.5) and 8.6, respectively, according to Lewis (1995). These schematic representations allow us to separately evaluate the two distinct surface albedo contributions using the spectral percentages of the direct and diffuse components as weight functions in Eq. (8.7). These black-sky and white-sky albedo contributions are differently subject to vary as a function of the numerous radiative parameters characterizing the atmospheric aerosols, such as the shape parameters of the particle number and mass size-distribution curves and complex refractive index.

Tomasi *et al.* (2013) showed that the instantaneous DARF term induced by airborne aerosols usually varies during the day, depending on (i) the time variations in AOT, measured at various visible and NIR wavelengths; (ii) the time variations in SSA, mainly depending on the particle size-distribution shape parameters and complex refractive index; (iii) the variations in the spectral and spatial characteristics of the underlying surface reflectance, which can be exhaustively represented by using the four OS non-Lambertian models proposed earlier to represent the ocean surface reflectance properties and the four VS models representing the vegetation-covered land surface reflectance; and (iv) the solar zenith angle ϑ_0, which allows us to determine the time variations of incoming solar radiation flux.

8.2.6
Dependence Features of Instantaneous DARF Terms on Aerosol Optical Parameters and Surface Reflectance

The scattering and absorption characteristics of a certain atmospheric aerosol content depend on particle concentration and size-distribution shape parameters, as well as on the particle origin and mass concentrations of absorbing and nonabsorbing substances. The higher is the columnar content of aerosol mass, the more marked extinction effects are produced by airborne particles along the atmospheric slant path of incoming solar radiation. This implies that the DARF effects occurring at the ToA level are expected to increase as the aerosol mass content becomes gradually higher and AOT at visible wavelengths correspondingly increases. It is also important to take into account that the aforementioned DARF term is strongly influenced by the aerosol SSA characteristics in relation to the local surface albedo conditions (Chylek and Coakley, 1974). Bearing in mind these dependence features, the DARF terms were calculated for the set of 43 aerosol extinction models and the 16 BRDF surface reflectance models described earlier to determine their mean dependence patterns on parameters AOT and SSA, surface reflectance properties in oceanic and land areas, and solar zenith angle ϑ_0 over its 0–80° range.

8.2.6.1 Dependence of Instantaneous DARF on Aerosol Optical Thickness

The instantaneous DARF term $\Delta F_{\mathrm{ToA}}(t)$, produced by aerosols through solar radiation scattering and absorption, varies almost linearly and without modifying the sign as AOT $\tau_a(0.55\,\mu\mathrm{m})$ increases in all cases with $\tau_a(0.55\,\mu\mathrm{m}) \leq 0.10$ (Chylek and Coakley, 1974). Conversely, nonlinear dependence features of $\Delta F_{\mathrm{ToA}}(t)$ are usually observed for $\tau_a(0.55\,\mu\mathrm{m})$ considerably greater than 0.10. In fact, $\Delta F_{\mathrm{ToA}}(t)$ is given in these cases by the combined radiative transfer effects associated with the atmospheric aerosol content and spectral optical characteristics (including those of SSA) and the spectral and geometrical surface reflectance characteristics, which can largely vary from one surface type to another. In the present analysis of the dependence features of instantaneous DARF terms on $\tau_a(\lambda)$, the DARF calculations were made for each of the aforementioned aerosol models and for (i) a null value of $\tau_a(0.55\,\mu\mathrm{m})$ considered to simulate the pristine atmosphere conditions and (ii) five values of $\tau_a(0.55\,\mu\mathrm{m})$ increasing in steps of 0.2 from 0.1 to 0.9 to represent the variety of atmospheric turbidity conditions observed in the real atmosphere.

The aforementioned choices allowed us to evaluate with good accuracy the dependence features of the instantaneous DARF terms at the ToA and BoA levels and in the atmosphere on $\tau_a(\lambda)$ for various airborne aerosol composition and optical characteristics and for largely different surface reflectance conditions. Figure 8.7 shows three examples of the dependence curves of instantaneous terms $\Delta F_{\mathrm{ToA}}(t)$, $\Delta F_{\mathrm{BoA}}(t)$, and $\Delta F_{\mathrm{Atm}}(t)$ on monochromatic AOT $\tau_a(0.55\,\mu\mathrm{m})$ obtained for three values of soar zenith angle ϑ_0, simulated over the OS3 BRDF sea surface and for the three M-1 (pure oceanic), M-8 (pure continental), and M-14 (heavy polluted anthropogenic) aerosol models determined by Tomasi *et al.* (2013) for the three values of weighted average SSA $\omega = 0.999$, $\omega = 0.855$, and $\omega = 0.651$, respectively. The BRDF non-Lambertian calculations of $\Delta F_{\mathrm{ToA}}(t)$ were found to be negative for all the three values of ϑ_0 and the three M-type aerosol models. The DARF values were found to decrease slowly as $\tau_a(0.55\,\mu\mathrm{m})$ increases, yielding slope coefficients which do not vary markedly with ϑ_0 and vary only slightly with the aerosol model, presenting the lowest absolute values for the more absorbing M-14 aerosols and the most marked negative slopes for the pure oceanic and

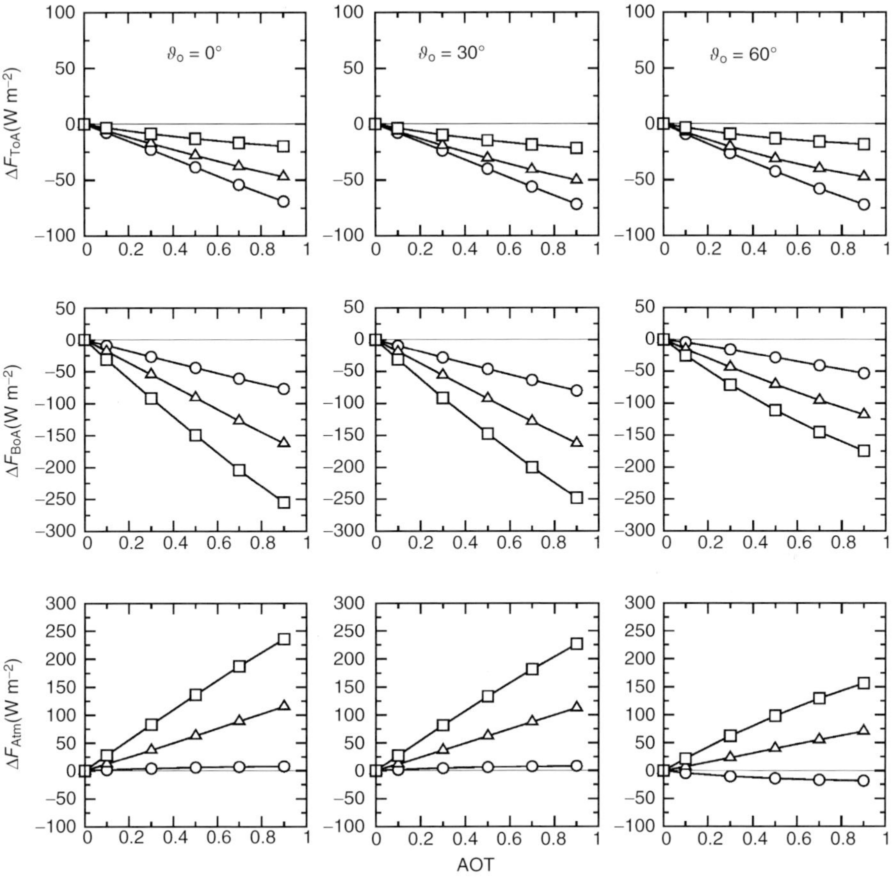

Figure 8.7 Dependence curves of instantaneous DARF terms $\Delta F_{ToA}(t)$ (upper part), $\Delta F_{BoA}(t)$ (middle part), and $\Delta F_{Atm}(t)$ (lower part) plotted as a function of aerosol optical thickness AOT at the 0.55 μm wavelength, as obtained for (i) the solar zenith angles $\vartheta_0 = 0°$ (left), $\vartheta_0 = 30°$ (middle), and $\vartheta_0 = 60°$ (right); (ii) the M-1 pure oceanic (open circles), M-8 pure continental (open triangles), and M-14 heavy polluted anthropogenic (open squares) aerosol models defined by Tomasi *et al.* (2013); and (iii) the OS3 BRDF oceanic surface reflectance model.

nonabsorbing M-1 aerosols. The values of $\Delta F_{BoA}(t)$ were found to decrease as a function of $\tau_a(0.55\,\mu m)$ with less marked slopes passing from $\vartheta_0 = 0°$ to $\vartheta_0 = 60°$, these negative slopes being most marked for the M-14 (absorbing) aerosol model and weakest for the M-1 (nonabsorbing) aerosol model.

Correspondingly, slightly convex patterns were described by $\Delta F_{Atm}(t)$ plotted versus $\tau_a(0.55\,\mu m)$ for all the three values of ϑ_0 but with average absolute slope coefficients estimated to appreciably decrease as ϑ_0 increases from 0° to 60°, presenting nearly null values for the M-1 oceanic model, most marked values for the

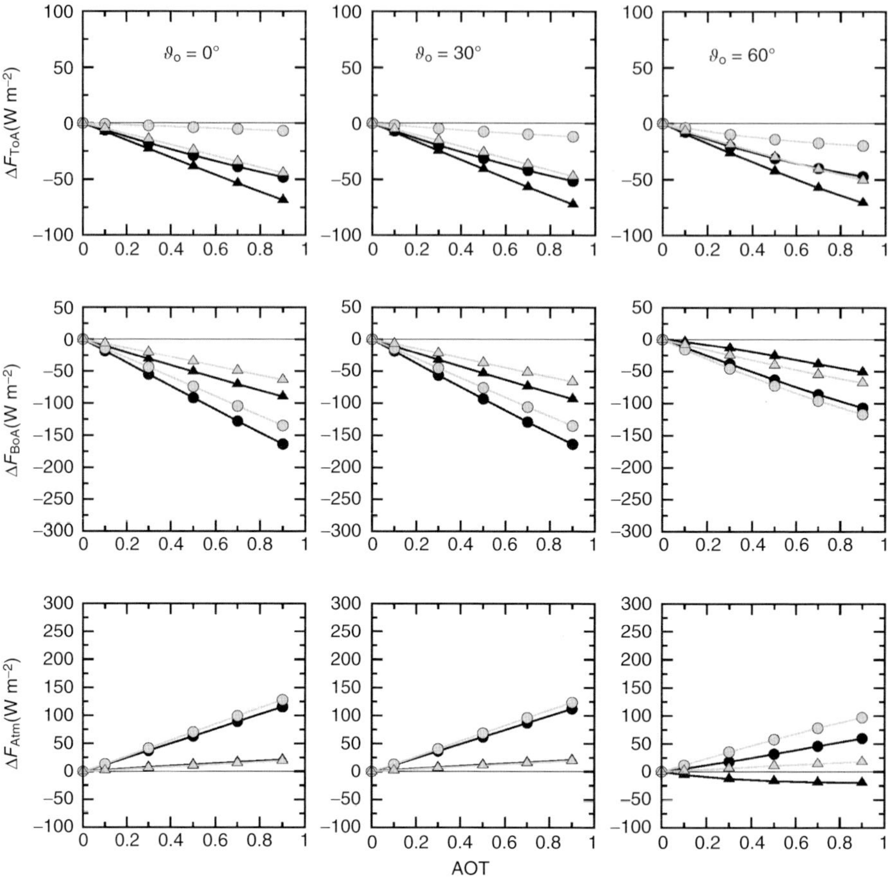

Figure 8.8 As in Figure 8.7, for the 6S-M maritime (circles) and 6S-C continental (triangles) and the OS2 BRDF sea surface (solid symbols) and VS2 BRDF vegetated land surface (gray symbols) reflectance models.

M-14 anthropogenic aerosol model, and intermediate values for the M-8 continental aerosol model.

In order to further study the dependence features of the three instantaneous DARF terms $\Delta F_{\mathrm{ToA}}(t)$, $\Delta F_{\mathrm{BoA}}(t)$, and $\Delta F_{\mathrm{Atm}}(t)$ on $\tau_a(0.55\,\mu\mathrm{m})$ for different values of ϑ_0 over both oceanic and vegetation-covered surfaces, the calculations of the three DARF terms shown in Figure 8.8 were made for (i) three values of solar zenith angle equal to 0°, 30°, and 60°; (ii) the 6S-M (maritime) and 6S-C (continental) aerosol models of Vermote *et al.* (1997a); and (iii) the BRDF non-Lambertian surface reflectance models OS2 (oceanic surface, with wind speed $\mathrm{WS} = 5\,\mathrm{m\,s^{-1}}$) and VS2 (corn field vegetated land surface). The linear dependence patterns obtained for all the three values of ϑ_0 and the 6S-C (continental) and 6S-M (maritime) aerosol models associated with the OS2 and VS2 models are shown in Figure 8.8.

The term $\Delta F_{\text{ToA}}(t)$ was estimated to vary in an approximately linear fashion as a function of AOT over both the surface reflectance models, yielding average negative slope coefficients which do not vary appreciably passing from $\vartheta_0 = 0°$ to $\vartheta_0 = 60°$, for both 6S-C and 6S-M aerosol models, and give more marked slopes over the OS2 surface. Instantaneous $\Delta F_{\text{BoA}}(t)$ was estimated to present nearly linear decreasing patterns, with gradually lower slopes passing from $\vartheta_0 = 0°$ to $\vartheta_0 = 60°$ and from the VS2 surface to the OS2 surface, until showing very similar features for $\vartheta_0 = 60°$, the average slopes being appreciably more marked for the 6S-C aerosol model than for the 6S-M aerosol model. The term $\Delta F_{\text{Atm}}(t)$ was estimated to be positive for $\vartheta_0 = 0°$ and $\vartheta_0 = 30°$, giving values for the 6S-M and 6S-C aerosol models which agree very closely over both surfaces. The values of $\Delta F_{\text{Atm}}(t)$ determined in the case with $\vartheta_0 = 60°$ were found to increase in Figure 8.8 as a function of $\tau_a(0.55\,\mu\text{m})$ for the 6S-C aerosol model over both surfaces, presenting a higher slope coefficient for the VS surface than for the OS2 surface. For the 6S-C continental aerosol model, very low slope coefficients were obtained over both surfaces, presenting positive signs for the VS2 surface and negative for the OS2 surface.

In general, it can be stated that the instantaneous DARF terms vary almost linearly as $\tau_a(0.55\,\mu\text{m})$ increases over its lower range for poorly absorbing aerosol particles, giving values of $\omega(0.55\,\mu\text{m}) > 0.95$. Measurements and modeling of the DARF effects induced by aerosols over the eastern coastal regions of the United States were performed by Russell, Hobbs, and Stowe (1999a); Russell *et al.* (1999b) during the TARFOX experiment, obtaining estimates of the instantaneous daytime upwelling shortwave flux changes ranging between +14 and +48 W m^{-2} for average values of τ_a over the visible spectral range varying between 0.20 and 0.55. These variations were found to be approximately proportional to τ_a for stable values of $\omega(0.55\,\mu\text{m})$ and to decrease considerably as $\omega(0.55\,\mu\text{m})$ decreases from 1.00 to 0.86. Bush and Valero (2002) followed the same procedure to evaluate the DARF occurring at the BoA level during the Indian Ocean Experiment (INDOEX) conducted at the Kaashidhoo Climate Observatory (KCO) (Republic of Maldives), obtaining a diurnal average value of ΔF_{BoA} over the sea surface equal to -72.2 ± 5.5 W m^{-2} per unit τ_a determined over the total solar broadband spectrum, corresponding to an average value of $\omega(0.50\,\mu\text{m}) = 0.874 \pm 0.028$, which was observed to vary appreciably as a function of SSA.

8.2.6.2 Dependence of Instantaneous DARF on Aerosol Single Scattering Albedo

The effects of aerosol SSA on the DARF term at the ToA level were studied in numerous articles (Russell, Kinne, and Bergstrom, 1997; Takemura *et al.*, 2002). To give an idea of these dependence features, a partial set of the aforementioned 43 aerosol models and some surface reflectance models described in the previous sections were used to evaluate the DARF terms at the ToA and BoA levels and within the atmosphere. Examining the spectral characteristics of numerous aerosol models, Tomasi *et al.* (2013) showed that airborne aerosols yield values of average SSA ω calculated over the 0.30–3.75 µm wavelength range, which vary between about 0.6–1.0 and, hence, cover the whole range of ω observable in the Earth's atmosphere. Correspondingly, the absorptance/reflectance ratio of airborne aerosols increases as the percent mass concentration of absorbing particulate components becomes higher. The results obtained by Chylek and Coakley (1974) indicate that aerosol particles causing cooling effects over relatively low-albedo surfaces, such as those measured over the offshore oceanic regions, can produce warming effects in the atmosphere when transported over high-albedo surfaces, such as those of the polar regions, consequently causing a change in the sign of ΔF_{ToA}.

To evaluate with good accuracy the variable dependence features of the three instantaneous DARF terms on average ω, the values of ΔF_{ToA}, ΔF_{BoA}, and ΔF_{Atm} obtained for 29 of the 43 aerosol models defined earlier are shown in Figure 8.9 as a function of parameter ω for the two cases with $\vartheta_0 = 30°$ and $\vartheta_0 = 60°$, over the OS1 oceanic surface (defined for calm wind conditions and surface-level wind speed WS = 2 m s^{-1}). Instantaneous ΔF_{ToA} was found to exhibit decreasing patterns over the 0.61 – 1.00 range of ω for both solar zenith angles, presenting comparable negative slopes of the two regression lines yielding ΔF_{ToA} varying from around −6 W m^{-2} to around −25 W m^{-2}. Instantaneous ΔF_{BoA} was estimated to be negative for $\omega < 0.75$, assuming values ranging from around −130 W m^{-2} to +5 W m^{-2} over the 0.61 – 1.00 range of ω. Figure 8.9 shows that the instantaneous ΔF_{BoA} increases almost linearly with ω, presenting some appreciable discrepancies between the values determined for the ten OPAC aerosol models and the other aerosol extinction models.

These findings indicate that aerosols of oceanic and continental origins, consisting mainly of poorly absorbing particulate matter giving values of $\omega > 0.95$, induce in general negative (cooling) effects on the surface – atmosphere system. Almost linear dependence features of instantaneous ΔF_{Atm} are also shown in Figure 8.9, presenting positive values for $\vartheta_0 = 30°$ for all the aerosol models over the total range of ω, which gradually decrease from more than +100 W m^{-2} to nearly null values as ω increases from 0.61 to 1.00. For $\vartheta_0 = 60°$, the values of ΔF_{Atm} were estimated to decrease almost linearly from more than +50 W m^{-2} to around −25 W m^{-2} over the 0.61 – 1.00 range of ω. This behavior can be reasonably explained considering that the soot content of particulate matter is so high for values of $\omega < 0.80$ as to produce marked aerosol absorption effects, which cause a moderate warming of the atmosphere and, hence, give a positive ΔF_{Atm} in each of these cases, exceeding the value of +50 W m^{-2} in the most polluted aerosol cases, associated with markedly negative (cooling) values of both ΔF_{ToA} and ΔF_{BoA}.

More significant results are shown in Figure 8.10, as obtained from the calculations made using the VS2 vegetation-covered surface reflectance model for the same aerosol models considered in Figure 8.9. Instantaneous ΔF_{ToA} was estimated to decrease gradually for both $\vartheta_0 = 30°$ and $\vartheta_0 = 60°$ as ω increases from ~0.61 to 1.00, giving in general positive values for $\omega < 0.80$ and negative values for $\omega > 0.80$. The values of ΔF_{ToA} determined using the 10 OPAC aerosol models were found to be considerably higher than those calculated for the other aerosol models chosen outside the OPAC set. In fact, the overall best-fit line drawn for the

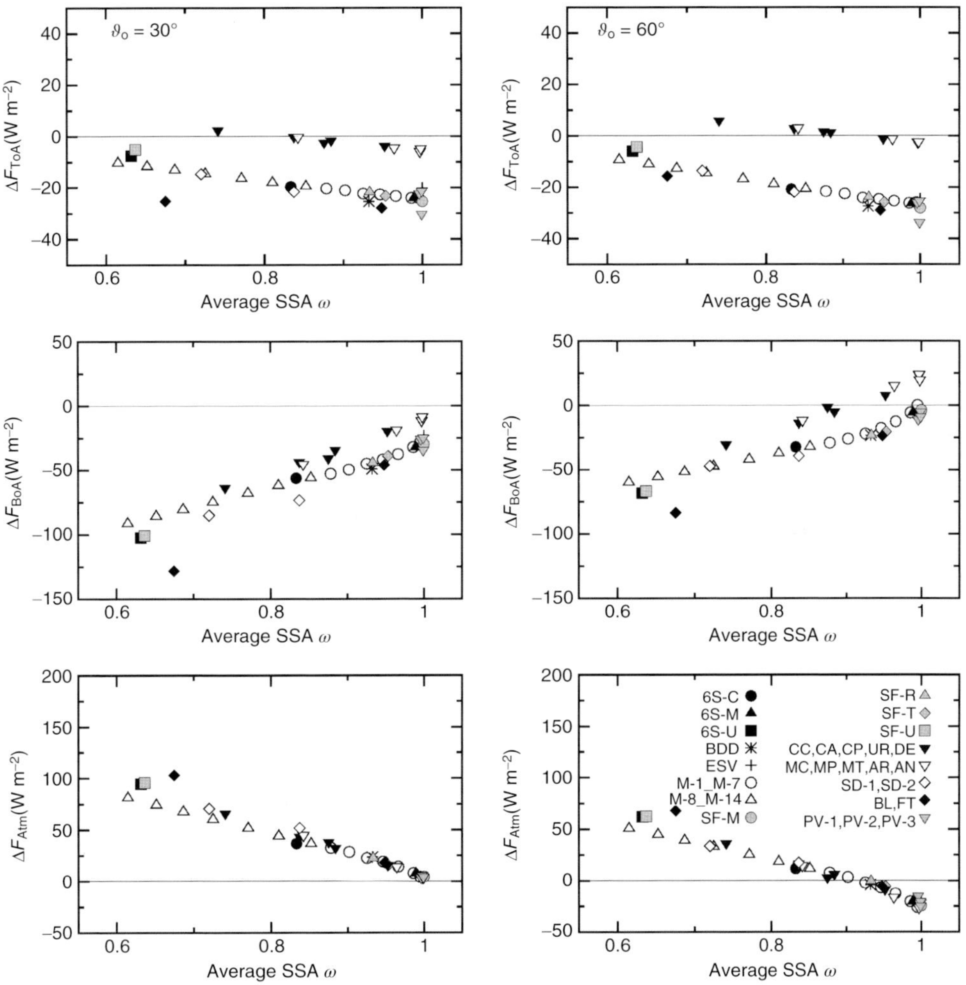

Figure 8.9 Scatter plots of instantaneous DARF terms $\Delta F_{ToA}(t)$ (upper part), $\Delta F_{BoA}(t)$ (middle part), and $\Delta F_{Atm}(t)$ (lower part) versus the average single scattering albedo (SSA) ω calculated over the OS1 BRDF sea surface for the solar zenith angles $\vartheta_0 = 30°$ (left) and $\vartheta_0 = 60°$ (right) and the following aerosol models: (i) the 6S-C (continental), 6S-M (maritime), and 6S-U (urban) of Vermote *et al.* (1997b); (ii) the background desert dust (BDD) and El Chichón stratospheric volcanic (ESV) aerosol of Tomasi *et al.* (2013); (iii) the M-1, M-7, M-8, and M-14 aerosol models of Tomasi *et al.* (2013); (iv) the 4 wet-air (RH = 50%) SF-M (maritime), SF-R (rural), SF-T (tropospheric), and SF-U (urban) aerosol models of Shettle and Fenn (1979); (v) the ten OPAC aerosol models of Hess, Koepke, and Schult (1998) defined for aerosols of different origins; and (vi) the seven additional aerosol models of Tomasi *et al.* (2013) for Saharan dust (SD-1 and SD-2), biomass burning smoke in free troposphere (FT) and atmospheric boundary layer (BL), the pre-Pinatubo (PV-1) background stratospheric particles, and the post-Pinatubo volcanic (PV-2, PV-3) particles in the low stratosphere.

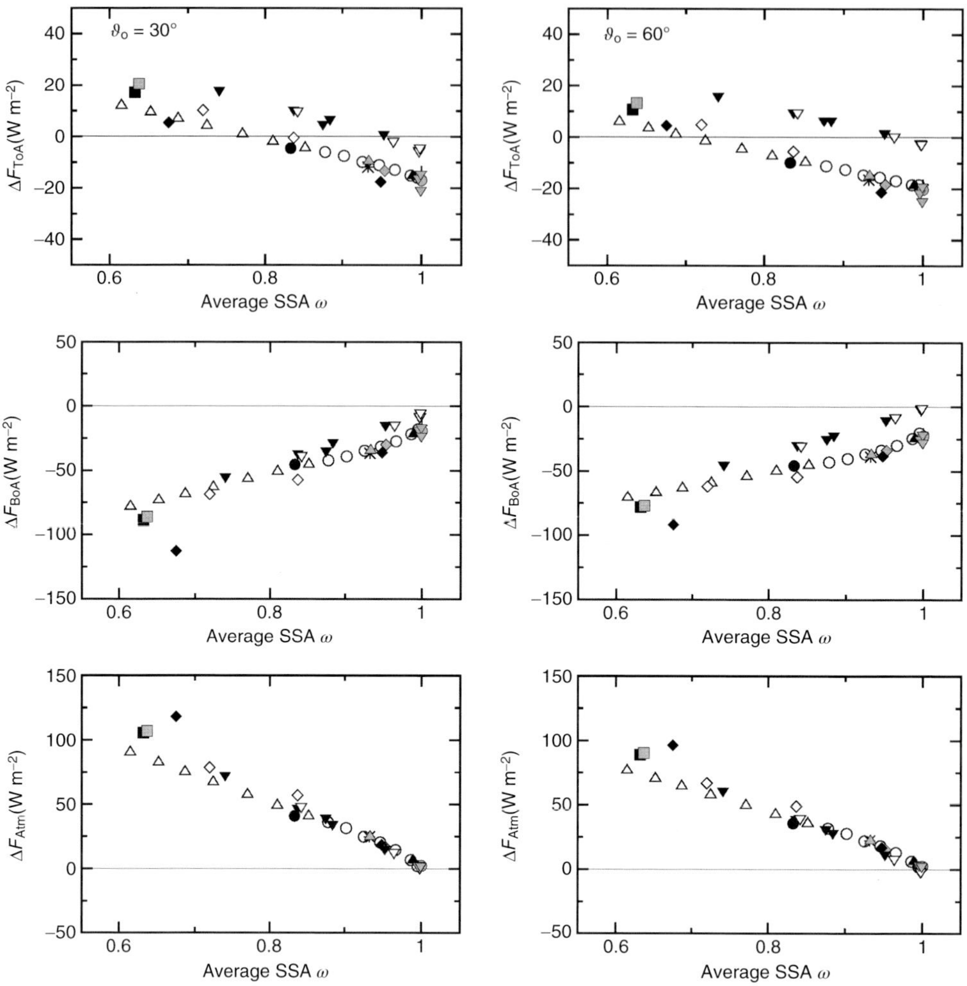

Figure 8.10 As in Figure 8.9, for (i) the solar zenith angles $\vartheta_0 = 30°$ (left) and $\vartheta_0 = 60°$ (right), (ii) the VS2 BRDF vegetation-covered (corn field) land surface reflectance model, and (iii) the twenty-nine aerosol models considered in Figure 8.9 and represented using the same symbols shown in the last right graph of Figure 8.9.

OPAC models yields slightly negative values of ΔF_{ToA} close to about $-5\,\text{W m}^{-2}$ for $\omega = 1.00$, while the best-fit line drawn for the non-OPAC aerosol models gives an intercept value of $\Delta F_{\text{ToA}} \approx -20\,\text{W m}^{-2}$ for unit ω in both cases with $\vartheta_0 = 30°$ and $\vartheta_0 = 60°$. Nearly linear dependence trends of instantaneous ΔF_{BoA} on ω were obtained for both solar zenith angles, yielding negative values increasing from less than $-80\,\text{W m}^{-2}$ for $\omega = 0.61$ to on average value between -25 and $-2\,\text{W m}^{-2}$ for unit ω, with slope coefficients not appreciably differing from the case with $\vartheta_0 = 30°$ to that with $\vartheta_0 = 60°$. Decreasing dependence patterns of instantaneous ΔF_{Atm} on average ω are shown in Figure 8.10, with only a small decrease in the evaluations made for $\vartheta_0 = 60°$ with respect to those determined for $\vartheta_0 = 30°$. The values of ΔF_{Atm} were found to decrease from around $+100\ \text{W m}^{-2}$ to nearly null values as ω increases from 0.61 to 1.00 at $\vartheta_0 = 30°$ and to decrease from around $+100\ \text{W m}^{-2}$ to nearly null values over the same range of ω in the case with $\vartheta_0 = 60°$.

8.2.6.3 Dependence of Instantaneous DARF on Underlying Surface Albedo

The previous results indicate clearly that the surface albedo characteristics play an important role in regulating the radiative transfer processes occurring within the surface–atmosphere system, due to the fact that the terrestrial surface (i) reflects upward the incoming solar radiation, which crosses again the atmosphere before reaching the outer space, being subject to further aerosol scattering and absorption processes influencing the DARF term ΔF_{ToA}, and (ii) absorbs part of the incoming solar radiation, thus subtracting such a fraction from the radiation balance of the atmosphere. Therefore, the surface reflectance causes important modifications in the spectrum of the solar radiation reflected upward and the spectral curve of the shortwave radiation absorbed by the atmosphere before reaching the ToA level. This suggests that surface albedo characteristics can have a strong impact on the instantaneous DARF term ΔF_{Atm}. The results shown in Figure 8.10 give a measure of the appreciably different evaluations of the instantaneous forcing terms ΔF_{ToA}, ΔF_{BoA}, and ΔF_{Atm} obtained for the surface albedo conditions of a vegetation-covered surface. The calculations of these instantaneous DARF terms made for $\vartheta_0 = 30°$ and $\vartheta_0 = 60°$, a set of fifteen aerosol models chosen among those described earlier, and the four OS BRDF surface reflectance models are plotted in Figure 8.11 versus the broadband surface albedo $A(\theta_o)$.

All the three terms ΔF_{ToA}, ΔF_{BoA}, and ΔF_{Atm} were estimated to vary largely as a function of the aerosol optical characteristics, especially for $\vartheta_0 = 60°$. Surface albedo $A(\vartheta_0 = 30°)$ varies from 0.035 to 0.051 passing from the OS1 to the OS4 surface reflectance models (see Table 8.2). Correspondingly, instantaneous ΔF_{ToA} varies from about $-24\ \text{W m}^{-2}$ to nearly $+4\ \text{W m}^{-2}$, (ii) ΔF_{BoA} varies from around -100 to $-10\ \text{W m}^{-2}$, and (iii) ΔF_{Atm} assumes increasing values from nearly null to $+100\ \text{W m}^{-2}$ for the various aerosol models. Surface albedo $A(\vartheta_0 = 60°)$ increases gradually from 0.105 for the OS4 surface reflectance model to 0.193 for the OS1 surface reflectance model, therefore covering a wider range than in the previous case (see Table 8.2). Instantaneous ΔF_{ToA} was estimated to vary from less than $-25\ \text{W m}^{-2}$ to nearly $+10\ \text{W m}^{-2}$ over such a larger range of A, while ΔF_{BoA} varies from less than $-80\ \text{W m}^{-2}$ to more than $+20\ \text{W m}^{-2}$ passing from the 6S-U aerosol model to the MC OPAC model, and ΔF_{Atm} varies from $-25\ \text{W m}^{-2}$ to more than $+75\ \text{W m}^{-2}$ passing from the MC OPAC model to the 6S-U model.

The results shown in Figure 8.11 can be compared with the analogous results obtained for the four VS BDRF surface reflectance models pertaining to vegetation-covered surfaces, which are shown in Figure 8.12.

Comparing the results of Figure 8.11 with those of Figure 8.12, it is evident that the four VS surface reflectance models provide a wider range of A than the OS models for both solar zenith angles $\vartheta_0 = 30°$ and $\vartheta_0 = 60°$, presenting a range

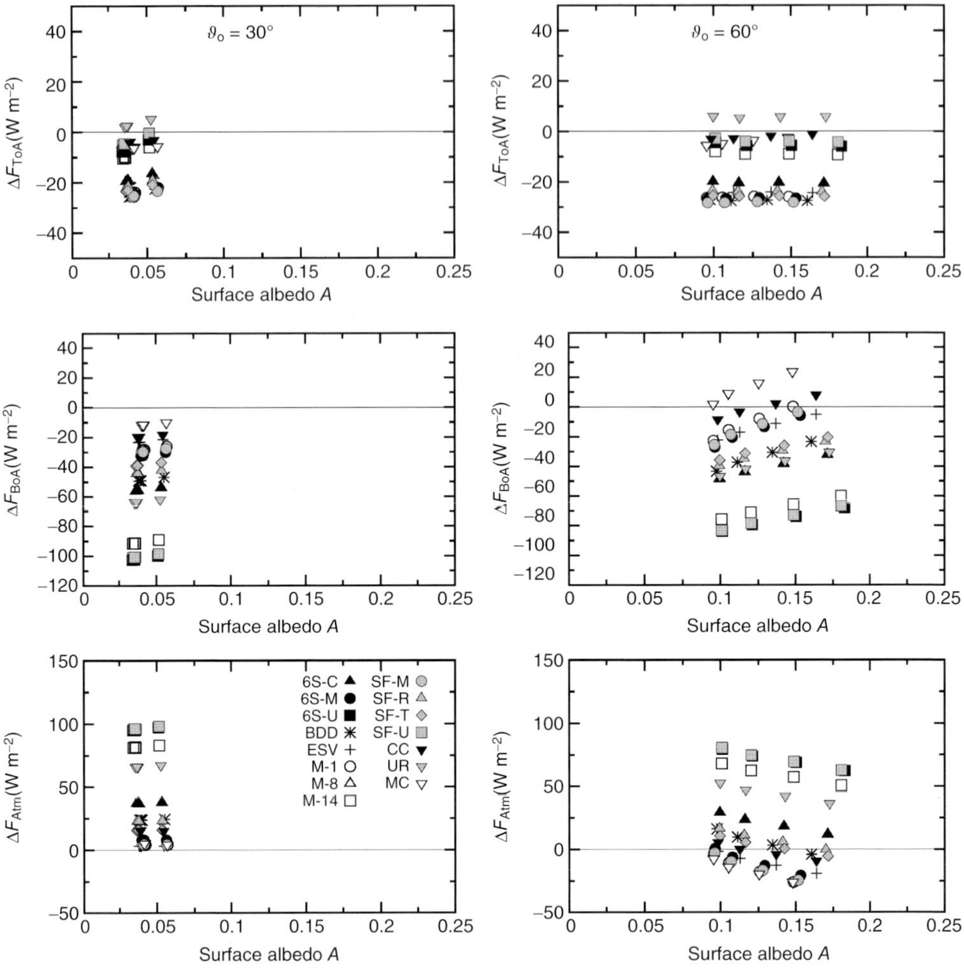

Figure 8.11 Scatter plots of instantaneous DARF terms $\Delta F_{ToA}(t)$ (upper part), $\Delta F_{BoA}(t)$ (middle part), and $\Delta F_{Atm}(t)$ (lower part) versus broadband surface albedo A, as obtained for (1) aerosol optical thickness τ_a(0:55 μm) = 0.30, (2) solar zenith angles $\vartheta_0 = 30°$ (left) and $\vartheta_0 = 60°$ (right), (3) over the four OS BRDF reflectance surfaces (with surface albedo gradually increasing from OS1 to OS4 models), and (4) the following aerosol models: (i) the 6S-C (continental), 6S-M (maritime), and 6S-U (urban) of Vermote *et al.* (1997a); (ii) the background desert dust (BDD) and El Chichón stratospheric volcanic (ESV) aerosol of Tomasi *et al.* (2013); (iii) the M-1, M-8, and M-14 aerosol models of Tomasi *et al.* (2013); (iv) the 4 wet-air (RH = 50%) SF-M (maritime), SF-R (rural), SF-T (tropospheric), and SF-U (urban) aerosol models of Shettle and Fenn (1979); and (v) the CC (continental clean), UR (urban), and MC (maritime clean) OPAC aerosol models of Hess, Koepke, and Schult (1998).

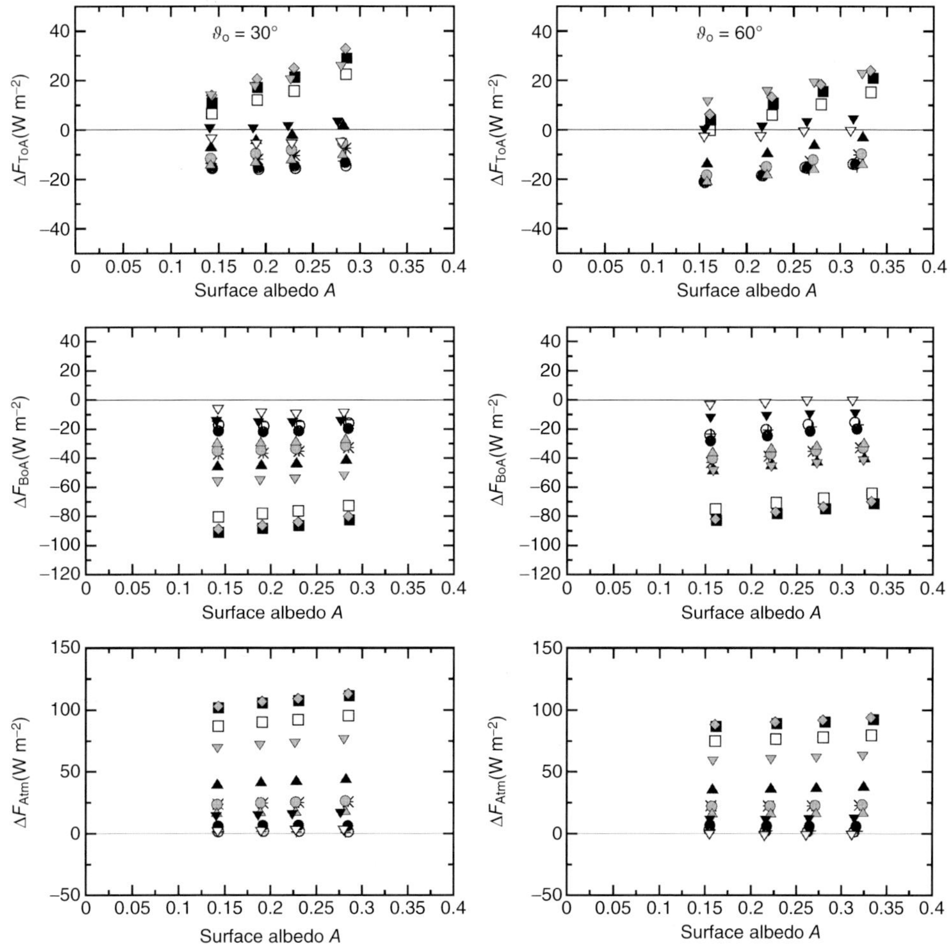

Figure 8.12 Scatter plots of instantaneous DARF terms $\Delta F_{ToA}(t)$ (upper part), $\Delta F_{BoA}(t)$ (middle part), and $\Delta F_{Atm}(t)$ (lower part) versus broadband surface albedo A, as obtained for (1) aerosol optical thickness $\tau_a(0.55\ \mu m) = 0.30$, (2) solar zenith angles $\vartheta_0 = 30°$ (left) and $\vartheta_0 = 60°$ (right), (3) over the four VS BRDF reflectance surfaces (with surface albedo gradually increasing from VS1 to VS4 models), and (4) the following aerosol models considered in this figure: (i) the 6S-C (continental), 6S-M (maritime), and 6S-U (urban) of Vermote *et al.* (1997a); (ii) the background desert dust (BDD) and El Chichón stratospheric volcanic (ESV) of Tomasi *et al.* (2013); (iii) the M-1, M-8, and M-14 aerosol models of Tomasi *et al.* (2013); (iv) the 4 wet-air (RH = 50%) SF-M (maritime), SF-R (rural), SF-T (tropospheric), and SF-U (urban) aerosol models of Shettle and Fenn (1979); and (v) the CC (continental clean), UR (urban), and MC (maritime clean) OPAC aerosol models of Hess, Koepke, and Schult (1998).

of $A(\vartheta_0 = 30°)$ from 0.139 (VS1) to 0.264 (VS4) and a range of $A(\vartheta_0 = 60°)$ from 0.155 (VS1) to 0.306 (VS4) (see Table 8.2). For surface albedo $A(\vartheta_0 = 30°)$ and the 15 selected aerosol models, instantaneous ΔF_{ToA} was found to vary from about $-18\,\mathrm{W\,m^{-2}}$ (6S-M aerosol model) to more than $+30\,\mathrm{W\,m^{-2}}$ (SF-T aerosol model), while ΔF_{BoA} assumes values ranging from around −90 (6S-U aerosol model) to $-5\,\mathrm{W\,m^{-2}}$ (MC OPAC model), and ΔF_{Atm} ranges from nearly null values (MC OPAC model) to more than $+110\,\mathrm{W\,m^{-2}}$ (6S-U aerosol model). For the case with $A(\vartheta_0) = 60°$, instantaneous ΔF_{ToA} was evaluated to vary from less than $-20\,\mathrm{W\,m^{-2}}$ (6S-M maritime aerosol model) to nearly $+10\,\mathrm{W\,m^{-2}}$ (UR OPAC aerosol model) over such a larger range of A, while ΔF_{BoA} increases from less than $-80\,\mathrm{W\,m^{-2}}$ (6S-U urban aerosol) to nearly null values (MC OPAC aerosol), and ΔF_{Atm} increases from nearly null values (MC OPAC model) to more than $+90\,\mathrm{W\,m^{-2}}$ (6S-U urban aerosol). The results provide evidence of the largely different DARF effects induced over the 0.15–0.30 surface albedo range by maritime (nonabsorbing) aerosols and urban-polluted (strongly absorbing) aerosols on the solar radiation fluxes.

8.2.6.4 **Dependence of Instantaneous DARF on Solar Zenith Angle**

A general picture is shown in Figure 8.13 of the variations of instantaneous DARF terms ΔF_{ToA}, ΔF_{BoA}, and ΔF_{Atm} determined over the $0° \le \vartheta_0 \le 80°$ range for a set of thirteen aerosol models chosen among those described earlier and the two surface reflectance cases represented by the OS2 and VS2 BRDF models. The dependence patterns shown in Figure 8.13 indicate that all the three instantaneous DARF terms vary greatly as a function of ϑ_0 for all the chosen aerosol models and both the surface reflectance models. It can be clearly seen that correct calculations of the largely variable instantaneous DARF terms as a function of ϑ_0 are of fundamental importance for the correct determination of the diurnally average DARF effects calculated over the whole 24 h insolation period, during which ϑ_0 varies regularly with time, presenting a daily minimum around noon. The results also indicate that the three instantaneous DARF terms can vary considerably as a function of ϑ_0, depending on local latitude and season. Evaluations of the variations in DARF due to such changes in ϑ_0 can be made on the basis of the dependence patterns of instantaneous ΔF_{ToA}, ΔF_{BoA}, and ΔF_{Atm} shown in Figure 8.13 for (i) $\tau_a(0.55\,\mu\mathrm{m}) = 0.30$, (ii) nine selected values of ϑ_0 taken in steps of 10° from 0° to 80°, (iii) a set of 13 aerosol models chosen among the 43 models described earlier, and (iv) the OS2 and VS2 surface reflectance models. It can be noted that instantaneous term ΔF_{ToA} exhibits rather stable patterns over both surfaces, presenting a wide and not pronounced minimum $\vartheta_0 \approx 60°$, with appreciable differences in the values of ΔF_{ToA} caused by the different surface reflectance conditions. Instantaneous term ΔF_{BoA} was found to assume rather stable values over the range $\vartheta_0 < 50°$, considerably differing from one aerosol model to another in both the OS2 and VS2 cases, and then to (i) rapidly increase as ϑ_0 increases from 50° to 80° over the OS2 surface until assuming positive values greater than $+100\,\mathrm{W\,m^{-2}}$ for $\vartheta_0 = 80°$ and (ii) remain more stable over the VS2 surface, slowly increasing with ϑ_0 until reaching slightly positive values for $\vartheta_0 = 80°$, only for the

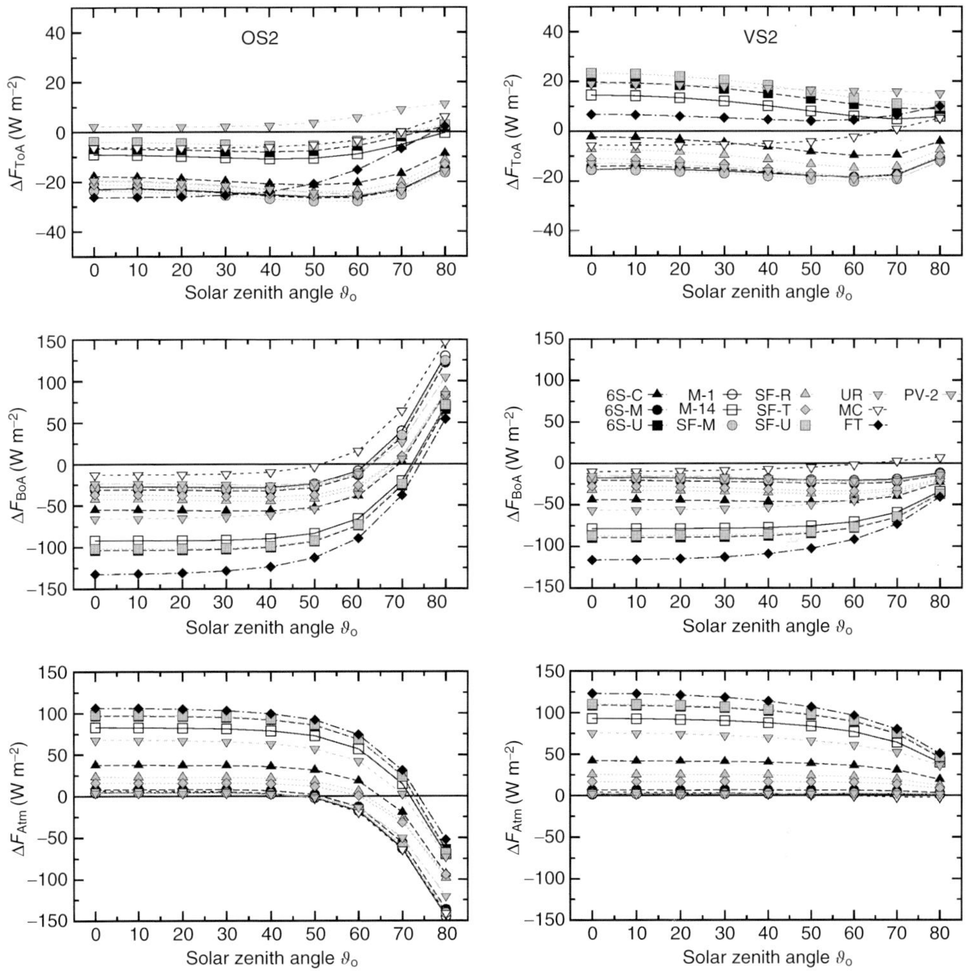

Figure 8.13 Instantaneous DARF terms $\Delta F_{ToA}(t)$ (upper part), $\Delta F_{BoA}(t)$ (middle part), and $\Delta F_{Atm}(t)$ (lower part) as a function of solar zenith angle ϑ_0 (determined in steps of 10° from 0° to 80°), as calculated for aerosol optical thickness $\tau_a(0.55\,\mu m) = 0.30$ and the BRDF surface reflectance models OS2 (left) and VS2 (right) using the following aerosol models: (i) the 6S-C (continental), 6S-M (maritime), and 6S-U (urban) of Vermote *et al.* (1997a); (ii) the M-1 (pure oceanic) and M-14 (heavy polluted) aerosol models of Tomasi *et al.* (2013); (iii) the 4 wet-air (RH = 50%) SF-M (maritime), SF-R (rural), SF-T (tropospheric), and SF-U (urban) aerosol models of Shettle and Fenn (1979); (iv) the UR (urban) and MC (maritime clean) OPAC aerosol models of Hess, Koepke, and Schult (1998); and (v) the free troposphere biomass burning smoke (FT) and the post-Pinatubo volcanic (PV-2) additional aerosol models of Tomasi *et al.* (2013), represented using the symbols given in the second right graph.

most absorbing aerosol models, such as the UR OPAC model. Correspondingly, instantaneous $\Delta F_{\rm Atm}$ was found to assume positive values over the $0° \leq \vartheta_0 \leq 50°$ range for both the OS2 and VS2 surface reflectance models, estimated to range from 0 to + 200 W m^{-2}, subsequently assume negative values for $\vartheta_0 > 50°$ over the OS2 surface, and exhibit positive values over the VS2 surface, gradually decreasing as ϑ_0 increases to assume values higher than 50°.

8.3 The Diurnally Average DARF Induced by Various Aerosol Types over Ocean and Land Surfaces

Using the instantaneous DARF evaluations of $\Delta F_{\rm ToA}(t)$ and $\Delta F_{\rm BoA}(t)$ calculated over the whole 24 h insolation period (from sunrise to sunset) in terms of Eqs. (8.12) and (8.13) at the ToA level and the BoA level, respectively, the so-called diurnally average aerosol forcing (hereinafter labeled with symbol ΔDF) can be calculated in terms of the following integrals:

$$\Delta{\rm DF}_{\rm ToA} = \int_{\rm sunrise}^{\rm sunset} \Delta F_{\rm ToA}(t)dt/24\,{\rm h}, \tag{8.18}$$

and

$$\Delta{\rm DF}_{\rm BoA} = \int_{\rm sunrise}^{\rm sunset} \Delta F_{\rm BoA}(t)dt/24\,{\rm h}, \tag{8.19}$$

at the ToA and BoA levels, respectively. It is well known that the sunlit period from sunrise to sunset varies considerably as a function of latitude and with season. Evaluations of $\Delta{\rm DF}_{\rm ToA}$ and $\Delta{\rm DF}_{\rm BoA}$ obtained from field measurements performed at a site very far from the equator may be subject to significant variations throughout the year. However, the concept underlying the definitions of $\Delta{\rm DF}_{\rm ToA}$ and $\Delta{\rm DF}_{\rm BoA}$ in terms of Eqs. (8.18) and (8.19) is correct, since these two optical quantities provide realistic estimates of RF associated with solar radiation budget change in the surface–atmosphere system on a certain day of the year and at a certain site.

The absolute magnitude of the DARF terms at the ToA level closely depends not only on the amount of radiation entering the atmosphere but also on the aerosol mass content and optical characteristics. An appropriate parameter suitable for defining the atmospheric content of aerosol particles is $\tau_a(\lambda)$, since it provides a measure of the solar radiation extinction produced by aerosols along the vertical atmospheric path. The DARF term at the ToA level is expected to be proportional to $\tau_a(\lambda)$ over a certain surface of known reflectance through a constant closely depending on the optical characteristics of atmospheric particles. In fact, $\Delta{\rm DF}_{\rm ToA}$ is given by the daily change in the upwelling flux of solar radiation reflected by the surface–atmosphere system toward space as a result of aerosol scattering and absorption processes. Therefore, it depends closely on the amount of aerosol particles extinguishing the incoming solar radiation, which is proportional to $\tau_a(\lambda)$. New calculations of $\Delta{\rm DF}_{\rm ToA}$ and $\Delta{\rm DF}_{\rm BoA}$ are given in this section, obtained from both field measurements of $\tau_a(\lambda)$ and airborne aerosol optical parameters and radiation transfer simulations. Particular attention was paid to define (i) the aerosol radiative parameters measured by means of ground-based remote sensing techniques and *in situ* sampling measurements and (ii) the spectral and directional characteristics of the surface reflectance BRDF functions, which are particularly important in order to achieve realistic estimates of the DARF terms (Chylek and Coakley, 1974). For this purpose, the BRDF non-Lambertian models determined by Tomasi *et al.* (2013) were used by Tomasi *et al.* (2015a) to represent the surface reflectance characteristics of the oceanic surface in the Atlantic Ocean and Mediterranean Sea and the land surface in European and North American continental regions and remote Himalayan, Arctic, and Antarctic areas.

Based on the previous concepts, the rate at which the surface–atmosphere system is forced at the ToA level per unit $\tau_a(\lambda)$ is known as the *aerosol forcing efficiency* E_{ToA} (according to the definitions given by Bush and Valero (2003)). This parameter can be evaluated over the whole 24 h period as the ratio between the aerosol RF $\Delta\text{DF}_{\text{ToA}}$ defined in Eq. (8.18) and the daily mean value of $\tau_a(\lambda)$ measured at wavelength $\lambda = 0.55\,\mu\text{m}$ (or at a different visible wavelength, in cases where $\tau_a(0.55\,\mu\text{m})$ is not directly derived from the sun-photometer measurements). Similarly, the aerosol forcing efficiency E_{BoA} at the BoA level can be calculated on each measurement day as the ratio between the diurnally average $\Delta\text{DF}_{\text{BoA}}$ determined in terms of Eq. (8.19) from field measurements and the daily mean value of $\tau_a(0.55\,\mu\text{m})$ calculated at this wavelength over the sunlit period or the average value of τ_a calculated over the 0.30–0.70 μm broadband spectral range, as made by Hignett *et al.* (1999) during the TARFOX experiment. On the basis of these remarks, we used the following equations:

$$E_{\text{ToA}} = \Delta\text{DF}_{\text{ToA}}/\tau_a(0.55\,\mu\text{m}) \tag{8.20}$$

and

$$E_{\text{BoA}} = \Delta\text{DF}_{\text{BoA}}/\tau_a(0.55\,\mu\text{m}) \tag{8.21}$$

to perform the present calculations. In addition, the DARF efficiency E_{Atm} in the atmosphere can be simply calculated as the difference

$$E_{\text{Atm}} = E_{\text{ToA}} - E_{\text{BoA}}, \tag{8.22}$$

between the values of E_{ToA} and E_{BoA}, this quantity being therefore given by ratio $\Delta\text{DF}_{\text{Atm}}/\tau_a(0.55\,\mu\text{m})$ and providing a measure of the atmospheric heating produced by aerosols.

The previous calculations of the DARF efficiency parameters are in general very useful to characterize the DARF effects occurring within the surface–atmosphere system, since the diurnal average DARF terms $\Delta\text{DF}_{\text{ToA}}$, $\Delta\text{DF}_{\text{BoA}}$, and $\Delta\text{DF}_{\text{Atm}} = \Delta\text{DF}_{\text{ToA}} - \Delta\text{DF}_{\text{BoA}}$ give a measure of aerosol RF at the ToA and BoA levels and of the amount of energy entering the atmosphere, respectively, which are all closely related to the atmospheric content and microphysical and radiative characteristics of airborne aerosols.

8.3.1 Description of the Calculation Method Based on the Field Measurements of Aerosol Optical Parameters

In order to calculate the DARF terms ΔDF_{ToA}, ΔDF_{BoA}, and ΔDF_{Atm} on the basis of the field measurements of the main aerosol radiative parameters conducted during various experimental campaigns, Tomasi *et al.* (2015a) followed a rigorous procedure (hereinafter referred to as DARF-PROC), subdivided into the following seven steps:

Analysis of Field Data to Determine the Columnar Aerosol Extinction Parameters

The spectral series of $\tau_a(\lambda)$ were derived at visible and NIR wavelengths from the direct solar irradiance measurements performed on cloudless-sky days using sophisticated sun photometers and sky/sun radiometers, such as the AERONET Cimel CE318 sun/sky radiometer (Holben *et al.*, 1998) and the SKYNET Prede POM-01 L and POM-02 L sun/sky radiometers (Nakajima *et al.*, 2003). The direct solar irradiance measurements taken at various wavelengths centered within the main windows of the atmospheric transmittance spectrum over the 0.38–2.20 μm wavelength range were examined in terms of the Bouguer–Lambert–Beer to determine the total atmospheric optical thickness $\tau(\lambda)$ at each wavelength, from which the values of $\tau_a(\lambda)$ were calculated after appropriate corrections for the Rayleigh scattering optical thickness (Tomasi *et al.*, 2005) and the partial optical thicknesses due to the absorption by minor gases (ozone, NO_2 and its dimer, H_2O, and oxygen dimer). To obtain a good coverage of the time variations in $\tau_a(\lambda)$ observed on each measurement day, the sun-photometric measurements were regularly taken on each clear-sky day with a frequency no lower than one spectral scanning every 20–30 min from sunrise to sunset, collecting on average more than 25 spectral scannings per day. The spectral values of $\tau_a(\lambda)$ were examined in terms of the Ångström (1964) formula to determine the atmospheric turbidity parameters α and β over the 0.38–0.87 μm spectral range for each sun-photometric scanning.

Determination of the Columnar Aerosol Refractive Index

Regular measurements of sky brightness in the almucantar were in general conducted using the AERONET and SKYNET sun/sky radiometers, with a frequency of at least one measurement per hour at some selected wavelengths from 0.40 to 1.10 μm, from which the main optical parameters of airborne aerosols were retrieved. The spectral values of real part $n(\lambda)$ and imaginary part $k(\lambda)$ of the complex particulate refractive index were derived at various measurement hours of each golden day (Holben *et al.*, 1998), finding the daily mean values of $n(0.50\ \mu m)$ and $k(0.50\ \mu m)$ given in Tables 8.4–8.11, which were obtained by Tomasi *et al.* (2015a) on several clear-sky days of seven field experimental campaigns (CLEARCOLUMN (ACE-2), PRIN-2004, AEROCLOUDS, Ev-K2-CNR, POLAR-AOD, Aerosols99, and DOE/ARM/AIOP).

Table 8.4 Daily mean values of aerosol optical thickness τ_a(0.55 μm) and single scattering albedo ω(0.55 μm) obtained on the 10 golden days of the CLEARCOLUMN (ACE-2) experiment conducted in summer 1997 at Sagres (Portugal) for different aerosol types and corresponding values of the diurnally average aerosol forcing terms ΔDF_{ToA} at the ToA level, ΔDF_{BoA} at the BoA level, and ΔDF_{Atm} within the atmosphere and the diurnally average DARF efficiencies E_{ToA}, E_{BoA}, and E_{Atm} determined by Tomasi *et al.* (2015a) using the BRDF surface reflectance models OS2, VS1, and VS4 defined in Tables 8.1 and 8.2.

Measurement day of 1997 and aerosol type	Daily mean aerosol optical thickness τ_a(0.55 μm)	Daily mean ω(0.55 μm)	Surface albedo model	Diurnally average DARF terms (W m^{-2})			Diurnally average DARF efficiencies (W m^{-2})		
				ΔDF_{ToA}	ΔDF_{BoA}	ΔDF_{Atm}	E_{ToA}	E_{BoA}	E_{Atm}
June 19 (maritime clean)	0.053	0.96	OS2	−5.2	−1.1	−4.2	−98.5	−19.9	−78.6
			VS1	−2.7	−4.5	+1.7	−51.8	−84.7	+32.9
			VS4	−1.7	−3.5	+1.8	−31.6	−66.1	+34.6
June 20 (maritime–continental)	0.067	0.81	OS2	−4.9	−4.0	−0.8	−72.4	−60.3	−12.1
			VS1	−2.0	−6.9	+4.9	−30.3	−103.6	+73.2
			VS4	−0.5	−5.7	+5.3	−6.8	−85.3	+78.4
June 23 (maritime–continental)	0.065	0.87	OS2	−4.7	−4.2	−0.5	−72.5	−64.8	−7.7
			VS1	−1.0	−7.7	+6.7	−14.9	−118.1	+103.2
			VS4	+1.8	−5.8	+7.6	+27.9	−89.5	+117.4
June 25 (maritime–continental)	0.086	0.90	OS2	−7.3	−10.3	+3.0	−85.1	−120.3	+35.2
			VS1	−3.1	−12.8	+9.7	−35.6	−148.4	+112.9
			VS4	−0.5	−11.0	+10.5	−6.1	−127.9	+121.8
July 5 (continental-polluted)	0.051	0.91	OS2	−3.7	−3.2	−0.5	−73.1	−63.0	−10.1
			VS1	−1.0	−5.6	+4.6	−19.1	−110.1	+91.0
			VS4	+1.0	−4.3	+5.3	+18.9	−84.8	+103.7
July 7 (continental-polluted)	0.132	0.80	OS2	−7.0	−9.3	+2.3	−52.8	−70.3	+17.5
			VS1	−2.3	−13.2	+10.8	−17.6	−99.7	+82.1
			VS4	+1.0	−11.0	+12.0	+7.5	−83.3	+90.8

(*continued overleaf*)

Table 8.4 (*continued*)

Measurement day of 1997 and aerosol type	Daily mean aerosol optical thickness τ_a(0.55 μm)	Daily mean ω(0.55 μm)	Surface albedo model	Diurnally average DARF terms (W m^{-2})			Diurnally average DARF efficiencies (W m^{-2})		
				ΔDF_{ToA}	ΔDF_{BoA}	ΔDF_{Atm}	E_{ToA}	E_{BoA}	E_{Atm}
July 10	0.233	0.85	OS2	−12.2	−19.0	+6.8	−52.5	−81.7	+29.1
(continental-			VS1	−5.8	−21.4	+15.6	−24.9	−92	+67.1
polluted)			VS4	−1.9	−18.8	+16.9	−8.1	−80.5	+72.4
July 11	0.217	0.89	OS2	−5.5	−5.9	+0.4	−25.4	−27.1	+1.7
(continental-			VS1	−2.5	−7.9	+5.4	−11.7	−36.5	+24.8
polluted)			VS4	−0.9	−6.7	+5.8	−4.3	−31.0	+26.7
July 18	0.298	0.90	OS2	−12.0	−13.5	+1.5	−40.1	−45.3	+5.2
(continental-			VS1	−5.7	−20.5	+14.8	−19.2	−68.9	+49.7
polluted)			VS4	−1.6	−17.5	+15.9	−5.5	−58.7	+53.2
July 20	0.291	0.84	OS2	−11.3	−14.9	+3.5	−39.0	−51.0	+12.0
(continental-			VS1	−5.4	−19.6	+14.2	−18.5	−67.4	+48.9
polluted)			VS4	−1.6	−16.9	+15.3	−5.5	−58.1	+52.6

Table 8.5 Daily mean values of aerosol optical thickness τ_a(0.55 μm) and single scattering albedo ω(0.55 μm) obtained on the 12 golden days of the PRIN-2004 campaign conducted from 25 April to 2 October 2003, at Lecce (Salento, Puglia, Southern Italy) (Lanconelli, 2007) for different aerosol types and corresponding values of the diurnally average aerosol forcing terms ΔDF_{ToA} at the ToA level, ΔDF_{BoA} at the BoA level, and ΔDF_{Atm} within the atmosphere and the diurnally average DARF efficiencies E_{ToA}, E_{BoA}, and E_{Atm} determined by Tomasi *et al.* (2015a) using the BRDF surface reflectance models OS3, OS4, VS1, and VS4 defined in Tables 8.1 and 8.2.

Measurement day of 1997 and aerosol type	Daily mean aerosol optical thickness τ_a(0.55 μm)	Daily mean ω(0.55 μm)	Surface albedo model	Diurnally average DARF terms ($W m^{-2}$)			Diurnally average DARF efficiencies ($W m^{-2}$)		
				ΔDF_{ToA}	ΔDF_{BoA}	ΔDF_{Atm}	E_{ToA}	E_{BoA}	E_{Atm}
April 25 (remote continental)	0.271	0.919	OS3	−11.2	−15.9	+4.6	−41.4	−58.6	+17.2
			OS4	−10	−17.7	+7.7	−36.7	−65.3	+28.6
			VS1	−6.4	−18.9	+12.5	−23.6	−69.7	+46.1
			VS4	−3.3	−16.6	+13.3	−12.2	−61.1	+48.9
April 30 (maritime −continental)	0.232	0.901	OS3	−12.5	−12.3	−0.1	−53.7	−53.1	−0.6
			OS4	−11.2	−14.9	+3.7	−48.1	−64.1	+16.0
			VS1	−7	−16	+9	−30.2	−68.9	+38.7
			VS4	−3.7	−13.3	+9.5	−16.1	−57.2	+41.1
May 1 (Saharan dust)	0.230	0.954	OS3	−12.3	−13.6	+1.3	−53.4	−59.0	+5.6
			OS4	−10.9	−16.4	+5.5	−47.3	−71.2	+23.9
			VS1	−6.4	−17.3	+10.9	−27.7	−75.3	+47.6
			VS4	−2.9	−14.4	+11.4	−12.7	−62.4	+49.7
May 20 (continental-polluted)	0.144	0.874	OS3	−6.9	−10.4	+3.5	−48.2	−72.3	+24.1
			OS4	−6	−11.7	+5.7	−42.0	−81.5	+39.5
			VS1	−3.3	−12.6	+9.3	−23.0	−87.5	+64.5
			VS4	−1	−10.9	+10	−6.6	−76.0	+69.4
May 29 (rural-continental)	0.186	0.938	OS3	−9.3	−9.4	+0.1	−49.8	−50.4	+0.6
			OS4	−8.3	−10.7	+2.4	−44.7	−57.8	+13.1
			VS1	−5.5	−11.7	+6.2	−29.7	−63.1	+33.4
			VS4	−3.6	−10.1	+6.6	−19.2	−54.5	+35.3

(*continued overleaf*)

Table 8.5 (*continued*)

Measurement day of 1997 and aerosol type	Daily mean aerosol optical thickness τ_a(0.55 μm)	Daily mean ω(0.55 μm)	Surface albedo model	Diurnally average DARF terms (W m^{-2})			Diurnally average DARF efficiencies (W m^{-2})		
				ΔDF_{ToA}	ΔDF_{BoA}	ΔDF_{Atm}	E_{ToA}	E_{BoA}	E_{Atm}
June 10 (rural-continental)	0.345	0.977	OS3	−16.3	−20.1	+3.8	−47.3	−58.2	+10.9
			OS4	−14.7	−23	+8.3	−42.6	−66.7	+24.1
			VS1	−9.8	−23.5	+13.7	−28.5	−68.3	+39.8
			VS4	−5.9	−20.5	+14.6	−17.1	−59.4	+42.3
July 17 (Saharan dust)	0.627	0.940	OS3	−31.4	−50.2	+18.8	−50.0	−80.0	+30.0
			OS4	−28.4	−55.7	+27.2	−45.3	−88.8	+43.5
			VS1	−17.8	−53.6	+35.8	−28.4	−85.5	+57.1
			VS4	−8	−45.2	+37.2	−12.8	−72.1	+59.3
July 24 (continental and Saharan dust)	0.402	0.919	OS3	−19.2	−26.5	+7.3	−47.9	−66.0	+18.1
			OS4	−17.2	−30.3	+13.1	−42.8	−75.4	+32.6
			VS1	−10.6	−31	+20.5	−26.3	−77.2	+50.9
			VS4	−4.9	−26.5	+21.6	−12.2	−65.9	+53.7
August 8 (continental-polluted)	0.540	0.939	OS3	−21.4	−27.8	+6.5	−39.6	−51.6	+12.0
			OS4	−19.4	−30.4	+11	−35.9	−56.3	+20.4
			VS1	−13.5	−31.5	+18	−25.0	−58.3	+33.3
			VS4	−8.4	−27.5	+19.1	−15.5	−50.9	+35.4
August 30 (continental-polluted)	0.582	0.938	OS3	−26.9	−24.9	−2	−46.2	−42.8	−3.4
			OS4	−24.8	−30.9	+6.1	−42.6	−53.1	+10.5
			VS1	−17.7	−31.5	+13.9	−30.4	−54.2	+23.8
			VS4	−11.5	−25.5	+14	−19.8	−43.9	+24.1
September 20 (rural-continental)	0.350	0.949	OS3	−13.9	−15.9	+2	−39.8	−45.6	+5.8
			OS4	−12.6	−19.3	+6.7	−35.9	−55.1	+19.2
			VS1	−8.7	−20.8	+12.1	−24.9	−59.4	+34.5
			VS4	−5.1	−18	+12.9	−14.6	−51.3	+36.7
October 2 (continental-polluted)	0.325	0.920	OS3	−15	−14.3	−0.7	−46.1	−44.0	−2.1
			OS4	−13.3	−19.7	+6.3	−41.0	−60.5	+19.5
			VS1	−8	−21.8	+13.9	−24.5	−67.1	+42.6
			VS4	−2.7	−17.3	+14.6	−8.4	−53.2	+44.8

Table 8.6 Daily mean values of aerosol optical thickness $\tau_a(0.55\ \mu m)$ and single scattering albedo $\omega(0.55\ \mu m)$ obtained on the 18 golden days of the AEROCLOUDS campaign conducted from May 2007 to March 2008 at San Pietro Capofiume (Po Valley, Italy) by Mazzola *et al.* (2010) for different aerosol types and corresponding values of the diurnally average aerosol forcing terms ΔDF_{ToA} at the ToA level, ΔDF_{BoA} at the BoA level, and ΔDF_{Atm} within the atmosphere and the diurnally average DARF efficiencies E_{ToA}, E_{BoA}, and E_{Atm} determined by Tomasi *et al.* (2015a) using the BRDF surface reflectance models VS1, VS2, VS3, and VS4 defined in Tables 8.1 and 8.2.

Measurement day of 1997 and aerosol type	Daily mean aerosol optical thickness $\tau_a(0.55\ \mu m)$	Daily mean $\omega(0.55\ \mu m)$	Surface albedo model	Diurnally average DARF terms ($W\ m^{-2}$)			Diurnally average DARF efficiencies ($W\ m^{-2}$)		
				ΔDF_{ToA}	ΔDF_{BoA}	ΔDF_{Atm}	E_{ToA}	E_{BoA}	E_{Atm}
May 18, 2007 (maritime –continental)	0.115	0.986	VS1	−5.1	−7.6	+2.5	−43.7	−65.2	+21.5
			VS2	−4.3	−7.0	+2.7	−37.1	−59.9	+22.8
			VS3	−4.0	−6.8	+2.8	−34.2	−58.3	+24.1
			VS4	−3.5	−6.4	+2.9	−29.7	−54.8	+25.1
May 24, 2007 (rural-continental)	0.290	0.980	VS1	−11.9	−15.6	+3.7	−42.7	−56.1	+13.4
			VS2	−10.9	−14.5	+3.6	−39.2	−52.3	+13.1
			VS3	−10.3	−14.1	+3.8	−37.1	−50.6	+13.5
			VS4	−9.6	−13.3	+3.7	−34.4	−47.8	+13.4
June 21, 2007 (continental-polluted)	0.349	0.902	VS1	−11.5	−29.0	+17.5	−33.6	−84.6	+51.0
			VS2	−9.5	−27.3	+17.9	−27.6	−79.7	+52.1
			VS3	−7.9	−26.3	+18.3	−23.1	−76.6	+53.5
			VS4	−6.0	−24.8	+18.8	−17.4	−72.2	+54.8
July 1, 2007 (maritime –continental)	0.318	0.952	VS1	−12.0	−17.3	+5.3	−41.4	−59.8	+18.4
			VS2	−10.5	−16.0	+5.5	−36.1	−55.2	+19.1
			VS3	−9.8	−15.5	+5.7	−33.8	−53.6	+19.8
			VS4	−8.7	−14.6	+5.9	−30.2	−50.6	+20.4
July 8, 2007 (maritime –continental)	0.156	0.968	VS1	−7.4	−10.5	+3.1	−47.8	−67.9	+20.1
			VS2	−6.5	−9.8	+3.3	−42.4	−63.7	+21.3
			VS3	−6.2	−9.6	+3.4	−40.0	−62.3	+22.3
			VS4	−5.5	−9.1	+3.6	−35.9	−59.0	+23.1

(*continued overleaf*)

Table 8.6 (*continued*)

Measurement day of 1997 and aerosol type	Daily mean aerosol optical thickness τ_a(0.55 μm)	Daily mean ω(0.55 μm)	Surface albedo model	Diurnally average DARF terms (W m^{-2})			Diurnally average DARF efficiencies (W m^{-2})		
				ΔDF_{ToA}	ΔDF_{BoA}	ΔDF_{Atm}	E_{ToA}	E_{BoA}	E_{Atm}
July 17, 2007 (continental-polluted)	0.292	0.901	VS1	−9.2	−22.3	+13.1	−32.0	−77.3	+45.3
			VS2	−7.3	−20.7	+13.4	−25.3	−71.7	+46.4
			VS3	−5.8	−19.6	+13.8	−20.3	−68.0	+47.7
			VS4	−4.3	−18.4	+14.1	−14.8	−63.6	+48.8
July 26, 2007 (maritime −continental)	0.153	0.932	VS1	−5.6	−11.2	+5.7	−36.7	−74.0	+37.3
			VS2	−4.4	−10.4	+6.1	−28.8	−68.9	+40.1
			VS3	−3.9	−10.2	+6.3	−25.4	−67.1	+41.7
			VS4	−3.0	−9.6	+6.6	−19.6	−63.1	+43.5
Aug. 2, 2007 (maritime −continental)	0.374	0.940	VS1	−13.5	−21.0	+7.4	−39.5	−61.2	+21.7
			VS2	−11.7	−19.4	+7.6	−34.2	−56.4	+22.2
			VS3	−10.5	−18.4	+7.8	−30.7	−53.5	+22.8
			VS4	−9.2	−17.2	+7.9	−27.0	−50.1	+23.1
Aug. 5, 2007 (continental-polluted)	0.132	0.976	VS1	−5.2	−9.2	+4.0	−38.2	−67.3	+29.1
			VS2	−4.2	−8.5	+4.3	−30.8	−61.9	+31.1
			VS3	−3.8	−8.2	+4.4	−27.8	−60.1	+32.3
			VS4	−3.1	−7.7	+4.6	−22.9	−56.4	+33.5
Aug. 19, 2007 (continental-polluted)	0.283	0.974	VS1	−9.2	−15.9	+6.8	−27.8	−48.4	+20.6
			VS2	−7.7	−14.6	+6.9	−23.3	−44.4	+21.1
			VS3	−7.0	−14.1	+7.1	−21.2	−42.8	+21.6
			VS4	−6.0	−13.3	+7.3	−18.3	−40.4	+22.1

Aug. 27, 2007 (continental-polluted)	0.256	0.935	VS1	−9.0	−15.5	+6.4	−34.9	−59.8	+24.9
			VS2	−7.6	−14.1	+6.5	−29.2	−54.4	+25.2
			VS3	−6.5	−13.2	+6.7	−25.3	−51.2	+25.9
			VS4	−5.6	−12.4	+6.8	−21.6	−47.9	+26.3
Sept. 9, 2007 (maritime −continental)	0.207	0.963	VS1	−7.3	−9.9	+2.6	−36.6	−49.8	+13.2
			VS2	−6.4	−9.0	+2.7	−32.2	−45.5	+13.3
			VS3	−6.0	−8.8	+2.8	−30.1	−44.2	+14.1
			VS4	−5.5	−8.3	+2.9	−27.5	−41.9	+14.4
Sept. 13, 2007 (maritime −continental)	0.084	0.910	VS1	−2.7	−8.0	+5.3	−35.5	−105.4	+69.9
			VS2	−1.5	−7.3	+5.8	−19.9	−96.5	+76.6
			VS3	−0.9	−7.0	+6.1	−12.2	−92.5	+80.3
			VS4	−0.1	−6.5	+6.4	−0.8	−85.7	+84.9
Oct. 15, 2007 (rural-continental)	0.296	0.928	VS1	−3.6	−8.8	+5.2	−24.9	−60.4	+35.5
			VS2	−2.5	−8.0	+5.4	−17.5	−54.7	+37.2
			VS3	−2.1	−7.7	+5.6	−14.2	−52.7	+38.5
			VS4	−1.5	−7.2	+5.8	−10.2	−49.8	+39.6
Dec. 19, 2007 (rural-continental)	0.147	0.937	VS1	−3.6	−4.7	+1.1	−26.9	−35.1	+8.2
			VS2	−3.0	−4.1	+1.1	−22.7	−30.6	+7.9
			VS3	−2.9	−4.1	+1.2	−21.7	−30.4	+8.7
			VS4	−2.7	−3.9	+1.2	−20.4	−29.2	+8.8
Jan. 24, 2008 (remote continental)	0.071	0.915	VS1	−1.2	−4.5	+3.3	−18.2	−66.8	+48.6
			VS2	−0.7	−4.1	+3.3	−11.0	−60.4	+49.4
			VS3	−0.6	−4.0	+3.4	−8.3	−59.1	+50.8
			VS4	−0.3	−3.8	+3.5	−4.8	−56.4	+51.6

(continued overleaf)

Table 8.6 (*continued*)

Measurement day of 1997 and aerosol type	Daily mean aerosol optical thickness $\tau_a(0.55\,\mu m)$	Daily mean $\omega(0.55\,\mu m)$	Surface albedo model	Diurnally average DARF terms ($W\,m^{-2}$)			Diurnally average DARF efficiencies ($W\,m^{-2}$)		
				ΔDF_{ToA}	ΔDF_{BoA}	ΔDF_{Atm}	E_{ToA}	E_{BoA}	E_{Atm}
Feb.12, 2008	0.245	0.945	VS1	−6.9	−10.4	+3.4	−28.9	−43.2	+14.3
(rural-			VS2	−5.9	−9.3	+3.4	−24.4	−38.6	+14.2
continental)			VS3	−5.4	−9.0	+3.6	−22.5	−37.3	+14.8
			VS4	−4.9	−8.5	+3.6	−20.5	−35.4	+14.9
Mar. 13, 2008	0.080	0.948	VS1	−2.8	−4.8	+2.0	−31.4	−54.3	+22.9
(maritime			VS2	−2.3	−4.4	+2.1	−25.8	−49.6	+23.8
−continental)			VS3	−2.1	−4.3	+2.2	−23.5	−48.5	+25.0
			VS4	−1.8	−4.1	+2.3	−20.4	−46.0	+25.6

Table 8.7 Daily mean values of aerosol optical thickness $\tau_a(0.55\ \mu m)$ and single scattering albedo $\omega(0.55\ \mu m)$ obtained during the Ev-K2-CNR campaigns of summer 1991 and summer 1992 employing two Volz sun photometers at the CNR Pyramid Laboratory (5050 m a.m.s.l.) in Himalaya (Nepal) and corresponding values of the diurnally average aerosol forcing terms ΔDF_{ToA} at the ToA level, ΔDF_{BoA} at the BoA level, and ΔDF_{Atm} within the atmosphere and the diurnally average DARF efficiencies E_{ToA}, E_{BoA}, and E_{Atm} determined by Tomasi *et al.* (2015a) using the BRDF surface reflectance models BS1, VS1, and PS4 defined in Tables 8.1 and 8.2 and chosen to represent the surface albedo characteristics of the Tibetan Plateau (TPR), subtropical broadleaf and coniferous forests (SBCF), and Himalayan mountain region (HMR), respectively.

Measurement period and aerosol type	Daily mean aerosol optical thickness $\tau_a(0.55\ \mu m)$	Daily mean $\omega(0.55\ \mu m)$	Surface albedo model	Diurnally average DARF terms ($W\ m^{-2}$)			Diurnally average DARF efficiencies ($W\ m^{-2}$)		
				ΔDF_{ToA}	ΔDF_{BoA}	ΔDF_{Atm}	E_{ToA}	E_{BoA}	E_{Atm}
July 24–July 26, 1991 (free troposphere)	0.116	0.945	BS1	−4.3	−16.2	+12.0	−37.1	−139.7	+103.4
			VS1	−3.9	−9.5	+5.7	−33.6	−81.9	+49.1
			PS4	−0.4	−6.6	+6.3	−3.4	−56.9	+54.3
July 28–Aug. 5, 1991 (stratos. volc. 2-month aged)	0.173	0.984	BS1	−4.3	−6.9	+2.6	−24.9	−39.9	+15.1
			VS1	−5.9	−8.5	+2.6	−34.1	−49.1	+15.0
			PS4	−2.8	−5.7	+2.9	−16.2	−32.9	+16.8
Sept. 19–Oct. 2, 1991 (stratos. volc. 3-month aged)	0.174	0.975	BS1	−4.3	−7.9	+3.6	−24.7	−45.4	+20.7
			VS1	−5.8	−9.3	+3.4	−33.3	−53.4	+19.5
			PS4	−1.8	−5.8	+4.0	−10.3	−33.3	+23.0
July 23–Aug. 11, 1992 (stratos. volc. 13-month aged)	0.138	0.975	BS1	−4.0	−6.8	+2.9	−29.0	−49.3	+21.0
			VS1	−5.5	−8.4	+2.8	−39.9	−60.9	+20.2
			PS4	−2.3	−5.5	+3.2	−16.7	−39.9	+23.2
March 1992 (stratos. volc. 9-month aged) (Kinne and Pueschel, 2001)	0.200	1.00	VS2	−4.7	−8.5	+3.8	−23.5	−42.5	+19.0

Table 8.8 Daily mean values of aerosol optical thickness τ_a(0.55 μm) and single scattering albedo ω(0.55 μm) obtained from the POLAR-AOD measurements conducted at various Arctic and Antarctic sites (Tomasi *et al.*, 2007, 2012) for different aerosol types and corresponding values of the diurnally average aerosol forcing terms ΔDF_{ToA} at the ToA level, ΔDF_{BoA} at the BoA level, and ΔDF_{Atm} within the atmosphere and the diurnally average DARF efficiencies E_{ToA}, E_{BoA}, and E_{Atm} determined by Tomasi *et al.* (2015a) using the BRDF surface reflectance models OS1, VS1, PS1, PS2, PS3, and PS4 defined in Tables 8.1 and 8.2.

Aerosol type and measurement site	Daily mean aerosol optical thickness τ_a(0.55 μm)	Daily mean ω(0.55 μm)	Surface albedo model	Diurnally average DARF terms (W m^{-2})			Diurnally average DARF efficiencies (W m^{-2})		
				ΔDF_{ToA}	ΔDF_{BoA}	ΔDF_{Atm}	E_{ToA}	E_{BoA}	E_{Atm}
Background Arctic summer aerosol (Barrow, Alaska)	0.076	0.978	OS1	−9.2	−1.9	−7.3	−121.0	−25.0	−96.0
			VS1	−4.2	−6.5	+2.3	−55.3	−85.5	+30.2
Background Arctic summer aerosol (Ny-Ålesund, Svalbard)	0.040	0.966	OS1	−7.2	+1.4	−8.6	−180.0	+35.0	−215.0
			PS4	+1.1	−2.3	+3.4	+27.5	−57.5	+85.0
Background Arctic summer aerosol (Summit, Greenland)	0.038	0.969	PS1	+2.2	−0.3	+2.5	+57.9	−7.9	+65.8
Background Arctic summer aerosol (Sodankylä, Finland)	0.058	0.965	OS1	−6.5	−1.6	−4.9	−112.1	−27.6	−84.5
			VS1	−3.0	−5.5	+2.5	−51.7	−94.8	+43.1
			PS2	+3.0	−0.4	+3.4	+51.7	−6.9	+58.6
Background Arctic summer aerosol (Tiksi, Siberia)	0.082	0.977	OS1	−10.3	−1.6	−8.7	−125.6	−19.5	+106.1
			VS1	−5.6	−9.0	+3.4	−68.3	−109.8	+41.5
Background Antarctic summer aerosol (Mario Zucchelli)	0.029	0.964	OS1	−6.2	+0.9	−7.1	−213.8	+31.0	−244.8
			PS2	+2.0	−0.1	+2.1	+69.0	−3.4	+72.4
Background Antarctic summer aerosol (Neumayer)	0.042	0.975	OS1	−8.2	+1.4	−9.6	−195.2	+33.4	−228.6
			PS1	+1.9	−0.2	+2.1	+45.2	−4.8	+50.0
			PS2	+2.1	0.0	+2.1	+50.0	+0.0	+50.0
Background Antarctic summer aerosol (Mirny)	0.021	0.985	PS1	+0.2	0.0	+0.2	+9.5	0.0	+9.5
			PS2	+0.6	0.0	+0.6	+28.6	0.0	+28.6

Background Antarctic summer aerosol (Dome C)	0.017	0.999	PS1	+0.4	+0.2	+0.2	+23.5	+11.7	+11.8
			PS2	+0.6	+0.4	+0.2	+35.3	+23.5	+11.8
Background Antarctic summer aerosol (South Pole)	0.016	0.988	PS1	+0.6	−0.1	+0.7	+37.5	−6.2	+43.7
			PS2	+0.7	−0.1	+0.8	+43.7	−6.3	+50.0
Arctic haze (Barrow, Alaska)	0.112	0.840	OS1	−5.5	+6.5	−12.0	−49.1	+58.0	−107.1
			PS1	+9.0	+0.1	+8.9	+80.4	+0.9	+79.5
			PS2	+8.8	+0.1	+8.7	+78.6	+0.9	+77.7
Arctic haze (Ny-Ålesund, Svalbard)	0.077	0.949	OS1	−5.3	−0.6	−4.7	−68.8	−7.8	−61.0
			PS3	+1.0	−0.4	+1.4	+13.0	−5.2	+18.2
Arctic haze (Sodankylä, Finland)	0.064	0.840	OS1	−7.7	−1.2	−6.5	−120.3	−18.7	−101.6
			VS1	−3.5	−5.3	+1.8	−54.7	−82.8	+28.1
			PS2	+2.2	−0.3	+2.5	+34.4	−4.7	+39.1
Dense Arctic summer aerosol (Ny-Ålesund, Svalbard)	0.12	0.852	OS1	−10.4	+4.2	−14.6	−86.7	+35.0	−121.7
			PS4	+13.0	−5.2	+18.2	+108.3	−43.4	+151.7
Asian dust (Barrow, Alaska)	0.20	0.858	OS1	−9.3	+16.0	−25.3	−46.5	+80.0	−126.5
			PS2	+18.5	+0.3	+18.2	+92.5	+1.5	+91.0
Boreal forest fire smoke (Barrow, Alaska)	0.30	0.758	OS1	−32.9	−6.7	−26.2	−109.7	−22.4	−87.3
			VS1	−20.2	−28.5	+8.3	−67.3	−95.0	+27.7

Table 8.9 Daily mean values of aerosol optical thickness $\tau_a(0.55\,\mu m)$ and single scattering albedo $\omega(0.55\,\mu m)$ obtained on three golden days of the Aerosols99 cruise conducted in the Atlantic Ocean (Voss *et al.*, 2001) for different aerosol types and corresponding values of the diurnally average aerosol forcing terms ΔDF_{ToA} at the ToA level, ΔDF_{BoA} at the BoA level, and ΔDF_{Atm} within the atmosphere and the diurnally average DARF efficiencies E_{ToA}, E_{BoA}, and E_{Atm} determined by Tomasi *et al.* (2015a) using the BRDF oceanic surface reflectance models OS2 and OS4 defined in Tables 8.1 and 8.2.

Measurement day and aerosol type	Daily mean aerosol optical thickness $\tau_a(0.55\,\mu m)$	Daily mean $\omega(0.55\,\mu m)$	Surface albedo model	Diurnally average DARF terms ($W\,m^{-2}$)			Diurnally average DARF efficiencies ($W\,m^{-2}$)		
				ΔDF_{ToA}	ΔDF_{BoA}	ΔDF_{Atm}	E_{ToA}	E_{BoA}	E_{Atm}
January 18, 1999 (maritime clean)	0.097	0.996	OS2	−4.4	+4.9	−9.3	−45.4	+50.5	−95.9
			OS4	−2.7	+3.0	−5.7	−27.8	+30.9	−58.7
January 25, 1999 (African dust)	0.280	0.883	OS2	−4.0	−0.1	−3.9	−14.3	−0.4	−13.9
			OS4	−1.9	−7.8	+5.9	−6.8	−27.9	+21.1
January 30, 1999 (biomass burning smoke)	0.331	0.950	OS2	−19.5	−18.1	−1.4	−58.9	−54.7	−4.2
			OS4	−17.3	−22.9	+5.6	−52.3	−69.2	+16.9

Table 8.10 Daily mean values of aerosol optical thickness $\tau_a(0.55\,\mu m)$ and single scattering albedo $\omega(0.55\,\mu m)$ evaluated for 19 clear-sky measurement days of the DOE/ARM/AIOP experiment field campaign conducted by Ferrare *et al.* (2006) in May 2003 over north-central Oklahoma (United States) for different aerosol types and corresponding values of the diurnally average aerosol forcing terms ΔDF_{ToA} at the ToA level, ΔDF_{BoA} at the BoA level, and ΔDF_{Atm} within the atmosphere and the diurnally average DARF efficiencies E_{ToA}, E_{BoA}, and E_{Atm} determined by Tomasi *et al.* (2015a) using the BRDF land surface reflectance models VS2 and VS3 defined in Tables 8.1 and 8.2. The values of $\omega(0.55\,\mu m)$ have been derived according to the measurements of $\omega(0.675\,\mu m)$ made by Strawa *et al.* (2006).

Measurement day and aerosol type	Daily mean aerosol optical thickness $\tau_a(0.55\,\mu m)$	Daily mean $\omega(0.55\,\mu m)$	Surface albedo model	Diurnally average DARF terms ($W\,m^{-2}$)			Diurnally average DARF efficiencies ($W\,m^{-2}$)		
				ΔDF_{ToA}	ΔDF_{BoA}	ΔDF_{Atm}	E_{ToA}	E_{BoA}	E_{Atm}
May 5	0.077	0.980	VS2	−3.2	−3.8	+0.4	−41.7	−47.3	+5.5
(rural-continental)			VS3	−3.1	−3.5	+0.5	−39.6	−45.7	+6.1
May 7	0.142	0.950	VS2	−5.5	−6.4	+0.9	−38.6	−44.8	+6.2
(continental-polluted)			VS3	−5.2	−6.1	+0.9	−36.7	−43.2	+6.5
May 8	0.473	0.985	VS2	−17.0	−20.5	+3.4	−36.0	−43.3	+7.3
(rural-continental)			VS3	−16.1	−19.6	+3.5	−34.1	−41.4	+7.4
May 9	0.367	0.975	VS2	−12.1	−14.9	+2.8	−33.0	−40.6	+7.5
(rural-continental)			VS3	−11.3	−14.1	+2.8	−30.8	−38.4	+7.7
May 10	0.438	0.980	VS2	−14.5	−17.5	+3.1	−33.0	−40.1	+7.1
(rural-continental)			VS3	−13.6	−16.7	+3.1	−31.0	−38.2	+7.2
May 11	0.068	0.920	VS2	−2.6	−3.0	+0.4	−37.9	−43.9	+6.0
(continental-polluted)			VS3	−2.5	−2.9	+0.4	−36.1	−42.7	+6.6
May 12	0.110	0.935	VS2	−4.3	−5.1	+0.7	−39.4	−46.2	+6.8
(continental-polluted)			VS3	−4.1	−4.9	+0.8	−37.3	−44.6	+7.3
May 14	0.315	0.925	VS2	−9.7	−11.3	+1.6	−30.7	−35.8	+5.1
(continental-polluted)			VS3	−9.2	−10.9	+1.6	−29.3	−34.6	+5.2
May 18	0.373	0.930	VS2	−11.7	−13.6	+1.9	−31.3	−36.5	+5.2
(continental-polluted)			VS3	−11.2	−13.1	+2.0	−29.9	−35.2	+5.3

(*continued overleaf*)

Table 8.10 *(continued)*

Measurement day and aerosol type	Daily mean aerosol optical thickness τ_a(0.55 μm)	Daily mean ω(0.55 μm)	Surface albedo model	Diurnally average DARF terms (W m^{-2})			Diurnally average DARF efficiencies (W m^{-2})		
				ΔDF_{ToA}	ΔDF_{BoA}	ΔDF_{Atm}	E_{ToA}	E_{BoA}	E_{Atm}
May 20	0.194	0.920	VS2	−5.9	−7.0	+1.1	−30.4	−36.3	+5.9
(continental-polluted)			VS3	−5.6	−6.8	+1.2	−29.0	−35.2	+5.2
May 21	0.245	0.915	VS2	−8.1	−9.5	+1.4	−33.0	−38.9	+5.9
(continental-polluted)			VS3	−7.7	−9.2	+1.5	−31.4	−37.5	+6.1
May 22	0.219	0.915	VS2	−7.1	−8.4	+1.3	−32.6	−38.4	+5.8
(continental-polluted)			VS3	−6.8	−8.2	+1.3	−31.2	−37.3	+6.1
May 24	0.129	0.920	VS2	−4.8	−5.4	+0.7	−37.0	−42.1	+5.1
(continental-polluted)			VS3	−4.6	−5.3	+0.7	−35.5	−40.9	+5.4
May 25 (continental	0.198	0.945	VS2	−7.4	−8.5	+1.0	−37.6	−42.7	+5.1
-polluted with SFFS[a])			VS3	−7.1	−8.2	+1.1	−36.1	−41.5	+5.4
May 26 (continental	0.159	0.940	VS2	−8.0	−6.8	+0.8	−37.9	−42.9	+5.0
-polluted with SFFS[a])			VS3	−5.8	−6.6	+0.8	−36.4	−41.7	+5.3
May 27 (continental	0.282	0.950	VS2	−9.1	−10.7	+1.6	−32.4	−37.9	+5.5
-polluted with SFFS[a])			VS3	−8.8	−10.4	+1.6	−31.1	−36.8	+5.7
May 28	0.383	0.928	VS2	−11.6	−13.9	+2.2	−30.4	−36.3	+5.9
(continental-polluted)			VS3	−11.2	−13.5	+2.3	−29.1	−35.2	+6.0
May 29	0.150	0.920	VS2	−5.4	−6.3	+0.9	−35.7	−41.9	+6.3
(continental-polluted)			VS3	−5.1	−6.1	+1.0	−34.1	−40.7	+6.6
May 30	0.160	0.910	VS2	−6.1	−7.3	+1.2	−37.9	−45.5	+7.6
(continental-polluted)			VS3	−5.8	−7.1	+1.3	−36.2	−44.1	+8.0

a) SFFS = Siberian forest fire smoke.

Table 8.11 Values of slope coefficient k_D, correlation coefficient c_{corD}, and standard error of estimate SEE_D determined for the regression lines of the diurnally average values of ΔDF_{ToA}, ΔDF_{BoA}, and ΔDF_{Atm} plotted versus the daily mean values of $\tau_a(0.55\ \mu m)$ obtained for the various aerosol types considered in the present study and evaluated separately over ocean and land surfaces.

Aerosol type	Surface	ΔDF_{ToA}			ΔDF_{BoA}			ΔDF_{Atm}		
	albedo	k_D (W m^{-2})	c_{corD}	SEE_D (W m^{-2})	k_D (W m^{-2})	c_{corD}	SEE_D (W m^{-2})	k_D (W m^{-2})	c_{corD}	SEE_D (W m^{-2})
Remote continental	Ocean	−39.1	—	—	−62.0	—	—	22.7	—	—
	Land	−21.0	−0.211	±30.8	−63.3	−0.608	±24.9	43.6	+0.324	±40.3
Rural-continental	Ocean	−34.8	−0.904	±6.2	−56.7	−0.872	±12.1	21.9	+0.543	±12.8
	Land	−25.8	−0.700	±40.7	−57.2	−0.785	±6.7	31.9	+0.518	±8.2
Free tropospheric	Land	12.7	+0.308	±19.7	−43.5	−0.433	±45.2	57.3	+0.688	±31.9
Continental-polluted	Ocean	38.6	−0.936	±3.1	−0.5	−0.842	±8.1	3.4	+0.060	±13.1
	Land	4.9	+0.175	±3.3	−91.6	−0.690	±11.2	103.4	+0.661	±14.3
Maritime clean	Ocean	37.5	+0.746	±19.3	114.8	+0.951	±21.6	−75.0	−0.727	±40.9
	Land	520.0	+0.949	±100.0	660.0	+0.968	±100.0	−150.0	−0.993	±10.0
Maritime–continental	Ocean	−15.9	−0.491	±3.5	−60.8	−0.558	±11.9	53.8	+0.733	±7.1
	Land	−32.1	−0.891	±2.2	−71.9	−0.864	±5.4	36.9	+0.735	±4.6
Desert dust	Ocean	−24.2	−0.519	±8.9	−11.8	−0.825	±19.1	97.1	+0.825	±16.1
	Land	−41.8	−0.632	±9.9	−99.3	−0.830	±13.4	68.9	+0.724	±13.4
Biomass burning	Ocean	6.4	+0.078	±31.2	28.1	+0.433	±29.3	16.0	+0.284	±26.9
	Land	−33.2	−0.488	±14.0	−127.6	−0.840	±20.0	91.0	+0.870	±12.5
Urban and industrial	Land	−49.2	−0.607	±21.5	−12.4	−0.754	±34.0	70.0	+0.843	±14.9
Polar	Ocean	−79.2	−0.837	±15.1	5.8	+0.081	±20.9	−85.8	−0.916	±10.8
	Land	−19.2	−0.188	±20.4	−68.6	−0.734	±13.2	54.3	+0.708	±11.3
Stratospheric volcanic	Land	−19.1	−0.331	±15.1	33.1	+0.336	±25.8	−53.7	−0.585	±20.7

Determination of the Size-Distribution Curves of Columnar Aerosol

The spectral series of $\tau_a(\lambda)$ measured at various AERONET and SKYNET sites on selected golden days were analyzed using the AERONET and SKYNET protocols to determine the Level 2.0 shape parameters of the multimodal aerosol size-distribution curves pertaining to the vertical atmospheric column. At other sites, the spectral series of $\tau_a(\lambda)$ were examined by Tomasi *et al.* (2015a) using the King (1982) inversion method for the daily mean values of complex refractive index determined at the previous step, retrieving the multimodal size-distribution curves over the 0.16–24.0 μm particle diameter range (Tomasi *et al.*, 2003). These size-distribution curves were obtained by Tomasi *et al.* (2015a) examining the spectral series of $\tau_a(\lambda)$ measured during the CREARCOLUMN/ACE-2 campaign (Vitale *et al.*, 2000), the PRIN-2004 campaign (Lanconelli, 2007; Tomasi *et al.*, 2015a), the AEROCLOUDS campaign (Mazzola *et al.*, 2010), the Aerosols99 cruise (Voss *et al.*, 2001), and the DOE/ARM/AIOP experiment (Ferrare *et al.*, 2006). In all these cases, the aerosol size-distribution curves of particle number density $N(a)$ and particle volume $V(a)$ were found to be bimodal (with fine and coarse particle modes) or trimodal (with nuclei, accumulation, and coarse particle modes).

Determination of the Columnar Aerosol Single Scattering Albedo

Additional measurements were conducted at the most important observational sites of the aforementioned sun-photometer networks using tropospheric LIDARs (to define the vertical profile of volume backscattering coefficient in the low troposphere), nephelometers, aethalometers, and particle soot absorption photometers (to carry out regular ground-level measurements of the volume scattering and absorption coefficients at visible wavelengths, from which average values of volume extinction coefficient $\beta_{ext}(0.550\,\mu m)$ and ground-level SSA parameter $\omega_o(0.550\,\mu m)$ were derived) and particle sampling measurements within various size ranges (to determine the chemical composition and microphysical characteristics of aerosols at the surface). In all cases where measurements of $\beta_{ext}(0.550\,\mu m)$ and $\beta_{sca}(0.550\,\mu m)$ and $\omega_o(0.550\,\mu m)$ were not conducted, the spectral curve of $\omega(\lambda)$ was calculated using the multimodal size-distribution curves retrieved at various hours of each golden day at the previous step and the values of $n(\lambda)$ and $k(\lambda)$ retrieved at step (2) from the sky-brightness measurements in the almucantar. Examining these data, the daily mean values of $\omega(0.55\,\mu m)$ were calculated for each golden day, finding values ranging from 0.94 to 0.98 for remote continental aerosol, from 0.92 to 0.98 for rural-continental aerosol, from 0.96 to 0.99 for maritime clean aerosol, from 0.81 to 0.90 for mixed maritime–continental aerosols, from 0.85 to 0.96 for continental-polluted aerosols, and from 0.80 to 0.90 for anthropogenic heavily polluted aerosol.

Choice of the Most Realistic Surface Reflectance Model

The most suitable non-Lambertian BRDF surface albedo model was chosen in order to represent the average characteristics of the underlying ocean or land surface around the measurement site. In general, the most realistic surface reflectance model was chosen among the 16 BRDF surface reflectance models described earlier and belonging to one of the OS, VS, BS, and PS classes by taking into account the average surface-level wind velocity conditions over the ocean or the most realistic surface albedo conditions of the land area under study, as defined by the set of MCD43C3 products derived from the MODIS Level 3.0 surface albedo data recorded during the observation period.

Calculation of the Daily Time Patterns of Instantaneous and Diurnally Average DARF Terms

The time patterns of instantaneous $\Delta F_{\mathrm{ToA}}(t)$ and $\Delta F_{\mathrm{BoA}}(t)$ were calculated for all the golden days of the various experimental campaigns using (i) the size-distribution curves retrieved at step (3) from the spectral series of $\tau_a(\lambda)$ determined at step (1) for the various solar zenith angles ϑ_o at which the sun-photometer measurements were conducted; (ii) the complex refractive index data defined at step (2), together with the main aerosol optical parameters calculated at step (4); and (iii) the surface albedo models individuated at step (5). The time patterns of $\Delta F_{\mathrm{ToA}}(t)$ and $\Delta F_{\mathrm{BoA}}(t)$ were subsequently integrated over the whole insolation period of each golden day, according to Eqs. (8.18) and (8.19), to determine the daily mean values of $\Delta \mathrm{DF}_{\mathrm{ToA}}$ and $\Delta \mathrm{DF}_{\mathrm{BoA}}$ for the selected BRDF surface reflectance model and calculate $\Delta \mathrm{DF}_{\mathrm{Atm}}$ according to Eq. (8.22). Such daily mean values of $\Delta \mathrm{DF}_{\mathrm{ToA}}$, $\Delta \mathrm{DF}_{\mathrm{BoA}}$, and $\Delta \mathrm{DF}_{\mathrm{Atm}}$ were then plotted versus the corresponding values of $\tau_a(0.55\,\mu\mathrm{m})$, separately for each BRDF surface reflectance model, to study the proportionality features existing between these DARF terms and the corresponding values of AOT.

Calculation of the Daily Mean Values of DARF Efficiency

The slope coefficients of the regression lines drawn in the previous scatter plots of $\Delta \mathrm{DF}_{\mathrm{ToA}}$, $\Delta \mathrm{DF}_{\mathrm{BoA}}$, and $\Delta \mathrm{DF}_{\mathrm{Atm}}$ versus $\tau_a(0.55\,\mu\mathrm{m})$ provide reliable average estimates of the corresponding efficiency terms, defined according to Eqs. (8.20–8.22). The daily mean values of E_{ToA}, E_{BoA}, and E_{Atm} obtained for the golden days of the various field experiments were subdivided into eleven aerosol classes and plotted versus $\omega(0.55\,\mu\mathrm{m})$ separately for each aerosol type and the ocean and land surfaces to give evidence of the dependence features of the efficiency parameters on the optical characteristics of airborne aerosols.

As mentioned earlier, the DARF-PROC procedure was applied by Tomasi *et al.* (2015a) to various sets of field measurements to determine the daily mean values of $\tau_a(0.55\,\mu\mathrm{m})$ and $\omega(0.55\,\mu\mathrm{m})$ and the diurnally average terms $\Delta \mathrm{DF}_{\mathrm{ToA}}$, $\Delta \mathrm{DF}_{\mathrm{BoA}}$, $\Delta \mathrm{DF}_{\mathrm{Atm}}$, E_{ToA}, E_{BoA}, and E_{Atm} calculated for the golden days of the following experimental campaigns:

A. The CLEARCOLUMN (ACE-2) experiment conducted in summer 1997 at Sagres (Portugal) by Vitale *et al.* (2000), for which the values of diurnally average DARF terms were calculated by Tomasi *et al.* (2015a) for 10 cloudless golden days characterized by maritime clean, mixed maritime–continental, and continental-polluted aerosol loads using the OS2, VS1, and VS4 surface reflectance models. The results are given in Table 8.4.

B. The PRIN-2004 field campaign conducted by Lanconelli (2007) from 25 April to 2 October 2003, at Lecce (Italy) in cooperation with the University of Salento group (Perrone *et al.*, 2005), for which the AERONET sun-photometer measurements conducted in this period were examined. A set of 12 golden days was selected by Tomasi *et al.* (2015a) to calculate the aforementioned DARF terms for the optical characteristics typical of atmospheric aerosol loads having different origins, such as remote continental, maritime–continental, continental-polluted, rural-continental, mixed continental-dust, and Saharan dust particles. The diurnally average values of the DARF terms were calculated by Tomasi *et al.* (2015a) for the data sets collected on the golden days (listed in Table 8.5, where the daily origins of aerosols are specifically indicated) and the OS3, OS4, VS1, and VS4 surface reflectance models. The results are given in Table 8.5 to illustrate the dependence features of the DARF terms on $\tau_a(0.55\,\mu m)$ and $\omega(0.55\,\mu m)$ for the various aerosol types.

C. The AEROCLOUDS field campaign conducted at the San Pietro Capofiume meteorological station in the Po Valley (Italy) from May 2007 to March 2008, for which the optical characteristics of various aerosol types were determined by means of spectral measurements of direct solar irradiance and sky-brightness measurements in the almucantar made using the SKYNET Prede POM-02 L sun/sky radiometer. Analyzing these measurements, the spectral values of $\tau_a(\lambda)$ and the aerosol phase function were determined at numerous visible and NIR wavelengths. Additional measurements were conducted with a portable tropospheric LIDAR working at $\lambda = 0.532\,\mu m$, the ground-based Radiance Research (RR) nephelometer (M903 model) measuring the volume aerosol scattering coefficient at $\lambda = 0.532\,\mu m$, and the particle soot absorption photometer (PSAP) measuring the ground-level volume aerosol absorption coefficient at $\lambda = 0.573\,\mu m$, from which the ground-level $\omega_o(0.550\,\mu m)$ were calculated. Various aerosol types were studied on eighteen golden days selected by Tomasi *et al.* (2015a) to calculate the DARF terms, of which eight days were characterized by maritime–continental aerosol, four days by rural-continental aerosol, five days by continental-polluted aerosol, and one day only by remote continental aerosol, as indicated in Table 8.6. The calculations of the three DARF terms were made by Tomasi *et al.* (2015a) for all the four aerosol types monitored on the chosen golden days and for the VS1, VS2, VS3, and VS4 surface reflectance models, obtaining the results given in Table 8.6.

D. The Ev-K2-CNR project field campaigns conducted in the summer months of 1991 and 1992, during which regular sun-photometer measurements of direct solar irradiance were carried out by Tomasi, Vitale, and Tarozzi (1997) at the CNR Pyramid Laboratory (27°57′N, 86°49′E, 5050 m a.m.s.l.), located at the foot of Mt. Everest (Himalaya, Nepal). During the two campaigns, the following measurements of $\tau_a(\lambda)$ were obtained: (i) from 24 to 26 July 1991, before the arrival of Mt. Pinatubo volcanic aerosol cloud over the CNR Pyramid Laboratory, for free tropospheric aerosol; (ii) from 28 July to 5 August 1991, for 2-month-aged volcanic aerosol suspended in the low stratosphere; (iii) from 19 September to 2 October 1991, for 3-month-aged volcanic aerosol at low stratospheric levels; and (d) from 23 July to 11 August 1992, for 13-month-aged volcanic aerosol particles in the low stratosphere. Table 8.7 provides the average values of $\tau_a(0.55\,\mu\text{m})$, $\omega(0.55\,\mu\text{m})$, $\Delta\text{DF}_{\text{ToA}}$, $\Delta\text{DF}_{\text{BoA}}$, $\Delta\text{DF}_{\text{Atm}}$, E_{ToA}, E_{BoA}, and E_{Atm} determined by Tomasi *et al.* (2015a) for the four measurement periods by using the BS1, VS1, and PS4 surface reflectance models to represent the surface albedo characteristics of the Tibetan Plateau (TPR), subtropical broadleaf and coniferous forests (SBCF), and Himalayan mountain region (HMR), respectively. The average values of aerosol optical parameters and DARF terms calculated by Kinne and Pueschel (2001) over a land surface were also considered for comparison, pertaining to a load of 9-month-aged volcanic aerosol present in the low stratosphere.

E. The POLAR-AOD sun-photometer measurements of $\tau_a(\lambda)$ conducted at various Arctic and Antarctic sites by Tomasi *et al.* (2007, 2012) for different aerosol types, together with measurements of $\omega(0.55\,\mu\text{m})$ and aerosol chemical composition determined from *in situ* measurements to characterize the optical properties of different polar aerosol types (background summer aerosol, AH, AD, BFFS, dense summer Arctic aerosol, background summer and winter Antarctic aerosols near the coasts, background summer aerosol over the Antarctic Plateau). The daily mean values of $\tau_a(0.55\,\mu\text{m})$, $\omega(0.55\,\mu\text{m})$, and the various DARF terms are given in Table 8.8 for all the aforementioned polar aerosol types using the OS1 surface reflectance model to represent the oceanic surface reflectance, the VS1 vegetation-covered surface reflectance model to simulate the tundra reflectance characteristics, and the four PS reflectance models to represent the snow- and ice-covered surface reflectance features.

F. The measurements of aerosol characteristics recorded in the Atlantic Ocean during the Aerosols99 cruise conducted by Voss *et al.* (2001) for maritime clean, African dust, and BBS, which were examined by Tomasi *et al.* (2015a) to calculate the average values of $\tau_a(\lambda)$, $\omega(0.55\,\mu\text{m})$ and DARF terms during three distinct periods of the cruise for the OS2 and OS4 BRDF oceanic surface reflectance models, obtaining the results given in Table 8.9.

G. The DOE/ARM/AIOP experiment field campaign conducted by Ferrare *et al.* (2006) in May 2003 over north-central Oklahoma (United States) for different aerosol types (rural-continental, continental-polluted, and continental-polluted mixed with BBS transported from Siberian areas). The calculations of $\tau_a(0.55\,\mu\text{m})$ and $\omega(0.55\,\mu\text{m})$ were made by Tomasi *et al.* (2015a) examining the local AERONET measurements recorded on nineteen clear-sky measurement days , while those of the DARF parameters were made by applying the DARF-PROC procedure to the field measurements for the VS2 and VS3 vegetated surface reflectance models chosen to represent the BDRF surface reflectance characteristics of the north-central Oklahoma agricultural area in spring.

8.3.2 Calculations of the Diurnally Average DARF Terms and Efficiency Parameters for Eleven Aerosol Types

The daily mean values of $\tau_a(0.55\,\mu m)$ and $\omega(0.55\,\mu m)$ and the corresponding diurnally average DARF and efficiency terms given in Tables 8.4–8.10 were examined together with analogous data found in the literature to provide a set of relationships between the three DARF terms estimated over the ocean and land surfaces and $\tau_a(0.55\,\mu m)$ and the corresponding relationships between the three DARF efficiencies and $\omega(0.55\,\mu m)$ for 11 different aerosol types. The dependence characteristics of the DARF terms on $\tau_a(0.55\,\mu m)$ and of the DARF efficiency parameters on monochromatic or average SSA in the visible have been defined separately for ocean and land surfaces for each of the following aerosol types.

8.3.2.1 Remote Continental Aerosols

Only one measurement day was found by us among the PRIN-2004 field measurements carried out on the selected golden days given in Table 8.5. It provides a set of reliable estimates of the DARF and efficiency parameters produced by remote continental aerosols over the ocean surface: for such an episode of pure continental aerosol transport from the Scandinavian and Baltic Sea regions across the eastern Europe and Balkans, observed on 25 April 2003, Tomasi *et al.* (2015a) determined the daily mean values of $\tau_a(0.55\,\mu m) = 0.271$ and $\omega(0.55\,\mu m) = 0.919$ over Lecce in southern Italy. The calculations of DARF and efficiency terms made for the OS3 and OS4 surface reflectance models provided the results reported in Table 8.5. Since only one value was found for each DARF term, so that the best-fit procedure cannot be correctly applied, we obtained the values of diurnally average DARF terms $\Delta DF_{ToA} = -39.1\ W\,m^{-2}$, $\Delta DF_{BoA} = -62.0\ W\,m^{-2}$, and $\Delta DF_{Atm} = +22.7\ W\,m^{-2}$ per unit AOT on that golden day, which can be reliably assumed as average estimates of efficiency parameters E_{ToA}, E_{BoA}, and E_{Atm}, respectively.

Conversely, a large set of DARF terms and efficiency evaluations were found in the literature for remote continental aerosol suspended over land surfaces, obtained from the studies carried out by (i) Tomasi *et al.* (2015a), examining the PRIN-2004 measurements conducted on 25 April 2003, over the VS1 and VS4 surface reflectance models (given in Table 8.5); (ii) Tomasi *et al.* (2015a), analyzing the AEROCLOUDS measurements performed on 24 January 2008, for the four VS1, VS2, VS3, and VS4 land surface reflectance models (given in Table 8.6); (iii) Le Blanc *et al.* (2012), analyzing the ARCTAS measurements conducted at Cold Lake (Canada) in June–July 2008 for the VS4 vegetated land surface reflectance model; and (iv) García *et al.* (2012), examining the AERONET data collected in Europe and North America during spring for land surfaces having albedo ranging from 0 to more than 0.50. The values of diurnally average DARF terms

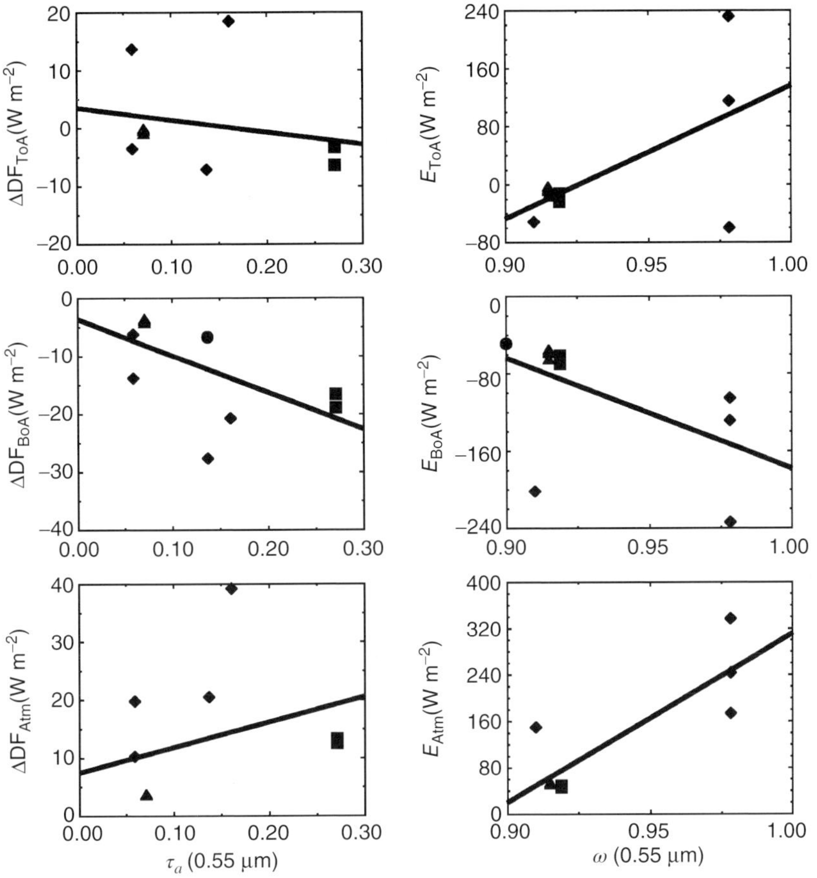

Figure 8.14 Left-hand side: scatter plots of the daily mean values of DARF terms ΔDF_{ToA} (upper part), ΔDF_{BoA} (middle part), and ΔDF_{Atm} (lower part) versus the corresponding daily mean values of aerosol optical thickness τ_a(0.55 μm), determined from various sets (*) of field measurements performed over land surfaces for atmospheric loads of remote continental aerosol particles. Right-hand side: corresponding scatter plots of the daily mean values of DARF efficiency parameters E_{ToA} at the ToA level, E_{BoA} at the BoA level, and E_{Atm} within the atmosphere versus the aerosol single scattering albedo ω(0.55 μm) over land surfaces for remote continental aerosol particles. (*) Tomasi *et al.* (2015a), PRIN-2004 (solid squares); Tomasi *et al.* (2015a), AEROCLOUDS (solid upward triangles); Le Blanc *et al.* (2012), ARCTAS (solid circles); Garcia *et al.* (2012), AERONET (solid diamonds).

ΔDF_{ToA}, ΔDF_{BoA}, and ΔDF_{Atm} obtained on the above golden days are plotted versus the corresponding daily mean values of τ_a(0.55 μm) in the left-hand side of Figure 8.14, showing that ΔDF_{ToA} ranges between about −10 and +20 W m^{-2}, ΔDF_{BoA} varies between −28 and −3 W m^{-2}, and ΔDF_{Atm} ranges between +3 and +40 W m^{-2}. Such largely spread characteristics of the results imply that the correlation coefficients of the regression lines are rather low and the standard errors of estimate are particularly great, as can be seen in Table 8.11. The best-fit values of the slope coefficients k_D determined for the regression lines are equal to −21.0, −63.3, and +43.6 W m^{-2}, respectively. These findings indicate that (i) more pronounced cooling effects are induced by remote continental aerosols at the ToA level over ocean than over land, (ii) DARF at the surface assumes negative (cooling) and comparable values over both ocean and land, and (iii) DARF in the atmosphere is positive in both cases, imply warming effects that are more marked over land than over sea. The scatter plots of the daily mean values of E_{ToA}, E_{BoA}, and E_{Atm} due to remote continental aerosols over land are shown versus the corresponding daily mean values of $\omega(0.55\,\mu\mathrm{m})$ in the right-hand side of Figure 8.14, providing the variations $\Delta E_{\mathrm{ToA}} = +185$ W m^{-2}, $\Delta E_{\mathrm{BoA}} = -115$ W m^{-2}, and $\Delta E_{\mathrm{Atm}} = +291$ W m^{-2} over the $0.90 \le \omega(0.55\,\mu\mathrm{m}) \le 1.00$ range.

8.3.2.2 Rural-Continental Aerosols

A reliable set of DARF forcing and efficiency evaluations induced by rural-continental aerosols over the oceanic surfaces was collected, including (i) the PRIN-2004 data set given in Table 8.5 for the field measurements conducted on 29 May, 10 June, and 20 September of 2003 and calculated by Tomasi *et al.* (2015a) for the OS3 and OS4 oceanic surface models, and (ii) the average data calculated by Tafuro *et al.* (2007) examining the AERONET measurements conducted at Lecce (Italy) from March 2003 to March 2004 and the radiative transfer calculations made for a mixed sea–land surface, assumed to consist of 75% sea surface and 25% land surface. In the left-hand side of Figure 8.15a, the daily values of $\Delta\mathrm{DF}_{\mathrm{ToA}}$, $\Delta\mathrm{DF}_{\mathrm{BoA}}$, and $\Delta\mathrm{DF}_{\mathrm{Atm}}$ are plotted versus the daily mean values of $\tau_a(0.55\,\mu\mathrm{m})$, showing that they vary almost linearly, providing a triplet of best-fit lines having the good correlation coefficients given in Table 8.11, and the best-fit values of k_D equal to −34.8, −56.7, and +21.9 W m^{-2}, respectively, all obtained with good correlation coefficients. The daily values of diurnally average efficiencies E_{ToA}, E_{BoA}, and E_{Atm} are plotted in the right-hand side of Figure 8.15a versus $\omega(0.55\,\mu\mathrm{m})$ over the 0.93–0.97 range, presenting (i) values of E_{ToA} varying between −50 and −30 W m^{-2}, (ii) values of E_{BoA} ranging between −68 and −45 W m^{-2}, and (iii) values of E_{Atm} varying between 0 and +28 W m^{-2}. The regression lines drawn for these scatter plots were found to yield changes of these efficiency parameters close to $\Delta E_{\mathrm{ToA}} = -4$ W m^{-2}, $\Delta E_{\mathrm{BoA}} = -15$ W m^{-2}, and $\Delta E_{\mathrm{Atm}} = +12$ W m^{-2} over the $0.90 \le \omega(0.55\,\mu\mathrm{m}) \le 1.00$ range.

A large number of field evaluations of the DARF effects was collected for rural-continental aerosols over land surfaces, derived from the following field measurements: (i) the PRIN-2004 measurements carried out by Tomasi *et al.* (2015a) on the aforementioned three golden days for the VS1 and VS4 BDRF surface reflectance models given in Table 8.3; (ii) the AEROCLOUDS measurements conducted by Tomasi *et al.* (2015a) on 24 May, 15 October, and 19 December of 2007 and on 12 February 2008, for the four VS1, VS2, VS3, and VS4 surface reflectance models given in Table 8.6; (iii) the DOE/ARM/AIOP field measurements performed by Tomasi *et al.* (2015a) on the four aforementioned golden days from 5 to 10 May 2003, reported in Table 8.10, which were examined following the DARF-PROC procedure and using the VS2 and VS3 land surface

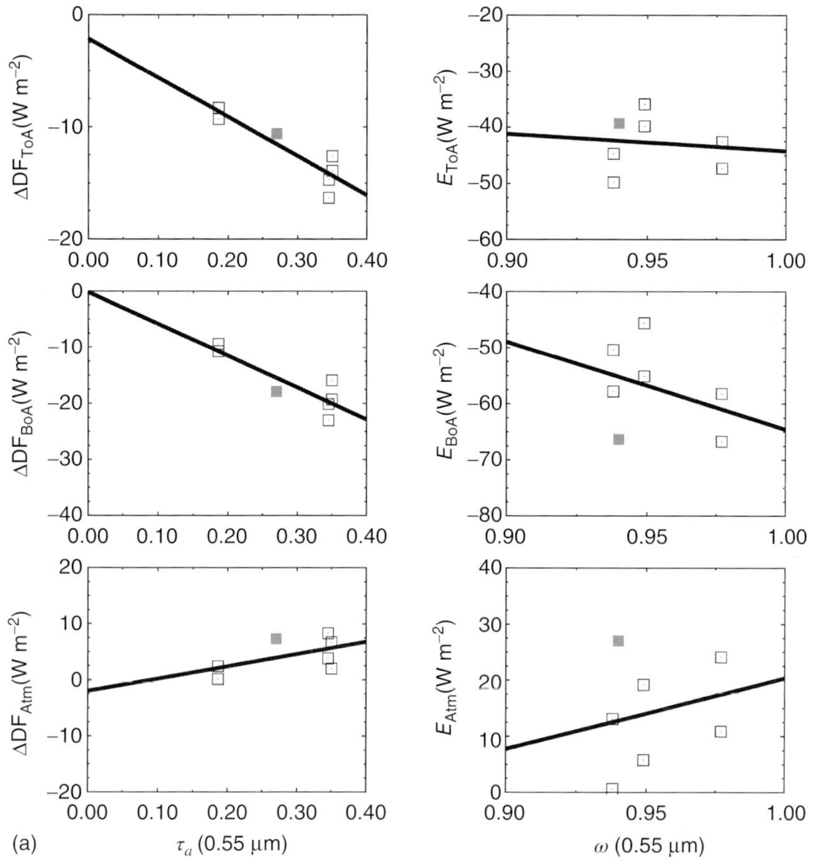

Figure 8.15a Left-hand side: scatter plots of the daily mean values of DARF terms ΔDF_{ToA} (upper part), ΔDF_{BoA} (middle part), and ΔDF_{Atm} (lower part) versus the corresponding daily mean values of aerosol optical thickness τ_a(0.55 μm), determined from two sets (*) of field measurements performed over ocean surfaces for atmospheric loads of rural-continental aerosol particles. Right-hand side: corresponding scatter plots of the daily mean values of DARF efficiency parameters E_{ToA} at the ToA level, E_{BoA} at the BoA level, and E_{Atm} within the atmosphere versus the aerosol single scattering albedo ω(0.55 μm) over ocean surfaces for rural-continental aerosol particles. (*) Tomasi *et al.* (2015a), PRIN-2004 (open squares); Tafuro *et al.* (2007), AERONET (gray squares).

reflectance models; (iv) the measurements conducted by Horvath *et al.* (2002) during the June 1999 field campaign conducted in the Almerían region (Spain), for which the DARF effects were calculated for vegetation-covered land surface reflectance conditions; (v) the NOAA/CMDL measurements carried out by Delene and Ogren (2002) at Bondville (Illinois) and Lamont (Oklahoma) from November 1994 to September 2000 over vegetation-covered land surfaces; (vi) the AERONET measurements and radiative transfer simulations performed by Yu *et al.* (2006) over the eastern United States and Europe in 2001–2010 for vegetation-covered land surfaces; (vii) the field measurements conducted by Xu *et al.* (2003) at Linan, in the rural region of the Yangtze delta (East China) during November 1999, as a part of the Joint US–China 1999 cooperative program; (ix) the EAST-AIRE field measurements conducted by Liu *et al.* (2007) at Xianghe in October 2004 and at Beijing (China) during the period from October 2004 to December 2005; (x) the EAST-AIRE measurements conducted by Xia *et al.* (2007b) at Taihu (China) from October 2004 to December 2005; and (xi) the EAST-AIRE measurements performed by Liu *et al.* (2011) at Lanzhou (China) from September 2005 to August 2006. The daily values of ΔDF_{ToA}, ΔDF_{BoA}, and ΔDF_{Atm} obtained on the various golden days are plotted versus $\tau_a(0.55\,\mu m)$ in Figure 8.15b.

The values of ΔDF_{ToA} and ΔDF_{BoA} provide a pair of not markedly spread scatter plots, while those of ΔDF_{Atm} are all positive, although more largely dispersed. The regression lines drawn for the three scatter plots have the coordinates given in Table 8.11 and are characterized by best-fit values of k_D equal to −25.8, −57.2, and +31.9 W m^{-2} at the ToA and BoA levels and in the atmosphere, respectively. These findings indicate that rural-continental aerosols cause (i) cooling effects at the ToA level, which are stronger over ocean than over sea; (ii) rather intense cooling effects at the BoA level, which are comparable over both ocean and land; and (iii) warming effects in the atmosphere, which turn out to be stronger over land than over ocean. The corresponding daily values of diurnal average DARF efficiencies E_{ToA}, E_{BoA}, and E_{Atm} over land are plotted in the right-hand side of Figure 8.15b versus the daily mean $\omega(0.55\,\mu m)$. The regression lines determined for these scatter plots yield the variations of the three aforementioned efficiencies equal to $\Delta E_{ToA} = -22$ W m^{-2}, $\Delta E_{BoA} = +22$ W m^{-2}, and $\Delta E_{Atm} = -48$ W m^{-2} over the $0.90 \leq \omega(0.55\,\mu m) \leq 1.00$ range.

8.3.2.3 Free Tropospheric Aerosols

Reliable evaluations of the DARF effects due to free tropospheric aerosols over ocean surfaces were not found by us in the literature, while a pair of data sets were derived from field measurements conducted over land surfaces by (i) Tomasi *et al.* (2015a) at the Ev-K2-CNR Pyramid Laboratory (5050 m a.m.s.l.) on the three days from 24 to 26 July 1991 (see Table 8.7), not affected by Mt. Pinatubo volcanic aerosols, for the BS1, VS1, and PS4 surface reflectance models defined in Tables 8.1 and 8.2, which were chosen to represent the surface albedo characteristics of the Tibetan Plateau (TPR), the subtropical broadleaf and coniferous forests (SBCF), and Himalayan mountain region (HMR), respectively; and (ii) García *et al.* (2012), who examined the AERONET measurements conducted at the Izaña (Canary Island, Spain) and Mauna Loa Observatory (Hawaii, United States) mountain stations for FT aerosol and average surface albedo ranging from 0.30 to 0.50. Such a few results indicate that (i) the diurnally average term ΔDF_{ToA} of these aerosol particles over land increases with $\tau_a(0.55\,\mu m)$, yielding $k_D = +12.7$ W m^{-2}; (ii) ΔDF_{BoA} decreases with $\tau_a(0.55\,\mu m)$, being $k_D = -43.5$ W m^{-2}; and (iii) ΔDF_{Atm} increases with $\tau_a(0.55\,\mu m)$, being $k_D = +57.3$ W m^{-2}.

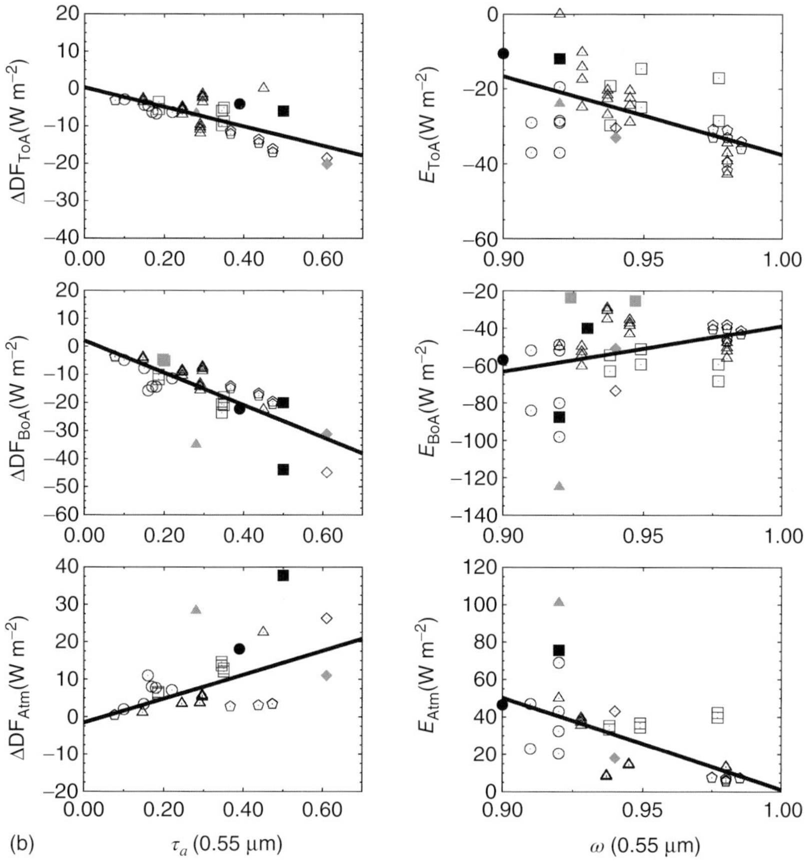

Figure 8.15b Left-hand side: scatter plots of the daily mean values of DARF terms ΔDF_{ToA} (upper part), ΔDF_{BoA} (middle part), and ΔDF_{Atm} (lower part) versus the corresponding daily mean values of aerosol optical thickness $\tau_a(0.55\,\mu m)$, determined from various sets (*) of field measurements performed over land surfaces for atmospheric loads of rural-continental aerosol particles. Right-hand side: corresponding scatter plots of the daily mean values of DARF efficiency parameters E_{ToA} at the ToA level, E_{BoA} at the BoA level, and E_{Atm} within the atmosphere versus the aerosol single scattering albedo $\omega(0.55\,\mu m)$ over land surfaces for rural-continental aerosol particles.

(*) Tomasi *et al.* (2015a), PRIN-2004 (open squares); Tomasi *et al.* (2015a), AEROCLOUDS (open upward triangles); Tomasi *et al.* (2015a), DOE/ARM/AIOP (open pentagons); Horvath *et al.* (2002), Granada University 1999 (solid circles); Delene and Ogren (2002), NOAA/CMDL campaign (gray squares); Yu *et al.* (2006), AERONET (open circles); Xu *et al.* (2003), Joint US–China 1999 project (open diamonds); Yu *et al.* (2006), Joint US–China 1999 project (gray diamonds); Liu *et al.* (2007), EAST-AIRE (solid squares); Xia *et al.* (2007b), EAST-AIRE (open downward triangles); Liu *et al.* (2011), EAST-AIRE (solid downward triangles).

8.3.2.4 **Continental-Polluted Aerosols**

A numerous set of DARF evaluations induced by continental-polluted aerosols over ocean surfaces was collected for (i) the CLEARCOLUMN measurements conducted by Tomasi *et al.* (2015a) on six golden days, from which the diurnal average DARF values were calculated for the OS2 surface reflectance model (see Table 8.4); (ii) the PRIN-2004 measurements conducted by Tomasi *et al.* (2015a) on the golden days of 20 May, 8 August, 30 August, and 2 October 2003, and analyzed using the OS3 and OS4 surface reflectance models, for which the DARF evaluations reported in Table 8.5 were obtained; (iii) the CLEARCOLUMN field measurements carried out on four clear-sky days from 5 to 20 July 1997, from which Tomasi *et al.* (2003) calculated the DARF effects over an oceanic surface reflectance model pertaining to clean water and calm winds; (iv) the TARFOX field measurements carried out by Hignett *et al.* (1999) in July 1996 over the North Atlantic coastal regions of United States for sea surface reflectance conditions; (v) the TARFOX field measurements of aerosol characteristics and ocean surface reflectance conditions, for which Russell *et al.* (1999b) evaluated the DARF effects over the sea; (vi) the TARFOX field data collected by Hignett *et al.* (1999) and analyzed by Kinne and Pueschel (2001); and (vii) the TARFOX data set collected by Russell *et al.* (1999b) and analyzed by Kinne and Pueschel (2001). The daily values of ΔDF_{ToA}, ΔDF_{BoA}, and ΔDF_{Atm} obtained on the various measurement days mentioned earlier are plotted in Figure 8.16a versus $\tau_a(0.55\,\mu m)$, presenting not markedly dispersed scatter plots, for which the regression lines were estimated to have the coordinates given in Table 8.11 (with correlation coefficients better than −0.93 at the ToA level and than −0.84 at the BoA level) and best-fit values of k_D equal to −38.6, −50.2, and +3.4 W m^{-2} at the ToA and BoA levels and in the atmosphere, respectively. The daily values of E_{ToA}, E_{BoA}, and E_{Atm} are plotted versus the daily mean $\omega(0.55\,\mu m)$ in the right-hand side of Figure 8.16a to determine the corresponding best-fit regression lines for such scatter plots characterized by rather good correlation coefficients, providing total variations of E_{ToA}, E_{BoA}, and E_{Atm} over the $0.80 \leq \omega(0.55\,\mu m) \leq 1.00$ range equal to $\Delta E_{ToA} = +9$ W m^{-2}, $\Delta E_{BoA} = +18$ W m^{-2}, and $\Delta E_{Atm} = -34$ W m^{-2}.

A large set of DARF evaluations was collected for the continental-polluted aerosols over land surfaces, obtained by analyzing the following sets of field measurements performed by (i) Tomasi *et al.* (2015a) on six CLEARCOLUMN golden days from 5 to 20 July 1997, for the VS1 and VS4 surface reflectance models defined in Table 8.3, obtaining the results given in Table 8.4; (ii) Tomasi *et al.* (2015a) for the four PRIN-2004 golden days from 20 May to 2 October 2003, listed in Table 8.5, using the VS1 and VS4 surface reflectance models; (iii) Tomasi *et al.* (2015a) for five AEROCLOUDS golden days from 21 June to 27 August 2007, obtaining the DARF evaluations reported in Table 8.6 for the four VS-type surface reflectance models defined in Table 8.3; (iv) Tomasi *et al.* (2015a) for the DOE/ARM/AIOP measurements performed on 12 clear-sky days from 7 to 30 May 2003, obtaining the DARF values given in Table 8.10 for the VS2 and VS3 surface reflectance models defined in Table 8.3; (v) Redemann *et al.* (2006), examining the INTEX-NA Phase A field measurements conducted in

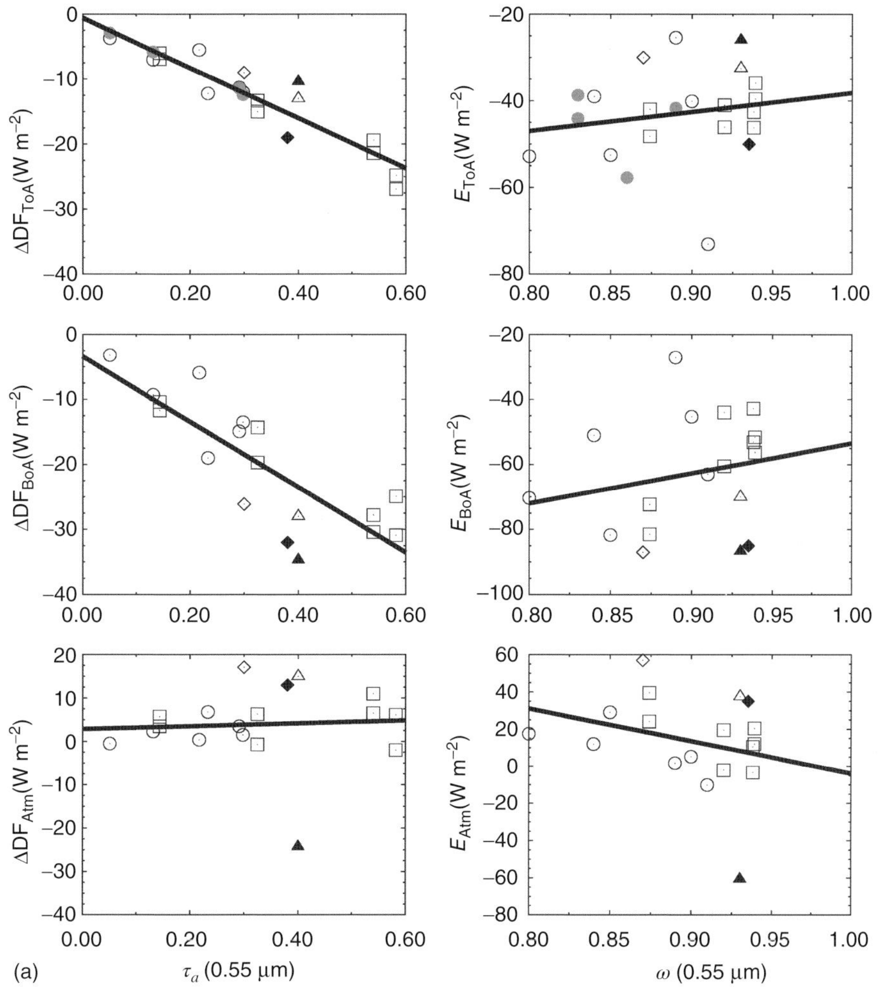

Figure 8.16a Left-hand side: scatter plots of the daily mean values of DARF terms ΔDF_{ToA} (upper part), ΔDF_{BoA} (middle part), and ΔDF_{Atm} (lower part) versus the corresponding daily mean values of aerosol optical thickness $\tau_a(0.55\,\mu m)$, determined from various sets (*) of field measurements performed over ocean surfaces for atmospheric loads of continental-polluted aerosol particles. Right-hand side: corresponding scatter plots of the daily mean values of DARF efficiency parameters E_{ToA} at the ToA level, E_{BoA} at the BoA level, and E_{Atm} within the atmosphere versus the aerosol single scattering albedo $\omega(0.55\,\mu m)$ over ocean surfaces for continental-polluted aerosol particles. (*) Tomasi *et al.* (2015a), CLEARCOLUMN/ACE-2 (open circles); Tomasi *et al.* (2015a), PRIN-2004 (open squares; Tomasi *et al.* (2003), CLEARCOLUMN/ACE-2 (gray circles); Hignett *et al.* (1999), TARFOX (open diamonds); Russell *et al.* (1999b), TARFOX (solid diamonds); Kinne and Pueschel (2001) and Hignett *et al.* (1999), TARFOX (solid upward triangles); Kinne and Pueschel (2001) and Russell *et al.* (1999b), TARFOX (open upward triangles).

June and July 2004 over the central and eastern US areas; (vi) Ramana *et al.* (2004), analyzing the atmospheric brown cloud (ABC) surface-level radiation measurements carried out at Kathmandu and Nagarkot (Nepal) in winter 2003; (vii) Ramanathan *et al.* (2007) using a Monte Carlo aerosol–cloud–radiation (MACR) model and the regional chemical transport model (STEM-2K) to analyze the ABC field measurements conducted over the megacity areas of South and East Asia and Indo-Gangetic Plains during the five-year period from 2003 to 2008; (viii) Li *et al.* (2007a), analyzing the EAST-AIRE measurements conducted at various sites in China during March and April 2005; (ix) Li *et al.* (2007a,2007b), examining the EAST-AIRE measurements conducted at Xianghe (China) from September 2004 to September 2005; (x) Xia *et al.* (2007c), analyzing the EAST-AIRE measurements carried out at Xianghe (China) in October 2004 for heavy polluted air conditions; and (xii) Liu *et al.* (2007), examining the EAST-AIRE measurements conducted at Beijing (China) in October 2004, for heavy anthropogenic haze conditions.

The scatter plots of the daily values of ΔDF_{ToA}, ΔDF_{BoA}, and ΔDF_{Atm} versus $\tau_a(0.55\,\mu m)$ are shown in the left-hand side of Figure 8.16b. The data appear to be quite dispersed, with ΔDF_{ToA} ranging between -20 and $+10\,W\,m^{-2}$, ΔDF_{BoA} varying mainly between -40 and $-2\,W\,m^{-2}$, and ΔDF_{Atm} ranging mainly between -10 and $+20\,W\,m^{-2}$. The regression lines were found to have the coordinates given in Table 8.11 and best-fit values of k_D equal to $+4.9$, -91.6, and $+103.4\,W\,m^{-2}$ at the ToA and BoA levels and in the atmosphere, respectively, for rather low values of their correlation coefficients. The daily values of E_{ToA}, E_{BoA}, and E_{Atm} are shown versus the corresponding daily mean values of $\omega(0.55\,\mu m)$ in the right-hand side of Figure 8.16b. For these largely dispersed values, the regression lines provide the overall variations $\Delta E_{ToA} = -31\,W\,m^{-2}$, $\Delta E_{BoA} = +7\,W\,m^{-2}$, and $\Delta E_{Atm} = -90\,W\,m^{-2}$ over the $0.80 \leq \omega(0.55\,\mu m) \leq 1.00$ range. The previous results indicate that rural-continental aerosols cause in general: (i) pronounced cooling effects at the ToA level over ocean and weak warming effects over land; (ii) rather intense cooling effects at the BoA level, slightly stronger over land than over ocean; and (iii) warming effects in the atmosphere, which are considerably stronger over land than over ocean.

8.3.2.5 Maritime Clean Aerosols

Only a few significant diurnally average estimates of ΔDF_{ToA}, ΔDF_{BoA}, and ΔDF_{Atm} and parameters E_{ToA}, E_{BoA}, and E_{Atm} induced by maritime clean aerosols over ocean surfaces were obtained by (i) Tomasi *et al.* (2015a), examining the CLEARCOLUMN data recorded on 19 June 1997 (see Table 8.4), using the OS2 surface reflectance model defined by Tomasi *et al.* (2013) (see Tables 8.1 and 8.2), and (ii) Tomasi *et al.* (2015a), analyzing the Aerosols99 cruise measurements conducted by Voss *et al.* (2001) on 18 January 1999, for which the DARF evaluations were made using the OS2 and OS4 oceanic surface reflectance models of Tomasi *et al.* (2013) described in Tables 8.1 and 8.2. These daily values of ΔDF_{ToA}, ΔDF_{BoA}, and ΔDF_{Atm} were plotted versus $\tau_a(0.55\,\mu m)$ to determine the regression lines giving the best-fit values of k_D equal to $+37.5$, $+114.8$, and $-75.0\,W\,m^{-2}$

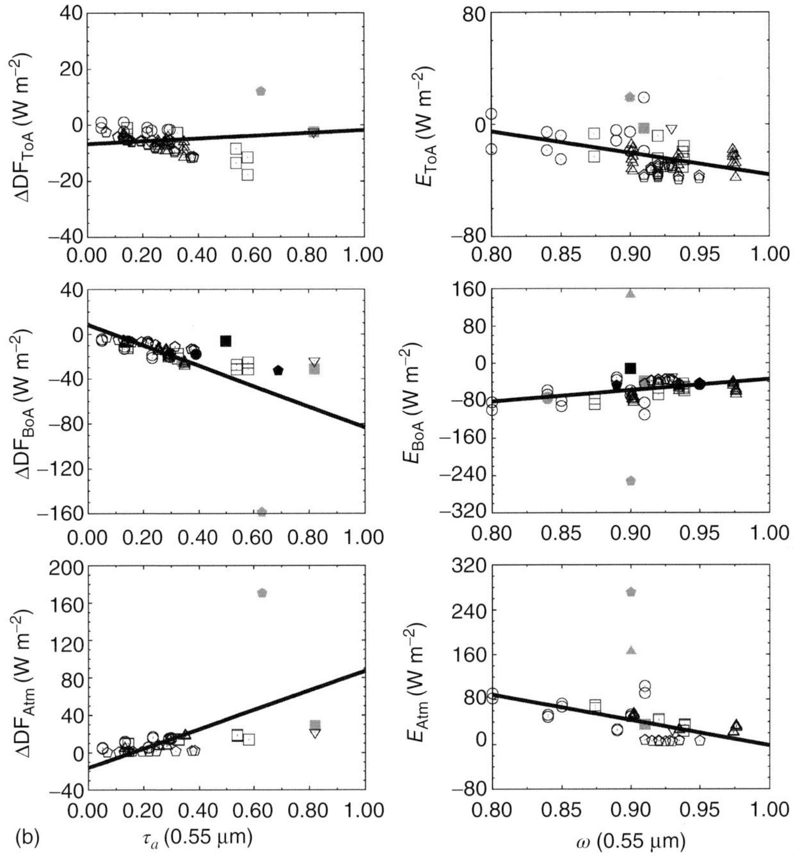

Figure 8.16b Left-hand side: scatter plots of the daily mean values of DARF terms ΔDF_{ToA} (upper part), ΔDF_{BoA} (middle part), and ΔDF_{Atm} (lower part) versus the corresponding daily mean values of aerosol optical thickness τ_a(0.55 μm), determined from various sets (*) of field measurements performed over land surfaces for atmospheric loads of continental-polluted aerosol particles. Right-hand side: corresponding scatter plots of the daily mean values of DARF efficiency parameters E_{ToA} at the ToA level, E_{BoA} at the BoA level, and E_{Atm} within the atmosphere versus the aerosol single scattering albedo ω(0.55 μm) over land surfaces for continental-polluted aerosol particles. (*) Tomasi *et al.* (2015a), CLEARCOLUMN/ACE-2 (open circles); Tomasi *et al.* (2015a), PRIN-2004 (open squares); Tomasi *et al.* (2015a), AEROCLOUDS (open upward triangles); Tomasi *et al.* (2015a), DOE/ARM/AIOP (open pentagons); Redemann *et al.* (2006), INTEX-NA Phase A) (solid circles); Ramana *et al.* (2004), ABC (gray circles); Ramanathan *et al.* (2007), ABC (solid squares); Li *et al.* (2007a), EAST-AIRE over the whole China (solid pentagons); Li *et al.* (2007a), EAST-AIRE, Xianghe (downward open triangles); Li *et al.* (2007b), EAST-AIRE, Xianghe (gray pentagons); Xia *et al.* (2007a), EAST-AIRE over the whole China (downward open triangles); Xia *et al.* (2007c), EAST-AIRE, Xianghe (gray squares); Liu *et al.* (2007), EAST-AIRE, Beijing (solid upward triangles).

at the ToA and BoA levels and in the atmosphere. The scatter plots of the daily values of E_{ToA}, E_{BoA}, and E_{Atm} versus $\omega(0.55\,\mu\text{m})$ over the narrow range from 0.95 to 1.00 provided the variations $\Delta E_{\text{ToA}} = +101\,\text{W m}^{-2}$, $\Delta E_{\text{BoA}} = +86\,\text{W m}^{-2}$, and $\Delta E_{\text{Atm}} = +2\,\text{W m}^{-2}$.

A set consisting of a few DARF evaluations was also collected for maritime clean aerosols over land by (i) Tomasi *et al.* (2015a), examining the CLEARCOLUMN measurements conducted on six golden days from 5 to 20 July 1997 (see Table 8.4), using the VS1 and VS4 surface reflectance models defined in Table 8.3, and (ii) Tomasi *et al.* (2015a), analyzing the PRIN-2004 measurements carried out on four golden days from 20 May to 2 October 2003 (see Table 8.5), using the VS1 and VS4 surface reflectance models of Tomasi *et al.* (2013). The daily values of $\Delta\text{DF}_{\text{ToA}}$, $\Delta\text{DF}_{\text{BoA}}$, and $\Delta\text{DF}_{\text{Atm}}$ were plotted versus $\tau_a(0.55\,\mu\text{m})$, yielding regression lines having slope coefficient k_D equal to +520, +660, and $-150\,\text{W m}^{-2}$ at the ToA and BoA levels and in the atmosphere, respectively. Correspondingly, the scatter plots of the daily values of E_{ToA}, E_{BoA}, and E_{Atm} versus $\omega(0.55\,\mu\text{m})$ yielded estimates of $\Delta E_{\text{ToA}} = -83\,\text{W m}^{-2}$, $\Delta E_{\text{BoA}} = -105\,\text{W m}^{-2}$, and $\Delta E_{\text{Atm}} = +24\,\text{W m}^{-2}$ over the $0.95 \leq \omega(0.55\,\mu\text{m}) \leq 1.00$ range. These results indicate that maritime clean aerosols induce in general more pronounced warming effects over land than over ocean at both ToA and BoA levels, while stronger cooling effects are induced in the atmosphere over land than over ocean.

8.3.2.6 Maritime–Continental Aerosols

A large set of daily mean values of $\Delta\text{DF}_{\text{ToA}}$, $\Delta\text{DF}_{\text{BoA}}$, and $\Delta\text{DF}_{\text{Atm}}$ was collected for maritime–continental aerosols suspended over ocean surfaces, obtained by (i) Tomasi *et al.* (2015a), examining the CLEARCOLUMN measurements conducted on 20, 23, and 25 June 1997, using the OS2 surface reflectance model; (ii) Tomasi *et al.* (2015a), analyzing the PRIN-2004 measurements performed on 30 April 2003, using the OS3 and OS4 models given in Table 8.5; (iii) Lelieveld *et al.* (2002); Markowicz *et al.* (2002), and Yu *et al.* (2006), examining the MINOS field measurements conducted in the Crete Island during July and August 2010; (iv) Tafuro *et al.* (2007), analyzing the AERONET measurements conducted at Lecce (Italy) in winter 2004 and assuming surface reflectance characteristics given by a mixed surface consisting of 75% sea surface and 25% land surface; (v) Schmidt *et al.* (2010), examining the MILAGRO/INTEX-B field measurements carried out over the Gulf of Mexico on 13 March 2006; (vi) Kinne and Pueschel (2001), analyzing (a) the ACE-2 field measurements conducted by Russell and Heintzenberg (2000) over the subtropical northeastern part of Atlantic Ocean in June–July 1997, (b) the INDOEX field measurements performed over the tropical Indian Ocean in winter 1998 and 1999, and (c) the so-called "Asia case 3" data derived from the NASA DC-8 aircraft measurements conducted over the Western Pacific Ocean off East Asia in early March 1994; (vii) Yu *et al.* (2006), examining (a) a set of ACE-2 field measurements, (b) a set of radiative transfer simulations made over the midlatitude part of northern Atlantic Ocean in summer 2001, and over the tropical part of northern Atlantic Ocean in winter and summer 2001, and (c) a set of MODIS, CERES-A, CERES-B, and MODIS-A satellite observations recorded over the midlatitude north Atlantic Ocean in summer 2001; (viii) Meywerk and Ramanathan (1999), analyzing the INDOEX field measurements conducted over the northern Arabian Sea in February 1998 and the tropical Indian Ocean in March 1998; (ix) Redemann *et al.* (2006), examining the INDOEX measurements conducted over the northern Arabian Sea in February 1998 and the tropical Indian Ocean (15°S latitude, ITCZ) in March 1998; (x) Satheesh *et al.* (1999), analyzing the radiative transfer simulations made over the tropical Indian Ocean and derived from the INDOEX measurements conducted at the KCO (Republic of Maldives) in February–March 1998; (xi) Podgorny *et al.* (2000), analyzing the INDOEX measurements conducted over the tropical Indian Ocean (in proximity of the Maldives Islands) in February–March 1998; (xii) Yu *et al.* (2006), analyzing (a) the INDOEX field measurements and radiative transfer simulations carried out at the KCO site in February–March 1998, (b) the INDOEX measurements carried out at the KCO site in February–March 1999, (c) the shipborne measurements conducted on board the R/V Sagar Kanya and the MODIS/Terra satellite-based observations made over the Bay of Bengal during three different periods of 2001 and 2003, (d) the MODIS, CERES-A, CERES-B, and MODIS-A satellite data collected over the Arabian Sea, northern Indian Ocean, and southern Asia in winter 2001, spring 2001, and autumn 2001, and (e) the radiative transfer simulations based on the CERES and MODIS satellite data collected over the northwestern Pacific Ocean in spring and autumn 2001; (xiii) Tahnk and Coakley (2002), analyzing the INDOEX data carried out on board the R/V Sagar Kanya and the MODIS and NOAA-14 AVHRR satellite data collected over the Arabian Sea, northern Indian Ocean, and Bay of Bengal in the winter monsoon season (January–March) from 1996 to 2000; (xiv) Podgorny and Ramanathan (2001), using the multiple scattering Monte Carlo radiation model for the INDOEX field measurements of aerosol optical parameters carried out at the KCO site during the winter monsoon season; (xv) Satheesh and Ramanathan (2000), examining the INDOEX field measurements performed at the KCO site in winter 1999; (xvi) Conant (2000), examining the INDOEX surface-level radiation measurements performed at the KCO site in February–March 1998; (xvii) Collins *et al.* (2002), performing radiative transfer simulations based on the observations made over the tropical Indian Ocean in January–March 1999; (xviii) Bush and Valero (2002), analyzing the INDOEX surface-level radiation measurements performed at the KCO site in February–March 1999; (xix) Rajeev and Ramanathan (2001), analyzing the satellite data collected over the tropical Indian Ocean (at latitudes from 25°S to 25°N) during the winter months from 1997 to 1999; (xx) Ramanathan *et al.* (2001a,2001b), examining the INDOEX field measurements conducted over the Arabian Sea and tropical Indian Ocean in January–April) 1998 and winter 1999; (xxi) Satheesh (2002) and Satheesh and Srinivasan (2002), analyzing a set of INDOEX surface-level measurements of aerosol optical parameters and a data set provided by the Indian Remote Sensing (IRS-P4) satellite over the Arabian Sea, tropical Indian Ocean, and Bay of Bengal in March 2001; (xxii) Sumanth *et al.* (2004), examining a set of radiation flux measurements performed on board the R/V Sagar Kanya and the MODIS/Terra satellite data set recorded over the Bay of Bengal during March 2001, February 2003, and October 2003; (xxiii) Markowicz *et al.* (2003), analyzing the ACE-Asia field measurements conducted over the Sea of Japan in March–April 2001; and (xxiv) Kim *et al.* (2005a), examining the ACE-Asia measurements performed at Anmyon and Gosan (South Korea) and Amami Ōshima (Japan) in April 2001. The daily values of ΔDF_{ToA}, ΔDF_{BoA}, and ΔDF_{Atm} determined earlier on the selected golden days were plotted versus τ_a(0.55 μm) in the left-hand side of Figure 8.17a.

The values of ΔDF_{ToA} were found to be all negative and higher than −16 W m^{-2}, those of ΔDF_{BoA} to be negative and mainly higher than −40 W m^{-2}, and those of ΔDF_{Atm} to range between −5 and +32 W m^{-2}. The regression lines were found to have the coordinates given in Table 8.11 and good correlation coefficients, providing best-fit values of k_D equal to −15.9, −60.8, and +53.8 W m^{-2} at the ToA and BoA levels and in the atmosphere, respectively. The daily mean values of E_{ToA}, E_{BoA}, and E_{Atm} were plotted versus the daily mean ω(0.55 μm) in the right-hand side of Figure 8.17a, yielding the average variations $\Delta E_{ToA} = -10$ W m^{-2}, $\Delta E_{BoA} = +30$ W m^{-2}, and $\Delta E_{Atm} = -16$ W m^{-2} over the $0.80 \leq \omega(0.55\ \mu m) \leq 1.00$ range.

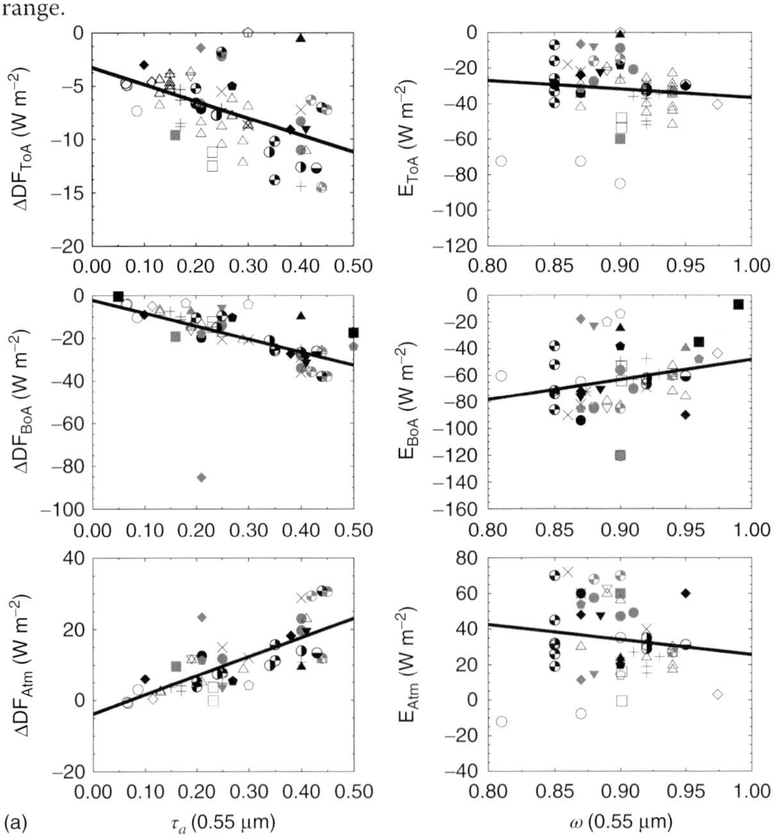

Figure 8.17a Left-hand side: scatter plots of the daily mean values of DARF terms ΔDF_{ToA} (upper part), ΔDF_{BoA} (middle part), and ΔDF_{Atm} (lower part) versus the corresponding daily mean values of aerosol optical thickness $\tau_a(0.55\,\mu m)$, determined from various sets (*) of field measurements performed over ocean surfaces for atmospheric loads of maritime–continental aerosol particles. Right-hand side: corresponding scatter plots of the daily mean values of DARF efficiency parameters E_{ToA} at the ToA level, E_{BoA} at the BoA level, and E_{Atm} within the atmosphere versus the aerosol single scattering albedo $\omega(0.55\,\mu m)$ over ocean surfaces for maritime–continental aerosol particles. (*) Tomasi *et al.* (2015a), CLEARCOLUMN/ACE-2 (open circles); Tomasi *et al.* (2015a), PRIN-2004 (open squares); Lelieveld *et al.* (2002), MINOS (gray diamonds); Markowicz *et al.* (2002), MINOS (gray pentagons); Yu *et al.* (2006), MINOS (solid circles); Tafuro *et al.* (2007), AERONET (open diamonds); Schmidt *et al.* (2010), MILAGRO (solid pentagons); Kinne and Pueschel (2001), ACE-2, INDOEX, and NASA DC-8 aircraft campaign (gray circles); Yu *et al.* (2006), ACE-2 (crosses (+)); Meywerk and Ramanathan (1999), INDOEX (solid squares); Redemann *et al.* (2006), INDOEX (gray pentagons); Satheesh *et al.* (1999), INDOEX (gray squares); Podgorny *et al.* (2000), INDOEX (downward open triangles); Yu *et al.* (2006), INDOEX (upward open triangles); Tahnk and Coakley (2002), R/V Sagar Kanya and MODIS and NOAA-14 AVHRR data (vertical half-solid circles); Podgorny and Ramanathan (2001), INDOEX (upward solid triangles); Satheesh and Ramanathan (2000), INDOEX (downward solid triangles); Conant (2000), INDOEX (upward gray triangles); Collins *et al.* (2002), INDOEX (downward gray triangles); Bush and Valero (2002), INDOEX (asterisks); Rajeev and Ramanathan (2001), INDOEX (solid diamonds); Ramanathan *et al.* (2001a,2001b), INDOEX, (open pentagons); Satheesh (2002) and Satheesh and Srinivasan (2002), INDOEX (slant hourglasses); Sumanth *et al.* (2004), INDOEX (gray slant hourglasses); Markowicz *et al.* (2003), ACE-Asia (horizontal half-solid circles); Kim *et al.* (2005a), ACE-Asia (crosses (×)).

A large set of daily values of ΔDF_{ToA}, ΔDF_{BoA}, and ΔDF_{Atm} and of E_{ToA}, E_{BoA}, and E_{Atm} was collected for maritime–continental aerosols over land surfaces, consisting of the evaluations of these parameters derived by (i) Tomasi *et al.* (2015a), examining the CLEARCOLUMN measurements conducted on the three golden days from 20 to 25 June 1997 (see Table 8.4), using the VS1 and VS4 surface reflectance models defined in Table 8.3; (ii) Tomasi *et al.* (2015a), analyzing the PRIN-2004 measurements carried out on 30 April 2003 (see Table 8.5), using the VS1 and VS4 surface reflectance models; (iii) Tomasi *et al.* (2015a), examining the AEROCLOUDS measurements conducted on eight golden days from 18 May 2007 to 13 March 2008, using the four VS surface reflectance models defined in Table 8.3; (iv) Delene and Ogren (2002), examining the NOAA/CMDL field measurements conducted at Sable Island (Nova Scotia, Canada) from November 1994 to September 2000 for vegetation-covered surface reflectance conditions; (v) Nakajima *et al.* (2003), analyzing the Asian Atmospheric Particle Environmental Change Studies (APEX) field measurements conducted in April 2001 over the East China Sea region by using four different methodologies to examine the data collected at Gosan (South Korea) and Amami Ōshima (Japan), based on (a) the aerosol chemical model SPRINTARS, (b) the mesoscale chemical model CFORS, (c) the ASD analysis procedure applied to satellite data, and (d) the AS-DARF analysis of surface-level DARF measurements; (vi) Bush and Valero (2003), examining the ACE-Asia field measurements conducted at

Gosan (South Korea) from 25 March to 4 May 2001; (vii) Won *et al.* (2004), analyzing the ACE-Asia field measurements carried out at Gosan (South Korea) on 15 April 2001; (viii) Kim *et al.* (2004), analyzing the ACE-Asia measurements of surface-level shortwave irradiance performed at Gosan in April 2001 and carrying out the corresponding radiative transfer simulations over land surfaces; (ix) Kim *et al.* (2005a), examining the ACE-Asia measurements conducted on seventeen clear-sky days of April 2001; (x) Kim *et al.* (2005b), analyzing the ACE-Asia measurements of sky radiation and surface solar fluxes performed at Sri Samrong (Thailand) in April 2001; (xi) Yu *et al.* (2006), examining the AERONET measurements and radiative transfer simulations made in spring and autumn 2001 over some land regions in East Asia; and (xii) Nakajima *et al.* (2007), analyzing the EAREX 2005 field measurements of surface-level solar radiation fluxes conducted at several sites of eastern Asia in March–April 2005. The scatter plots of the daily values of ΔDF_{ToA}, ΔDF_{BoA}, and ΔDF_{Atm} versus $\tau_a(0.55\,\mu m)$ are shown in the left-hand side of Figure 8.17b, providing the regression lines having the coordinates given in Table 8.11, all obtained with very good correlation coefficients, and best-fit values of k_D equal to −32.1, −71.9, and +36.9 W m^{-2} at the ToA and BoA levels and in the atmosphere, respectively. The daily values of E_{ToA}, E_{BoA}, and E_{Atm} were plotted versus the daily mean $\omega(0.55\,\mu m)$ in the right-hand side of Figure 8.17b, obtaining the variations $\Delta E_{ToA} = -30$ W m^{-2}, $\Delta E_{BoA} = +77$ W m^{-2}, and $\Delta E_{Atm} = -106$ W m^{-2} over the $0.80 \leq \omega(0.55\,\mu m) \leq 1.00$ range. The comparison between the results shown in Figures 8.17a and 8.17b for maritime–continental aerosols indicates that (i) considerably more pronounced cooling effects are induced at the ToA level over land than over ocean, (ii) appreciably more marked cooling effects are produced at the BoA level over land than over ocean, and (iii) more pronounced warming effects are caused on average in the atmosphere by these particles over ocean than over land.

8.3.2.7 **Desert Dust**

A large set of the six DARF parameters was calculated for desert dust over ocean surfaces by (i) Tomasi *et al.* (2015a), examining the PRIN-2004 field measurements carried out on the three golden days of 1 May, 17 July, and 24 July 2003, characterized by Saharan dust transport over southern Italy and reported in Table 8.5, using the OS3 and OS4 surface reflectance models; (ii) Tomasi *et al.* (2015a), analyzing the Aerosols99 measurements conducted by Voss *et al.* (2001) on 25 January 1999, during an African dust transport episode over the Atlantic Ocean for the OS2 and OS4 surface reflectance models; (iii) Yu *et al.* (2006), examining the ACE-Asia field measurements conducted on board the NOAA R/V *Ronald H. Brown* and a C-130 aircraft over the Sea of Japan from 5 to 15 April 2001, for average reflectance characteristics of the ocean surface; (iv) Kim *et al.* (2005a), analyzing the ACE-Asia field measurements of aerosol optical parameters and surface-level solar and sky radiation fluxes conducted at Anmyon and Gosan (South Korea) and Amami Ōshima (Japan) in April 2001 during AD transport events over the ocean; (v) Christopher *et al.* (2003), examining the Puerto Rico Dust Experiment (PRIDE) field measurements performed over the Caribbean region during a 26-day period

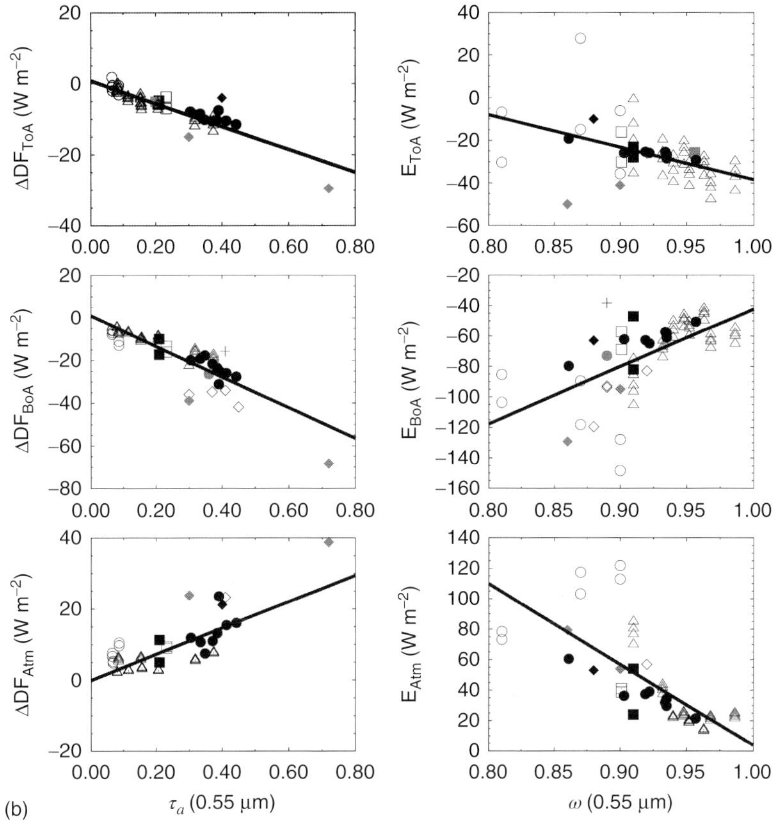

Figure 8.17b Left-hand side: scatter plots of the daily mean values of DARF terms ΔDF_{ToA} (upper part), ΔDF_{BoA} (middle part), and ΔDF_{Atm} (lower part) versus the corresponding daily mean values of aerosol optical thickness τ_a(0.55 μm), determined from various sets (*) of field measurements performed over land surfaces for atmospheric loads of maritime–continental aerosol particles. Right-hand side: corresponding scatter plots of the daily mean values of DARF efficiency parameters E_{ToA} at the ToA level, E_{BoA} at the BoA level, and E_{Atm} within the atmosphere versus the aerosol single scattering albedo ω(0.55 μm) over land surfaces for maritime–continental aerosol particles.

(*) Tomasi *et al.* (2015a), CLEARCOLUMN/ACE-2 (open circles); Tomasi *et al.* (2015a), PRIN-2004 (open squares); Tomasi *et al.* (2015a), AEROCLOUDS (open upward triangles); Delene and Ogren (2002), NOAA/CMDL campaign (gray squares); Nakajima *et al.* (2003), APEX (solid circles); Bush and Valero (2003), ACE-Asia (gray circles); Won *et al.* (2004), ACE-Asia (gray diamonds); Kim *et al.* (2004), ACE-Asia (solid diamonds); Kim *et al.* (2005a), ACE-Asia (open diamonds); Kim *et al.* (2005b), ACE-Asia (crosses (+)); Yu *et al.* (2006), AERONET (solid squares); Nakajima *et al.* (2007), EAREX 2005 (open upward triangles).

from mid-June to late July 2000 and the upwelling radiation data recorded by CERES during some mineral dust transport episodes from North Africa; (vi) Li, Vogelmann, and Ramanathan (2004), analyzing two sets of CERES and MODIS satellite-based observations made over the midlatitude part of northern Atlantic Ocean in the winter months from November 2000 to January 2001 and in the summer months from June 2000 to August 2001; (vii) Yu *et al.* (2006), examining some sets of MODIS, CERES-A, CERES-B, and MODIS-A data recorded over the tropical north Atlantic Ocean in summer 2001 during some Saharan dust transport episodes; and (viii) Markowicz *et al.* (2008), examining the United Arab Emirates Unified Aerosol Experiment (UAE)[2] measurements conducted over the Persian Gulf and Arabian Peninsula in August–September 2004. The daily values of ΔDF_{ToA}, ΔDF_{BoA}, and ΔDF_{Atm} over ocean were plotted versus $\tau_a(0.55\,\mu m)$ in the left-hand side of Figure 8.18a to determine the regression lines found to have the coordinates given in Table 8.11.

The values of ΔDF_{ToA} were all negative, ranging from -32 to $-2\,W\,m^{-2}$, those of ΔDF_{BoA} varied between -75 and $0\,W\,m^{-2}$, and those of ΔDF_{Atm} from -4 to $+66\,W\,m^{-2}$, being characterized by rather good correlation features. The best-fit lines provided values of k_D equal to -24.2, -118.4, and $+97.1\,W\,m^{-2}$ at the ToA and BoA levels and in the atmosphere, respectively. The daily values of E_{ToA}, E_{BoA}, and E_{Atm} were plotted versus the daily mean $\omega(0.55\,\mu m)$ in the right-hand side of Figure 8.18a, yielding the average variations $\Delta E_{ToA} = -45\,W\,m^{-2}$, $\Delta E_{BoA} = +22\,W\,m^{-2}$, and $\Delta E_{Atm} = -70\,W\,m^{-2}$ over the $0.80 \leq \omega(0.55\,\mu m) \leq 1.00$ range.

The set of DARF parameters pertaining to desert dust over land was calculated by (i) Tomasi *et al.* (2015a), analyzing the PRIN-2004 measurements performed on the three aforementioned golden days using the VS1 and VS4 surface reflectance models; (ii) García *et al.* (2012), examining the AERONET measurements conducted during some mineral dust transport episodes in North and West Africa, Arabian Peninsula, and Asia for different surface reflectance characteristics; (iii) Won *et al.* (2004), analyzing the ACE-Asia radiative transfer simulations made at Gosan (South Korea) during the AD episode observed on 13 April 2001; (iv) Seinfeld *et al.* (2004), examining the ACE-Asia field measurements conducted on board the NOAA R/V Ronald H. Brown and a C-130 aircraft over the Sea of Japan from 5 to 15 April 2001; (v) Kim *et al.* (2005a), examining the ACE-Asia solar and sky radiation ground-based measurements conducted at Dunhuang (China) in April–July from 1998 to 2000; (vi) Kim *et al.* (2005b), analyzing the ACE-Asia ground-based measurements performed at Gosan (South Korea) on 10 April 2001; (vii) Liu *et al.* (2011), analyzing the EAST-AIRE measurements performed at Lanzhou (China) in March–May 2009 during some dust events; (viii) Xia *et al.* (2007b), examining the EAST-AIRE measurements carried out at Taihu (China) from September 2005 to August 2008 during some AD transport episodes; (ix) Ge *et al.* (2010, 2011), examining the ARM/AAF/SMART-COMMIT 2008 field measurements conducted at Zhangye and Minqin in northwestern China in April–June 2008 during some dust transport episodes from the Gobi Desert; (x) Hsu, Herman, and Weaver (2000),

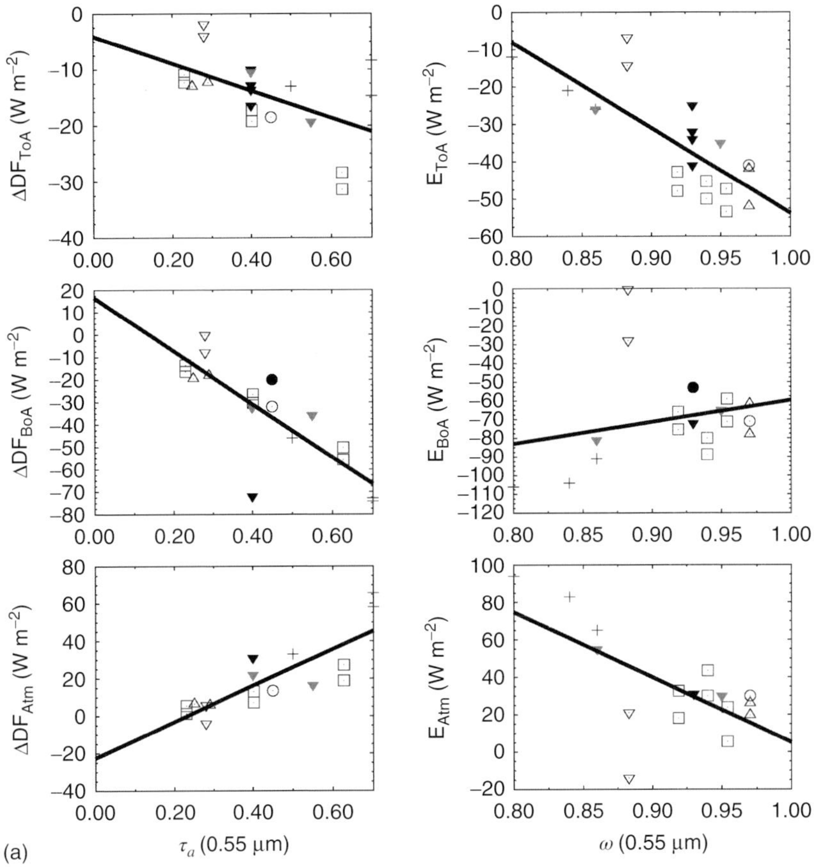

Figure 8.18a Left-hand side: scatter plots of the daily mean values of DARF terms ΔDF_{ToA} (upper part), ΔDF_{BoA} (middle part), and ΔDF_{Atm} (lower part) versus the corresponding daily mean values of aerosol optical thickness τ_a(0.55 μm), determined from various sets (*) of field measurements performed over ocean surfaces for atmospheric loads of desert dust particles. Right-hand side: corresponding scatter plots of the daily mean values of DARF efficiency parameters E_{ToA} at the ToA level, E_{BoA} at the BoA level, and E_{Atm} within the atmosphere versus the aerosol single scattering albedo ω(0.55 μm) over ocean surfaces for desert dust particles. (*) Tomasi *et al.* (2015a), PRIN-2004 (open squares); Tomasi *et al.* (2015a), Aerosols99 (downward open triangles); Yu *et al.* (2006), ACE-Asia (open circles); Kim *et al.* (2005b), ACE-Asia ((crosses (+)); Christopher and Zhang (2002) and Christopher *et al.* (2003), PRIDE (upward open triangles); Li, Vogelmann, and Ramanathan (2004), CERES and MODIS (downward gray triangles); Yu *et al.* (2006), CERES and MODIS (downward solid triangles); Markowicz *et al.* (2008), UAE[2] (solid circles).

analyzing the Nimbus-7 TOMS and NOAA-9 ERBE satellite data recorded over (a) Cape Verde in February 1985 and (b) the Bidi Bahn land area in July 1985 during a few Saharan dust long-range transport episodes; and (xi) Tanré *et al.* (2003), examining the AERONET and SHADE field measurements conducted at Cape Verde from 19 to 29 September 2000. All the daily values of ΔDF_{ToA}, ΔDF_{BoA}, and ΔDF_{Atm} obtained previously on the selected golden days were plotted versus $\tau_a(0.55\,\mu m)$ in the left-hand side of Figure 8.18b to determine the regression lines found to have the coordinates and correlation coefficients given in Table 8.11 and values of k_D equal to −41.8, −99.3, and +68.9 W m^{-2} at the ToA and BoA levels and in the atmosphere, respectively.

The daily values of E_{ToA}, E_{BoA}, and E_{Atm} were plotted versus the daily mean $\omega(0.55\,\mu m)$ in the right-hand side of Figure 8.18b, determining the average variations $\Delta E_{ToA} = -55$ W m^{-2}, $\Delta E_{BoA} = +40$ W m^{-2}, and $\Delta E_{Atm} = -90$ W m^{-2} over the range $0.80 \leq \omega(0.55\,\mu m) \leq 1.00$. Comparing the results shown in Figure 8.18a for desert dust over ocean with those of Figure 8.18b pertaining to desert dust particles over land, it is evident that these wind-forced aerosol particles induce considerably stronger cooling effects at the ToA and BoA levels over land (about twice) than over ocean and more pronounced warming effects in the atmosphere over ocean than over land at parity of $\tau_a(0.55\,\mu m)$.

8.3.2.8 Biomass Burning Aerosols

An interesting set of DARF evaluations was collected for some episodes of BBS transport over the ocean, including the estimates made by (i) Tomasi *et al.* (2015a), examining the Aerosols99 measurements collected on 30 January 1999 (Voss *et al.*, 2001), and given in Table 8.9 for the OS2 and OS4 surface reflectance models; (ii) Ross, Hobbs, and Holben (1998), analyzing the SCAR-B aerosol measurements conducted over the southern Atlantic Ocean in August–September 1995 for BBS transported from the Brazilian Amazon and Cerrado regions; (iii) Podgorny, Li, and Ramanathan (2003), analyzing the satellite-based observations and Monte Carlo radiative transfer simulations made over the tropical Indian Ocean during episodes of Indonesian forest fire smoke transport in September–November 1997; and (iv) Yu *et al.* (2006), examining the MODIS and CERES satellite data recorded over the tropical Atlantic Ocean in winter and autumn 2001.These daily values of ΔDF_{ToA}, ΔDF_{BoA}, and ΔDF_{Atm} were plotted versus $\tau_a(0.55\,\mu m)$ in the left-hand side of Figure 8.19a, showing values of ΔDF_{ToA} ranging between −20 and −3 W m^{-2}, values of ΔDF_{BoA} between −24 and −11 W m^{-2}, and values of ΔDF_{Atm} between −1 and +9 W m^{-2}. The regression lines were estimated to have the coordinates given in Table 8.11, providing best-fit values of k_D equal to +6.4, +28.1, and +16.0 W m^{-2} at the ToA and BoA levels and in the atmosphere, respectively. The daily values of E_{ToA}, E_{BoA}, and E_{Atm} were plotted versus the daily mean $\omega(0.55\,\mu m)$ in the right-hand side of Figure 8.19a, yielding the average variations $\Delta E_{ToA} = -28$ W m^{-2}, $\Delta E_{BoA} = -41$ W m^{-2}, and $\Delta E_{Atm} = -29$ W m^{-2} over the 0.85–1.00 range.

The evaluations of DARF parameters pertaining to BBS over land were provided by (i) Tomasi *et al.* (2015a), examining the DOE/ARM/AIOP field

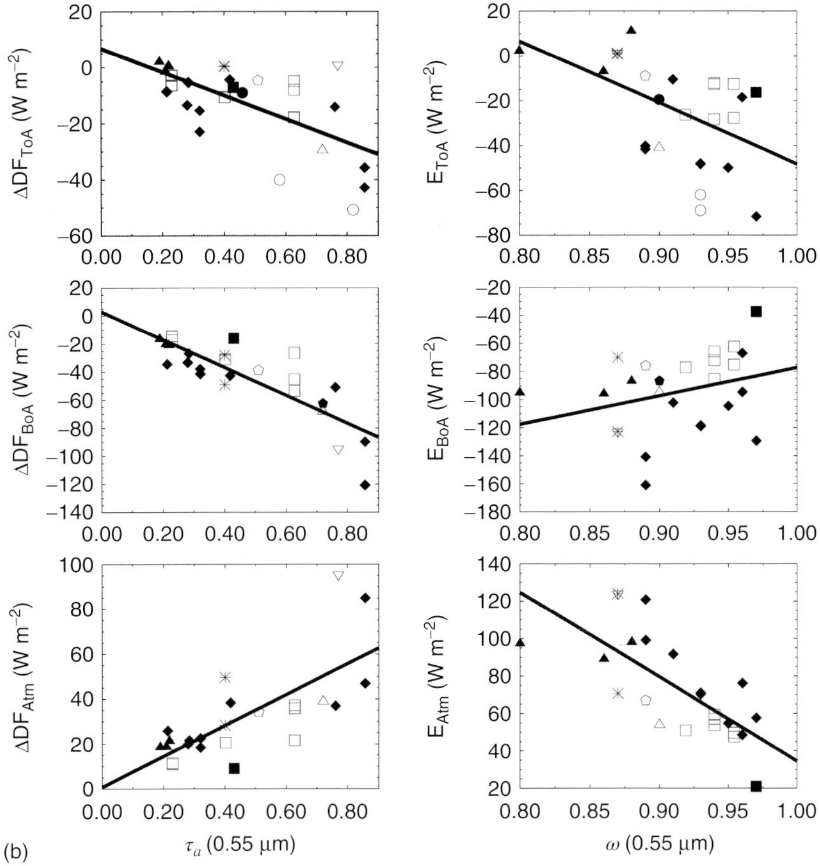

Figure 8.18b Left-hand side: scatter plots of the daily mean values of DARF terms ΔDF_{ToA} (upper part), ΔDF_{BoA} (middle part), and ΔDF_{Atm} (lower part) versus the corresponding daily mean values of aerosol optical thickness τ_a(0.55 μm), determined from various sets (*) of field measurements performed over land surfaces for atmospheric loads of desert dust particles. Right-hand side: corresponding scatter plots of the daily mean values of DARF efficiency parameters E_{ToA} at the ToA level, E_{BoA} at the BoA level, and E_{Atm} within the atmosphere versus the aerosol single scattering albedo ω(0.55 μm) over land surfaces for desert dust particles.

(*) Tomasi *et al.* (2015a), PRIN-2004 (open squares); García *et al.* (2012), AERONET (solid diamonds); Won *et al.* (2004), ACE-Asia (gray squares); Seinfeld *et al.* (2004), ACE-Asia (solid squares); Kim *et al.* (2005a), ACE-Asia (open pentagons); Kim *et al.* (2005b), ACE-Asia (open pentagons); Liu *et al.* (2011), EAST-AIRE (asterisks); Xia *et al.* (2007b), EAST-AIRE (downward open triangles); Ge *et al.* (2010, 2011), ARM/AAF/SMART-COMMIT 2008 (solid upward triangles); Hsu, Herman, and Weaver (2000), Cape Verde and Bidi Bahn, satellite-borne observations (open circles); Tanré *et al.* (2003), Cape Verde, AERONET and SHADE data (solid circles).

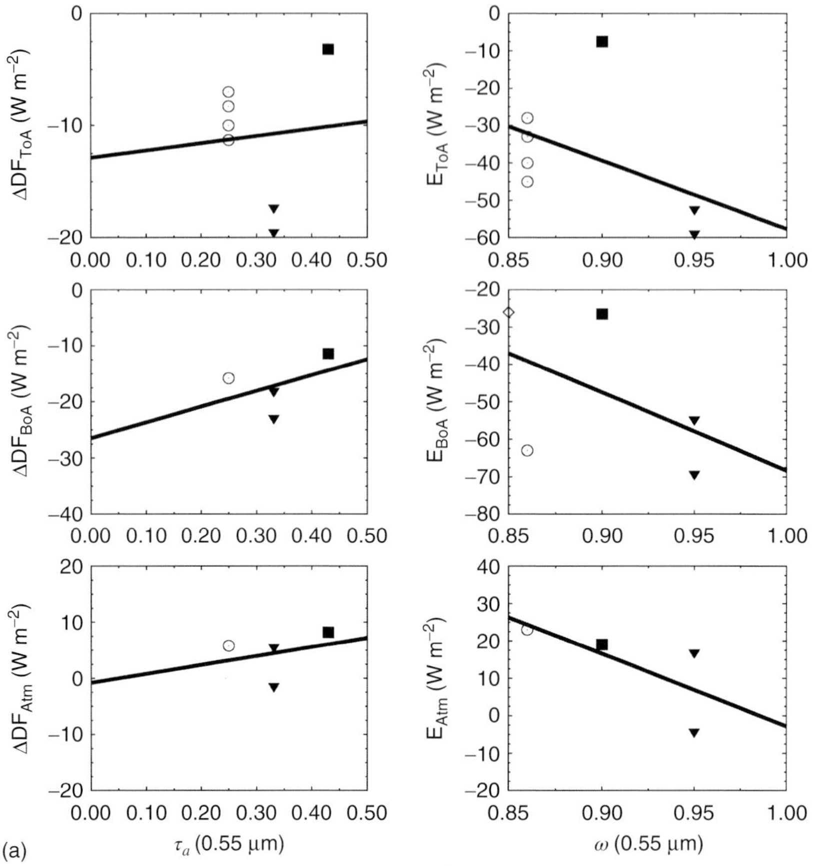

Figure 8.19a Left-hand side: scatter plots of the daily mean values of DARF terms ΔDF_{ToA} (upper part), ΔDF_{BoA} (middle part), and ΔDF_{Atm} (lower part) versus the corresponding daily mean values of aerosol optical thickness τ_a(0.55 μm), determined from various sets (*) of field measurements performed over ocean surfaces for atmospheric loads of biomass burning smoke particles. Right-hand side: corresponding scatter plots of the daily mean values of DARF efficiency parameters E_{ToA} at the ToA level, E_{BoA} at the BoA level, and E_{Atm} within the atmosphere versus the aerosol single scattering albedo ω(0.55 μm) over ocean surfaces for biomass burning smoke particles. (*) Tomasi *et al.* (2015a), Aerosols99 (downward solid triangles); Ross, Hobbs, and Holben (1998), SCAR-B (open diamonds); Podgorny, Li, and Ramanathan (2003), satellite data (solid squares); Yu *et al.* (2006), CERES and MODIS data (open circles).

measurements carried out from 25 to 27 May 2003 (see Table 8.10), during which continental-polluted aerosols mixed with Siberian forest fire smoke were transported over north-central Oklahoma (United States); (ii) García *et al.* (2012), analyzing the AERONET measurements conducted in various continental areas during summer and autumn periods from 1993 to 2006; (iii) Ross, Hobbs, and Holben (1998), examining the SCAR-B field measurements performed in the

Brazilian Amazon and Cerrado regions in August–September 1995; (iv) Kinne and Pueschel (2001), analyzing the SAFARI-92 field measurements conducted in Zambia during September and October 1992; (v) Kinne and Pueschel (2001), analyzing the SCAR-B measurements carried out over the Amazon forest in August–September 1995; (vi) Christopher and Zhang (2002), examining the GOES-8 and CERES data recorded over the Amazon forests (Brazil) in late July and August 1998; (vii) Yu *et al.* (2006), analyzing the SCAR-B field measurements collected over the Amazon forest fires in August–September 1995; and (viii) Yu *et al.* (2006), examining the AERONET measurements and radiative transfer simulations performed over southern America in autumn 2001. The daily values of ΔDF_{ToA}, ΔDF_{BoA}, and ΔDF_{Atm} were plotted versus $\tau_a(0.55\,\mu m)$ in the left-hand side of Figure 8.19b. The values of ΔDF_{ToA} were found to range from -60 to $0\,W\,m^{-2}$, those of ΔDF_{BoA} from -140 to around $-10\,W\,m^{-2}$, and those of ΔDF_{Atm} from nearly 0 to more than $+100\,W\,m^{-2}$. The regression lines were estimated to have the coordinates defined in Table 8.11 and slope coefficients k_D equal to -33.2, -127.6, and $+91.0\,W\,m^{-2}$ at the ToA and BoA levels and in the atmosphere, respectively. The daily values of E_{ToA}, E_{BoA}, and E_{Atm} were plotted versus the daily mean $\omega(0.55\,\mu m)$ in the right-hand side of Figure 8.19b, obtaining the average overall variations $\Delta E_{ToA} = -34\,W\,m^{-2}$, $\Delta E_{BoA} = -13\,W\,m^{-2}$, and $\Delta E_{Atm} = -104\,W\,m^{-2}$ over the 0.8–1.00 range of $\omega(0.55\,\mu m)$.

The comparison between the results shown in Figures 8.19a and 8.19b over ocean and land, respectively, clearly indicates that these combustion particles induce (i) considerably strong cooling effects at the ToA level over land and weak warming effects on average over ocean, (ii) rather moderate warming effects at the BoA level over ocean and very strong cooling effects over land, and (iii) weak warming effects in the atmosphere over oceanand more intense warming effects over land.

8.3.2.9 Urban and Industrial Aerosols

Reliable evaluations of the DARF effects were not found by us for urban and industrial aerosols over ocean surfaces. Conversely, a considerable number of DARF evaluations were found for this aerosol type over land, determined by (i) García *et al.* (2012), examining the AERONET measurements performed at various sites in Europe, North America, and Asia from 1998 to 2008; (ii) Kinne and Pueschel (2001), analyzing the SCAR-A field measurements conducted in the eastern United States during July 1993; (iii) Horvath *et al.* (2002), examining the surface-level measurements conducted during the two field campaigns carried out by the University of Vienna group over the Vienna city urban area in June 1998 and 1999; (iv) Schmidt *et al.* (2010), examining the MILAGRO/INTEX-B field measurements conducted over the Mexico City area on 19 March 2006; (v) Le Blanc *et al.* (2012), analyzing the CALNEX measurements carried out over Los Angeles (California, United States) in May 2010; and (vi) Kim *et al.* (2005a), examining the ACE-Asia field measurements performed at Yinchuan (China) in the September–December months from 1997 to 2000.

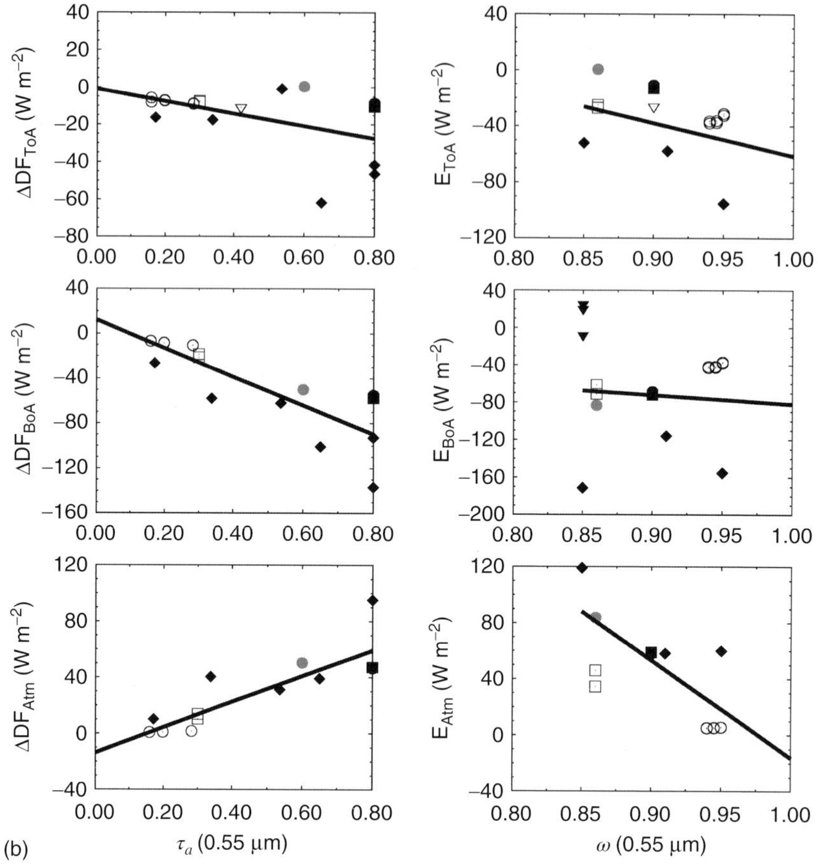

Figure 8.19b Left-hand side: scatter plots of the daily mean values of DARF terms ΔDF_{ToA} (upper part), ΔDF_{BoA} (middle part), and ΔDF_{Atm} (lower part) versus the corresponding daily mean values of aerosol optical thickness $\tau_a(0.55\,\mu m)$, determined from various sets (*) of field measurements performed over land surfaces for atmospheric loads of biomass burning smoke particles. Right-hand side: corresponding scatter plots of the daily mean values of DARF efficiency parameters E_{ToA} at the ToA level, E_{BoA} at the BoA level, and E_{Atm} within the atmosphere versus the aerosol single scattering albedo $\omega(0.55\,\mu m)$ over land surfaces for biomass burning smoke particles. (*) Tomasi *et al.* (2015a), DOE/ARM/AIOP (open circles); García *et al.* (2012), AERONET (solid diamonds); Ross, Hobbs, and Holben (1998), SCAR-B (downward solid triangles); Kinne and Pueschel (2001), SAFARI-92 (gray circles); Kinne and Pueschel (2001), SCAR-B (solid circles); Christopher and Zhang (2002), GOES-8 and CERES data (downward open triangles); Yu *et al.* (2006), SCAR-B (solid squares); Yu *et al.* (2006), AERONET (open squares).

The daily values of ΔDF_{ToA}, ΔDF_{BoA}, and ΔDF_{Atm} were plotted versus $\tau_a(0.55\,\mu m)$ in the left-hand side of Figure 8.20. They were found to vary in an almost linear fashion with $\tau_a(0.55\,\mu m)$, providing the regression lines having the coordinates given in Table 8.11 and values of k_D equal to −49.2, −12.4, and +70.0 W m^{-2} at the ToA and BoA levels and in the atmosphere, respectively.

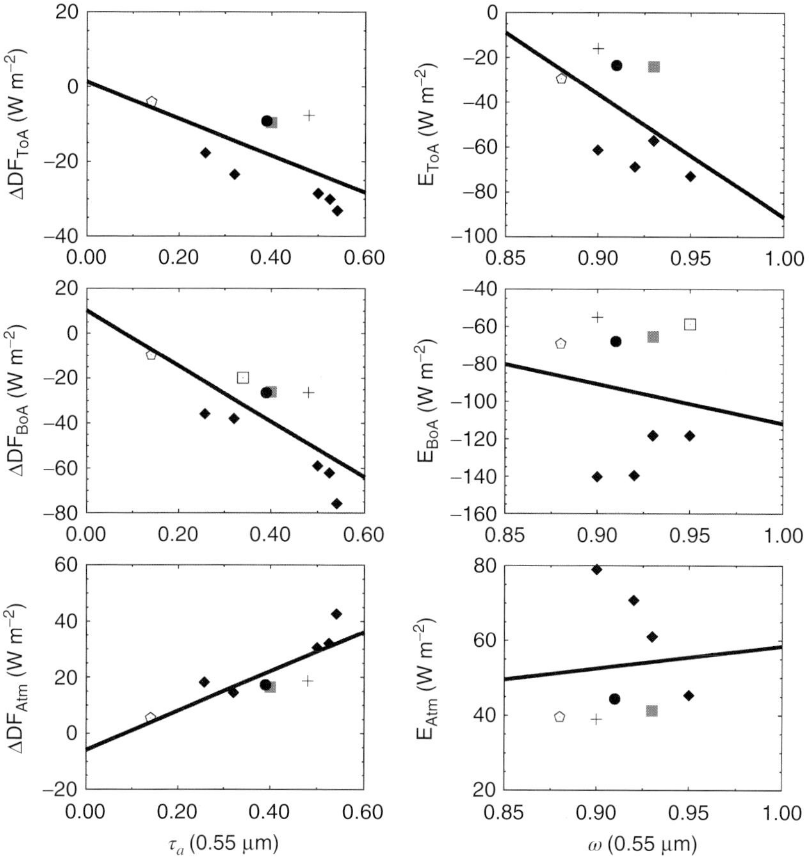

Figure 8.20 Left-hand side: scatter plots of the daily mean values of DARF terms ΔDF_{ToA} (upper part), ΔDF_{BoA} (middle part), and ΔDF_{Atm} (lower part) versus the corresponding daily mean values of aerosol optical thickness τ_a(0.55 μm), determined from various sets (*) of field measurements performed over land surfaces for atmospheric loads of urban and industrial aerosol particles. Right-hand side: corresponding scatter plots of the daily mean values of DARF efficiency parameters E_{ToA} at the ToA level, E_{BoA} at the BoA level, and E_{Atm} within the atmosphere versus the aerosol single scattering albedo ω(0.55 μm) over land surfaces for urban and industrial aerosol particles. (*) García *et al.* (2012), AERONET (solid diamonds); Kinne and Pueschel (2001), SCAR-A (gray squares); Horvath *et al.* (2002), Vienna City (Austria) (solid circles); Schmidt *et al.* (2010), MILAGRO (open pentagons); Le Blanc *et al.* (2012), CALNEX (open squares); Kim *et al.* (2005a), ACE-Asia (crosses (+)).

The daily values of E_{ToA}, E_{BoA}, and E_{Atm} were plotted versus the daily mean $\omega(0.55\,\mu m)$ in the right-hand side of Figure 8.20, determining the average variations $\Delta E_{ToA} = -81\ W\,m^{-2}$, $\Delta E_{BoA} = -32\ W\,m^{-2}$, and $\Delta E_{Atm} = +9\ W\,m^{-2}$ over the $0.85 \le \omega(0.55\,\mu m) \le 1.00$ range.

8.3.2.10 Polar Aerosols

Calculations of the DARF effects induced by different polar aerosols over the ocean were made by Tomasi *et al.* (2015a) using the OS1 surface reflectance model to represent the calm wind oceanic surface reflectance characteristics. The analysis was performed for the POLAR-AOD field measurements conducted for the six aerosol types listed in Table 8.8: (a) background Arctic summer aerosol, measured at Barrow (Alaska), Ny-Ålesund (Spitsbergen, Svalbard), Summit (Greenland), Sodankylä (Finland), and Tiksi (north-central Siberia); (b) AH, measured at Barrow, Ny-Ålesund, and Sodankylä; (c) dense Arctic summer aerosol, measured at Ny-Ålesund; (d) AD, measured at Barrow; (e) BFFS, measured at Barrow; and (f) background Antarctic summer aerosol, measured at the Mario Zucchelli (Italy) and Neumayer III (Germany) stations.

The daily values of ΔDF_{ToA}, ΔDF_{BoA}, and ΔDF_{Atm} obtained by examining the POLAR-AOD measurements for the OS1 surface reflectance characteristics were plotted versus $\tau_a(0.55\,\mu m)$ in the left-hand side of Figure 8.21a to determine the regression lines having the coordinates given in Table 8.11 and the best-fit values of k_D equal to −79.2, +5.8, and −85.8 W m^{-2} at the ToA and BoA levels and in the atmosphere, respectively. The daily values of E_{ToA}, E_{BoA}, and E_{Atm} were plotted versus the daily mean $\omega(0.55\,\mu m)$ in the right-hand side of Figure 8.21a, obtaining the average variations $\Delta E_{ToA} = -93$ W m^{-2}, $\Delta E_{BoA} = -18$ W m^{-2}, and $\Delta E_{Atm} = -17$ W m^{-2} over the 0.75–1.00 range of $\omega(0.55\,\mu m)$.

The set of DARF evaluations for polar aerosols over land was collected from the field measurements conducted by (i) Delene and Ogren (2002), examining the NOAA/CMDL sun-photometer measurements conducted at Barrow from November 1994 to September 2000 for land surface characteristics similar to those of model VS1, and (ii) Tomasi *et al.* (2015a), analyzing the POLAR-AOD field measurements performed for the following aerosol types: (a) background Arctic summer aerosol, measured at Barrow (for the VS1 surface), Ny-Ålesund (PS4 surface), Summit (PS1 surface), Sodankylä (VS1 and PS2 surfaces), and Tiksi (VS1 surface); (b) AH, measured at Barrow (PS1 and PS2 surfaces), Ny-Ålesund (PS3 surface), and Sodankylä (VS1 and PS2 surfaces); (c) dense Arctic summer aerosol, measured at Ny-Ålesund (PS4 surface); (d) AD, measured at Barrow (PS2 surface); (e) BFFS, measured at Barrow (VS1 surface); and (f) background Antarctic summer aerosol, measured at Mario Zucchelli (PS2 surface), Neumayer III (PS1 and PS2 surfaces), Mirny (Russia) (PS1 and PS2 surfaces), Dome C (France/Italy) (PS2 surface), and South Pole (PS1 and PS2 surfaces).

The daily values of ΔDF_{ToA}, ΔDF_{BoA}, and ΔDF_{Atm} calculated for the aforementioned field measurements and the previously chosen surface reflectance models were plotted versus $\tau_a(0.55\,\mu m)$ in the left-hand side of Figure 8.21b, obtaining the regression lines having the coordinates defined in Table 8.11 and the values of k_D equal to −19.2, −68.6, and +54.3 W m^{-2} at the ToA and BoA levels and in the atmosphere, respectively. The daily mean values of E_{ToA}, E_{BoA}, and E_{Atm} were

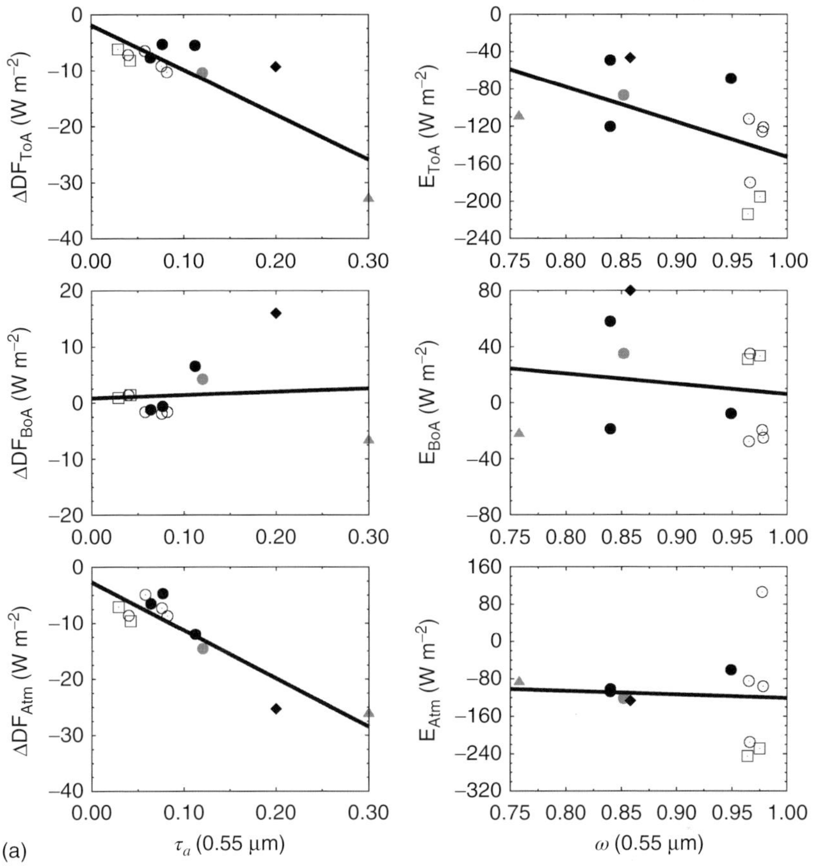

Figure 8.21a Left-hand side: scatter plots of the daily mean values of DARF terms ΔDF_{ToA} (upper part), ΔDF_{BoA} (middle part), and ΔDF_{Atm} (lower part) versus the corresponding daily mean values of aerosol optical thickness $\tau_a(0.55\,\mu m)$, determined from various sets (*) of field measurements performed by over ocean surfaces for atmospheric loads of polar aerosol particles. Right-hand side: corresponding scatter plots of the daily mean values of DARF efficiency parameters E_{ToA} at the ToA level, E_{BoA} at the BoA level, and E_{Atm} within the atmosphere versus the aerosol single scattering albedo $\omega(0.55\,\mu m)$ over ocean surfaces for polar aerosol particles. (*) Tomasi *et al.* (2015a), POLAR-AOD: background Arctic summer aerosol (open circles); background Antarctic summer aerosol (open squares); Arctic haze (solid circles); dense Arctic summer aerosol (gray circles); Asian dust (solid diamonds); Boreal forest fire smoke (upward gray triangles).

plotted versus the corresponding daily mean values of $\omega(0.55\,\mu m)$ in the right-hand side of Figure 8.21b, obtaining the average variations $\Delta E_{ToA} = -9\,W\,m^{-2}$, $\Delta E_{BoA} = +34\,W\,m^{-2}$, and $\Delta E_{Atm} = -40\,W\,m^{-2}$ over the $0.75 \leq \omega(0.55\,\mu m) \leq 1.00$ range.

The comparison between the results shown in Figure 8.21a over ocean and in Figure 8.21b over land clearly indicates that polar aerosols induce in general (i)

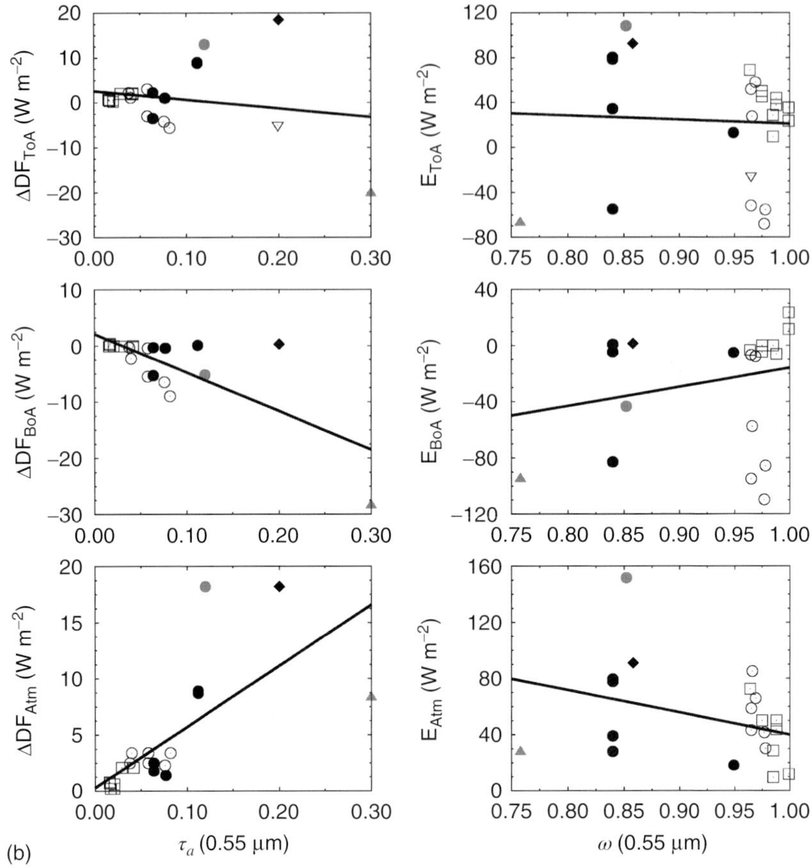

Figure 8.21b Left-hand side: scatter plots of the daily mean values of DARF terms ΔDF_{ToA} (upper part), ΔDF_{BoA} (middle part), and ΔDF_{Atm} (lower part) versus the corresponding daily mean values of aerosol optical thickness τ_a(0.55 μm), determined from various sets (*) of field measurements performed over land surfaces for atmospheric loads of polar aerosol particles. Right-hand side: corresponding scatter plots of the daily mean values of DARF efficiency parameters E_{ToA} at the ToA level, E_{BoA} at the BoA level, and E_{Atm} within the atmosphere versus the aerosol single scattering albedo ω(0.55 μm) over land surfaces for polar aerosol particles. (*) Delene and Ogren (2002), NOAA/CMDL measurements at Barrow (downward open triangles); Tomasi *et al.* (2015a), POLAR-AOD: background Arctic summer aerosol (open circles); background Antarctic summer aerosol (open squares); Arctic haze (solid circles); dense Arctic summer aerosol (gray circles); Asian dust (solid diamonds); boreal forest fire smoke (upward gray triangles).

cooling effects at the ToA level, which are considerably stronger over ocean than over land; (ii) weak warming effects at the BoA level over ocean and very strong cooling effects over land; and (iii) strong cooling effects in the atmosphere over ocean, while very intense warming effects in the atmosphere are usually produced over land.

8.3.2.11 Stratospheric Volcanic Aerosols

Useful evaluations of the DARF effects produced by volcanic aerosol layers in the low stratosphere over ocean surfaces were not calculated in the present study over ocean surfaces. Conversely, two sets of DARF evaluations over land were obtained by analyzing the following data: (i) those determined by Kinne and Pueschel (2001), analyzing the Mt. Pinatubo volcanic aerosol measurements performed on board an airplane flying at stratospheric altitudes in March 1992, 9 months after the violent eruption of June 1991, for surface reflectance characteristics similar to those represented with the VS1 surface reflectance model described in Table 8.3, and (ii) those obtained by Tomasi *et al.* (2015a), examining the field measurements carried out at the CNR Pyramid Laboratory (5050 m a.m.s.l.) in Himalaya (Nepal) from (a) 28 July to 5 August 1991, for Mt. Pinatubo 2-month volcanic aerosols, (b) 19 September to 2 October 1991, for 3-month volcanic aerosols, and (c) 23 July to 11 August 1992, for 13-month volcanic aerosols. For these measurements, the DARF evaluations given in Table 8.7 were obtained for surface reflectance conditions represented using the BS1, VS1, and PS4 surface reflectance models. The daily values of ΔDF_{ToA}, ΔDF_{BoA}, and ΔDF_{Atm} calculated for the previous measurements and the vegetation-covered surface reflectance conditions were plotted versus $\tau_a(0.55\,\mu m)$ in the left-hand side of Figure 8.22, obtaining the regression lines having the coordinates given in Table 8.11 and the values of k_D equal to −19.1, +33.1, and −53.7 W m^{-2} at the ToA and BoA levels and in the atmosphere, respectively. The daily values of E_{ToA}, E_{BoA}, and E_{Atm} plotted versus the daily mean $\omega(0.55\,\mu m)$ were found to be largely dispersed, as shown in the right-hand side of Figure 8.22, yielding the average variations $\Delta E_{ToA} = +1$ W m^{-2}, $\Delta E_{BoA} = +118$ W m^{-2}, and $\Delta E_{Atm} = -118$ W m^{-2} over the $0.90 \leq \omega(0.55\,\mu m) \leq 1.00$ range.

8.4 Variations of DARF Efficiency as a Function of Aerosol Single Scattering Albedo

The regression lines drawn in Figures 8.14–8.22 for the scatter plots of the daily values of E_{ToA}, E_{BoA}, and E_{Atm} versus the corresponding daily mean values of $\omega(0.55\,\mu m)$ were defined obtaining in general rather good correlation coefficients over the ranges of $\omega(0.55\,\mu m)$ given in Table 8.12 for the following types of aerosols suspended over land and ocean surfaces: (i) remote continental aerosols over land, (ii) rural-continental aerosols over land, (iii) free tropospheric aerosols over land, (iv) maritime clean aerosols over ocean and over land, (v) maritime–continental aerosols over land, (vi) desert dust over ocean and over land, and (vii) biomass burning particles over ocean. Conversely, worse correlation coefficients were obtained over the ranges of $\omega(0.55\,\mu m)$ given in Table 8.12 for the other aerosol types, such as (a) rural-continental aerosols over ocean, (b) continental-polluted aerosols over ocean, (c) continental-polluted aerosols over land, (d) maritime–continental aerosol over ocean, (e) biomass burning particles over land, (f) urban and industrial aerosols over land, (g) polar aerosols

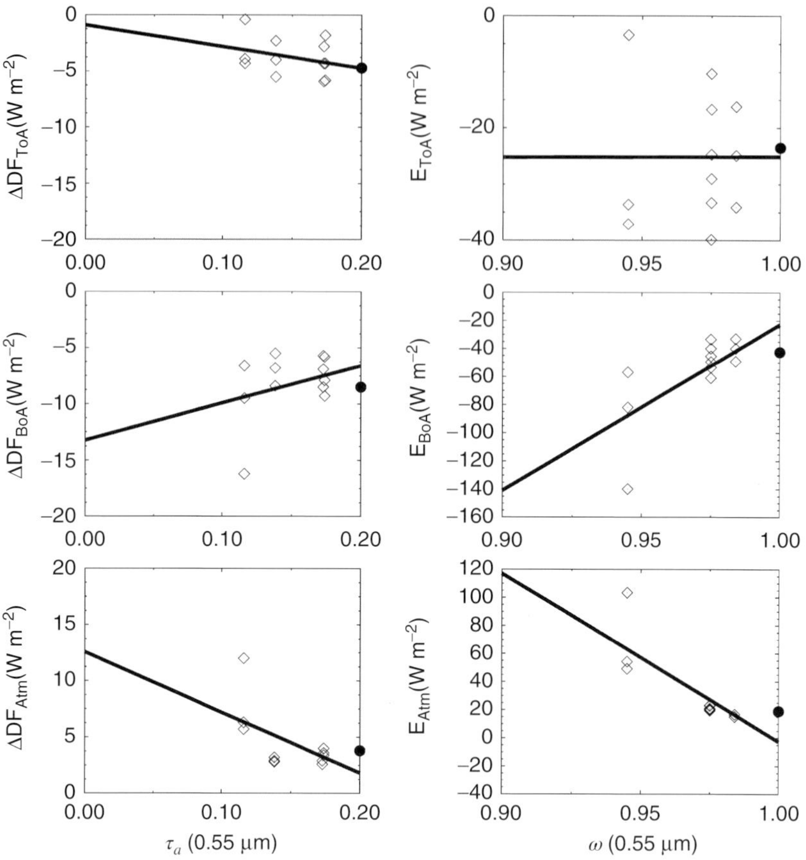

Figure 8.22 Left-hand side: scatter plots of the daily mean values of DARF terms ΔDF_{ToA} (upper part), ΔDF_{BoA} (middle part), and ΔDF_{Atm} (lower part) versus the corresponding daily mean values of aerosol optical thickness τ_a(0.55 μm), determined from two sets (*) of field measurements performed over land surfaces for atmospheric loads of stratospheric volcanic aerosol particles. Right-hand side: corresponding scatter plots of the daily mean values of DARF efficiency parameters E_{ToA} at the ToA level, E_{BoA} at the BoA level, and E_{Atm} within the atmosphere versus the aerosol single scattering albedo ω(0.55 μm) over land surfaces for stratospheric volcanic aerosol particles. (*) Tomasi *et al.* (2015a), Ev-K2-CNR (open diamonds); Kinne and Pueschel (2001), Mauna Loa Observatory measurements (solid circle).

over ocean, (h) polar aerosols over land, and (i) stratospheric volcanic aerosols over land.

The values of slope coefficient k_E were found to vary largely from one aerosol type to another and from ocean to land, due to the large scatter of the experimental data determined for the different measurement areas and atmospheric turbidity conditions, being in general characterized by the high values of the standard errors of estimate.

Table 8.12 Values of slope coefficient k_E and correlation coefficient c_{corE} determined for the regression lines of the diurnally average values of efficiencies E_{ToA}, E_{BoA}, and E_{Atm} plotted versus the daily mean values of $\omega(0.55\ \mu m)$, as obtained for the various aerosol types considered in the present study over the respective experimental ranges of $\omega(0.55\ \mu m)$, separately over ocean and land surfaces.

Aerosol type	Surface albedo	Experimental range of $\omega(0.55\ \mu m)$	E_{ToA}		E_{BoA}		E_{Atm}	
			k_E (W m^{-2})	c_{corE}	k_E (W m^{-2})	c_{corE}	k_E (W m^{-2})	c_{corE}
Remote continental	Land	0.90–1.00	+1834.0	−0.622	−1144.0	−0.551	+2916.0	+0.855
Rural-continental	Ocean	0.90–1.00	−30.4	−0.108	−156.2	−0.350	+125.8	+0.227
	Land	0.90–1.00	−208.8	−0.563	+244.6	+0.317	−491.5	−0.593
Free tropospheric	Land	0.90–1.00	−8153.0	−0.971	−10850	−0.920	+2644.0	+0.630
Continental-polluted	Ocean	0.80–1.00	+43.98	+0.175	+92.5	+0.214	−175.1	−0.282
	Land	0.80–1.00	−153.8	−0.406	+238.0	+0.237	−451.2	−0.400
Maritime clean	Ocean	0.90–1.00	+1719.0	−0.971	+1683.0	+0.963	+36.1	+0.040
	Land	0.90–1.00	+1647.0	−0.959	−2131.0	+0.979	+481.9	+0.996
Maritime–continental	Ocean	0.80–1.00	−47.9	−0.100	+150.6	+0.239	−84.5	−0.134
	Land	0.80–1.00	−151.9	−0.473	+37.7	+0.617	−530.0	−0.736
Desert dust	Ocean	0.80–1.00	−228.7	−0.777	+116.4	+0.231	−348.0	−0.666
	Land	0.80–1.00	−274.0	−0.490	+201.2	+0.302	−451.7	−0.717
Biomass burning	Ocean	0.85–1.00	−182.7	−0.456	−208.8	−0.488	−193.9	−0.694
	Land	0.85–1.00	−235.6	−0.360	+0.97.7	−0.071	−695.0	−0.730
Urban and industrial	Land	0.85–1.00	−547.8	−0.528	−211.2	−0.138	+59.1	+0.084
Polar	Ocean	0.75–1.00	−375.0	−0.522	−72.9	−0.149	−76.6	−0.063
	Land	0.75–1.00	−36.9	−0.050	+135.7	+0.238	−158.8	−0.346
Stratospheric volcanic	Land	0.90–1.00	+1.1	+0.002	+1179.0	+0.705	−1197.0	−0.810

The regression lines were found to yield the following average trends of E_{ToA}, E_{BoA}, and E_{Atm}:

i. E_{ToA} varying from −48 to + 137 W m^{-2}, E_{BoA} from −63 to −178 W m^{-2}, and E_{Atm} from +20 to +311 W m^{-2} for remote continental aerosols over land as ω(0.55 μm) increases from 0.90 to 1.00
ii. E_{ToA} varying from −40 to −44 W m^{-2}, E_{BoA} from −49 to −64 W m^{-2}, and E_{Atm} from +8 to +20 W m^{-2} for rural-continental aerosols over ocean as ω(0.55 μm) increases from 0.90 to 1.00
iii. E_{ToA} varying from −16 to −38 W m^{-2}, E_{BoA} from −62 to −40 W m^{-2}, and E_{Atm} from +50 to +2 W m^{-2} for rural-continental aerosols over land as ω(0.55 μm) increases from 0.90 to 1.00
iv. E_{ToA} varying from 0 to −150 W m^{-2}, E_{BoA} from −50 to −250 W m^{-2}, and E_{Atm} from +45 to +110 W m^{-2} for free tropospheric aerosols over land as ω(0.55 μm) increases from 0.90 to 1.00
v. E_{ToA} varying from −47 to −38 W m^{-2}, E_{BoA} from −71 to −53 W m^{-2}, and E_{Atm} from +30 to −4 W m^{-2} for continental-polluted aerosols over ocean as ω(0.55 μm) increases from 0.80 to 1.00
vi. E_{ToA} varying from −5 to −36 W m^{-2}, E_{BoA} from −80 to −73 W m^{-2}, and E_{Atm} from −90 to 0 W m^{-2} for continental-polluted aerosols over land as ω(0.55 μm) increases from 0.80 to 1.00
vii. E_{ToA} varying from −127 to −28 W m^{-2}, E_{BoA} from −38 to +48 W m^{-2}, and E_{Atm} from −79 to −77 W m^{-2} for maritime clean aerosols over ocean as ω(0.55 μm) increases from 0.90 to 1.00
viii. E_{ToA} varying from −25 to −108 W m^{-2}, E_{BoA} from −56 to −161 W m^{-2}, and E_{Atm} from +29 to +53 W m^{-2} for maritime clean aerosols over land as ω(0.55 μm) increases from 0.90 to 1.00
ix. E_{ToA} varying from −27 to −37 W m^{-2}, E_{BoA} from −79 to −49 W m^{-2}, and E_{Atm} from +42 to +26 W m^{-2} for maritime−continental aerosols over ocean as ω(0.55 μm) increases from 0.80 to 1.00
x. E_{ToA} varying from −9 to −39 W m^{-2}, E_{BoA} from −119 to −42 W m^{-2}, and E_{Atm} from +110 to +4 W m^{-2} for maritime−continental aerosols over land as ω(0.55 μm) increases from 0.80 to 1.00
xi. E_{ToA} varying from −8 to −53 W m^{-2}, E_{BoA} from −82 to −60 W m^{-2}, and E_{Atm} from −75 to +5 W m^{-2} for desert dust over ocean as ω(0.55 μm) increases from 0.80 to 1.00
xii. E_{ToA} varying from +7 to −48 W m^{-2}, E_{BoA} from −118 to −78 W m^{-2}, and E_{Atm} from −125 to −35 W m^{-2} for desert dust over land as ω(0.55 μm) increases from 0.80 to 1.00
xiii. E_{ToA} varying from −30 to −58 W m^{-2}, E_{BoA} from −37 to −78 W m^{-2}, and E_{Atm} from +26 to −3 W m^{-2} for biomass burning particles over ocean as ω(0.55 μm) increases from 0.85 to 1.00
xiv. E_{ToA} varying from −26 to −60 W m^{-2}, E_{BoA} from −68 to −81 W m^{-2}, and E_{Atm} from +88 to −16 W m^{-2} for biomass burning particles over land as ω(0.55 μm) increases from 0.85 to 1.00

xv. E_{ToA} varying from −10 to −91 W m^{-2}, E_{BoA} from −80 to −112 W m^{-2}, and E_{Atm} from +50 to +59 W m^{-2} for urban and industrial aerosols over land as ω(0.55 μm) increases from 0.85 to 1.00

xvi. E_{ToA} varying from −59 to −152 W m^{-2}, E_{BoA} from +24 to +6 W m^{-2}, and E_{Atm} from −103 to −120 W m^{-2} for polar aerosols over ocean as ω(0.55 μm) increases from 0.75 to 1.00

xvii. E_{ToA} varying from +30 to +21 W m^{-2}, E_{BoA} from −50 to −16 W m^{-2}, and E_{Atm} from +80 to +40 W m^{-2} for polar aerosols over land as ω(0.55 μm) increases from 0.75 to 1.00

xviii. E_{ToA} varying from −26 to −25 W m^{-2}, E_{BoA} from −140 to −22 W m^{-2}, and E_{Atm} from +116 to −2 W m^{-2} for stratospheric volcanic aerosols over land as ω(0.55 μm) increases from 0.90 to 1.00.

The results described previously and shown in Figures 8.14–8.22 provide considerably high values of efficiency terms E_{ToA}, E_{BoA}, and E_{Atm} in most cases, which can deceive the reader if he/she does not take into account that efficiency is calculated per unit AOT τ_a(0.55 μm), while this quantity exhibits relatively low values in the real cases over the large part of the Earth. The analysis made by Takemura *et al.* (2005) for a large aerosol data set obtained using the SPRINTARS chemical model shows that:

a. Sulfate aerosols provide annual mean values of τ_a(0.55 μm) ranging (i) from 0.07 to 0.70 in the most polluted areas of Europe and China and (ii) from 0.02 to 0.07 in the midlatitude land and oceanic areas and lower than 0.02 in the remote mid- and high-latitude oceanic regions.
b. Sea-salt aerosols yield annual mean values of τ_a(0.55 μm) lower than 0.02 over the continental regions and ranging mainly from 0.02 to 0.06 over the oceans.
c. BC + OC (carbonaceous) aerosols exhibit annual mean values of τ_a(0.55 μm) ranging (i) from 0.07 to more than unit in the continental areas most frequently involved by forest fires (like those of tropical Africa), (ii) from 0.03 to 0.20 in the continental areas of Asia and southern America, and (iii) from 0.01 to 0.05 over the oceans, with the higher values in regions not far from the continental particle sources.
d. Soil dust produces annual mean values of τ_a(0.55 μm) ranging (i) from 0.10 to 0.70 over the Saharan, Middle Eastern, and Persian Gulf regions, (ii) from 0.03 to 0.07 over the continental regions not far from desert areas, and (iii) from 0.01 to 0.02 over the major part of the oceans.

8.5 Concluding Remarks on the DARF Effects over the Global Scale

The calculations made in the present study indicate that both natural and anthropogenic aerosols induce important direct effects on the radiation budget of the land–atmosphere–ocean system. These effects exert a powerful influence on the Earth's climate mainly through the strong scattering and absorption processes of shortwave radiation and less intensively through weak scattering and absorption of the longwave radiation emitted upward by the terrestrial surface and both downward and toward outer space by the atmosphere. These radiative effects have been studied exhaustively over the past 40 years through field measurements, satellite-based observations, and radiative transfer model simulations (McCormick and Ludwig, 1967; Charlson, Langner, and Rodhe, 1990; Charlson *et al.,* 1992; Schwartz and Andreae, 1996; Ramanathan *et al.,* 2001b; Kaufman, Tanré, and Boucher, 2002), showing the complexity of these processes and the large uncertainties affecting the evaluations of the DARF effects. The cited studies have shown that the DARF effects are particularly intense in the midlatitude industrial and more densely populated regions of the planet, where the anthropogenic aerosol emissions are particularly strong (Charlson *et al.,* 1991; Chin *et al.,* 2007), whereas they are considerably weaker in remote oceanic areas (Takemura *et al.,* 2005) and polar regions (Blanchet, 1989; Tomasi *et al.,* 2007), where the background number and mass concentrations of particles are usually much lower than in midlatitude populated regions.

The interactions of aerosols with the incident solar radiation generally cause a marked decrease in the direct solar radiation flux density reaching the surface, leading to an increase in the solar radiation fraction reflected back to space and an increase in the overall albedo of the climate system (Haywood and Boucher, 2000). Scattering by aerosols causes also an increase in the diffuse solar radiation flux reaching the surface, in this way contributing to enhance the surface–atmosphere system albedo. Absorption of incoming solar radiation by airborne aerosols is particularly strong in cases where significant contents of soot substances (containing BC and/or EC) are present in the airborne particulate matter (Andreae and Gelencsér, 2006), inducing appreciable warming effects in the lower part of the troposphere (Kaufman, 1987; Bellouin *et al.,* 2003). In addition, scattering and absorption of longwave terrestrial radiation can modify appreciably the cooling rate of the atmospheric boundary layer (Lubin *et al.,* 2002), especially in the presence of dense haze particle layers (Grassl, 1973; Yu, Liu, and Dickinson, 2002) or Saharan dust layers (Zhang and Christopher, 2003). However, aerosols are generally estimated to scatter and absorb the IR radiation rather weakly on the global scale, only slightly enhancing the warming processes that are generated in the cloudless atmosphere by the thermal radiation absorption by water vapor, carbon dioxide, methane, nitrous oxide, and many other atmospheric greenhouse gases (Hansen, Sato, and Ruedy, 1997). Due to the irregular distribution of aerosol concentration over the planet, marked variations in time and discontinuities in space characterize the regional distribution features of the DARF effects, which are estimated to exceed or be comparable in magnitude to the greenhouse warming effects occurring in some areas of the Earth, often inducing opposite warming and cooling effects in different regions (King *et al.,* 1999; Takemura *et al.,* 2002). Therefore, the interactions between aerosols and incoming solar radiation are among the main sources of uncertainty in modeling climate changes within the global circulation models (Hansen *et al.,* 1998).

There are still large knowledge gaps on this matter, for instance, regarding the dependence of DARF effects on surface reflectance for aerosol types presenting largely different physicochemical properties (Charlson *et al.,* 1992). Recent estimates of the average DARF effects over the global scale have been assessed by Boucher *et al.* (2013) to be $-0.45\ \mathrm{W\,m^{-2}}$, obtained over the range from -0.95 to $+0.05\ \mathrm{W\,m^{-2}}$, while these effects were assessed to be $-0.35\ \mathrm{W\,m^{-2}}$ (-0.85 to $+0.15$) $\mathrm{W\,m^{-2}}$ in the polar regions, over surfaces covered by snow and ice. In particular, they estimated that (i) fossil fuel and biofuel emissions contribute via sulfate aerosol to yield an average DARF of $-0.4\ \mathrm{W\,m^{-2}}$ (ranging from -0.6 to $-0.2\ \mathrm{W\,m^{-2}}$), (ii) BC aerosols provide an average DARF of $+0.4\ \mathrm{W\,m^{-2}}$ (over the range from $+0.05$ to $+0.8\ \mathrm{W\,m^{-2}}$), (iii) primary and secondary organic aerosols yield an average DARF of $-0.12\ \mathrm{W\,m^{-2}}$ (over the range from -0.4 to $+0.1\ \mathrm{W\,m^{-2}}$), (iv) biomass burning particles (in which BC and OC aerosol changes offset each other) give an average null DARF (over the range from -0.2 to $+0.2\ \mathrm{W\,m^{-2}}$), (v) nitrate aerosols yield an average DARF of $-0.11\ \mathrm{W\,m^{-2}}$) (over the range from -0.3 to $-0.03\ \mathrm{W\,m^{-2}}$), and (vi) mineral dust (in part due to natural processes and in part to anthropogenic changes in land use and water use) provides an average DARF of $-0.1\ \mathrm{W\,m^{-2}}$) (over the range from -0.3 to $+0.1\ \mathrm{W\,m^{-2}}$).

8.6
On the Indirect Aerosol Effects Acting in the Earth's Climate System

For relatively high number concentrations, airborne aerosols can also exert important indirect effects on the terrestrial climate system through their interactions with surrounding clouds. In fact, aerosols act as cloud condensation nuclei (CCNs) and change the cloud droplet concentration and size-distribution parameters, thus modifying the cloud amount and, hence, influencing the radiation balance and hydrology through their impact on cloud microphysical processes. Aerosols exert also a semidirect effect, in which the heating by aerosol particles (mainly due to solar radiation absorption by the soot components of particulate matter) results in a decrease of cloud amount. Therefore, both natural and anthropogenic aerosols can change significantly the cloud optical properties by causing a substantial increase in the cloud albedo and modifying the average cloudiness conditions in numerous areas of the Earth (Ackerman, Toon, and Hobbs, 1994; Ackerman *et al.,* 2000, 2004; Kaufman and Fraser, 1997; Jiang *et al.,* 2006; Koren *et al.,* 2008). In these processes, aerosols can redistribute solar energy as thermal energy inside cloud layers and enhance the liquid water content of clouds, thus altering the cloud lifetime and strongly influencing the heterogeneous chemistry of the atmosphere. The IPCC TAR (2001) report stated that "the indirect effect is the mechanism by which aerosols modify the microphysical and, hence, the radiative properties, amount and lifetime of clouds." Key parameter for determining the indirect effect is the effectiveness of an aerosol particle to act as a CCN, which is a function of the size, chemical composition, mixing state, and ambient environment (e.g., Penner *et al.,* 2001).

Consequently, aerosols serve as CCNs upon which liquid droplets can form and therefore can alter both the size-distribution shape parameters of cloud droplets and the lifetime of cloud droplets. Through these processes, the following pair of indirect effects can occur:

1. The microphysically induced effect on the cloud droplet number concentration and, hence, cloud droplet size, with the liquid water content held fixed, which is commonly called the *first indirect effect* (Ramaswamy *et al.*, 2001), or the *cloud albedo effect* (Lohmann and Feichter, 2005), or the *Twomey effect* (Twomey, 1977), in which more aerosols result in a larger concentration of smaller droplets, leading to the formation of a brighter cloud. In this process, a higher concentration of polluted aerosols near the ground may cause more polluted low-level clouds, which exhibit in general lower brightness features.
2. The microphysically induced effect on the liquid water content, cloud height, and lifetime of clouds, which is called the *second indirect effect* (Ramaswamy *et al.*, 2001), or the *cloud lifetime effect* (Lohmann and Feichter, 2005), or the *Albrecht effect* (Albrecht, 1989). However, there are many other possible aerosol–cloud–precipitation processes which may amplify or dampen the second indirect effect.

The IPCC TAR (2001) report classified the indirect effects into the two aforementioned types, denoted as *cloud albedo effect* and *cloud lifetime effect*, respectively. These terms are more descriptive of the microphysical processes acting on the climate system than the other aforementioned definitions: the cloud albedo effect was considered to be a purely RF mechanism, leading to the distribution of the same cloud liquid water content over a larger space (and, hence, leading to higher cloud reflectivity), whereas the cloud lifetime effect was considered to be given by the combination of various physical processes regulating the evolutionary patterns of a cloud. To investigate these aspects, global model calculations were performed to describe the influence of increased aerosol concentration on the cloud optical properties for a fixed liquid water content of the cloud. This effect is considered to be a key uncertainty source in the RF of climate. For this reason, a well-defined estimate of this RF effect was not provided by IPCC TAR (2001), indicating only its range from 0 to $-2\ \mathrm{W\ m^{-2}}$ for liquid water clouds. Other different indirect effects were not considered in the IPCC TAR (2001) report to generate additional RF processes. It was assumed that the hydrological cycle is invariably altered through feedback processes in suppressing drizzle, increasing the cloud height, and modifying the cloud lifetime in atmospheric models, without providing evidence of radiative effects. Among these processes, the impact of anthropogenic aerosols on the formation and modification of the physical and radiative properties of ice clouds was also considered (Penner *et al.*, 2001), although such an RF effect was not accurately quantified because of the uncertainties and unknowns on the ice cloud nucleation process. It is worth mentioning that the IPCC TAR (2001) report did not include any assessment of the semidirect effect, which can be defined as the mechanism by which absorption of shortwave radiation by tropospheric aerosols leads to the heating of the troposphere and, hence, changes the relative humidity conditions and stability of the troposphere, thereby influencing cloud formation and lifetime (Hansen, Sato, and Ruedy, 1997; Ackerman *et al.*, 2000; Cook and Highwood, 2004; Johnson, Shine, and Forster, 2004).

In the Earth's climate system, the direct and indirect aerosol-induced effects are accomplished with numerous other important radiation budget changes arising from the variability of sea surface temperature, the selective absorption of solar radiation by atmospheric water vapor and other minor gases, and the numerous variations in the cloud coverage characteristics occurring at various tropospheric altitudes as a result of the complex atmospheric circulation processes. The combination of all such effects depicts a very complicate representation of the radiative behavior of the surface–atmosphere system (Waliser and Graham, 1993; Ramanathan *et al.*, 2001a; Takemura *et al.*, 2005). The indirect effects constitute a significant source of such uncertainties due to the complexity and variety of atmospheric interactions involved and the wide range of scales on which these interactions take place. In this complex picture, the irregular distribution of the aerosol radiative effects on the global scale is enhanced by the marked emissions of anthropogenic aerosols in the industrialized and most populated regions of the planet, which tend to render even more irregular the spatial distribution of atmospheric heat released by aerosols after their absorption of solar radiation. On the basis of these remarks, the IPCC Climate Change (2007) report described the following not direct effects:

1. The *cloud albedo* indirect effect, which regards all the cloud types: in these processes, the increase in the smaller cloud particle concentration leads to less precipitation and in cloud optical thickness causes a decrease in the solar radiation flux reaching the surface. In fact, for the same cloud water or ice content, more but smaller cloud particles reflect more efficiently the solar radiation, inducing a negative change (of medium magnitude) in the net radiative flux at the ToA level and producing a negative change (of medium magnitude) in the global mean net shortwave radiation at the surface.
2. The *cloud lifetime* indirect effect, which involves all the cloud types and in which smaller cloud particles decrease the precipitation efficiency, thereby presumably prolonging cloud lifetime, inducing a negative medium change in the net radiative flux at the ToA level, and causing a negative small change in the global mean net shortwave radiation at the surface.
3. The *semidirect* effect, which regards all the cloud types and in which reflecting aerosols tend to brighten clouds and make them last longer, while aerosols containing BC from soot can have the opposite effect contributing to warming the surrounding atmosphere and causing cloud droplets to evaporate. The absorption of solar radiation by absorbing aerosols produces heating of the atmosphere and affects the static stability characteristics of the atmosphere as well as the surface energy budget and may lead to an evaporation of cloud particles (with shrinking of clouds), inducing positive or negative small changes in the net radiative flux at the ToA level and negative large changes in the global mean net shortwave radiation at the surface.

4. The *glaciation* indirect effect, which involves the mixed-phase clouds and in which an increase in the ice nuclei (IN) concentration results in a rapid glaciation of a supercooled liquid water cloud, due to the difference in vapor pressure over ice and water. Unlike cloud droplets, these ice crystals grow in an environment of high supersaturation with respect to ice, quickly reaching precipitation size, with the potential to turn a nonprecipitating cloud into a precipitating cloud. Therefore, an increase in the IN concentration leads to an increase of precipitation efficiency and causes (i) a positive medium change in the net radiative flux at the ToA level and (ii) a positive medium change in the global mean net shortwave radiation at the surface. Thus, these processes delay freezing and favor the formation of higher and colder clouds, causing more precipitation.
5. The *thermodynamic* effect, which refers to a delay in freezing by the smaller droplets, causing supercooled clouds to extend to colder temperatures. These processes generate positive or negative medium changes in both the net radiative flux at the ToA level and the global mean net shortwave radiation at the surface.

The scientific understanding of the aforementioned effects was estimated to be low in the first case and very low in the remaining four cases, since they involve various feedbacks in the climate system (see IPCC Climate Change, 2007). The cloud albedo effect cannot be easily separated from the other effects, since the processes yielding a decrease in the cloud droplet size per given liquid water content also cause a decrease in the precipitation formation, thus prolonging the cloud lifetime. The cloud feedbacks remain the largest source of uncertainty in climate sensitivity estimates, and the relatively poor simulation of boundary layer clouds in the present climate is an additional reason for such uncertainties. On the global scale, the indirect effects described previously are in opposition to the warming by greenhouse gases, since they cause in general a cooling of the atmosphere. While greenhouse gases disperse widely in the troposphere and have a fairly consistent impact from region to region, aerosol effects are less consistent, partly because of how the particles affect clouds. In fact, aerosols are distributed around the planet differently than greenhouse gases, causing effects which do not simply cancel through combination with the greenhouse effect. On the whole, clouds are thought to cool Earth's surface by shading about 60 percent of the planet at any one time and by increasing the reflectivity of the atmosphere. On this matter, it is worth noting that just a 5% increase in cloud reflectivity can compensate for the entire increase in the global average warming effects produced by greenhouse gases from the modern industrial era. This clearly indicates that long-term variations in cloudiness could have a major impact on Earth's climate.

The previous remarks clearly indicate that clouds and aerosols contribute the largest uncertainty to estimates and interpretations of the Earth's changing energy budget. Focusing on process understanding and considering all together observations, theory, and models to assess how clouds and aerosols contribute and respond to climate change, Boucher *et al.* (2013) pointed out that many of the cloudiness and humidity changes simulated by climate models in warmer climates are now understood as responses to large-scale circulation changes that do not appear to depend strongly on subgrid-scale model processes, therefore increasing confidence in climate change. For example, multiple lines of evidence now indicate positive feedback contributions from circulation-driven changes in both the height of high clouds and the latitudinal distribution of clouds (medium to high confidence). However, some aspects of the overall cloud response vary substantially among models and appear to depend strongly on subgrid-scale processes in which there is less confidence. Climate-relevant aerosol processes are currently better understood, and climate-relevant aerosol properties better observed, than in the IPCC Climate Change (2007) report. The representation of relevant processes varies greatly in global aerosol and climate models, while it remains unclear what level of sophistication is required to model their effect on climate. Globally, between 20% and 40% of AOT (medium confidence) and between one quarter and two thirds of CCN concentrations (low confidence) are of anthropogenic origin.

Recent studies have clarified the importance of distinguishing forcing (defined as the instantaneous change in the radiative budget) and rapid adjustments (which modify the radiative budget indirectly through fast atmospheric and surface changes) from feedbacks (which operate through changes in climate variables that are mediated by a change in surface temperature). Furthermore, it seems realistic to distinguish between the traditional concept of RF and the relatively new concept of effective radiative forcing (ERF), which also includes rapid adjustments. With regard to aerosols, one can further distinguish forcing processes arising from aerosol–radiation interactions (using the suffix *ari*) and aerosol–cloud interactions (using the suffix *aci*).

The quantification of cloud and convective effects in models and of aerosol–cloud interactions continues to be a challenge. Climate models are incorporating more of the relevant processes than those considered in the IPCC Climate Change (2007) report, but confidence in the representation of these processes remains weak. Cloud and aerosol properties vary at scales significantly smaller than those resolved in climate models, and cloud-scale processes respond to aerosol in nuanced ways at these scales. Until subgrid-scale parameterizations of clouds and aerosol–cloud interactions are able to address these issues, model estimates of aerosol–cloud interactions and their radiative effects will carry large uncertainties. Satellite-based estimates of aerosol–cloud interactions remain sensitive to the treatment of meteorological influences on clouds and assumptions on what constitutes preindustrial conditions.

Precipitation and evaporation are expected to not only increase on average in a warmer climate but also undergo global and regional adjustments to carbon dioxide (CO_2) and other forcing mechanisms that differ from their warming responses. Moreover, there is high confidence that extreme precipitation rates on daily time scales will increase faster than the time average, as climate warms. Changes in average precipitation must remain consistent with variations in the net rate of cooling of the troposphere, which is affected not only by its temperature but also by greenhouse gases and aerosols. Consequently, while the increase in global mean precipitation due to surface temperature change alone would be 1.5–3.5% per degree Celsius, warming caused by CO_2 and absorbing aerosols results in a smaller sensitivity, even more so if it is partially offset by albedo increases. The complexity of land surface and atmospheric processes turns out to limit the confidence in regional projections of precipitation change, especially over land, although there is a component of a "wet-get-wetter" and "dry-get-drier" response over oceans at the large scale. Changes in local extremes on daily and subdaily time scales are strongly influenced by lower tropospheric water vapor concentrations and on average will increase by roughly 5–10% per degree Celsius of warming (medium confidence). Aerosol–cloud interactions can influence the character of individual storms, but evidence for a systematic aerosol effect on storm or precipitation intensity is still more limited and ambiguous.

The net feedback from water vapor and lapse rate changes combined, as traditionally defined, is likely evaluated to be extremely positive (amplifying global climate changes). The sign of the net radiative feedback due to all cloud types is not only less certain but also likely positive. Uncertainty in the sign and magnitude of the cloud feedback is primarily attributed to continuing uncertainty in the impact of warming on low clouds.

Boucher *et al.* (2013) estimated the water vapor plus lapse rate feedback as equal to +1.1 (+0.9 to +1.3) $W\,m^{-2}\,°C^{-1}$ and the cloud feedback from all cloud types to be +0.6 (−0.2 to +2.0) $W\,m^{-2}\,°C^{-1}$. These ranges are broader than those of climate models to account for additional uncertainty associated with processes that may not have been accounted for in those models. The mean values and ranges in climate models are essentially unchanged since IPCC Climate Change (2007) but are now supported by stronger indirect observational evidence and better process understanding, especially for water vapor. Conversely, the behavior of low clouds (which contribute positive feedback in most models) is at present not well understood.

Aerosol–climate feedbacks occur mainly through changes in the source strength of natural aerosols or changes in the sink efficiency of natural and anthropogenic aerosols: a limited number of modeling studies have estimated the feedback parameter within +0.2 $W\,m^{-2}\,°C^{-1}$ with low confidence. ERF due to aerosol–radiation interactions that takes rapid adjustments into account (ERF_{ari}) is assessed to be −0.45 (−0.95 to +0.05) $W\,m^{-2}$. The RF produced by absorbing aerosol on snow and ice is assessed separately to be +0.04 (+0.02 to +0.09) $W\,m^{-2}$. Prior to adjustments taking place, the RF due to aerosol–radiation interactions (RF_{ari}) is assessed to be −0.35 (−0.85 to +0.15) $W\,m^{-2}$. This forcing term encompasses radiative effects from anthropogenic aerosols before any adjustment takes place and corresponds to what is usually referred to as the aerosol direct effect (DARF). Such an estimate made by Boucher *et al.* (2013) is less negative than that determined in the IPCC Climate Change (2007) report because of a reevaluation of aerosol absorption. Its uncertainty estimate is wider but more robust, based on multiple lines of evidence from models, remotely sensed data, and ground-based measurements: (i) fossil fuel and biofuel emissions contribute to RF_{ari} via sulfate aerosol yielding an average effect of −0.4 (−0.6

to −0.2) W m^{-2}, BC aerosol provides an average contribution of +0.4 (+0.05 to +0.8) W m^{-2}, and primary and secondary organic aerosols yield an average effect of −0.12 (−0.4 to +0.1) W m^{-2}. Additional RF_{ari} contributions occur via biomass burning emissions (equal to +0.0 (−0.2 to +0.2) W m^{-2}), nitrate aerosol (equal to −0.11 (−0.3 to −0.03) W m^{-2}), and mineral dust (equal to −0.1 (−0.3 to +0.1) W m^{-2}), although the latter may not be entirely of anthropogenic origin. While there is robust evidence for the existence of rapid adjustment of clouds in response to aerosol absorption, these effects are multiple and not well represented in climate models, leading to large uncertainty. Unlike in the last IPCC assessment, the RF from BC on snow and ice includes the effects on sea ice, accounts for more physical processes, and incorporates evidence from both models and observations. The RF_{ari} term has a 2–4 times larger global mean surface temperature change per unit forcing than a change in CO_2 concentration. The RF_{aci} terms is due to aerosol–cloud interactions and refers to the instantaneous effect on cloud albedo caused by changes in the CCN and IN concentrations, also known as the Twomey effect.

All the subsequent changes to the cloud lifetime and thermodynamics are rapid adjustments, which contribute to the ERF from (i) aerosol–radiation interactions (abbreviated as ERF_{ari} and including what has been referred to as the semidirect effect in the IPCC Climate Change (2007) report) and (ii) aerosol–cloud interactions (abbreviated as ERF_{aci}), which is a theoretical construct that is not easy to separate from other aerosol–cloud interactions and was therefore not quantified by Boucher *et al.* (2013). The total ERF due to aerosols ($ERF_{ari} + ERF_{aci}$), excluding the effect of absorbing aerosol on snow and ice, is assessed to be −0.9 (−1.9 to −0.1) W m^{-2} with medium confidence. The total ERF estimate includes rapid adjustments, such as changes to the cloud lifetime and aerosol microphysical effects on mixed-phase, ice, and convective clouds. This range was obtained from expert judgment guided by climate models that include aerosol effects on mixed-phase and convective clouds in addition to liquid clouds, satellite studies, and models that allow cloud-scale responses. This forcing can be much larger regionally, but the global mean value is consistent with several new lines of evidence, suggesting less negative estimates for the ERF due to aerosol–cloud interactions than in the IPCC Climate Change (2007) report. Persistent contrails from aviation contribute an RF of +0.01 (+0.005 to +0.03) W m^{-2} for year 2011, and the combined contrail and contrail-cirrus ERF from aviation is assessed to be +0.05 (+0.02 to +0.15) W m^{-2}. This forcing can be much larger regionally, but there is now medium confidence that it does not produce observable regional effects on either the mean or diurnal xrange of surface temperature.

Abbreviations

AOT	aerosol optical thickness
BC	black carbon
BoA	bottom of atmosphere

BRDF	bidirectional reflectance distribution functions
BS	bare soil
CCN	cloud condensation nucleus
CFCs	chlorofluorocarbons
DARF	direct aerosol-induced radiative forcing
EC	elemental carbon
ERF	effective radiative forcing
IN	ice nucleus
LAI	leaf area index
NIR	near-infrared
OC	organic carbon
OPAC	optical properties of aerosols and clouds
OS	ocean surface
PS	polar surface
RF	radiative forcing
SSA	single-scattering albedo
ToA	top of atmosphere
UV	ultraviolet
VS	vegetated surface
WS	wind speed

List of Symbols

Latin

a	diameter of a spherical particle
$A(\vartheta_0)$	broadband albedo represented as a function of solar zenith angle
AOT	aerosol optical thickness
C_{AB}	chlorophyll content of leaf coverage
c_{corrD}	correlation coefficient of the regression line drawn through the scatter plots of diurnally average values of ΔDF_{ToA}, ΔDF_{BoA}, and ΔDF_{Atm} versus $\tau_a(0.55\ \mu m)$
c_{corrE}	correlation coefficient of the regression line drawn through the scatter plots of diurnally average efficiency parameters E_{ToA}, E_{BoA}, and E_{Atm} versus $\omega(0.\ 55\ \mu m)$
C_{ext},	extinction cross section
C_{abs}	absorption cross section
C_{sca}	scattering cross section
C_{lw}	leaf water equivalent thickness
C_n	ratio of refractive indices of the leaf surface wax and internal material
C_p	seawater pigment concentration (measured in mg m^{-3})
C_{ss}	seawater salt concentration
$D\downarrow$	diffuse fraction of the incoming solar irradiance

$D\downarrow(\lambda)$	spectral curve of the diffuse fraction of downwelling (global) solar radiation $I\downarrow(\lambda)$ at surface level
$dF\downarrow(\vartheta_0, \phi_0)$	incident solar irradiance coming from the direction individuated by the pair of solar zenith angle ϑ_0 and solar azimuthal angle ϕ_0
$dL\uparrow(\eta_{pz}, \phi)$	solar radiance reflected upward by the surface in the direction given by the pair of polar angles η_{pz} and ϕ
D_o	intercept value of the regression line of the scatter plots of diurnally average values of ΔDF_{ToA}, ΔDF_{BoA}, and ΔDF_{Atm} versus $\tau_a(0.55\ \mu m)$ obtained for $\tau_a(0.55\ \mu m) = 0.10$
E_{Atm}	diurnally average DARF efficiency within the atmosphere
E_{BoA}	diurnally average DARF efficiency at the bottom-of-atmosphere level
E_{ell}	elliptical eccentricity of the leaf angle distribution
E_o	intercept value of the regression line of the scatter plots of diurnally average values of E_{ToA}, E_{BoA}, and E_{Atm} versus $\omega(0.\ 55\ \mu m)$ obtained for unit $\omega(0.\ 55\ \mu m)$
ERF_{aci}	effective radiative forcing processes arising from aerosol–cloud interactions
ERF_{ari}	effective radiative forcing processes arising from aerosol–radiation interactions
E_{ToA}	diurnally average DARF efficiency at the top-of-atmosphere level
$F\uparrow(\eta_{pz}, \phi)$	solar irradiance reflected by the surface in a certain upwelling direction
f_e	ratio between solar radiance $dL\uparrow(\eta, \phi)$ and incident solar radiation $dF\downarrow(\vartheta_0, \phi_0)$
$g(\lambda)$	asymmetry factor of an aerosol particle population at wavelength λ
$I^*(\lambda)$	spectral curve of incoming solar radiation at the surface for solar zenith angle $\vartheta_0 = 60°$ in the U. S. Standard Atmosphere 1976 of Anderson *et al.* (1986)
$I_{BoA\downarrow}(\lambda)$	solar irradiance reaching the Earth's surface after its passage through the atmosphere
$I_{BoA\uparrow}(\lambda)$	solar irradiance reflected upward by the Earth's surface
$I_{ToA\downarrow}(\lambda)$	spectral curve of incoming solar radiation at the top level of the atmosphere
$I_d\downarrow(\lambda, \vartheta_0)$	incoming solar diffuse radiation flux at wavelength λ for solar zenith angle ϑ_0
k_D	slope coefficient of the regression line drawn through the scatter plots of diurnally average values of ΔDF_{ToA}, ΔDF_{BoA}, and ΔDF_{Atm} versus $\tau_a(0.55\ \mu m)$
k_E	slope coefficient of the regression line drawn through the scatter plots of diurnally average values of E_{ToA}, E_{BoA}, and E_{Atm} versus $\omega(0.\ 55\ \mu m)$

$K_{ext}(a, \lambda)$, $K_{sca}(a, \lambda)$, $K_{abs}(a, \lambda)$	extinction, scattering, and absorption efficiency factors of an aerosol particle as a function of particle diameter a and wavelength λ
$m(\lambda)$	complex refractive index of particulate matter at wavelength λ, equal to $n(\lambda) - i\chi(\lambda)$)
$n(\lambda)$	real part of particulate matter refractive index at wavelength λ
N_{eff}	effective number of elementary layers inside a leaf
N_{tot}	total particle number concentration of aerosol particles per unit cubic centimeter of air
Q_m	modal inclination of the leaf distribution
$R(\lambda, \vartheta_0, \phi_o, \eta_{pz}, \phi)$	bidirectional reflectance
$R_{bs}(\lambda, \vartheta_0)$	spectral directional hemispherical reflectance, commonly called "black-sky albedo"
$R_{bs}(\vartheta_0 = 0°)$	black-sky albedo for null zenith illumination angle
$R_{ws}(\lambda)$	spectral bihemispherical reflectance called "white-sky albedo"
$R_L(\lambda, \theta_o)$	spectral curve of surface albedo (Lewis, 1995)
RF_{aci}	radiative forcing processes arising from aerosol–cloud interactions
RF_{ari}	radiative forcing processes arising from aerosol–radiation interactions
S_1	weight of the first Price (1990) function for the soil reflectance
SEE_D	standard error of estimate obtained for the regression line drawn through the scatter plots of diurnally average values of ΔDF_{ToA}, ΔDF_{BoA}, and ΔDF_{Atm} versus $\tau_a(0.55\ \mu m)$
SEE_E	standard error of estimate obtained for the regression line drawn through the scatter plots of diurnally average values of E_{ToA}, E_{BoA}, and E_{Atm} versus $\omega(0.55\ \mu m)$
S_L	relative leaf size with respect to the canopy depth
T_e	effective absolute blackbody emission temperature of the Sun
WD	wind direction
WS	wind speed (usually measured in $m\ s^{-1}$)
x	Mie size parameter equal to $\pi a/\lambda$
z	vertical coordinate or altitude above mean sea level

Greek

$\beta_{ext}(\lambda)$	volume extinction coefficient of an aerosol particle population at wavelength λ
$\beta_{sca}(\lambda)$	volume scattering coefficient of an aerosol particle population at wavelength λ

ϕ_0	solar azimuth angle
ϑ_0	solar zenith angle
ϕ	viewing azimuthal angle of the observer
α	Ångström's exponent calculated for the values of $\beta_{ext}(\lambda)$ determined over the visible and near-infrared wavelength range (usually from 0.40 to 0.86 μm)
ω	mean single-scattering albedo of a dry-air aerosol particle population calculated over the 0.30–3.75 μm wavelength range
θ	scattering angle
λ	wavelength (usually measured in μm)
$\chi(\lambda)$	imaginary part of particulate matter refractive index at wavelength λ
$\omega(\lambda)$	single-scattering albedo of an aerosol particle population at wavelength λ
$\omega(0.55\ \mu m)$	aerosol single-scattering albedo at wavelength $\lambda = 0.55\ \mu m$
$\omega(0.40-3.75\ \mu m)$	mean single-scattering albedo of the 10 OPAC (RH = 50%) aerosol models and SF aerosol models calculated over the 0.40–3.75 μm wavelength range
ω^*	weighted average single-scattering albedo of a dry-air aerosol particle population calculated using as weight function of the spectral curve $I^*(\lambda)$ of incoming direct solar irradiance at the surface
μ, m	relative optical air mass
ϕ'	difference between the two azimuthal angles ϕ_0 and ϕ
$\tau_a(\lambda)$	aerosol optical thickness at wavelength λ
$\tau_a(0.55\ \mu m)$	aerosol optical thickness at wavelength $\lambda = 0.55\ \mu m$
$\beta_{abs}(\lambda)$	volume absorption coefficient of an aerosol particle population at wavelength λ
ρ_{am}	air mass density
$\beta_{atm}(z)$	atmospheric optical density as a function of altitude z
ΔDF_{Atm}	diurnally average DARF term within the atmosphere
ΔDF_{BoA}	diurnally average DARF term at the bottom-of-atmosphere level
ΔDF_{ToA}	diurnally average DARF term at the top-of-atmosphere level
σ_{ext}	areal extinction cross section of a single aerosol particle
η_{pz}	polar zenith angle

References

Ackerman, A.S., Kirkpatrick, M.P., Stevens, D.E., and Toon, O.B. (2004) The impact of humidity above stratiform clouds on indirect aerosol climate forcing. *Nature*, **432**, 1014–1017. doi: 10.1038/nature03174

Ackerman, A.S., Toon, O.B., and Hobbs, P.V. (1994) Reassessing the dependence of cloud condensation nucleus concentration on formation rate. *Nature*, **367**, 445–447. doi: 10.1038/367445a0

Ackerman, A.S., Toon, O.B., Stevens, D.E., Heymsfield, A.J., Ramanathan, V., and Welton, E.J. (2000) Reduction of tropical cloudiness by soot. *Science*, **288**, 1042–1047. doi: 10.1126/science.288.5468.1042

Albrecht, B. (1989) Aerosols, cloud microphysics and fractional cloudiness. *Science*, **245**, 1227–1230. doi: 10.1126/science.245.4923.1227

Anderson, G.P., Clough, S.A., Kneizys, F.X., Chetwind, J.H., and Shettle, E.P. (1986) AFGL Atmospheric Constituent Profiles (0–120 km). Environmental Research Papers, No. 954, AFGL-TR-86-0110. Optical Physics Division (Air Force Geophysics Laboratory), Hanscom Air Force Base, MA.

Andreae, M.O. and Gelencsér, A. (2006) Black carbon or brown carbon? The nature of light-absorbing carbonaceous aerosols. *Atmos. Chem. Phys.*, **6**, 3131–3148. doi: 10.5194/acp-6-3131-2006

Ångström, A. (1964) The parameters of atmospheric turbidity. *Tellus*, **16** (1), 64–75. doi: 10.1111/j.2153-3490.1964.tb00144.x

Bellouin, N., Boucher, O., Tanré, D., and Dubovik, O. (2003) Aerosol absorption over the clear-sky oceans deduced from POLDER-1 and AERONET observations. *Geophys. Res. Lett.*, **30**, 1748. doi: 10.1029/2003GL017121

Blanchet, J.P. (1989) Toward estimation of climatic effects due to arctic aerosols. *Atmos. Environ.*, **23**, 2609–2625. doi: 10.1016/0004-6981(89)90269-2

Born, M. and Wolf, E. (1975) *Principles of Optics, Electromagnetic Theory of Propagation, Interference and Direction of Light*, 5th edn, Pergamon Press, Oxford.

Boucher, O., Randall, D. *et al.* (2013) Clouds and aerosols. Climate Change 2013: The Physical Science Basis, IPCC Working Group I Contribution to the Fifth Assessment Report of the Intergovernmental Panel on Climate Change, Chapter 7, 573–657 pp.

Bush, B.C. and Valero, F.P.J. (2002) Spectral aerosol radiative forcing at the surface during the Indian Ocean Experiment (INDOEX). *J. Geophys. Res.*, **107** (D19), 8003. doi: 10.1029/2000JD000020

Bush, B.C. and Valero, F.P.J. (2003) Surface aerosol radiative forcing at Gosan during the ACE-Asia campaign. *J. Geophys. Res.*, **108** (D23), 8660. doi: 10.1029/2002JD003233

Carbone, C., Decesari, S., Mircea, M., Giulianelli, L., Finessi, E., Rinaldi, M., Fuzzi, S., Marinoni, A., Duchi, R., Perrino, C., Sargolini, T., Vardè, M., Sprovieri, F., Gobbi, G.P., Angelini, F., and Facchini, M.C. (2010) Size-resolved aerosol chemical composition over the Italian peninsula during typical summer and winter conditions. *Atmos. Environ.*, **44** (39) 5269–5278. doi: 10.1016/j.atmosenv.2010.08.008.

Carr, S.B. (2005) The Aerosol Models in MODTRAN: Incorporating Selected Measurements from Northern Australia. Technical Report of the Defence Science and Technology Organisation, No. DSTO-TR-1803, Edinburgh, South Australia, 67 pp, see http://www.ewp.rpi.edu/hartford/~brazw/Project/Other/Research/Soot/Carr2005_AerosolModelsInMODTRAN.pdf (accessed 14 June 2016).

Charlson, R.J., Langner, J., and Rodhe, H. (1990) Sulphate aerosol and climate. *Nature*, **348**, 22–26. doi: 10.1038/348022a0

Charlson, R.J., Langner, J., Rodhe, H., Leovy, C.B., and Warren, S.G. (1991) Perturbation of the northern hemisphere radiative balance by backscattering from anthropogenic sulfate aerosols. *Tellus*, **43 B**, 152–163. doi: 10.1034/j.1600-0870.1991.00013.x

Charlson, R.J., Schwartz, S.E., Hales, J.M., Cess, R.D., Coakley, J.A. Jr., Hansen, J.E., and Hofmann, D.J. (1992) Climate forcing by anthropogenic aerosols. *Science*, **255**, 423–430. doi: 10.1126/science.255.5043.423

Chin, M., Diehl, T., Ginoux, P., and Malm, W. (2007) Intercontinental transport of pollution and dust aerosols: implications for regional air quality. *Atmos. Chem. Phys.*, **7**, 5501–5517. doi: 10.5194/acp-7-5501-2007

Christopher, S.A. and Zhang, J. (2002) Daytime variation of shortwave direct radiative forcing of biomass burning aerosols from GOES-8 Imager. *J. Atmos. Sci.*, **59** (3), 681–691. doi: 10.1175/1520-0469(2002)059<0681:DVOSDR>2.0CO:2

Christopher, S.A., Wang, J., Ji, Q., and Tsay, S.-C. (2003) Estimation of diurnal shortwave dust aerosol radiative forcing during PRIDE. *J. Geophys. Res.*, **108** (D19), 8956. doi: 10.1029/2002JD002787

Chylek, P. and Coakley, J.A. Jr., (1974) Aerosols and climate. *Science*, **183** (4120), 75–77. doi: 10.1126/science.183.4120.75

Coakley, J.A. Jr., and Chylek, P. (1975) The two-stream approximation in radiative transfer: including the angle of the incident radiation. *J. Atmos. Sci.*, **32** (2), 409–418. doi: 10.1175/1520-0469(1975)032 <0409:TTSAIR>2.0.CO;2

Collins, W.D., Rasch, P.J., Eaton, B.E., Fillmore, D.W., Kiehl, J.T., Beck, C.T., and Zender, C.S. (2002) Simulation of aerosol distributions and radiative forcing for INDOEX: regional climate impacts. *J. Geophys. Res.*, **107** (D19), 8028. doi: 10.1029/2000JD000032

Conant, W.C. (2000) An observational approach for determining aerosol surface radiative forcing: results from the first field phase of INDOEX. *J. Geophys. Res.*, **105** (D12), 15347–15360. doi: 10.1029/1999JD901166

Cook, J. and Highwood, E.J. (2004) Climate response to tropospheric absorbing aerosols in an intermediate general-circulation model. *Q. J. R. Meteorol. Soc.*, **130** (596), 175–191. doi: 10.1256/qj.03.64

Cox, C. and Munk, W. (1954) Statistics of the sea surface derived from sun glitter. *J. Mar. Res.*, **13** (2), 198–227.

Delene, D.J. and Ogren, J.A. (2002) Variability of aerosol optical properties at four North American surface monitoring sites. *J. Atmos. Sci.*, **59** (6), 1135–1150. doi: 10.1175/1520-0469(2002)059 <1135:VOAOPA>2.0.CO;2

Dubin, M., Sissenwine, N., and Teweles, S. (1966) U. S. Standard Atmosphere Supplements, *1966*, vol. **20402**, Environmental Science Services Administration, NASA, U. S. Air Force, Washington, DC, p. 289.

Ferrare, R., Feingold, G., Ghan, S., Ogren, J., Schmid, B., Schwartz, S.E., and Sheridan, P. (2006) Preface to special section: Atmospheric Radiation Measurement Program May 2003 Intensive Operations Period examining aerosol properties and radiative influences. *J. Geophys. Res.*, **111**, D05S01. doi: 10.1029/2005JD006908

Fröhlich, C. (2013) Solar constant and total solar irradiance variations, in *Solar Energy* (eds C. Richter, D. Lincot, and C.A. Gueymard), Springer-Verlag, pp. 399–416. doi: 10.1007/978-1-4614-5806-7_443

García, O.E., Díaz, J.P., Expósito, F.J., Díaz, A.M., Dubovik, O., Derimian, Y., Dubuisson, P., and Roger, J.-C. (2012) Shortwave radiative forcing and efficiency of key aerosol types using AERONET data. *Atmos. Chem. Phys.*, **12** (11), 5129–5145. doi: 10.5194/acp-12-5129-2012

Ge, J.M., Huang, J.P., Su, J., Bi, J.R., and Fu, Q. (2011) Shortwave radiative closure experiment and direct forcing of dust aerosol over northwestern China. *Geophys. Res. Lett.*, **38**, L24803. doi: 10.1029/2011GL049571

Ge, J.M., Su, J., Ackerman, T.P., Fu, Q., Huang, J.P., and Shi, J.S. (2010) Dust aerosol optical properties retrieval and radiative forcing over northwestern China during the 2008 China - U.S. joint field experiment. *J. Geophys. Res.*, **115**, D00K12. doi: 10.1029/2009JD013263

Gras, J.L. (1991) Southern hemisphere tropospheric aerosol microphysics. *J. Geophys. Res.*, **96** (D3), 5345–5356. doi: 10.1029/89JD00429

Grassl, H. (1973) Aerosol influence on radiative cooling. *Tellus*, **25** (4), 386–395.

Hansen, J.E., Sato, M., Lacis, A., Ruedy, R., Tegen, I., and Matthews, E. (1998) Climate forcings in the industrial era. *Proc. Natl. Acad. Sci. U.S.A.*, **95** (12), 12,753–12,758.

Hansen, J., Sato, M., and Ruedy, R. (1997) Radiative forcing and climate response. *J. Geophys. Res.*, **102** (D6), 6831–6864. doi: 10.1029/96JD03436

Haywood, J. and Boucher, O. (2000) Estimates of the direct and indirect radiative forcing due to tropospheric aerosols: a review. *Rev. Geophys.*, **38** (4), 513–543. doi: 10.1029/1999RG000078

Heintzenberg, J. (1994) The life cycle of the atmospheric aerosol, in *Topics in Atmospheric and Interstellar Physics and Chemistry*, Vol. 1, Chapter XII (ed. F. Boutron), Les Editions de Physique, Sciences, Les Ulis, France ERCA, pp. 251–270.

Hess, M., Koepke, P., and Schult, I. (1998) Optical properties of aerosols and clouds: the software package OPAC. *Bull.*

Am. Meteorol. Soc., **79** (5), 831–844. doi: 10.1175/1520-0477(1998)079<0831:OPOAAC>2.0.CO;2

Hignett, P., Taylor, J.P., Francis, P.N., and Glew, M.D. (1999) Comparison of observed and modeled direct aerosol forcing during TARFOX. *J. Geophys. Res.*, **104** (D2), 2279–2287. doi: 10.1029/98JD02021

Holben, B.N., Eck, T.F., Slutsker, I., Tanré, D., Buis, J.P., Setzer, A., Vermote, E., Reagan, J.A., Kaufman, Y.J., Nakajima, T., Lavenu, F., Jankowiak, I., and Smirnov, A. (1998) AERONET – a federated instrument network and data archive for aerosol characterization. *Remote Sens. Environ.*, **66** (1), 1–16. doi: S0034-4257(98)00031-5

Horvath, H., Arboledas, L.A., Olmo, F.J., Jovanovic, O., Gangl, M., Kaller, W., Sanchez, C., Sauerzopf, H., and Seidl, S. (2002) Optical properties of the aerosol in Spain and Austria and its effect on radiative forcing. *J. Geophys. Res.*, **107** (D19), 4386. doi: 10.1029/2001JD001472

Hsu, N.C., Herman, J.R., and Weaver, C.J. (2000) Determination of radiative forcing of Saharan dust using combined TOMS and ERBE data. *J. Geophys. Res.*, **105** (D16), 20649–20661. doi: 10.1029/2000JD900150

IPCC (Intergovernmental Panel on Climate Change) (1996) Climate Change 1995, The Science of Climate Change (eds J.T. Houghton, L.G. Meira Filho, B.A. Callander, N. Harris, A. Kattenberg, and K. Maskell), Cambridge University Press, Cambridge, p. 572.

IPCC Climate Change (2007) *Climate Change 2007: Synthesis Report, Contribution of Working Groups I, II and III to the Fourth Assessment Report of the Intergovernmental Panel on Climate Change* (eds R.K. Pachauri and A. Reisinger), IPCC, Geneva, p. 104.

IPCC TAR (2001) *Third Assessment Report, Climate Change 2001, Working Group I: The Scientific Basis* (eds J.T. Houghton, Y. Ding, D.J. Griggs, M. Noguer, P.J. van der Linden, X. Dai, K. Maskell, and C.A. Johnson), Cambridge University Press, Cambridge and New York, p. 881.

Iqbal, M. (1983) *An Introduction to Solar Radiation*, Academic Press, Toronto, p. 390.

Jacquemoud, S. and Baret, F. (1990) PROSPECT: a model of leaf optical properties spectra. *Remote Sens. Environ.*, **34** (2), 75–91. doi: 10.1016/0034-4257(90)90100-Z

Jaenicke, R. (1993) Tropospheric aerosols, in *Aerosol-Cloud-Climate Interactions* (ed. P.V. Hobbs), Academic Press, San Diego, CA, pp. 1–31.

Jiang, H., Xue, H., Teller, A., Feingold, G., and Levin, Z. (2006) Aerosol effects on the lifetime of shallow cumulus. *Geophys. Res. Lett.*, **33**, L14806. doi: 10.1029/2006GL026024

Johnson, B.T., Shine, K.P., and Forster, P.M. (2004) The semi-direct aerosol effect: impact of absorbing aerosols on marine stratocumulus. *Q. J. R. Meteorol. Soc.*, **130**, 1407–1422. doi: 10.1256/qj.03.61

Kaufman, Y.J. (1987) Satellite sensing of aerosol absorption. *J. Geophys. Res.*, **92** (D4), 4307–4317. doi: 10.1029/JD092iD04p04307

Kaufman, Y.J. and Fraser, R.S. (1997) The effect of smoke particles on clouds and climate forcing. *Science*, **277**, 1636–1639. doi: 10.1126/science.277.5332.1636

Kaufman, Y.J., Tanré, D., and Boucher, O. (2002) A satellite view of aerosols in the climate system. *Nature*, **419** (6903), 215–223. doi: 10.1038/nature01091

Kim, S.-W., Jefferson, A., Yoon, S.-C., Dutton, E.G., Ogren, J.A., Valero, F.P.J., Kim, J., and Holben, B.N. (2005a) Comparisons of aerosol optical depth and surface shortwave irradiance and their effect on the aerosol surface radiative forcing estimation. *J. Geophys. Res.*, **110**, D07204. doi: 10.1029/2004JD004989

Kim, D.-H., Sohn, B.-J., Nakajima, T., and Takamura, T. (2005b) Aerosol radiative forcing over east Asia determined from ground-based solar radiation measurements. *J. Geophys. Res.*, **110**, D10S22. doi: 10.1029/2004JD004678

Kim, D.-H., Sohn, B.-J., Nakajima, T., Takamura, T., Takemura, T., Choi, B.-C., and Yoon, S.-C. (2004) Aerosol optical properties over east Asia determined from ground-based sky radiation measurements. *J. Geophys. Res.*, **109**, D02209. doi: 10.1029/2003JD003387

King, M.D. (1982) Sensitivity of constrained linear inversions to the selection of the Lagrange multiplier. *J. Atmos. Sci.*, **39** (6), 1356–1369. doi: 10.1175/1520-0469(1982)039 <1356:SOCLIT>2.0.CO;2

King, M.D., Harshvardhan, and Arking, A. (1984) A model of the radiative properties of the El Chichón stratospheric aerosol layer. *J. Clim. Appl. Meteorol.*, **23** (7), 1121–1137. doi: 10.1175/1520-0450(1984)023<1121:AMOTRP>2.0.CO;2

King, M.D., Kaufman, Y.J., Tanré, D., and Nakajima, T. (1999) Remote sensing of tropospheric aerosols from space: past, present and future. *Bull. Am. Meteorol. Soc.*, **80** (11), 2229–2259. doi: 10.1175/1520-0477(1999)080<2229:RSOTAF>2.0.CO;2

Kinne, S. and Pueschel, R. (2001) Aerosol radiative forcing for Asian continental outflow. *Atmos. Environ.*, **35** (30), 5019–5028. doi: 10.1016/S1352-2310(01)00329-6

Koepke, P. (1984) Effective reflectance of oceanic whitecaps. *Appl. Opt.*, **23** (11), 1816–1824. doi: 10.1364/AO.23.001816

Kokhanovsky, A.A. (2004) Spectral reflectance of whitecaps. *J. Geophys. Res.*, **109**, C05021. doi: 10.1029/2003JC002177

Kokhanovsky, A.A. and Breon, F.-M. (2012) Validation of an analytical snow BRDF model using PARASOL multi-angular and multispectral observations. *IEEE Geosci. Remote Sens. Lett.*, **9** (5), 928–932. doi: 10.1109/LGRS.2012.2185775

Koren, I., Martins, J.V., Remer, L.A., and Afargan, H. (2008) Smoke invigoration versus inhibition of clouds over the Amazon. *Science*, **321**, 5891. doi: 10.1126/science.1159185

Kriebel, K.T. (1978) Measured spectral bidirectional reflection properties of four vegetated surfaces. *Appl. Opt.*, **17** (2), 253–259. doi: 10.1364/AO.17.000253

Kuusk, A. (1994) A multispectral canopy reflectance model. *Remote Sens. Environ.*, **50** (2), 75–82. doi: 10.1016/0034-4257(94)90035-3

Lanconelli, C. (2007) Evaluation of the effects produced by surface reflectance anisotropy in the direct radiative forcing induced by atmospheric aerosols. PhD thesis. University of Ferrara, 19th cycle of the Research Doctorate in Physics, February 5, 2007, 131 pp.

Le Blanc, S.E., Schmidt, K.S., Pilewskie, P., Redemann, J., Hostetler, C., Ferrare, R., Hair, J., Langridge, J.M., and Lack, D.A. (2012) Spectral aerosol direct radiative forcing from airborne radiative measurements during CalNex and ARCTAS. *J. Geophys. Res.*, **117**, D00V20. doi: 10.1029/2012JD018106

Lelieveld, J., Berresheim, H., Borrmann, S., Crutzen, P.J., Dentener, F.J., Fischer, H., Feichter, J., Flatau, P.J., Heland, J., Holzinger, R., Korrmann, R., Lawrence, M.G., Levin, Z., Markowicz, K.M., Mihalopoulos, N., Minikin, A., Ramanathan, V., de Reus, M., Roelofs, G.J., Scheeren, H.A., Sciare, J., Schlager, H., Schultz, M., Siegmund, P., Steil, B., Stephanou, E.G., Stier, P., Traub, M., Warneke, C., Williams, J., and Ziereis, H. (2002) Global air pollution crossroads over the Mediterranean. *Science*, **298** (5594), 794–799. doi: 10.1126/science.1075457

Lewis, P. (1995) On the implementation of linear kernel-driven BRDF models. Proceedings of Annual Conference of the Remote Sensing Society '95, "Remote Sensing in Action", Southampton, September 11–14, 1995, pp. 333–340.

Lewis, P. and Barnsley, M.J. (1994) Influence of the sky radiance distribution on various formulations of the Earth surface albedo. Proceedings of the 6th International Symposium on Physical Measurements and Signatures in Remote Sensing, Val d'Isere, France, January 17–21, 1994, pp. 707–715.

Li, Z., Chen, H., Cribb, M., Dickerson, R., Holben, B., Li, C., Lu, D., Luo, Y., Maring, H., Shi, G., Tsay, S.-C., Wang, P., Wang, Y., Xia, X., Zheng, Y., Yuan, T., and Zhao, F. (2007a) Preface to special section on East Asian studies of tropospheric aerosols: an international regional experiment (EAST-AIRE). *J. Geophys. Res.*, **112**, D22S00. doi: 10.0129/2007JD008853

Li, Z., Xia, X., Cribb, M., Mi, W., Holben, B., Wang, P., Chen, H., Tsay, S.-C., Eck, T.F., Zhao, F., Dutton, E.G., and Dickerson, R.R. (2007b) Aerosol optical properties and their radiative effects in northern China. *J. Geophys. Res.*, **112**, D22S01. doi: 10.1029/2006JD007382

Li, F., Vogelmann, A.M., and Ramanathan, V. (2004) Dust aerosol radiative forcing measured from space over the Western Africa. *J. Clim.*, **17** (13), 2558–2571.

doi: 10.1175/1520-0442(2004)017 < 2558:SDARFM > 2.0.CO;2

Liu, Y., Huang, J., Shi, G., Takamura, T., Khatri, P., Bi, J., Shi, J., Wang, T., Wang, X., and Zhang, B. (2011) Aerosol optical properties and radiative effect determined from sky-radiometer over Loess Plateau of Northwest China. *Atmos. Chem. Phys.*, **11**, 11455–11463. doi: 10.5194/acp-11-11455-2011

Liu, J., Xia, X., Wang, P., Li, Z., Zheng, Y., Cribb, M., and Chen, H. (2007) Significant aerosol direct radiative effects during a pollution episode in northern China. *Geophys. Res. Lett.*, **34**, L23808. doi: 10.1029/2007GL030953

Lohmann, U. and Feichter, J. (2005) Global indirect aerosol effects: a review. *Atmos. Chem. Phys.*, **5**, 715–737. doi: 10.5194/acp-5-715-2005

Lubin, D., Satheesh, S.K., McFarquar, G., and Heymsfield, A.J. (2002) Long-wave radiative forcing of Indian Ocean tropospheric aerosol. *J. Geophys. Res.*, **107** (D19), 8004. doi: 10.1029/2001JD001183

McCormick, R.A. and Ludwig, J.H. (1967) Climate modification by atmospheric aerosols. *Science*, **156** (3780), 1358–1359. doi: 10.1126/science.156.3780.1358

Markowicz, K.M., Flatau, P.J., Quinn, P.K., Carrico, C.M., Flatau, M.K., Vogelmann, A.M., Bates, D., Liu, M., and Rood, M.J. (2003) Influence of relative humidity on aerosol radiative forcing: an ACE-Asia experiment perspective. *J. Geophys. Res.*, **108** (D23), 8662. doi: 10.1029/2002JD003066

Markowicz, K.M., Flatau, P.J., Ramana, R.V., Crutzen, P.J., and Ramanathan, V. (2002) Absorbing mediterranean aerosols lead to a large reduction in the solar radiation at the surface. *Geophys. Res. Lett.*, **29** (1968), 2002. doi: 10.1029/2002GL015767

Markowicz, K.M., Flatau, P.J., Remiszewska, J., Witek, M., Reid, E.A., Reid, J.S., Bucholtz, A., and Holben, B. (2008) Observations and modeling of the surface aerosol radiative forcing during UAE2. *J. Atmos. Sci.*, **65** (9), 2877–2891. doi: 10.1175/2007JAS2555.1

Mazzola, M., Lanconelli, C., Lupi, A., Busetto, M., Vitale, V., and Tomasi, C. (2010) Columnar aerosol optical properties in the Po Valley, Italy, from MFRSR data. *J. Geophys. Res.*, **115**, D17206. doi: 10.10129/2009JD013310

Meywerk, J. and Ramanathan, V. (1999) Observations of the spectral clear-sky aerosol forcing over the tropical Indian Ocean. *J. Geophys. Res.*, **104** (D20), 24359–24370. doi: 10.1029/1999JD900502

Mie, G. (1908) Beiträge zur Optik trüber Medien, speziell kolloidaler Metallösungen. *Ann. Phys., Vierte Folge*, **25** (3), 377–445.

Morel, A. (1988) Optical modeling of the upper ocean in relation to its biogenous matter content (Case I waters). *J. Geophys. Res.*, **93** (C9), 10749–10768. doi: 10.1029/JC093iC09p10749

Nakajima, T., Sekiguchi, M., Takemura, T., Uno, I., Higurashi, A., Kim, D., Sohn, B.J., Oh, S.-N., Nakajima, T.Y., Ohta, S., Okada, I., Takemura, T., and Kawamoto, K. (2003) Significance of direct and indirect radiative forcings of aerosols in the East China Sea region. *J. Geophys. Res.*, **108** (D23), 8658. doi: 10.1029/2002JD003261

Nakajima, T., Yoon, S.-C., Ramanathan, V., Shi, G.-Y., Takemura, T., Higurashi, A., Takamura, T., Aoki, K., Sohn, B.-J., Kim, S.-W., Tsuruta, H., Sugimoto, N., Shimizu, A., Tanimoto, H., Sawa, Y., Lin, N.-H., Lee, C.-T., Goto, D., and Schutgens, N. (2007) Overview of the Atmospheric Brown Cloud East Asian Regional Experiment 2005 and a study of the aerosol direct radiative forcing in east Asia. *J. Geophys. Res.*, **112**, D24. doi: 10.1029/2007JD009009

Nilson, T. and Kuusk, A. (1989) A reflectance model for the homogeneous plant canopy and its inversion. *Remote Sens. Environ.*, **27** (2), 157–167. doi: 10.1016/0034-4257(89)90015-1

Penner, J.E., Andreae, M., Annergarn, H., Barrie, L., Feichter, J., Hegg, D., Jayaraman, A., Leaitch, R., Murphy, D., Nganga, J., and Pitari, G. (2001) Aerosols, their direct and indirect effects, in *Climate Change 2001: The Scientific Basis. Contribution of Working Group I to the Third Assessment Report of the Intergovernmental Panel on Climate Change* (eds J.T. Houghton, Y. Ding, D.J. Griggs, M. Noguer, P.J. van der Linden, X. Dai, K. Maskell, and C.A. Johnson), Cambridge University Press, Cambridge and New York, pp. 289–348.

Perrone, M.R., Santese, M., Tafuro, A.M., Holben, B., and Smirnov, A.

(2005) Aerosol load characterization over south-east Italy by one year of AERONET sun-photometer measurements. *Atmos. Res.*, **75** (1-2), 111–133. doi: 10.1016/j.atmosres.2004.12.003

Podgorny, I.A. and Ramanathan, V. (2001) A modeling study of the direct effect of aerosols over the tropical Indian Ocean. *J. Geophys. Res.*, **106** (D20), 24097–24105. doi: 10.1029/2001JD900214

Podgorny, I.A., Conant, W.C., Ramanathan, V., and Satheesh, S.K. (2000) Aerosol modulation of atmospheric and solar heating over the tropical Indian Ocean. *Tellus*, **52B** (3), 947–958. doi: 10.1034/j.1600-0889.2000.d01-4.x

Podgorny, I.A., Li, F., and Ramanathan, V. (2003) Large aerosol radiative forcing due to the 1997 Indonesian forest fire. *Geophys. Res. Lett.*, **30** (1), 1028. doi: 10.1029/2002GL015979

Price, J.C. (1990) On the information content of soil reflectance spectra. *Remote Sens. Environ.*, **33** (2), 113–121. doi: 10.1016/0034-4257(90)90037-M

Pueschel, R.F., Kinne, S.A., Russell, P.B., Snetsinger, K.G., and Livingston, J.M. (1993) Effects of the 1991 Pinatubo volcanic eruption on the physical and radiative properties of stratospheric aerosols, in IRS '92: Current Problems in Atmospheric Radiation, Proceedings of International Radiation Symposium, *Tallinn (Estonia), 2–8 August, 1992* (eds S. Keevallik and O. Kärner), A. Deepak Publishing, Hampton, VA, pp. 183–186.

Rahman, H., Pinty, B., and Verstraete, M.M. (1993) Coupled surface-atmosphere reflectance (CSAR) model 2. Semiempirical surface model usable with NOAA Advanced Very High Resolution Radiometer data. *J. Geophys. Res.*, **98** (D11), 20791–20801. doi: 10.1029/93JD02072

Rajeev, K. and Ramanathan, V. (2001) Direct observations of clear-sky aerosol radiative forcing from space during the Indian Ocean Experiment. *J. Geophys. Res.*, **106** (D15), 17221–17235. doi: 10.1029/2000JD900723

Ramana, M.V., Ramanathan, V., Podgorny, I.A., Pradhan, B.B., and Shrestha, B. (2004) The direct observations of large aerosol radiative forcing in Himalayan region. *Geophys. Res. Lett.*, **31**, L05111. doi: 10.1029/2003GL018824

Ramanathan, V., Crutzen, P.J., Lelieveld, J., Mitra, A.P., Althausen, D., Anderson, J., Andreae, M.O., Cantrell, W., Cass, G.R., Chung, C.E., Clarke, A.D., Coakley, J.A., Collins, W.D., Conant, W.C., Dulac, F., Heintzenberg, J., Heymsfield, A.J., Holben, B., Howell, S., Hudson, J., Jayaraman, A., Kiehl, J.T., Krishnamurti, T.N., Lubin, D., McFarquhar, G., Novakov, T., Ogren, J.A., Podgorny, I.A., Prather, K., Priestley, K., Prospero, J.M., Quinn, P.K., Rajeev, K., Rasch, P., Rupert, S., Sadourny, R., Satheesh, S.K., Shaw, G.E., Sheridan, P., and Valero, F.P.J. (2001a) Indian Ocean Experiment: an integrated analysis of the climate forcing and effects of the great Indo-Asian haze. *J. Geophys. Res.*, **106** (D22), 28371–28398. doi: 10.1029/2001JD900133

Ramanathan, V., Crutzen, P.J., Kiehl, J.T., and Rosenfeld, D. (2001b) Aerosols, climate and the hydrological cycle. *Science*, **294** (5549), 2119–2124. doi: 10.1126/science.1064034

Ramanathan, V., Li, F., Ramana, M.V., Praveen, P.S., Kim, D., Corrigan, C.E., Nguyen, H., Stone, E.A., Schauer, J.J., Carmichael, G.R., Adhikary, B., and Yoo, S.C. (2007) Atmospheric brown clouds: hemispherical and regional variations in long-range transport, absorption, and radiative forcing. *J. Geophys. Res.*, **112**, D22S21. doi: 10.1029/2006JD008124

Ramaswamy, V., Boucher, O., Haigh, J., Hauglustaine, D., Haywood, J., Myhre, G., Nakajima, T., Shi, G.Y., and Solomon, S. (2001) Radiative forcing of climate change, in *Climate Change 2001: The Scientific Basis. Contribution of Working Group I to the Third Assessment Report of the Intergovernmental Panel on Climate Change* (eds J.T. Houghton, Y. Ding, D.J. Griggs, M. Noguer, P.J. van der Linden, X. Dai, K. Maskell, and C.A. Johnson), Cambridge University Press, Cambridge and New York, pp. 349–416.

Redemann, J., Pilewskie, P., Russell, P.B., Livingston, J.M., Howard, S., Schmid, B., Pommier, J., Gore, W., Eilers, J., and Wendisch, M. (2006) Airborne measurements of spectral direct aerosol radiative forcing in the Intercontinental chemical

Transport Experiment/Intercontinental Transport and Chemical Transformation of anthropogenic pollution, 2004. *J. Geophys. Res.*, **111**, D14210. doi: 10.1029/2005JD006812

Remer, L.A. and Kaufman, Y.J. (2006) Aerosol direct radiative effect at the top of the atmosphere over cloud free ocean derived from four years of MODIS data. *Atmos. Chem. Phys.*, **6** (1), 237–253. doi: 10.5194/acp-6-237-2006

Remer, L.A., Kaufman, Y.J., Holben, B.N., Thompson, A.M., and McNamara, D. (1998) Biomass burning aerosol size distribution and modeled optical properties. *J. Geophys. Res.*, **103** (D24), 31879–31891. doi: 10.1029/98JD00271

Román, M.O., Schaaf, C.B., Lewis, P., Gao, F., Anderson, G.P., Privette, J.L., Strahler, A.H., Woodcock, C.E., and Barnsley, M. (2010) Assessing the coupling between surface albedo derived from MODIS and the fraction of diffuse skylight over spatially-characterized landscapes. *Remote Sens. Environ.*, **114** (4), 738–760. doi: 10.1016/j.rse.2009.11.014

Ross, J.L., Hobbs, P.V., and Holben, B. (1998) Radiative characteristics of regional hazes dominated by smoke from biomass burning in Brazil: closure tests and direct radiative forcing. *J. Geophys. Res.*, **103** (D24), 31925–31941. doi: 10.1029/97JD03677

Russell, P.B. and Heintzenberg, J. (2000) An overview of the ACE 2 clear sky column closure experiment (CLEARCOLUMN). *Tellus B*, **52** (2), 463–483. doi: 10.1034/j.1600-0889.2000.00013.x

Russell, P.B., Hobbs, P.V., and Stowe, L.L. (1999a) Aerosol properties and radiative effects in the United States East Coast haze plume: an overview of the Tropospheric Aerosol Radiative Forcing Observational Experiment (TARFOX). *J. Geophys. Res.*, **104** (D2), 2213–2222. doi: 10.1029/1998JD200028

Russell, P.B., Livingston, J.M., Hignett, P., Kinne, S., Wong, J., Chien, A., Bergstrom, R., Durkee, P., and Hobbs, P.V. (1999b) Aerosol-induced radiative flux changes off the United States mid-Atlantic coast: comparison of values calculated from sunphotometer and in situ data with those measured by airborne pyranometer. *J. Geophys. Res.*, **104** (D2), 2289–2307. doi: 10.1029/1998JD200025

Russell, P.B., Kinne, S.A., and Bergstrom, R.W. (1997) Aerosol climate effects: local radiative forcing and column closure experiments. *J. Geophys. Res.*, **102** (D8), 9397–9407. doi: 10.1029/97JD00112

Satheesh, S.K. (2002) Radiative forcing by aerosols over Bay of Bengal region. *Geophys. Res. Lett.*, **29** (22), 2083. doi: 10.1029/2002GL015334

Satheesh, S.K. and Ramanathan, V. (2000) Large differences in tropical aerosol forcing at the top of the atmosphere and Earth's surface. *Nature*, **405** (6782), 60–63. doi: 10.1038/35011039

Satheesh, S.K. and Srinivasan, J. (2002) Enhanced aerosol loading over Arabian Sea during the pre-monsoon season: natural or anthropogenic? *Geophys. Res. Lett.*, **29** (18), 1874. doi: 10.1029/2002GL015687

Satheesh, S.K., Ramanathan, V., Li-Jones, X., Lobert, J.M., Podgorny, I.A., Prospero, J.M., Holben, B.N., and Loeb, N.G. (1999) A model for the natural and anthropogenic aerosols over the tropical Indian Ocean derived from Indian Ocean Experiment data. *J. Geophys. Res.*, **104** (D22), 27421–27440. doi: 10.1029/1999JD900478

Schmidt, K.S., Pilewskie, P., Bergstrom, R., Coddington, O., Redemann, J., Livingston, J., Russell, P., Bierwirth, E., Wendisch, M., Gore, W., Dubey, M.K., and Mazzoleni, C. (2010) A new method for deriving aerosol solar radiative forcing and its first application within MILAGRO/INTEX-B. *Atmos. Chem. Phys.*, **10** (16), 7829–7843. doi: 10.5194/acp-10-7829-2010

Schwartz, S.E. and Andreae, M.O. (1996) Uncertainty in climate change caused by aerosols. *Science*, **272** (5265), 1121–1122. doi: 10.1126/science.272.5265.1121

Seinfeld, J.H. and Pandis, S.N. (2006) *Atmospheric Chemistry and Physics, from Air Pollution to Climate Change*, 2nd edn, John Wiley & Sons, Inc., Hoboken, NJ, p. 1225.

Seinfeld, J.H., Carmichael, G.R., Arimoto, R., Conant, W.C., Brechtel, F.J., Bates, T.S., Cahill, T.A., Clarke, A.D., Doherty, S.J., Flatau, P.J., Huebert, B.J., Kim, J., Markowicz, K.M., Quinn, P.K., Russell, L.M., Russell, P.B., Shimizu, A., Shinozuka, Y., Song, C.H., Tang, Y., Uno,

I., Vogelmann, A.M., Weber, R.J., Woo, J.-H., and Zhang, X.Y. (2004) ACE-Asia: regional climatic and atmospheric chemical effects of Asian dust and pollution. *Bull. Am. Meteorol. Soc.*, **85** (3), 367–380. doi: 10.1175/BAMS-85-3-367

Shettle, E.P. (1984) Optical and radiative properties of a desert aerosol model, in IRS '84: Current Problems in Atmospheric Radiation, Proceedings of International Radiation Symposium, *Perugia, Italy, 21–28 August, 1984* (ed. G. Fiocco), A. Deepak Publishing, Hampton, VA, pp. 74–77.

Shettle, E.P. and Fenn, R.W. (1979) Models for the Aerosols of the Lower Atmosphere and the Effects of Humidity Variations on their Optical Properties. Environmental Research Papers, No. 676. Air Force Geophysics Laboratory, AFGL-Technical Report 79-0214. Hanscom AFB, MA, 94 pp.

Strawa, A.W., Elleman, R., Hallar, A.G., Covert, D., Ricci, K., Provencal, R., Owano, T.W., Jonsson, H.H., Schmid, B., Luu, A.P., Bokarius, K., and Andrews, E. (2006) Comparison of in situ aerosol extinction and scattering coefficient measurements made during the Aerosol Intensive Operating Period. *J. Geophys. Res.*, **111**, D05S03. doi: 10.1029/2005JD006056

Sumanth, E., Mallikarjuna, K., Stephen, J., Moole, M., Vinoj, V., Satheesh, S.K., and Moorthy, K.K. (2004) Measurements of aerosol optical depths and black carbon over Bay of Bengal during post-monsoon season. *Geophys. Res. Lett.*, **31**, L16115. doi: 10.1029/2004GL020681

Tafuro, A.M., Kinne, S., De Tomasi, F., and Perrone, M.R. (2007) Annual cycle of aerosol direct radiative effect over southeast Italy and sensitivity studies. *J. Geophys. Res.*, **112**, D20202. doi: 10.1029/2006JD008265

Tahnk, W.R. and Coakley, J.A. Jr., (2002) Aerosol optical depth and direct radiative forcing for INDOEX derived from AVHRR: observations, January–March 1996–2000. *J. Geophys. Res.*, **107** (D19), 8010. doi: 10.1029/2000JD000183

Takemura, T., Nakajima, T., Dubovik, O., Holben, B.N., and Kinne, S. (2002) Single-scattering albedo and radiative forcing of various aerosol species with a global three-dimensional model. *J. Clim.*, **15** (4), 333–352. doi: 10.1175/1520-0442(2002)015<0333:SSAARF>2.0.CO;2

Takemura, T., Nozawa, T., Emori, S., Nakajima, T.Y., and Nakajima, T. (2005) Simulation of climate response to aerosol direct and indirect effects with aerosol transport-radiation model. *J. Geophys. Res.*, **110**, D02202. doi: 10.1029/2004JD005029

Tanré, D., Haywood, J., Pelon, J., Léon, J.F., Chatenet, B., Formenti, P., Francis, P., Goloub, P., Highwood, E.J., and Myhre, G. (2003) Measurement and modeling of the Saharan dust radiative impact: overview of the Saharan Dust Experiment (SHADE). *J. Geophys. Res.*, **108** (D18), 8574. doi: 10.1029/2002JD003273

Tomasi, C. (1982) Features of the scale height for particulate extinction in hazy atmospheres. *J. Appl. Meteorol.*, **21** (7), 931–944. doi: 10.1175/1520-0450(1982)021<0931:FOTSHF>2.0.CO;2

Tomasi, C., Lanconelli, C., Lupi, A., and Mazzola, M. (2015a) Diurnally averaged direct aerosol-induced radiative forcing from clear-sky field measurements performed during seven regional experiments, in Light Scattering Reviews, *Light Scattering and Radiative Transfer*, Springer-Praxis Books in Environmental Sciences, vol. **9**, Chapter 8 (ed. A.A. Kokhanovsky), Springer-Verlag, Berlin, Heidelberg, pp. 297–425. doi: 10.1007/978-3-642-37985-7

Tomasi, C., Kokhanovsky, A.A., Lupi, A., Ritter, C., Smirnov, A., O'Neill, N.T., Stone, R.S., Holben, B.N., Nyeki, S., Wehrli, C., Stohl, A., Mazzola, M., Lanconelli, C., Vitale, V., Stebel, K., Aaltonen, V., de Leeuw, G., Rodriguez, E., Herber, A.B., Radionov, V.F., Zielinski, T., Petelski, T., Sakerin, S.M., Kabanov, D.M., Xue, Y., Mei, L., Istomina, L., Wagener, R., McArthur, B., Sobolewski, P.S., Kivi, R., Courcoux, Y., Larouche, P., Broccardo, S., and Piketh, S.J. (2015b) Aerosol remote sensing in polar regions. *Earth Sci. Rev.*, **140**, 108–157. doi: 10.1016/j.earscirev.2014.11.001

Tomasi, C., Lanconelli, C., Lupi, A., and Mazzola, M. (2013) Dependence of direct aerosol radiative forcing on the optical properties of atmospheric aerosol and underlying surface, in Light Scattering Reviews, *Radiative Transfer and Light*

Scattering, Springer-Praxis Books in Environmental Sciences, vol. **8**, Chapter 11, (ed. A.A. Kokhanovsky), Springer-Verlag, Berlin, Heidelberg, pp. 505–626. doi: 10.1007/978-3-642-32106-1

Tomasi, C., Lupi, A., Mazzola, M., Stone, R.S., Dutton, E.G., Herber, A., Radionov, V.F., Holben, B.N., Sorokin, M.G., Sakerin, S.M., Terpugova, S.A., Sobolewski, P.S., Lanconelli, C., Petkov, B.H., Busetto, M., and Vitale, V. (2012) An update on polar aerosol optical properties using POLAR-AOD and other measurements performed during the International Polar Year. *Atmos. Environ.*, **52** (C), 29–47. doi: 10.1016/j.atmosenv.2012.02.055

Tomasi, C., Prodi, F., and Tampieri, F. (1979) Atmospheric turbidity variations caused by layers of Sahara dust particles. *Beitr. Phys. Atmos.*, **52** (3), 215–228.

Tomasi, C., Vitale, V., Lupi, A., Cacciari, A., Marani, S., and Bonafè, U. (2003) Marine and continental aerosol effects on the upwelling solar radiation flux in Southern Portugal during the ACE-2 experiment. *Ann. Geophys.*, **46** (2), 467–479.

Tomasi, C., Vitale, V., Lupi, A., Di Carmine, C., Campanelli, M., Herber, A., Treffeisen, R., Stone, R.S., Andrews, E., Sharma, S., Radionov, V., von Hoyningen-Huene, W., Stebel, K., Hansen, G.H., Myhre, C.L., Wehrli, C., Aaltonen, V., Lihavainen, H., Virkkula, A., Hillamo, R., Ström, J., Toledano, C., Cachorro, V., Ortiz, P., de Frutos, A., Blindheim, S., Frioud, M., Gausa, M., Zielinski, T., Petelski, T., and Yamanouchi, T. (2007) Aerosols in polar regions: a historical overview based on optical depth and in situ observations. *J. Geophys. Res.*, **112**, D16205. doi: 10.1029/2007JD008432

Tomasi, C., Vitale, V., Petkov, B., Lupi, A., and Cacciari, A. (2005) Improved algorithm for calculations of Rayleigh-scattering optical depth in standard atmospheres. *Appl. Opt.*, **44** (16), 3320–3341. doi: 10.1364/AO.44.003320

Tomasi, C., Vitale, V., and Tarozzi, L. (1997) Sun-photometric measurements of atmospheric turbidity variations caused by the Pinatubo aerosol cloud in the Himalayan region during the summer periods of 1991 and 1992. *Il Nuovo Cimento C*, **20** (1), 61–88.

Twomey, S.A. (1977) The influence of pollution on the shortwave albedo of clouds. *J. Atmos. Sci.*, **34** (7), 1149–1152.

Valero, F.P.J. and Bush, B.C. (1999) Measured and calculated clear-sky solar radiative fluxes during the Subsonic Aircraft Contrail and Cloud Effects Special Study (SUCCESS). *J. Geophys. Res.*, **104** (D22), 27387–27398. doi: 10.1029/1999JD900947

Vermote, E.F., Tanré, D., Deuzé, J.L., Herman, M., and Morcrette, J.-J. (1997a) Second simulation of the satellite signal in the solar spectrum (6S): an overview. *IEEE Trans. Geosci. Remote Sens.*, **35** (3), 675–686. doi: 10.1109/36.581987

Vermote, E., Tanré, D., Deuzé, J.L., Herman, M., and Morcrette, J.J. (1997b) Second Simulation of the Satellite Signal in the Solar Spectrum *(6S), 6S User Guide Version 2, July 1997*, Universitè de Lille, France, p. 218.

Vitale, V., Tomasi, C., Lupi, A., Cacciari, A., and Marani, S. (2000) Retrieval of columnar aerosol size-distributions and radiative forcing evaluations from sun-photometric measurements taken during the CLEARCOLUMN (ACE 2) Experiment. *Atmos. Environ.*, **34** (29-30), 5095–5105. doi: 10.1016/S1352-2310(00)00269-7

Voss, K.J., Welton, E.J., Quinn, P.K., Frouin, R., Miller, M., and Reynolds, R.M. (2001) Aerosol optical depth measurements during the Aerosols99 experiment. *J. Geophys. Res.*, **106** (D18), 20811–20819. doi: 10.1029/2000JD900783

Waliser, D.E. and Graham, N.E. (1993) Convective cloud systems and warm-pool sea surface temperatures: coupled interactions and self-regulation. *J. Geophys. Res.*, **98** (D7), 12881–12893. doi: 10.1029/93JD00872

Warren, S.G. (1984) Optical constants of ice from the ultraviolet to the microwave. *Appl. Opt.*, **23** (8), 1206–1225. doi: 10.1364/AO.23.001206

Warren, S.G. and Wiscombe, W.J. (1980) A model for the spectral albedo of snow. II: Snow containing atmospheric aerosols. *J. Atmos. Sci.*, **37** (12), 2734–2745. doi: 10.1175/1520-0469(1980)037<2734:AMFTSA>2.0.CO;2

Wiscombe, W.J. and Warren, S.G. (1980) A model for the spectral albedo of snow. I: Pure snow. *J. Atmos. Sci.*, **37** (12), 2712–2733. doi: 10.1175/1520-0469(1980)037<2712:AMFTSA>2.0.CO;2 DOI:10.1175/1520-0469(1980)037%3c2712:AMFTSA%3e2.0.CO;2

Won, J.-G., Yoon, S.-C., Kim, S.-W., Jefferson, A., Dutton, E.G., and Holben, B.N. (2004) Estimation of direct radiative forcing of Asian dust aerosols with Sun/sky radiometer and lidar measurements at Gosan, Korea. *J. Meteorol. Soc. Jpn.*, **82** (1), 115–130.

Xia, X., Chen, H., Li, Z., Wang, P., and Wang, J. (2007a) Significant reduction of surface solar irradiance induced by aerosols in a suburban region in northeastern China. *J. Geophys. Res.*, **112**, D22S02. doi: 10.1029/2006JD007562

Xia, X., Li, Z., Holben, B., Wang, P., Eck, T., Chen, H., Cribb, M., and Zhao, Y. (2007b) Aerosol optical properties and radiative effects in the Yangtze Delta region of China. *J. Geophys. Res.*, **112**, D22S12. doi: 10.1029/2007JD008859

Xia, X., Li, Z., Wang, P., Chen, H., and Cribb, M. (2007c) Estimation of aerosol effects on surface irradiance based on measurements and radiative transfer model simulations in northern China. *J. Geophys. Res.*, **112**, D22S10. doi: 10.1029/2006JD008337

Xu, J., Bergin, M.H., Greenwald, R., and Russell, P.B. (2003) Direct aerosol radiative forcing in the Yangtze delta region of China: observation and model estimation. *J. Geophys. Res.*, **108** (D2), 4060. doi: 10.1029/2002JD002550

Yu, H., Kaufman, Y.J., Chin, M., Feingold, G., Remer, L.A., Anderson, T.L., Balkanski, Y., Bellouin, N., Boucher, O., Christopher, S., DeCola, P., Kahn, R., Koch, D., Loeb, N., Reddy, M.S., Schulz, M., Takemura, T., and Zhou, M. (2006) A review of measurement-based assessments of the aerosol direct radiative effect and forcing. *Atmos. Chem. Phys.*, **6** (3), 613–666. doi: 10.5194/acp-6-613-2006

Yu, H., Liu, S.C., and Dickinson, R.E. (2002) Radiative effects of aerosols on the evolution of the atmospheric boundary layer. *J. Geophys. Res.*, **107**, 4142. doi: 10.1029/2001JD000754

Zhang, J. and Christopher, S.A. (2003) Longwave radiative forcing of Saharan dust aerosols estimated from MODIS, MISR, and CERES observations on Terra. *Geophys. Res. Lett.*, **30**, 2188. doi: 10.1029/2003GL018479

Zhao, T.X.-P., Yu, H., Laszlo, I., Chin, M., and Conant, W.C. (2008) Derivation of component aerosol direct radiative forcing at the top of atmosphere for clear-sky oceans. *J. Quant. Spectrosc. Radiat. Transfer*, **109** (7), 1162–1186. doi: 10.1016/j.jqsrt.2007.10.006

9
Aerosol and Air Quality

Sandro Fuzzi, Stefania Gilardoni, Alexander A. Kokhanovsky, Walter Di Nicolantonio, Sonoyo Mukai, Itaru Sano, Makiko Nakata, Claudio Tomasi, and Christian Lanconelli

9.1
Introduction

Clean air is considered a basic requirement for human health, and air pollution has become a serious problem worldwide. According to the World Health Organization (WHO), an estimated 7 million people per year die from air pollution-related diseases. These include stroke and heart diseases, respiratory illness, and cancer. Globally, air pollution is the second leading risk factor for the global burden of disease in 2010, behind high blood pressure, and together with tobacco smoking. In 2010, outdoor aerosol air pollution alone was responsible for 3.2 million premature deaths, with an additional 3.5 million premature deaths due to indoor aerosol air pollution (Lim *et al.*, 2012). Indoor pollution goes beyond the scope of this chapter, where we will only discuss outdoor pollution.

9.1.1
Aerosol Air Pollution

People living in low- and middle-income countries disproportionately experience the burden of outdoor air pollution with 88% (of the 3.2 million premature deaths) occurring in low- and middle-income countries and the greatest burden in the Western Pacific and Southeast Asian regions (CCAC, 2014, see Figure 9.1).

9.1.2
Aerosol Sources and Size Distribution in Relation to Human Health Effects

The aerosol mass size distribution is characterized by two main modes: the accumulation (or fine) and the coarse modes (Figure 9.2). These modes have different chemical composition and sources. The fine mode originates from combustion-related sources and contains most of the particulate matter (PM) formed chemically in the atmosphere (secondary aerosols). The coarse mode

Atmospheric Aerosols: Life Cycles and Effects on Air Quality and Climate, First Edition.
Edited by Claudio Tomasi, Sandro Fuzzi, and Alexander Kokhanovsky.

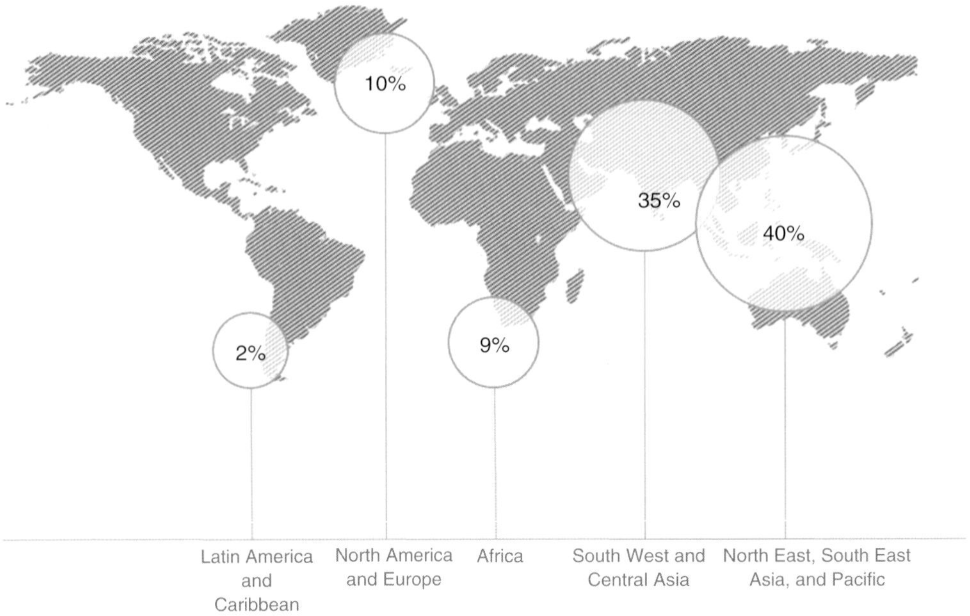

Figure 9.1 Approximate share of premature deaths from air pollution (both outdoor and indoor) (CCAC, 2014).

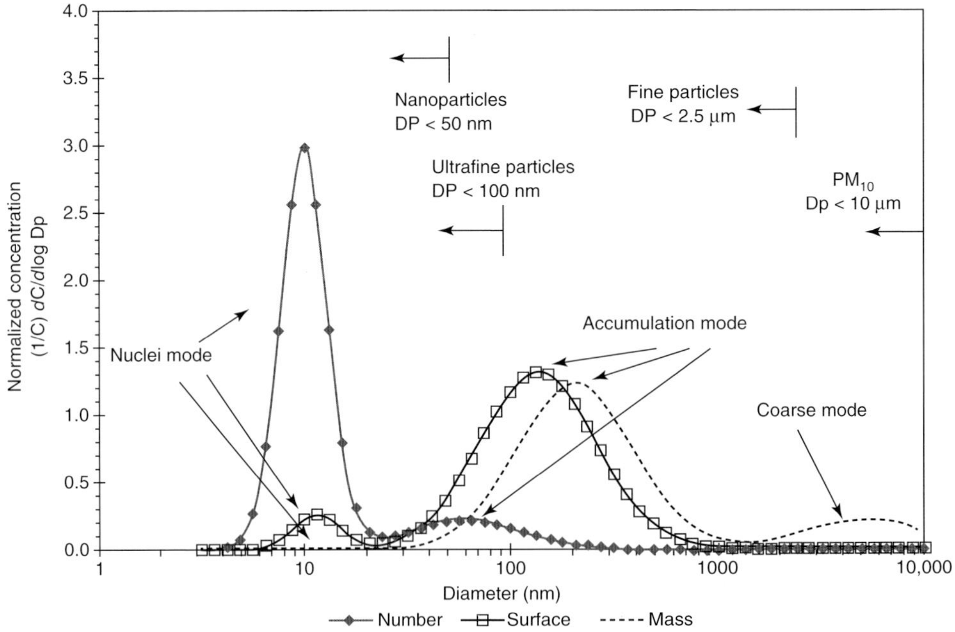

Figure 9.2 Normalized size distribution of particle number, surface, and mass for a typical roadside PM (HEI, 2010).

contains mainly resuspended dust, sea salt, and other mechanically generated particles.

With respect to the health effects of fine aerosols (conventionally known in the air quality (AQ) literature as $PM_{2.5}$), these particles penetrate into the lungs more efficiently than the larger coarse particles (known as PM_{10}) that deposit mainly in the large conducting airways. Fine particle mass mainly comprises secondary inorganic species (ammonium, nitrate, and sulfate), elemental (or black) carbon, and organic compounds, which can have primary or secondary origin. The contribution of the inorganic and organic fractions varies spatially and seasonally. Coarse particle mass is dominated by inorganic species, such as crustal minerals and sea salt. In addition to major components, fine and coarse aerosols contain also minor components that might be relevant in defining aerosol sources and toxicological properties, such as black carbon (BC) transition metals and aromatic or halogenated organic species (Fuzzi *et al.*, 2015).

A few studies highlight a correlation of mortality and morbidity in urban areas with ultrafine particle (UFP) concentration (e.g., Sioutas, Delfino, and Singh, 2005). UFP (particles with aerodynamic diameter smaller than 100 nm) represent less than 10% of fine aerosol mass in most urban locations but comprise more than 90% of particle number concentration. UFP in urban atmosphere are mainly emitted by vehicular road traffic, through incomplete combustion of petrol and diesel fuel, and are mainly composed by elemental carbon (EC) and organic compounds.

9.1.3 Aerosol Chemical Composition and Health Effects

Both fine and coarse aerosol particles are linked to health effects, although on different timescale exposure: fine particle effects are seen for both long- and short-term exposure, while coarse particles have been shown to have effects so far only from short-term exposure (Fuzzi *et al.*, 2013).

Toxicological and epidemiological studies begin to reveal potential mechanisms responsible for aerosol health effects (WHO, 2007), such as (i) obstruction of respiratory conducts and asthma, (ii) carcinogenic and mutagenic response, (iii) oxidative stress and consequent tissue inflammation, and (iv) translocation of nanoparticles (with sizes $a < 50$ nm) from the lungs to the circulatory system and from them to a number of organs.

The main chemical classes contributing to the health effects are expected to be soot (WHO, 2012), transition metals (Duvall, 2008), aromatic and halogenated compounds (Delfino *et al.*, 2010), organic carbon (Hu *et al.*, 2008), and primary biogenic aerosol particles (PBAPs) (Polymenakou *et al.*, 2008).

Soot is a component of fine PM and is emitted by combustion engines (especially diesel engines), residential heating (including wood and coal burning), energy production by heavy oil and coal combustion, waste burning, and forest fires. Depending on the methodologies used for its determination, soot is quantified as BC (through optical measurements) or EC (through thermooptical measurements).

Within aerosol particles, soot is associated with a large variety of organic species and metals coemitted during combustion. Reviews of available epidemiological studies reveal a link between short-term variation of BC concentration and short-term health effects, including all-cause and cardiovascular mortality and hospital admissions (WHO, 2012). Cohort studies show a relation between BC exposure and long-term health effects, such as all-cause and cardiopulmonary mortality (Figure 9.3). The link between traffic-related health effects and BC concentration is more robust than with PM concentration, indicating that BC is a good indicator of traffic emission harmfulness. Nevertheless, a review of toxicological studies shows that BC (or EC) might not be the major toxic component of PM, but may act as a carrier for more toxic substances emitted during combustion (WHO, 2012). Laboratory experiments have shown that organic compounds in both gasoline fuel and diesel engine exhaust can induce oxidative response in cell or are associated with oxidative potential (Hu *et al.*, 2008). In particular volatile organic species, likely secondary organic aerosol from condensation process, are characterized by the highest oxidative potential (Biswas *et al.*, 2009).

A few epidemiological studies investigated the health effects of aerosol-bound transition metals. These studies observed a correlation between some transition metals (e.g., Ni, V, Cr, and As) and hospitalization for cardiovascular and respiratory disease and cardiovascular mortality. Toxicological investigations

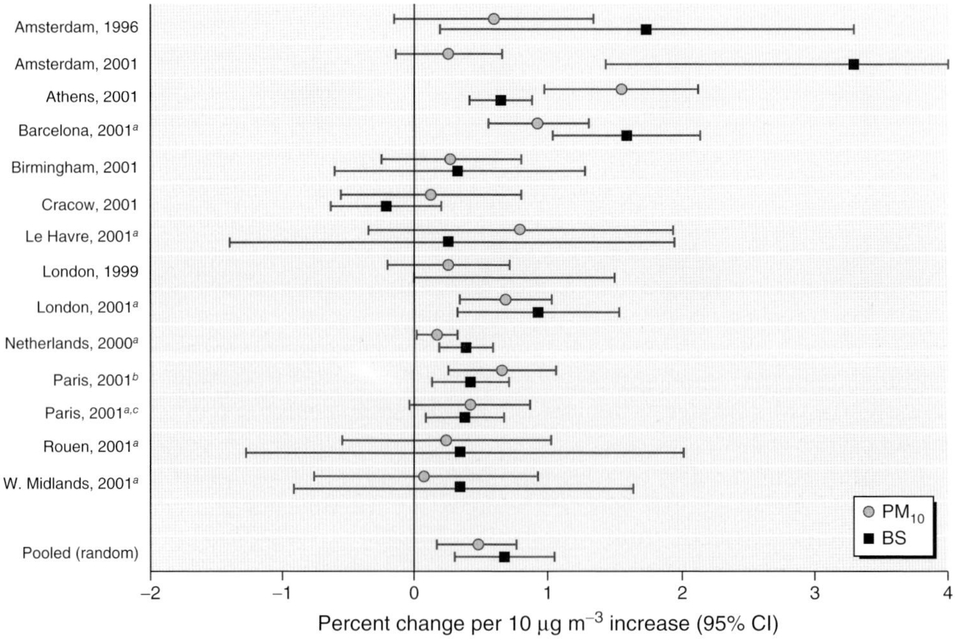

Figure 9.3 Single-city estimates of all-cause mortality for PM_{10} and black smoke (BS) (10 μg m^{-3} of BS correspond about to 1 μg m^{-3} of BC). Year indicates year of publication. (Janssen *et al.* (2011). Reproduced with permission from ProQuest LLC.)

link metals to increased particle oxidative potential and carcinogenic properties (Duvall, 2008). Finally, the health impact of metals was observed to depend on their solubility and bioavailability, which in turn is affected by particle age and processing (Harrison and Yin, 2000). Aromatic compounds, such as polycyclic aromatic hydrocarbons (PAHs), have been extensively investigated for their toxicity, mutagenic, and carcinogenic effects. In addition, aromatic compounds can be converted into oxygenated species, like quinones, which induce oxidative stress response in cells.

PBAPs, such as viruses, bacteria, pollen, spores, fungi, plant, and animal tissue debris, are associated with negative health impact, such as some allergies and diseases. PBAPs contribute to both fine and coarse PM and are present in rural and urban areas. Long-distance transport of PBAPs, allergens, and pathogens has been proposed, for example, during Saharan dust outbreaks over Europe (Polymenakou *et al.*, 2008). Combined exposure to pollen and air pollutants enhances allergic sensitization, whereas it is still unclear whether this sensitization arises from the impact of air pollution on the pollen on the exposed individuals.

Attention has also been dedicated to secondary inorganic aerosols (predominantly nitrate and sulfate salts), which make up a sizable fraction of $PM_{2.5}$ in many areas (Rohr and Wyzga, 2012; and references therein). Epidemiological studies have frequently found these to be associated with adverse health effects. This is in contrast to findings from toxicological laboratory studies where exposure to higher than ambient concentrations of sulfates and nitrates generally fails to elicit much of a biological effect (Kelly and Fussell, 2012). In ambient air, nitrates and sulfates possibly alter aerosol hygroscopicity, which may increase exposure to soluble transition metals and other toxic aerosol components.

9.1.4
Atmospheric Aerosols, Air Pollution, and Climate Change

Many aerosol pollutants, such as light-absorbing species (BC) and light-scattering species (sulfate and nitrate), not only deteriorate AQ with detrimental effects on human health but also contribute, with greenhouse gases, to climate change. Since the main sources of air pollutants are also sources of greenhouse gases, a synergic strategy should be adopted to mitigate climate change and improve AQ. Such a strategy should balance reduction of cooling and warming aerosol components, together with reduction of greenhouse gas emissions in order to meet climate targets. A wide range of policy options for reducing emissions of air pollutants and greenhouse gases in urban areas has been investigated. Concerning specific aerosol components, an important target is BC since it has both a clear warming climate effect and a negative impact on health. Recent studies demonstrate that for climate, BC is a target as important as greenhouse gases, like CH_4 and CO_2. In addition, BC deposition on ice and snow in the Arctic is responsible for positive radiative forcing and Arctic atmospheric warming. The recent program Climate and Clean Air Coalition (http://www.ccacoalition.org), initiated by United

Nations Environment Programme (UNEP) and World Meteorological Organization (WMO), is investigating and implementing emission reduction measures for BC and ozone that can lead to an improvement in AQ while also benefiting climate (Shindell *et al.*, 2012).

9.1.5
Aerosol Load in Different Areas of the World

Aerosol measurements have revealed air pollution problems in many of the major urban areas of the world. Some typical ranges of concentrations of the PM indicator found in a selection of cities around the world are summarized in Table 9.1. The highest concentrations of PM_{10} are found in Africa, Asia, and Latin America. Trends in AQ development differ in respect to the indicator pollutants. In Europe, PM_{10} levels had decreased by the end of last century but have tended to rise again, which may be partially explained by changing weather conditions. Even though large Asian cities have seen a slight reduction in PM_{10} levels over the last few decades, PM_{10} and $PM_{2.5}$ are still the major air pollutants in Asia. Many of the large cities in Latin America, such as Mexico City, still experience high levels of PM.

Air pollution in megacities around the world has been an issue for several years, since large conurbation exposes a high number of inhabitants to particularly elevated pollutant concentrations. There are 24 megacities, each with more than 10 million inhabitants, whose population is still increasing. Twelve of these megacities are located in Asia, four in Latin America, and two in Africa. According to the World Bank, some African cities are growing by more than 10% annually. In many megacities, such as Beijing, Calcutta, Mexico City, Rio de Janeiro, and Cairo, high levels of PM constitute a major problem. The serious consequences of exposure to high levels of urban ambient air pollution were made clear in the mid-twentieth century, when cities in Europe and the United States experienced air pollution episodes. Subsequent clean air legislation and actions reduced ambient air pollution in many regions. The winter smog problems associated with coal combustion

Table 9.1 Ranges of PM_{10} annual average concentration in a selection of cities in different regions of the world.

Region	PM_{10} (annual average concentration) ($\mu g\ m^{-3}$)
Africa	40–150
Asia	35–220
Australia/New Zealand	28–127
Canada/the United States	20–60
Europe	20–70
Latin America	30–129

that were common in some cities during the 1980s and early 1990s have been eradicated, and it is now mainly emissions from traffic that pose the main threat to good AQ. The previously frequent winter smog comprising a mixture of sulfurous compounds and particles (soot) has in this way changed over the years. Suspended particles, and especially submicron particles, combined with secondary pollutants such as oxides of nitrogen and ozone, have become a major problem in the large urban areas around the world. At the same time, the populations of the rapidly expanding megacities of Asia, Africa, and Latin America are increasingly exposed to levels of ambient air pollution that rival and often exceed those experienced in industrialized countries in the first half of the twentieth century (Krzyzanowski and Schwela, 1999).

Air pollution levels, including PM concentrations, are normally higher in developing countries than in highly developed industrialized countries. PM_{10} concentrations have been reported from countries such as India and Pakistan to be four to five times as high as the international standard values for AQ. Figure 9.4 shows the total suspended particle (TSP) and PM_{10} mass concentration in regional background sites located in developing countries.

Regional background concentrations in Asia can be as high as $100\,\mu g\,m^{-3}$ and are significantly higher compared to regional background concentrations in Africa and South America (Gilardoni, Vignati, and Cavalli, 2007). Asia experiences relatively high background concentrations owing to forest fires, local emissions of particles from the use of poor-quality fuels, and dust transport. A well-known springtime meteorological phenomenon throughout East Asia, causing the Asian dust, originates from windblown dust from the arid region of Mongolia and China and adds to the general level of PM in the region. Chinese cities experience very high airborne particle concentrations due to primary particles emitted from coal and biomass combustion and motor vehicle exhaust, as well as secondary sulfates formed by atmospheric chemical reaction of the sulfur dioxide emitted from coal combustion. Fine particles are responsible for most visibility problems in Asia. Recent measurements in Beijing show $PM_{2.5}$ concentrations averaging just over $100\,\mu g\,m^{-3}$. The monthly average concentrations varied between 61 and $139\,\mu g\,m^{-3}$. During air

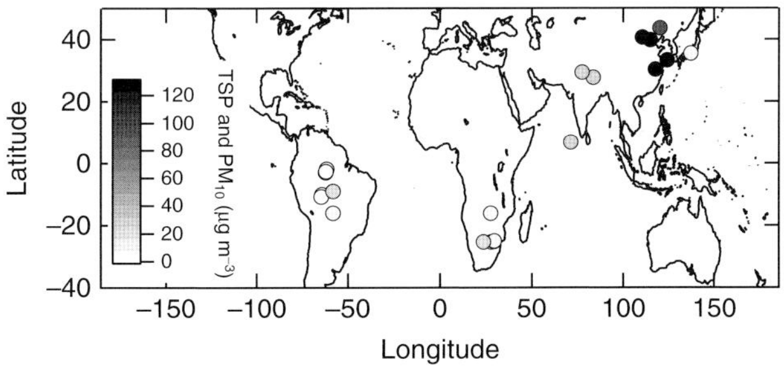

Figure 9.4 TSP and PM_{10} concentration in developing countries regional background sites.

pollution episodes, daily mean $PM_{2.5}$ values can reach 300 μg m^{-3} (Zheng *et al.*, 2005).

9.2
Aerosol Load as Derived from Satellite-Based Measurements

It is well known that AQ monitoring and guideline strongly depend on economic, political, and social factors. That is why satellite-based measurements of aerosol pollution are useful. In fact, satellite observations make possible to monitor the Earth on a global scale from a viewpoint of equal standard. Unique and outstanding space sensors with multidirectional observation of state of polarization, as well as radiation intensity, are planned for not only aerosol remote sensing but also understanding of air pollution from space (Kokhanovsky *et al.*, 2015).

9.2.1
VIS/NIR/SWIR Multispectral Satellite Observations for Evaluating PM Concentrations: An Example over the Northern Italy Area

In the recent past, the use of satellite observations for the assessment of AQ has advanced significantly (e.g., Wang and Christopher, 2003; Engel-Cox *et al.*, 2004; Al-Saadi *et al.*, 2005; van Donkelaar, Martin, and Park, 2006; Di Nicolantonio *et al.*, 2007; Gupta and Christopher, 2008; Di Nicolantonio, Cacciari, and Tomasi, 2009a; Di Nicolantonio and Cacciari, 2011). The increase in the capabilities of Earth observation satellites in remote sensing retrieval of anthropogenic and natural aerosols can, in fact, provide very useful information for improving estimates of spatial distribution and transport of PM and, therefore, for estimating population exposure to aerosols. Satellite data can be used to define maps of aerosol optical thickness (AOT) over large areas, which can be related to ground-level concentration of PM. Thus, data coming from satellite sensors can add very useful information, providing a direct estimate of PM concentrations. The importance of these evaluations is linked to the negative effects on human health of anthropogenic aerosol (respiratory problems and cardiovascular and lung diseases), which is mainly due to the exposure to fine particles (Pope, Ezzati, and Dockery, 2009). PM concentration satellite-based maps highlight areas affected by significant levels of pollution and thus those possibly exceeding legislation limits. Pollutant concentration can also be estimated in areas where no sampling sites are available.

In the next sections, method and results on the estimates of $PM_{2.5}$ concentrations based on moderate-resolution imaging spectroradiometer (MODIS) satellite aerosol optical properties (Level 2 data) retrieved from observations in the VIS/near-infrared (NIR)/SWIR spectral channels over the whole Northern Italy area are described. The method was implemented in a downstream preoperational service developed and demonstrated by CGS SpA Compagnia Generale per lo Spazio in the frame of MACC-I (Monitoring Atmospheric Composition

& Climate, 2009–2011) and Promote Air quality Services integrating Observations – Development Of Basic Localised information for Europe (PASODOBLE; 2011–2013) EC FP7 projects.

9.2.1.1 MODIS-Based PM Concentration Estimates at the Surface

The capability of MODIS sensors, on board Terra and Aqua/NASA platforms to retrieve aerosol optical properties (Levy *et al.*, 2007), is used in a semiempirical approach to estimate $PM_{2.5}$ content at the ground over a domain containing whole Northern Italy. In particular, this approach fuses MODIS/Terra and Aqua data, meteorological simulations, and ground-based data in order to provide maps of $PM_{2.5}$ concentration at the ground at the resolution of $10 \times 10\,km^2$ (Di Nicolantonio *et al.*, 2007). The analysis domain includes Po Valley, which is one of the most polluted and densely populated European areas. The particular orographic conditions of Northern Italy, characterized by the Alps on the western and northern sides and the Apennines on the southern side, often favor pollutant trapping processes within the ground layer of the atmosphere.

Bearing in mind these particular conditions, estimates of AOT over land can be regarded as being mainly due to PM atmospheric loading from anthropogenic activities (Bäumer, Rinke, and Vogel, 2008), especially in heavily polluted areas and in the absence of natural aerosol transport (such as desert dust) toward the monitored area. The relationship between AOT and PM at the surface varies with region and season as well as with vertical profiles and microphysical properties of aerosols.

Satellite data provide a measure of the abundance of aerosols in terms of AOT, as the integral of aerosol extinction from the ground to the top of atmosphere (TOA). Adopting the concept of aerosol scale height, H_a (m), defined as the ratio between AOT and the corresponding volume extinction coefficient determined at ground level, the aerosol mass concentration ρ (0) ($\mu g\, m^{-3}$) measured at the ground level for ambient relative humidity rh can be expressed as follows:

$$\rho(0) = \frac{\mathrm{AOT}(0.55\,\mu m)}{\sigma_{ext}(0) H_a}, \tag{9.1}$$

where $\sigma_{ext}(0)$ ($m^2\,\mu g^{-1}$) is the aerosol extinction cross section per unit mass at surface.

Focusing over Northern Italy, vertical aerosol features are described by two structures, which can be shown by aerosol profiles in the Po Valley area. In the first case, the aerosol is considered well mixed within the mixing layer, that is, up to mixing layer height H_{mix}, while in the second case, an exponential profile is characterized by a scale height H_a that can be well approximated by H_{mix}. Furthermore, it is important to take into account that PM concentrations are usually sampled at values of relative humidity rh < 50% and that aerosol optical properties differ considerably from those of dry air only in cases where ambient rh is above 60–70%, also as a function of aerosol chemical composition (Hänel, 1976; Hänel and Bullrich, 1978). Thus, the aerosol hygroscopic growth factor at surface, $F_{hyg}(rh; F_H)$, has to be also considered in estimating the aerosol extinction cross section at the surface, depending upon the increasing or decreasing phase of rh as well as on the Hänel growth factor F_H, varying with aerosol type (see, for instance, Randriamiarisoa *et al.*, 2006).

Applying a linear regression approach to Eq. (9.1) at reference wavelength $\lambda = 550$ nm, provided that (i) PM_f samplings (where f identifies PM type and in this analysis it stands for 2.5 or 10 μm) are available in different ground sites within the study area, (ii) the fraction of AOT due to fine particles can be set to be equal to the product $\eta_f \times$ AOT for $PM_{2.5}$ (being η_f the MODIS fine mode fraction (Levy *et al.*, 2007)), and (iii) H_{mix} and rh data are also available as corresponding data in spatiotemporal coincidence, the following expression can be assumed:

$$PM_f(i,j) = A_f(i,j)\frac{AOT(0.55\,\mu m)}{F_{hyg}(rh; F_H)H_{mix}} + B_f(i,j), \tag{9.2}$$

where $A_f(i,j)$ (μg m^{-2}) and $B_f(i,j)$ (μg m^{-3}) can be considered calibration parameters for the ith sampling site, jth month, and PM_f aerosol type. The slope coefficient parameter $A_f(i,j)$ in Eq. (9.2) is the inverse of the aerosol extinction cross section, varying with the chemical composition and microphysical properties of dry mass particles, and can be assumed as remaining almost constant for a fixed site and a certain month. Intercept parameter $B_f(i,j)$ is related to the nonlinear relationships between the physical and radiative properties of columnar and ground-level aerosols. Thus, $A_f(i,j)$ and $B_f(i,j)$ usually vary from one site to another and from one period of the year to another. They are estimated for each sampling site and each month through a procedure based on the selection of the best-fit regression cases characterized by (i) values of regression coefficient better than 0.7 and (ii) corresponding intercept $B_f(i,j)$ absolute values lower than 10% of the sampled PM_f monthly mean.

Once the calibration parameters for the sampling sites, months, and the PM analyzed are determined, a spatial interpolation function – based on an average weighted by the inverse of the square of the distance between each sampling site and a given geolocation (x, y) – is applied over the domain defining two calibration maps, $A_f(x, y; j)$ and $B_f(x, y; j)$, for each jth month. Employing these monthly calibration maps, estimates of $PM_f(x, y)$ concentration can be carried out using also AOT, H_{mix}, and rh, the first defined from the MODIS-derived data and the others simulated by the MM5 meteorological model (Grell, Dudhia, and Stauffer, 1994).

Summarizing, the method allowing to derive maps of MODIS-based PM concentration at the ground is characterized by two main stages. The first stage is the calibration, and it is aimed at the determination of monthly maps of calibration coefficients over the domain of interest. The following stage estimation generates the daily maps of satellite-based PM mass concentration at the ground corresponding to MODIS observations.

9.2.1.2 Data Set and Results

In the following it presented the data set – relative to Northern Italy and to 2004–2009 time period – to which the semiempirical method described in the previous section has been applied. Use is made of MODIS aerosol level 2 data (C005) provided by NASA at $10 \times 10\,km^2$ resolution both from Terra and Aqua observations. In particular, aerosol optical properties as AOT and η_f (which defines the contribution to AOT given by the fine-dominated (nondust) mode) are employed. The temporal corresponding meteorological characterization of the domain – at $12 \times 12\,km^2$ resolution – is also in input, defined by hourly fields of mixing layer height H_{mix} and rh at the ground as simulated by MM5 employing ECMWF analysis as boundary and initial conditions.

Furthermore, samplings of PM concentration at selected sites within the domain are employed to infer aerosol optical properties to PM conversion parameters on a monthly basis and, subsequently, to compare and validate the test-case performances. $PM_{2.5}$ concentration data set is relative to samplings carried out in whole 2004, summer 2007, and winter 2008 for a total of 25 sampling sites. Together with these, further data were taken into account relative to samplings in whole 2009 in 23 sites located in the Emilia–Romagna (ER) region. PM_{10} concentrations data set has the same features of the $PM_{2.5}$ data set, except for 2009 for which measurements from 49 sampling sites have been collected.

Each PM_f concentration data set was then parted into two subsets: the first employed for the determination of the calibration parameters and the second one used for comparing satellite-based results.

Then, on a monthly basis, for each PM type and for each calibration site, the corresponding spatiotemporal data of three data sets (satellite aerosol data, meteorological simulations, and PM samplings) were grouped to determine the best regression parameters $A_f(i,j)$ and $B_f(i,j)$ in terms of Eq. (9.2). As reported in the previous section, the spatial interpolation of calibration parameters over the whole region, on a monthly basis, leads to 12 pairs of maps of the $A_f(x,y;j)$ and $B_f(x,y;j)$ conversion coefficients for each jth month.

Examples of daily maps of MODIS-/Terra-based $PM_{2.5}$ concentrations for 17 July 2007, and for 10 February 2008, are presented in Figure 9.5. As it is typically expected, with respect to a summer day, during winter days, higher values of PM concentrations are registered together to a major number of missing MODIS aerosol data due to the higher presence of clouds.

Daily maps of satellite-based PM concentrations are calculated by averaging the MODIS-based PM estimates derived from observations coming both from Terra and Aqua platforms for the same day. The daily maps were then monthly averaged over the whole analyzed time period. Monthly averages from May to October 2008 are presented in Figure 9.6. A common feature in the spatial distribution of $PM_{2.5}$ concentration can be easily recognized in these maps and consisting in higher values in the area located north of the Apennines with respect to southern areas and confirming that the Po Valley is a polluted area also during summer months. Furthermore, passing from June to August, a decrease in particle concentration

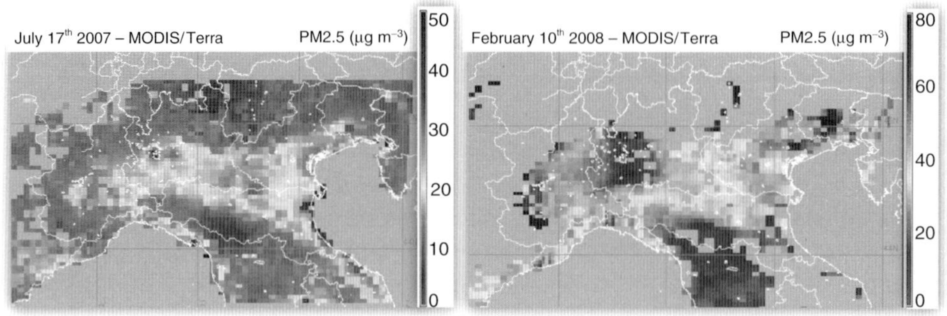

Figure 9.5 Example of daily MODIS/Terra-based $PM_{2.5}$ concentrations estimate for a summer day, 17 July 2007, and for a winter day, 10 February 2008. (Di Nicolantonio *et al.* (2009b).)

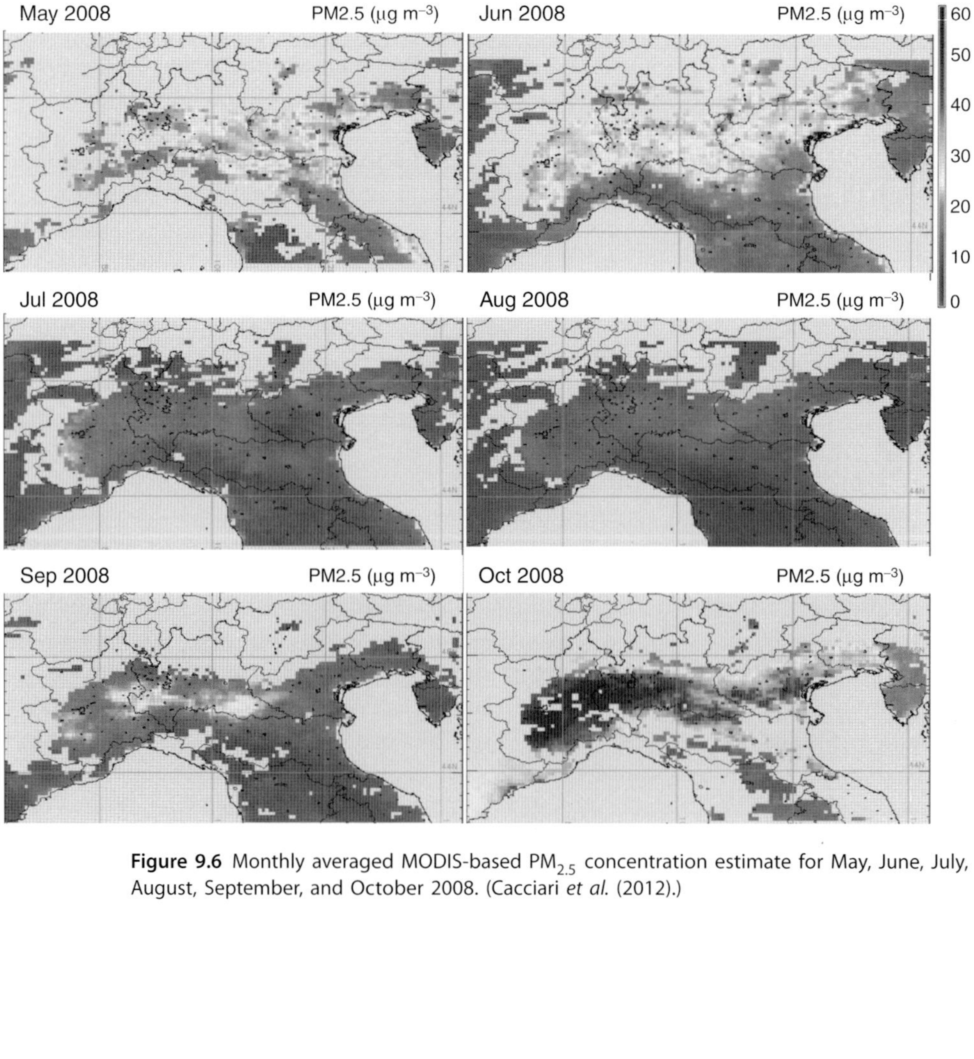

Figure 9.6 Monthly averaged MODIS-based $PM_{2.5}$ concentration estimate for May, June, July, August, September, and October 2008. (Cacciari *et al.* (2012).)

values can be seen over the whole domain followed by an increase in September and October. This can be ascribed to the main holiday season period and the consequent lowering of anthropogenic activities during summer (Minguzzi *et al.*, 2011).

As stated, two subsets of $PM_{2.5}$ and PM_{10} samplings were parted from the initial data set so as to provide the benchmarking data set. The comparison between MODIS-based PM values and PM concentration sampled at the ground has been made in terms of a scatter plot of daily values where MODIS-derived PM concentration estimates are set against corresponding *in situ* sampled PM concentration values. An example of the comparison for daily $PM_{2.5}$ is illustrated in the scatter plot of Figure 9.7. Employed data refer to 2008–2009 period and to 12 ARPA ground sites located in Lombardy and ER regions. The number of available corresponding daily samplings, 187, refers to the independent data set not employed in the calibration phase. A good agreement is found in this comparison: the linear best fit presents the Pearson product-moment linear correlation coefficient, slope, and intercept equal to 0.75, 0.986, and −1.61 μm m^{-3}, respectively.

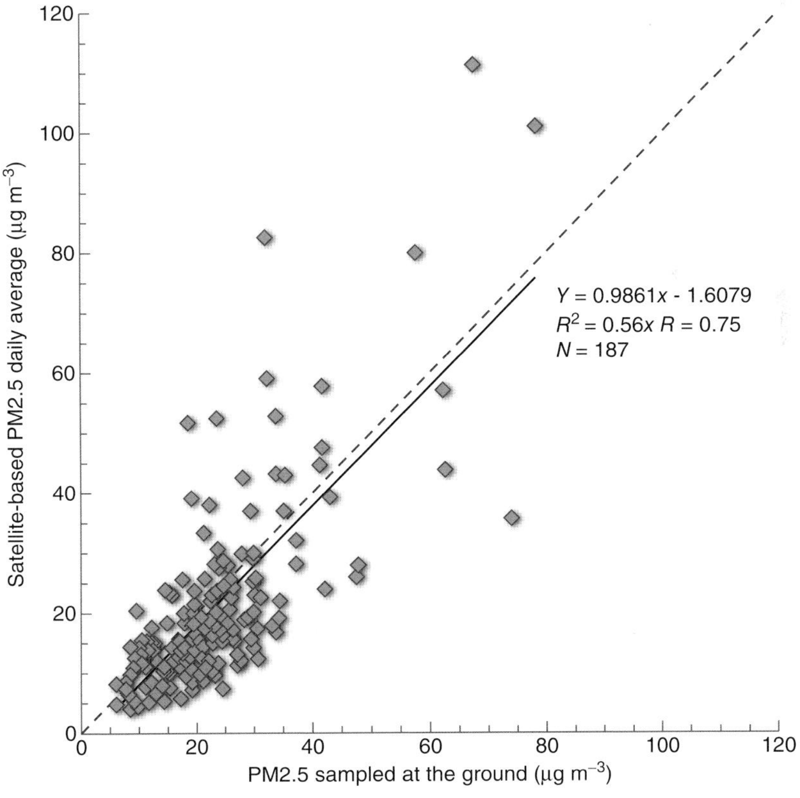

Figure 9.7 Comparison between daily MODIS-based (average of Terra and Aqua estimates) and ground-based $PM_{2.5}$ concentrations. The dashed line is the $y = x$ line.

Table 9.2 Results of the regression analysis between MODIS-derived $PM_{2.5}/PM_{10}$ concentration estimates and *in situ* $PM_{2.5}/PM_{10}$ and samplings (N_c indicates the number of samples available for the analysis).

PM	Comparison type	N_c	Slope	Intercept ($\mu g\,m^{-3}$)	Pearson's coefficient
$PM_{2.5}$	Daily	187	0.98	−1.6	0.75
	Monthly	112	1.05	2.2	0.82
PM_{10}	Daily	806	0.52	14.4	0.47
	Monthly	391	0.68	10.5	0.59

In addition to this, a comparison has been performed also for monthly averaged values. In Table 9.2, results of these comparisons are summarized in terms of regression parameters as slope, intercept, and Pearson's R coefficient both for $PM_{2.5}$ and PM_{10}. In the case of PM_{10}, 38 ARPA ground sites located in Lombardy and ER regions, for the same 2-year period, 2008–2009, were employed.

As can be noted from the parameters presented in Table 9.2, a good agreement is found both for the daily and monthly comparison in the $PM_{2.5}$ case, giving evidence of a significant improvement with respect to the previous performances of the method reported in Di Nicolantonio, Cacciari, and Tomasi (2009a) which was characterized by a tendency of satellite-based values to underestimate the $PM_{2.5}$ daily and monthly values by nearly 20%. This improvement can be ascribed to the use of ECMWF analysis as boundary conditions in simulating MM5 meteorological hourly field both for the calibration and processing phases, while in the previous analysis NCEP and ECMWF analysis were employed in the two different phases. Moreover, AOT to $PM_{2.5}$ calibration coefficient have been here determined making use of a wide data set, taking into account $PM_{2.5}$ samplings for the whole 2009.

Concerning PM_{10}, it should be taken into account here that these estimates represent the first and preliminary results obtained for MODIS-based PM_{10} concentration values. Results presented in Table 9.2 turned out to provide an underestimation of the MODIS-based values higher than about 30% both for daily and monthly values.

9.2.1.3 Satellite PM Multiannual Monitoring: Looking for Compliance to European Air Quality Directive

Daily MODIS-based maps related to the 2004–2009 period, both for $PM_{2.5}$ and PM_{10}, were then averaged on a yearly basis in the frame of a possible compliance monitoring of the legislation limit on the particle pollutant concentration set by the Directive 2008/50/EC on ambient AQ and cleaner air for Europe (EC Directive, 2008).

For monitoring spatial distributions of PM concentration, satellite-based averaged maps derived in this analysis could be very useful. For instance, they can be employed to locate areas characterized by high values of PM concentrations in

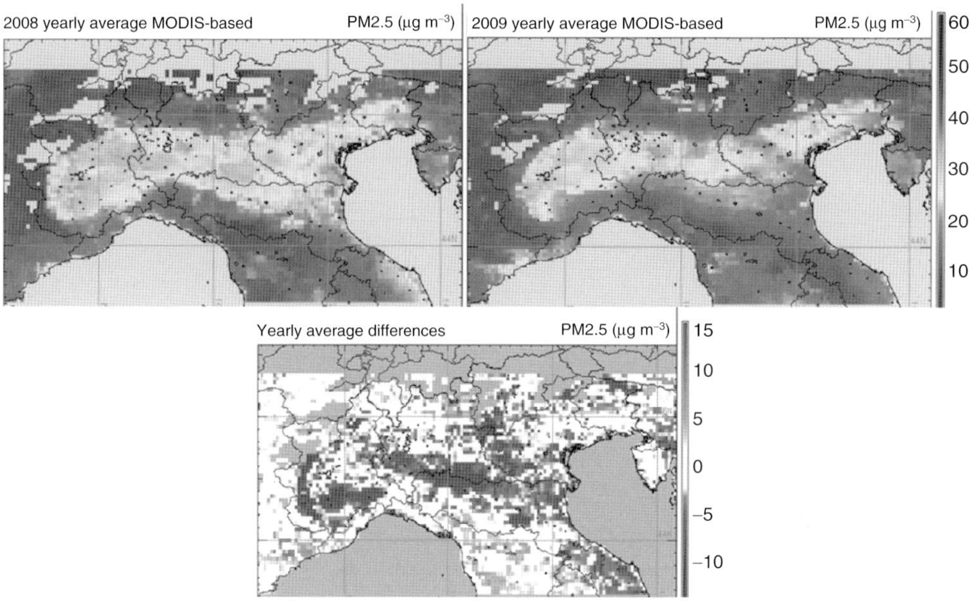

Figure 9.8 Annual averaged MODIS-based $PM_{2.5}$ concentrations estimate for 2008 and 2009 and (at bottom) map of relative difference (2009–2008). (Adapted from a graph of Cacciari, Di Nicolantonio, and Tiesi (2012).)

different periods of the year, as well as to evaluate the year-to-year variations in PM concentration.

With regard to the last aspect, the yearly averaged maps of MODIS-based $PM_{2.5}$ concentrations are illustrated in Figure 9.8 for years 2008 and 2009 together with the difference map between 2009 and 2008, providing the possibility to analyze the eventual annual trend. The difference map presents negative values in the Po Valley area, indicating that on a yearly basis 2009 PM concentrations in this area are lower than corresponding 2008 values, in accordance with the findings reported in the regional report on AQ written by the regional environmental agency ARPA Emilia Romagna (2012). In the maps shown in the upper part of Figure 9.8, locations showing yearly average values of $PM_{2.5}$ concentrations higher than the legal limit value of European directive, 25 or 20 $\mu g\,m^{-3}$, can be recognized.

An example of how the satellite-based PM_{10} maps may be used in support of AQ monitoring for analysis concerning the compliance to the EC directive is presented in Figure 9.9. The map in Figure 9.9a provides the 2007 yearly averages of PM_{10} MODIS-based values, with values ranging between 5 and 58 $\mu g\,m^{-3}$. In these maps each black pixel represents a location with PM_{10} concentration above or equal to the yearly legal limit of 40 $\mu g\,m^{-3}$.

The maps in Figure 9.9b show instead the distribution over the Northern Italy of the percentage of days over 1 year (2007) for which MODIS-based PM_{10} concentration exceeded the daily limit value of 50 $\mu g\,m^{-3}$. Values lower than

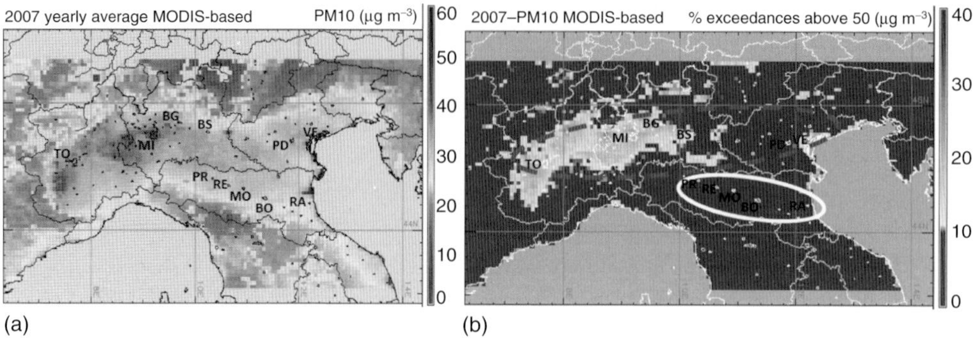

Figure 9.9 PM_{10} MODIS-based maps related to 2007 compliance to EC directive. (a) Yearly averaged values (legal limit: 40 μg m^{-3}). (b) Percentage of number of days over 1 year exceeding the daily limit concentration equal to 50 μg m^{-3}. The legal limit for the PM_{10} is 35 days per year, which corresponds to about 10% of the days in 1 year. (Adapted from a graph of Cacciari, Di Nicolantonio, and Tiesi (2012).)

10% (white in the map in Figure 9.9b) highlight areas for which the number of days exceeding the threshold concentration value is lower than the legal limit of 35 days in the whole year. Critical areas are instead represented by values higher than 10%, with colors, in the maps, ranging from black to gray as the percentage increases.

The percentage of days exceeding the PM_{10} daily limit calculated for 2007 from MODIS data (Figure 9.9) is compared to the percentage number of days exceeding the 50 μg m^{-3} limit value in PM_{10} concentration sampled at ground sites for several cities located in the Po Valley, as derived from the AirBase database managed by the European Environment Agency (http://www.eea.europa.eu/themes/air/airbase). Results of this comparison are presented in Figure 9.10.

Ground-based and satellite-based data turn out to provide a similar behavior presenting lower values for cities located in the middle of the Po Valley in the ER region (Parma, PR; Reggio Emilia, RE; Modena, MO; Bologna, BO; Ravenna, RA) and higher values for cities located in the northern part of the Po Valley (Torino, TO; Milano, MI; Bergamo, BG; Brescia, BS; Padova, PD; Venezia, VE; see also Figure 9.9). The difference between the percentage exceedances of these two Po Valley areas can be due, in particular, to the different (i) orography, being the cited northern cities located closer to the Alps which favor air mass persistence, and (ii) population density with significant anthropogenic activities, which are higher in the Lombardy and Veneto regions with respect to the ER region.

In this analysis it must be also taken into account that, with respect to the summer season, as the number of missing MODIS data is higher in winter when the PM ground concentrations are more often above the legal threshold, the satellite-based PM data generally underestimate the number of days registering the above threshold concentration values.

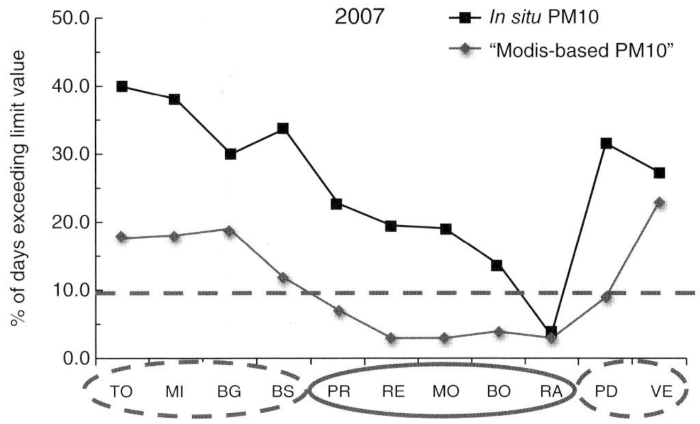

Figure 9.10 Annual percentage of exceedances over the 50 μg m^{-3} limit value in PM_{10} daily concentration sampled at the ground (solid squares) compared to corresponding MODIS-based PM_{10} annual percentage of exceedances (gray diamonds) for 11 cities located in the Po Valley. Satellite-derived data are extracted from the map of Figure 9.5b. (Adapted from a graph of Cacciari, Di Nicolantonio, and Tiesi (2012).)

9.2.2 PM Estimations over Osaka (Japan) Based on Satellite Observations

9.2.2.1 Introduction

Figure 9.11 presents two kinds of pictures taken in clear condition of atmosphere (low aerosol loading) and on a severe polluted day (high aerosol loading) in Osaka, Japan. A mountain illustrated in Figure 9.11a is clearly visible because the distance to the mountain from the view position is around 4 km away. Figure 9.11b shows the polluted case, and the mountain is difficult to see because aerosol particles scatter incident light. The pictures imply that light scattering is

(a) (b)

Figure 9.11 Pictures of clear and polluted days in Osaka, Japan (a view of Mt Ikoma from Kindai University).

important to understand the aerosol pollution and vice versa; we can retrieve the aerosol properties through the atmospheric scattering measurements.

Mie theory deals with light-scattering phenomena by a spherical particle. Van de Hulst (1957) did pioneer work to present the solution for Mie scattering from the numerical computational point of view. Further Deirmendjian (1969) took into account the condition of polydisperse particles. Multiple light scattering by particles is also considered. This process is called radiative transfer problem. The calculations of radiative transfer take into account the Rayleigh scattering by molecules and Mie scattering by aerosols in the Earth atmosphere model. Solving radiative transfer equation is required to understand the coupled Earth atmosphere–surface system. Chandrasekhar (1950) showed the basic equations, and van de Hulst (1980) gave the varieties of computational methods in his works, for example, method of successive orders, method based on Ambartsumian's principle of invariance, adding–doubling method, and so on. For understanding the aerosol pollution based on the ground and spaceborne measurements, a single scattering pattern of aerosols from Mie theory and radiative transfer simulations in the Earth atmosphere model are necessary. Poor AQ occurs when aerosol pollution reaches high enough concentrations, which are called aerosol pollution episodes associated with excessive loading of aerosols in the atmosphere. In the case of aerosol episode, the radiative transfer simulations become to be important. It is known that the extreme concentrations of aerosols in the atmosphere during episodes such as dust storms and biomass burning plumes can prevent aerosol monitoring with surface-level sun/sky photometers, whereas satellites can still be used to observe the Earth atmosphere from the space. This is why aerosol remote sensing with satellites is well known to be useful and effective. Figures 9.12 and 9.13 respectively present space-based images in the case of aerosol pollutions observed by SeaWiFS and Aqua/MODIS over East Asia.

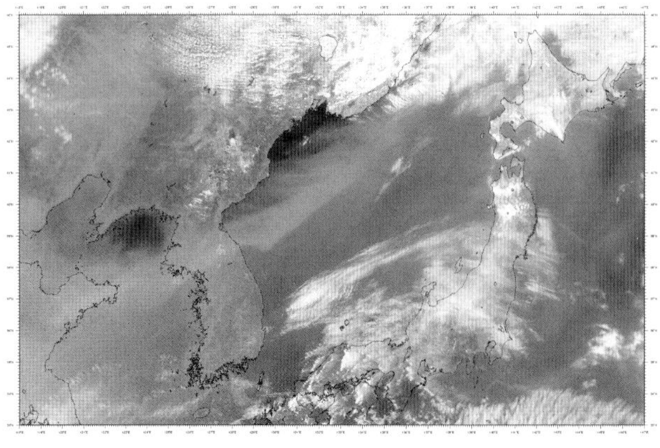

Figure 9.12 Sandstorm observed by SeaWIFS on 20 March 2001 over East Asia. (Cited from NASA Goddard Space Flight Center.)

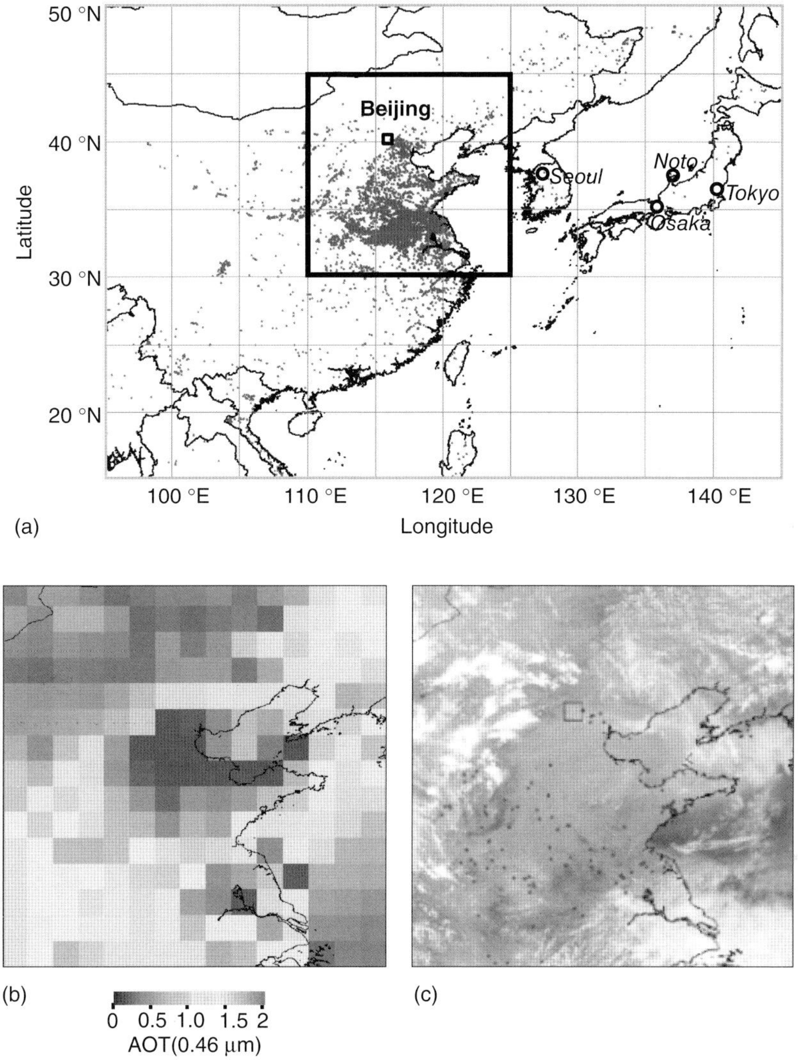

Figure 9.13 (a) Distribution of hot spots accumulated over June in 2012 in the East Asia (MCD14 Collection 5.1). (b) Monthly averaged AOT(0.46 μm) in June of 2012 over the East Central China (MYD08 Collection 5.1). (c) Biomass burning plume on 19 June, 2012 over the East Central China.

9.2.2.2 **Aerosol Remote Sensing**

Satellites are useful tools to perform Earth observations on a global scale with different spatial, spectral, and temporal resolutions. Retrieval of aerosol properties derived from spaceborne measurements has been discussed in many works. Some of the algorithms engaged by the author themselves are briefly shown here in order. Aerosol characteristics play important roles in all kinds of remote sensing products. At first atmospheric correction for the space-based ocean color imagery has been dealt with. It has been known that almost all of the satellite measurements are contaminated with scattered lights by molecular gases and aerosols in the atmosphere. As a result the observed signals are changed from the origins in all observing wavelengths. In other words, spectral information of radiation is modified due to light scattering by atmospheric particles. Then removal of the light-scattering effects from satellite data is necessary for the first step of satellite data usage. This process is called atmospheric correction. In the atmospheric correction, appropriate aerosol models should be set for radiative transfer simulations. Mukai *et al.* (1992) employed a method to estimate aerosol characteristics as comparison of lookup table (LUT) that stored the precalculated upward radiance at the TOA with database of satellite measurements (DB) (refer to Figure 9.14). There the coastal zone color scanner (CZCS) data in April and July of 1981 on Nimbus-7 satellite are examined. The results show the existence of dense aerosol particles around Japan especially in April. This result suggests such a possibility that the water-soluble aerosols have transported from the China continent in this seasonal pressure system, although those particles might be emitted within Japan.

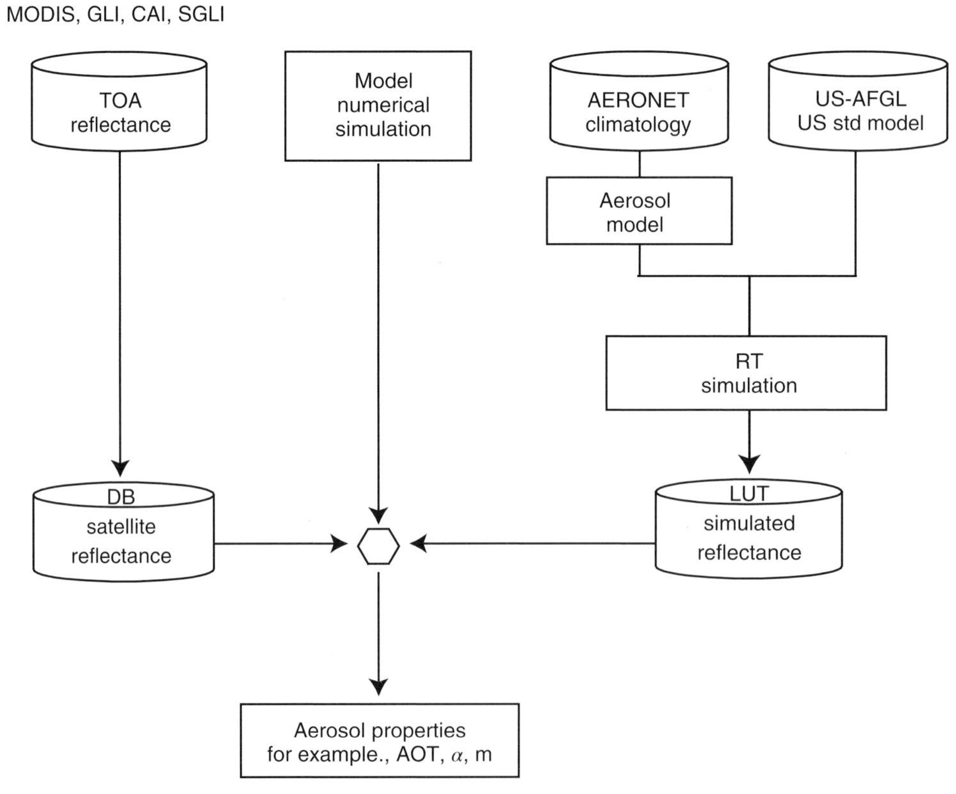

Figure 9.14 Block diagram for satellite-based aerosol retrieval.

In July, the Pacific high-pressure system has extended over the Japan islands, and hence clean air consisting of oceanic-type aerosols covers around Japan. These facts have been also shown by Higurashi and Nakajima (1999). They successfully retrieved two kinds of aerosol properties, that is, AOT and Ångström exponent (α) as size information on a global scale from the advanced very high-resolution radiometer (AVHRR) aboard NOAA satellite. The observation period is 1990, but the same trend as that from CZCS is confirmed.

It has been already mentioned earlier that the aerosol retrieval over ocean is performed by using NIR wavelength measurements because the radiance from the ocean itself (i.e., water leaving radiance) is small in this spectral region. On the contrary, the land surface is complicated, and the radiance reflected from the land is much brighter in visible and NIR wavelengths than that from the ocean. These facts lead to be difficult for aerosol retrieval over land. Herman *et al.* (1997) have proposed a new method to the retrieval of aerosols over land. They adopt two UV channel data of Nimbus-7/total ozone mapping spectrometer(TOMS) in order to extract the absorbing aerosol properties such as biomass burning plume and also mineral dust events. The French group has developed a new sensor, POLarization and Directionality of the Earth's Reflectances(POLDER), which is available for polarization measurements from space. It provides us with three Stokes parameters (I, Q, and U) at blue-, red-, and near-IR observing channels (Deschamps *et al.*, 1994). The polarized reflectance at TOA is composed of scattered light by atmospheric particles, that is, aerosols and molecules, and reflected light from the Earth's surface. The polarization signal from the ground is much smaller than that from the atmosphere. Deuzé *et al.* (2001) and Sano (2004) have retrieved the aerosol properties on a global scale by using POLDER polarization data. They showed that polarization information is suitable to detect the anthropogenic aerosols in particular. Most of these aerosols are due to the fire events or industrial activities. Dubovik *et al.* (2011) have developed a new inversion technique by using all of the advantages of POLDER measurements as multidirection, spectral reflectance, and polarization information to estimate various aerosol properties, that is, optical thickness, size distribution, complex refractive index, single scattering albedo, and so on. The retrieved results governed with this method provide excellent performances in comparison with others (Kokhanovsky *et al.*, 2010).

It is of interest to mention that large-scale air pollution is likely to continue because air pollution becomes severe due to both the increasing emissions of anthropogenic aerosols and the complicated behavior of natural aerosols. Mukai *et al.* (2012) and Mukai, Yasumoto, and Nakata (2014) propose that aerosol retrieval in the hazy atmosphere is achieved based on radiation simulation method of successive order of scattering (MSOS). They have shown that MSOS is available for such dense aerosol episodes as yellow sand storms or biomass burning plumes.

Figure 9.14 shows an example of flow diagram for aerosol retrieval as derived from satellite-based measurements. The retrieval process shown in Figure 9.14 is divided into three parts: satellite data analysis (S), model simulations (M), and radiative transfer calculations (R). Aerosol properties such as the AOT, refractive index (m), and size are estimated by comparing satellite measurements with the numerical values of radiation simulations in the Earth atmosphere–surface model. The model of the Earth's atmosphere is based on the AFGL code, which provides the aerosol and molecule distributions with height (Kneizys *et al.*, 1989). The multiple scattering calculations of radiative transfer take Rayleigh scattering by molecules and Mie scattering by aerosols in the atmosphere into account. The aerosol models are estimated from NASA/AERONET data (Dubovik *et al.*, 2002; Omar *et al.*, 2005).

Worldwide ground-based sun/sky photometry network, called AERONET, managed by NASA, has been established since early 1990s (Holben *et al.*, 1998). Nowadays, nearly 400 AERONET stations collect the aerosol information throughout a year. The network uses Cimel sun/sky photometers which take the direct sun measurements as well as sky radiance measurements by changing azimuth angles with fixed elevation angle. The accuracy of the measurements is maintained based on the comparison with the master instrument. Several master instruments are setting at Mauna Loa Observatory (MLO) in Hawaii to get the calibration constants. One of those instruments is shipped to the Goddard Space Flight Center (GSFC) of NASA in New York and transferred the obtained calibration constants into other instruments. Sky radiance sensor is also calibrated by using integrated sphere. The calibration procedure of each instrument is usually performed at least once a year. Moreover each instrument is checked and maintained in detail. For example, the blocking performance of each interference filter is evaluated and replaced if the performance is bad. The filter number has been registered in the internal AERONET database. Thus the accuracy of AOT at visible wavelengths is achieved, which is less than 0.01. The observed results at each AERONET site are transferred to NASA/GSFC. Then the data are processed which include inversion process to obtain the aerosol properties, for example, volume size distribution, complex refractive index, single scattering albedo, and many other parameters, based on the optimized algorithms (Dubovik and King, 2000). All of the measurements of aerosol properties are archived and released to the general public on the website of GSFC (http://aeronet.gsfc.nasa.gov).

9.2.2.3 Estimation of PM from Satellite-Based AOT

It is shown that AERONET ground-based sun photometric measurements give accurate aerosol information, particularly AOT. Smirnov *et al.* (2000) compared the columnar AOT measured by sun photometry with dust concentration by high volume air sampler which was measured at Barbados Island. Their results suggest a linear relationship between AOT and concentration of dust. This idea is promptly applied for estimation of the PM concentration from AOT measurements, because we have to consider health risk appraisal of PM then. Particulate air pollutants comprise material in solid or liquid phase suspended in the atmosphere. Such particles can be either primary or secondary and cover a wide range of sizes. Newly formed secondary particles can be as small as 1–2 nm in diameter ($1\,\text{nm} = 10^{-9}\,\text{m}$), while coarse dust and sea salt particles can be as large as

100 μm (1 μm = 10^{-6} m) or 0.1 mm in diameter. However, the very large particles have a short atmospheric existence, tending to fall out rapidly through gravity and wind-driven impaction processes. Thus in practice there are few particles in the atmosphere exceeding 20 μm in diameter, except in areas very close to sources of emission. PM can be separated from atmospheric gases by drawing air through a filter fine enough to retain the particles or by accelerating air through a jet that fires them at a fixed plate, onto which the particles impact and are collected. Particulate air pollutants have very diverse chemical compositions that are highly dependent on their source. They are also diverse in terms of particle size. PM_{10}, $PM_{2.5}$, and UFP fractions are typically those measured within the atmosphere for the purposes of health effects studies; the first two fractions are also used for compliance monitoring.

Wang and Christopher (2003) tried to derive the AQ index and $PM_{2.5}$ maps over Jefferson County in Alabama, United States, based on the AOT product derived from MODIS spectroradiometer observations made onboard the Terra and Aqua platforms. The correlations must vary with air humidity condition and vertical distribution of aerosols because the size of particles is increased by water uptake in the humid condition and decreased by water evaporation in the dry condition. The changing of particle size leads to the change of extinction cross section of each particle. As a result AOT values could be changed. van Donkelaar, Martin, and Park (2006) also estimate $PM_{2.5}$ on a global scale based on the MODIS AOT and also using calculated aerosol properties, which is a correction factor of AOT to $PM_{2.5}$ conversion, by chemical transport model (GEOS-CHEM) driven with assimilate weather model (NASA's GEOS-3). Model-based approach is available for global-scale applications. Thus global distribution of PM has to be considered.

Since the relationship between AOT and $PM_{2.5}$ has been also investigated by authors' group (Mukai *et al.*, 2007), the method and results are briefly introduced here. We have such an advantage that both instruments of AERONET and $PM_{2.5}$ sampler are set up at the roof of 50 m building and collocated within a 10 m^2 area. Then the measurements of AOT and $PM_{2.5}$ are made with the same ambient airs. The $PM_{2.5}$ measurements are taken by the method of β-ray gauge that gives the 1 h high temporal information against the daily averaged information. We can describe the linear relationship based on the classification by the first derivative of AOT with respect to wavelength (yielding the Ångström exponent α) and the second derivative (called α'). Three kinds of classification give the more accurate $PM_{2.5}$ estimation from AOT data. Further Sano *et al.* (2010) simplified the algorithm as two kinds of classes for applying the satellite AOT information, which takes into account the effect of dust particles in the $PM_{2.5}$ because some dust particles are smaller than the diameter of 2.5 μm. Also the dust is usually transported in high altitude; therefore coefficient of linear trend equation should be treated. Vertical profile of aerosol extinction is provided by the light detection and ranging (LIDAR) instrument. Sano *et al.* (2008) correct the aerosol's vertical profile effect by using NIES 2ch LIDAR measurements and demonstrate that the relationship between near-surface AOT and $PM_{2.5}$ gives better correlation than that without LIDAR measurements.

Here $PM_{2.5}$ retrieval over Osaka in Japan is demonstrated (Sano *et al.*, 2012). Osaka including large cities such as Kobe, Kyoto, and others is the second largest (15 million peoples) area in Japan. The cloud aerosol imager (CAI) dataset on board the greenhouse gases observing satellite (GOSAT) is chosen for practice. It observes the TOA reflectance at wavelengths of 380, 670 and 870 nm and 1.6 μm. The measurement at near UV is much darker than the visible wavelength, thus it is suitable for aerosol remote sensing over the land. Accurate ground surface reflectance is obtained from the time series data of atmospheric corrected reflectance based on the observational results of AERONET ground sun/sky photometry. This means that observed spectral AOT values and complex refractive index are used to correct the whole images. Note that we adopt the specific points of aerosol properties as representative of target region. In this case, the aerosol properties may change within the target area. So the minimum reflectance of the target region based on time series of atmospheric corrected reflectance is chosen to avoid the problem. Now the aerosol retrieval is performed based on the CAI measurements of 0.38 and 0.67 μm dataset, which gives AOT at 0.55 μm and Ångström exponent (α). The ground LIDAR measurement of 0.532 μm is used to apply the aerosol height correction of retrieved AOT. Then linear correlation of AOT and $PM_{2.5}$ is used to estimate the surface-level $PM_{2.5}$ concentration. Figure 9.15 shows the distribution of $PM_{2.5}$ concentration over Osaka from June of 2009 to January of 2010. Note the retrieval target area is only urbanized area. The concentration of $PM_{2.5}$ is low on 1 June and 17 September 2009. A high concentration appeared at center of Osaka City on 20 January 2010. Lightly polluted condition appeared from 23 October 2009, to 18 January 2010.

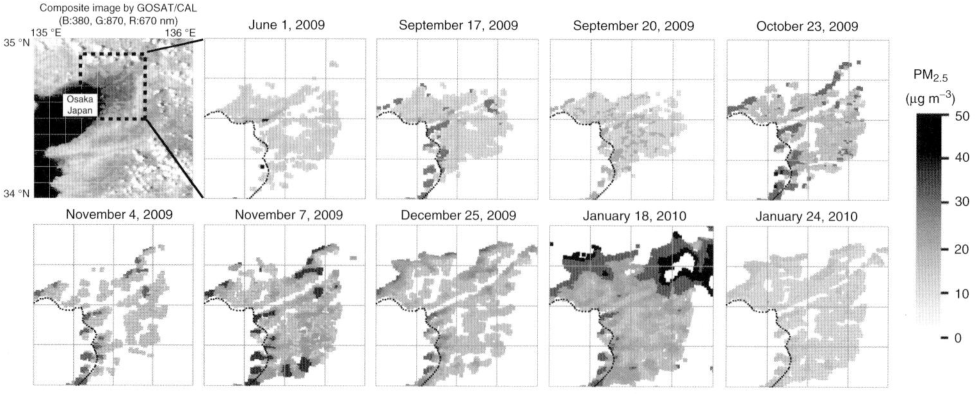

Figure 9.15 Contrast-enhanced composite image of GOSAT/CAI (B: 380, G: 870, R: 670 nm) in the upper-right figure and distribution of the retrieved concentration of ground-level $PM_{2.5}$ in Osaka, Japan. Target pixels are selected with the condition of low NDVI value <0.2 for processing Osaka City area. Most of the sky conditions are clear except on 18 January 2010, when the upper region is covered with some thin clouds.

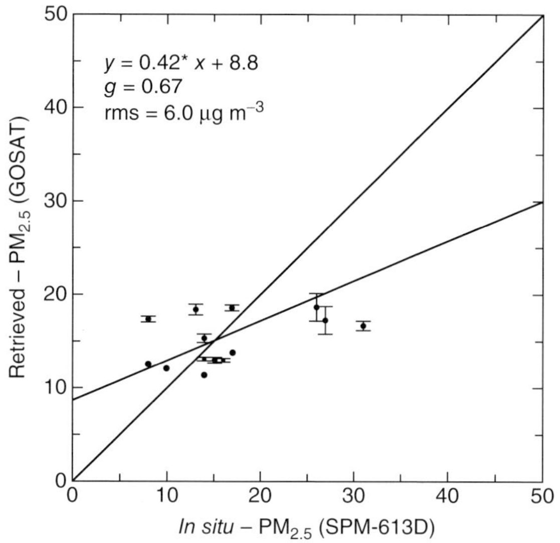

Figure 9.16 Comparison of retrieved $PM_{2.5}$ concentration derived from GOSAT/CAI with *in situ* instrument which is installed at ~50 m high roof from ground level.

In addition, note that the upper region is covered with some thin clouds on 18 January 2010.

Comparison of retrieved $PM_{2.5}$ concentration against the ground measurements is shown in Figure 9.16. The result indicates that the correlation coefficient is ~0.67.

9.3
Characterization of Mass Concentration and Optical Properties of Desert Dust in Different Areas of the Earth

Natural aerosol can contribute appreciably to the deterioration of AQ, producing not negligible effects on the human health. The most obvious example of these effects is given by the transport of desert dust, which can seriously affect AQ also at locations very far from the desert sources. The examination of the optical characteristics of desert dust made in various areas of the Earth clearly indicates that the aerosol microphysical properties can vary largely from one region to another, being closely related not only to the different morphological and mineralogical features of the mobilization areas but also to the distance from the source regions and the contamination of the original mineral dust during its transport with other natural (maritime and continental) and anthropogenic aerosol particles generated by biomass burning and industrial/urban activities.

Examining the AERONET ground-based measurements (Holben *et al.*, 1998, 2001) and the data retrieved from the AVHRR satellite observations

(Higurashi *et al.*, 2000), Takemura *et al.* (2002) determined the annual average global maps of monochromatic AOT(0.55 μm) produced by various aerosol types (including desert dust) and defined the dependence features of AOT(0.55 μm) on Ångström (1964) wavelength exponent (α) and aerosol single scattering albedo (ω) characteristics. These findings are in reasonable agreement with the spatial representations of the aerosol mass concentration derived from SPRINTARS[1)] data (Takemura *et al.*, 2000, 2002) and the results provided by different global aerosol transport models and satellite retrievals (Kinne *et al.*, 2003). The desert dust map drawn by Takemura *et al.* (2005) showed that high mineral dust mass concentrations are mobilized in the Saharan region and then transported toward the tropical Atlantic Ocean until reaching the coasts of Eastern United States. Important loads of Saharan dust are also episodically transported toward the Mediterranean Sea and Southern Europe under favorable meteorological conditions, more frequently in late winter and spring. High mass concentrations of desert dust were often monitored in the Arabian Peninsula and the most arid regions of Middle East. Strong desert dust transport events are sometimes observed over large areas of China, Korean peninsula, and Japan. All the aforementioned desert dust transport episodes can exert a strong influence on the ground-level characteristics of AQ, since wind-borne dust consists in general of various mineral aerosol components, often combined with water-soluble substances in cases where the aerosol particles are transported over long distances across oceanic and land regions (Hess, Koepke, and Schult, 1998).

9.3.1
Dust Storms in the Southwestern United States

Dust storm episodes frequently occur in the Southwestern United States (Arizona, New Mexico, and Texas), especially in summer during thunderstorm events, when winds move in a direction opposite to the storm's travel and from all directions into the thunderstorm. To improve the knowledge of these dust storms, Lei and Wang (2014) analyzed the meteorological characteristics of a set of historical dust transport episodes measured by means of AQ and satellite observations, finding that the dust mass concentrations captured by *in situ* measurements are consistent with the variations of AOT(0.55 μm) retrieved from local ground-based radiometer and satellite observations. Four typical dust storm cases were considered: (i) case D1, regarding a cold front-induced dust storm characterized by rapid processes and strong dust emissions; (ii) case D2, caused by meso- to small-scale weather systems and presenting a very high emission level; (iii) case D3, regarding a dust storm caused by tropical disturbances and yielding a stronger particle concentration of dust for an emission

1) SPRINTARS is a global aerosol transport-radiation model named "Spectral Radiation-Transport Model for Aerosol Species," which is coupled with an atmospheric general circulation model including a mixed layer ocean and simulates changes in the meteorological field due to the aerosol direct and indirect radiative effects produced by the main components of tropospheric aerosol.

time appreciably longer than those of the D1 and D2 cases; and (iv) case D4, pertaining to a dust storm triggered by very long cyclogenesis. Considering the hazardous effects of polluted aerosols on the human health, rigorous limits of $PM_{2.5}$ and PM_{10} were adopted by the National Ambient Air Quality Standards (NAAQS) of the US Environmental Protection Agency (EPA) for the particulate mass concentration measured at the AQ monitoring stations, equal to $35\,\mu g\,m^{-3}$ for the daily average $PM_{2.5}$ and $150\,\mu g\,m^{-3}$ for the daily average PM_{10}. The field measurements conducted in the previous four cases indicated that (i) case D1 was rather strong, yielding daily average values of $PM_{2.5} \approx 15\,\mu g\,m^{-3}$ and PM_{10} slightly higher than $100\,\mu g\,m^{-3}$, associated with a peak value of $AOT(0.55\,\mu m) \approx 0.30$; (ii) case D2 provided a value of $PM_{2.5} > 40\,\mu g\,m^{-3}$, therefore slightly exceeding the NAAQS/EPA limit and a value of $PM_{10} \approx 230\,\mu g\,m^{-3}$, which also exceeded the NAAQS/EPA limit, presenting a daily maximum value of $AOT(0.55\,\mu m) = 0.96$; (iii) case D3 was relatively weak, with $PM_{2.5} \approx 7\,\mu g\,m^{-3}$, $PM_{10} \approx 52\,\mu g\,m^{-3}$, and daily average $AOT(0.55\,\mu m) \approx 0.22$; and (iv) case D4 provided daily average values of $PM_{2.5} \approx 22\,\mu g\,m^{-3}$, $PM_{10} \approx 40\,\mu g\,m^{-3}$, and $AOT(0.55\,\mu m) \approx 0.29$. Considering only the three cases D1, D3, and D4, we obtained values of ratio $PM_{2.5}/AOT(0.55\,\mu m)$ ranging from 32 to $77\,\mu g\,m^{-3}$, which provided a daily average value of $(52 \pm 23)\,\mu g\,m^{-3}$, suggesting that the $PM_{2.5}$ limit established by NAAQS/EPA is most plausibly reached for $AOT(0.55\,\mu m) > 0.67$. Following a similar procedure, the ratio $PM_{10}/AOT(0.55\,\mu m)$ was estimated to vary from 140 to $325\,\mu g\,m^{-3}$, finding an average value equal to $234\,\mu g\,m^{-3}$, for which the NAAQS/EPA limit of PM_{10} is plausibly exceeded for $AOT(0.55\,\mu m) = 0.64$.

9.3.2
Saharan Dust Transport over the Southeastern United States and the Caribbean Region

Analyzing continuous daily aerosol samplings carried out at Miami (Florida, United States) over a 23-year period, Prospero (1999) showed that large amounts of African mineral dust are periodically carried in summer over the Southeastern United States, causing daily average surface-level particulate mass concentrations ranging usually from 10 to more than $100\,\mu g\,m^{-3}$. During these events, important loads of mineral dust are transported from the African source regions toward the downwind areas thousands of miles away, such as the wide Caribbean region. These dust transport episodes typically last several days or longer, being most intense in July. They were estimated to yield a monthly mean value of dust mass concentration equal to $16.3\,\mu g\,m^{-3}$ throughout the year, with relatively high monthly mean concentrations of $8.4\,\mu g\,m^{-3}$ in June and $9.8\,\mu g\,m^{-3}$ in August, such a considerable year-to-year variability being apparently linked to various meteorological factors including also climate conditions in North Africa associated with drought periods. Satellite data show that African dust incursions are synoptic-scale events, which can impact a large region of the Southern and Eastern United States. Coupled with inputs from local emissions, the dust transport events can produce important effects on the AQ parameters over the whole Atlantic Ocean. Half-hourly measurements of $AOT(0.67\,\mu m)$

were retrieved from the geostationary satellite imager GOES 8 data during the Puerto Rico Dust Experiment (PRIDE) conducted at Puerto Rico from 27 June to 24 July 2000, by Christopher *et al.* (2003), which obtained a monthly mean value of AOT(0.67 μm) equal to 0.26 ± 0.13. This estimate agrees very well with the simultaneous AERONET measurements conducted in this area by Holben *et al.* (1998), providing a daily mean value of α varying from 0.18 to 0.31, such rather low values being due to the predominant extinction by Saharan coarse particles.

9.3.3
Saharan Dust Transport over the Tropical Atlantic Ocean and the Western Coast of Africa

Regular measurements of AOT(0.50 μm) were conducted by Voss *et al.* (2001) using a ship-borne sun photometer during the Aerosols99 cruise conducted from 14 January to 8 February 1999, across the whole Atlantic Ocean from Norfolk (Virginia) to Cape Town (South Africa). Average values of AOT(0.50 μm) $= 0.10 \pm 0.03$ and $\alpha = 0.30 \pm 0.30$ were measured during the first phase of the cruise for maritime clean aerosol conditions without dust. On 25 January 1999, during an African dust event, average daily mean values of AOT(0.50 μm) $= 0.29 \pm 0.05$ and $\alpha = 0.36 \pm 0.13$ were measured, indicating that the average contribution given by the African dust load to the overall value of AOT(0.50 μm) was equal to around 0.19, while only weak variations of parameter α were produced by the dust load with respect to maritime aerosols. The location, the intensity, and the frequency of the dust transport events over the Atlantic Ocean vary largely with season and meteorological conditions occurring over the mobilization areas. As confirmed by the regular AERONET observations performed at Sal Island, in the Cape Verde archipelago, the largest dust concentration over the equatorial Atlantic Ocean is generally measured in the summer months (Holben *et al.*, 2001). Analyzing the AERONET datasets collected at Sal Island over the 6-year period from 1994 to 1999, Tanré *et al.* (2003) found that (i) AOT(0.55 μm) assumed relatively high monthly mean values throughout the year, varying from 0.4 to 0.7 in May–September and from 0.3 to 0.5 in January–February and (ii) exponent α assumed monthly mean values gradually varying from 0.25 to 0.45 in January–February to less than 0.2 in May–September, closely depending on the percentage coarse particle mass content. The combination of high values of AOT(0.55 μm) in summer with corresponding low values of α clearly indicated that the aerosol regime was dominated by the occurrence of frequent Saharan dust outbreaks and very weak extinction effects due to biomass burning aerosols. Conversely, the relatively high aerosol content observed in winter and associated with intermediate values of α can still be attributed to the combination of dust with significant contents of sea salt particles, sometimes mixed with polluted sulfate aerosols transported from the urban and industrial regions of Northern Africa and Southern Europe. Through an accurate study of the Saharan dust optical properties conducted during the Saharan Dust Experiment (SHADE) from 19 to 29 September 2000, in the Cape Verde Islands area near the west coast of

Africa and at the AERONET sites located at Sal Island (Cape Verde) and M'Bour (Senegal, 80 km south of Dakar), Tanré *et al.* (2003) measured the microphysical, chemical, and radiative parameters of aerosols using different *in situ* and airborne radiometric techniques. The daily mean values of AOT(0.67 μm) measured in Sal Island were found to slowly increase from September 23 to 25, until reaching a value close to 0.50 on September 25, and present a marked maximum of 1.70 in the early morning of September 26, followed by a sharp decrease to ~1.10 in the middle afternoon. On the same days, AOT(0.67 μm) was found to exhibit a daily mean value of ~0.40 on September 24 in M'Bour and to increase to ~1.50 at noon of September 25 and decrease to 1.20 in the late afternoon. The analysis of these particulate composition data showed that aerosols were only weakly contaminated by anthropogenic pollution and biomass burning fires, whereas the size distribution curves showed bimodal characteristics in most cases, with mode diameters a_c equal to around 0.40 and 1.5–2.0 μm, respectively.

9.3.4
Saharan Dust Transport Toward Southern Europe

Intrusions of Saharan dust northward to Europe are frequently observed during the period of the year from late winter to early summer. Analyzing an 11-year set of particulate samples collected from 1968 to 1978 at the Plateau Rosá mountain station (3480 a.m.s.l.) in the Western Alps, Prodi and Fea (1978) identified an overall number of 34 dust transport cases from the Saharan desert region. They were classified separately for three main transport mechanisms: (i) straight transport (25 cases), characterized by the south-to-north transport direction, a cold front acting on the mobilization area, and a warm advection moving from the south directly toward the Italian peninsula; (ii) transport over the Mediterranean with anticyclonic rotation from the Libyan coasts across the Eastern Mediterranean Sea, Aegean Sea, and Hellenic peninsula (six cases); and (iii) diverted transport with cyclonic rotation from the Western Saharan desert across the western part of the Mediterranean Sea and/or Iberian peninsula (three cases). These findings indicate that Saharan dust incursions take place most frequently across the Italian peninsula, due to the closer proximity of Italy to the most frequent mobilization areas, and that the orographic effects of the Apennines and the Alps may influence appreciably the circulation patterns enhancing cloud formation and precipitation scavenging by rain during the dust transport. The analysis of a pair of dust samples deposited by rains and collected by Tomasi, Prodi, and Tampieri (1979) at Sestola on the Apennines (Northern Italy) provided evidence of the multimodal characteristics of the Saharan dust size distribution curves and was found to consist of a first mode of nuclei/accumulation particles with mode diameter a_c ranging from 0.10 to 0.24 μm, a second mode of accumulation/coarse particles with a_c ranging from 0.90 to about 1.10 μm, and a third mode of coarse particles having mode diameter a_c ranging mainly from 3 to 4 μm.

Regular measurements of daily particulate mass concentration $PM_{2.5}$ were more recently conducted at the ARPA ER meteorological station, located at San Pietro

Capofiume (SPC) (Italy) during the two AEROCLOUDS project campaigns performed in summer 2007 (from May 22 to July 24) and winter 2008 (from February 6 to March 17). The time patterns of the daily values of $PM_{2.5}$ measured by the ARPA ER technicians during the two field campaigns are shown in Figure 9.17 and compared with the threshold value of $PM_{2.5} = 25\,\mu g\,m^{-3}$ established by the EC Directive 2008/50/(2008). The daily values of $PM_{2.5}$ determined during the summer 2007 campaign exceeded slightly the EC threshold value only on a few days of late May, in concomitance with the presence of Saharan dust layers over the Po Valley, and were higher than $25\,\mu g\,m^{-3}$ only in early June, because of strong anthropogenic aerosol emissions.

During the remaining weeks of June and July, the daily values of $PM_{2.5}$ varied mainly from 10 to $20\,\mu g\,m^{-3}$, in spite of some weak Saharan dust transport episodes recorded from mid-June to mid-July at the Mt. Cimone Observatory (MCO) (2165 m a.m.s.l.), located on the top of the Northern Apennines. Conversely, very high daily values of $PM_{2.5}$ were measured during the winter 2008 campaign, ranging from less than $10\,\mu g\,m^{-3}$ to more than $70\,\mu g\,m^{-3}$, mainly associated with emissions of anthropogenic/urban particles. Only on February 12, 17, and 29, traces of Saharan dust were sampled at the MCO station. Correspondingly, values of $PM_{2.5}$ equal to 35.0, 18.2, and $62.5\,\mu g\,m^{-3}$ were measured at the SPC station. On the other days of February and March, the daily values of $PM_{2.5}$ were found to exceed very often the EC limit, because of the local emissions of anthropogenic polluted aerosol.

Level 2.0 spectral measurements of AOT were routinely performed during the two AEROCLOUDS field campaigns by using a PREDE POM-02L sun/sky radiometer of the SKYNET network, equipped with seven narrow-band interference filters having peak wavelengths ranging from 0.340 to 1.020 μm. The best-fit values of exponent α were calculated over the 0.400–0.870 μm spectral range. The time patterns of the daily mean values of AOT(0.50 μm) and α(0.40–0.87 μm) measured during the two AEROCLOUDS field campaigns are shown in Figure 9.17 for comparison with the $PM_{2.5}$ data. The daily mean values of AOT(0.50 μm) measured in late May (with Saharan dust) varied between 0.30 and 0.34 and presented a peak value of 0.84 on May 26, while exponent α ranged between 1.01 and 1.32, presenting a minimum equal to 0.80 on May 26. The time patterns of AOT(0.50 μm) measured in June and July were characterized by sharp day-to-day variations due to Saharan dust transport events or anthropogenic pollution episodes giving daily mean values ranging from less than 0.10 in early July to more than 0.30 on July 18 (with Saharan dust). Correspondingly, α varied largely from about 0.5 (for Saharan dust) to more than 1.4 (for predominant anthropogenic pollution). During the winter 2008 field campaign, the daily mean values of AOT(0.50 μm) were found to increase gradually in February until reaching a value of 0.75 on 23 February 2008, and to vary largely in the subsequent weeks, from less than 0.10 to more than 0.50. In particular, values of AOT(0.50 μm) ranging from 0.16 to 0.57 were measured on February 12, 17, and 29 in the presence of Saharan dust over the Po Valley, while α assumed daily mean values equal to 1.41, 1.47, and 0.66, respectively.

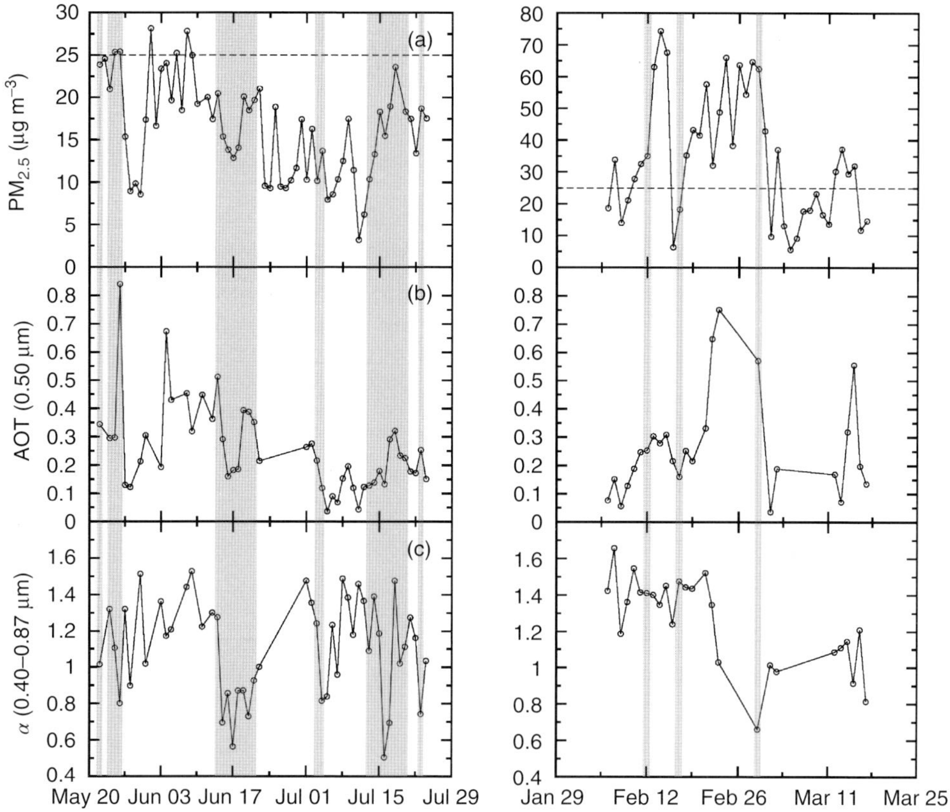

Figure 9.17 Left-hand side: In the upper part (a), time patterns of the daily values of particulate mass concentration $PM_{2.5}$ of fine particles measured by ARPA ER at the SPC meteorological station during the AEROCLOUDS experimental campaign conducted in summer 2007 (from May 22 to July 24). The horizontal dashed line marks the EC threshold value of 25 μg m^{-3}. In the central part (b), time patterns of the daily mean values of aerosol optical thickness (AOT) (0.50 μm) measured with the PREDE POM-02L sun/sky radiometer of the SKYNET network during the AEROCLOUDS summer campaign. In the lower part (c), time patterns of the daily mean values of the Ångström exponent α(0.40–0.87 μm) derived from the PREDE POM-02L spectral measurements. Right-hand side: as in the left-hand side, for the measurements conducted during the AEROCLOUDS experimental campaign in winter 2008 (from February 6 to March 17). The measurement days characterized by the presence of Saharan dust in the low atmosphere during the two field campaigns (revealed through the *in situ* sampling observations conducted at the MCO station on the Apennines) are marked with shaded thin vertical strips.

The daily values of $PM_{2.5}$ measured during the two AEROCLOUDS field campaigns are plotted versus the corresponding daily mean values of AOT(0.50 μm) to provide evidence of the different atmospheric extinction characteristics observed during the two campaigns. In summer 2007, about 40% of the measurement days were affected by Saharan dust transport. Thus, the least square line drawn to fit the scatter plot of the daily values of $PM_{2.5}$ versus AOT(0.50 μm) does not differ considerably from that obtained for the Saharan dust days only, the slope coefficient of the overall best-fit line being equal to $(54.5 \pm 6.6)\ \mu g\ m^{-3}$, and that of the Saharan dust line being equal to $(52.5 \pm 6.9)\ \mu g\ m^{-3}$. These two values of the slope coefficient imply that a value of AOT(0.50 μm) = 0.46 corresponds to the EC threshold value of $PM_{2.5} = 25\ \mu g\ m^{-3}$, as clearly shown in Figure 9.18, while a value of AOT(0.50 μm) equal to nearly 0.48 corresponds to the EC threshold value of $PM_{2.5}$ for Saharan dust particles only.

The least square line drawn for the scatter plot of daily values of $PM_{2.5}$ versus those of AOT(0.50 μm) obtained during the winter 2008 campaign was found to yield a best-fit slope coefficient equal to $(104.8 \pm 6.6)\ \mu g\ m^{-3}$ for this set of measurements conducted for a prevailing number of cases without Saharan dust transport episodes. Such a high value of the slope coefficient clearly indicates that the EC threshold value of $PM_{2.5} = 25\ \mu g\ m^{-3}$ is frequently exceeded in cases with AOT(0.50 μm) > 0.22 for atmospheric turbidity conditions dominated by anthropogenic urban and rural aerosols.

9.3.5
Saharan Dust Transport Toward the Middle Eastern and the Persian Gulf

East Mediterranean trajectories of dust-carrying storms from the Sahara were studied over several years by Yaalon and Ganor (1979), who estimated that about 25 Mtons of dust follow trajectories originated from the Libyan, Egyptian, Sinai, and Negev deserts to annually reach the East Mediterranean Basin mostly settling into the Mediterranean Sea. The Arabian Peninsula is also a very important source of atmospheric dust, since about 15–20 dust outbreaks per year are generally observed in this region. Examining the storm that occurred from 18 to 20 March 2012, Prakash *et al.* (2015) estimated that a very large area was involved by the storm, including the entire Middle East, from Northeastern Africa to Afghanistan and Pakistan and from the river valleys of the Tigris and Euphrates to the Red Sea and the Persian Gulf coasts. It is worth mentioning that strong interactions of polluted aerosols (from local refineries, factories, and fossil fuel combustion, as well as from the Indian subcontinent after transport over long distances) with mineral dust can often occur in the Persian Gulf area in summer, forming dust particles mixed with BC.

9.3.6
Asian Dust Transport Over Central Asia and China

Asian dust transport is a sporadic meteorological phenomenon involving large regions of Asia during the spring months. Dust particles originate mainly in the deserts of Mongolia, Northern China, and Kazakhstan, where high-speed surface winds mobilize large amounts of fine and coarse dry soil particles during intense dust storms. These aerosols are then carried by prevailing winds and transported over the Russian Far East, China, Korean peninsula, and Japan. In the last

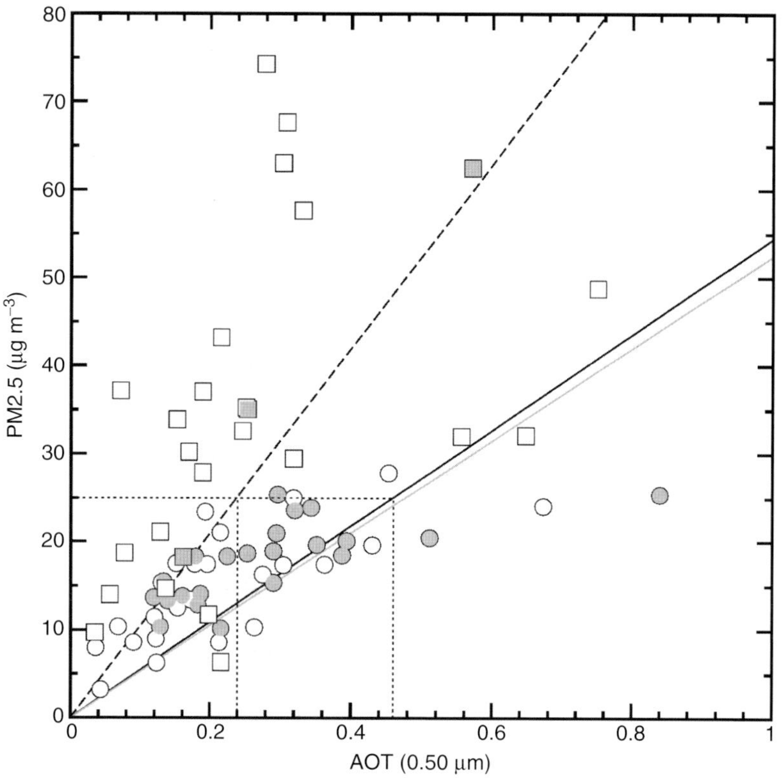

Figure 9.18 Scatter plot of the daily values of particulate mass concentration $PM_{2.5}$ versus those of aerosol optical thickness (AOT) (0.50 μm) measured by us at the ARPA ER meteorological station located at the SPC station during the two AEROCLOUDS experimental campaigns conducted in summer 2007 (circles) and winter 2008 (squares) on days without evidence of Saharan dust transport (open symbols) or on days in which Saharan dust transport episodes were monitored at the MCO station on the Apennines (marked with gray symbols). Three least square lines passing through the origin are drawn in the previous graph: (i) the solid line, determined for the overall set of the summer 2007 campaign data and found to have a slope coefficient equal to $(54.5 \pm 3.3)\,\mu g\,m^{-3}$; (ii) the dashed line, determined for the overall set of winter 2008 campaign measurement data and found to have a slope coefficient equal to $(104.8 \pm 12.7)\,\mu g\,m^{-3}$; and (iii) the gray line, calculated for the 25 days of the summer 2007 campaign on which Saharan dust transport was observed and found to have a slope coefficient equal to $(52.5 \pm 4.5)\,\mu g\,m^{-3}$. The horizontal dotted lines are drawn to intercept the dashed and solid lines and determine the threshold values of AOT(0.50 μm) equal to 0.24 and 0.46, respectively.

decades, Asian dust has become a serious problem due to the increase of industrial pollutants contained in the dust and intensified desertification in China, causing longer and more frequent dust mobilization occurrences. Such an intensification was clearly observed in the last few decades, when the Aral Sea of Kazakhstan and Uzbekistan started drying up due to the diversion of the Amu River and Syr

River following a Russian agricultural program to irrigate Central Asian deserts, mainly for cotton plantations. For this reason, a series of field experiments were recently conducted in Central Asia and China over the last 10 years in order to study the optical and microphysical characteristics of dust particles and evaluate their radiative effects on climate.

Regular sun photometer measurements of AOT(0.50 μm) were performed by Chen *et al.* (2013) at Bishkek (42.68N, 74.69E, 1750 m a.m.s.l.) and the so-called LIDAR (not far from Karakol, 42.47N, 78.53E, 2000 m a.m.s.l.) stations in Northern Tien-Shan (Kyrgyzstan) from July 2008 to June 2009. The measurements were analyzed in order to establish statistical relationships between AOT(0.50 μm) and $PM_{2.5}$ and PM_{10} mass concentrations. A portable Microtops II sun photometer was used to measure AOT(0.50 μm), while routine measurements of continuous hourly $PM_{2.5}$ concentration were conducted together with daily sampling measurements of $PM_{2.5}$ and PM_{10}. A cloud screening procedure based on LIDAR observations was applied to the sun photometer data in order to remove the measurements taken on cloud-impacted days. A good correlation was found between $PM_{2.5}$ and AOT(0.50 μm) in spring and autumn, associated with the transport of Asian dust from various source areas, among which the most important is the Taklimakan Desert in Northwestern China. Yearly average values of AOT(0.50 μm) equal to 0.195 ± 0.05 and 0.185 ± 0.05 were measured at Bishkek and LIDAR sites, respectively, while the annual mean $PM_{2.5}$ was found to be equal to $11.65 \pm 3.7\ \mu g\ m^{-3}$ at Bishkek and $9.95 \pm 3.4\ \mu g\ m^{-3}$ at the LIDAR site. Typical values of AOT(0.50 μm) and $PM_{2.5}$ varied during the warm season from 0.1 to 0.2 and from 5 to 25 $\mu g\ m^{-3}$, respectively. The monthly mean values of $PM_{2.5}$, PM_{10}, and AOT(0.50 μm) measured from July 2008 to June 2009 at the two sites presented very similar annual variations with pronounced maxima occurring at the end of summer and in early autumn, mainly due to seasonal factors and the strong impact of long-range desert dust transport. Field measurements of the particle size distributions showed that Asian dust contains a large fraction of particles having sizes ranging from 1 to 20 μm and forming a significantly skewed mode toward the supermicron sizes. Coarse particles not only dominate the size distribution during dust transport events but also influence appreciably the $PM_{2.5}$ concentration.

A retrieval method of the main aerosol optical parameters from simultaneous ground-based measurements of sky radiation and solar irradiance was applied by Kim *et al.* (2005) to the multiyear data sets collected at the Mandalgovi, Dunhuang, and Yinchuan sites of the SKYNET sun/sky radiometer network in China. The spectral values of AOT obtained from these measurements were large compared with those generally measured in North America or Europe, since the overall atmospheric aerosol extinction was produced by aerosols of various origins, such as urban pollution, biomass burning aerosols, and desert dust. Significant temporal variations were also observed throughout the year at the three SKYNET sites, especially at Dunhuang, where large extinction effects were caused by the dust particles transported downwind from the arid/desert source regions in Northern China and Mongolia. In particular, it was found in Dunhuang that the fine particle mode is very weak throughout the whole year, where the coarse mode is considerably more marked due to the continuous transport of dust particles from the nearby eastern edge of the Taklimakan Desert. The analysis of the previous SKYNET data showed that single scattering albedo ω presents lower average values than those evaluated for the same aerosol types in other areas of the world. In fact, ω was estimated to be equal to 0.89 for Asian dust in Dunhuang and 0.90 for urban aerosols in Yinchuan, clearly indicating that aerosols over East Asia absorb comparatively more solar radiation than in the other areas of the Earth.

The Atmosphere Radiation Measurements Program's Ancillary Facility (AAF/SMART-COMMIT) was deployed during the period of late April to mid-June in 2008 at Zhangye (39.08N, 100.28E, 1461 m a.m.s.l.), located in a semidesert area of Northwestern China, at the south edge of Gobi Desert. Field measurements of dust AOT, Ångström exponent α, size distribution curves, and spectral dependence features of single scattering albedo $\omega(\lambda)$ and asymmetry factor $g(\lambda)$ were conducted by Ge *et al.* (2010) using a multifilter rotating shadow-band radiometer during some dust transport episodes from the Gobi Desert, where dust and loess particles are widely spread. They are mobilized in the late winter and early spring of each year, when strong winds caused by cold air outbreaks lift surface crustal material into the free troposphere and transport the dust particles to downwind regions, from the Asian continent to West Pacific Ocean. Eleven cases of dust transport were observed, yielding values of AOT(0.67 μm) ranging from 0.07 to 0.25, exponent α ranging from 0.34 to 0.93, $\omega(\lambda)$ increasing from 0.76 ± 0.02 at wavelength $\lambda = 0.415$ μm to 0.86 ± 0.01 at $\lambda = 0.870$ μm, and $g(\lambda)$ decreasing from 0.74 ± 0.04 to 0.70 ± 0.02 over the same spectral range. In particular, the range of the values of AOT(0.67 μm) measured in Zhangye was from 0.08 to 3.10 during the experiment, the highest values (>1.0) being caused by severe dust storm outbreaks and the major part of data (75%) being in the range from 0.08 to 0.48, mainly due to anthropogenic pollution.

The aerosol optical properties and their associated radiative effects were investigated by Liu *et al.* (2011) by means of systematic sky radiometer and surface solar radiation measurements conducted at the Semi-Arid Climate and Environment Observatory of Lanzhou (SACOL) University (36.05N, 103.86E) in the Gansu province (central China) over the March–May 2009 period. The measurements provided a seasonal mean value of AOT(0.50 μm) = 0.40, the monthly mean values being equal to 0.50, 0.42, and 0.28 in March, April, and May, respectively. The largest value recorded in March is due to the frequent dust events observed over the Loess Plateau. Single scattering albedo ω(0.50 μm) was found to vary significantly during the spring, ranging from 0.82 to 0.98, the average value for background aerosol being equal to 0.90 and that measured during the dust event outbreak period being 0.87 which is slightly lower because of the predominant scattering effects by coarse particles.

As a fast-developing country covering a large territory, China is experiencing rapid environmental changes. As mentioned earlier, the East Asian Study of Tropospheric Aerosols: An International Regional Experiment(EAST-AIRE) experiment was conducted in order to characterize the physical, optical, and chemical properties of the aerosols over China and study their effects on climate. The measurements carried out by Li *et al.* (2007) provided some preliminary results using continuous high-quality measurements of aerosol, cloud, and radiative quantities made at the first EAST-AIRE project baseline station at Xianghe, about 70 km east of Beijing over a period of 1 year (September 2004 to September 2005). The region was found to be often covered by a thick layer of haze, with yearly mean AOT(0.50 μm) = 0.82 and a daily mean maximum greater than 4, primarily due to strong anthropogenic aerosol emissions. For anthropogenic pollution conditions similar to those observed at Xianghe, $PM_{2.5}$ samples were collected by Yu *et al.* (2011) at the Beijing urban site during a 3-year period from spring 2005 to spring 2008, with the purpose of explaining the meteorological and source-driven dynamics of $PM_{2.5}$. Special attention was paid to seven well-identified episodes of extremely high $PM_{2.5}$ concentrations, in which anthropogenic pollution was superimposed by dust storms occurring especially in spring. Nearly 60% of the high-pollution $PM_{2.5}$ cases were measured more frequently in spring and winter and occurred during typical dust intrusion episodes, which invaded Beijing from March to May. However, high levels of $PM_{2.5}$ were also observed in summer and winter, because of strong emissions from primarily anthropogenic sources, which exceeded the value of 75 $\mu g\,m^{-3}$ over 35% of time.

9.3.7
Asian Dust Transport Over Korea and Japan

In East Asia, a mixture of industrial pollution and mineral dust influences large areas of Korean peninsula, Japan islands, and Northern Pacific Ocean and may even reach the North American continent, especially during the spring season. Kim *et al.* (2005) analyzed the SKYNET measurements regularly conducted at the Anmyon and Gosan sites in South Korea and at Amami Oshima in Japan to define the aerosol optical characteristics and found that the average single scattering albedo ω of Asian dust decreases gradually during its transport from the East Asian source regions toward Korea and Japan. The measurements taken in April provided values of ω equal to 0.86 at Anmyon, 0.84 at Gosan, and 0.80 at Amami Oshima, which are all appreciably lower than the mean value of 0.89 measured in Dunhuang. These results obtained over the Korean peninsula and the East China Sea suggest that Asian dust becomes gradually more absorbing during its transport phase because of its mixing with soot particles generated in the industrial/urban areas of China.

The transport of mineral dust, its vertical distribution, and the associated meteorological conditions were deeply investigated by Kim *et al.* (2010) during an Asian dust event which occurred in Seoul on May 29–31, 2008. The field measurements performed with ground-based and spaceborne two-wavelength polarization LIDARs, particulate mass sampling instruments, and surface meteorological stations showed that the dust particles were transported from the Inner Mongolia, Man-Ju, and Ordos source regions toward Korea passing over the northeastern sector of China. The numerous observations of PM_{10} and meteorological parameters made at some Korean sites revealed that the major fraction of the heavy dust particles transported across South Korea move most frequently from the northwest to southeast and are characterized by the prevailing presence of spherical and relatively small particles.

9.3.8
Desert Dust Transport Over Oceanic Areas

Multiple satellite data sets in conjunction with the simulations made using the Monte Carlo Aerosol–Cloud Radiation (MACR) model were employed by Zhu *et al.* (2007) to determine the climatological distributions of dust plumes over the Pacific, Indian, and Atlantic oceans and evaluate their radiative impacts. Three target regions (Yellow Sea, Arabian Sea, and Saharan Coast) were examined for quantitative comparisons of the dust properties. Twenty-year averages of AVHRR AOT data clearly showed that the peak dust season was March–April–May for the dust plumes over the Pacific Ocean and Yellow Sea and June–July–August for those over the Arabian Sea and Saharan coasts. Evaluations of the dust fraction with respect to the maritime aerosol content were obtained using the Goddard Global Ozone Chemistry Aerosol Radiation and Transport (GOCART) model and MODIS data. The vertical distribution of dust concentration was studied by examining the Stratospheric Aerosol and Gas Experiment (SAGE) II aerosol extinction data, revealing the existence of a pronounced peak at around 4 km altitude over the Arabian Sea, Indian Ocean, and Saharan coasts of the Atlantic Ocean, while a multilayered dust plume structure was often observed over the Yellow Sea.

Abbreviations

AQ	air quality
ARPA ER	Regional Agency for Prevention and Environment of Emilia Romagna (Italy)
AVHRR	advanced very high resolution radiometer aboard NOAA satellite
BC	black carbon
BS	black smoke
CZCS	coastal zone color scanner aboard the Nimbus-7 satellite
EC	elemental carbon
MCO	Mt. Cimone Observatory
PAHs	polycyclic aromatic hydrocarbons
PBAPs	primary biogenic aerosol particles
PM	particulate matter
POLDER	POLarization and Directionality of the Earth's Reflectances satellite sensor
SPC	San Pietro Capofiume meteorological station
TOA	top level of the atmosphere
TOMS	Total Ozone Mapping Spectrometer aboard the Nimbus-7 satellite

List of Symbols

Latin

a	aerosol particle size (or diameter)
a_c	mode diameter of an unimodal size-distribution curve of aerosol particles
AOT(λ)	monochromatic aerosol optical thickness
AOT(0.67 μm)	aerosol optical thickness retrieved from the geostationary satellite imager GOES 8 data recorded using the 0.67 μm spectral channel
AOT(0.46 μm)	monthly average aerosol optical thickness calculated in June 2012 over the East Central China examining the MYD08 Collection 5.1 data set
AOT(0.50 μm)	aerosol optical thickness measured at wavelength $\lambda = 0.50$ μm using an example of sun photometer (PFR and Microtops II models) or sun/sky radiometer (like the PREDE POM-01 and POM-02L models of the SKYNET network and the CIMEL CE 318 models of the AERONET network)
AOT(0.55 μm)	aerosol optical thickness measured at the 0.55 μm wavelength by means of CIMEL CE 318 sun/sky radiometers of the AERONET network or derived from satellite-based AVHRR imager observations
$A_f(i,j)$	slope coefficient adopted in Eq. (2.2) as calibration parameter for the ith sampling site, the jth month, and PM_f type of particulate matter
$B_f(i,j)$	intercept parameter adopted in Eq. (2.2) for the ith sampling site, the jth month, and PM_f type of particulate matter
$F_{hyg}(rh; F_H)$	aerosol hygroscopic growth factor at the surface
F_H	Hänel growth factor
$g(\lambda)$	asymmetry factor of aerosol particles as a function of wavelength
H_a	aerosol scale height (measured in meter or kilometer)
H_{mix}	mixing layer height (measured in meter or kilometer)
PM_{10}	24-hour average mass concentration of particles having aerodynamic diameter <10 μm
$PM_{2.5}$	24-hour average mass concentration of particles with aerodynamic diameter <2.5 μm
$PM_f(i,j)$	particulate matter 24-hour average mass concentration of aerosol particles with sizes $a > f$ for the ith sampling site, the jth month, and PM_f type of particulate matter
rh	relative humidity of air

Greek

α	Ångström wavelength exponent in the visible and near-infrared

$\alpha(0.40-0.87\ \mu m)$	Ångström wavelength exponent measured over the 0.40–0.87 μm wavelength range
η_f	MODIS fine particle mode fraction
λ	wavelength (measured in μm)
$\rho(0)$	aerosol mass concentration measured at the ground level (in $\mu g\ m^{-3}$)
$\sigma_{ext}(0)$	surface-level aerosol extinction cross section per unit mass (measured in $m^2\ \mu g^{-1}$)
ω	average single-scattering albedo of an aerosol particle polydispersion over the visible wavelength range
$\omega(\lambda)$	single-scattering albedo of aerosol particles at wavelength λ
$\omega(0.50\ \mu m)$	monochromatic single-scattering albedo at the 0.50 μm wavelength

References

Al-Saadi, J., Szykman, J., Pierce, R.B., Kittaka, C., Neil, D., Chu, D.A., Remer, L., Gumley, L., Prins, E., Weinstock, L., MacDonald, C., Wayland, R., Dimmick, F., and Fishman, J. (2005), Improving national air quality forecasts with satellite aerosol observations. *Bull. Am. Meteorol. Soc.*, **86**, 1249–1261. doi: 10.1175/BAMS-86-9-1249

Ångström, A. (1964) The parameters of atmospheric turbidity. *Tellus*, **16** (1), 64–75. doi: 10.1111/j.2153-3490.1964.tb00144.x

ARPA Emilia Romagna (2012) La Qualità dell'aria in Emilia Romagna, Edizione 2011, 2012, a cura di Regione Emilia Romagna – Direzione Generale Ambiente e Difesa del Suolo e della Costa.

Bäumer, D., Rinke, E., and Vogel, B. (2008) Weekly periodicities of aerosol optical thickness over Central Europe – Evidence of an anthropogenic direct aerosol effect. *Atmos. Chem. Phys.*, **8**, 39–90. doi: 10.5194/acp-8-83-2008

Biswas, S., Verma, V., Schauer, J.J., Cassee, F.R., Cho, A.K., and Sioutas, C. (2009) Oxidative potential of semi-volatile and non volatile particulate matter (PM) from heavy-duty vehicles retrofitted with emission control technologies. *Environ. Sci. Technol.*, **43**, 3905–3912. doi: 10.1021/es9000592

Cacciari, A., Di Nicolantonio, W., and Tiesi, A. (2012) Satellite-based PM annual compliance monitoring for Northern Italy: a PASODOBLE downstream sub-service. Proceedings of Advances in Atmospheric Science and Applications, ESA SP-708, Bruges, Belgium, June 18–22, 2012.

CCAC (Climate and Clean Air Coalition to Reduce Short-Lived Climate Pollutants) (2014) Time to Act to Reduce Short-Lived Climate Pollutants. UNEP (United Nations Environmental Programme) CCAC Secretariat, Paris (France).

Chandrasekhar, S. (1950) *Radiative Transfer*, Dover Publications, New York, 393 pp.

Chen, B.B., Sverdlik, L.G., Imashev, S.A., Solomon, P.A., Lantz, J., Schauer, J.J., Shafer, M.M., Artamonova, M.S., and Larmichael, G. (2013), Empirical relationship between particulate matter and aerosol optical depth over Northern Tien-Shan, Central Asia. *Air Qual. Atmos. Health*, **6** (2), 385–396. doi: 10.1007/s11869-012-0192-5

Christopher, S.A., Wang, J., Ji, Q., and Tsay, S.-C. (2003) Estimation of diurnal shortwave dust aerosol radiative forcing during PRIDE. *J. Geophys. Res.*, **108** (D19), 8956. doi: 10.1029/2002JD002787

Deirmendjian, D. (1969) *Electromagnetic Scattering on Spherical Polydispersions*, American Elsevier Publ. Co., New York, 312 pp.

Delfino, R.J., Staimer, N., Tjoa, T., Arhami, M., Polidori, A., Gillen, D.L., Kleinman, M.T., Schauer, J.J., and Sioutas, C. (2010) Association of biomarkers of systemic inflammation with organic components and source tracers in quasi-ultrafine particles. *Environ. Health Perspect.*, **118** (6), 756–762. doi: 10.1289/ehp.0901407

Deschamps, P.Y., Bréon, F.M., Leroy, M., Podaire, A., Bricaud, A., Buriez, J.-C., and Sèze, G. (1994) The POLDER mission: Instrument characteristics and scientific objectives. *IEEE Trans. Geosci. Remote Sens.*, **32** (3), 598–615. doi: 10.1109/36.297978

Deuzé, J.L., Bréon, F.M., Devaux, C., Goloub, P., Herman, M., Lafrance, B., Maignan, F., Marchand, A., Nadal, F., Perry, G., and Tanré, D. (2001) Remote sensing of aerosols over land surfaces from POLDER ADEOS-1 polarized measurements. *J. Geophys. Res.*, **106** (D5), 4913–4926. doi: 10.1029/2000JD900364

Di Nicolantonio, W. and Cacciari, A. (2011) MODIS multiannual observations in support of air quality monitoring in Northern Italy. *Ital. J. Remote Sens.*, **43** (3), 97–109. doi: 10.5721/ItJRS20114337

Di Nicolantonio, W., Cacciari, A., Bolzacchini, F., Ferrero, L., Volta, L., and Pisoni, E. (2007) MODIS aerosol optical properties over North Italy for estimating surface-level $PM_{2.5}$. Proceedings of the Envisat Symposium, Montreux, Switzerland, April 23–27, 2007, ESA SP-636, July 2007.

Di Nicolantonio, W., Cacciari, A., and Tomasi, C. (2009a) Particulate matter at surface: Northern Italy monitoring based on satellite remote sensing, meteorological fields, and in-situ samplings. *IEEE J. Sel. Top. Appl. Earth Obs. Remote Sens.*, **2** (4), 284–292. doi: 10.1109/JSTARS.2009.2033948

Di Nicolantonio, W., Cacciari, A., Petritoli, A., Carnevale, C., Pisoni, E., Volta, M.L., Stocchi, P., Curci, G., Bolzacchini, E., Ferrero, L., Ananasso, C., and Tomasi, C. (2009b) MODIS and OMI satellite observations supporting air quality monitoring. *Radiat. Prot. Dosim.*, **137**, 3–4. doi: 10.1093/rpd/ncp231

van Donkelaar, A., Martin, R.V., and Park, R.J. (2006) Estimating ground-level $PM_{2.5}$ using aerosol optical depth determined from satellite remote sensing. *J. Geophys. Res.*, **111**, D21201. doi: 10.1029/2005JD006996.

Dubovik, O., Herman, M., Holdak, A., Lapyonok, T., Tanré, D., Deuzé, J.L., Ducos, F., Sinyuk, A., and Lopatin, A. (2011) Statistically optimized inversion algorithm for enhanced retrieval of aerosol properties from spectral multi-angle polarimetric satellite observations. *Atmos. Meas. Tech.*, **4**, 975–1018. doi: 10.5194/amt-4-975-2011

Dubovik, O., Holben, B., Eck, T.F., Smirnov, A., Kaufman, Y.J., King, M.D., Tanré, D., and Slutsker, I. (2002) Variability of absorption and optical properties of key aerosol types observed in worldwide locations. *J. Atmos. Sci.*, **59** (3), 590–608. doi: 10.1175/1520-0469(2002)059<0590:VOAAOP>2.0.CO;2

Dubovik, O. and King, M.D. (2000) A flexible inversion algorithm for retrieval of aerosol optical properties from Sun and sky radiance measurements. *J. Geophys. Res.*, **105** (D16), 20673–20696. doi: 10.1029/2000JD900282

Duvall, R.M. (2008) Source apportionment of particulate matter in the U.S. and associations with lung inflammatory markers. *Inhal. Toxicol.*, **20** (7), 671–683. doi: 10.1080/08958370801935117

EC Directive (2008) Directive 2008/50/EC of the European Parliament and of the Council of 21 May 2008 on "Ambient air quality and cleaner air for Europe". *Off. J.*, **L152**, 1–44.

Engel-Cox, J.A., Holloman, C.H., Coutant, B.W., and Hoff, R.M. (2004) Qualitative and quantitative evaluation of MODIS satellite sensor data for regional and urban scale air quality. *Atmos. Environ.*, **38**, 2495–2509. doi: 10.1016/j.atmosenv.2004.01.039

Fuzzi, S., Baltensperger, U., Carslaw, K., Decesari, S., Denier van der Gon, H., Facchini, M.C., Fowler, D., Koren, I., Langford, B., Lohmann, U., Nemitz, E., Pandis, S., Riipinen, I., Rudich, Y., Schaap, M., Slowik, J.G., Spracklen, D.V., Vignati, E., Wild, M., Williams, M., and Gilardoni, S. (2015) Particulate matter, air quality and climate: lessons learned and future needs. *Atmos. Chem. Phys.*, **15**, 8217–8299. doi: 10.5194/acp-15-8217-2015

Fuzzi, S., Gilardoni, S., Baklanov, A., Brunekreef, B., Kulmala, M., Pandis, S., Sokhi, R., Querol, X., Vignati, E., and

Williams, M. (2013) in *Research Findings in Support of the EU Air Quality Review* (eds M. Maione and S. Fuzzi), European Commission, Luxembourg, pp. 29–42.

Ge, J.M., Su, J., Ackerman, T.P., Fu, Q., Huang, J.P., and Shi, J.S. (2010) Dust aerosol optical properties retrieval and radiative forcing over northwestern China during the 2008 China – U.S. joint field experiment. *J. Geophys. Res.*, **115**, D00K12. doi: 10.1029/2009JD013263

Gilardoni, S., Vignati, E., and Cavalli, F. (2007) Regional background measurements of elemental and organic carbon in East Asia, India, Africa, and South America, in *Proceedings of 2007 EUCAARI Annual Meeting*, Yliopistopaino, pp. 98–100, ISBN/ISSN: 978-952-5027-86-0.

Grell, G.A., Dudhia, J., and Stauffer, D.R. (1994) A Description of the Fifth-Generation Penn State/NCAR Mesoscale Model (MM5). NCAR Technical Note TN-398, STR National Center for Atmospheric Research, Boulder, CO.

Gupta, P. and Christopher, S.A. (2008) Seven year particulate matter air quality assessment from surface and satellite measurements. *Atmos. Chem. Phys.*, **8**, 3311–3324. doi: 10.5194/acp-8-3311-2008

Hänel, G. (1976) The properties of atmospheric aerosol particles as functions of the relative humidity at thermodynamic equilibrium with the surrounding moist air. *Adv. Geophys.*, **19**, 73–188.

Hänel, G. and Bullrich, K. (1978) Physico-chemical property models of tropospheric aerosol particles. *Beitr. Phys. Atmos.*, **51**, 129–138.

Harrison, R.M. and Yin, J. (2000) Particulate matter in the atmosphere: which particle properties are important for its effects on health? *Sci. Total Environ.*, **249** (1–3), 85–101. doi: 10.1016/S0048-9697(99)00513-6

HEI (2010) Panel on the Health Effects of Traffic-Related Air Pollution – A Critical Review of the Literature on Emissions, Exposure, and Health Effects. Technical Report no: 17, Health Effect Institute, http://pubs.healtheffects.org/view.php?id=334 (accessed 13 June 2016).

Herman, J.R., Bhartia, P.K., Torres, O., Hsu, C., Seftor, C., and Celarier, E. (1997) Global distribution of UV-absorbing aerosols from Nimbus 7/TOMS data. *J. Geophys. Res.*, **102**, 16911–16922. doi: 10.1029/96JD03680

Hess, M., Koepke, P., and Schult, I. (1998) Optical properties of aerosols and clouds: the software package OPAC. *Bull. Am. Meteorol. Soc.*, **79** (5), 831–844. doi: 10.1175/1520-0477(1998)079<0831:OPOAAC>2.0.CO;2

Higurashi, A. and Nakajima, T. (1999) Development of a two channel aerosol retrieval algorithm on global scale using NOAA/AVHRR. *J. Atmos. Sci.*, **56**, 924–941. doi: 10.1175/1520-0469(1999)056<0924:DOATCA>2.0.CO;2

Higurashi, A., Nakajima, T., Holben, B.N., Smirnov, A., Frouin, R., and Chatenet, B. (2000) A study of global aerosol optical climatology with two channel AVHRR remote sensing. *J. Clim.*, **13** (12), 2011–2027. doi: 10.1175/1520-0442(2000)013<2011:ASOGAO>2.0.CO;2

Holben, B.N., Eck, T.F., Slutsker, I., Tanre, D., Buis, J.P., Setzer, A., Vermote, E., Reagan, J.A., Kaufman, Y.J., Nakajima, T., Lavenu, F., Jankowiak, I., and Smirnov, A. (1998) AERONET – A federated instrument network and data archive for aerosol characterization. *Remote Sens. Environ.*, **66**, 1–16.

Holben, B.N., Tanré, D., Smirnov, A., Eck, T.F., Slutsker, I., Abuhassan, N., Newcomb, W.W., Schafer, J.S., Chatenet, B., Lavenu, F., Kaufman, Y.J., Vande Castle, J., Setzer, A., Markham, B., Clark, D., Frouin, R., Halthore, R., Karneli, A., O'Neill, N.T., Pietras, C., Pinker, R.T., Voss, K., and Zibordi, G. (2001) An emerging ground-based aerosol climatology: aerosol optical depth from AERONET. *J. Geophys. Res.*, **106** (D11), 12067–12097. doi: 10.1029/2001JD900014

Hu, S., Polidori, A., Arhami, M., Shafer, M.M., Schauer, J.J., Cho, A., and Sioutas, C. (2008) Redox activity and chemical speciation of size fractioned PM in the communities of the Los Angeles-Long Beach harbor. *Atmos. Chem. Phys.*, **8**, 6439–6451. doi: 10.5194/acp-8-6439-2008

van de Hulst, H.C. (1957) *Light Scattering by Small Particles*, John Wiley & Sons, Inc., New York and London, 470 pp.

van de Hulst, H.C. (1980) *Multiple Light Scattering: Tables, Formulas, and*

Applications, 2 volumes (vol. 1 of 336 pp., and vol. of 464 pp), Academic Press, New York.

Janssen, N.A.H., Hoek, G., Simic-Lawson, M., Fischer, P., van Bree, L., ten Brink, H., Keuken, M., Atkinson, R.W., Anderson, H.R., Brunekreef, B., and Cassee, F.R. (2011) Black carbon as an additional indicator of the adverse health effects of airborne particles compared with PM_{10} and $PM_{2.5}$. *Environ. Health Perspect.*, **119**, 1691–1699. doi: 10.1289/ehp.1003369

Kelly, F.J. and Fussell, J.C. (2012) Size, source and chemical composition as determinants of toxicity attributable to ambient particulate matter. *Atmos. Environ.*, **60**, 504–526. doi: 10.1016/j.atmosenv.2012.06.039

Kim, D.-H., Sohn, B.-J., Nakajima, T., and Takamura, T. (2005) Aerosol radiative forcing over east Asia determined from ground-based solar radiation measurements. *J. Geophys. Res.*, **110**, D10S22. doi: 10.1029/2004JD004678

Kim, S.-W., Yoon, S.-C., Kim, J., Kang, J.-Y., and Sugimoto, N. (2010) Asian dust event observed in Seoul, Korea, during 29–31 May 2008: analysis of transport and vertical distribution of dust particles from lidar and surface measurements. *Sci. Total Environ.*, **408**, 1707–1718. doi: 10.1016/j.scitotenv.2009.12.018

Kinne, S., Lohmann, U., Feichter, J., Schulz, M., Timmreck, C., Ghan, S., Easter, R., Chin, M., Ginoux, P., Takemura, T., Tegen, I., Koch, D., Herzog, M., Penner, J., Pitari, G., Holben, B., Eck, T., Smirnov, A., Dubovik, O., Slutsker, I., Tanrè, D., Torres, O., Mishchenko, M., Geogdzhayev, I., Chu, D., and Kaufman, Y. (2003) Monthly averages of aerosol properties: a global comparison among models, satellite, and AERONET ground data. *J. Geophys. Res.*, **108** (D20), 4634. doi: 10.1029/2001JD001253

Kneizys, F.X., Shettle, E.P., Abreu, L.W., Chetwynd, J.H., Anderson, G.P., Gallery, W.O., Selby, J.E.A., and Clough, S.A. (1989) Users guide to LOWTRAN 7, AFGL-TR-88-0177, Environmental Research Paper No. 1010, Air Force Geophysical Laboratory, Hanscom AFB, MA.

Kokhanovsky, A.A., Davis, A.B., Cairns, B., Dubovik, O., Hasekamp, O.P., Sano, I., Mukai, S., Rozanov, V.V., Litvinov, P., Lapyonok, T., Kolomiets, I.S., Oberemok, Y.A., Savenkov, S., Martin, W., Wasilewski, A., Di Noia, A., Stap, F.A., Rietjens, J., Xu, F., Natraj, V., Duan, M., Cheng, K.T., and Munro, R. (2015) Space-based remote sensing of atmospheric aerosols: the multi-angle spectro-polarimetric frontier. *Earth Sci. Rev.*, **145**, 85–116. doi: 10.1016/j.earscirev.2015.01.012

Kokhanovsky, A.A., Deuzé, J.L., Diner, D.J., Dubovik, O., Ducos, F., Emde, C., Garay, M.J., Grainger, R.G., Heckel, A., Herman, M., Katsev, I.L., Keller, J., Levy, R., North, P.R.J., Prikhach, A.S., Rozanov, V.V., Sayer, A.M., Ota, Y., Tanré, D., Thomas, G.E., and Zege, E.P. (2010) The inter-comparison of major satellite aerosol retrieval algorithms using simulated intensity and polarization characteristics of reflected light. *Atmos. Meas. Tech.*, **3**, 909–932. doi: 10.5194/amt-3-909-2010

Krzyzanowski, M. and Schwela, D. (1999) in *Air Pollution and Health* (eds S.T. Holgate *et al.*), Academic Press, San Diego, CA, pp. 105–113.

Lei, H. and Wang, J.X.L. (2014) Observed characteristics of dust storm events over the western United States using meteorological, satellite, and air quality measurements. *Atmos. Chem. Phys.*, **14**, 7847–7857. doi: 10.5194/acp-14-7847-2014

Levy, R.C., Remer, L.A., Mattoo, S., Vermote, E.F., and Kaufman, Y.J. (2007) Second generation operational algorithm: retrieval of aerosol properties over land from inversion of Moderate Resolution Imaging Spectroradiometer spectral reflectance. *J. Geophys. Res.*, **112**, D13211. doi: 10.1029/2006JD007811

Li, Z., Xia, X., Cribb, M., Mi, W., Holben, B., Wang, P., Chen, H., Tsay, S.-C., Eck, T.F., Zhao, F., Dutton, E.G., and Dickerson, R.E. (2007) Aerosol optical properties and their radiative effects in northern China. *J. Geophys. Res.*, **112**, D22S01. doi: 10.1029/2006JD007382

Lim, S.S. *et al.* (2012) A comparative risk assessment of burden of disease and injury attributable to 67 risk factors clusters in 21 regions, 1990–2010: a systematic analysis for the global burden of disease 2010. *Lancet*, **380**, 2224–2260. doi: 10.1016/S0140-6736(12)61766-8

Liu, Y., Huang, J., Shi, G., Takamura, T., Khatri, P., Bi, J., Shi, J., Wang, T., Wang, X., and Zhang, B. (2011) Aerosol optical properties and radiative effect determined from sky-radiometer over Loess Plateau of Northwest China. *Atmos. Chem. Phys.*, **11**, 11455–11463. doi: 10.5194/acp-11-11455-2011

Minguzzi, E., Ballista, G., Bonafè, G., Cacciari, A., Deserti, M., Di Nicolantonio, W., Stortini, M., and Tiesi, A. (2011) Multi annual comparison between satellite-based observations and CTM estimates of surface aerosol concentrations in Northern Italy. Proceedings of the 14th International Conference on Harmonisation within Atmospheric Dispersion Modelling for Regulatory Purposes (HARMO).

Mukai, S., Nishina, M., Sano, I., Nakata (Mukai), M., Iguchi, N., and Mizobuchi, S. (2007) Suspended particulate matter sampling at an urban AERONET site in Japan, part 1: clustering analysis of aerosols. *J. Appl. Remote Sens.*, **1**, 013518. doi: 10.1117/1.2762202

Mukai, S., Sano, I., Masuda, K., and Takashima, T. (1992) Atmospheric correction for ocean color remote sensing: optical properties of aerosols derived from CZCS imagery. *IEEE Trans. Geosci. Remote Sens.*, **30**, 818–824.

Mukai, S., Yasumoto, M., and Nakata, M. (2014) Estimation of biomass burning influence on air pollution around Beijing from an aerosol retrieval model. *Sci. World J.*, **2014**, 10, Article ID 649648, doi: 10.1155/2014/649648

Mukai, S., Yokomae, T., Sano, I., Nakata, M., and Kokhanovsky, A. (2012) Multiple scattering in a dense aerosol atmosphere. *Atmos. Meas. Tech. Discuss.*, **5** (1), 881–907. doi: 10.5194/amtd-5-881-2012

Omar, A.H., Won, J.-G., Winker, D.M., Yoon, S.-C., Dubovik, O., and McCormick, M.P. (2005) Development of global aerosol models using cluster analysis of Aerosol Robotic Network (AERONET) measurements. *J. Geophys. Res.*, **110** (D10S14), 1–14. doi: 10.1029/2004JD004874

Polymenakou, P., Mandalakis, M., Stephanou, E.G., and Tselepides, A. (2008) Particle size distribution of airborne microorganisms and pathogens during an intense African dust event in the eastern Mediterranean. *Environ. Health Perspect.*, **116** (3), 292–296. doi: 10.1289/ehp.10684

Pope, C.A. III,, Ezzati, M., and Dockery, D.W. (2009) Fine-particulate air pollution and life expectancy in the United States. *N. Engl. J. Med.*, **360** (4), 376–386. doi: 10.1056/NEJMsa0805646

Prakash, P.J., Stenchikov, G., Kalenderski, S., Osipov, S., and Bangalath, H. (2015) The impact of dust storms on the Arabian Peninsula and the Red Sea. *Atmos. Chem. Phys.*, **15**, 199–222. doi: 10.5194/acp-15-199-2015

Prodi, F. and Fea, G. (1978) Transport and deposition of Saharan dust over the Alps. Verhandlungen des Fünfzehnten Internationalen Tagung für Alpine Meteorologie, Grindelwald, Switzerland, September 19–23, 1978, pp. 179–182.

Prospero, J.M. (1999) Long-term measurements of the transport of African mineral dust to the southeastern United States: implications for regional air quality. *J. Geophys. Res.*, **104** (D13), 15917–15927. doi: 10.1029/1999JD900072

Randriamiarisoa, H., Chazette, P., Couvert, P., Sanak, J., and Mègie, G. (2006) Relative humidity impact on aerosol parameters in a Paris suburban area. *Atmos. Chem. Phys.*, **6**, 1289–1407. doi: 10.5194/acp-6-1389-2006

Rohr, A.C. and Wyzga, R.E. (2012) Attributing health effects to individual particulate matter constituents. *Atmos. Environ.*, **62**, 130–152. doi: 10.1016/j.atmosenv.2012.07.036

Sano, I. (2004) Optical thickness and Ångström exponent of aerosols over the land and ocean from space-borne polarimetric data. *Adv. Space Res.*, **34**, 833–837. doi: 10.1016/j.asr.2003.06.039

Sano, I., Mukai, M., Iguchi, N., and Mukai, S. (2010) Suspended particulate matter sampling at an urban AERONET site in Japan, part 2: relationship between column aerosol optical thickness and $PM_{2.5}$ concentration. *J. Appl. Remote Sens.*, **4** (1), 043504. doi: 10.1117/1.3327930

Sano, I., Mukai, S., Nakata, M., Sugimoto, N., and Holben, B. (2012) Local distribution of PM2.5 concentration over Osaka based on space and ground measurements. *Proc. SPIE*, **8534**, 85340C-1–85340C-7. doi: 10.1117/12.974434

Sano, I., Mukai, M., Okada, Y., Mukai, S., Sugimoto, N., Matsui, I., and Shimizu, A. (2008) Improvement of $PM_{2.5}$ analysis by using AOT and LIDAR data. *Proc. SPIE*, **7152**, 7152M. doi: 10.1117/12.804903

Shindell, D., Kuylenstierna, J.C.I., Vignati, E., van Dingenen, R., Amann, M., Klimont, Z., Anenberg, S.C., Muller, N., Janssens-Maenhout, G., Raes, F., Schwartz, J., Faluvegi, G., Pozzoli, L., Kupiainen, K., Höglund-Isaksson, L., Emberson, L., Streets, D., Ramanathan, V., Hicks, K., Oanh, N.T.K., Milly, G., Williams, M., Demkine, V., and Fowler, D. (2012) Simultaneously mitigating near-term climate change and improving human health and food security. *Science*, **335**, 183–189. doi: 10.1126/science.1210026

Sioutas, C., Delfino, R.J., and Singh, M. (2005) Exposure assessment for atmospheric ultrafine particles (UFPs) and implications in epidemiologic research. *Environ. Health Perspect.*, **113** (8), 947–955. doi: 10.1289/ehp.7939

Smirnov, A., Holben, B., Savoie, D., Prospero, J.M., Kaufman, Y.J., Tanré, D., Eck, T.F., and Slutsker, I. (2000) Relationship between column aerosol optical thickness and in situ ground based dust concentrations over Barbados. *Geophys. Res. Lett.*, **27** (11), 1643–1646. doi: 10.1029/1999GL011336

Takemura, T., Nakajima, T., Dubovik, O., Holben, B.N., and Kinne, S. (2002) Single scattering albedo and radiative forcing of various aerosol species with a global three-dimensional model. *J. Clim.*, **15** (4), 333–352. doi: 10.1175/1520-0442(2002)015<0333:SSAARF>2.0.CO;2

Takemura, T., Nozawa, T., Emori, S., Nakajima, T.Y., and Nakajima, T. (2005) Simulation of climate response to aerosol direct and indirect effects with aerosol transport-radiation model. *J. Geophys. Res.*, **110**, D02202. doi: 10.1029/2004JD005029

Takemura, T., Okamoto, H., Maruyama, Y., Numaguti, A., Higurashi, A., and Nakajima, T. (2000) Global three-dimensional simulation of aerosol optical thickness distribution of various origins. *J. Geophys. Res.*, **105** (D14), 17853–17873. doi: 10.1029/2000JD900265

Tanré, D., Haywood, J., Pelon, J., Léon, J.F., Chatenet, B., Formenti, P., Francis, P., Goloub, P., Highwood, E.J., and Myhre, G. (2003) Measurement and modeling of the Saharan dust radiative impact: overview of the Saharan Dust Experiment (SHADE). *J. Geophys. Res.*, **108** (D18), 8574. doi: 10.1029/2002JD003273

Tomasi, C., Prodi, F., and Tampieri, F. (1979) Atmospheric turbidity variations caused by layers of Sahara dust particles. *Contr. Atmos. Phys.*, **52** (3), 215–228.

Voss, K.J., Welton, E.J., Quinn, P.K., Frouin, R., Miller, M., and Reynolds, R.M. (2001) Aerosol optical depth measurements during the Aerosols99 experiment. *J. Geophys. Res.*, **106** (D18), 20811–20819. doi: 10.1029/2000JD900783

Wang, J. and Christopher, S.A. (2003) Intercomparison between satellite-derived aerosol optical thickness and $PM_{2.5}$ mass: implications for air quality studies. *Geophys. Res. Lett.*, **30**, 2095. doi: 10.1029/2003GL018174

WHO (2007) Health Relevance of Particulate Matter from Various Sources, Report, Bonn, Germany, March 26–27, 2007, http://www.euro.who.int/__data/assets/pdf_file/0007/78658/E90672.pdf (accessed 14 June 2016).

WHO (2012) Health Effects of Black Carbon, WHO Regional Office for Europe, Copenhagen, Denmark, http://www.euro.who.int/en/health-topics/environment-and-health/air-quality/publications/2012/health-effects-of-black-carbon-2012.

Yaalon, D.H. and Ganor, E. (1979) in *SCOPE 14, Saharan Dust* (ed. C. Morales), John Wiley &Sons, Inc., New York, pp. 187–193.

Yu, Y., Schleicher, N., Norra, S., Fricker, M., Dietze, V., Kaminski, U., Cena, K., and Stüben, D. (2011) Dynamics and origin of $PM_{2.5}$ during a three-year sampling period in Beijing, China. *J. Environ. Monit.*, **13**, 334–346. doi: 10.1039/C0EM00467G

Zheng, M., Salmon, L.G., Schauer, J.J., Zeng, L., Kiang, C.S., Zhang, Y., and Cass, G.R. (2005) Seasonal trends in $PM_{2.5}$ source contributions in Beijing, China. *Atmos. Environ.*, **39** (22), 3967–3976. doi: 10.1016/j.atmosenv.2005.03.036

Zhu, A., Ramanathan, V., Li, F., and Kim, D. (2007) Dust plumes over the Pacific, Indian, and Atlantic oceans: climatology and radiative impact. *J. Geophys. Res.*, **112**, D16208. doi: 10.1029/2007JD008427

10 Impact of the Airborne Particulate Matter on the Human Health

Marina Camatini, Maurizio Gualtieri, and Giulio Sancini

10.1 Introduction

Particulate matter (PM) is a mixture of many different components with local and regional variations. PM can be characterized by origin (anthropogenic or natural, primary or secondary), by source (combustion processes and traffic or erosion and degradation), or by physicochemical properties, such as solubility, organic carbon content, or metals. PM is characterized by the size of particles (aerodynamic diameter). Total suspended particles (TSPs) include particles of any size present in the air; even particles larger than 30–70 μm may remain suspended for a very short period before deposition. In general, the smaller and lighter a particle is, the longer it will remain suspended in the air. Larger particles tend to settle to the ground by gravity in a matter of hours, whereas the smallest particles can stay in the atmosphere for weeks and are mostly removed by precipitation. PM_{10} and $PM_{2.5}$ are composed by particles which have respectively a mean aerodynamic diameter less than 10 and 2.5 μm; ultrafine particles (UFP) have a diameter less than 0.1 μm, and they are referred as $PM_{0.1}$. $PM_{2.5}$ particles are defined as "fine particles" (FPs), while PM_{10} particles, comprised between 10 and 2.5 μm, are named "coarse particles."

The fate of particles into the lungs helps to address PM fraction: thoracic fraction (PM_{10}) is the one which penetrates the upper respiratory tree, while the respirable fraction ($PM_{2.5}$) is the one which reaches the alveoli, where the gases exchange takes place.

PM is a heterogeneous and highly variable mixture of liquid or solid particles, differing in size and chemical composition, in which two main components can be identified. Particles are called "primary" if they are in the same chemical form in which they are emitted in the atmosphere, while they are called "secondary" if they are formed by chemical reactions from gaseous precursors. "Primary" PM is mainly originated by combustion processes, as well as mechanical processes, like abrasions of surfaces, suspension of soil, and biological materials. "Secondary" PM

Atmospheric Aerosols: Life Cycles and Effects on Air Quality and Climate, First Edition.
Edited by Claudio Tomasi, Sandro Fuzzi, and Alexander Kokhanovsky.

is generated in the atmosphere by chemical reactions of free, adsorbed, or dissolved gases such as nitrogen oxide (NO) or volatile organic compounds (VOCs). The mentioned variability suggests that PM is regarded as a particular contaminant whose morphological, chemical, physical, and thermodynamic properties significantly depend on specific properties of the sampling site and on the sampling campaign.

Current European air quality standards for human health and environment protection for PM (EC, 2008) are still above the World Health Organization Air Quality Guidelines (WHO-AQG) that aim to protect public health. In Europe, annual mean PM_{10} should not exceed $40\,\mu g\,m^{-3}$ (limit value set in 2005), and member states are requested to reduce exposure to $PM_{2.5}$ in urban areas below $20\,\mu g\,m^{-3}$ by 2015 (legally binding value). The WHO-AQG for PM, chosen as the lowest levels at which total, cardiopulmonary, and lung cancer mortality has been shown to significantly increase in response to long-term exposure to PM, are set as an annual mean of $20\,\mu g\,m^{-3}$ for PM_{10} and $10\,\mu g\,m^{-3}$ for $PM_{2.5}$ (Beelen *et al.*, 2014). Stakeholders drafting policies to reduce air pollution must take into account not only the human health but also considerations on the economic and social constraints, political orientations, and urban planning.

Traffic is a major source of PM_{10}, $PM_{2.5}$, and UFP; traffic intensity contributes significantly to PM-related health effects. It is one of the most important causes of PM concentrations, and its contribution varies among size fractions. Comparison of PM levels in several European cities demonstrated a positive correlation between PM concentrations and traffic density. Recently attention has been paid to urban PM_1 concentrations (50% of PM_{10}) and its elemental composition, and it has been demonstrated that traffic intensity influences the PM physicochemical characteristics, particularly with regard to concentrations of polycyclic aromatic hydrocarbon (PAH), metals, and radical-generating capacity. PAHs and PAH derivatives are formed during combustion in fossil fuel-driven engines, whereas metals (including transition metals) originate from car catalysts. Since no threshold for PM-induced adverse health effects has been clearly established, there is a general consensus that ambient PM levels should be reduced as much as possible.

Although an overall reduction of PM concentrations is likely to result in reduction of PM-associated health risks, emission reduction strategies have to be considered as priority. In order to develop such environmental health policies, relationships between PM emission sources, physicochemical characteristics, and its toxicity have received a lot of attention. Over the last decades, a great effort has been made to study its adverse health effects using many different research designs and approaches, including large epidemiological studies (Hamra *et al.*, 2014), animal experiments (Kelly and Fussell, 2012), and *in vitro* toxicological assays (Gualtieri *et al.*, 2010) to establish mutagenicity, cytotoxicity, and DNA reactivity in relation to PM chemical composition and specific size fractions.

Health endpoints, often studied in relationship to PM, are mortality and morbidity, hospital admissions for cardiovascular (CV) respiratory diseases, lung function, and functional endpoints, such as heart rate variability (HRV). Particularly

$PM_{2.5}$ and the even smaller UFP penetrate deeply into the lung and reach the alveoli and are thus more likely to induce adverse health effects than larger particles, such as PM_{10} and TSP. Furthermore, it is generally acknowledged that specific chemicals present in PM, such as PAHs and their derivatives, biological material, acid aerosols, and metals, attached to a carbonaceous core, determine the toxic potency to a large extent.

The PM adverse health effects may relate to its physicochemical characteristics, including mass, size, number, surface area, composition, concentration, and source. PM has a great complexity for its physicochemical properties, and the possible variability of the chemicals adsorbed on inhalable particles may change its properties and must be taken into account. Some studies provide evidence on the role of PM components and their associated adverse health effects, but a consensus within the scientific community to which specific components are the most significant determinants of the toxicological response is still lacking. For this reason toxicological studies on both animals and cellular systems have to be further analyzed in order to uncover the fine mechanisms of effects activated by PM on the lungs and other target organs (Figure 10.1).

The WHO (2013) and IARC on October 17, 2013 (Loomis *et al.*, 2013), designates PM a group 1 carcinogen. Particulates are among the deadliest forms of air pollution due to its ability to penetrate deep into the lungs and bloodstreams unfiltered, causing permanent DNA mutations, heart attacks, and premature death. In 2013, a study involving 312 944 people in nine European countries revealed that

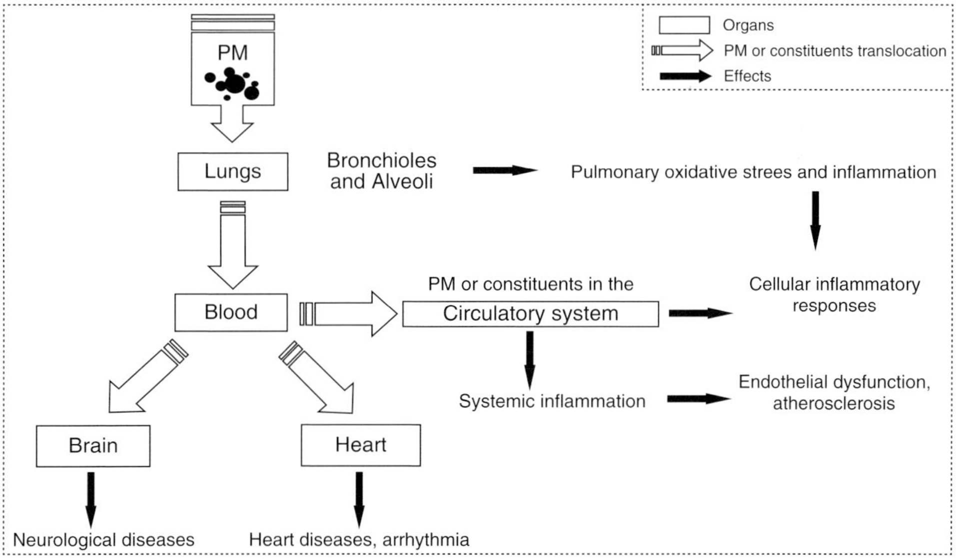

Figure 10.1 Pathways linking PM exposure and biological effects. Fine particles reach the alveoli and promote pulmonary oxidative stress and inflammation. PM or its components may translocate to the circulatory system and impact on other organs such as heart and brain.

there was no safe level of PM and that for every increase of 10 μg m^{-3} in PM_{10}, the lung cancer rate is 22%. The smaller $PM_{2.5}$ was particularly deadly, with a 36% increase in lung cancer per 10 μg m^{-3} inhaled since it can penetrate deeper into the lungs.

In summary epidemiological approaches are relevant, and numerous data have been furnished by the literature from 1970: time-series analyses and cohort studies, case-crossover studies, cross-sectional studies, panel studies, and case–control studies are the records present. These studies have demonstrated that exposure to urban PM is associated with several adverse health effects. Long-term exposure to high concentrations of PM increases the risk of lung cancer, respiratory diseases, and arteriosclerosis, whereas short-term exposure peaks can cause exacerbation of several forms of respiratory diseases, including bronchitis and asthma, as well as changes in HRV.

Despite the correlation between pollution and health, which emerged in the epidemiological studies, there are still some unsolved questions with respect to PM health effects:

- Is there a PM fraction that is more health relevant?
- What is the correlation between PM exposure and biological effects?
- Is PM a causative agent, an indicator, or a container?

When toxicological experiments will parallel clinical and epidemiological studies, the fine biological processes activated by particles and particle components will be clarified.

10.2 Epidemiological Evidences

The most direct evidence for linking health effects to air pollution exposure is provided by epidemiological studies, having the advantage of accurately representing conditions in a population (Englert, 2004; Schwartz *et al.*, 2008). They also present an opportunity to study vulnerable effects, even though these studies are complicated by uncertainties over exposure causality and confounding factors.

In the past, epidemiological studies routinely adopted central pollution monitoring, which assumes that every person in a study area is exposed to the same recorded pollution levels. Research is now progressing to methods that measure personal exposure or that better approximate actual exposure to highly variant local emissions.

An association between high levels of air pollutants and human illnesses has been known for more than half a century. A few episodes of markedly increased mortality rates during extreme elevations in urban pollution, such as in the Meuse Valley, Belgium (December 1930), and the United Kingdom (the London fog incident of 1952), sparked the initial epidemiological research (Lippmann, 2014). As a result during several decades, long effort has been done to reduce air pollution culminated in the Clean Air Act legislation of 1970. Despite improvements in air quality over the past few decades, associations between current ambient pollution levels and excess morbidity and mortality have been consistently detected.

There are several hundred published epidemiological studies linking air pollution with human illnesses. Although many pollutants may cause disease individually or in combination (e.g., O_3, SO_2, and NO_2), over the past decade, PM has become a major focus of research. During the past 15 years, the magnitude of evidence and the number of studies linking air pollution to cardiovascular diseases (CVDs) have grown substantially. In general, studies can be separated into those that have investigated the health effects of acute (Englert, 2004) or chronic (Lepeule *et al.*, 2012) air pollution exposure. Mortality is the most important health effect and has been studied the longest. The earliest evidence of mortality relates back to smog episodes with PM concentrations in the range of thousands of micrograms per cubic meter (Lippmann, 2014). Even though the air quality has improved markedly in most developed countries and concentrations have fallen with one to two orders of magnitude, it is still possible to discern short-term temporal associations with daily mortality at the present levels of air pollution. It is estimated that across the EU, PM_{10} is causing more than 350 000 premature deaths per year. Another early observation was that mortality was higher in more polluted areas. This has been confirmed by modern cohort studies that account for other potential explanations for such associations, and there does not appear to be a threshold of effect within the range of concentrations. Advances in the understanding of air pollution and mortality have been driven by the combined development of methods and biomedical concepts. The most significant methodological developments have been in time-series techniques and in the establishment of large cohort studies. Throughout the history of air pollution epidemiology, mortality studies have been central, and they will continue for the availability of mortality data on a large population scale and the weight of mortality in estimating impacts for policy development. Environmental epidemiology aims to understand the relationship between environment and health in human populations to use this knowledge to set up rules able to improve public health.

The epidemiological approach comprises two stages. The first is the use of statistical methods to establish whether or not the hazard is associated with health after accounting for chance and known sources of bias or confounding. The second stage is a causal implication; even this is problematic since most epidemiological evidence depends on observations rather than experiments. Causal judgment brings to nonepidemiological evidence, since the most important observations come from clinical and laboratory experiments. The epidemiology of air pollution and mortality is further constrained by the necessity to use opportunist data on mortality and air pollution collected for some other purpose (Lepeule *et al.*, 2012). One of the major difficulties of air pollution epidemiology is that the health risks of pollution levels are extremely small when compared to the traditional risks, such as cigarette smoking (Pope *et al.*, 2011) (Figure 10.2).

As a consequence, it is difficult to distinguish the effects of air pollution from the noise of natural fluctuations in health events, such as hospital admissions and

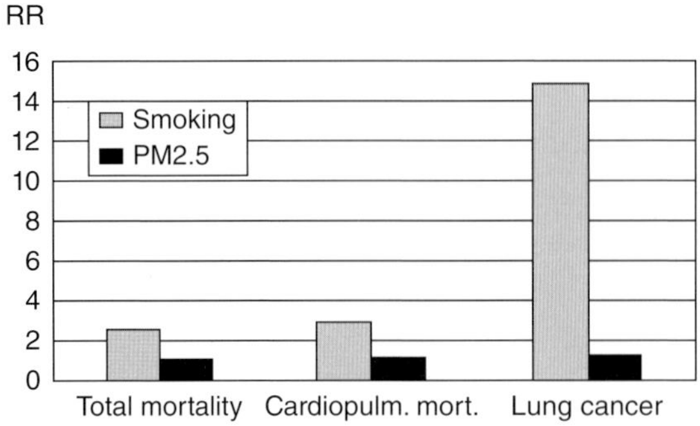

Figure 10.2 Comparison of health effects (relative risk (RR) ratio) associated with exposure to ambient $PM_{2.5}$ and smoking (data from Pope *et al.*, 2002) calculated for the difference in mean $PM_{2.5}$ concentrations between the most polluted city and the least polluted city (24.5 µg m^{-3}) and for an average current smokers (22 cigarettes per day for 33.5 years, with initiation before 18 years).

deaths attributable to other factors. At present refined statistical approaches have been developed to extrapolate the "weather" signal from the noise.

10.2.1
Exacerbation of Lung Diseases

The effects of PM on respiratory health are universally acknowledged for the results of various studies examining respiratory symptoms, exacerbation of respiratory diseases, decrease in pulmonary function, and mortality in patients with chronic respiratory diseases. The short-term effects of PM_{10} with a 50% cutoff on daily mortality have been estimated to be in the range of 0.3–1.5% for 10 µg m^{-3} PM_{10} (Pope *et al.*, 2011).

Although the association between air particles and respiratory mortality has been known for several years, many aspects of PM_{10}–respiratory mortality relationship, such as the specific shape of the exposure–response curve, the latency interval of the effect, and the individual characteristics that can modify the effect, deserve further investigation. A greater effect of PM_{10} on respiratory mortality in respect to all natural mortality causes has been reported. The effect is stronger in the warm than in the cold season. The most consistent effect, not heterogeneous across cities, has been observed for 4-day exposure, and the effect is not influenced by a contemporaneous increase of NO_2 or O_3 concentration. The effect increased linearly with PM_{10} concentration without a threshold. Females and subjects with a previously diagnosed chronic condition are likely the most susceptible people. Most previous studies in Europe and China, analyzing the short-term effect of increasing levels on cause-specific mortality, have reported a greater effect on respiratory mortality than on natural or CV mortality. Also more recent studies in Europe (Raaschou-Nielsen *et al.*, 2013), in the United States (Correia *et al.*, 2013), and in Southeast Asia (Shang *et al.*, 2013) report that PM effect on respiratory mortality is twice in comparison with natural or CV mortality.

The recent results reported (Hamra *et al.*, 2014) and the decision of the IARC working group to classify outdoor air pollution within group 1 of carcinogenic substances further justify efforts to reduce exposures to air pollutants that can arise from many sources. The Global Burden of Disease in 2010 has estimated that approximately 3.22 million deaths are caused by exposure to air pollution, an increase from 2.91 million deaths attributed to air pollution in 1990. Cancers of the trachea, bronchus, or lung (Hamra *et al.*, 2014) represent approximately 7% of total mortality attributable to $PM_{2.5}$ in 2010. The group 1 classification raises questions regarding individual components in the air pollution mixture, such as the carcinogenic potential of each component as well as through what pathways they may contribute to cancer risk (Table 10.1).

10.2.2 Effects on the Cardiovascular System

Numerous data have concluded that short-term exposure to PM contributes to acute CV morbidity and mortality and that exposure to elevated PM levels over the long term can reduce life expectancy by a few years (Pope *et al.*, 2011). The literature data corroborate several credible biological pathways through which PM could produce acute CVDs and promote chronic disease (Brook *et al.*, 2010). There is increasing evidence documenting the epidemiological studies on the elevated risk for CV events associated with exposure to UFP (Lippmann, 2014). In fact $PM_{2.5}$ has been associated with increased risks of myocardial infarction, stroke, arrhythmia, and heart failure exacerbation within hours to days of exposure in susceptible individuals. Several new studies have also demonstrated that residing in locations with higher long-term average PM levels elevates the risk for CV morbidity and mortality. Moreover recent evidence also involves other size fractions, such as UFP, gaseous pollutants (O_3 and NO_x), and specific sources such as traffic, which are able to produce CVD. Air pollutants have been linked with endothelial dysfunction and vasoconstriction, increased blood pressure, coagulant changes, arrhythmias, and a progression of atherosclerosis (Figure 10.1). The US Environmental Protection Agency (EPA) has completed its updated "Air Quality Criteria for Particulate Matter." The National Ambient Air Quality Standard (NAQS) for daily $PM_{2.5}$ levels was 65 $\mu g\, m^{-3}$, and from 2006 the threshold value is 35 $\mu g\, m^{-3}$ (US Environmental Protection Agency, 2008).

Numerous epidemiological studies have demonstrated consistent associations between short-term elevations in PM and increase in daily CV morbidity and mortality. Several studies have also reported adverse CV outcomes in relation to long-term PM exposure. At present the PM components responsible for mediating these effects remain to be clarified, even though there is experimental evidence suggesting biological mechanisms involved in promoting and exacerbating CVD. One pathway involves the initiation of pulmonary and systemic oxidative stress

Table 10.1 Summary of studies included in the meta-analyses of lung cancer risk associated with exposure to PM.

Continent	ID	Source	Number of events (mortality or incidence)	Total population	Study period	Exposure assessment method	Exposure distribution (mean ± SD)	Study title
North America								
California, United States	01	Beeson *et al.* (1998)	16 (incidence)	6 338	1977–1992	Fixed site monitor	PM_{10}: 51.0 ± 16.5	Adventist Health Study on Smog (AHSMOG)
California, United States	02	McDonnell *et al.* 2000	13 (mortality)	3 769	1977–1992	Fixed site monitor	$PM_{2.5}$: 31.9 ± 10.7	AHSMOG
United States	03	Pope *et al.* 2002	n/a	415 000	1982–1998	Fixed site monitor	PM_{10}: 28.8 ± 5.9	American Cancer Society's Cancer Prevention Study II
United States	04	Krewski *et al.* (2009)	16 615 (mortality)	499 968	1982–2000	Fixed site monitor	$PM_{2.5}$: 21.2 ± 10.8 (1979–1983)	American Cancer Society's Cancer Prevention Study II
$PM_{2.5}$: 14.0 ± 9.1 (1999–2000)								
Los Angeles, CA, United States	05	Jerrett *et al.* (2013)	1 481 (mortality)	73 711	1982–2000	Land-use regression	$PM_{2.5}$: 14.1 ± 12.4	American Cancer Society's Cancer Prevention Study II
United States	06	Hart *et al.* (2011)	800 (mortality)	53 814	1985–2000	Inverse distance weighting ($PM_{2.5}$)/ spatiotemporal (PM_{10})	$PM_{2.5}$: 14.1 ± 4.0 PM_{10}: 26.8 ± 6.0	Trucking Industry Particle Study (TrIPS)
California, United States	07	Lipsett *et al.* (2011)	275 (PM_{10}), 234 ($PM_{2.5}$) (mortality)	101 784 (PM_{10}), 73 489 ($PM_{2.5}$)	1997–2005	Inverse distance weighting	$PM_{2.5}$: 15.6 ± 4.5 PM_{10}: 29.2 ± 9.8	California Teachers Study
United States	08	Lepeule *et al.* (2012)	632 (mortality)	8 096	1975–2009	Fixed site monitor	$PM_{2.5}$: 15.9[a)]	Harvard Six Cities Study

Canada	09	Hystad *et al.* (2013)	2 390 (incidence)	5 897	1994–1997	Spatiotemporal model	$PM_{2.5}$: 11.9 ± 3.0	National Enhanced Cancer Surveillance System Case–Control study
United States	10	Puett *et al.* (2013)	1 648 (incidence)	97 565	1998–2010	Spatiotemporal model	PM_{10}: 21.6 ± 6.0	Nurses' Health Study
Europe								
Netherlands	11	Beelen *et al.* (2008)	1 888 (mortality)	120 852	1986–1997	Land-use regression	PM_{10}: 21.6 ± 6.0	Netherlands Cohort Study of Diet and Cancer
United Kingdom	12	Carey *et al.* (2013)	5 273 (mortality)	830 842	2003–2007	Air dispersion	$PM_{2.5}$: 12.9 ± 1.4 PM_{10}: 19.7 ± 2.3	Clinical Practice Research Datalink
Italy	13	Cesaroni *et al.* (2013)	12 208 (mortality)	1 265 058	2001–2010	Air dispersion	$PM_{2.5}$: 23.0 ± 4.4	Rome Longitudinal Study
Germany	14	Heinreich *et al.* (2013)	41 (mortality)	4 752	1980–2008	Fixed site monitor	PM_{10}: 43.7[a)]	German Women's Health Study
European Union	15	Raaschou-Neilsen *et al.* (2013)	2 095 (incidence)	312 944	1990s[a]	Land-use regression	$PM_{2.5}$: 13.4 ± 1.2 PM_{10}: 21.3 ± 2.7	European Study of Cohorts for Air Pollution Effects (ESCAPE)
Other								
China	16	Cao *et al.* (2011)	624 (mortality)	70 947	1991–2000	Fixed site monitor	$PM_{2.5}$: [a]	China National Hypertension Follow-Up Survey
Japan	17	Katanoda *et al.* (2011)	421 (mortality)	63 520	1983–1995	Fixed site monitor	$PM_{2.5}$: 28.8[a)]	Three Prefecture Cohort Study
New Zealand	18	Hales *et al.* (2012)	1 686 (mortality)	1 050 222	1996–1999	Land-use regression	PM_{10}: 8.3 ± 8.4	New Zealand Census Mortality Study

The numbers reported represent the range of exposure estimated by converting TSP to $PM_{2.5}$ with a 3 : 1 ratio (Hamra *et al.*, 2014).
a) Standard deviation not reported.
Source: Hamra *et al.* (2014).

and inflammation by components within PM; subsequently, a cascade of physiological responses may follow. These include alterations in blood and cardiac rhythmus and vascular dysfunction (Figures 10.1 and 10.8).

The most recent analysis of the National Morbidity, Mortality, and Air Pollution Study, based on data from 90 of the largest cities in the United States, estimates that daily total and cardiopulmonary mortality increased in the short term by 0.21% and 0.31%, respectively, for each 10 $\mu g\,m^{-3}$ increase in PM_{10}. The American Cancer Society has estimated that for each 10 $\mu g\,m^{-3}$ increase in annual average exposure to $PM_{2.5}$, long-term all-cause, cardiopulmonary, and lung cancer mortality increased by approximately 4%, 6%, and 8%, respectively. A statistically significant association between $PM_{2.5}$ and the overall CV mortality has been confirmed for a 10 $\mu g\,m^{-3}$ increase in long-term exposure. There is no evidence for excess mortality in the entire cohort due to other reasons.

Thus the epidemiological data suggest some conclusions:

- Exposure to $PM_{2.5}$ over a few hours to weeks can trigger CVD-related mortality and nonfatal events.
- Longer-term exposure (a few years) increases the risk for CV mortality to an even greater extent than exposures over a few days and reduces life expectancy of the population by several months to a few years.
- Reductions in PM levels are associated with decreases in CV mortality within a time frame as short as a few years.
- Many credible biological mechanisms have been elucidated that can explain these findings (Figures 10.5, 10.7, and 10.8).

The overall evidence is consistent with a causal relationship between $PM_{2.5}$ exposure and CV morbidity and mortality.

10.2.3 Life Expectancy and PM Concentration

There are numerous studies taking into consideration the life expectancy in association with the effects of PM levels on human health. The studies performed on European (Pascal *et al.*, 2013) and US (Correia *et al.*, 2013; Pope, Ezzati, and Dockery, 2013) cities deserve attention as well as the consideration made on some countries in Asia and Africa (Shang *et al.*, 2013), where air pollution becomes an increasing concern.

Urban air quality represents a major public health burden and is a long-standing concern to European citizens. Despite a major decrease in the pollutant levels in Europe since 1950 and the implementation in the 2008 (EC, 2008) of the first European Commission directive on ambient air quality in 1980, important disparities in exposure to air pollution within European countries remain (Pascal *et al.*, 2013). Air pollution still has a major public health impact in European cities, and life expectancy and monetary benefits increase significantly when levels of FP are further reduced. The total impact of air pollution on European's health is confirmed by the numerous studies pointing on the reduction of PM_{10} and $PM_{2.5}$, which would significantly decrease mortality and morbidity, improving the quality of life. The Clean Air for Europe estimated that PM was responsible for about 348 000 premature deaths and 100 000 hospitalizations for respiratory and CV causes. The associated economic valuation ranges between 276 and 790 billion euros. A decrease by 5 $\mu g\,m^{-3}$ in the cities with the highest pollution levels would result in smaller even visible benefits. Chronic effects largely exceed acute effects for mortality, as the long-term monetary benefits associated with a reduction of $PM_{2.5}$ are more than 200 times greater than the monetary benefits associated with a reduction of PM_{10}, whatever the scenario is (Table 10.2).

This result certainly deserves the attention of public decision makers. Linking the reduction of air pollution with a reduction of traffic-related noise or a reduction of greenhouse gas (GHG) emissions would result in potentially larger benefits, providing additional arguments to promote more ambitious air quality targets in Europe. Similarly, more knowledge on the chronic effects of air pollution in Europe is needed.

In US cities, ambient levels of $PM_{2.5}$ have continued to decline from 2000 to 2007 as a result of interventions, but the decline has been at a slower rate than during 1980–2000 (Correia *et al.*, 2013). Whether these more recent and slower declines of $PM_{2.5}$ levels continue to improve life expectancy and whether they benefit all populations equally are unknown. A decrease of 10 $\mu g\,m^{-3}$ in the concentration of $PM_{2.5}$ has been associated with an increase in mean life expectancy of 0.35 years. This association is stronger in more urban and densely populated counties. Reductions in $PM_{2.5}$ have been associated with improvements in life expectancy for the period from 2000 to 2007, and air pollution control in the last decade has continued to have a positive impact on public health.

Since 1970 the increase in air quality controls has led to improvements in air quality in the United States at costs that the EPA has estimated to be $25 billion per year. Air pollutant concentrations have been generally decreased in the United

Table 10.2 Health outcome and associated concentration–response functions, expressed as relative risk (RR) for 10 $\mu g\,m^{-3}$ increase in each pollutant used in the health impact assessment.

Health outcome	Ages	RR	References
Short-term impacts of PM_{10}			
Respiratory hospitalizations	All	1.011 [1.006–1.017]	Medina *et al.* (2005)
Cardiac hospitalizations	All	1.006 [1.003–1.009]	Medina *et al.* (2005)
Long-term impacts of $PM_{2.5}$			
Total (including external) mortality	>30	1.06 [1.02–1.11]	Pope *et al.* (2002)
Cardiovascular mortality	>30	1.12 [1.08–1.15]	Pope *et al.* (2004)

Source: Pascal *et al.* (2013). Reproduced with the permission from Elsevier.

States, with substantial differences in reductions across metropolitan areas. Levels of $PM_{2.5}$ remain relatively high in some areas. In 2010 the EPA estimated that 62 US countries, accounting for 26% of the total population, had $PM_{2.5}$ concentrations exceeding the NAQS (US Environmental Protection Agency, 2008).

Reductions in PM are associated with reductions in both cardiopulmonary and overall mortality. In the mid-1990s, the American Cancer Society reported associations of cardiopulmonary mortality risk with chronic exposure to $PM_{2.5}$. These studies have confirmed that $PM_{2.5}$ is an important independent environmental risk factor for cardiopulmonary disease and mortality. Moreover cohort studies, population-based studies, and short-term time-series studies have also shown associations between reduction in air pollution and reduction in mortality, demonstrating an association between $PM_{2.5}$ and life expectancy. An interesting observation may be posed: these findings point to the role of air pollution in contributing to the problem of avoidable premature mortality. But despite this observation, much of the world's population remains exposed to PM concentrations one or two orders of magnitude higher than those in the United States during the years of 2000–2007 (Table 10.3). Moreover exposure levels in many countries reflect high levels of outdoor air pollution from vehicles, industry, and biomass burning. Outdoor air pollution receives little regulatory attention in many countries; thus control of air pollution must be taken into account as economic development drives up pollution emissions.

For the years between 1981 and 2000, northern China's air contained a daily average of 551.6 μgm^{-3} TSPs, whereas the south had an average of 354 $\mu g\,m^{-3}$

Table 10.3 Estimated regression coefficients for the association between the changes in $PM_{2.5}$ and changes in life expectancy for 545 countries for the years 2000–2007 for selected regression models are summarized.

Variable	Mean (SD)
Life expectancy (years)	
2000	76.7 (1.7)
2007	77.5 (2.0)
Change	0.8 (0.6)
$PM_{2.5}$ ($\mu g\,m^{-3}$)	
2000	13.2 (3.4)
2007	11.6 (2.8)
Reduction	1.6 (1.5)
Per capita income (in thousands of $)	
2000	27.9 (7.4)
2007	30.4 (7.9)
Change	2.5 (2.3)
Population (in hundreds of thousands)	
2000	3.5 (6.3)
2007	3.8 (6.6)
Change	0.3 (0.6)

Source: Correia *et al.* (2013). Reproduced with permission from Correia.

(Shang *et al.*, 2013). In the United States the legally required standard for air quality from 1971 to 1987 was 75 μg m^{-3}. In China between 1981 and 2001, concentrations of TSPs were more than double the country's earlier average of 200 μg m^{-3} and more than five times the average amount in the United States. Besides these observations and these elevated numbers, it must be considered that pollution problem is not restricted to the outdoor air, as families in developing countries are exposed to pollution inside the home too: almost half the world prepares food with wood, charcoal, or coal, all common sources of air pollutants. It is important to keep in mind that developing countries are trying to find the right balance between economic growth and environmental quality of life and the existing studies and all the certainties acquired will help them in finding a proper relationship between pollution and health.

10.3 Toxicological Evidences

10.3.1 Particle Dosimetry, Particle Deposition, and Real Exposure

The dimension of particles determines their fate into the lungs: this is a matter of fact. Even if with slight differences among various models and reports, PM aerodynamic classes relate to possible deposition at specific lung regions. However the deposition in the airways of each particle is determined by physical processes which are impaction, settling, diffusion, and interception (Figure 10.3).

Impaction determines the deposition of particles at bifurcation, or when the air stream makes a turn, in this case a particle's inertia maintains particle in its original path and impact on the lung epithelium. Since inertia depends on the velocity and mass of particle, this mechanism is important when particle size/mass and air velocity are quite high, and therefore impaction determines deposition mainly in bronchial regions.

Settling becomes important when air velocity is low and other forces govern the path of the particles. In the distal regions of the lung (terminal bronchioles), the airways are almost horizontal, and gravity exerts its effect on particles. The combined effect of low speed of airstream, airway orientation, and particle mass determines the settling.

Diffusion is important for UFP or particles with aerodynamic diameter up to 0.5 μm. Deposition in the small airway, where residence time of particles is long, is dominated by their random movements (Brownian motion), which impact the lung epithelium.

Interception is related to the real shape of particles, and it is a relevant mechanism when a particle follows the airstream without deviation and for its size and shape enters in contact with airway surface. This mechanism has little importance in human lungs except for fibers.

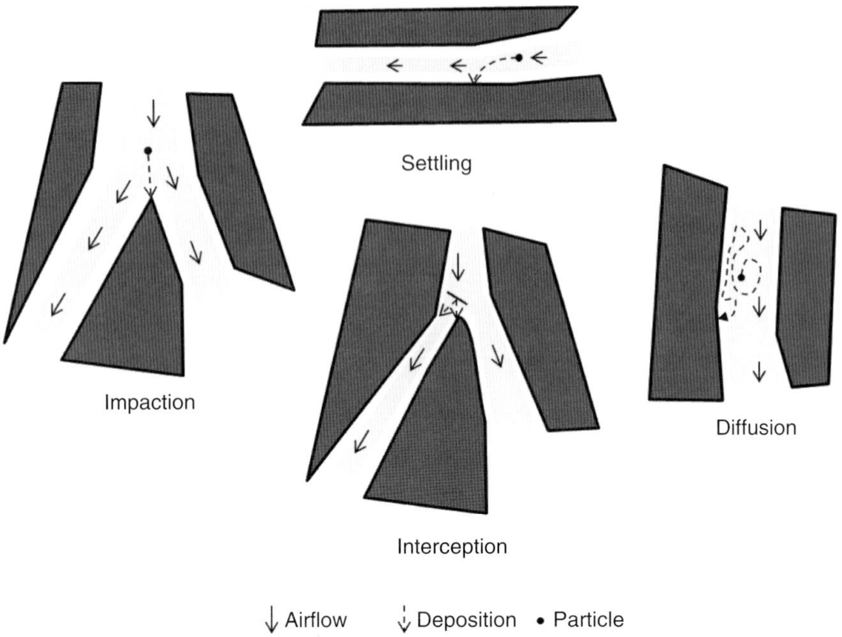

Figure 10.3 Processes determining particle deposition in the lung.

The site of particle deposition determines the activation of specific clearance mechanisms, and combined deposition and clearance define the residence time of particles in the airways. Particles with aerodynamic diameter higher than 10 μm are trapped and directly removed by nasal airways. Particles lower than 10 μm (PM_{10}) pass the nose and enter the upper lungs; but while particles with aerodynamic diameter higher than 1 μm are trapped at the extrathoracic and thoracic level (which is the region comprising the trachea to the terminal bronchioles), UFP enter deeply, deposit at alveolar level and may cross the air–blood barrier (ABB), and enter the blood circulation.

Lung clearance mechanisms play therefore a fundamental role in maintaining the physiological condition of the respiratory system. Inhaled particles may be cleared by different mechanisms in relation to the site of deposition and interaction with lung epithelia.

Mucociliary clearance is the main mechanism of clearance of thoracic airways. This process relies on the presence of mucus over the lung epithelium, formed by cells with cilia at their top. The mucus is a viscous fluid in which particles remain trapped. Then the cilia moving together in an orchestrated fashion, beating at defined interval, propel the mucus up way till the larynx where it is swallowed into the gastrointestinal tract or the lymphatic system. Mucociliary clearance is a fast process, within few hours for PM_{10}, while it becomes less efficient when diameter of particles is smaller than 3 μm ($PM_{2.5}$).

Particles may become soluble in the mucus or trapped by macrophages, specialized cells engulfing and degrading them; thus these defense mechanisms protect the respiratory system. Macrophage activity is thus the another main mechanism of clearance, and it is fundamental in the alveoli where mucus is not present and cells are not ciliated.

This rapid overview on deposition and clearance mechanism makes clear the complexity behind the definition of real exposure. However the possibility to know exposure levels is essential for sustaining epidemiological evidences of the effects promoted by PM exposure. Similarly toxicologists need to know environmental exposure to set up experiments to provide the biological explanation of the epidemiological effects. Furthermore toxicological studies need to define appropriate particle dosimetry to correlate biological outcomes with a defined PM measure unit. So far PM mass has been used as reference dosimetry. However it is clear that PM mass does not represent the ideal reference unit. For example, coarse PM is comprised for a large part of salts with minimal biological effects. On the other hand, trace compounds (such as PAHs) on the total PM mass are well-known inducers of serious adverse biological outcomes. Therefore, although current environmental standards for PM_{10} and $PM_{2.5}$ are set up on a mass basis, researchers are working to define one (or more) dosimetry able to better describer for the PM toxic effects.

At present particle number is considered a good candidate to substitute mass. Since most of the active toxicological compounds, adsorbed on PM surface, derive from combustion processes and since combustion processes emit a higher number of really tiny particles, in which contribution to total mass is low, particle number has a heavy role. The research, based on this parameter, evaluates the contribution of those PF and UFP, which vehicle the majority of biological active compounds for their high surface area. Another possible dosimetry for PM exposure is the surface area. Since PM is a complex mixture of coarse particle, FP, and UFP, their surface area will be always affected by this variability. Another valid alternative to mass metrics is the oxidative potential of particles. Since reactive oxygen species (ROS) are formed through numerous mechanisms derived by PM physicochemical properties, the evaluation of ROS might give a parameter able to integrate different properties.

A common PM dosimetry will be essential for future epidemiological and toxicological studies. These latter are furthermore dealing with other technical and methodological challenges. For example, *in vivo* (see Section 10.3.2) and *in vitro* systems (see Section 10.3.3) can be exposed to PM samples in different ways. A researcher may use exposure chambers, which mimic somehow the real exposure; they may use instruments for particle instillation in animals or make particles inhalable through nose deposition. In *in vitro* experiments exposure may be performed in the classic submerged conditions using a dispersion of particles in cell culture medium or using more complex air–liquid interface systems by which particles are deposited directly on the cells. In all these cases however the question is whether the nominal dose used to describe the experimental plan will be really representative of the ambient exposure dose. For example, submerged condition for *in vitro* exposure may underestimate the contribution of UFP, in which deposition at the cell membrane is slow in comparison with the larger particles. PM intratracheal instillation in *in vivo* systems is not exactly homogeneous in all airways, determining local higher concentration of PM.

Numerical models are helping in the description of each of these exposure conditions, but the steps to take are still numerous and will require in the next years a considerable effort from researchers of different disciplines.

10.3.2
In Vivo Evidences

The biological mechanisms behind the association between PM and health effects are not fully understood; even the results of *in vitro* toxicological research have shown that PM induces several types of adverse cellular effects, including cytotoxicity, mutagenicity, DNA damage, and stimulation of proinflammatory cytokine production. Because traffic is an important source of PM emission, it seems obvious that traffic intensity has an important impact on both quantitative and qualitative aspects of ambient PM, including its chemicophysical and toxicological characteristics. This evaluation shows that, in general, the smaller PM fractions have the highest toxicity, contain higher concentrations of extractable organic matter (comprising a wide spectrum of chemical substances), and possess a relatively high radical-generating capacity. Also, associations between chemical characteristics and PM toxicity tend to be stronger for the smaller PM size fractions. Most importantly, traffic intensity does not always explain local differences in PM toxicity, and these differences are not necessarily related to PM mass concentrations.

It has just been outlined in Section 10.2 that there is an increase in the association between respiratory disease and levels of PM_{10}, especially among susceptible populations. Current research suggests that this fraction may contribute predominantly to possible toxic and proinflammatory effects by its constituents, which include biological agents (Camatini *et al.*, 2012), metals, and organic compounds.

Recently, attention has been focused on the fraction constituted of particles with an aerodynamic diameter lower than 2.5 μm ($PM_{2.5}$). These particles, which are usually present in high number in PM samples, may be more harmful than larger ones, as they are more efficiently retained in the alveolar lung. As a general rule, the primary biological targets of inhaled particles are cells of the pulmonary epithelium as well as resident macrophages. Among these cells, the alveolar epithelium seems to be affected in a number of respiratory diseases. Fine fraction is frequently suggested to be responsible for cardiac disorders and acute CV events, such as myocardial infarction, as well as inflammatory pathologies. Increased plasma viscosity and other changes in blood-related parameters have been detected after inhalation of fine PM. However, the biological mechanisms of such events are still unclear; even the pulmonary endothelium appears to be the initial target site for UFP, which can translocate into the blood vessels and reach other organs, as illustrated in Figure 10.1.

10.3.2.1 Lung Inflammation

Major concerns for human health from exposure to PM include effects on breathing and respiratory systems, damage to lung tissue, cancer, and premature death. The elderly, children, and people with chronic lung disease, influenza, or asthma are especially sensitive to PM. To protect human health, the US EPA has established the NAQS to regulate "criteria" of air pollutants: O_3, PM, NO_x, SO_x, lead, and carbon monoxide (CO). Regulation of these criteria through the Clean Air Act is based on the scientific evidence that exposure to air pollution is associated with the development (long-term effect) or exacerbation (short-term effect) of a variety of respiratory conditions (Plummer, Smiley-Jewell, and Pinkerton, 2012). Acute and chronic exposures to atmospheric gases and PM are associated with increased emergency room visits and hospital admissions for exacerbations of pulmonary conditions, including asthma and chronic obstructive pulmonary disease. Adverse human health effects from these key pollutants may still occur at low levels of exposure. The potential impacts of adverse health effects of low levels of air pollution are of particular concern for susceptible populations, including children, the elderly, and those with preexisting respiratory conditions. Inhaled pollutants initiate rapid and efficient pulmonary defense mechanisms following interaction with airway cells as part of the immune response. These cellular and biochemical defense mechanisms function to prevent excessive tissue damage and are orchestrated through numerous redundant inflammatory and oxidative stress pathways. These effects may result from direct interaction of air pollutants with various receptors or via interactions with secondary mediators within these inflammatory pathways. Numerous investigations of air pollution-induced health effects are investigating the mechanism behind the inflammatory response to inhaled air pollutants using human, laboratory, and cellular systems (Plummer, Smiley-Jewell, and Pinkerton, 2012). It has just been reported (see Section 10.2) that PM physicochemical nature is influenced by many factors such as particle size, emission season, and site; in such a way it seems difficult to understand which is the contribution promoting the inflammatory pathways. PM inhalation induces a local response in the lung that is initiated by alveolar macrophages and airway epithelial cells. The macrophages are several times more potent in producing proinflammatory mediators that contribute to the local inflammatory response; moreover they promote also the subsequent systemic inflammatory response (Hogg and Van Eeden, 2009). Recently different sized PM collected in a very polluted city and their effects on mice lungs and extra pulmonary tissues have been investigated (Figure 10.4; Table 10.4) (Farina *et al.*, 2013).

Different proinflammatory, cytotoxic, and procarcinogenic markers and TLR4 receptor activation pathway have been evaluated in mice bronchoalveolar lavage fluid (BALf), lungs, blood, heart, and brain, both at 24 h after one PM_{10} intratracheal instillations and 1 week after three instillation (Table 10.4).

Gene upregulation (HMOX1, Cyp1B1, IL-1β, MIP-2, MPO) and miR-21 have been investigated in the lungs and blood. Inflammation in the respiratory tract of PM_{10}-treated mice has been evidenced in BALf and lung parenchyma by increase of defense cells and cytokine levels.

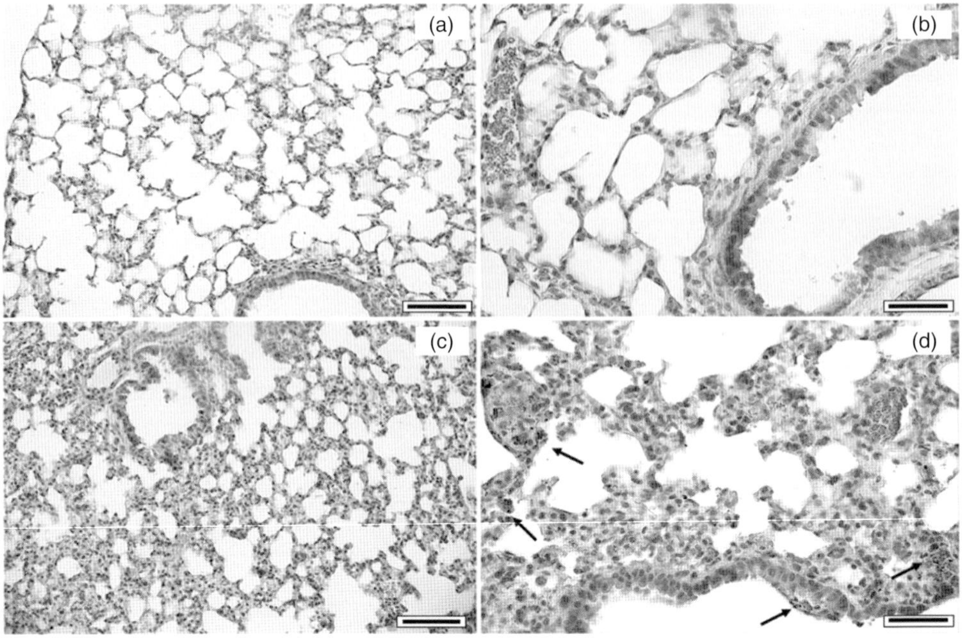

Figure 10.4 Histology of lung tissue 24 h after the third intratracheal instillation. (a, b) Lung parenchyma of sham mice. (c, d) PM_{10} sum-treated lung parenchyma, showing inflammatory cells recruitment and AM infiltration (arrows) in the connective surrounding terminal bronchioles. Each figure represents the status evidenced examining 6 sham and 6 PM_{10} sum-treated mice. a and c bars = 150 μm; b and d bars = 50 μm. (From Farina *et al.* (2013).)

The blood endothelium too was activated as demonstrated by a significant increase of plasma proteins. Furthermore PM_{10} determined the heart endothelial activation and caused an increase of defense enzymes in the brain (Figure 10.5).

These results demonstrate the translocation of inflammation mediators from the lungs to the bloodstream, thus triggering a systemic reaction, which involves heart and brain and provides additional insight toward the toxicity of inhaled PM. Prolonged exposure to antigens via the airways can cause severe pulmonary arterial remodeling; in fact vascular changes have been well documented in asthma, a typical airway disease. These experimental and clinical findings link the responses in the airways with those in the lung's vasculature. Consequently PM can cause, or exacerbate, diseases in the pulmonary circulation. Evidences on the PM health effects on the lung's vasculature with emphasis on the lung's inflammatory responses to the particle deposition and pulmonary hypertension have been described (Grunig *et al.*, 2014). The American Thoracic Society (Leary *et al.*, 2013) has demonstrated that living in proximity to roadways, where exposure to high $PM_{2.5}$ levels and NO_x is very high, is associated with a greater right ventricular mass and changes in right ventricular function. Right ventricular mass is a measure of the damage, which can be caused by molecular changes

Table 10.4 Immunoblotting analysis in the lung (a), heart (b), and brain (c) parenchyma from sham ($n = 6$) and PM_{10} sum-treated mice ($n = 6$), 24 h after the third intratracheal instillation.

		Sham ($n = 6$)		PM_{10} sum ($n = 6$)		
		Mean	± SE	Mean	± SE	p
(a) Lung	ET-1	1	0.1	1.69	0.52	*
	HO-1	1	0.14	3.07	0.64	*
	Hsp70	1	0.19	1.68	0.23	*
	Cyp1B1	1	0.26	0.39	0.11	*
	iNOS	1	0.14	0.75	0.08	
	MPO	1	0.24	1.32	0.32	
	Casp8-p18	1	0.24	1.04	0.15	
	Casp3-p17	1	0.32	1.1	0.2	
	pIRAK1/IRAK1	1	0.16	0.66	0.09	
	pTAK1/TAK1	1	0.24	1.21	0.12	
	pIKBα/IKBα	1	0.18	1.05	0.08	
(b) Heart	ET-1	1	0.24	2.41	0.58	*
	Cyp1B1	1	0.37	2.65	0.18	*
	Hsp70	1	0.33	2.04	0.11	*
	HO-1	1	0.44	1.5	0.27	
	MPO	1	0.26	2.45	0.11	
	Casp8-p18	1	0.27	1.02	0.14	
	Casp3-p17	1	0.34	0.72	0.22	
(c) Brain	ET-1	1	0.2	2.86	0.86	*
	HO-1	1	0.22	2.3	0.72	*
	Cyp1B1	1	0.31	0.68	0.25	
	CD68	1	1.14	4.5	1.06	
	Soluble TNF-α	1	0.24	2.22	0.36	
	Memb TNF-α	1	0.2	1.19	0.41	
	Hsp70	1	0.2	1.17	0.28	
	GFAP	1	0.29	1.03	0.22	
	Casp8-p18	1	0.31	1.35	0.52	
	Casp3-p17	1	0.29	1.65	0.42	

The proteins have been normalized to β-actin, and each protein in PM_{10}-treated group has been normalized onto respective sham group. All the data are expressed as mean 6 SE sham versus PM_{10} sum-treated: $^*p = 0.05$.
Source: From Farina *et al.* (2013).

in the right ventricle or changes in the pulmonary circulation, in turn causing limitations in the blood flow. These evidences even need a further confirmation indicating that air pollution can exacerbate parameters of the pulmonary hypertension. A deeper understanding of the health implications of PM exposure will furnish suitable indication to manage lung diseases, including those involving pulmonary circulation.

10.3.2.2 Cardiovascular Damages

Two decades of research have shown that air pollution can trigger heart attacks, strokes, and irregular heart rhythms, particularly in people already at risk for these

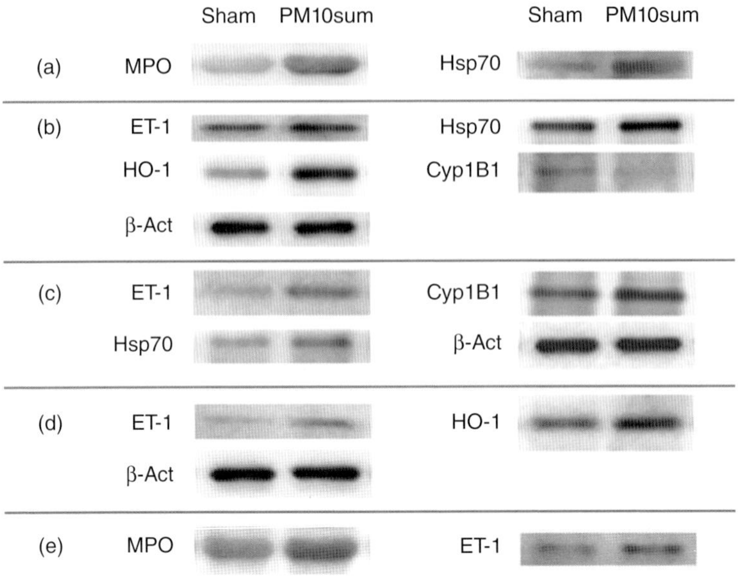

Figure 10.5 Western blots displaying proteins in BALf (a), lung (b), heart (c), brain (d), and plasma (e) in control (sham) and PM_{10} sum-treated mice, 24 h after the third intratracheal instillation. Reported blots are representative of 6 sham and 6 PM_{10} sum-treated mice. (From Farina *et al.* (2013)).

conditions. The relationship between PM and the CVD has received public attention only recently. Several reviews of these studies conclude that air pollution poses a risk to heart health (Gold and Mittleman, 2013). The EPA, American Heart Association (AHA), American Stroke Association, and American College of Cardiology (ACC) have published a concise leaflet on "Heart Disease, Stroke and Outdoor Air Pollution." AHA has published also "Guidelines for Primary Prevention of Cardiovascular Disease and Stroke" as an aid to healthcare professionals and their patients without established coronary artery disease or other atherosclerotic diseases. The statement intends to complement the AHA/ACC "Guidelines for Preventing Heart Attack and Death in Patients with Atherosclerotic Cardiovascular Disease." These recommendations emphasize multifactorial interventions with suggested measures to modify individual CV risk factors with diet, drugs, exercise, weight management, and complete smoking cessation. Over the last decade, however, a growing body of epidemiological, clinical, and experimental evidence has led to a heightened concern about the relation between PM and heart disease. Besides PM fractions, attention is actually also dedicated to CO, NO_x, SO_2, O_3, and lead. All these pollutants are associated with increased hospitalization and mortality due to CVDs especially in persons with congestive heart failure or frequent arrhythmias. The well-established causal associations between active and passive smoking with heart disease and stroke support the plausibility of an adverse effect of PM on the CV system. As reported in Section 10.2.3, the deleterious health effect of an increment of 10 $\mu g\, m^{-3}$ has been detected when a range of maximum city-specific 24-h PM_{10} concentrations was from 26 to 534 μg m^{-3}. Besides ambient exposure, measures have been performed for job exposure too. An increasing risk of myocardial infarction has been found in above 3000 Swedish workers with cumulative exposure to products from nonvehicular combustion processes. Short-term exposure to PM can increase blood pressure and cause arterial vasoconstriction and changes in vascular tone. These immediate effects are probably driven by the changes in activation of the autonomic nervous system, as indicated by the coincident changes in HRV (Figure 10.1). At the same time, blood becomes more prothrombotic following exposure, which may accelerate the formation of a thrombus in the circulatory system. Because of impaired endothelial vasomotor responses, arteries cannot dilate to compensate for the impending obstruction, and endogenous systems – which normally limit coagulation progression – are also impaired so limiting the ability of the vascular system to overpass this damage. The end result of this cascade of effects is occlusion of the coronary artery and acute myocardial infarction. Furthermore, because of changes in the autonomic control of the heart, exposed persons may become more prone to developing ventricular arrhythmias in conjunction with myocardial infarction.

It is difficult to attribute this vascular endothelial dysfunction either to activation of the autonomic nervous system, as these changes appear to be immediate and transient, or to systemic inflammation, as studies have failed to demonstrate any changes in the inflammatory cascade for at least 18 h following exposure to pollutants. This suggests that a third mechanism may be responsible for these effects. It has been proposed that $PM_{2.5}$ and UFP, which are of a similar physical size to low-density lipoprotein (LDL) cholesterol particles, are able to translocate from the lungs into the systemic circulation. Although only a small fraction of the inhaled mass may enter the circulation, this represents a huge number of particles that may have a direct impact on the vascular endothelium. Besides all these evidences, it has to be outlined that for most people, air pollution poses only a small risk as a trigger for heart attacks, strokes, or irregular heart rhythm. Nevertheless, some people with known pathologies are at higher risk, and these are the ones who have had a heart attack or angioplasty or angina, heart failure, some types of heart rhythm problems, or diabetes. Studies suggest that hot and cold weather may increase the risk of heart attacks and these atmospheric weather changes together with high pollution may be particularly dangerous for those at risk (Gold and Mittleman, 2013). Short-term exposure to a high amount of PM_{10} leads to higher hospitalization rates for CVDs and increases risk for myocardial infarction and ischemic stroke (IS), even 24 h after peak of pollution. Acute exposure to traffic-derived PM has been linked to the triggering of acute myocardial infarction, suggesting a direct causal link. It has been proposed that inhaled particles may trigger an inflammatory response within the alveoli, with a subsequent secondary systemic inflammatory effect resulting in CVD.

Alternatively, the UFP for their small size may translocate across the alveolar membrane, gaining access to the bloodstream and directly influencing the vascular endothelium. Other pathways may also contribute, including translocation and stimulation of sensory receptors in the upper airways and nasal epithelium, translocation across the gastrointestinal tract after mucociliary clearance, and accumulation of particle constituents. There is good evidence for all the main hypotheses and each pathway may play an important role in the overall CV response.

Oxidative stress is involved in all the mechanisms suggested (Figure 10.11) (Miller, Shaw, and Langrish, 2012). Several different types of PM have been shown to exert an oxidative insult alone in cell cultures, in isolated tissues, or in animal models, and evidence is now beginning to emerge in man too. Even the exact mechanism remains to be established, a combined oxidative insult from PM itself, through stimulation of ROS-generating enzymes in the vascular endothelium and indirectly through activation of inflammatory cells, is the main responsible. Certainly PM-induced oxidative stress will exacerbate its detrimental CV effects. Identification and removal of the PM constituents able to generate free radicals may represent a way to limit the harmful effects of air pollution. Although the absolute increase in risk associated with PM exposure is modest in comparison with other risk factors, the ubiquitous nature of air pollution exposure means that the whole population is affected (Miller, Shaw, and Langrish, 2012). Even if the causality and biological mechanisms need further investigation, air pollution is a pervasive public health issue with major CV and healthcare economic consequences presenting a key target for intervention (Shah *et al.*, 2013).

Genetic susceptibility is likely to play a role in response to air pollution; therefore, gene–environment interaction studies can be a tool for exploring the mechanisms and the importance of the pathway in the association between air pollution and CV outcome (Zanobetti, Baccarelli, and Schwartz, 2011).

10.3.2.3 Brain and Other Target Organs

Accumulating evidence suggests that PM may have a significant impact on central nervous system (CNS) (Block and Calderón-Garcidueñas, 2009). About a decade ago, the CNS has also been proposed to be a target organ for the detrimental effects of airborne pollutants (Oberdörster and Utell, 2002). Emerging evidence from recent epidemiological, clinical, and experimental studies suggests that certain neurological diseases, such as Alzheimer's disease (AD), Parkinson's disease (PD), and stroke, may be strongly associated with ambient air pollution (Tonelli and Postolache, 2012). Air pollutants affect the CNS either directly by transport of UFP into the CNS or through systemic inflammations. Either of the effects can be caused either by the physical characteristics of the particle itself or by toxic compounds adsorbed on them. Although the exact mechanisms underlying brain pathology are not fully understood, several lines of current evidence point out that neuroinflammation, oxidative stress, glial activation, and cerebrovascular damage might be the primary pathways activated by inhaled UFP. A major risk of UFP arises from the fact that they are able to penetrate deep into the lung, where they eventually enter the blood circulation and can get distributed throughout the body. Increasing evidence links air pollution to CNS diseases, behavior

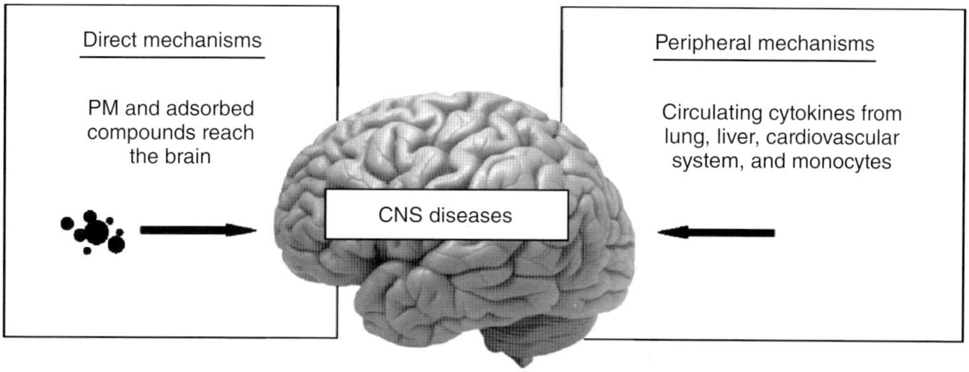

Figure 10.6 Air pollution impacts the brain through multiple pathways; PM may cause diverse CNS pathology through several interrelated mechanisms. Some CNS effects have been attributed to specific PM components; even actually a clear pathway responsible for CNS damage has yet to be identified.

deficits, neuroinflammation, and neuropathology in human and animal studies (Figure 10.6).

Cerebrovascular disease, common in the elderly, precedes and accompanies age-related cognitive dysfunction and neurodegeneration (Zlokovic, 2011) and is a risk factor not only for vascular dementia (VD) but also for AD. AD and VD often coexist in people suffering from dementia. Accumulation of amyloid beta (Aβ) in the vessel wall triggers cerebrovascular structure changes, vascular fibrosis, and inflammation. The blood–brain barrier (BBB) microvasculature, having a crucial role in the regulation of cerebral blood flow (Zlokovic, 2011) which guides the entry into the brain of molecules triggering inflammation, oxidative stress, and cytotoxicity, plays a pivotal role in AD and AD-like pathogenesis. Soluble factors, released from endothelial cells, glial cells, neurons, circulating cells, and platelets, after UFP-induced damages, might influence different intracellular pathways, finally converging on brain damage and neuronal death mechanisms. Among soluble factors released after vessel damages, vascular NO production and activity appear to be an important contributor to AD. NO is a regulatory molecule for the host defense that plays a fundamental role in the CV, immune, and nervous systems. NO is synthesized by the enzyme NO synthase (NOS). Recent evidence supports that this bioactive molecule affects the brain and vascular homeostasis, influencing pathology typifying AD. Inhalation of diesel exhaust particle (DEP) impairs vascular function in *in vivo* animals and in man too by a mechanism still under investigation. It has been hypothesized that exposure to DEP and UFP would cause endothelial dysfunction as a consequence of systemic inflammation. It is well known that UFP can generate ROS that in turn can produce or regulate apoptosis typically through caspase-3 activation (Figure 10.7). *In vitro* studies have demonstrated that DEP, UFP, and their organic extracts can induce HO-1 expression in various cell lines. Previous studies of DEP have shown that organic compounds play

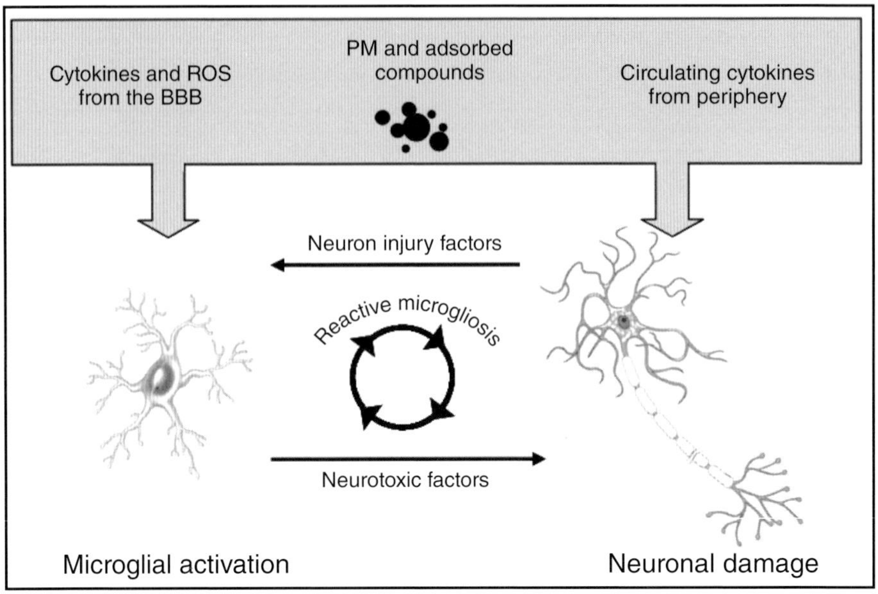

Figure 10.7 Microglia may be activated by proinflammatory stimuli, or in response to neuronal damage, air pollution can contribute to toxic microglial activation through different mechanisms.

an important role in their prooxidative and proinflammatory effects. PAH and their oxygenated derivatives (e.g., quinones), which are able to redox cycle and generate ROS, may target cell populations such as cerebrovascular endothelial cells and neurons. HO-1 expression is dependent on the function of antioxidant response element in its promoter and is typically induced by the DEP extract. At high doses DEP activate intracellular pathways responsible for proinflammatory effects. Activation of these cascades may constitute the principal mechanism by which UFP exerts adjuvant effects in AD-like diseases. Thus in this scenario inflammation and oxidative stress may increase Aβ generation, inducing further neuroinflammation in AD patients. Therefore the Aβ-increasing intracellular deposits might promote the HO-1 upregulation and the intracellular oxidative stress modifying the mitochondrial membrane permeability. Consequently the Aβ accumulation within cells may trigger event which sustains the persistent upregulation of HO-1, thus disclosing its cytotoxic rather than its protective role.

In the presence of relatively high concentrations of UFP, HO-1 may undergo to a transition from cytoprotective to cytotoxic effects. Thus the mechanisms contributing to oxidative damage and mitochondrial insufficiency in brain aging and AD may include air pollution as a further promoting agent. Neuropathological analyses of postmortem tissues from humans living in highly polluted areas have shown elevations in particular proteins.

All these data then suggest that exposure to air pollution has the potential to determine neurodegenerative disease pathways. However, at present the evidence is still insufficient to clearly demonstrate PM involvement to neurons damage, either directly or through inflammation. Some of the individual constituents or characteristics of specific components have already been implicated in neurotoxicity and nervous system disease independently. Recent reports reveal that controlled inhalation exposure to ambient PM, O_3, and DEP results in elevated cytokine expression and oxidative stress in the brain. At present, the mechanisms responsible for the initiation of neuroinflammation in response to UFP are poorly understood and may be exposure specific. One theory implicates activated brain microglia in neuroinflammation (Campbell, 2004). Microglia are the resident innate immune cells in the brain and are predominant regulators of neuroinflammation, evidenced by their production of proinflammatory proteins: IL-1β, TNF-α, and ROS. Although microglia are necessary for normal brain function, excessive and chronic activation can result in neurotoxicity, such as the initiation and/or amplification of neuronal damage. Microglia are activated in neurodegenerative diseases, as indicated by analysis of postmortem brains from AD and PD patients, and have been implicated in the progressive nature of each disease (Figure 10.7).

The systemic inflammatory response is characterized by mobilization of inflammatory cells from the bone marrow into the circulatory system, followed by their activation, as well as the production of acute-phase proteins by the liver, and an increase in circulating inflammatory mediators. Over the last decade an increasing number of studies have provided data showing a relationship between PM exposure and IS. Currently available data suggest that PM exposure is a risk factor for IS, especially in patients with preexisting illnesses. With regard to the mechanisms leading to PM-dependent vascular damage, proinflammatory, prooxidative, and proatherogenic pathways have been suggested to be involved. Public policies and clinical guidelines emphasize lifestyle modifications, including healthy environments and social context, that facilitate appropriate caloric consumption and physical activity. In 2011, the US National Toxicology Program conducted an international workshop review to comprehensively evaluate the epidemiologic and experimental evidence on the relationship of environmental chemicals with obesity, diabetes, and metabolic syndrome. Despite differences in the pollutants, age groups, and study designs, most of the studies showed weak evidence of an association between diabetes and air pollution. Particles originating from human activities have existed for millennia, but recently the anthropogenic particulate pollution has markedly increased. Significantly, technological advancement (Euro 4 and 5 new engines) has also changed the PM character, increasing the proportion of UFP and expanding the variety of chemical compositions. It remains to tackle the issue of the effect of the exposure to UFP on the CNS homeostasis by the synergistic use of *in vivo* and *in vitro* approaches combined with a strong bioinformatics support in order to be able to indicate early molecular markers specifically referable to UFP action mode within the inner molecular pathways, leading toward neuroinflammation and neurodegenerative process.

10.3.3 *In Vitro* Evidences

The difficulty of investigating PM toxicity lies in the variability and complexity of this pollutant, whose heterogeneity in structure and composition can lead to different biological outcomes. PM is indeed a mixture of different chemical components, organic compounds, metals, ions, and elemental carbon. Toxicological studies are thus fundamental for understanding the causal relation so to explain the epidemiological findings. *In vitro* toxicology has the advantages to provide controlled and simplified systems that can help to clarify the PM mechanisms of action and the molecular pathways involved. Moreover, the repeatability of the *in vitro* systems helps in comparative analyses of different PM classes, while the reproducibility of experiments helps to associate specific biological effects with particular physicochemical characteristics. *In vitro* systems are therefore a fundamental step to uncover the correlation between particles properties and biological outcomes.

In vitro experiments are an important alternative to animal testing along with *in silico* simulation. The main reasons for alternatives to *in vivo* systems are summarized in the 3Rs: reduction, refinement, and replacement. Cell cultures are primary alternatives when considering reduction and replacement of animals. Indeed *in vitro* systems may be important tools for diminishing the number of animal sacrificed for testing or completely avoiding their use when the goal is testing molecular or cellular mechanisms.

In vitro systems are widely applied in PM toxicological studies since they allow the analysis of molecular pathways activated by PM exposure at cellular level and therefore add fundamental information for explaining the fine mechanisms triggering the biological outcomes observed in humans. The *in vitro* systems used for PM toxicity analysis are cell cultures of various cell types. Most of them are alveolar and bronchial epithelial cell lines, representative of lung epithelia; monocyte/macrophage cells, used to mimic the immune system defenses; and endothelial cell lines, for understanding the activation of the endothelium in response to PM. These cell lines are also used in cocultured systems, that is, culturing two different cell lines together, in order to assess the crosstalk among different cell types in determining significant alteration in tissues. Recently cell lines from other organs have received attention since UFP are able to cross the ABB to reach other organs.

The following paragraphs report the evidences resulting from *in vitro* studies describing different cellular mechanisms related to PM exposure.

10.3.3.1 Inflammatory Response

Airway inflammatory response is a normal function of lung tissues, activated mainly by lung immune system cells in response to different stimuli. It consists in the identification of a possible threat; the activation of biochemical messengers which signals the presence of the threat; the recruitment of macrophages (scavenger cells which are able to engulf, internalize, and degrade adverse elements, such as bacteria); and the final elimination of these elements. During recognition, signaling, and elimination processes, a number of biochemical signaling molecules (cytokines, chemokines, ROS, etc.) are released by epithelial cells. The resulting inflammatory response is a central process of the organism against adverse stimuli. When it occurs with a high intensity or for a prolonged period, it promotes airway inflammation and lung diseases such as chronic obstructive pulmonary disease and asthma.

Cytokines are signaling proteins involved in the regulation of many physiological events; in the inflammatory process, they can act as pro- or anti-inflammatory mediators. Chemoattractant cytokines (chemokine) are produced and released in the case of injury, or other stimuli, to recruit and activate the immune cells. Interleukins 8 and 6 (IL-8 and IL-6) are well-known and widely studied proinflammatory cytokines; despite some functional redundancy, these proteins present specific functions. IL-8 is fast released after recognition of activating stimuli and immediately elicits the recruitment of immune cells (in particular neutrophils, basophils, and macrophages) at the site of threat; IL-8 therefore represents the first route of cellular defense since its effect is rapid and long lasting.

On the contrary, IL-6 is mostly involved in the regulation of acute-phase response to injury, and it induces the final maturation of cells to antibody-producing defense cells.

Complex molecular pathways perform the regulation of cytokine release. The different pathways may involve the activation of cell membrane receptors, such as Toll-like receptors (TLRs) or the activation of other stress mediators including tumor necrosis factor (TNF) and interleukin-1β (IL-1β).

TLRs are membrane proteins involved in the recognition of specific molecules and disseminating the signal to intracellular transcription factors; most of TLRs are responsible for the activation of cell pathways that respond to danger stimuli such as the presence of bacterial fragments or other molecules. In general these membrane receptors recognize both conserved microbial ligands, termed pathogen-associated molecular patterns, and endogenous ligands derived from stressed cells, termed damage-associated molecular patterns.

Upon activation of TLRs a cascade of intracellular signals triggers the transcription of DNA into RNA which will be in turn translated to proteins, which are the final effectors of cell responses. Usually TLRs activate the transcription of DNA region encoding IL proteins, and they are therefore considered one of the main pathways leading to the proinflammatory responses.

The relation between PM exposure and short-term lung inflammatory effect in humans has been largely accepted. However, local lung inflammation may result in a more complex and systemic response that can contribute also to CV effects and increase the risk of stroke (McColl, Rothwell, and Allan, 2007). The reason underlying the propagation of inflammation from the lung to other organs is linked to the presence of UFP, soluble PM components, or inflammatory mediators themselves, which may translocate from the lung to the circulatory system. Once in the bloodstream, biochemical messengers and/or pollutants may exert their action interacting with other cell types (Mills *et al.*, 2007). The possibility to uncover these systemic effects requires *in vivo* experiments (see Section 10.3.2) which allow the evaluation of the effects of inhaled PM toward other organs. The *in vitro* experiments too, when based on coculture systems of lung epithelia and endothelial cells, may provide useful information on the ability of PM or biochemical messengers to cross the ABB.

In this respect, IL-1β may be a major contributor to the systemic responses induced by PM inhalation. IL-1β is a proinflammatory cytokine involved in many cellular functions, such as proliferation and activation of the immune response through the induction of other interleukins, and, noteworthy, it has a critical role in the systemic phase of inflammation (McColl, Rothwell, and Allan, 2007). Several *in vitro* experiments have demonstrated the PM ability to increase significantly the levels of IL-1β, suggesting also the possible activation mechanisms for this response.

PM inflammatory potential has been mainly related to PM_{10}, while $PM_{2.5}$ and UFP have been found to be less effective. This observation has been linked to a higher content of endotoxins in PM_{10}. Endotoxins, such as lipopolysaccharides (LPS), are components of Gram-negative bacteria wall, which are known to induce inflammation in cells through the activation of specific TLRs.

Actually many studies on PM effects suggest that endotoxin is not the sole factor eliciting PM-induced inflammation; metals, contained in particles, have been related to the PM inflammatory potential (Veranth *et al.*, 2006). The metal-related inflammatory effect has been partly ascribed to their capability to induce ROS formation determining a state of oxidative stress (Figure 10.8).

PM for its complex composition may activate a fine regulated cell pathway. The release of IL-1β requires the activation of membrane TLRs as well as the internalization of particles, promoting ROS production. IL-1β release is associated with lung diseases such as silicosis and asbestosis, and some *in vitro* evidences show that PM_{10} is able to determine a significant release of the interleukin at low exposure doses.

In vitro studies have demonstrated that PM inflammatory potential depends on the chemical composition of particles rather than their dimension. PM_{10} is a potent inducer of inflammation due to the presence of specific components which are not present, or only in small quantity, in fine PM. The relation "PM composition–source–size" is thus a parameter much more significant for the understanding of PM effects than particle size only.

10.3.3.2 **Oxidative Stress**

In simple terms oxidative stress is the overwhelming production of free radical over cell antioxidant defense mechanisms. Reactive species, mainly oxygen and nitrogen ones, are commonly generated during a number of cell physiological processes. Several enzymes catalyze oxygen reactions, the most important of which is oxidative phosphorylation, which takes place in mitochondria allowing ATP production and cell survival. During oxidative phosphorylation, a number of ROS are formed, but the rate and frequency of formation are low. In normal conditions cells are able to cope with ROS through a number of antioxidant molecules, such as

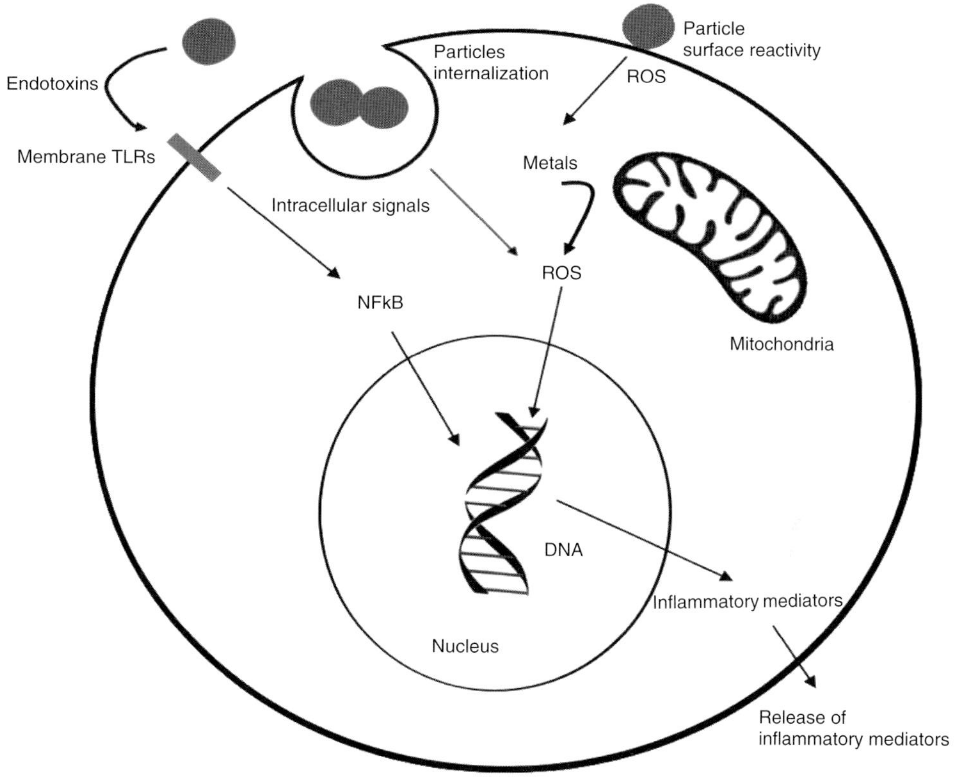

Figure 10.8 Hypothesis of inflammation mechanism in cells exposed to PM.

glutathione (GSH), superoxide dismutase (SOD), and others, which scavenge free radicals. However, external stimuli may determine abundant and/or abnormal formation of ROS, which overcome cell antioxidant capability, promoting oxidative stress. The decrease of cellular GSH/glutathione disulfide (GSSG) ratio is normally considered a parameter for evaluating the redox disequilibrium of cells and oxidative stress conditions. At low levels ROS determine a slight disequilibrium in the GSH/GSSG ratio; in response several oxidative stress response pathways are activated, protecting cellular homeostasis and increasing the ability of cells to scavenge free radicals. However, if these responses fail to provide sufficient protection, ROS are not completely scavenged, and their levels may determine the activation of inflammatory pathways. In this case the response is mediated by molecules sensitive to reactive species that activate other effector molecules able to induce the release of inflammatory mediators. If the level of intracellular ROS is too high, it may cause cytotoxic effects. Mainly these effects are mediated by the release of factors inducing cell death by damaged mitochondria (see Section 10.3.3.4) or activated by genotoxic stress as a consequence of DNA damages. In addition, ROS may interact with phospholipids of cell membranes. The

ROS reaction with lipids determines chemical and functional variations of the membrane and may also activate cell death pathways.

PM-induced ROS are formed through numerous mechanisms including the Fenton reaction of soluble metals, metabolism of organic compounds such as PAHs, quinone redox cycling, cell membrane interactions caused by physicochemical surface properties of particles, and NADPH oxidase activation. ROS may be formed also as a consequence of secondary effects related to mitochondrial damages or particle phagocytosis and consequent oxidative burst in lysosomes, organelles able to degrade external agents, such as bacteria.

10.3.3.3 **DNA Damage**

Toxicological studies have widely described the PM genotoxic effect in humans. These results, together with the epidemiological evidences of lung cancer mortality for long-term exposure to $PM_{2.5}$, suggest a role of DNA damage in PM-related health effects (EPA, 2009). $PM_{2.5}$ has a higher genotoxic potential compared to PM_{10} in *in vitro* systems, and these results agree with the epidemiological studies. The fine fraction from industrialized, traffic, and wood/biomass combustion areas has been found to exert a stronger effect on DNA. The high genotoxicity of these PMs is due to the particle composition and source, since particles derived from combustion processes are rich in metals and organic compounds, such as PAHs (Mehta *et al.*, 2008).

Actually, different mechanisms have been reported for PM-induced DNA damage, which include oxidative stress, organic metabolite formation, and direct interaction of particle organic compounds with the DNA.

Generation of ROS and consequent oxidative stress is widely recognized as an important toxicological mechanism in PM-induced DNA double-strand breaks (DSBs)/oxidation and possibly lung cancer. Besides ROS formation, the metabolic activation of PAHs by P450 cytochromes (CYP1A1/1B1) forms reactive molecules, epoxides, that may form bulky adducts to the DNA. The formation of such adducts is a well-known process in cells exposed to PAHs, such as benzo[*a*]pyrene (BaP) (Park *et al.*, 2009). BaP is one of the organic molecules present in fine PM, and this organic fraction released from particles could activate such processes. Interestingly, it has been recently demonstrated that transition metals (Cr, Cd, Ni, and arsenic) and aldehydes present in PM can inhibit DNA damage repair mechanism, and this effect may enhance the PM genotoxic effect and contribute to lung carcinogenesis (Mehta *et al.*, 2008). In fact, when DNA damage is present, the cell activates different repair mechanisms whose initial step is the DNA lesion recognition and binding by specific sensors such as ataxia telangiectasia mutated (ATM) and AT and Rad3 related (ATR); these proteins recognize quite different kind of damages, ATM being more specific for DSBs and ATR for single-strand breaks (SSBs) and stalled replication forks (Paulsen and Cimprich, 2007). Both the sensors phosphorylate the histone 2AX, a protein which is present in the nucleosome and induces chromatin changes around the lesion site, leading to the recruitment of many checkpoint and repair factors. However, upon DNA DSB formation, phosphorylated H2AX (γH2AX) appears in chromatin as discrete foci, with each focus representing a single point of DSB. Thus, the presence of γH2AX foci has been adopted as clear and consistent quantitative marker for DSBs (Dickey *et al.*, 2009), the most dangerous DNA lesion that can result in genetic abnormalities leading to cancers. The DNA damage detection is then followed by the activation of transcription factors, which induce the expression of genes involved in DNA repair, cell cycle arrest, and cell death. DNA-integrity checkpoints, when DNA is damaged, function to delay the cell cycle progression providing time to repair the lesions.

Cell cycle is a sequence of events which allow replicating cells to growth and finally divide into two daughter cells. Cell cycle is conventionally divided in four phases (gap 1 (G1), synthesis (S), gap 2 (G2), and mitosis (M)) which together form the cell cycle. A number of molecules regulate their cycle, and the passage from a phase to the subsequent requires the satisfaction of specific checkpoints. Checkpoints are mechanisms that verify the proper completion of the phase allowing the continuity of DNA replication and correct cell division (Figure 10.6). If the requirements of a checkpoint are not satisfied, a series of signaling pathways are activated, and these led to cell cycle arrest for repairing the mechanism. One of main sources of checkpoint activation is DNA damage since it is fundamental that cell proliferation is stopped in their presence in order to avoid mutation in DNA of daughter cells (Figure 10.9).

10.3.3.4 **Cell Death**

The evaluation of cell death and/or cell proliferation is a rapid test to determine the PM potential toxic effects. Cytotoxic effects, such as cell death, are usually determined by PM-induced damages too high to prevent the recovery through

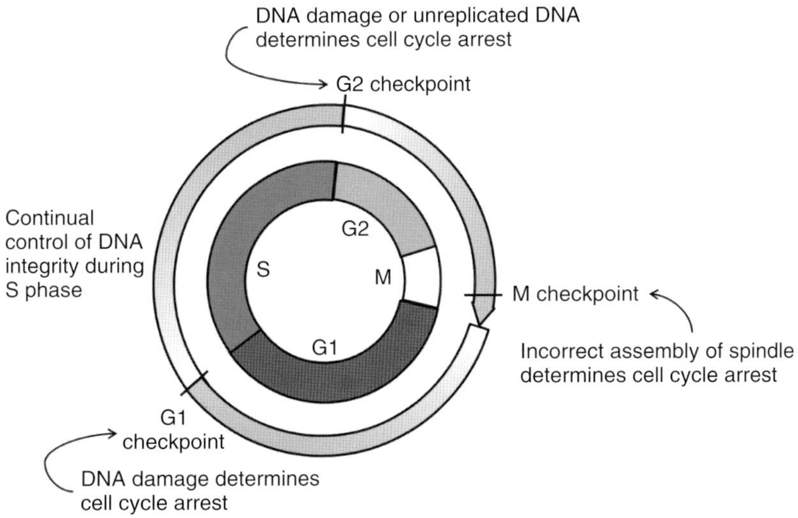

Figure 10.9 Cell cycle, DNA damage, and checkpoint activation in normal proliferating cells.

repair mechanisms. Death can occur in cells roughly by two distinct mechanisms: one is necrosis, caused by factors external to the cells, and the other is apoptosis or generally programmed cell death (PCD), depending on the activation of death pathways within the cells. The specific mechanisms triggered by PM depend not only on the chemical properties of the different fractions but also on the exposure dose; in fact the same stimuli which induce apoptosis at low doses can result in necrosis at higher doses (Elmore, 2007). Necrosis is a passive death process characterized by the failure of the cell to maintain homeostasis, subsequent cell lysis, and release of the cytoplasmic contents into the extracellular space.

In vivo this event contributes to the inflammation and the development of both acute health effects and chronic lung diseases; in fact, the release of inflammatory mediators from the damaged epithelium sustains and prolongs the inflammatory reaction and possibly leads to chronic effects.

Apoptosis is a form of PCD caspase mediated and characterized by specific biochemical and morphological features. The caspases related to apoptosis are protease with specific targets and can be divided in initiators, such as caspases 8 and 9, and effectors, such as caspase 3, of the process. The activation of effector caspases produces the cleavage of DNA and cytoskeleton proteins that leads to the typical morphological features of apoptosis: nuclear fragmentation, cell shrinkage, membrane blebbing, chromatin condensation, formation of apoptotic bodies with intact membrane, and exposure of phosphatidylserine, molecules targeting cell corpses for phagocytosis. Two main apoptotic pathways have been identified:

(i) The extrinsic or death receptor pathway is TNF-α related, since high levels of this protein provoke activation of both initiator and effector caspases. This pathway is partly related to the inflammatory process and ROS formation since both of these events induce high levels of TNF-α (Figure 10.10).
(ii) The intrinsic or mitochondrial pathway involves the release into the cytosol of a specific protein, cytochrome *c*, that is normally confined in the mitochondria and that is able to activate a cascade of events in the cytoplasm; cytochrome *c* released from mitochondria converts an inactive form of a initiator caspase into its active form which then promotes apoptosis through other effector caspases. This mechanism involves a complex of proteins including tumor suppressor proteins and the activity of pro- and antiapoptotic proteins, which govern mitochondrial membrane permeability. This pathway is related to DNA damage, which provokes the activation of the proapoptotic proteins responsible of cytochrome *c* release (Figure 10.10).

However, also high levels of ROS may activate the apoptotic mitochondrial pathway directly causing mitochondrial damage. A particular feature of apoptosis is mitotic catastrophe, a type of cell death that occurs during mitosis in the presence of DNA and/or cellular damage. Cell death occurs during the M/A transition and is characterized by the activation of a specific caspase, in response

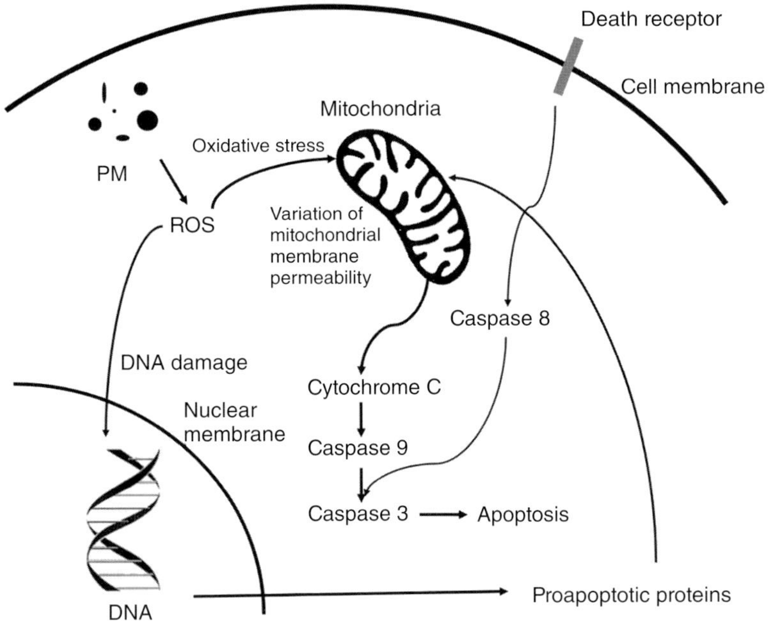

Figure 10.10 Simplified scheme of apoptosis pathways and PM exposure.

to DNA damage, and/or mitochondrial membrane permeability with the release of cytochrome *c* and subsequent activation of caspases.

In cells exposed to PM, both necrotic and apoptotic processes have been widely reported; necrosis is often related to a severe inflammatory status or oxidative stress characterized by high levels of released cytokines, ROS formation, and various metabolites which contribute to epithelial damage. Intrinsic and extrinsic apoptotic pathways have been described especially for $PM_{2.5}$; this fraction has been found to induce apoptosis through the TNF-α-dependent pathway and by the mitochondrial pathway.

Both pathways have been related to severe DNA damage and ROS formation (Figure 10.10). PM-induced apoptosis and necrosis have also been reported in primary human AM, and apoptotic process has been related to the PM organic components.

In *in vitro* systems, it is the PM dose that determines necrosis or apoptosis. For instance, the DEP organic compounds induce apoptosis in cells through the mitochondrial pathway; even higher doses induce necrotic events. On the other hand, PM organic compounds, such as BaP, are known to alter cell proliferation instead of causing cell death. The alteration of genes involved in cell cycle and cell cycle control systems is a possible cause of cell transformation. In the unfortunate event of transformation, cells start to proliferate uncontrollably, eventually determining tumor initiation.

The balance between cell death and cell proliferative signals following PM exposure is of great importance in determining extremely different cascade of events. The research is focusing now on unraveling the mechanisms by which PM may trigger biological effects as demonstrated by epidemiological data and *in vivo* experiments.

10.4 Mechanism of Effects

10.4.1 The Inflammatory Paradigm

The prevalence of asthma, a chronic disease characterized by swelling and inflammation of lung tissues, is increasing worldwide, and its relation with PM levels has been recently described (Guarnieri and Balmes, 2014). Also increases in lung respiratory infections in at-risk groups such as children and elderly (Brugha and Grigg, 2014) are reported.

The balance between immune response inflammation and the ability of the immune system to tolerate stress signals is fundamental for lung tissue homeostasis. Uncontrolled response of lung cells to external stimuli and inflammation lead to lung disease, such as asthma. On the other hand, increased tolerance of the immune system to external stimuli leads to higher possibility of infection for the lowered capability of immune system to cope in due time with pathogens. Inflammatory response is the primary and key response of lung immune cells to constant impact from daily inhaled air.

Epidemiological and clinical studies have clearly shown the relation among PM air concentration, inflammatory response, and lung diseases. The ability of PM to alter the fine regulated inflammatory balance is therefore at the base of a number of health outcomes. *In vitro* and *in vivo* studies have described similar PM capability, from different areas of the world, to activate biological pathway related to inflammation.

On one side, *in vitro* and *in vivo* experiments have demonstrated that DEP, the major source of pollution in highly urbanized cities, are able to increase the ability of pathogen bacteria, such as *Streptococcus pneumoniae*, to bind to cells, to reduce the ability of pulmonary clearance, and to determine a state of infection. Furthermore, PM is also able to alter the growth and basal response of immune cells, mining at the base the ability of the immune systems to respond to other stress factors, such as viruses causing influenza and other diseases.

The mechanisms determining the increased sensibility of PM-exposed lung to infections have not been completely clarified. This may depend on the fact that epidemiological studies supporting this relation cannot completely separate the contributions from different air pollutants (i.e., PM, O_3, NO_x). Thus *in vitro* and *in vivo* data have not clarified the complex response pathways activated by PM or

DEP, even if they have provided good evidences of causal relation between PM and immune system responses.

The ability of PM to increase levels of cytokines is largely reported, as well as the ability of certain PM components to interact with TLRs and other cell receptors (Figure 10.10). The presence of biogenic components (such as pollens, bacterial wall endotoxins, viruses, and fungal spores) in PM_{10} is related to inflammatory responses *in vitro* and *in vivo*. In addition, PM may act as sensitizer, therefore increasing the ability of cellular receptors to react in response to other ligands commonly inhaled and determining in the end an alteration of inflammatory response of lung tissues. The exact mechanism of this sensitization is not fully understood, but *in vitro* experiments have demonstrated an alteration of the expression of receptors in response to DEP.

One big challenge of the research focuses on the understanding of the mechanisms elicited by PM and other pollutants in triggering lung inflammatory diseases such as asthma. Increasing evidences link PM exposure to epigenetic modifications, that is, changes in gene expression, heritable during mitosis, which are not originally coded in the DNA. DNA methylation or modifications of DNA-related proteins (histone acetylation) can alter gene expression related to different cellular mechanisms, such as inflammatory response.

If lung inflammation is well documented and cellular mechanisms are under investigation, increasing evidences relate PM exposure and inflammatory responses to other organs. The CNS is receiving particular attention due to the number of reports pointing out chain of events relating air pollutants to CNS adverse effects. So far two main mechanisms relate PM exposure and CNS inflammation: the first focuses on the transfer of systemic inflammatory signals to the brain, and the second implies a direct interaction between PM or PM components (UFP or soluble components) and CNS cells. Activation of lung inflammatory response determines the transfer in circulation of cytokines such as IL-1β and TNF-α, which are known to induce CNS inflammation. At the same time, a number of studies have demonstrated that the finest PM particles or soluble components normally adsorbed on PM may reach the brain and induce inflammatory responses, although the mechanisms are still unclear; direct cell–particle interaction may activate cellular inflammatory responses as answer to ROS formation or cell receptor activation. However, some evidence suggest that particles act as Trojan horses, since they transport and release a number of toxic compounds once they reach the brain.

Inflammation plays a central role in PM-related health effects, and its activation determines the alterations of the physiology of different organs. The mechanisms underlying such alterations are not fully understood; even the role of proinflammatory protein and particle translocation is of primary relevance. Lungs are the primary target of PM, but an increasing number of reports show significant inflammatory effect to other organs.

10.4.2
The Reactive Oxygen Species

ROS are generated in response to both endogenous and exogenous stimuli. Under physiological concentrations, ROS act as signaling molecules mediating cell growth, migration, and differentiation, whereas at higher concentrations, they induce cell death, apoptosis, and senescence. Excessive production of ROS exceeds the detoxification capacity of cell antioxidant defenses, thereby triggering a cascade of events closely associated with inflammation.

ROS generate oxidative stress, which can be removed by the body's defenses. Oxygen-derived free radicals such as superoxide (O_2^-) and hydroxyl free radicals, as well as other oxidants, such as hydrogen peroxide, are the key mediators that drive oxidative stress and impair cellular function. Free radical generation is a normal by-product of cellular respiration, and redox regulation of proteins is essential for normal cell function. However, depletion of antioxidants can lead to the formation of free radicals able to contribute to disease processes (Ziech *et al.*, 2010). High levels of ROS cause a change in the redox status of the cell, triggering a cascade of events associated with inflammation and, at higher concentrations, apoptosis. ROS are formed through the reduction of oxygen by the reducing agents NADH and NADPH, with redox-active chemical species such as organic chemicals and metals.

Numerous types of PM may generate oxygen free radicals. The oxidative capacity of PM varies with its physiochemical characteristics. *In vitro* studies have demonstrated a clear link between the biological actions of PM and levels of free radicals (Longhin *et al.*, 2013). *In vivo* studies consistently find evidence for increased levels of oxidative stress in PM-exposed animals (Farina *et al.*, 2013). At present it is unclear if particle-derived oxidative stress mediates the CV actions of inhaled particles.

Urban air contains a mixture of oxidizing gases, PAHs, and VOCs, arising from power plants, motor vehicles, wildfires, and waste incinerators. Evidence suggests that chronic exposure to polluted air induces irreversible damage to cellular macromolecules (DNA, proteins, lipids, etc.) via ROS production and their accumulation in cells and tissues.

In order to counterbalance ROS-mediated injury, endogenous antioxidant defense systems exist and function by quenching and clearing intracellular ROS activity and accumulation and maintaining redox equilibrium. This is of great importance in the case of an intracellular imbalance resulting from chronic oxidative stress, where the occurring damage to DNA, lipids, and proteins will result in irreversible modifications of cellular viability and function. The antioxidant defenses such as SOD, glutathione peroxidase, and catalase can counterbalance oxidative stress by chelating superoxide and various other peroxides. Thus ROS can be viewed as critical determinants of intracellular redox states and may be considered an important cell regulatory mechanism both for health and disease (Figure 10.11).

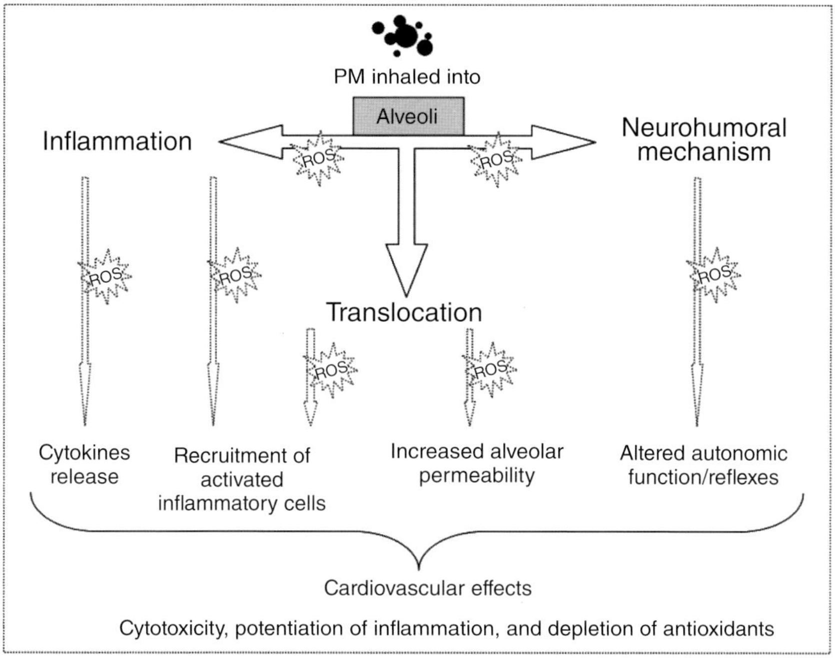

Figure 10.11 Inhaled particles induce an inflammatory response in the lungs, leading to the release of cytokines and other mediators that enter the systemic circulation. UFP translocate across the alveolar wall and directly interact with the cardiovascular system. Particles may activate the autonomic nervous system through sensory receptors on the alveolar surface. The ROS could play a role in exacerbating the stages of each pathway, as well as in promoting interactions between pathways (e.g., in conjunction with inflammation).

The presence of soluble transition metals in PM enhances the inflammatory responses via increased oxidative stress, which occurs after exposure to UFP and DEP, $PM_{2.5}$, and cigarette smoke. After 2 h exposure of rats to concentrated ambient $PM_{2.5}$, the quantity of ROS produced is duplicated in hearts and lungs. This may occur in response to a variety of transition metals or free radical components, present within $PM_{2.5}$ as a result of atmospheric chemical reactions. DEP or its organic extracts may induce apoptosis or necrosis on the cells of the respiratory system for the oxidant effects on mitochondria, decreasing the defenses to respiratory infection or increasing airway reactivity.

There is strong evidence on the potential role of ROS to the pathways linking the pulmonary and CV effects (Figures 10.1 and 10.11). The oxidative stress emerges at numerous points of the three possible hypotheses: particles inhaled into the lungs may trigger an inflammatory response within the alveoli, with a subsequent secondary systemic inflammatory effect, resulting in CV effects; UFP may allow their translocation across the alveolar epithelium, entering the bloodstream and influencing the vascular endothelium; and particles may trigger sensory receptors on the alveolar surface, leading to indirect alterations to CV function (Figure 10.11). Thus at present attention is focused on ROS as the principal signal to PM impact.

10.4.3 Translocation of Particles: If Yes Then Where

UFP are constituents of urban ambient air aerosol, generally traffic linked and of anthropogenic origin, and they are a health risk factor for urban subjects. UFP occur in single and agglomerate forms; their number concentration shows tens of thousand times greater levels in urban aerosol than in nonurban aerosol. These pollutants seem to have more aggressive implications than other respirable fractions of urban aerosol. In literature, it is hypothesized that a chronic exposure to their high number concentrations and their vast surface area, transporting various toxicants, injures tissues or cells and induces inflammation and many adverse health effects. UFP, deposited deep in the tissues, translocate and skip the innate clearance mechanisms. After retention for a long time, they can permeate the lung cells. The molecular changes are proposed to be the subclinical effects that manifest disease exacerbations or the predisposition of subjects to pathologies after exposure to UFP (Kumar, Verma, and Srivastava, 2013). Upon inhalation, airborne UFP come into contact with a series of structural and functional barriers that protect the respiratory system against harmful material. This is important as the internal surface area of the lungs is vast (alveoli and airways have a surface of above 140 m^2), facilitating efficient access to the lung tissue (Figures 10.1 and 10.11). The physiological barriers of the respiratory system may not be effective to protect the body from UFP. Deposition as well as the subsequent fate of inhaled UFP is different from that of larger particles, as it has been described in Section 10.3.1. It has been shown that titanium dioxide (TiO_2) particles with a diameter of less than 0.1 μm cross cellular membranes of a rat lung with a mechanism different from a phagocytic process and that a small fraction of these particles are rapidly transported from the airway lumen to the connective tissue and subsequently released into the systemic circulation. As these particles were also found inside pulmonary capillary erythrocytes, it is not surprising that UFP too have been localized in many other organs of the body, including the liver, heart, and the nervous system, within a few hours after deposition in the respiratory system. Studies of UFP on *in vitro* and *in vivo* systems clearly show that their interactions with the organism, cells, and tissues can be very different from those of larger particles. Moreover apparently it is the particle composition that promotes translocation across the ABB. In fact 240 nm polystyrene particles deposited in the alveolar region were unable to translocate into the blood circulation. When coated with lecithin, these particles translocate and are present in blood monocytes. Thus, it seems that the small dimension is not the only parameter consenting UFP translocation, but their surface properties have an important role. Thus, effects such as inflammation, oxidative stress, and molecular cell activation are likely to occur not only in the primary organ of entry but also in secondary target organs. Such effects are unlikely to occur with larger particles, except under lung particle overload conditions (silicosis of the liver, spleen, and bone marrow) and in the case of asbestos-induced mesothelioma. UFP deposited in the upper and lower respiratory tract may translocate directly to the blood compartment or via the nasal and lung lymphatic circulation (Figures 10.1 and 10.6). Based on studies published so far, UFP translocation rates into the blood circulation appear to be very low. UFP translocation across the BBB is more likely to occur at sites where this barrier is less well developed or injured. Although translocation rates of UFP from the portal of entry to the secondary organs may be very low, a continuous exposure may result in significant accumulation of particles in a secondary target organ (Oberdörster *et al.*, 2009). There are a significant systematic review and statistical analysis of particle translocation from the respiratory system (Nakane, 2012). A categorical regression analysis based on currently available data showed that all of the effects of particle size, particle material, animal species, and exposure route were statistically significant. The effects were large for particle size and particle material and small for exposure route and animal species. In humans, the literature on the translocation of UFP from the lungs into the blood circulation is still conflicting.

There seem to be significant differences in inhaled PM and UFP biokinetics and translocation between rodents and human lungs. Macrophage-mediated clearance kinetics in humans and canine lungs has been shown to be one order of magnitude slower than in rodents.

Further evidence for species differences in particle relocation and accumulation in pulmonary tissues comes from a long-term exposure study to various dusts in monkeys and rats: DEP, coal, and a combination of both. Though a similar retention pattern as in canine pulmonary tissues has been observed in monkeys, particle retention in rats is much more prominent on the alveolar epithelium. While the coal dust was micron sized (<7 μm), the DEP consisted of agglomerated submicron particles. Both these particles were retained similarly in pulmonary tissues of monkeys, thus demonstrating the access of particles of any size to pulmonary tissues. These data suggest that the UFP relocation toward pulmonary tissues is similar in rodents, dogs, and monkeys and most likely in man, even though no detailed long-term clearance kinetics data are known for humans. Yet the kinetics of macrophage-mediated clearance seems to be independent of the location of the retained particle as what the rodent data showed. UFP in rodent lungs are retained predominantly in the long term in interstitial spaces of the alveolar region, and there is only limited translocation toward the circulation.

Particles do not necessarily remain on their site of deposition after inhalation but undergo numerous transport processes within the various sites of the lungs and out of them toward other organs. Hence, these transport processes and their kinetics may differ among the animal models studied. Moreover, the transport processes are significantly affected by particle size, dosimetry, and toxicology; consequently the differences within species and the tested particle size may affect the evaluation of their "translocation properties."

10.4.4 Dimension versus Composition: Two Heads of the "PM Hydra"

If the dimension of particles or their chemical composition is the most important property in determining PM biological effects, researchers are still debating on how and whether to give priority to one over the other. The importance of particle dimensions has been already described (Section 10.3.1) in terms of potential deposition in the airways and possible clearance mechanisms. However, the dimension of particles determines also the activation of cellular mechanisms that have great influence on the final outcomes. Particles with small diameter are rapidly and easily internalized in epithelial cells, while larger particles remain in the extracellular space. Internalization process may promote the activation of specific chemical component adsorbed on PM; in addition larger particles may facilitate the activation of membrane receptors (Figure 10.8). The importance of particle size in determining biological effects is also supported by the notion that the effects observed in biological systems after PM exposure are not easily explained by the chemical components alone. However it is well accepted that chemical and biogenic compounds adsorbed on PM have a fundamental role in their biological effects.

Organic compounds (PAHs, quinones, dioxin-like molecules, etc.) have been associated with DNA damaging effects, mutagenicity, oxidative stress, and cell death. Metals play as well a central role in oxidative stress and inflammation, while biogenic components are pivotal triggers of inflammatory responses.

It is therefore likely that these two PM faces, size and chemical composition, play an important role on biological effects. PM_{10} inflammatory effects are largely related to biogenic components which activate surface receptors, but at the same time metals and crustal elements promote the release of mediators of inflammation. Metals on the other hand when adsorbed on fine or ultrafine PM fractions, which are easily internalized, increase intracellular ROS, promoting rupture of lysosomes (intracellular vesicles with the function to degrade and digest waste cellular debris and dangerous biological compounds).

The relation between size and composition is complex, but the PM hydra has another head that makes the system much more complex; this third parameter is the surface area. Surface area has been indeed related per se to biological outcomes. UFP with their high surface area are able to interact at a larger extent with biological components, such as cellular membrane, and induce local damages. Several experiments, conducted with particles of same composition but different surface area, have shown a relation among surface area, ROS formation, and biological effects.

$PM_{2.5}$ and, above all, UFP have in general a higher surface area than PM_{10}, and this relation link size and surface area. In addition, higher surface area means higher adsorption and absorption potential; therefore, $PM_{2.5}$ may carry more compounds on the surface than PM_{10}. As a consequence there is a link also between surface area and chemical composition. The correlation among size, chemical composition, surface area, and biological effects of PM is extremely complex. However, in this last decade several reports are using a multidisciplinary approach to uncover clear relation, at least, between particles from well-defined emission sources, therefore easily characterized in terms of physicochemical properties and potential biological outcomes. In the future the efforts have to be dedicated to analyze systematically the contribution of selected sources of primary PM and their biological effects.

10.5 Conclusions

PM creates a substantial burden of disease, reducing life expectancy. Numerous people live in cities with levels of PM exceeding WHO-AQG, and only a lightly decreasing trend in average concentrations has been observed in the EU countries over the past decade. Moreover, there is no evidence of a safe level of PM exposure or a threshold below which no adverse health effects occur. Air quality management becomes imperative to achieve WHO-AQG levels and reduce health risks to a minimum, since even at low concentrations, air pollution has a big impact on health.

Exposure to air pollutants cannot base its control to individuals, since it requires action by public authorities at regional, national, and international levels. Measures to reduce air pollution and the associated risk to human health need a multidirected approach, engaging transport, housing, energy production, and industry. The EU directive of 2008 on ambient air quality and cleaner air for Europe explicitly states that "emissions of harmful air pollutants should be avoided, prevented or reduced and appropriate objectives set for ambient air quality taking into account relevant World Health Organization standards, guidelines and program." The year 2013 has been declared the EU year of air by the European Commission, and the scientific evidence on the health effects of air pollutants has been reviewed. Thus evidence not only on the health effects of the major air pollutants (PM, O_3, and NO_x) but also on emissions from individual metals and PAHs has been considered. This activity has been performed in view of a substantial review of EU's air quality policies.

At present the potential adverse health effects of air pollution, particularly among susceptible subpopulations, continue to be of concern, not only in rapidly expanding cities of developing countries where levels of air pollution have increased significantly but also in North American and western European cities where levels have been decreasing. Beside the hundreds of research papers investigating the association between PM and various health endpoints appeared in the literature and the evident relationship between air pollution and health, not enough attention is dedicated to such complex aspect by the Institutions. Even though air pollution rises a myriad of methodological problems, it deserves a more deep consideration, since its impact is directed not only to health but also to less visible damages, such as to materials and ecosystems. The main indirect impact is climate change. The biomass and fossil fuels that cause air pollution have caused the warming of the Earth's atmosphere resulting from the release of GHGs. Recently US EPA has summarized the climate impacts of air pollution as well as the air quality impacts of climate change. The 2013 report of the European Environment Agency has emphasized also that "air quality and climate change can be cost effectively tackled together by using an integrated approach when defining policies and measures." However, the EU policies for controlling air-polluting substances and for mitigating climate change remain separate.

Abbreviations

ABB	air–blood barrier
ACC	American College of Cardiology
AD	Alzheimer's disease
AHA	American Heart Association
ATM	ataxia telangiectasia mutated
ATP	adenosine triphosphate
ATR	ATM and Rad3 related
Aβ	amyloid beta
BALf	bronchoalveolar lavage fluid
BaP	benzo[*a*]pyrene
BBB	blood–brain barrier
CNS	central nervous system
CV	cardiovascular
CVD	cardiovascular disease
DEP	diesel exhaust particle
DNA	deoxyribonucleic acid
DSBs	double-strand breaks
US EPA	US Environmental Protection Agency
FP	fine particles
GSH	glutathione
GSSG	glutathione disulfide
HDL	high-density lipoprotein
HO-1	heme oxygenase-1
HRV	heart rate variability
IARC	International Agency for Research on Cancer
IL-1β	interleukin-1β
LDL	low-density lipoprotein
LPS	lipopolysaccharide
NAD	nicotinamide adenine dinucleotide (coenzyme)
NADPH	nicotinamide adenine dinucleotide phosphate
NAQS	National Ambient Air Quality Standards
NOS	enzyme NO synthase
PAHs	polycyclic aromatic hydrocarbons
PCD	programmed cell death

PD	Parkinson's disease
PM	particulate matter
RNA	ribonucleic acid
ROS	reactive oxygen specie
RR	relative risk
SD	standard deviation
SOD	superoxide dismutase
SSBs	single-strand breaks
TLRs	toll-like receptors
TNF	tumor necrosis factor
TSP	total suspended particles
UFPs	ultrafine particles
VD	vascular dementia
VOCs	volatile organic compounds
WHO-AQG	World Health Organization Air Quality Guidelines

List of Symbols

PM_{10}	particulate matter consisting of particles having mean aerodynamic diameter <10 μm
$PM_{2.5}$	particulate matter consisting of particles having mean aerodynamic diameter <2.5 μm
PM_1	particulate matter consisting of fine particles having mean aerodynamic diameter <1 μm
$PM_{0.1}$	ultrafine particles (UFP) having mean aerodynamic diameter <0.1 μm

References

Beelen, R., Raaschou-Nielsen, O., Stafoggia, M., Andersen, Z.J., Weinmayr, G., Hoffmann, B. *et al.* (2014) Effects of long-term exposure to air pollution on natural cause mortality: an analysis of 22 European cohorts within the multicenter ESCAPE project. *Lancet*. 2014, **383** (9919), 785–795. doi: 10.1016/S0140-6736(13)62158-3.

Beelen, R., Hoek, G., van den Brandt, P.A., Goldbohm, R.A., Fischer, P., Schouten, L.J., Jerrett, M., Hughes, E., Armstrong, B., and Brunekreef, B. (2008) Long-term effects of traffic-related air pollution on mortality in a Dutch cohort (NLCS-AIR study). *Environ. Health Perspect*, **116**, 196–202. doi: 10.1289/ehp.10767

Beeson, W.L., Abbey, D.E., and Knutsen, S.F. (1998) Long-term concentrations of ambient air pollutants and incident lung cancer in California adults: results from the AHSMOG study. *Environ. Health Perspect*, **106**, 813–822.

Block, M.L. and Calderón-Garcidueñas, L. (2009) Air pollution: mechanisms of neuroinflammation and CNS disease. *Trends Neurosci.*, **32** (9), 506–516. doi: 10.1016/j.tins.2009.05.009

Brook, R.D., Rajagopalan, S., Pope, C.A., Brook, J.R., Bhatnagar, A., Diez-Roux, A.V., Holguin, F., Hong, Y., Luepker, R.V., Mittleman, M.A., Peters, A., Siscovick, D., Smith, S.C. Jr.,, Whitsel, L., and Kaufman, J.D. (2010) Particulate matter air pollution and cardiovascular disease: an update to the scientific statement from the American Heart Association.

Circulation, **121** (21), 2331–2378. doi: 10.1161/CIR.0b013e3181dbece1

Brugha, R. and Grigg, J. (2014) Urban air pollution and respiratory infections. *Paediatr. Respir. Rev.*, **15** (2), 194–199. doi: 10.1016/j.prrv.2014.03.001

Camatini, M., Corvaja, V., Pezzolato, E., Mantecca, P., and Gualtieri, M. (2012) PM10-biogenic fraction drives the seasonal variation of proinflammatory response in A549 cells. *Environ. Toxicol.*, **27** (2), 63–73. doi: 10.1002/tox.20611

Campbell, A. (2004) Inflammation, neurodegenerative diseases, and environmental exposures. *Ann. N. Y. Acad. Sci.*, **1035**, 117–132. doi: 10.1196/annals.1332.008

Carey, I.M., Atkinson, R.W., Kent, A.J., van Staa, T., Cook, D.G., and Anderson, H.R. (2013) Mortality associations with long-term exposure to outdoor air pollution in a national English cohort. *Am. J. Respir. Crit. Care Med.*, **187**, 1226–1233.

Cao, J., Yang, C.X., Li, J.X., Chen, R.J., Chen, B.H., Gu, D., and Kan, H. (2011) Association between long-term exposure to outdoor air pollution and mortality in china: a cohort study. *J Hazard Mater*, **186** (2-3), 1594–1600.

Cesaroni, G., Badaloni , G. Gariazzo, C., Stafoggia, M., Sozzi, R., Davoli, M., and Forastiere, F. (2013) Long-term exposure to urban air pollution and mortality in a cohort of more than a million adults in Rome. *Environ Health Perspect*, **121**, 324–331. doi: 10.1289/ehp.1205862.

Correia, A.W., Pope, C.A. III, Dockery, D.W., Wang, Y., Ezzati, M., and Dominici, F. (2013) Effect of air pollution control on life expectancy in the United States: an analysis of 545 U.S. counties for the period 2000 to 2007. *Epidemiology*, **24** (1), 23–31. doi: 10.1097/EDE.0b013e3182770237

Dickey, J.S., Redon, C.E., Nakamura, A.J., Baird, B.J., Sedelnikova, O.A., and Bonner, W.M. (2009) H2AX: functional roles and potential applications. *Chromosoma*, **118** (6), 683–692. doi: 10.1007/s00412-009-0234-4

EC (2008) Directive on Ambient Air Quality and Cleaner air for Europe (Directive 2008/50/EC), http://eur-lex.europa.eu/LexUriServ/LexUriServ.do?uri=OJ:L:2008:152:0001:0044:en:PDF.

Elmore, S. (2007) Apoptosis: a review of programmed cell death. *Toxicol. Pathol.*, **35** (4), 495–516. doi: 10.1080/01926230701320337

Englert, N. (2004) Fine particles and human health: a review of epidemiological studies. *Toxicol. Lett.*, **149** (1-3), 235–242.

EPA (2009) Integrated Science Assessment for Particulate Matter. U.S. Environmental Protection Agency, FRL-9090-9; Docket ID No. EPA-HQ-ORD-2007-0517.

Farina, F., Sancini, G., Battaglia, C., Tinaglia, V., Mantecca, P., Camatini, M., and Palestini, P. (2013) Milano summer particulate matter (PM10) triggers lung inflammation and extra pulmonary adverse events in mice. *PLoS One*, **8** (2), e56636. doi: 10.1371/journal.pone.0056636

Gold, D.R. and Mittleman, M.A. (2013) New insights into pollution and the cardiovascular system: 2010 to 2012. *Circulation*, **127** (18), 1903–1913. doi: 10.1161/CIRCULATIONAHA.111.064337

Grunig, G., Marsh, L.M., Esmaeil, N., Jackson, K., Gordon, T., Reibman, J., Kwapiszewska, G., and Park, S.H. (2014) Perspective: ambient air pollution: inflammatory response and effects on the lung's vasculature. *Pulm. Circ.*, **4** (1), 25–35. doi: 10.1086/674902

Gualtieri, M., Øvrevik, J., Holme, J.A., Perrone, M.G., Bolzacchini, E., Schwarze, P.E., and Camatini, M. (2010) Differences in cytotoxicity versus pro-inflammatory potency of different PM fractions in human epithelial lung cells. *Toxicol. in Vitro*, **24** (1), 29–39. doi: 10.1016/j.tiv.2009.09.013

Guarnieri, M. and Balmes, J.R. (2014) Outdoor air pollution and asthma. *Lancet*, **383** (9928), 1581–1592. doi: 10.1016/S0140-6736(14)60617-6

Hales, S., Blakely, T., and Woodward, A. (2012) Air pollution and mortality in New Zealand: cohort study. *J Epidemiol Community Health*, **66**, 468–473.

Hamra, G.B., Guha, N., Cohen, A., Laden, F., Raaschou-Nielsen, O., Samet, J.M., Vineis, P., Forastiere, F., Saldiva, P., Yorifuji, T., and Loomis, D. (2014) Outdoor particulate matter exposure and lung cancer: a systematic review and meta-analysis. *Environ. Health Perspect.*, **122** (9), 906–911. doi: 10.1289/ehp.1408092

Hart, J.E., Garshick, E., Dockery, D.W., Smith, T.J., Ryan, L., and Laden, F. (2011) Long-term ambient multipollutant exposures and mortality. *Am. J. Respir. Crit. Care Med.*, **183**, 73–78.

Heinreich , J., Thiering, E., Rzehak, P., Kramer, U., Hochadel, M., Rauchfuss, K.M., Gehring, U., and Wichmann, H.E. (2013) Long-term exposure to NO_2 and PM_{10} and all-cause and cause-specific mortality in a prospective cohort of women. *Occup. Environ. Med.*, **70**, 179–186.

Hogg, J.C., and Van Eeden, S. (2009) Pulmonary and systemic response to atmospheric pollution. *Respirology*, **14** (3), 336–346.

Hystad, P., Demers, P.A., Johnson, K.C., Carpiano, R.M., and Brauer, M. (2013) Long-term residential exposure to air pollution and lung cancer risk. *Epidemiology*, **24**, 762–772.

Jerrett, M., Burnett, R.T., Beckerman, B.S., Turner, M.C., Krewski, D., Thurston, G., Martin, R.V., van Donkelaar, A., Hughes, E., Shi, Y., Gapstur, S.M., Thun, M.J., and Pope, C.A. III. (2013) Spatial analysis of air pollution and mortality in California. *Am. J. Respir. Crit. Care Med.*, **188**, 593–599.

Katanoda, K., Sobue, T., Satoh, H., Tajima, K., Suzuki, T., Nakatsuka, H., Takezaki, T., Nakayama, T., Nitta, H., Tanabe, K., and Tominaga, S. (2011) An association between long-term exposure to ambient air pollution and mortality from lung cancer and respiratory diseases in Japan. *J. Epidemiol.*, **21**, 132–143.

Kelly, F.J. and Fussell, J.C. (2012) Size, source and chemical composition as determinants of toxicity attributable to ambient particulate matter. *Atmos. Environ.*, **60**, 504–526. doi: 10.1016/j.atmosenv.2012.06.039

Krewski, D., Jerrett, M., Burnett, R.T., Ma, R., Hughes, E., Shi, Y., Turner, M.C., Pope, C.A., III, Thurston, G., Calle, E.E., Thun, M.J., Beckerman, B., DeLuca, P., Finkelstein, N., Ito, K., Moore, D.K., Newbold, K.B., Ramsay, T., Ross, Z., Shin, H., and Tempalski, B. (2009) Extended follow-up and spatial analysis of the American Cancer Society study linking particulate air pollution and mortality. *Res. Rep. Health Eff. Inst*, **140**, 5–114.

Kumar, S., Verma, M.K., and Srivastava, A.K. (2013) Ultrafine particles in urban ambient air and their health perspectives. *Rev. Environ. Health*, **28** (2-3), 117–128. doi: 10.1515/reveh-2013-0008

Leary, P.J., Barr, R.G., Bluemke, D.A., Hough, C.L., Kaufman, J.D., Szpiro, A.A., Kawut, S.M., and Van Hee, V.C. (2013) The relationship of roadway proximity and NOx with right ventricular structure and function: the MESA-Right Ventricle and MESA-Air studies. *Am. J. Respir. Crit. Care Med.*, **187**, A3976.

Lepeule, J., Laden, F., Dockery, D., and Schwartz, J. (2012) Chronic exposure to fine particles and mortality: An extended follow-up of the Harvard six cities study from 1974 to 2009. *Environ. Health Perspect.*, **120** (7), 965–970. doi: 10.1289/ehp.1104660

Lippmann, M. (2014) Toxicological and epidemiological studies of cardiovascular effects of ambient air fine particulate matter (PM2.5) and its chemical components: coherence and public health implications. *Crit. Rev. Toxicol.*, **44** (4), 299–347. doi: 10.3109/10408444.2013.861796

Lipsett, M.J., Ostro, B.D., Reynolds, P., Goldberg, D., Hertz, A., Jerrett, M., Smith, D.F., Garcia, C., Chang, E.T., and Bernstein, L. (2011) Long-term exposure to air pollution and cardiorespiratory disease in the California Teachers Study cohort. *Am. J. Respir. Crit. Care Med.*, **184**, 828–835.

Longhin, E., Holme, J.A., Gutzkow, K.B., Arlt, V.M., Kucab, J.E., Camatini, M., and Gualtieri, M. (2013) Cell cycle alterations induced by urban PM2.5 in bronchial epithelial cells: characterization of the process and possible mechanisms involved. *Part. Fibre Toxicol.*, **10**, 63. doi: 10.1186/1743-8977-10-63

Loomis, D., Grosse, Y., Lauby-Secretan, B., El Ghissassi, F., Bouvard, V., Benbrahim-Tallaa, L., Guha, N., Baan, R., Mattock, H., Straif, K. (2013) The carcinogenicity of outdoor air pollution. *Lancet Oncol*, **14**, 1262–1263. doi: 10.1016/S1470-2045(13)70487-X

McColl, B.W., Rothwell, N.J., and Allan, S.M. (2007) Systemic inflammatory stimulus potentiates the acute phase and CXC chemokine responses to experimental stroke and exacerbates brain damage via interleukin-1- and neutrophil-dependent mechanisms. *J. Neurosci.*, **27** (16), 4403–4412. doi: 10.1523/JNEUROSCI.5376-06.2007

McDonnell, W.F., Nishino-Ishikawa, N., Petersen, F.F., Chen, L.H., and Abbey, D.E. (2000) Relationships of mortality with the fine and coarse fractions of long-term ambient PM10 concentrations in non-smokers. *J Expo. Anal. Environ. Epidemiol*, **10**, 427–436.

Medina, S., Boldo, E., Krzyzanowski, M., Niciu, E.M., Mueke, H.G., and Atkinson, R. (2005) and the Aphesis working group. APHEIS. Health impact assessment of air pollution and communication strategy. Third year report. -200. Saint-Maurice (Fra): Institut de veille sanitaire. [30-9-2011. Ref Type: Report].

Mehta, M., Chen, L.C., Gordon, T., Rom, W., and Tang, M.S. (2008) Particulate matter inhibits DNA repair and enhances mutagenesis. *Mutat. Res.*, **657** (2), 116–121. doi: 10.1016/j.mrgentox.2008.08.015

Miller, M.R., Shaw, C.A., and Langrish, J.P. (2012) From particles to patients: oxidative stress and the cardiovascular effects of air pollution. *Future Cardiol.*, **8** (4), 577–602. doi: 10.2217/fca.12.43

Mills, N.L., Törnqvist, H., Gonzalez, M.C., Vink, E., Robinson, S.D., Söderberg, S., Boon, N.A., Donaldson, K., Sandström, T., Blomberg, A., and Newby, D.E. (2007) Ischemic and thrombotic effects of dilute diesel-exhaust inhalation in men with coronary heart disease. *N. Engl. J. Med.*, **357** (11), 1075–1082. doi: 10.1056/NEJMoa066314

Nakane, H. (2012) Translocation of particles deposited in the respiratory system: a systematic review and statistical analysis. *Environ. Health Preventative Med.*, **17** (4), 263–274. doi: 10.1007/s12199-011-0252-8

Oberdörster, G. and Utell, M.J. (2002) Ultrafine particles in the urban air: to the respiratory tract--and beyond? *Environ. Health Perspect.*, **110** (8), A440–A441.

Oberdörster, G., Elder, A., and Rinderknecht, A. (2009) Nanoparticles and the brain: cause for concern? *J. Nanosci. Nanotechnol.*, **9** (8), 4996–5007.

Park, J.H., Mangal, D., Frey, A.J., Harvey, R.G., Blair, I.A., and Penning, T.M. (2009) Aryl hydrocarbon receptor facilitates DNA strand breaks and 8-oxo-2'-deoxyguanosine formation by the aldo-keto reductase product benzo[*a*]pyrene-7,8-dione. *J. Biol. Chem.*, **284** (43), 29725–29734. doi: 10.1074/jbc.M109.042143

Pascal, M., Corso, M., Chanel, O., Declercq, C., Badaloni, C., Cesaroni, G., Henschel, S., Meister, K., Haluza, D., Martin-Olmedo, P., Medina, S., and Aphekom Group (2013) Assessing the public health impacts of urban air pollution in 25 European cities: results of the Aphekom project. *Sci. Total Environ.*, **449**, 390–400. doi: 10.1016/j.scitotenv.2013.01.077

Paulsen, R.D. and Cimprich, K.A. (2007) The ATR pathway: fine-tuning the fork. *DNA Repair*, **6** (7), 953–966. doi: 10.1016/j.dnarep.2007.02.015

Plummer, L.E., Smiley-Jewell, S., and Pinkerton, K.E. (2012) Impact of air pollution on lung inflammation and the role of Toll-like receptors. *Int. J. Interferon, Cytokine Mediator Res.*, **4**, 43–57. doi: 10.2147/IJICMR.S29352

Pope, C.A. III,, Burnett, R.T., Turner, M.C., Cohen, A., Krewski, D., Jerrett, M., Gapstur, S.M., and Thun, M.J. (2011) Lung cancer and cardiovascular disease mortality associated with ambient air pollution and cigarette smoke: shape of the exposure–response relationships. *Environ. Health Perspect.*, **119** (11), 1616–1621. doi: 10.1289/ehp.1103639

Pope, C.A. III,, Ezzati, M., and Dockery, D.W. (2013) Fine particulate air pollution and life expectancies in the United States: the role of influential observations. *Air Waste Manag. Assoc.*, **63** (2), 129–132. doi: 10.1080/10962247.2013.760353

Pope, C.A., III, Burnett, R.T., Thun, M.J., Calle, E.E., Krewski, D., Ito, K., and Thurston, G.D. (2002) Lung cancer, cardiopulmonary mortality, and long-term exposure to fine particulate air pollution. *JAMA*, **287** (9), 1132–1141.

Pope, C.A., III, Burnett, R.T., Thurston, G.D., Thun, M.J., Calle, E.E., Krewski, D., and Godleski, J.J. (2004) Cardiovascular mortality and long-term exposure to particulate air pollution: epidemiological evidence of general pathophysiological pathways of disease. *Circulation*, **109**, 71–7.

Puett, R.C., Hart, J.E., Yanosky, J.D., Spiegelman, D., Wang, M., Fisher, J.A., Hong, B., and Laden, F. (2014) Particulate matter air pollution exposure, distance to road, and incident lung cancer in the Nurses' Health Study cohort. *Environ. Healt. Perspect*, **122**, 926–932. doi: 10.1289/ehp.1307490

Raaschou-Nielsen, O., Andersen, Z.J., Beelen, R., Samoli, E., Stafoggia, M., Weinmayr, G. *et al.* (2013) Air pollution and lung cancer incidence in 17 European cohorts: prospective analyses from the European Study of Cohorts for Air Pollution Effects (ESCAPE). *Lancet Oncol.*, **14** (9), 813–822. doi: 10.1016/S1470-2045(13)70279-1

Schwartz, J., Coull, B., Laden, F., and Ryan, L. (2008) The effect of dose and timing of dose on the association between airborne particles and survival. *Environ. Health Perspect.*, **116** (1), 64–69. doi: 10.1289/ehp.9955

Shah, A.S.V., Langrish, J.P., Nair, H., McAllister, D.A., Hunter, A.L., Donaldson, K., Newby, D.E., and Mills, N.L. (2013) Global association of air pollution and heart failure: a systematic review and meta-analysis. *Lancet*, **382** (9897), 1039–1048. doi: 10.1016/S0140-6736(13)60898-3

Shang, Y., Sun, Z., Cao, J., Wang, X., Zhong, L., Bi, X., Li, H., Liu, W., Zhu, T., and Huang, W. (2013) Systematic review of Chinese studies of short-term exposure to air pollution and daily mortality. *Environ. Int.*, **54**, 100–111. doi: 10.1016/j.envint.2013.01.010

Tonelli, L.H. and Postolache, T.T. (2012) Airborne inflammatory factors: "from the nose to the brain". *Front. Biosci. (Schol Ed.)*, **2**, 135–152.

US Environmental Protection Agency (2008) National Ambient Air Quality Criteria Standards, http://www.epa.gov/air/criteria.html (accessed 1 September 2008).

Veranth, J.M., Moss, T.A., Chow, J.C., Labban, R., Nichols, W.K., Walton, J.C., Watson, J.G., and Yost, G.S. (2006) Correlation of in vitro cytokine responses with the chemical composition of soil-derived particulate matter. *Environ. Health Perspect.*, **114** (3), 341–349.

WHO (World Health Organization) (2013) Evidence on Health Aspects of Air Pollution to Review EU Policies–the REVIHAAP Project.

Zanobetti, A., Baccarelli, A., and Schwartz, J. (2011) Gene-air pollution interaction and cardiovascular disease: a review. *Prog. Cardiovasc. Dis.*, **53** (5), 344–352. doi: 10.1016/j.pcad.2011.01.001

Ziech, D., Franco, R., Georgakilas, A.G., Georgakila, S., Malamou-Mitsi, V., Schoneveld, O., Pappa, A., and Panayiotidis, M.I. (2010) The role of reactive oxygen species and oxidative stress in environmental carcinogenesis and biomarker development. *Chem. Biol. Interact.*, **188** (2), 334–339. doi: 10.1016/j.cbi.2010.07.010

Zlokovic, B.V. (2011) Neurovascular pathways to neurodegeneration in Alzheimer's disease and other disorders. *Nat. Rev. Neurosci.*, **12** (12), 723–738. doi: 10.1038/nrn3114

11
Aerosol Impact on Cultural Heritage: Deterioration Processes and Strategies for Preventive Conservation

Alessandra Bonazza, Paola De Nuntiis, Paolo Mandrioli, and Cristina Sabbioni

11.1
Introduction

Studies over the past few decades have explored the mechanisms that cause damage to heritage materials, both indoors and outdoors located, exposed to various climates. These have typically been related to deterioration processes occurring on monumental complexes, archeological sites, and heritage objects in urban and remote areas and highlighted air pollutants, biological attack, and a number of meteorological parameters relevant to changing the rate and typology of damage. This chapter focuses on the deterioration phenomena occurring on cultural heritage due to aerosol impact and presents a methodological approach for monitoring climate and pollution parameters suitable for cultural assets preservation with the aim of identifying the environmental parameters to measure, assessing their spatial and temporal variability, and defining the number of measurements required. Appropriated analyses to perform on aerosol sampled are also presented.

Thanks to the experience gained by the authors, particular attention is devoted to the explanation of how to perform specific monitoring campaigns for the conservation of cultural heritage outdoors, and relating EN standards for environmental monitoring are summarized. Finally guidelines for managing the conservation of cultural heritage in urban areas toward pollution impact are proposed.

11.2
Monitoring for Cultural Heritage Conservation

Since the nineteenth century, environmental changes caused by anthropogenic activities have had a major impact on the conservation of indoor and outdoor cultural heritage sites. The link between heritage conservation and environmental conditions had been introduced by British conservationists during the Second

Atmospheric Aerosols: Life Cycles and Effects on Air Quality and Climate, First Edition.
Edited by Claudio Tomasi, Sandro Fuzzi, and Alexander Kokhanovsky.

World War. Brandi (1977) subsequently renewed and extended the concept of restoration, historically linked to cultural heritage, to include "the protection, removal of danger and assurance of favorable conditions." In practice, the catastrophic floods that occurred in Florence and Venice in the late 1960s gave rise to a new discipline in Italy: the science of conservation. In the 1970s, the International Center for the Study of the Preservation and Restoration of Cultural Property (ICCROM) defined preventive conservation as a series of measures aimed at enhancing the state of conservation of a monument over time. It was only in 2000 at the "Preventive Conservation Strategy" meeting held in Finland that a document containing the recommendations and guidelines for the shared implementation of preventive conservation strategies in Europe was finally approved (Menegazzi, 2003).

As cultural heritage belongs to the category of "unique and nonrenewable" resources, conservation is a priority. Artistic heritage deteriorates naturally and the capacity of materials to adapt to changing environmental parameters diminishes over time. Decay is therefore a cumulative process developing in an irreversible nonlinear fashion. The causes of deterioration stemming from unfavorable environmental conditions have different effects depending on the object in question and its history. Preventive conservation in the field of cultural heritage means providing an artwork with a suitable environment designed to increase its life expectancy (Guichen, 1999) by implementing indirect measures globally (De Nuntiis, 2013), rather than on individual artifacts, on both intact and damaged collections.

The type of building material, microclimatic conditions, and the chemical and biological pollution of the premises in which they are stored can all trigger processes leading to irreversible damage to works of art. Air pollution, precipitation, wind, biological contamination, and physical stress, resulting from temperature and relative humidity cycling, are among the main factors responsible for the surface deterioration of artifacts (Nugari, 2003; Bonanni *et al.*, 2006; Camuffo, 2013). The microclimate resulting from the set of physical factors such as temperature, solar radiation, and humidity has a major impact on all artifacts and their surrounding environment (Camuffo, 2013). The microclimate has an ongoing action and its long-term effects are caused by successive characteristic stages like the alternation of day and night and the seasons and by the location of the artifact in an open, closed, or underground environment. In addition to its direct impact on materials, the microclimate has an indirect effect in creating a habitat favorable to forms of biological decay or undesirable chemical reactions, especially in the presence of air pollutants. The effect of pollutants depends on the relative humidity (RH) of the air: high RH levels ($RH > 70\%$) in a nonhygroscopic material lead to the formation of a surface water film transforming soluble gaseous pollutants and particulate matter and favoring chemical reactions causing deterioration. In addition, synergic effects are common as the action of certain pollutants is exacerbated by factors like solar radiation acting as a photooxidation catalyst.

To be meaningful, microclimate monitoring must be long lasting, especially in areas of the world with seasonal variability where it is advisable to monitor environmental variables continuously for at least a year to collect enough data for analysis. Monitoring will provide data sets useful not only to assess the risk factors for the conservation of a given artifact but above all to check on the management solutions adopted over time. Therefore, microclimate control is a key undertaking requiring reliable instruments as incorrect measurements may leave the object exposed to inappropriate conditions or, worse, trigger unnecessary and potentially hazardous actions (e.g., covering an outdoor site or activating an indoor dehumidification system).

The Italian, European, and international standards (UNI, EN, ISO) provide indications on the specific instrumentation to use, setting levels of accuracy, range, uncertainty, resolution, stability, periodic checkups, maintenance, calibration, and measurement recommendations (e.g., UNI EN 16242: 2012 "Conservation of cultural heritage. Procedures and instruments for measuring humidity in the air and moisture exchanges between air and cultural property"). The standards also provide the measuring methods for specific samples (UNI 11526: 2014 Beni culturali – Materiali lapidei naturali ed artificiali – Cianobatteri ed alghe verdi: tecniche colturali, Cultural Heritage – Natural and Artificial Stone – Cyanobacteria and Green Algae: Techniques of Cultivation). Table 11.1 summarizes the EN standards currently in force for monitoring cultural heritage in the outdoor environment.

Table 11.1 EN standards in force recommended for outdoor environmental monitoring.

European Standard – EN	Title
EN 15801:2009	Conservation of cultural property – Test methods – Determination of water absorption by capillarity
EN 15757:2010	Conservation of cultural property – Specifications for temperature and relative humidity to limit climate-induced mechanical damage in organic hygroscopic materials
EN 15758:2010	Conservation of cultural property – Procedures and instruments for measuring temperatures of the air and the surfaces of objects
EN 15898:2011	Conservation of cultural property – Main general terms and definitions
EN 16085:2012	Conservation of cultural property – Methodology for sampling from materials of cultural property – General rules
EN 16096:2012	Conservation of cultural property – Condition survey and report of built cultural heritage
EN 16242:2012	Conservation of cultural heritage – Procedures and instruments for measuring humidity in the air and moisture exchanges between air and cultural property
EN 16302:2013	Conservation of cultural heritage – Test methods – Measurement of water absorption by pipe method
EN 16322:2013	Conservation of cultural heritage – Test methods – Determination of drying properties

Since microclimate cannot be considered the only cause for all deterioration processes (physical, chemical, and biological), microclimate monitoring alone will not guarantee conservation. Thus, an integrated environmental monitoring referring to a set of sensor-based measurements designed to identify the main deterioration processes occurring over time is here proposed. The integrated methodology aims to:

- Determine the conditions to which a work of art is subject over time
- Describe the risk factors on the basis of qualitative and quantitative data and assess the synergic effects
- Devise maintenance programs
- Identify or suggest accurate and specific heritage conservation conditions
- Recommend management solutions
- Identify sustainable use
- Reassess management solutions *a posteriori* to continue or change direction
- Devise catastrophe risk management strategies

The integrated environmental monitoring plan outlined below focuses on outdoor heritage largely made of building materials like stone, mortar, brick, and plaster. Many well-documented museum monitoring experiences have been reported, whereas outdoor systems are less common and vary in relation to fluctuating weather conditions and seasonal changes more easily controlled in the indoor environment. The environmental monitoring plan was devised, tested, and implemented by the authors in different settings in the last 10 years and includes the megalithic temples of Hagar Qim e Mnajdra (Mandrioli *et al.*, 2006) commissioned to CNR-ISAC by Heritage Malta (Cassar *et al.*, 2011), the Camposanto Monumentale in Pisa required by Opera Primaziale Pisana as part of a project to relocate the frescoes in the four-sided portico (Sabbioni *et al.*, 2008; Mandrioli *et al.*, 2008), and the Baptistery of Florence for understanding the aerosol impact on the different sides of the building. The specific aim of an integrated environmental monitoring plan is to monitor and assess the environmental conditions around the site of an individual artifact, monumental complex, church, or archeological site. An accurate measurement protocol should be based on a thorough site inspection with preliminary sampling to determine the macroscopic processes responsible for deterioration and the correlated parameters to measure, assessing the environmental variability and defining the number of measurements required for continuous monitoring and seasonal campaigns (De Nuntiis *et al.*, 2007).

The *physical parameters* to be monitored continuously by micrometeorological stations for *microclimatic characterization* of the outdoor environment are air temperature (T, °C), relative humidity (RH, %), rainfall (rain, mm min^{-1}), wind speed and direction (WS, m s^{-1}, *WD*, degrees), and solar radiation (I_o, W m^{-2}). Surface temperature (T_{surf}, °C), dew point of the air (T_d, °C), and surface condensation should also be included in continuous monitoring as different atmosphere–material interactions occur during the year. Services currently available for automatic micrometeorological measurements are based on the installation of a local network of sensors with wireless communications technology. The system involves automatic data handling through an intranet network exchanging data from the site connected to the Data Management Centre (DMC) using a GPS mobile phone. All the data from the remote sites are transmitted in real time to a dynamic database installed at the DMC providing analyses and forecasts on the trend of the environmental conditions. Continuous monitoring of physical parameters is associated with measurements conducted for short-term campaigns as outlined in the following.

An assessment of the risk to cultural heritage due to air pollution, including the impact of aerosol, must identify potential stationary (such as industry, domestic heating, power plants) and mobile combustion sources (road traffic, railway) to characterize the environment surrounding the works of art. Data from air quality networks can be used in such assessments, but they tend to lack detail and may not be available. Environmental monitoring of the seasonal concentration and composition of gaseous pollutants and aerosols suggested as a tool to assess risks and to make a realistic prognosis of the forthcoming deterioration processes. A good risk definition will justify conservation investments designed to reduce the risks.

Gaseous pollutants and aerosols emitted by mobile combustion sources are currently considered the chief causes of surface blackening and damage layer formation on cultural and built heritage located in European cities. Despite the ever-growing energy demand for fossil fuels in the past, improvements to industrial plants, the utilization of low-sulfur fuels, and mitigation policies have dramatically reduced the industrial air pollution. Today, emissions from mobile combustion sources are among the main agents responsible for the pollution encountered in most western European cities, and the negative effects of traffic are expected to continue and multiply in the future. Because of the rise in vehicle emissions, the physicochemical characteristics of the particles emitted into the atmosphere have changed substantially. Mainly constituted by organic compounds, they are submicronic in size and characterized by a fractal morphology, which implies a larger specific surface and therefore a greater interaction with gaseous pollutants.

As shown in Section 11.3, the quantification of carbon in damage layers is particularly important to discern the deposition deriving from anthropogenic sources. Along with carbonate carbon (CC), often linked to the building material itself, organic carbon (OC) must be discriminated from elemental carbon (EC). EC originates solely from particles emitted into the atmosphere by combustion processes. The OC found in black crusts is more difficult to understand because it originates not only from the deposition of primary and secondary pollutants but also from the biological activity of microorganisms colonizing building surfaces and the protective treatments applied to monuments in the past (Ghedini *et al.*, 2006). OC also contains a considerable fraction of formates, acetates, and oxalates, anions that are always present in damage layers (Sabbioni, 2003). Primary and secondary air pollutants are the main sources of formates and acetates, while oxalates are principally the product of biological activity and the oxidative deterioration of organic treatments applied to building surfaces, although they also derive from reactions occurring in the atmosphere or on the surfaces of primary pollutants.

To evaluate the effects of emissions from mobile and stationary sources on the cultural heritage, the results obtained so far analyzing damage layers from monuments in urban centers have shown that it is important not only to quantify the particulate mass emitted but also to gain a knowledge of its composition by measuring its carbonaceous (EC and OC) components and soluble salt content (sulfates, nitrates, carbonates, chlorides, formates, acetates, and oxalates). This requires seasonal *aerosol sampling* campaigns at the site of interest in order to study the interaction between the air and heritage materials.

Total suspended particles (TSPs) are sampled using quartz fiber filters to collect the whole size range of atmospheric particulate matter as all particles are deposited on and interact with materials. Gravimetric analysis of the TSP mass ($\mu g\, m^{-3}$) and ion chromatography (IC) measurement of the concentration of soluble anions and cations ($\mu g\, m^{-3}$) and carbon fractions (OC and EC) serve to determine TSP origin and possible sources (e.g., sulfates from industrial activity, nitrates and organic compounds from motor vehicle traffic, and chlorides of marine origin).

Viable bioaerosol is sampled by aspirating a known volume of air with a surface air system sampler that deposits the particles on a Petri dish for selective culture (De Nuntiis *et al.*, 2003). Laboratory culture methods and light microscopy (LM) are used to determine the atmospheric concentration (total microbial, fungal, and bacterial load, $CFU\, m^{-3}$) of the potentially biodeteriogenic air-dispersed viable component (Figure 11.1) once it has settled in favorable environmental conditions (e.g., available liquid water) to resume its vegetative activity. A Hirst-type impactor (Lanzoni, Bologna VPPS1000, VPPS2000) is used to sample biological particles like pollen grains or fungal spores (Hirst, 1995). Subsequent LM analysis yields qualitative and quantitative measurements of the air-dispersed *non-viable* biological particulate which is useful as a biological air mass tracer or to determine the contribution of spores, propagules, and seeds from remote or local biological sources (Figure 11.2).

Integrated environmental monitoring must include diagnostic tests on the material components of an artwork and any layers of deterioration already present. Most outdoor heritage is composed of natural (granite, porphyry, limestone, travertine, sandstone, marble, slate) or artificial (ceramic materials, bricks, plaster coating, stucco, concrete) stone. The petrographic characterization of the stone is done on fragments of material sampled if possible directly on the work of art by successive observations under transmitted and reflected LM and scanning electron microscopy (SEM), X-ray diffraction analysis (EDX) to identify the crystalline phases and IC analysis to determine the amount of soluble salts. The stone (silicate or carbonate) composition of an artwork will determine its resistance or sensitivity to air pollutants. All chemical compounds able to lower the pH of an aqueous solution play a role in corroding calcium carbonate-based materials. The more porous the materials are, the more pollution-sensitive they

Figure 11.1 Petri dish with colonization of the viable bioaerosol grown on specific cultural media.

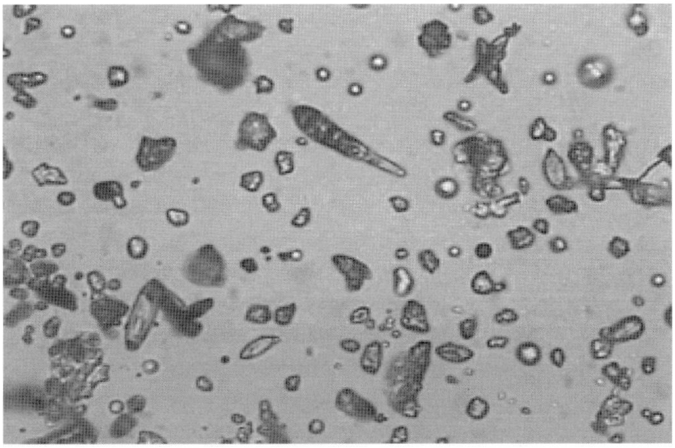

Figure 11.2 Optical microscope image of non-viable bioaerosol (fungal spores, pollen grains) sampled with impactor.

will be. In the case of outdoor cultural heritage, restoration, protection, and consolidation schemes often contribute additional inorganic and especially organic material that not only interacts with the environment but can also provide nutrients for microorganisms. An analysis of deterioration must include detailed information collected from written and iconographic sources. Examples of deterioration of outdoor artifacts include blackening, black crust formation, powdering, soluble salt-induced efflorescence, alveolization, exfoliation, deposits of particulate matter, biological colonization and biodegradation.

Surface monitoring yields information not only on the esthetic impact of pollutants (blackening color loss) but also on deterioration processes (alveolization, corrosion, etc.) or degradation caused by organisms using the artwork as a substrate (lichen growth). While many techniques and instruments are used in this sector, this chapter briefly describes those aimed at general deterioration processes. *In situ* colorimetric analysis and spectrophotometric data based on a set of numbers provide information on the color of an artifact, while measurements repeated over time will define its state of conservation or any deterioration underway. Sampling of surface deposits on artifacts in the form of incoherent material or fragments will allow an analysis of the effects due to the microclimate-dependent deposition of gases and aerosols. Deposited airborne particles are identified by SEM-EDX, while ion chromatography (Dionex 4500i) is used to characterize water-soluble ions (i.e., sulfate, nitrate, chloride, acetate, formate, oxalate). Contact Petri dishes (RODAC type), nitrocellulose, nylon membranes, and microbiological analysis of viable bioaerosol deposited on surfaces will identify any biodeteriogenic microbial components. The results of surface monitoring are expressed in CFU cm^{-2}. The biological monitoring of cyanobacteria and chlorophyceae (green algae) on stone is based on the instructions provided in standard UNI 11526: 2014. Infrared thermography is an example of a nondestructive and noninvasive diagnostic technique producing thermal maps of an artifact from single images showing the infrared radiation emitted by the object during heat exchange cycles, for example, day–night. The infrared measurements can disclose any thermal inhomogeneity or discontinuity in the structure below the surface, also revealing processes like thermoclastism (Bonazza *et al.*, 2009; Mandrioli *et al.*, 2006) and water-related deterioration such as the capillary rise of damp in walls or the formation of water vapor condensation under certain seasonal and meteorological conditions (Sabbioni *et al.*, 2008; Mandrioli *et al.*, 2012).

The protection of cultural heritage not only has an economic impact, but it is also a quality of life priority meriting national and European support as the basis for sustainable development. An integrated environmental monitoring system offers concrete support to decision makers and sector operators alike.

11.3
Damage and Black Crusts Formation on Building Materials

Many literature studies on the degradation of the built heritage in urban areas have shown that the most important deterioration process underway on architectural surfaces is black crusts formation (Leysen, Roekensand, and Van Grieken, 1989; Sabbioni and Zappia, 1992; Sabbioni, 1995; Rossval, 1998; Saiz-Jimenez, 2003).

Figure 11.3 Black crusts on Candoglia marble, Milan Cathedral (Italy).

Black crusts are mainly observed on stones wetted by rain but sheltered from intense rainwater washout (Figure 11.3). Their formation is due almost exclusively to the deposition effect of SO_2 and particles emitted by stationary and mobile combustion sources. Although major advances have been made in the knowledge of black crust composition and formation, attention has mainly focused on atmospheric sulfur impact, whereas the role of aerosol and its composition in causing deterioration have only been addressed by recent research (Bonazza, Sabbioni, and Ghedini, 2005; Ghedini *et al.*, 2006; Ozga *et al.*, 2013).

Carbonaceous particles in the atmosphere primarily derive from various combustion processes, and their concentration in urban areas has increased exponentially over the last century. All combustion processes produce abundant amounts of carbonaceous species, which contain both OC and EC. While the atmospheric concentrations of sulfur dioxide have been drastically reduced in recent years, the air content of carbonaceous particles has continued to rise. There is no doubt that these changes will have an impact on the blackening and surface deterioration of historic monuments and buildings located in urban centers.

11.3.1 Damage to Carbonate Stone

The earliest and most extensive studies on the impact of aerosol particles on cultural heritage concern monuments mainly built in marble and compact limestone. Firstly these materials have been widely used all over the world in statuary and building construction. Secondly they represent a simpler system than other stones on account of their chemical homogeneity, being almost entirely composed of calcium carbonate (or, more rarely, Ca and Mg carbonates) with low porosity, which permits the identification of a clearly defined interaction interface between the atmosphere and the undamaged material. In addition to chemical composition, porosity has been recognized as a further driving feature favoring the formation and pattern distribution of crusts on building facades.

In urban sites, degradation typologies externally observed on monuments built of marble and compact limestone are largely related to the way rainwater wets surfaces. The geometry of a building is important in determining the final distribution of blackening and soiling. White areas occur in those parts of the building that are unsheltered and are associated with dominating rainwater runoff, leading to dissolution of carbonate rock, so that the original color of the material is evident. Black crusts are observed on stone surfaces wetted by rainwater but sheltered from intensive runoff, which is where pollutants accumulate, along with products of reactions between stone and gases and atmospheric aerosols (Sabbioni, 2003). These are the areas on a building most exposed to pollution-induced damage. Protection from leaching by rain washout favors the formation of crusts, which are black in color due to the atmospheric particles embedded during their growth.

In the case of carbonate porous stones, such as sandstone and calcarenite, the weathering pattern is not so directly linked to the geometry of exposure to environmental agents. Here, the blackening is homogeneous due to high porosity and superficial roughness, which favor rain wetting and particle matter capture (Sabbioni and Zappia, 1992).

Carbonate stones, both metamorphic and sedimentary lithotypes, are in general more prone to pollution impact leading to the formation of crusts and damage layers than silicate rocks, like granite and sandstones with siliceous cement. Crusts on carbonate stones with high and low porosity are mainly composed of gypsum (i.e., calcium sulfate dihydrate) due to the SO_2 reaction with calcite, $CaCO_3$, the main mineralogical component of the substrate, following the well-known chemical reaction:

$$CaCO_3 + H_2SO_4 + H_2O{\cdot}CaSO_4 \rightarrow 2H_2O + CO_2.$$

This deterioration process, basically consisting of the transformation of calcite in gypsum, is known as sulfation (Sabbioni, 2003). The black color of the gypsum layers is determined by the carbonaceous particles, which become embedded in the layers during their formation. Such particles are the main agents responsible for the soiling of architectural surfaces and, hence, the esthetic impairment of monuments (Grossi and Brimblecombe, 2004).

Black crusts have long disfigured monument and building façades located in city centers and industrial areas. From a mineralogical standpoint, the crust composition has been studied in depth by optical microscopy observations of thin sections in transmitted light and X-ray diffraction and SEM analysis, which have disclosed the important role of aerosol in determining crust formation (Camuffo *et al.*, 1982; Del Monte *et al.*, 2001; Amoroso, 2002; Fassina *et al.*, 2004; Bonazza, Sabbioni, and Ghedini, 2005; Perez-Monserrat *et al.*, 2013; Ozga *et al.*, 2013). The crusts are

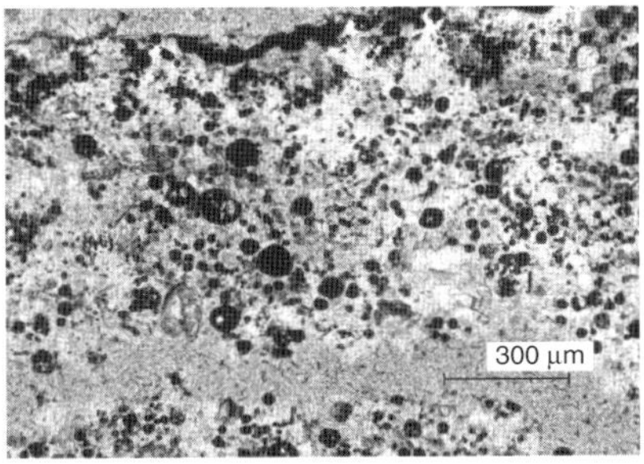

(a)

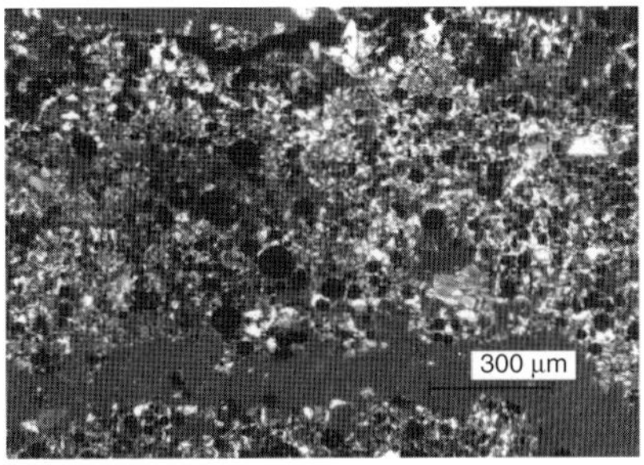

(b)

Figure 11.4 Micrographs showing black crust on Candoglia marble from Milan Cathedral (Italy): (a) planned-polarized light, black carbonaceous particles can be seen embedded in the damage layer; (b) cross-polarized light, needle-shaped crystals of gypsum are recognizable.

mainly composed of (80–90%) gypsum with lamellar/needle shape and/or microcrystalline habit due to dry and wet atmospheric SO_2 deposition. Calcite, linked to the carbonate substrate, is generally present as fragments of the underlying material embedded in crust layers or due to dissolution and secondary crystallization (Bonazza, Sabbioni, and Ghedini, 2005; Ozga *et al.*, 2013). Quartz, feldspars, clay minerals, and iron oxides, deriving from soil dust deposition, are also present. Generally layers have a dendritic and inhomogeneous morphology and embedded soot, with characteristic carbonaceous particles with sizes reaching 50–60 μm causing blackening and surface darkening (Figure 11.4a and b).

11.3.2 Damage to Silicate Stone

In comparison to damage layers on carbonate stone, limited research is available on damage formation and composition on silicate stones. Although gypsum crystallization on granitic building stone has also been ascribed to sulfur dioxide, the decay mechanisms due to air pollution impact on stone with a low Ca content remain unsettled. Among silicate stones, granites are widely present in the built heritage in Europe. This rock is more durable than marble and limestone toward environmental pollution due to its chemical composition and low porosity.

Schiavon (2007) and Schiavon *et al.* (1994, 1995) has studied air pollution impact and damage formation on granite, arguing that in humid mesothermic climates, SO_2 from anthropogenic emissions is responsible for accelerated kaolinization of feldspars in granites. Sulfur dioxide is suggested to play a dual role in the damage of silicate stones being responsible for the sulfation of Ca-bearing minerals (e.g., plagioclase) leading to gypsum formation and weathering of feldspar to form kaolinite. Both processes weaken the stone leading to spalling, powdering, and exfoliation. This was confirmed by Perez-Monserrat *et al.* (2013) who studied in detail the decay mechanisms of building materials, including granite, at the Royal

Palace of Madrid. Carbonaceous particles are often found embedded in surface deterioration layers, albeit in smaller amounts than those encountered in crusts on carbonate stone.

11.3.3 Anthropogenic Aerosol in Crusts

Carbonaceous particles deriving from combustion processes embedded in crusts and damage layers are responsible for the darkening of architectural surfaces. Several studies have characterized atmospheric particles on the basis of their morphology and elemental composition using optical and SEM with energy dispersive X-ray microanalyzer to identify the emission sources. Such techniques have identified different types of particles, which fall into three general categories:

- Spherical/subspherical carbonaceous particles with circular or irregular pores, emitted by oil or distilled oil combustion in domestic heating systems and electric power plants (Sabbioni, 1995), with diameters ranging between 10 and 60 μm (Del Monte, Sabbioni, and Vittori, 1981), and by vehicle exhaust (Chabas and Lefevre, 2000) (Figure 11.5a);
- Smooth non-porous spherical aluminosilicate (Figures 11.5b) particles produced by coal combustion (Sabbioni, 1995), with diameters ranging between 5 and 30 μm (Del Monte, Sabbioni, and Vittori, 1981), and by wood burning (Ausset *et al.*, 1998);
- Rare smooth metal particles, mainly composed of iron or titanium oxides, emitted by coal combustion processes (Sabbioni, 1995).

Whether atmospheric particles simply incur esthetic damage or play a role in the actual formation of the gypsum crust has been the subject of debate. Simulations of environmental damage processes on marble and limestone performed in a flow chamber, with controlled temperature, relative humidity, and SO_2 concentration, demonstrated that carbonaceous particles produced by combustion of oil-based fuels contribute significantly to the sulfation of carbonate stone (Sabbioni,

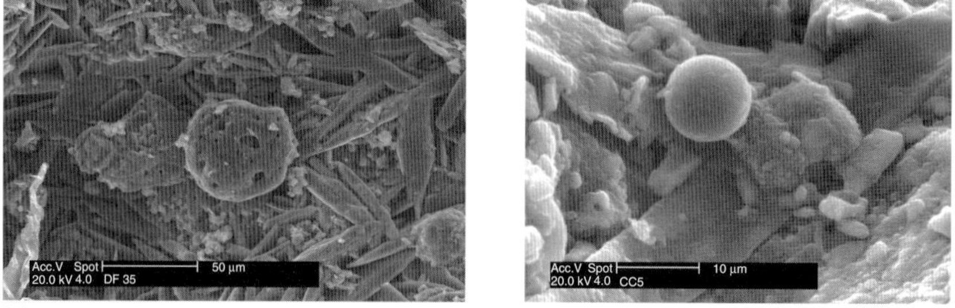

(a) (b)

Figure 11.5 (a) Carbonaceous particle with a spongy structure embedded in black crust from S. Maria del Fiore Cathedral in Florence. (b) Aluminosilicate particle in a damage layer from Cologne Cathedral.

Zappia, and Gobbi, 1996). The data obtained indicated that atmospheric carbonaceous particles increase the amount of SO_4^{2-} and favor the oxidation of sulfite into sulfate. The heavy metal content of carbonaceous particles, particularly iron, has been found to play a driving role in the heterogeneous oxidation of SO_2 into SO_4^{2-}, while their specific surface was demonstrated to have a limited effect. By contrast, other experiments using fly ash on carbonate stone reached the conclusion that carbonaceous particles play no active role in the sulfation process and actually mask the surface, providing some degree of protection (Hutchinson *et al.*, 1992). Additionally, laboratory experiments have shown how in controlled conditions (high relative humidity, absence of SO_2 and NO_x) carbonaceous particles produced by the combustion of fuel oil in electric power plants nucleate different mineralogical species, such as gypsum, associated with vanadates and amorphous particles of opal (i.e., $SiO_2{\cdot}nH_2O$) (Del Monte and Sabbioni, 1984). Furthermore, fly ash, also emitted by coal-fired power plants, has shown crystallite growth of gypsum in laboratory conditions during storage (Fisher, Chang, and Brummer, 1976; Del Monte and Sabbioni, 1984). The formation of gypsum is attributed to the reaction of calcium and sulfur, which are particularly enriched in the silicate matrix constituting the fly ash. Nevertheless the gypsum nucleated by the particles themselves is negligible compared to the gypsum due to atmospheric SO_2 deposition, as shown by the simulation experiments (Sabbioni, Zappia, and Gobbi, 1996).

11.3.4
Organic and Elemental Carbon

In addition to sulfur, carbon is important element of anthropogenic origin whose measurement can yield valuable input for the identification of pollution sources causing crust formation on building facades.

The analysis of carbon compounds in black crusts carried out to date have shown that the total carbon (TC) content in patinas is composed of two main fractions: carbonate (CC) and noncarbonate carbon (NCC):

$$\mathrm{TC} = \mathrm{CC} + \mathrm{NCC}.$$

The NCC fraction includes two different components, OC of biogenic and anthropic origin and EC, which is mainly a product of combustion processes:

$$\mathrm{NCC} = \mathrm{OC} + \mathrm{EC}.$$

Carbon included in the damage layers on monuments may have four different origins:

1. Calcium carbonate, deriving almost exclusively from the original building materials, both stone and mortars.
2. Deposition of atmospheric particles containing elemental and OC as primary and secondary pollutants.

3. Biological weathering due to the action of organisms: among the products of their metabolism, formic, acetic, and oxalic acids react with the underlying materials, leading to the formation of calcium formate, acetate, and oxalate.
4. Surface treatments, such as oils, waxes, proteins, and so on, frequently used in the past to protect monuments and historic buildings.

Since NCC is a large and important fraction of the TC present in black crusts, its measurement and discrimination into EC and OC is essential for a complete identification and quantification of the main components of the damage layers on monuments and of their origin (Ghedini *et al.*, 2006). Ghedini *et al.* (2006) have devised a chemical–thermal method to discriminate and quantify carbon fractions, including OC and EC. All of the experimental measurements of carbon in damage layers are carried out by combustion analysis using a carbon–hydrogen–nitrogen–sulfur–oxygen analyzer (CHNSO EA 1108 FISONS Instruments). For a complete analysis, the analytical methodology requires about 1 g of material and consists of three steps. First, TC is determined by burning the bulk sample, and then NCC is measured by burning the bulk sample after eliminating its CC fraction by treatment with HCl vapors. CC is then calculated according to the relation:

$$CC = TC - NCC.$$

The third step determines the EC content after eliminating the organic and inorganic species through a series of alternating basic and acid treatments at about 120 °C and applying CHNSO thermal analysis on the residual part. Finally OC is calculated according to the relation:

$$OC = NCC - EC.$$

Following results published by Bonazza, Sabbioni, and Ghedini (2005), Bonazza *et al.* (2007), and Ozga *et al.* (2013), the organic fraction in crusts tends to predominate over the elemental component (Figure 11.6). The OC concentration was particularly high in damage layers collected from monument sides exposed directly to traffic-induced air pollution, which is known to be the current main cause of urban pollution in Western Europe, emitting submicronic particles mainly composed of organic compounds (Cachier *et al.*, 2004).

11.3.5 Damage to Coastal Areas

As illustrated in the previous sections, atmospheric pollutants such as SO_2, NO_x, particles, and acid rain have a strong impact on the deterioration of building surfaces in urban environments. Blackening and soiling of surfaces are typical degradation effects formed at urban sites, closely connected to local pollution sources, for example, traffic near the heritage site.

Because of sea spray deposition, heritage sites located in urban coastal environments are particularly prone to damage such as fracture and delamination of

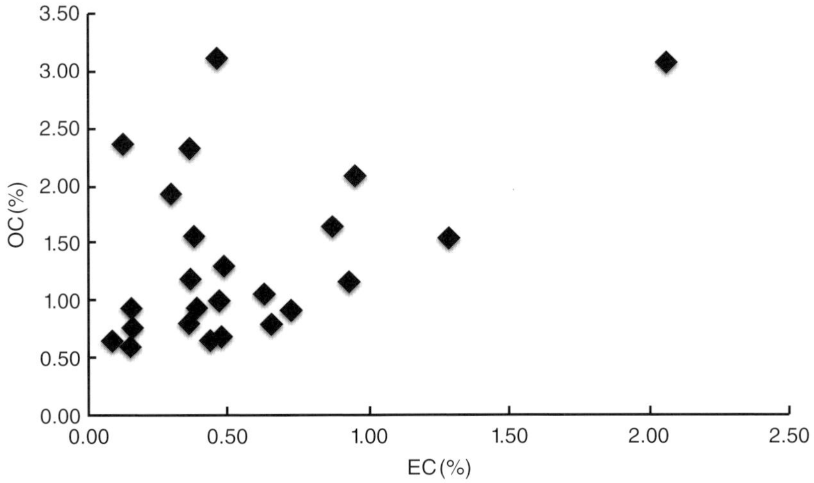

Figure 11.6 Organic (OC) and elemental (EC) carbonate concentration in black crusts from monuments in carbonate stone, such as marble and limestone.

porous materials due to salt crystallization and metal corrosion caused by chloride deposition. Marine aerosols, primarily composed of seawater along with particles naturally generated by the action of the wind on the seawater surface, introduce ionic species into the atmosphere, principally chlorides and sulfates. Chlorides are particularly destructive to porous materials as their high solubility allows them to penetrate the porous network and crystallize inside the material. This creates pressure that leads to the formation of microcracks (Cardell *et al.*, 2003). The corrosion rate at exposure sites strongly depends on their distance from the coast (Corvo *et al.*, 2008).

11.4 Bioaerosol Effects on Cultural Heritage

As widely discussed elsewhere in this book, airborne biological material is composed of particles generated from natural sources and resuspended in the atmosphere, often aggregated with each other or with other solid or liquid particles in suspension. Bioaerosol can be defined as a suspension of aerosols or particulate matter of microbial, plant, or animal origin and may consist of pathogenic or nonpathogenic, live or dead bacteria and fungi, algae, viruses, pollen, and plant fibers, high-molecular weight allergens, bacterial endotoxins, mycotoxins, peptidoglycans, or glucans (Douwes *et al*, 2003) passively carried by air (insects are excluded) (Cox and Wathes, 1995). Bioaerosol, also known as primary biological aerosol particles (PBAPs), vary widely in size and morphology and can be seen with a magnifying glass or microscope. Their aerodynamic diameter may range from molecular size to large and giant particle size (Jaenicke, 2005; Hinds, 1999; Pöschl, 2005), for example, viruses (0.05–0.15 μm), bacteria (0.1–4 μm), fungal spores (0.5–15 μm), algae (<10 μm), and pollen (10–100 μm) (Després *et al.*, 2012; Fuzzi *et al.*, 2015).

Although PBAP concentration in the atmosphere varies with the season and location, it has been estimated to constitute up to 25% of total aerosol mass and is sometimes numerically close to 50% of all aerosol particles on a global basis (Jones and Harrison, 2004; Jaenicke, 2005). Recent studies implementing molecular techniques demonstrated that the fungal spectrum suspended in the air is much richer than previously known (Fröhlich-Nowoisky *et al.*, 2009; Pashley *et al.*, 2012).

The atmosphere also contains vegetative forms and many forms of resistance like bacteria and fungal spores due to adverse environmental conditions such as dehydration, UV radiation, and temperature that may be lethal or detrimental for their reproduction. When microorganisms are incorporated into fog droplets for many hours, the environmental conditions are favorable for rapid growth, giving rise to secondary biological aerosol particles, commonly called SBAPs, as already described in this work (Fuzzi, Mandrioli, and Perfetto, 1997; Després *et al.*, 2012; Fuzzi *et al.*, 2015).

Once they are emitted by natural or/and anthropogenic sources (see Chapter 1), bioaerosol is dispersed in the atmosphere primarily with air currents. Small particles, ranging in size from 1.0 to 5.0 μm, remain suspended in the air for a longer time, whereas larger particles tend to settle more quickly on surfaces due to their increased mass. Bioaerosol can be transported in the atmosphere for long distances (Gregory, 1973) and a longer time due to its vicinity to particle sources and resuspension of deposited particles (Tampieri, Mandrioli, and Puppi, 1977; Mandrioli *et al.*, 1980, 1984; Rantio-Lehtimäiki, 1994). Kellogg and Griffin (2006) identified the global transport of desert dust as the main mechanism responsible for the transport of aerosol microbiota – pollens, fungi, and bacteria (Figure 11.7). Modern aerobiological studies claim that pollen transported by Saharan dust from North Africa has reached Southwestern Spain (Cordóba) (Cariñanos *et al.*, 2004). In fact, pollen from nonnative plants such as *Cannabis*, *Cupressus*, *Pinus*, *Platanus*, and *Sambucus* was detected exclusively on days when dust clouds were present.

The average residence time of biological particles in the atmosphere can range from less than a day to a few weeks, depending on their size and aerodynamic properties (Després *et al.*, 2012). Particles are usually removed from the air by sedimentation and deposition onto all surfaces, not only horizontal planes. Alternatively, they are washed out by precipitation, the most efficient removal mechanism for particles 0.1–10 μm in diameter.

Bioaerosol research started, thanks to the interest of the health sector, mainly related to allergology, and agriculture, namely, phytopathology. The cultural heritage sector is just one of many areas (pharmaceutical plants, food processing plants, waste treatment plants, electronic and space components factories, spacecraft, hospitals, agriculture, and offices) in which bioaerosol can cause injury to persons and/or objects, with major economic and other consequences.

Figure 11.7 Saharan dust from North Africa (by http://eoimages.gsfc.nasa.gov/images/).

The biodegradation of cultural heritage therefore has a cultural, scientific, and economic impact. Studies on biodeterioration are a constantly evolving field for scientific and technological research. They currently focus not only on bioaerosol sampling and identification methods but also on transport and deposition mechanisms and above all on the ecology of the species involved. For example, it has recently been demonstrated that the presence of lichen spores in the atmosphere is closely correlated with rainfall events (Favero-Longo *et al.,* 2014).

Bioaerosol on objects/artworks is only harmful for preservation in concomitance with other factors: microclimatic conditions, the nature of the object, its state of preservation, and the chemical and physical degradation processes already in place. Biological degradation is seldom caused by a single microorganism but is produced by complex communities, real ecosystems that develop on the article. Under favorable microclimatic conditions (RH, T, and light), the bioaerosol deposited on the surface can grow and reproduce using the substrate as a nutrient (heterotrophic) or support (autotrophic), causing damage to the material component (Hueck, 2001) of the cultural heritage whether it consists of historic artifacts and/or modern materials like polymers (Sabev *et al.,* 2006). The substrate may be a statue, painting, building facade, archeological site, or fresco made of a single material or polymaterials, organic (wood, paper, etc.), or inorganic (stone, metal, etc.). The organisms causing damage to works of art are called biodeteriogens but do not correspond to all the bioaerosol deposits generally found on works of art. Hueck (2001) defined biodeterioration as "any undesirable change in the properties of a material caused by the vital activities of organisms." Some of these particles may be viable but not culturable (VBNC): culturable, meaning the ability of a microorganism to form a colony on a nutrient agar plate under certain growth conditions (time, temperature, and nutrients). Many bioaerosol particles cannot be cultivated on conventional media (they do not form colonies on solid media), but their existence can be proved using other methods (Năşcuţiu, 2010). A number of different factors, which can operate cumulatively, can "stress" a microorganism and affect its viability – the organism species, growth conditions for the culture, method of aerosol generation, sampler, and the airborne environment. Desiccation, radiation, oxygen, ozone, and its reaction products and various pollutants can affect the viability of microorganisms (Griffith and De Cosemo, 1994). In air, biological spores survive better than vegetative cells as the humidity in the air is a very important factor of potential microorganism stress. The indoor environment (museum, gallery, archives, hypogeum) can be particularly suitable for microbial growth as it protects the microorganisms themselves from extreme variations in outside temperature and UV rays that can damage the bioaerosol.

The Mediterranean climate and topography foster the growth of numerous vegetal species whose flowering period is longer than in other bioclimatic zones, leading to the world's highest incidence of bioaerosol. The processes responsible for physical or mechanical failure, delamination/surface shrinkage, fragmentation, and loss of parts are mainly caused by the penetration of biological structures into the material (hyphae, roots, and rhizines). Fungi are among the most harmful organisms associated with the biodeterioration of organic and inorganic materials (Sterflinger, 2010). The most common fungal colonizers in rock are the so-called black meristematic fungi as they contain melanin in the cell wall. They are also called rock-inhabiting fungi (commonly called RIF) and are the most damaging organisms penetrating the surfaces of stone monuments (Urzì *et al.*, 2000; Cappitelli *et al.*, 2007; Sterflinger and Piñar, 2013; De Leo and Urzì, 2014; Onofri *et al.*, 2014).

Many organisms excrete waste metabolic products, including pigmented or acid compounds, that may disfigure materials, altering their colors or causing mechanical damage. Esthetic biodeterioration should not be underestimated as it severely alters the perception of beauty and the legibility of artworks (algae, fungi, mosses, and black lichens) although generally it is less aggressive toward the materials. For example, the fungi present on different types of materials can determine the formation of visible films, spots, exfoliations, disruptions, and pitting. Chemical processes lead to the transformation, alteration, and decomposition of the substrate and are much more common than in the past. In addition, the pores and fractures caused by chemical and mechanical interactions can host further biodeteriogens (Urzì *et al.*, 2000). A chemical action may be attributed both to assimilation processes (when the organisms use the material as nourishment by means of extracellular enzyme activity or ion exchange) and to the excretion of metabolic intermediates or substances having an inhibitory or waste function (such as acid, alkaline and complexing substances, and pigments). In physical biodeterioration, the organism breaks or simply deforms the material with growth or movement. For instance, the expansion of fungal hyphae between rock layers may lead to cracking and significant loss of material. Lithophytic lichens can penetrate rock faces down to many millimeters (Piervittori, Favero-Longo, and Gazzano, 2009).

Biodeterioration is more readily triggered on indoor and outdoor bioaccumulation (metabolic products, droppings, and dead organisms). This can determine the subsequent colonization of the surface dirt by other microorganisms such as fungi and algae (e.g., black patinas or microbial biofilms can grow on guano in archeological sites) and even vascular plants (on buildings and monuments, algae can produce patches or extended biofilms that facilitate subsequent colonization by vascular plants) such as *Parietaria* spp., *Capparis* spp., and *Ficus carica.* Colonies of lichens, mosses, and small plants adhere better on rough walls. Preventive conservation activities must always include both indoor and outdoor maintenance protocols providing regular and adequate cleaning of surfaces and objects (e.g., periodic dusting).

The indoor conservation of artworks (museums, gallery, archives, etc.) is closely correlated to the condition of the building fabric as poor maintenance of buildings can greatly contribute to increased indoor microbial exposure. In addition, the use of water-sensitive materials in areas with hot and humid climates can lead to indoor mould growth. Another element to highlight is the air conditioning systems frequently present in places of cultural heritage conservation. On the one hand, these systems treat the incoming air, while on the other, if they are not properly maintained, they can become a cause of contamination and microbial amplification. This has raised international concern about the impact of bioaerosol on works of art inside building structures and not least on the health of people working in the cultural heritage sector.

The study of bioaerosol dispersion also serves to understand the role of biological organisms, especially lichens and fungi, in the formation of patinas of calcium oxalate (whewellite or weddellite), thanks to the accumulation of metabolism products and their reaction with the calcium carbonate of the substrate. This type of decay is very common in monuments of different types, origin, and ages (Caneva *et al.,* 2003). Due to the wide variety of materials, substrates, types of works, and exposure conditions, monuments and archeological sites offer diverse environments that determine the colonization of very rich micro- and macroflora. Photosynthetic organisms, such as cyanobacteria and lichens, are resistant to water stress, heat, and nourishment and are considered pioneers in the colonization of environments, including works of art. Algae, mosses, and liverworts are particularly abundant in conditions of high soil moisture, while mushrooms colonize organic matter. Climactic succession also affects ruderal herbaceous plants, woody shrubs, and trees.

Another important aspect is the indirect relationship between bioaerosol and outdoor cultural heritage. Here biological particles are involved in atmospheric precipitation and climate due to their optical properties and because they act as ice nuclei (IN) or cloud condensation nuclei (CCN) (Rantio-Lehtimäiki, 1994; Després *et al.*, 2012). Moreover, a number of authors have noted that biogenic volatile organic compounds (BVOCs) attributable to emission by local urban vegetation determine the abundant ozone and aerosol formation in the urban atmosphere that contributes to climate change (Kanakidou *et al.*, 2005; Noe, Penuelas, and Niinemets, 2008; Calfapietra *et al.*, 2013).

In some cases, microorganisms like lichens (Ariño *et al.*, 1995; Carballal, Paz-Bermúdez, and Sánchez-Biezma, 2001) can be considered bioprotective rather than a potential danger (St. Clair and Seaward, 2004) since they enhance the preservation of a manufactured article, for instance, from atmospheric pollution (role of a network of lichenized fungal hyphae in the protection of stone). Moreover, there is a growing interest in the development of environmentally friendly biological technologies (neutral pH) that do not require the use of toxic materials. Biotechnology has been applied with success in different domains, such as biocleaning or bioremediation of ornamental stone or corrosion control of metal artworks in the BIOBRUSH (2002–2005, http://cordis.europa.eu/result/rcn/84151_en.html) and BAHAMAS (2010–2012, http://www2.unine.ch/lamun/cultural_heritage) research projects.

11.5
Guidelines for the Preventive Conservation of Cultural Heritage in Urban Areas

In view of the development of short- and long-term strategies for preventive conservation and management of cultural heritage toward aerosol impact, the results concerning the research in the field carried out up to now and discussed in this chapter can be generalized into the following guidelines (Ghedini *et al.*, 2011):

- Define risks of cultural heritage against pollutants and climate parameters by adopting continuous environmental monitoring close to monuments as a valuable tool. Definition of risks allows heritage managers to single out the priorities for greater investment in preservation and funding of actions to reduce the causes.
- Identify and prioritize pollutants and climatic variables causing deterioration, depending also on the sensitivity of materials and the environmental context in which monuments are located. The monitoring of aerosol (including bioaerosol) concentrations and composition with at least seasonal resolution is suggested as the correct methodological approach in environmental assessment for cultural heritage protection.
- Promote actions aimed at reducing exposure of cultural heritage to anthropogenic pollutants, in particular soluble and carbon fraction of particulate matter.

- Mitigate the negative causes by regulation at regional and local levels aimed at the reduction of pollutants, especially in developing urban plans for a sustainable mobility, which envisages the impact on cultural heritage.
- Plan creation and modification of public green in urban areas including vegetation species with low impact in terms of pollen production and BVOC emission.
- Develop management strategies for tourism fruition ensuring controlled flow in the surrounding areas and in the access to the cultural assets, with the aim of reducing atmospheric particle resuspension.
- Evaluate new cultural asset location also to take account of environmental condition.
- Favor research results transfer to urban planners and cultural heritage owners and managers.

Abbreviations

BVOC	biogenic volatile organic compounds
CC	carbonate carbon
CCN	cloud condensation nuclei
CFU	colony forming unit
DMC	Data Management Centre
EC	elemental carbon
EDX	X-ray diffraction analysis
IC	ion chromatography
IN	ice nuclei
LM	light microscopy
NCC	noncarbonate carbon
OC	organic carbon
PBAP	primary biological aerosol particles
SEM	scanning electron microscopy
TC	total carbon
TSP	total suspended particles
UNI, EN, ISO	Italian Organization for Standardization, European Standard, International Organization for Standardization

List of Symbols

RH	air relative humidity
T	air temperature
T_{surf}	surface temperature
T_d	dew point
WS	wind speed
WD	wind direction
I_o	solar radiation intensity

References

Amoroso, G. (2002) *Trattato di Scienza della Conservazione dei Monumenti. Etica della conservazione, degrado dei monumenti, interventi conservativi, consolidanti e protettivi*, ALINEA Editrice, Firenze, 416 pp.

Ariño, X., Ortega-Calvo, J.J., Gomez-Bolea, A., and Saiz-Jimenez, C. (1995) Lichen colonization of the Roman pavement at Baelo Claudia (Cadiz, Spain): biodeterioration vs. bioprotection. *Sci. Total Environ.*, **67**, 353–363. doi: 10.1016/0048-9697(95)04595-R

Ausset, P., Bannery, F., Del Monte, M., and Lefevre, R.A. (1998) Recording of pre-industrial atmospheric environment by ancient crusts on stone monuments. *Atmos. Environ.*, **32** (16), 2859–2863. doi: 10.1016/S1352-2310(98)00063-6

Bonanni, P., Daffinà, R., Gaddi, R., Silli, V., Cirillo, M., Cacace, C., and Giovagnoli, A. (2006) I risultati dello studio sperimentale APAT-ICR per la valutazione del danno dei beni storico-artistici esposti all'aperto. *Gazzetta dell'ambiente*, **2**, 63–74, edizioni Colombo, Roma.

Bonazza, A., Brimblecombe, P., Grossi, C.M., and Sabbioni, C. (2007) Carbon in black crusts from the Tower of London. *Environ. Sci. Technol.*, **41** (12), 4199–4204. doi: 10.1021/es062417w

Bonazza, A., Sabbioni, C., and Ghedini, N. (2005) Quantitative data on carbon fractions in interpretation of black crusts and soiling on European built heritage. *Atmos. Environ.*, **39** (14), 2607–2618. doi: 10.1016/j.atmosenv.2005.01.040

Bonazza, A., Sabbioni, C., Messina, P., Guaraldi, C., and De Nuntiis, P. (2009) Climate change impact: mapping thermal stress on Carrara marble in Europe. *Sci. Total Environ.*, **407** (15), 4506–4512. doi: 10.1016/j.scitotenv.2009.04.008

Brandi, C. (1977) *Teoria del restauro*, Einaudi, Torino, (I ed. Roma, Edizioni di Storia e Letteratura, 1963), **80** pp.

Cachier, H., Sarda-Esteve, R., Oikonomou, K., Sciare, J., Bonazza, A., Sabbioni, C., Greco, M., Rejes, J., Hermosin, B., and Saiz-Jimenez, C. (2004) *Air Pollution and Cultural Heritage* (ed. C. Saiz-Jimenez), A.A. Balkema Publishers, London, pp. 3–12.

Calfapietra, C., Fares, S., Manes, F., Morani, A., Sgrigna, G., and Loreto, F. (2013) Role of Biogenic Volatile Organic Compounds (BVOC) emitted by urban trees on ozone concentration in cities: a review, *Environ. Pollut.*, **183**, SI, 71–80, doi: 10.1016/j.envpol.2013.03.012

Camuffo, D. (2013) *Microclimate for Cultural Heritage – Conservation and Restoration of Indoor and Outdoor Monuments*, 2nd edn, vol. **23**, Elsevier Science, New York, 560 pp.

Camuffo, D., Del Monte, M., Sabbioni, C., and Vittori, O. (1982) Wetting, deterioration and visual features of stone surfaces in an urban area. *Atmos. Environ.*, **16** (9), 2253–2259. doi: 10.1016/0004-6981(82)90296-7

Caneva, G., Maggi, O., Nugari, M.P., Pietrini, A.M., Piervittori, R., Ricci, S. and Roccardi, A. (2003) *Cultural Heritage and Aerobiology* (Methods and measurement techniques for biodeterioration monitoring) (eds P. Mandrioli, G. Caneva, and C. Sabbioni), Kluwer Academic Publishers, Dordrecht, pp. 3–29.

Cappitelli, F., Nosanchuk, J.D., Casadevall, A., Toniolo, L., Brusetti, L., Florio, S., Principi, P., Borin, S., and Sorlini, C. (2007) Synthetic consolidants attacked by melanin-producing fungi: case study of the biodeterioration of Milan (Italy) cathedral marble treated with acrylics. *Appl. Environ. Microbiol.*, **73** (1), 271–277. doi: 10.1128/AEM.02220-06

Carballal, R., Paz-Bermúdez, G., and Sánchez-Biezma, M.J. (2001) Lichen colonization of coastal churches in Galicia: biodeterioration implications. *Int. Biodeterior. Biodegrad.*, **47** (3), 157–163. doi: 10.1016/S0964-8305(01)00044-0

Cardell, C., Delalieux, F., Roumpopoulos, K., Moropoulou, A., Auger, F., and Van Grieken, R. (2003) Salt-induced decay in calcareous stone monuments and buildings in a marine environment in SW France. *Constr. Build. Mater.*, **17** (3), 165–179. doi: 10.1016/S0950-0618(02)00104-6

Cariñanos, P., Galán, C., Alcázar, P., and Domínguez, E. (2004) Analysis of the particles transported with dust-clouds reaching Córdoba, Southwestern Spain. *Arch. Environ. Contam. Toxicol.*, **46** (2), 141–146. doi: 10.1007/s00244-003-2273-9

Cassar, J., Galea, M., Grima, R., Stroud, K., and Torpiano, A. (2011) Shelters over the Megalithic Temples of Malta: debate, design and implementation. *Environ. Earth Sci.*, **63** (7-8), 1849–1860. doi: 10.1007/s12665-010-0735-8

Chabas, A. and Lefevre, R.A. (2000) Chemistry and microscopy of atmospheric particulates at Delos (Cyclades-Greece). *Atmos. Environ.*, **34** (2), 225–238. doi: 10.1016/S1352-2310(99)00255-1

Corvo, C., Perez, T., Martin, Y., Reyes, J., Dzib, L.R., Gonzalez, J., and Castaneda, A. (2008) Time of wetness in tropical climate: considerations on the estimation of TOW according to ISO 9223 standard. *Corros. Sci.*, **50** (1), 206–219. doi: 10.1016/j.corsci.2007.06.012

Cox, C.S. and Wathes, C.M. (Eds.) (1995) *Bioaerosols handbook*, CRC Press, 656 pp.

De Leo, F. and Urzì, C. (2014) *Fungi from Different Substrates* (eds J.K. Misra, J.P. Tewari, S.K. Deshmukh, and C. Vágvölgyi), CRC Press, New York, pp. 144–158.

De Nuntiis, P. (2013) *Science and Conservation for Museum Collections* (ed. B. Fabbri), Nardini Editore – Kermes Quaderni, Firenze, pp. 1–42.

De Nuntiis, P., Guaraldi, C., Mandrioli, P., Monco, A. (2007) *Oggetti nel tempo. Principi e tecniche di conservazione preventiva*, Regione Emilia Romagna, Istituto per i Beni artistici, culturali e naturali, CLUEB, Bologna, 258 pp.

De Nuntiis, P., Maggi, O., Mandrioli, P., Ranalli, G., and Sorlini, C. (2003) *Cultural Heritage and Aerobiology* (Methods and measurement techniques for biodeterioration monitoring) (eds P. Mandrioli, G. Caneva, and C. Sabbioni), Kluwer Academic Publishers, Dordrecht, pp. 107–144.

Del Monte, M., Ausset, P., Forti, P., Lefevre, R.A., and Tolomelli, M. (2001) Air pollution records on selenite in the urban environment. *Atmos. Environ.*, **35** (22), 3885–3896. doi: 10.1016/S1352-2310(01)00143-1

Del Monte, M. and Sabbioni, C. (1984) Morphology and mineralogy of fly ash from coal-fuelled power plant. *Arch. Meteorol. Geophys. Biochem.*, **B35** (1-2), 93–104. doi: 10.1007/BF02269412

Del Monte, M., Sabbioni, C., and Vittori, O. (1981) Airborne carbon particles and marble deterioration. *Atmos. Environ.*, **15** (5), 645–652. doi: 10.1016/0004-6981(81)90269-9

Després, V.R., Huffman, J.A., Burrows, S.M., Hoose, C., Safatov, A.S., Buryak, G., Frohlich-Nowoisky, J., Elbert, W., Andreae, M.O., Pöschl, U., and Jaenicke, R. (2012) Primary biological aerosol particles in the atmosphere: a review. *Tellus Ser. B*, **64**, 15598. doi: 10.3402/tellusb.v64i0.15598

Douwes, J., Thorne, P., Pearce, N., and Heederik, D. (2003) Bioaerosol Health Effects and Exposure Assessment: Progress and Prospects. *Ann Occup Hyg.*, **47**, 187–200, doi: 10.1093/annhyg/meg032

Fassina, V., Schilling, M.R., Keeney, J., and Khanijan, H. (2004) *Air Pollution and Cultural Heritage* (ed. C. Saiz-Jimenez), A.A. Balkema Publishers, London, pp. 117–125.

Favero-Longo, S.E., Sandrone, S., Matteucci, E., Appolonia, L., and Piervittori, R. (2014) Spores of lichen-forming fungi in the mycoaerosol and their relationships with climate factors. *Sci. Total Environ.*, **466**, 26–33. doi: 10.1016/j.scitotenv.2013.06.057

Fisher, G.L., Chang, D.P.Y., and Brummer, M. (1976) Fly-ash collected from electrostatic precipitators: microcrystalline structures and mystery of spheres. *Science*, **192** (4239), 53–555. doi: 10.1126/science.192.4239.553

Fröhlich-Nowoisky, J., Pickersgill, D.A., Després, V.R., and Poschl, U. (2009) High diversity of fungi in air particulate matter. *Proc. Natl. Acad. Sci. U.S.A.*, **106**, 12814–12819. doi: 10.1073/pnas.0811003106

Fuzzi, S., Baltensperger, U., Carslaw, K., Decesari, S., Denier van der Gon, H., Facchini, M.C., Fowler, D., Koren, I., Langford, B., Lohmann, U., Nemitz, E., Pandis, S., Riipinen, I., Rudich, Y., Schaap, M., Slowik, J.G., Spracklen, D.V., Vignati, E., Wild, M., Williams, M., and Gilardoni, S. (2015) Particulate matter, air quality and climate: lessons learned and future needs. *Atmos. Chem. Phys.*, **15** (14), 8217–8299. doi: 10.5194/acp-15-8217-2015

Fuzzi, S., Mandrioli, P., and Perfetto, A. (1997) Fog droplets – an atmospheric source of secondary biological aerosol particles. *Atmos. Environ.*, **31** (2), 287–290. doi: 10.1016/1352-2310(96)00160-4

Ghedini, N., Ozga, I., Bonazza, A., Dilillo, M., Cachièr, H., and Sabbioni, C. (2011) Atmospheric aerosol monitoring as a strategy for the preventive conservation of urban monumental heritage: the Florence Baptistery. *Atmos. Environ.*, **45** (33), 5979–5987. doi: 10.1016/j.atmosenv.2011.08.001

Ghedini, N., Sabbioni, C., Bonazza, A., and Gobbi, G. (2006) Chemical-thermal quantitative methodology for carbon speciation in damage layers on building surfaces. *Environ. Sci. Technol.*, **40** (3), 939–944. doi: 10.1021/es050164l

Gregory, P.H. (1973) *The Microbiology of the Atmosphere*, 2nd edn, Leonard Hall, Aylesbury, 377 pp.

Griffith, W.D. and De Cosemo, G.A.L. (1994) The assessment of bioaerosols: a critical review. *J. Aerosol Sci.*, **25** (8), 1425–1458. doi: 10.1016/0021-8502(94)90218-6

Grossi, C.M. and Brimblecombe, P. (2004) Aesthetics of simulated soiling patterns on architecture. *Environ. Sci. Technol.*, **38** (14), 3971–3976. doi: 10.1021/es0353762

Guichen, G.D. (1999) Preventive conservation: a mere fad or far-reaching change? (In museums). *Mus. Int.*, **51** (1), 4–6. doi: 10.1111/1468-0033.00183

Hinds, W.C. (1999) *Aerosol Technology: Properties, Behavior, and Measurement of Airborne Particles*, 2nd edn, Wiley-Interscience, 504 pp.

Hirst, J.M. (1995) Bioaerosols: Introduction, Retrospect and Prospect. In *Bioaerosols Handbook* (eds C.S. Cox and C.M. Wathes), CRC Press, pp. 5–14.

Hueck, H.J. (2001) The biodeterioration of materials: an appraisal (Reprinted). *Int. Biodeterior. Biodegrad.*, **48** (1-4), 5–11. doi: 10.1016/S0964-8305(01)00061-0

Hutchinson, A.J., Johnson, J.B., Thompson, G.E., Wood, G.C., Sage, P.W., and Cooke, M.J. (1992) The role of fly-ash particulate material and other catalysts in stone degradation. *Atmos. Environ.*, **26** (15), 2795–2803. doi: 10.1016/0960-1686(92)90017-F

Jaenicke, R. (2005) Abundance of cellular material and proteins in the atmosphere. *Science*, **308** (5718), 73. doi: 10.1126/science.1106335

Jones, A.M. and Harrison, R.M. (2004) The effects of meteorological factors on atmospheric bioaerosol concentrations, a review. *Sci. Total Environ.*, **326** (1–3), 151–180. doi: 10.1016/j.scitotenv.2003.11.021

Kanakidou, M., Seinfeld, J.H., Pandis, S.N., Barnes, I., Dentener, F.J., Facchini, M.C., Van Dingenen, R., Ervens, B., Nenes, A., Nielsen, C.J., Swietlicki, E., Putaud, J.P., Balkanski, Y., Fuzzi, S., Horth, J., Moortgat, G.K., Winterhalter, R., Myhre, C.E.L., Tsigaridis, K., Vignati, E., Stephanou, E.G., and Wilson, J. (2005) Organic aerosol and global climate modelling: a review. *Atmos. Chem. Phys.*, **5** (4), 1053–1123. doi: 10.5194/acp-5-1053-2005

Kellogg, C.A. and Griffin, D.W. (2006) Aerobiology and the global transport of desert dust. *Trends Ecol. Evol.*, **21** (11), 638–644. doi: 10.1016/j.tree.2006.07.004

Leysen, L., Roekensand, E., and Van Grieken, R. (1989) Air pollution induced chemical decay of a sandy-limestone cathedral in Belgium. *Sci. Total Environ.*, **78**, 263–287. doi: 10.1016/0048-9697(89)90038-7

Mandrioli, P., De Nuntiis, P., Guaraldi, C., Monco, A., Sabbioni, C., Bonazza, A., and Bernardi, A. (2006) Environmental Monitoring at Hagar Qim and Mnajdra Temples. Final Report, Tender CT 2592/04, ISAC-CNR, Bologna, 99 pp.

Mandrioli, P., D. Fernandez, P. De Nuntiis and L. Branzanti (2012), *Prima, durante … invece del restauro, Proceedings of the 6° International Conference: Color and Conservation,* CESMAR7, (eds C. Lodi and C. Sburlino), pp. 77–85.

Mandrioli, P., Negrini, M.G., Cesari, G., and Morgan, G. (1984) Evidence for long range transport of biological and antropogenic aerosol particles in the atmosphere. *Grana*, **23** (1), 43–53. doi: 10.1080/00173138409428876

Mandrioli, P., Negrini, M.G., Scarani, C., Tampieri, F., and Trombetti, F. (1980) Mesoscale transport of *Corylus* pollen grains in winter atmosphere. *Grana*, **19** (3), 227–233. doi: 10.1080/00173138009425007

Mandrioli, P., Sabbioni, C., Bonazza, A., De Nuntiis, P., Guaraldi, C., Ozga, I., and Pili, F. (2008) *Il Camposanto di Pisa. Un progetto di restauro integrato, Conference Proceedings*, Pisa, Italy, Auditorium G. Toniolo Nuovi Laboratori di Campaldo Marzo 6–8, 2008, Opera Primaziale Pisana, Pisa, pp. 60–64.

Menegazzi, C. (2003) *Servizi e professionalità nuove per la tutela: La Conservazione Preventiva delle Raccolte Museali, Proceedings of the International Conference Restauro'99*, Ferrara, Italy, Marzo 27, 1999, Nardini, Coll (eds C. Menegazzi and I. Silvestri), Kermesquaderni, Firenze, pp. 9–11.

Năşcuţiu, A.M. (2010) Viable non-culturable bacteria. *Bacteriol. Virusol. Parazitol. Epidemiol.*, **55** (1), 11–18.

Noe, S.M., Penuelas, J., and Niinemets, U. (2008) Monoterpene emissions from ornamental trees in urban areas: a case study of Barcelona, Spain. *Plant Biol.*, **10** (1), 163–169. doi: 10.1111/j.1438-8677.2007.00014.x

Nugari, M.P. (2003) *Biological Problems in Cultural heritage*, Coalition Advanced Course, Florence, Italy, 8–9 November, 2002, (ed. C. Saiz-Jimenez), Coalition Newsletter, **6** (2), SI, pp. 8–10.

Onofri, S., Zucconi, L., Isola, D., and Selbmann, L. (2014) Rock-inhabiting fungi and their role in deterioration of stone monuments in the Mediterranean area. *Plant Biosyst.*, **148** (2), 384–391. doi: 10.1080/11263504.2013.877533

Ozga, I., Bonazza, A., Ait Lyazidi, S., Haddad, M., Ben-Ncerc, A., Ghedini, N., and Sabbioni, C. (2013) Pollution impact on the ancient ramparts of the Moroccan city Salé. *J. Cult. Heritage*, **14** (3), S25–S33. doi: 10.1016/j.culher.2012.10.018

Pashley, C.H., Fairs, A., Free, R.C., and Wardlaw, A.J. (2012) DNA analysis of outdoor air reveals a high degree of fungal diversity, temporal variability, and genera not seen by spore morphology. *Fungal Biol.*, **116** (2), 214–224. doi: 10.1016/j.funbio.2011.11.004

Perez-Monserrat, E.M., Alvarez de Buergo, M., Gomez-Heras, M., Varas Muriel, M.J., and Fort Gonzalez, R. (2013) An urban geomonumental route focusing on the petrological and decay features of traditional building stones used in Madrid, Spain. *Environ. Earth Sci.*, **69** (4), 1071–1084. doi: 10.1007/s12665-012-2164-3

Piervittori, R., S.E., Favero-Longo, C. Gazzano (2009), Lichens and biodeterioration of stonework: a review, *Chim. Oggi*, **27**, 6, 8-11, ISSN 0392-839X.

Pöschl, U. (2005) Atmospheric aerosols: composition, transformation, climate and health effects. *Angew. Chem. Int. Ed.*, **44** (46), 7520–7540. doi: 10.1002/anie.200501122

Rantio-Lehtimäiki, A. (1994) Short, medium and long range transported airborne particles in viability and antigenicity analyses. *Aerobiologia*, **10** (2), 175–181. doi: 10.1007/BF02459233

Rossval, J. (1998) *Air Pollution and Conservation*, Elsevier, Amsterdam.

Sabbioni, C. (1995) Contribution of atmospheric deposition to the formation of damage layers. *Sci. Total Environ.*, **167**, 49–56. doi: 10.1016/0048-9697(95)04568-L

Sabbioni, C. (2003) *The Effects of Air Pollution on the Built Environment* (ed. P. Brimblecombe), Imperial College Press, Singapore, pp. 63–106.

Sabbioni, C., Mandrioli, P., Bonazza, A., De Nuntiis, P., and Guaraldi, C. (2008) Indagini diagnostiche finalizzate alla identificazione delle cause di degrado degli affreschi del camposanto monumentale di Pisa, Report tecnico contratto OPA Pisa – ISAC CNR, Bologna, 154 pp.

Sabbioni, C. and Zappia, G. (1992) Decay of sandstone in urban areas correlated with atmospheric aerosol. *Water Air Soil Pollut.*, **63** (3-4), 305–316. doi: 10.1007/BF00475497

Sabbioni, C., Zappia, G., and Gobbi, G. (1996) Carbonaceous particles and stone damage in a laboratory exposure system. *J. Geophys. Res.*, **101** (D14), 19621–19627. doi: 10.1029/95JD03755

Sabev, H.A., Barratt, S.R., Greenhalgh, M., Handley, P.S., and Robson, G.D. (2006) *Fungi in Biogeochemical Cycles* (ed. G.M. Gadd), Cambridge University Press, pp. 212–235. doi: 10.1017/CBO9780511550522.010

Saiz-Jimenez, C. (2003) *The Effects of Air Pollution on the Built Environment* (ed. P. Brimblecombe), Imperial College Press, Singapore, pp. 183–225.

Schiavon, N. (2007) Kaolinisation of granite in an urban environment. *Environ. Geol.*, **52** (2), 333–341. doi: 10.1007/s00254-006-0473-0

Schiavon, N., G. Chiavari, D. Fabbri, and G. Schiavon (1994) In Stone and Monuments: Methodologies for the Analyses of Weathering and Conservation, Proceedings of the 3rd International Symposium on the Conservation of Monuments in the Mediterranean Basin, Venice, Italy, June 22–25, 1994 (eds V. Fassina, H. Ott, and F. Zezza), Balkema, Venice, pp. 93–99.

Schiavon, N., Chiavari, G., Schiavon, G., and Fabbri, D. (1995) Nature and decay effects of urban soiling on granitic building stones. *Sci. Total Environ.*, **167**, 87–101. doi: 10.1016/0048-9697(95)04572-I

St. Clair, L. and Seaward, M.R.D. (2004) *Biodeterioration of Stone Surfaces: Lichens and Biofilms as Weathering Agents of Rocks and Cultural Heritage*, Kluwer Academic Publishers, Dordrecht, London, 77 pp.

Sterflinger, K. (2010) Fungi: their role in the deterioration of cultural heritage. *Fungal Biol. Rev.*, **24** (1–2), 47–55. doi: 10.1016/j.fbr.2010.03.003

Sterflinger, K. and Piñar, G. (2013) Microbial deterioration of cultural heritage and works of art-tilting at windmills? *Appl. Microbiol. Biotechnol.*, **97** (22), 9637–9646. doi: 10.1007/s00253-013-5283-1

Tampieri, F., Mandrioli, P., and Puppi, G.L. (1977) Medium range transport of airborne pollen. *Agric. Meteorol.*, **18** (1), 9–20. doi: 10.1016/0002-1571(77)90023-1

Urzì, C., De Leo, F., De Hoog, G.S., and Sterflinger, K. (2000) *Of Microbes and Art. The Role of Microbial Communities in the Degradation and Protection of Cultural Heritage* (eds O. Ciferri, P. Tiano, and G. Mastromei), Kluwer Academic Publisher, New York, pp. 3–19. doi: 10.1007/978-1-4615-4239-1

Index

Atmospheric Aerosols: Life Cycles and Effects on Air Quality and Climate, First Edition.
Edited by Claudio Tomasi, Sandro Fuzzi, and Alexander Kokhanovsky.